HANDBOOK OF LIVESTOCK MANAGEMENT

HANDBOOK OF LIVESTOCK MANAGEMENT

FOURTH EDITION

Richard A. Battaglia

Animal and Veterinary Science Department
University of Idaho

PEARSON
Prentice
Hall

Upper Saddle River, New Jersey
Columbus, Ohio

Library of Congress Cataloging-in-Publication Data

Battaglia, Richard A.
    Handbook of livestock management / Richard A. Battaglia.—4th ed.
      p. cm.
    Includes index
    ISBN 0-13-118933-6
     1. Livestock—Handbooks, manuals, etc. I. Title II. Title: Livestock management.

SF65.2 .B38 2007
636—dc22

2006045719

Editor in Chief: Vernon Anthony
Associate Editor: Jill Jones-Renger
Editorial Assistant: Yvette Schlarman
Production Coordination: Carlisle Publishers Services
Production Editor: Holly Shufeldt
Design Coordinator: Diane Ernsberger
Cover Designer: Ali Mohtman
Cover photo: Getty One
Text Illustrator: Marge Battaglia
Production Manager: Matt Ottenweller
Marketing Manager: Jimmy Stephens
Senior Marketing Coordinator: Liz Farrell
Marketing Assistant: Alicia Dysert

Photograph credits are listed on page vi.
This book was set in New Baskerville by Carlisle Communications, Ltd. It was printed and bound by
R. R. Donnelly & Sons Company. The cover was printed by Coral Graphic Services, Inc.

Pearson Education Ltd.   Pearson Education Australia Pty. Limited
Pearson Education Singapore Pte. Ltd.   Pearson Education North Asia Ltd.
Pearson Education Canada, Ltd.   Pearson Educación de Mexico, S. A. de C. V.
Pearson Education—Japan   Pearson Education Malaysia Pte. Ltd.

10 9 8 7 6 5 4 3
ISBN 0-13-118933-6

To my family . . . without their continuing support, this revision could not have happened, nor would there have been a purpose. To past and present authors and editors . . . their contributions to the body of livestock management knowledge enabled the entire process. To the friends whom this work has developed over the past 25 years.

| PHOTO LIST |

Chapter 2   Beef Cattle Management Techniques: Beef Cattle Breeds

Charolais Bull   Courtesy of American International Charolais Association

Limousin Bull   Photo courtesy of The Limousin World Magazine. Used with permission.

Simmental Bull   Courtesy of American Simmental Association

Shorthorn Bull   Courtesy of American Shorthorn Association

Beefmaster Bull   Courtesy of Lasater Ranch

Angus Bull   Courtesy of Department of Animal Science, Oklahoma State University

Hereford Bull   Courtesy of Department of Animal Science, Oklahoma State University

Brahman Bull   Courtesy of Department of Animal Science, Oklahoma State University

Chapter 3   Cattle Management Techniques: Beef Cow-Calf Pairs

Black Angus   Courtesy of Department of Animal Science, Oklahoma State University

Beefmaster-Lasater   Courtesy of Lasater Ranch

Salers   Courtesy of Department of Animal Science, Oklahoma State University

Charolais   Courtesy of American International Charolais Association

Limousin   Photo courtesy of The Limousin World Magazine. Used with permission.

Shorthorn   Courtesy of Department of Animal Science, Oklahoma State University

Simmental   Courtesy of American Simmental Association

Red Angus   Courtesy of Department of Animal Science, Oklahoma State University

Chapter 4 Dairy   Cattle Management Techniques: Dairy Cattle Breeds

Jersey Cow   Reprinted with permission from Hoard's Dairyman.

Holstein Cow   Reprinted with permission from Hoard's Dairyman.

Guernsey Cow   Reprinted with permission from Hoard's Dairyman.

Brown Swiss Cow   Reprinted with permission from Hoard's Dairyman.

Ayrshire Cow   Reprinted with permission from Hoard's Dairyman.

Red and White Cow   Courtesy of Red & White Dairy Cattle Association

Milking Shorthorn Cow   Courtesy of Department of Animal Science, Oklahoma State University

Milking Devon Cow   Courtesy of Department of Animal Science, Oklahoma State University

Chapter 5   Swine Management Techniques: Swine Breeds

Chester White Boar   Courtesy of Department of Animal Science, Oklahoma Slate University

Duroc Boar   Courtesy of National Swine Registry

Hampshire Boar   Courtesy of National Swine Registry

Landrace Boar   Courtesy of National Swine Registry

Poland China Boar   Courtesy of Department of Animal Science, Oklahoma State University

Spots Boar   Courtesy of Department of Animal Science, Oklahoma State University

Yorkshire Boar   Courtesy of National Swine Registry

Tamworth Boar   Courtesy of Department of Animal Science, Oklahoma State University

Chapter 6   Horse Management Techniques: Horse Breeds

American Quarter Horse   Reprinted with permission from the ©American Quarter Horse Association.

Appaloosa   Courtesy of Appaloosa Horse Club Inc.

Arabian Horse   Courtesy of Arabian Horse Association, photographed by Jeff Janson.

American Paint Horse   Courtesy of American Paint Horse Association .

Lipizzan   Courtesy of Department of Animal Science, Oklahoma State University

Morgan   Courtesy of Department of Animal Science, Oklahoma State University

Thoroughbreds Racing   Courtesy of Department of Animal Science, Oklahoma State University

Standard Bred Trotter   Courtesy of United States Trotting Association

Chapter 7   Sheep Management Techniques: Sheep Breeds

Suffolk   Courtesy of American Sheep Industry Association

Columbia   Courtesy of American Sheep Industry Association

Dorset   Courtesy of American Sheep Industry Association

Hampshire   Courtesy of American Sheep Industry Association

Rambouillet   Courtesy of American Sheep Industry Association

Polypay   Courtesy of American Sheep Industry Association

Montadale   Courtesy of American Sheep Industry Association

Dorper   Courtesy of American Sheep Industry Association

Chapter 8   Goat Management Techniques: Goat Breeds

Alpine   Courtesy of American Dairy Goat Association

La Mancha   Courtesy of American Dairy Goat Association

Nubian   Courtesy of American Dairy Goat Association

Saanen   Courtesy of American Dairy Goat Association

Oberhasli   Courtesy of American Dairy Goat Association

Toggenburg   Courtesy of American Dairy Goat Association

Angora   Courtesy of Texas Sheep & Goat Raisers' Association

Boer   Courtesy of Texas Sheep & Goat Raisers' Association

Chapter 9   Poultry Management Techniques: Chicken and Turkey Breeds

Barred Rock   Courtesy of WATT Publishing Company. Prints may be purchased at www.waltcountrystore.com.

Black Orpington   Courtesy of WATT Publishing Company. Prints may be purchased at www.wattcountrystore.com.

Buff Cochin   Courtesy of WATT Publishing Company. Prints may be purchased at www.wattcountrystore.com.

Rhode Island Red   Courtesy of WATT Publishing Company. Prints may be purchased at www.wattcountrystore.com.

White Wyandotte   Courtesy of WATT Publishing Company. Prints may be purchased at www.wattcountrystore.com.

Black Turkey   Courtesy of WATT Publishing Company. Prints may be purchased at www.wattcountrystore.com.

Broad-Breasted Bronze   Courtesy of WATT Publishing Company. Prints may be purchased at www.wattcountrystore.com.

White Holland Turkey   Courtesy of WATT Publishing Company. Prints may be purchased at www.wattcountrystore.com.

| BRIEF CONTENTS |

|CONTENTS|

Chapter 5   Swine Management Techniques

Chapter 6   Horse Management Techniques

Chapter 7   Sheep Management Techniques

Chapter 8 Goat Management Techniques

Chapter 9 Poultry Management Techniques

Chapter 10 Animal Health Management

Appendices

| PREFACE |

The first edition of *Handbook of Livestock Management* was designed to fill the need for a book that described in detail the skills and techniques needed by those who managed livestock. We, the authors and contributors to that first edition, thought it important that the teacher, student, and producer utilize sound and safe procedures in the management of farm animals, thereby maximizing animal well-being and productivity as well as enjoyment for the manager. The book was intended as a comprehensive text and reference for students of animal science, veterinary science and technology, and vocational agriculture, and as a handbook and reference for livestock producers and technicians.

Today, just as when we completed the first three editions, the need is the same—a text, a guidebook, that provides the details of useful, complete, accurate, and safe procedures for performing the absolutely necessary livestock management techniques. This book is intended to fill in the gaps between the classroom and the laboratory. The material is not theoretical; it is primarily hands-on. It describes not only what to do but how to do it.

In reality, the need for the *Handbook* is greater today than it was 25 years ago. There are fewer of us growing up on farms, there is a greater awareness of the need to provide for an animal's well-being, and there is less time to teach (and learn) hands-on manipulative skills because of the ever-increasing body of technical and cognitive skills that must be taught to (and learned by) our students in the classroom.

Just as in the first three editions, included in this edition of the *Handbook* are chapters on the management of beef cattle, dairy cattle, swine, horses, sheep, goats, and poultry, along with chapters on livestock restraint and herd health. The presentation of the techniques in each chapter follows the same format. A brief introduction to each technique is presented along with a listing of necessary equipment, a discussion of required restraint, a detailed step-by-step procedure with cautionary notes at appropriate danger points, a description of the normal recovery sequence, and a discussion of necessary postprocedural management. At every opportunity, the verbal directions are supported by illustrations, of which there are over 800.

Parts of the animal, with terms accepted by livestock professionals, are included in Appendixes A–M. Labeled drawings of the skeletal structures of the various animals are presented in the appendices. A useful glossary and a comprehensive, cross-listed index complete the educational package.

No specialized knowledge is required for an understanding of the text. Many of the techniques can be self-taught. The book is written and organized in such a manner that the student can learn by doing. When it is commonly recognized that there are several ways of performing a technique, the one presented is preferred by the author and recommended as best for both the livestock manager and the animal.

The material presented in *Handbook of Livestock Management* reflects the years of experience and livestock production backgrounds of the author and contributors. It is unique in its approach and in its completeness. The author hopes that this book will serve as a comprehensive, useful text and reference work for those engaged in teaching livestock management to young men and women and for those actively working in production agriculture careers.

Moscow, Idaho. RAB

ACKNOWLEDGMENTS

Just as in the first edition, there have been many people who have given generously of their counsel, energy, and talents to the construction of this fourth edition. Most certainly, I hope that I have thanked each official and unofficial contributor directly and thoughtfully, but I would be remiss to not also thank him or her publicly.

I continue to be thankful for and appreciative of the undergraduate classroom students, whose questions and concerns about relevancy and usefulness continually identified the route along which to proceed. I need to thank the livestock, dairy, poultry, and horse producers who contributed significantly by supplying requested information to me and by demanding from me answers to questions of concern to them.

I am particularly grateful to past chapter editors. Their reviews, verbal comments, and written materials, when it was necessary to add new techniques, have been and continue to be critical to the successful completion of successive editions of the *Handbook*. These people deserve special thanks and acknowledgment: Mark Boggess and Hobe Jones (swine), Ed Fiez and Norb Moeller (dairy cattle), Steve Maki and Jim Outhouse (sheep and goats), Pat Momont and Ron Lemenager (beef cattle and general cattle), Ed Steele and Jim Carson (poultry), and Brad Williams and Wayne Kirkham (animal health). Their materials, their professionalism, and their genuine knowledge of livestock management enabled the entire process. Because they did such a fine job of identifying the important "core" and nearly timeless management techniques to include in the first three editions, this revision has been far easier than it might have been. Particular thanks are due to Jonathan Barrett, Ohio Central School System, and Brian G. Bolt, Clemson University, who reviewed the manuscript for this edition.

Vern Mayrose, my coauthor for the first edition, and his spouse, Sandy, are good friends, and Vern is as fine a professional cohort as a person could wish for. When asked if he'd collaborate again, after the first edition, Vern responded in his typical direct style: something to the extent that he had retired, enjoyed a no-deadline type of lifestyle, was never going to meet another deadline . . . thanks, but no thanks! Vern and Sandy, it has not been the same without you. Thank you for all you have done!

Finally, I wish to thank my artist for the second, third, and fourth editions, Marge Battaglia. In addition to being appreciated and recognized as a talented artist, Marge should also receive a meritorious service/hazardous duty award for serving simultaneously as artist for the project and spouse of the author.

ABOUT THE AUTHOR

Richard A. Battaglia, Professor and Head of the Department of Animal and Veterinary Science, University of Idaho, Moscow

Dr. Battaglia was raised in a farm setting (mixed livestock and grain) near Belleville, Illinois. He graduated from Southern Illinois University, Carbondale, with a B.S. degree in Animal Industries in 1966. He received the M.S. degree in 1968 and Ph.D. in 1969, both from Virginia Polytechnic Institute and State University in Blacksburg, Virginia. He worked for 2 years as Animal Scientist (Reproductive Physiologist) at Oklahoma State University; 3 years as Animal Production Program and Horse Program Coordinator at the University of Minnesota Technical College for Agriculture in Waseca, Minnesota; 10 years as an Animal Scientist and Horse Extension Specialist at Purdue University in West Lafayette, Indiana; and 8 years at South Dakota State University, where he served as Program Leader for Agriculture, Director of Cooperative Extension Service, Dean of the College of Agriculture and Biological Sciences, and Development Officer for the University Foundation. In 1991, he joined the College of Agriculture at the University of Idaho as Professor and Head of the Department of Animal and Veterinary Science. He is a member of the American Society of Animal Science, American Registry of Professional Animal Scientists, American Dairy Science Association, Equine Nutrition and Physiology Society, and the National Association of Colleges and Teachers of Agriculture.

chapter ONE

Livestock Restraint Techniques

1.1 INTRODUCTION

Some degree of control over an animal's movement and activity is required for every technique that the livestock producer performs. Restraint can vary from the psychological control that a handler's voice may exert over his animals to the complete restriction of activity and total immobilization that chemical agents can provide. With large, potentially dangerous animals, a combination of psychological, physical, and chemical restraint is often employed.

Restraint practices became important thousands of years ago when man first domesticated animals for food and fiber and as beasts of burden. This domestication altered the natural lifestyle of these animals, and forced man to be responsible for the animal's needs. Managing these animals for human purposes necessitated control of the beasts; there is evidence of early restraint in crude fencelike enclosures. Not until many years later was there any concern about the appropriateness of the restraint for the task at hand.

Today, we realize that it is our responsibility, since we have total control over the animal's life, to be concerned for its welfare, its sensation of pain, and its psychological well-being as they are affected by our production systems and our management techniques. However, we should not overemphasize the sentiment of not causing the animal *any* discomfort and lose sight of the fact that animals are housed, maintained, and ministered to for the production of food, fiber, and pleasure. Certain management techniques must be performed to achieve those ends. Some of these techniques will cause a degree of pain.

Pain is a necessary phenomenon of nature that signals to the animal that something out of the ordinary is happening to its body. Without this sensation of pain, noxious factors could destroy an animal's body without its knowing it. The animal manager's responsibility is to perform his tasks of caring for the animals in the most currently appropriate manner, while inflicting the least amount of pain and causing the least amount of psychological upset (fright). When the animal is restrained and the manager is performing necessary management techniques, there is no escape from the pain for the animal, no relief from the fright. Anyone failing to realize this or failing to do everything within his power to alleviate the pain and fright should not be allowed to manipulate the lives of animals.

When it is your responsibility to select the restraint method to be used in performing a given management technique, you must ask yourself the following questions: (1) Will the restraint method minimize the danger to the handler? (2) Will the restraint method minimize the danger to the animal? (3) Will the method cause unnecessary pain or fright? (4) Will the restraint method allow the management technique to be completed as necessary? If any one of the questions is answered negatively, an alternative method must be chosen.

Whatever the method of restraint selected, it is the manager's responsibility to become proficient

1

at it. Since the majority of people are not farm-raised today, proficiency in restraining animals is not passed from one generation of stockmen to the next, as it was in times past. Diligent study of the available restraint methods, a thorough understanding of the animal's anatomy, physiology, and psychology, demonstrations from correctly experienced livestock managers, and then practice on your own are the only ways to acquire the expertise necessary to perform a restraint method and a management technique safely, correctly, quickly, and as painlessly as possible.

1.2 TYPES OF RESTRAINT

There are five categories of restraint: (1) psychological, (2) sensory diminishment, (3) use of confining chutes, alleys, and barriers, (4) use of tools and physical force, and (5) chemical sedation or immobilization. Each of these has its advantages and disadvantages, depending upon the species of livestock involved and the management technique to be performed.

Psychological restraint depends upon the manager having a thorough working knowledge of the behavior patterns of the species to be restrained. With this knowledge, the manager can take steps to either make use of or offset the animal's natural behavioral tendencies. For example, when working a group of sheep, the experienced shepherd appreciates the futility of blindly trying to rush them through an opening and, instead, calmly "works" them until one of the flock starts through. At that point, the others will follow with a minimum of coercion. Another example, this time using a combination of psychological restraint and a tool, would be the proper use of a hog snare to restrain a pig. It is the natural tendency of the pig to pull backward against the pull of the snout snare. The knowledgeable manager would allow the pig to pull backward until it has positioned itself into a corner, thus maneuvering itself into being restrained from both ends.

Sometimes the human voice can actually be used as a restraint tool, depending on the previous conditioning of the animals. Authority (or the lack of it), confidence (or fear), and a coincident soothing (or exciting) effect can all be transmitted in a voice. Animals can readily perceive this and they do respond to it.

More than likely, the animal is actually responding to the combination of voice and mannerisms of the manager. The manager must move confidently and quickly, but not with the false bravado of shouting, frantic arm waving, and jumping about. Self-confidence in one's ability to get the job done comes through naturally, and if it's there, the animal will respond accordingly. The only things that

you can do to develop this self-confidence are to study the animal's behavior, anatomy, and physiology, watch the techniques being performed properly by others, practice on your own, and then believe in your ability.

Sensory diminishment as a way of restraining animals usually involves blindfolding. Under certain conditions, such as when animals must be maintained in ultranoisy surroundings that continually excite them, plugging the ears with cotton often has a quieting effect.

Blindfolding sometimes works, but be cautioned and aware that just because a horse is blindfolded does not mean that it will blindly enter a trailer that it has resisted for hours! Neither will the horse suddenly allow you to clip its ears with a buzzing electric clipper just because you blindfold it. (Inserting large cotton balls into the ear, however, may just be the extra edge you need to make the horse submit to clipping.) Blindfolding is probably best utilized as an adjunct restraint to assist you in quieting and controlling domestic animals that are resisting the primary restraint in too violent a manner.

The use of *confining alleys, chutes,* and *barriers* is one of the most common ways to restrain domestic livestock, especially cattle and sheep. However, before the manager makes the decision to use this method as his primary restraint, he must once again call upon his knowledge of animal behavior, anatomy, and physiology, and upon his own common sense. For example: 40-pound lambs do not need a squeeze chute for vaccinating; hogs are more easily worked with a snare for blood testing than with a headgate; horses would "blow up" if squeezed in a chute or if their heads were "taken away" from them in a headgate, but cattle should be placed in a headgate with a nose bar for dehorning. There is no substitute for knowledge, experience, and common sense when making these decisions.

Alleys and chutes speak for themselves about how they are used to assist the manager in handling animals. The use of barriers as an aid to performing a management technique should also be explored. Not all livestock operations are of a large enough scale to warrant their having a full set of (expensive) handling equipment. Barriers can be temporary (perhaps even bales of hay), inexpensive, serve multiple purposes (such as gate panels), or they may already exist (a stall, pen door, or side of building). Once again, if you are proficient at the technique to be performed, understand the behavior of the animal in question, and know the anatomy and physiology involved, you may be able to make do with what you have, instead of purchasing new equipment for a handful of animals. Perhaps the most obvious examples of barrier use are (1) placing bales of straw between you and the animal to prevent being kicked and (2) using a gate panel to squeeze an animal against a fence or shed for control.

The use of *tools* and *physical force* to restrain animals is as old as man and the domestication of animals. The tools are actually used as magnifiers and extenders of the physical force that man can exert on an animal. That such tools are necessary should be obvious because the size, strength, and agility of the animal are oftentimes 10 times that of the man.

The types of tools, all handheld, that are commonly used to control animals are ropes, snares, leg hooks, sheep crooks, nose tongs, nose rings, bull staffs, sorting poles, canes, electric (shock) prods, whips, and slappers. Each of these can assist the manager to safely, efficiently, and responsibly control and restrain the animal so that the management technique to be performed can be completed properly. It is also possible to misuse any one of these tools. Things move quickly and can become hectic when dealing with livestock. Most of the time, not every detail goes as you planned it, and it is easy to lose your temper. Before any person can effectively control and manage livestock, he must be capable of controlling himself!

Chemical restraint is the strongest method of controlling the activity of an animal. With chemicals, it is possible to totally immobilize an animal and have it lie recumbent before you, as is necessary for surgery. It is also possible to take the edge off an animal's unruly, or at least uncooperative, behavior by selecting a different drug, altering a dosage level, or using an alternate route of administration.

This sounds like a tailor-made method of control—no hassle, no sweat, no pain to either the animal or the manager. Unfortunately, this is not the case. There are many disadvantages to using chemical restraint, including the following. The use of chemicals implies a knowledge of anatomy, physiology, and pharmacology (drug effects and drug interaction effects) usually beyond that of the livestock producer. Any drug that can affect the nervous or muscular system to the extent of total immobilization is dangerous, and it is easy to overdose, causing death. Also, most animals will need to be restrained in order to give them the injection in the first place; so it hardly makes sense to hassle the animal with a "shot" when all you want to do is pierce an ear for identification. The cost, the time spent in waiting for the drug to take effect, and the risk involved can be out of proportion to the potential benefits!

Perhaps the greatest hazard of the promiscuous use of chemicals for restraint is the eventual deterioration of the livestock manager's skills. It will become too easy to attempt to solve every restraint (or training or management technique) problem with

a needle and syringe. Chemicals for the most part should remain a valuable tool, a type of adjunct restraint, for the manager—not a panacea!

1.3 WORKING WITH ROPE

Rope Types and Construction

Rope is one of the tools most often used by the livestock producer. No matter how mechanized the operation, how sophisticated the management plan, there are times in the management cycle of all species that a knowledge of rope, knots, and hitches is indispensable.

Traditionally, rope was constructed of natural plant fibers. Cotton, manila (or hemp), and sisal were the fibers of choice because of the manageability, strength, and low cost that each respectively brought to the rope-manufacturing process. Manmade, chemical-based fibers came upon the scene near the end of World War II. They brought with them ease of manufacture and greatly increased rope strength due to fibers, filaments, and yarns that ran the full length of the rope, as compared to fibers that could be only as long as the parent plant was tall.

Man-made or synthetic materials used to manufacture rope include nylon, polyester, polypropylene, polyethylene, and the newest and strongest of all, aramid (Kevlar) fibers. For the most part, manmade fibers have the advantage of being stronger than plant fibers, more resistant to rot and mildew, easier and cheaper to manufacture, resistant to all sorts of chemicals and sunlight that wreak havoc with plant-fiber rope, and can be made in a variety of brilliant colors. The latter attribute may not seem too important until show day arrives. Then, the manager may find that the color-coordinated lead rope, exactly matching the farm colors, is the necessary psychological edge that is needed for his son or daughter to feel confident enough to compete well in the show ring or arena.

One final consideration is the twist, or lack of twist, in the rope. There are plaited ropes and braided ropes on the market that may have a purpose on the farm or ranch. They do have the advantage of being quite strong for their diameter, and they are often smooth and easy to the touch. However, therein lies their inherent disadvantage for knots, hitches, and splices—they are so smooth that they hold knots very poorly. In the final analysis, the livestock manager should select the rope type that most closely meets the intended purpose. See Table 1.1 for further recommendations.

For the most part, the livestock manager will be well served by selecting and using a three-strand

TABLE 1.1 Types of Rope Available for Restraining Livestock

Type of Fiber	Breaking Strength (pounds)					Advantages	Disadvantages	Suggested Uses
	⅜"	½"	⅝"	¾"	1"			
Cotton	890	1450	2150	3100	5100	Soft, flexible Least likely to cause rope burn Intermediate in cost	Least strong Low abrasion resistance Will rot and deteriorate	Tying of animal limbs Neck ropes Hobbles Lead ropes, if ⅝"or larger
Sisal	1080	2120	3250	4320	7200	Same as manila, except that sisal is 75–80% the strength of manila		
Manila	1350	2650	4400	5400	9000	Good strength for natural fiber rope Has good "grass" rope "feel"	Subject to rotting Harsh on hands Likely to cause rope burn	Some use as a lariate—not suggested for tying animal limbs (legs) Lead ropes, if ½"or larger
Polypropylene	2650	4200	5700	8200	14000	Very strong, second only to nylon and dacron Will not rot Resistant to barnyard acids and bases	Rope burn is likely Flame or heat will cause melting	Good for lead ropes Excellent for slinging and total restraint
Nylon	4000	7100	10500	14200	24600	Highest strength of any rope Will not rot from water or mildew	Will stretch Very likely to cause rope burn Flame or heat will cause melting	Strongest lead rope available Excellent for slinging and total restraint

braided rope for the majority of activities. The discussion and illustrations that follow are based upon the manager using three-strand twisted rope of either natural or man-made fibers.

Three-strand rope is constructed of plant or man-made *fibers* that are twisted together to form a type of *yarn*. The yarn is then twisted (in our case, three yarns) into a *strand*. *Rope* is formed by twisting three strands together. The tightness of the rope strands and the overall flexibility of the rope depend upon the amount of twist imparted to the fibers, yarns, and strands at each step of the operation. This twisting and reverse twisting is called "laying" the rope. A hard-lay rope is a stiff, hard-feeling rope that keeps a loop open quite well for throwing but is a poor choice for knot tying. On the other hand, a soft-lay rope does not hold a loop well but is convenient for knot tying because of its flexibility. A medium-lay rope, depending upon the manufacturer, is somewhere between hard and soft and is intermediate in its usefulness to you as a throwing and tying rope. Not all rope is of the three-strand, twisted variety that we use for lariats, hay ropes, most adjustable rope halters, and most lead ropes. Machines can also be used to braid fibers into a variety of shapes, sizes, and types of rope. Perhaps

the most well known of the braided ropes are clothesline and the towrope used by water-skiers. The three-stranded rope is most suitable for livestock restraint.

The type of rope chosen by the livestock manager will depend upon the use to which the rope will be put and the amount of strength required. If the livestock producer's primary need is to catch and hold calves for range management techniques, an entirely different type of rope should be selected than the one used to cast and tie a horse. Consult Table 1.1 before making these decisions.

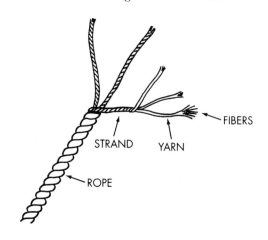

Preparation of Rope

The following is a brief glossary of terms. These terms must be learned well, because they are basic to all rope work.

1. The *standing part* of the rope is the portion that is not being worked with, except perhaps to wrap the running end or working end of the rope around, as in forming a knot, hitch, or loop.

2. The *working end* (running end) or *working part* (running part) of the rope is the portion that is being worked with or being used to form a knot, hitch, or loop.

3. A *bight* or bend is a turn of the rope that does not cross itself.

4. A *loop* is a turn of the rope that does cross itself. If the working end of a bight is crossed over its standing part, a loop is formed.

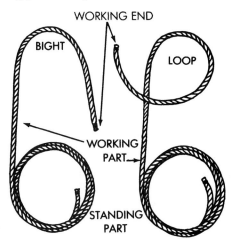

When a rope is purchased, it is sized to the requested length and then cut from the parent coil. If this length of rope is then taken back to the livestock enterprise and used with no further preparation, it will cease to be of value in short order because both raw or cut ends will begin to unravel. To prevent this, the cut ends must be whipped, crowned, ferruled, dipped, or burned (melted), depending upon the type of rope. An overhand or figure-eight knot can also be used to prevent unraveling.

Whipping

Step-by-Step Procedure

1. Place the rope to be whipped in front of you, the standing part on your left, with the working end directed to your right.

2. Form a bight in the working end of the whipping cord (30- to 60-lb-test braided nylon baitcasting line works well for ⅜" to ⅝" rope), and place the bight lengthwise onto the first 2" of the rope. The bend of the bight should be directed toward the working end of the rope.

3. Begin wrapping the whipping cord around the rope in a clockwise rotation. Start at the open end of the bight and move outward toward the working end of the rope. As you begin to wrap the whipping cord around the rope (and around its own bight), be certain to leave a 1" or longer portion of the whipping cord protruding from below the wrapping.

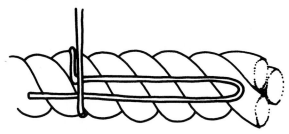

4. Continue whipping up the rope, being certain to wrap the whipping cord about the rope tightly and place each successive wrap snugly against the preceding one.

5. Stop whipping when you are approximately ½ inch from the end of the rope. At this point, the bight of the whipping cord should be located and opened to accept the working end of the whipping cord as it completes its last wrap of the rope.

6. Insert the working end of the whipping cord into the whipping cord bight and pull the working end until the last whipping wrap is as snug as the preceding ones.

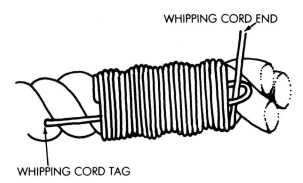

7. Hold the whipping cord end against the end of the rope with your right thumb and index finger. With your left thumb and index finger, grasp the tag of whipping cord that is protruding from the other end of the whipping on the rope and begin to pull it toward the standing end of your rope.

8. As you pull on the tag of the whipping cord, the bight at the other end of the whipping will begin to be drawn under the whipping wraps. As this occurs, the working end of the whipping cord will also be drawn down and under the wraps. Maintain thumb and index finger pressure on the working end of the whipping cord so that when it is in position, it will still be tight.

9. Continue pulling on the end of the whipping cord until the interlocking bights are near the center of the whipped area. You can feel for the bulge to be certain.

10. Closely snip off the ends of whipping cord from both ends of the whipped area.

Temporary Whipping

Masking, friction, adhesive, plastic electrician's, vinyl, and duct tape can all be used as temporary whipping. On seldom-used rope, they perform satisfactorily; however, if the rope will see heavy use, a permanent whip should be selected. If a type of tape is used, draw it down very tightly and wrap it around the rope three or four times. In general, tape does not adhere well to synthetic-fiber ropes.

Crowning (Crown Knot and Backsplice)

The crown knot has one feature—it produces a bulge that is double the diameter of the parent rope—that is either an advantage or disadvantage, depending upon the purposes for which the rope will be used. It does provide a convenient handhold and an alert that the end of the rope has arrived, but it is a nuisance and perhaps unusable if the rope must be threaded through a tightly sized pulley.

Step-by-Step Procedure

1. Place the working end of the rope to be crowned into your left hand and unwind about 4 inches of it. Before forming the knot and backsplicing, the end of each strand of the rope must be "finished" to prevent its unraveling during the splicing. With polypropylene, nylon, and dacron ropes, this is most easily done by using heat or flame to fuse the ends together. Cotton, manila, or sisal rope should be wrapped with masking tape.

2. Place your thumb on the front of the rope and your fingers on the rear. Your thumb and index finger should be pinching the rope strands and preventing further unraveling.

3. Arrange the strands so that two of them come across the top of the rope from a lower left to an upper right direction. The third strand appears to come from behind the front two, in a lower right to upper left direction. This arrangement is essential if the crown is to be properly constructed.

4. With your right hand, take the uppermost of the front strands (strand 1) and bend it over to the

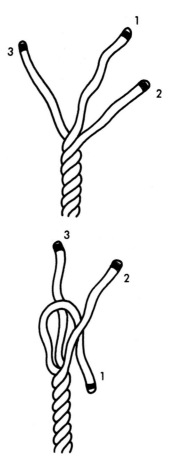

right forming a bight. Be certain that the bight goes behind strand 2, the second of the rope strands. Secure the end of strand 1 between the index and middle fingers of your left hand.

5. With your right hand, take strand 2, the remaining strand of the two that originally came across the "top" of the rope, and bend it around the bight in strand 1 that you are holding in your left hand. The wrap must be taken around the working end of the bight.

6. After making this wrap, place the end of strand 2 between the standing end of the bight in strand 1 and strand 3. Strand 3 is the only strand left untouched at this point.

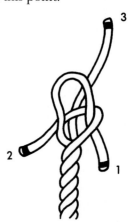

7. Secure the end of strand 2 between the index finger and thumb of your left hand and the rope.

8. With your right hand, take the end of strand 3 and place it under the bight in strand 1 and over all parts of strand 2. Study this arrangement for a moment and you will notice that each strand locks and is in turn locked by another.

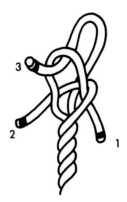

9. Release your grip on the strand ends. Start with any of the three strands and tug on it to begin tightening the crown knot. Do not attempt to pull one strand totally tight before beginning another. Take each up alternately, a little at a time, until the crown is tight.

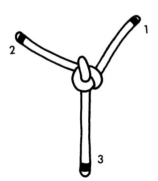

At this point, the crown knot is complete and the backsplicing must be performed to finish the process of "crowning" to prevent unraveling of the rope.

10. Hold the end of the rope with the crown knot in your left hand. Select any of the strands sticking out from the crown and grasp it with your right hand. Notice that it passes under a strand of the crown and then lies on or passes over a strand of the end of the rope being crowned. Study the whole crown and each strand so that you are aware that this "under–over" arrangement is correct for each strand.

11. Select a strand, call it strand 1, sticking out from under the crown knot and begin the backsplicing at that point by placing your right thumb partially under it and at the same time upon the strand that it is passing over or lying upon. Grip the rest of the crown knot with the tips of your index and middle fingers.

12. With your left index finger and thumb grasp the strand immediately below the one your right thumb is holding. This is the strand directly below the one that the working part of strand 1 passes over.

13. With your right hand, twist the crown to the right (clockwise) while twisting the standing end of the rope to the left (counterclockwise) with your left hand. This will open the rope and enable you to isolate the second strand below the point where strand 1 exits from under the crown.

14. Keep this strand isolated, and with your right hand place the end of strand 1 under it and pull it through until it is pulling against the crown itself.

15. Retighten the crown by holding it in your left hand and pulling each crowned strand downward with a clockwise twist.

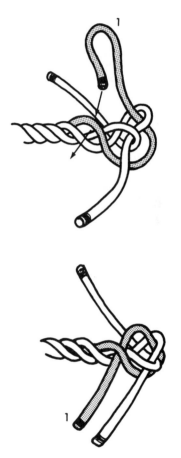

16. Move to the next strand to be spliced back into the parent rope by twisting the rope approximately one-third turn in either direction until the next strand coming out from under the crown knot is reached. This is strand 2.

17. As with the first over and under, place your right thumb under strand 2 and upon the strand it is passing over. Grip the rest of the crown knot with the tips of your index and middle fingers.

18. Study the crowning at this point and keep in mind the over-and-under principle. The crowned strand you are now working with, strand 2, is lying upon the strand it will pass over. You must take the strand of the parent rope next in line below this and isolate it between the thumb and index finger of your left hand. This is accomplished by twisting the crown to the right and the standing end of the rope to the left.

19. Keep this strand isolated, and with your right hand place the end of strand 2 under it and pull it through until it is pulling against the crown itself.

20. Retighten the crown as before by pulling each crowned strand downward with a clockwise twist.

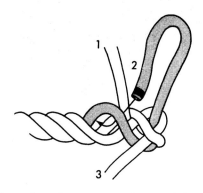

21. There is only one strand, strand 3, remaining, and it too must go through the over-and-under process. As before, there should be no difficulty in identifying the strand to be passed over, because strand 3 is lying upon it.

22. Grasp the knot in your right hand exactly as you did before for the first and second strands. Now, before you begin twisting, identify the strand of the standing end of the rope to pass the crowned strand under. Once again, it is the strand immediately below the strand being passed over. Since this is your last strand, the crowning is becoming "crowded" and it is easy to make a mistake. Keep in mind that *only one strand is passed over at a time,* that *only one strand can pass below another,* and that *only the strands of the standing end itself can be the ones passed over and gone under.* With this in mind, grasp the standing end of the rope in your left hand, twist the strands open as before, and insert the last crowned strand.

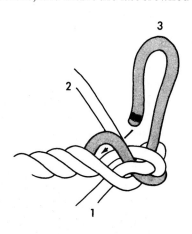

23. Retighten the crowning as before by pulling each strand downward with a clockwise twist.

24. To continue the crowning or backsplicing, repeat the preceding steps for as many rounds of splice as you desire. Except for your wishes, there is no need to repeat the process for more than three total rounds (over–under sequences).

25. Finish the crowning or backsplicing with a final retightening and a rolling of the entire crown between the palms of your hands. Cut off the ends of the crowned strands approximately ¼" from the last strand they went under. Taper the ends when cutting so that the crown is less rough on your hands. The crown will not unravel and is now a permanent feature of the rope.

Ferrules to Prevent Unraveling

Sometimes called "whipping with a ferrule," this is the easiest and most rapid method available to prevent unraveling. It consists of placing an appropriately sized metal split ring or band over the working end of the rope and then hammering or pinching it tightly about the rope. The best types of bands are split at an angle across their width and are about ³⁄₈" to ½" wide. Hog rings, applied with the pliers-like hog ringer, will also serve as the whipping ferrule.

Care must be taken with any of the metal whippings to be certain that no sharp edges are present. Gloves, clothing, or flesh can easily be torn as this sharp edge is pulled through your hand.

Dipping

Dipping the ends of all types of rope, except polypropylene, works quite well to prevent unraveling. Usually only an inch or two is dipped, although no harm is done if an additional portion is treated. Some dipping agents weaken the rope, so take care not to soak a part of the rope that will be used to form a knot or otherwise carry the stress of a load.

The reason that dipping is unsuitable for polypropylene ropes is because the plasticlike fibers of polypropylene strands do not absorb common dipping agents, nor do they provide a porous surface to which the dips can fuse.

Common dipping agents for nylon, cotton, manila, and other fiber ropes are enamel, lacquer, varnish, shellac, liquid vinyl, cold vulcanizing compounds, and plastic pipe solvent. Each of these is flammable, highly volatile, and gives off potentially hazardous fumes until dry. Use them cautiously.

There are three disadvantages to this type of unraveling prevention: (1) when the dip has dried, it produces a very hard surface that some managers object to; (2) after dipping, the rope is unavailable for use for up to 24 hours; (3) the end of the rope is discolored (although this may be objectionable to some, others consider it a plus because it identifies ownership).

Heating

Heating the ends of nylon, dacron, and polypropylene rope can be a satisfactory way to prevent the unraveling of the strands. It is of no value when dealing with natural-fiber ropes such as cotton and manila.

There are two methods of fusing (melting together) the ends of rope with heat or flame. One involves an open flame as the source of heat; the other utilizes a hot iron or blade.

Open-flame fusing (matches, lighter, propane torch) is satisfactory if the flame is held far enough below the rope end (start with an inch or two) so that the rope heats uniformly instead of only the outer fibers of each strand being heated. When this is done properly, the fibers melt and flow together until the flame is withdrawn, at which point they reharden in their new fused-together shape. The drawback to this is that the new shape may be a large glob or ragged, sharp-edged mass at the end. Do not attempt to shape this mass with your fingers as it is very hot, clings to your flesh, and imparts a severe burn!

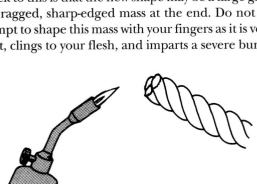

Hot-iron fusing makes use of the same chemistry and physics of melting and reshaping man-made fibers but avoids the disadvantage of forming a glob at the rope's end. To fuse with a hot iron or blade, select a discarded knife blade, hot-iron docking tool, or a commercially made electric docking iron and heat it to a red-hot condition. Taking care to avoid burning yourself, draw the rope across the hot edge. Cutting to size and fusing of the fibers occur simultaneously. If the cut edge is too sharp for your wishes, it can be rounded by rotating it while holding it at an angle against the hot iron.

Stopper Knots

Stopper knots can be used to prevent the end of a rope from unraveling or pulling through an eye, hole, or loop of another knot (as in a honda knot). There are three important, often-used stopper knots: the overhand knot, figure-eight knot, and blood knot (also known as the multiple overhand knot).

Overhand Knot

1. Position the rope with the standing part to your left and the working end running to your right.

2. Position the working part of the rope, approximately 12" from its end, in your left hand. Grasp the running end of the rope and form a loop with it by placing it behind the standing part of the rope.

3. Wrap the running end of the rope around the standing part and run it through the loop from front to back.

4. Pull the running end with your left hand and the standing part with your right hand to form and firm up the knot.

5. Position the overhand knot as a fray stopper by "sliding" it, before you pull it tight, to wherever you need to have it on the parent rope.

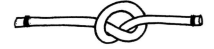

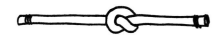

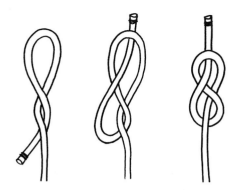

Figure-Eight Knot

1. Form a bight near the end of the rope. Allow the running end to hang to the left, with the standing end to the right.

2. Wrap the running end around the standing part, starting around the front first. Bring the running end out from behind and insert it into the loop held in your left hand from front to rear.

3. Grasp the running end with your right hand and pull while pulling on the standing part with your left hand.

4. Slide, shape, and position the knot as necessary.

Blood Knot

1. Start by forming a simple overhand knot as detailed below.

2. Make three or four additional turns, through the loop, with the working end of the rope. Keep them snugly wrapped around the standing end of the rope, so that the knot will draw up properly in step 3.

3. Draw the knot up by pulling on both ends of the rope at the same time. There are two ways to do this—both work. Use the one that works for you most often. One method is to pull gently on both ends while twisting the ends in opposite directions. The other method is to "jerk" the ends sharply apart. The quick jerk method works well with "slick" ropes such as nylon braid.

4. Form the knot by pushing the coils together. When the knot is tied correctly, the appearance is that of a single line of rope running through a series of wraps or coils.

Storage and Care of Rope

All types of rope must be kept dry, free of chemicals and chemical vapors, and untangled between uses if they are to maintain their strength and be instantly available for your use.

Freedom from moisture and chemical agents will assure you of continued strength if you have not nicked the fibers with a sharp instrument or otherwise damaged the strands. Freedom from moisture implies proper drying after use, before coiling and storage, and shelter from moisture during storage.

Rope, contrary to what appears certain at times, does not tangle itself up during storage. If you cannot peel it apart freely after storage, you did not coil or "hank" it properly prior to storage. Rope should always be coiled in the direction of its lay; i.e., if the strands are right-handed (running from lower left to upper right) the coils must be built in a right-handed or clockwise rotation. The reverse is true for the rare left-handed rope. Lariats and any other hard-twist rope are always coiled, never hanked. The size of the coil depends on the diameter of the rope. If you wish, the coils can be held together by wrapping the coiled rope at two or three places with a strand of twine or piece of tape.

Hanking a Rope

Large-diameter ropes—especially the medium-to-soft-lay, ⅝"-and-larger-diameter, natural-fiber ropes—are hanked for storage.

1. To hank a rope, coil it as you would any other rope, forming a coil that is approximately 24" from top to bottom.

2. When you are finished coiling, take the working end of the rope and wrap it entirely around the coils at least six or seven times, starting near the center and working upward.

3. Secure this working end of the wrap in one of three ways:

   a. As the final wrap comes around from back to front, insert the working end into the top bights of the rope from front to back, wrap it around the left arm of the bight, bring it around to the front, cross both arms of the bights, carry the working end around to the

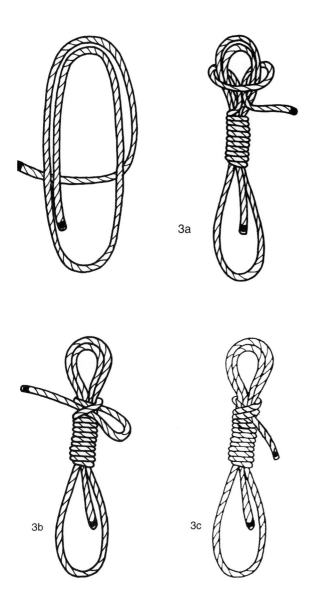

3a

3b   3c

back, insert the working end into the bight from back to front, and pass it under the crossing part of the working end of the rope. Pull the working end to snug the wraps.

b. As the final wrap comes around from back to front, form a bight with the working end of the rope and insert the bight under the top wrap. Pull the bight to snug the wraps.

c. After the final wrap has crossed the front, insert the working end of the rope into the top bights formed by the hanked rope, from back to front. Continue the working end through the bights and insert it under the last wrap or two. Pull the working end of the rope to snug the wraps.

The total care and protection of rope involve more than coiling it in some manner and then hanging it over a stob.

A rope to be used in animal restraint is of value to you only if it does not break in the midst of the

procedure. The following are *guidelines for the care and preservation of rope.*

1. When a rope is new and kinky (has a strong coil memory), it should be dragged for several minutes over level, nonrocky ground before use. This will help remove the kinkiness and improve the overall performance of the rope.

2. When a rope is coiled after use, it should always be coiled in the direction of its lay; i.e., a right-handed rope should be coiled to the right, or clockwise, and vice versa for the left-handed rope.

3. All rope should be protected from oils, paints, and other man-made chemicals. Urine and manure should be rinsed from the rope with clear water. Do not use soaps or detergents.

4. Do not coil and store any rope before it is thoroughly dried. The mildew fungus seriously weakens rope. Dry the rope by suspending it above the ground.

5. Look for worn or frayed spots on the outside of the rope strands and periodically inspect the insides of the rope by untwisting it (twist it against its lay) and exposing the inner strands. They must be new-looking, bright, and unspotted for the rope strength to be normal.

6. As you use the rope, prevent kinks, which can cause permanent damage (weakening) by overtaking the fibers at the point of the bend. A kink is much like a poorly constructed knot—it can reduce rope strength by as much as 55%.

7. To hang the hanked rope, use a 2" × 4" stob or other broad support to prevent a sharp kink or crease in the rope. The temptation to hang it over a nail should be avoided as this nearly always results in a sharp crease being placed in the coils.

8. Keep the rope clean. When rope becomes dirty, some of the dirt particles become embedded into the fibers and act as abrasives. Over a surprisingly short time, these abrasives can weaken a rope. When a rope does become dirty, wash it with water and drape it in a cool, dry place until it is thoroughly dried.

Knots, Hitches, and Splices

Knots are used to join ropes together, to attach ropes to a post or rail, or to attach ropes to an animal. *Hitches* are used to attach a rope to a post or rail. They are not truly knots, in that the only thing securing the rope to the post is the pressure of one rope coil wrapping upon the others. *Splices* are used to permanently join ropes to one another. They differ from knots in that, in a splice, the individual strands from each rope are interwoven with the individual strands from the other.

All knots, hitches, and splices weaken the ropes with which they are formed. This is because they

TABLE 1.2 Strengths of Knots, Hitches, and Splices

Type of Knot, Hitch, or Splice	% Efficiency*
Fresh, dry, undamaged, unknotted rope	100
Short splice	80
Double bowline knot	70
Half hitch, timber hitch	65
Bowline knot, slipknot, clove hitch, quick-release knot	60
Square knot, sheet bend knot	50
Overhand knot	45

*This is the percentage of remaining strength, compared to new unknotted ropes, after the knot, hitch, or splice has been formed.

form a bend or "Nip" in the fibers that distributes the stress on the fibers unequally. The sharper the bend or Nip on the fibers, the weaker the knot, hitch, or splice formed by it.

Table 1.2 illustrates the efficiency of several of the common knots used in animal restraint.

Several of these knots, hitches, and splices will be detailed and illustrated. At first, they may seem complicated and their construction clumsy. Just as with any other skill, practice is necessary and will make the formation of the knots second nature. It is important that you practice the knots enough to arrive at this "second-nature" status. When you are in the midst of tying a full-size, frightened farm animal, you cannot stop to think about knot formation.

Overhand Knot

The overhand knot is the simplest of all knots to tie, but is the least useful of the common knots when used by itself. It is the first step in the formation of more complex and more useful knots. It serves very nicely as a temporary "whipping" to prevent unraveling of a rope end. It also provides a convenient handhold at rope's end (or at intervals for climbing).

Step-by-Step Procedure

1. Position the rope with the standing part to your left and the working end running to your right.

2. Position the working part of the rope, approximately 12" from its end, in your left hand. Grasp the running end of the rope and form a loop with it by placing it behind the standing part of the rope.

3. Wrap the running end of the rope around the standing part and run it through the loop from front to back.

4. Pull the running end with your left hand and the standing part with your right to form the knot.

5. As your knot-tying skill and confidence increase, you will realize that this knot can be tied from either the right or left side and from either the front or back.

Square Knot (or Reef Knot)

In essence, the square knot is but two overhand knots—one tied on top of the other. Tied correctly, it is an excellent knot for joining two pieces of equal or nearly equal size rope or for tying the ends of a single rope together to form a loop. A variation of the square knot—the surgeon's knot (or suture knot)—is the basic knot of surgery. In animal restraint, the major use of square knots or surgeon's knots is to tie or secure the gates of cages.

Step-by-Step Procedure

1. Grasp the two ends of rope to be fastened together, one in each hand, about 4" from their running ends.

2. Lay the end of the right-hand rope across the end of the left-hand rope near the point where you are holding them.

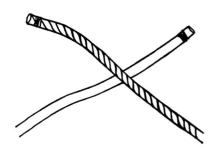

3. At the point where the two ropes cross, hold them in position between the thumb and index finger of your right hand.

4. Using the fingers of your left hand, wrap the end of the original right-hand rope over, around, and under the left-hand rope.

5. Now take the new left-hand end and cross it over, around, and under the new right-hand end.

6. Grasp an end in each hand and pull to tighten and secure. The end result should look and act like two interlocking bights. If the knot does not look or act correctly, it is very likely that you have tied a granny knot. The cause of this is a mix-up on which strand was used to start the second overhand knot.

7. A true square knot can be built every time if you remember the sequence, "right over left—left over right" or the exact reverse, "left over right—right over left." If the left–right switch is not made, the granny knot will result.

8. The granny knot is undesirable because it will slip under tension. The square knot or its variations, such as the surgeon's knot and reefer's knot, will not slip regardless of the tension.

9. The square knot is best suited to join together the two ends of a single rope. It is especially appropriate to use the square knot to tie a cloth or gauze bandage around the limb of an injured animal. If the square knot is used to join the ends of two ropes together, it can become unsafe. If it is used to fasten or join the ends of two different ropes, it can "upset" or untie if either of the working or protruding ends of the ropes is pulled sharply.

Thief Knot

There is an interesting knot, with a most intriguing bit of legend attached to it, that closely resembles the reef knot in name, appearance, and utility. Its name is the *thief knot*. Legend has it that sailors, leaving their quarters for a watch tour, used to tie their seabags shut with a thief knot. A burglar, often in a hurry and always distracted with the task at hand, would, following a quick glance, open what appeared to be a reef knot securing the about-to-be-purloined contents of the seabag. When the dishonest soul would retie the bag with the reef knot, the sailors could immediately tell that their bags had been pilfered by a thief . . . hence the name of the knot. While similar in all other aspects, the thief knot ties entirely differently from the reef knot.

Step-by-Step Procedure

1. Form a bight near the working end of one rope. Secure this bight, near its lower end, in one of your hands.

2. Take note of on which side of the bight the running end of the rope is positioned. Insert the running end of the second rope into the loop of the first bight . . . from the bottom to the top.

3. Wrap this running end around the lower end of the first bight. Be certain to start the wrap toward the side of the bight formed by the running end of the first rope. Continue around the bottom of the first bight, go around the back, come back out to the top, and insert the running end back into the first bight from top to bottom.

4. Pull on both ropes to firm up the knot.

5. Note that the thief knot and the reef knot are nearly identical in appearance, except for the fact that in the reef knot the tag ends of the bights are on the same side of the knot, while in the thief knot the tag ends are on opposite sides.

6. For the purposes of illustration, we have used two ropes to form the knot. The thief knot, like the reef knot, can also be used to secure the ends of a single rope . . . as around a seabag!

REEF KNOT

THIEF KNOT

Surgeon's Knot

The surgeon's knot is a variation of the square knot, and as such it shares the ease of tying and nonslip characteristics of the parent knot. It has the added advantage of having the first part of the knot hold while the second is being tied. Imagine this advantage while trying to tie shut a gate or door in a windstorm, securing an overstuffed box, or for its intended purpose of suturing.

Step-by-Step Procedure

1. Grasp the two ends of rope to be fastened together, one in each hand, about 6" from their running ends.

2. Lay the end of the right-hand rope across the end of the left-hand rope near the point where you are holding them.

3. At the point where the two ropes cross, hold them in position between the thumb and index finger of your right hand.

4. Using the fingers of your left hand, form the first of the overhand knots as you did for the square knot, going original right over original left.

5. Now, simply make one more wrap with the original right end. The end result of this step is to form a twice-wrapped overhand knot instead of a singly wrapped knot.

6. Now take the new left-hand end and wrap it over, around, and under the new right-hand end.

7. Grasp an end in each hand and pull to tighten and secure.

8. Like the square knot, the sequence can be reversed. The surgeon's knot can be tied correctly by double-wrapping right over left, left over right, or double-wrapping left over right, right over left.

9. There is another subtle but useful variation that can be used in tying this knot: a second wrap can be taken in the top part of the knot, similar to step 6. This will result in a bit bulkier knot, but will be less prone to slip under tension, especially with "slick" fibers such as nylon.

10. Grasp an end in each hand and pull to tighten and secure the knot.

Reefer's Knot

The reefer's knot is also known as the *bowknot* or *quick-release square knot*. Like the square knot, it is a good nonslip knot with which to tie ends of rope together. It has the added advantage that it can be untied under tension—a most important feature for any knot used to restrain livestock.

Step-by-Step Procedure

1. The first four steps are identical to those used in tying the square and surgeon's knots: a simple overhand knot, coming from right over left, is formed.

2. Now, begin to tie the second overhand knot, coming from left to right, by laying the new left-hand strand over the new right-hand strand.

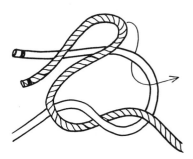

3. Instead of inserting the running end of the new left-hand strand into the loop formed by the crossing strands, form a bight in the new left-hand strand and insert it into the loop.

4. Grasp the bight with the thumb and index finger of your right hand and pull it partway through the loop.

5. Grasp the left-hand strand and left working end in your left hand and the right-hand strand in your right hand. Pull to shape and secure the knot. Be certain that the end of the bight is "trapped" in the center of the knot.

6. In an emergency, the free end of the bight can be pulled sharply, immediately releasing the knot.

7. The sequence is reversible; i.e., instead of initially right over left, then left over right, it can be tied left over right, right over left.

Bowline Knot

Knot users, from stockmen to seafarers, consider this the most useful of all knots. It is a nonslip knot, and as such can be used to form a loop or slip that will not tighten or draw down when placed around an animal's body or a post. Tied small, the loop formed makes a usable honda for a lariat. It is relatively easy to untie.

Step-by-Step Procedure

1. Position the rope so that the standing part is to your left, the working end to your right.

2. Form a right-hand loop by passing the working end of the rope over the standing part.

3. Secure the loop by positioning the strands where they cross between the thumb and index finger of your left hand.

4. Insert the working end of the rope into the loop from the back.

5. Cross the working end over the top of the standing part and wrap it around the rear of the standing part.

6. Reinsert the working end into the loop from the front.

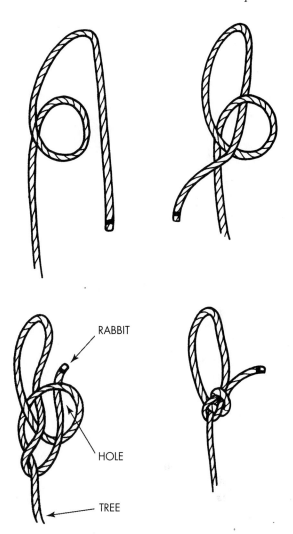

7. There is a short little "ditty" that helps us remember the sequence and direction for tying this knot. For the purposes of the ditty, the working end of the rope is the "rabbit," the loop formed in the rope is the "hole," and the standing end of the rope is the "tree." It goes like this: The rabbit comes out of the hole, goes around the tree, and goes back into the hole again.

8. Grasp the working end of the rope and the right-hand strand of the loop in your right hand, the standing part of the rope in your left hand.

9. Pull to shape and secure the knot.

10. The size of the loop formed is dependent upon the amount of working end originally allowed for use.

11. It makes no difference whether this knot is tied in the hand or about an animal's body—the steps are the same.

Doubled Bowline Knot (Three-Loop Bowline)

The doubled bowline knot is just what it says—it is a bowline knot tied into a doubled rope. It is sometimes described as being a bowline-on-a-bight because it is formed by a bight of rope, not a single

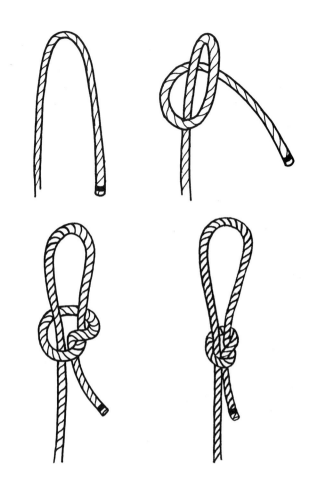

strand. It differs from the true bowline-on-a-bight only slightly. Many tyers do not appreciate the fine distinction, and so the doubled bowline (three-loop bowline) will likely always be misnamed the bowline-on-a-bight. Whatever its proper name, the knot is particularly useful whenever you must form a loop in the center of a long rope. It is a comfortable loop for the animal because it is two strands wide and therefore only one-half as irritating to the hide.

The steps followed in tying the doubled bowline knot are exactly the same as those for the single bowline, with the exception of the initial doubling of the rope and consequent formation of the bight in what becomes the working end of the rope.

Slipknot

The slipknot is not a particularly useful knot in animal restraint for the following reasons: it slips under tension, draws down upon the body of the animal restrained, and does not release itself when the tension ceases; it is next to impossible to untie this knot while it is under tension, or for that matter after it has been under tension; and last, it is one of the least efficient of knots (weakest). Its primary value appears to be that of providing a starting point for tying the quick-release knot. Be certain that the slipknot is never placed about the neck or body of an animal.

Step-by-Step Procedure

1. Start with the rope before you. Place the standing part to your left, the working end to your right.

2. Grasp the standing part in your left hand, about 24" from the working end.

3. Form a bight in the working end of the rope by placing the working part between the left thumb and index finger next to the standing part of the rope.

4. With your right hand, wrap the working end over the top and around the back of the original bight. Complete this wrap by inserting the running

end of the working part of the rope into the loop just formed—from top to bottom (front to back).

5. Grasp the standing part of the rope and the running end of the rope in your left hand and the loop in your right. Pull to shape and secure the knot.

6. Mention has been made of the many weaknesses and faults of this knot. One of those faults, the fact that the knot slips under tension, even to the point of becoming completely undone, can be remedied by securing an overhand knot near the

end of the working end of the rope. The slipknot will still slip, but only to the point where the overhand knot is jammed against the primary knot.

7. *There are better ways to do whatever you had planned to accomplish with the slipknot.*

Quick-Release Knot

This knot is known as the *mooring knot* among yachtsmen and as the *halter knot* in animal circles. Whatever the circumstances and the name of the moment, this handy knot is the standard way to tie an animal or bouncing, bobbing boat to a post or bollard and be able to release it very quickly even while it is under tension. It is a simple variation of the slipknot.

Step-by-Step Procedure

1. As with the slipknot, start out by forming a bight near the end of the rope end securing it between the thumb and index finger of your left hand. Leave 10" to 12" of the working end below your left hand.

2. With your right hand, wrap the working end over the top and around the back of the original bight, forming a loop.

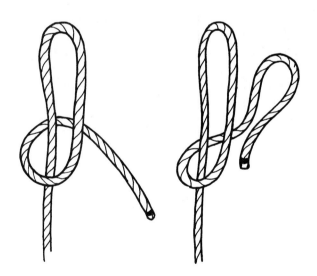

3. Now, instead of completing this wrap by inserting the running end of the rope into the loop just formed (and thereby making a slipknot), form a bight in the running or working part and insert it into the loop.

4. Grasp the standing part of the rope and the bend of the bight in your left hand and the loop in your right. Pull to shape and secure the knot.

5. Horses have the notorious habit of biting on the knots restraining them and ultimately freeing themselves. To prevent this, insert the running end of the rope into the bight.

6. This knot can be tied in the hand or around a post or tree. It should never be placed upon the neck or body of an animal.

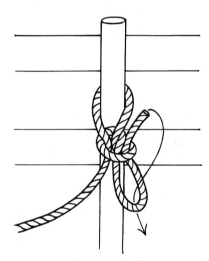

Sheet Bend Knot

The sheet bend is a friction knot used primarily to join together two ropes of unequal size. If the ropes are of similar size, the easier-to-tie square knot should be used. Ropes of unequal size do not bind together well and keep the true square knot orientation, so the sheet bend is the knot of choice for this purpose.

Step-by-Step Procedure

1. Form a bight in the working end of the heavier of the ropes and hold it in position with the thumb and fingers of your left hand.

2. Run the working end of the lighter rope into the bight of the heavier rope from below.

3. Run the working end of the lighter rope over, under, and around the bight in the heavier rope and finish by bringing it over the bight and inserting it between its own standing part and the strands of the bight.

4. Shape and secure the knot by pulling on the heavy rope bight with your left hand and on the standing part of the light rope with your right.

Learning one more step in the process of tying sheet bend knots can save a lot of frustration. Recall that the sheet bend is a friction knot; i.e., it holds securely only when under tension. This same tension that makes it a worthy knot makes it most difficult to untie. This can be remedied by an additional step that converts the sheet bend to a "slipped" sheet bend.

5. Take the working end of the lighter rope that you have just brought over, under, and around the heavier rope bight, and insert it, as in step 3, between its own standing part and the bight of the heavy rope. But this time, before you tighten the bend, form a bight in the running end of the light rope and insert the tag end of it under the standing part of the light rope. Now secure the bend by pulling on the bight in the heavy rope and on the standing part of the light rope.

6. When it is time to release this knot, under tension or not, a quick jerk on the tag end of the bight in the light rope is all that is necessary.

Double Sheet Bend Knot

Because the sheet bend knot holds well only when under tension, an improved form called the *double*

sheet bend has been developed to hold securely, even without tension on the knot. This knot is formed in the same way as the sheet bend, except that an additional wrap is taken with the working end around the strands of the bight.

Step-by-Step Procedure

1. Form a bight in the heavy rope and run the working end of the lightweight rope into the bight from below as with the sheet bend knot.

2. Run the working end of the lighter rope over, under, up around, and then over, under, and up around once again before inserting it under its own standing part where it has entered the bight.

3. Shape and secure the knot by pulling on the heavy rope bight with your left hand and on the standing part of the light rope with your right.

Figure-Eight Bend

Earlier, the figure-eight knot was illustrated. Its usefulness is primarily as a stopper knot in the end of a single rope, although it is an attractive and enjoyable knot to work with. When there are two ropes to be joined, the figure-eight principle can be expanded into the figure-eight bend. The figure-eight bend is also known as the *Flemish knot*. It is one of the strongest bends, simple to tie, and holds 1/4" cord as well as 1" rope.

1. Form a bight near the end of one of the ropes to be joined. Allow the running end to hang to the left, with the standing end to the right. Hold the bend of the bight in your left hand.

2. Wrap the running end of the bight around the standing part, coming over the front and around to the back. Bring the running end out from behind and insert it into the loop held in your left hand.

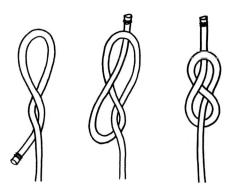

3. Stop at this point, before you pull the ends of the rope and firm up the knot. Arrange the figure-eight knot so that each of the loops is open and well formed.

4. Now, take the running end of a second rope to be joined to the original one (with the figure-eight in the end of it), and insert it into the original figure-eight knot by sliding it along the tag (working) end of the first rope which is sticking out of the figure-eight knot. Continue to place the second rope into the figure-eight knot. The coils of the two ropes should lie against one another through each turn of the figure eight.

5. As you complete this bend, the standing end of rope 1 and the tag end of rope 2 will be exiting the figure-eight bend and the standing end of rope 2 and the tag end of rope 1 will be exiting to the opposite direction.

6. At this point, position and firm up the knot.

Half Hitch

The half hitch is a very quick knot or hitch to form and use. It is a useful method of temporarily holding a rope in position around a post or tree in an emergency situation (such as an animal escaping). Any hitch is a tension knot; i.e., it must be held under constant pressure for it to remain tight. The half hitch is particularly vulnerable in this regard in that the working end must be held by the handler for the knot to remain secure. It is particularly useful as a leverage hitch to keep any ground gained while trying to move an animal closer to a post, trailer, or chute.

Step-by-Step Procedure

1. Position the standing part of the rope to your left and grasp the working end of the rope in your right hand.

2. Pass the running end (working end) of the rope over or around the post.

3. Bring the running end over the standing part of the rope, under it, and then insert it into the loop (the one around the post) from the bottom.

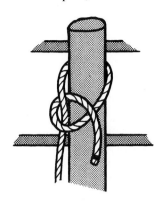

4. The knot is held by handler tension on the running end and animal tension on the standing end.

Double Half Hitch (Two Half Hitches)

The double half hitch is an extremely useful knot for the handler of livestock. It is quick, easy to tie, acts like a slipknot, and provides a convenient way to tie up the end of a rope when no other knot seems appropriate.

Step-by-Step Procedure

1. Position the standing part of the rope to your left and grasp the working end of the rope in your right hand.

2. Pass the running end of the rope over or around the post.

3. Bring the running end over the standing part of the rope, under it, and then insert it into the loop (the one around the post) from the bottom.

4. Repeat step 3 to form the second half hitch; i.e., take the running end of the rope and pass it over the standing part, under it once again, and then insert it into the loop just formed below the first half hitch.

5. Shape and secure the knot by pulling on the running end of the rope. Animal tension will provide the pull on the standing end of the rope.

Round Turn and Double Half Hitch

This knot is a fine substitute for the clove hitch and an extremely practical knot. The practicality involves using the wrap around the post, which was probably used to winch the animal into position, as the start of the knot, and then finishing it off with the easy-to-remember, easy-to-tie-under-tension double half hitch.

Step-by-Step Procedure

1. Start the knot by wrapping the working end of the rope around a post in a clockwise direction. As it comes back to the front, pass it either above or below the standing part of the rope and continue around the post once more. Stop as the running end of the rope is ready to pass over the top of the standing part.

2. At this point, a double half hitch is formed exactly as before (steps 1, 2, 3, 4, and 5).

Clove Hitch/Miller's Knot

The clove hitch is a knot that can be tied around a post or leg, or it can be preformed and dropped over the top of a post. It is not a secure knot and it is not particularly easy to tie around a leg or post. Better choices are the ring knot for dropping over a post and the round turn with double half hitch for tying around a post.

Drop-Over Clove Hitch

Step-by-Step Procedure

1. Form a loop in a section of rope by laying the working part of the rope over the standing part— refer to this as the loop to the front.

2. Keeping this front loop formed, make a second loop, this time a loop to the rear, by again placing the working part of the rope over the standing part.

3. Hold the front loop in your right hand and the rear loop in your left. Position the two loops atop one another by sliding the front loop in your right hand behind the rear loop in your left hand.

4. Insert fingers from both hands into the double loops and pull to size the knot *before* attaching it to the animal. After dropping the double loop over a post or placing it over a leg, pull on the ends to shape and tighten the knot.

Wraparound Clove Hitch

Step-by-Step Procedure

1. Grasp the working end of the rope in your right hand and wrap it around the post or leg to be secured, starting from the right side of the post or leg.

2. Cross the working end under the standing end of the rope in front of the post or leg. Pinch this point of crossing between your thumb and index finger.

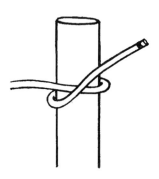

3. Wrap the working end about the post or leg a second time, this time above the first loop, and as the working end comes around to the front, bring it between the pole or leg and itself.

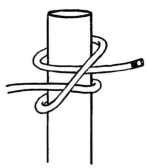

4. Slide the loops together. Shape and secure the knot by pulling on the working end and standing part of the rope.

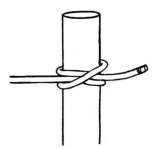

Ring Knot

The ring knot is also known as a *girth knot* or *girth hitch*. It is a superior substitute for the clove hitch in that it does not slip and is more easily tied. Just as with the clove hitch, it can be preformed to drop over a post or tied about the post or leg.

Drop-Over Ring Knot

The drop-over ring knot is a friction knot, like the clove hitch, but uses a bight formed in the rope to "drop over" a post or stob. The bight for the drop-over ring knot is typically formed near the end of the rope; however, it can be formed at any point in the rope.

Step-by-Step Procedure

1. Form a bight at the desired position in the rope by grasping the two strands in your right hand.

2. Insert your left hand into the bight and grasp the strands just in front of where your right hand is securing them. Remove your right hand from the strands.

3. With your right hand, grasp the bend of the bight and pull it over your left hand and down onto the standing parts of the rope.

4. The double loops now formed are the ring knot. Shape and size them in your hands to fit your needs as dictated by post or leg.

Wraparound Ring Knot

Step-by-Step Procedure

(If the knot is to be tied to a ring, this is the only method of forming it.)

1. Form a tight bight at the desired position in the rope by grasping it with your left hand.

2. Wrap the bight around the post or leg or insert it into the ring. As it comes around from behind the leg, grasp it in your right hand.

3. Take the doubled strands of the standing part of the rope and place them into this bight. Switch hands, so that the left hand is now controlling the bight while the right hand pulls the standing end of the rope into and through the bight.

4. After the entire lengths of the standing ends have been drawn through the bight, release the bight and tighten the knot by pulling with the right hand.

Honda Knot

A honda is a small loop secured into the working end of a rope through which the standing part of the rope passes as it forms a much larger loop. Most lariats come with the honda knot pretied into an end. A few are manufactured with a quick-release honda tied into the end. Bulk rope or broken lariats must have honda knots retied into their ends.

Step-by-Step Procedure

1. Tie an overhand knot tightly into the end of the rope.

2. Approximately 8' below this, tie another overhand knot, only this time leave it in the loosened state.

3. Grasp the loose overhand knot in your hands and study it until you have determined how to orient it so that the working end of the rope comes out from the loop and toward you. From there it runs upward to the end knot.

4. Grasp the running end of the rope and bend it so that it lies over the bend of the overhand knot loop.

5. Insert it into the overhand knot loop *between* the bend of the loop and its own standing part. Study the diagram carefully because it is easy to place the running end improperly.

6. Shape and secure the knot by pulling on the loop and standing part of the rope.

Sheepshank Knot

There will be times, as you work around a farming operation, when you must secure something with a rope or otherwise perform work with a rope that you wish were several feet shorter. Cutting is not the answer because there are tasks where the greater length of rope is valuable. The sheepshank knot allows you to temporarily "shorten" a rope. Since it holds only when there is a strain on it, it is unsuitable for livestock handling.

Step-by-Step Procedure

1. Position the rope in a Z configuration with the top and bottom bars of the Z about 12" in length and lying nearly on top of one another (the diagonal leg of the Z only about 2" tall).

2. Slide your left hand to the right side of the Z, forming a bight in the rope and holding it upon the bottom leg of the Z.

3. Form a loop to the rear in the lower leg of the Z, twist it in a counterclockwise motion and place it over the bight being held in your left hand. Snug it around the bight, being certain that about 2" of the bight protrude from the loop.

4. Slide your right hand to the other side of the Z, forming a bight in the rope and holding it up against the top leg of the Z.

5. This time, form a loop to the front in the upper leg of the Z, twist it counterclockwise, and place it over the bight held in your right hand. Snug and position it as before.

6. Tension must be kept on the two ends of the rope to keep the sheepshank secure.

Splices

Splicing becomes necessary when two ropes are to be joined or when an eye (honda) must be permanently installed in the end of a rope. Splicing the ends of rope together is necessary to repair a break or to create a longer rope from two shorter lengths. While the break or short ropes could be joined together with knots, splices are preferred because they are stronger than knots, less bulky, and permanent.

Short Splice

Step-by-Step Procedure

1. Unlay about 6" of the ends of the two ropes to be joined. With $1/2$" to $5/8$" rope, this will allow for three complete over–under sequences. Experience has shown that three sequences provide maximum strength.

2. Whip, tape, or melt (nylon and dacron) the unlaid end of each strand of both ropes. This is absolutely necessary if you wish to avoid a hopeless frayed-fiber mess after only the first over–under sequence.

3. Position the two ropes to be joined so that one lies to your left, the other to your right. Grasp the working end of the left rope in your left hand, the working end of the right in your right hand.

4. Bring the two ends together, alternating the strands; i.e., arrange it so that no two unlaid strands from the same rope lie next to one another without a strand from the other rope between them.

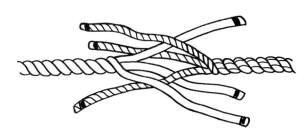

5. Push the ropes together snugly, twisting them slightly back and forth. Tie or tape the strands of one rope onto the standing end of the other rope. This is to prevent further unraveling of the rope during the splicing sequence.

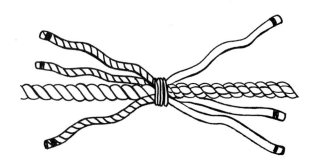

6. Study the junction of the ropes and the unlaid ends. You will notice that each of the unlaid strands lies directly over (upon) a strand of the intact rope. This arrangement is assured if step 4 was carefully followed.

7. Each of the unlaid strands must go *over* the rope strand that it is lying upon and *under* the one immediately to its left if you are splicing in a right-to-left direction.

8. Select any of the strands and begin your splice at that point by placing your right thumb partially under it and at the same time upon the strand that it is passing over or lying upon. Grip the junction area of the ropes to be spliced with the rest of your right hand.

9. With the index finger and thumb of your left hand, grasp the strand to be gone under—this is the one immediately adjacent to the one that the unraveled strand is passing over.

10. With your right hand, twist the junction to the right (clockwise) while twisting the rope being spliced to the left (counterclockwise) with your left hand. This will open the rope and enable you to isolate the strand under which to pass the end of the unlaid strand.

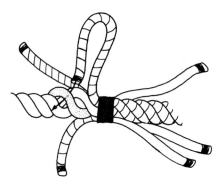

11. Keep this strand isolated, and with your right hand place the end of the unlaid strand under it and pull it through until it is pulling on the body of its parent rope.

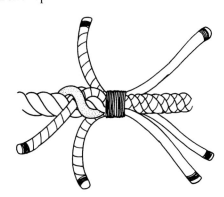

12. Rotate the rope in a clockwise direction, and repeat steps 8 through 11 for the second and third strands.

13. After you have completed the first series of over-and-under tucking, the work should be tightened. Grasp each individual unlaid strand, one at a time, and while pulling it into the rope being spliced, give it a strong twist in a clockwise direction.

14. Repeat steps 8 through 13 for the second and third sequences of over-and-under tucking. From this point onward, the over–under splicing is the same as for the backsplice.

15. Remove the tape or string that previously held the other set of unlaid strands and repeat the entire sequence (steps 6 through 14) for these strands and the part of the rope to be spliced into.

16. After the strands have been tucked over and under for three sequences, the entire splice should be stretched and faired (rolled on the floor under your foot). Clip off the excess unlaid strands approximately ¹/₄" from the last strand they went under.

Eye Splice

Step-by-Step Procedure

1. Unlay about 6" to 8" of the end of the rope to receive the eye. With ¹/₂" to ⁵/₈" rope, this will allow for three over–under sequences. Just as with the short splice, three courses have proven to be the most efficient number to use.

2. Whip, tape, or melt the unlaid end of each strand of the rope. This is absolutely necessary if you wish to avoid a hopeless frayed-fiber mess after only the first over–under sequence.

3. Hold the standing end of the rope in your left hand and the unraveled strands in your right. Arrange the strands so that two of them come across the top of the rope from a lower left to an upper right direction. The remaining strand appears to come from behind the front two, in a lower right to upper left direction.

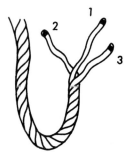

4. Form a loop the size appropriate to your needs by placing the unlaid strands against the standing part of the rope in the following manner. The

strand coming from behind in a lower right to upper left direction, and the topmost of the strands lying in the lower left to upper right direction go on top of the standing part, while the remaining lower-left- to upper-right-lying strand goes beneath this standing part of the rope. Push these into position snugly and hold them with your left hand.

5. Using both of your hands, while maintaining the loop (or eye) positioning, twist open the standing part of the rope where the unlaid strands pass over it. Take the center one of the three unlaid strands, identify it as strand 1, and insert it under any strand of the twisted-open rope. Take care not to lose the original orientation of the unlaid strands.

6. Now study the eye and rope. Note that unlaid strand 1 has just gone under a rope strand. Unlaid strand 2 is lying upon the rope strand that unlaid strand 1 went under.

7. Unlaid strand 2, the other strand on top, is now inserted over the rope strand that 1 is under and under the rope strand immediately adjacent to it.

8. If you look closely, you will note that strands 1 and 2 are touching and that strand 2 enters where strand 1 came out.

9. For the remaining unlaid strand, look to the back side of the eye and study the eye and rope. You will note that there is only one rope strand that does not have an unlaid strand under it. This is the strand under which strand 3 must go.

10. Strand 3 must be placed *over the top* of this remaining rope strand and come back under it in a left-to-right direction. At this point, strand 3 goes under where strand 2 comes out, and 3 comes out where 1 goes under.

11. Tighten and shape the eye by pulling and twisting in a clockwise direction on the unlaid strands. If you have performed the eye formation properly, the unraveled strands will come out at the same height and at equal spacing from one another.

12. From this point on, the unlaid strands are tucked twice more in the familiar over-and-under sequence detailed in the discussions of crowning and short splicing.

13. The strands are also finished as detailed in the discussions of those splice types.

Adjustable Rope Halter

A supply of low-cost, easy-to-make, adjustable rope halters is a necessity on farms or ranches where beef and dairy cattle are kept. Larger breeds of sheep and goats are also more easily handled in a rope halter than by the traditional methods.

Adjustable rope halters are used to teach cattle, sheep, and goats to lead, to routinely handle them in day-to-day activities, and to restrain them as necessary. The halters' low cost allows the producer to make extras and place them about the farmstead so that they will be handy when needed.

1. Select a 12' to 15' length of $1/2$" three-strand rope. Any of the rope types from cotton through nylon will work. The choice will depend upon the strength needed, durability required, and cost. Rope of $1/4$" to $3/8$" diameter is suitable for sheep and goats.

2. Finish one end of the rope by whipping it, clamping it with a ferrule, dipping, or heat-treating. (The method selected will depend upon your wishes and the type of rope selected.) If you do not finish the rope ends, they will unlay and fray. Temporarily finish the other end of the rope with tape or string. A crown knot will be formed into this end after the halter is constructed.

3. Mark a point with your hand about 12" to 15" from the whipped end of the rope. Refer to this 12" to 15" length as the *short end* of the rope—the remaining length is the *long end.*

4. Place the short end to your right, the long end to your left. Grasp the rope at the 12" to 15" mark between the thumb and first two fingers of both hands. Separate your right and left hands by about 2". Rotate the rope clockwise with your right hand and counterclockwise with your left. This will open the strands of the rope between your hands.

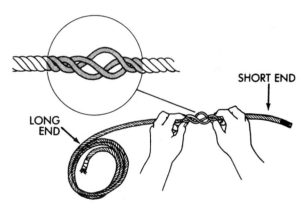

5. Isolate any one of the opened strands with the thumb and index fingers of your left hand. Use your right hand to insert the whipped end of the short end of the rope under this strand opening until the loop formed has an eye opening of about double the rope diameter.

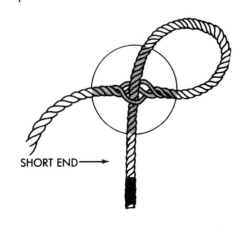

6. Now position the rope so that the eye loop is in your left hand with the short end pointing toward three o'clock and the long end exiting toward six o'clock. Grasp the eye loop and the single strand running across the short end of the rope between your left thumb and index finger. With your right thumb and index finger, grasp the short end of the rope at a point near the eye loop. Twist the eye loop and short end of the rope with your hands until you have isolated two strands between your right thumb and index finger.

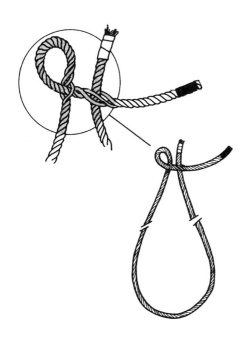

7. Use your left hand to insert the long end of the rope from bottom to top under and through these two strands. Pull it completely through until all the slack is gone. If done properly, one side of the loop will show three strands lying smoothly side by side. This is important because they will be positioned against the animal's face.

8. With the eye loop to your right, grasp the short end of the rope between your left thumb and index finger about 2" from the whipped end. Two inches farther from the whipped end, grasp it in the same manner in your right hand. Open the strands by twisting clockwise with your right and counterclockwise with your left hand. When the strands are opened wide, push your hands together. This will cause the strands to buckle and fold over, forming three loops.

9. Line these three loops up in order and work into them a sharpened stick of diameter equal to that of the rope. Use your right hand to feed the long end of the rope into the loops, starting at the one closest to the eye loop. Remove the stick from one loop at a time as you run the long end of the halter through them.

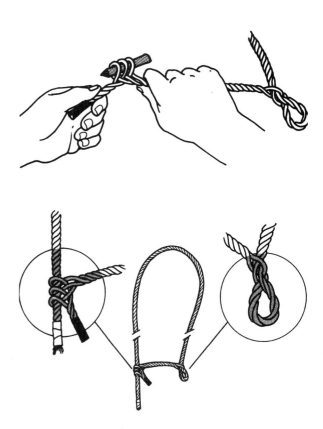

10. Run the long end of the rope into and through the eye of the loop. This completes the halter.

11. Permanently finish the long end of the rope in the chosen manner. Consider crowning the end because a crown knot and backsplice creates a convenient handle. Do not use a hog ring to finish the end because it could catch and tear the skin of your hand.

12. Always place the halter on the animal so that the eye loop is on the left side. Lead from the left.

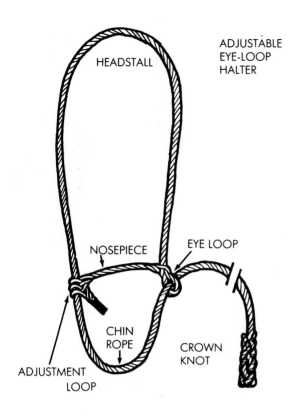

HEADSTALL

ADJUSTABLE EYE-LOOP HALTER

NOSEPIECE

EYE LOOP

ADJUSTMENT LOOP

CHIN ROPE

CROWN KNOT

1.4 CATTLE RESTRAINT

To be able to safely and correctly perform the day-to-day management techniques necessary for efficient cattle production, a thorough working knowledge of cattle-restraint methods is necessary. Efficient cattle restraint is more than "whoopin' and hollerin'" and "ropin' and tyin'." In fact, the more whooping, hollering, roping, and tying involved, the more inefficient the whole process becomes.

Cattle are large and strong, and while they are not particularly quick by animal standards, they can move rapidly enough to evade man's brute-force efforts. If you are going to be successful in restraining cattle, you must (1) have the mechanics of the restraint method well practiced so that you do not have to stop and think about what step is next, (2) study and try to understand their behavior so that you can outsmart them instead of trying to out-muscle them, and (3) have strong, appropriately sized, well-cared-for equipment that you thoroughly understand how to use. Some of the restraint methods discussed in this section may not be required on dairy operations, since dairy cattle are generally not as difficult to restrain as beef cattle. Common stanchion facilities, halters, and pen systems are all that are usually required for dairy cattle. In contrast, beef cattle often require working chutes and/or squeeze chutes to partially immobilize the animals.

Working Chute

Working chutes can be used effectively as the only restraint method for many of the management techniques discussed in Chapters 2 and 3. Some of the techniques, such as vaccinations, spraying, and applying pour-on insecticides, can be done while the animals are crowded head-to-tail in the chute. For more rigid confinement, and for other techniques, a squeeze chute and headgate, a scale, or a loading chute can be placed at the end of the working chute.

Many commercial and homemade designs are available, but the best of them share the following characteristics: (1) they have V-shaped sides or an adjustable side (18 to 30 inches) so that large or small cattle can be handled; (2) the "V" measures 14 to 17 inches wide at the bottom and remains so for the first 2 feet from the bottom, at which point it flares to approximately 30 inches at hip and horn height. Straight-sided permanent chutes measure 26 to 28 inches in width; (3) the sides should be solid so that the cattle cannot see through and become distracted; (4) the crowding pen and working chute should be bent or curved so that the cattle do not sense and resist the tunnel effect; and (5) the working-chute exit gate–squeeze-chute entry gate is usually made of bars (not solid) to give the illusion to the cattle of being able to escape, thereby causing them to enter.

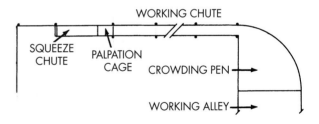

WORKING CHUTE

SQUEEZE CHUTE

PALPATION CAGE

CROWDING PEN

WORKING ALLEY

When constructing the working chute, build an access gate into it, adjacent to the squeeze-chute end, so that you can enter the working chute to perform palpations and castrations.

Step-by-Step Procedure

1. Move the cattle to the working alley and crowding pen that funnels into the working chute.

2. Start the cattle into the working chute. It will be more efficient to get one singled-out animal started down the chute instead of trying to drive all of them at once. If the cattle will simply not cooperate, it may be necessary to enter the crowding pen and force the cattle with the sparing use of broom, whip, cane, or tail twist.

> **CAUTION:** With several large animals in a relatively small area, you risk being squeezed, stepped on, or kicked. Avoid using the electric prod or "hot shot." It will make the cattle nervous and less cooperative, increasing the risk of injury not only while you work cattle this time but also when they are reworked later.

3. Keep the cattle moving down the working chute by walking along the outside of it and talking to them, slapping them on the rump with your hand and tail-twisting as necessary.

4. Place a bar completely across the working chute behind the last animal and also behind the first animal in the chute. This should be placed just above the hock and will prevent the animal from backing up.

5. Open the gate to the squeeze chute and allow the first animal in the working chute to enter.

Squeeze Chute and Headgate

The combination of squeeze chute and headgate can be used to advantage on any type of farm or ranch for such management techniques as dehorning, castrating, branding, implanting, ear tagging, stomach tubing, artificial insemination, and blood testing. It is made even more valuable when it is positioned at the terminal end of a working chute.

There are many commercial and homemade designs available. A workable combination should consist of the following: a squeeze mechanism, a headgate with head and nose bars, a tailgate, removable solid side panels measuring approximately 24 inches from the ground, and removable side bars for easy access to the animal's side.

There are three basic headgate designs used in beef production: straight-bar headgate, positive-type headgate, and curved-bar headgate. The *straight-bar headgate* generally is designed to automatically catch an animal as it walks through the chute and fits well into purebred and commercial cow–calf operations. The greatest advantage of the straight-bar headgate is its protection against choking the animal. The main disadvantage is that the animal can move its head up and down easily, which can create some problems in certain techniques that require head immobilization.

The *positive-type headgate* operates something like a guillotine. The main advantage of this headgate is almost complete head control, both sideways and up and down. The headgate almost completely immobilizes the head without the necessity of a head and nose bar. This type of headgate is popular among feedlot operators.

The *curved-bar headgate* combines most of the safety of the straight-bar headgate and some of the

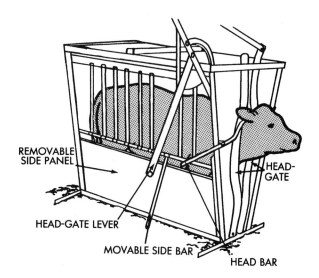

REMOVABLE SIDE PANEL

HEADGATE

HEAD-GATE LEVER

MOVABLE SIDE BAR

HEAD BAR

> **CAUTION:** It is absolutely necessary that the headgate be adjusted to the correct height of the animal being worked to prevent choking. In addition, it is mandatory that the animal's head be released from the headgate before the gate is opened for the animal to exit. If this is not done, the animal can lunge forward and break its neck or choke to death.
>
> Regardless of the headgate design used, it is important that the headgate be adjusted for the size of animals being worked to prevent choking or escape.

up-and-down head restraint of the positive-type headgate. This headgate is an excellent compromise for the cow–calf feedlot operator.

Step-by-Step Procedure

1. Position the squeeze chute at the end of the working chute. Be sure they fit squarely and snugly together. In permanent squeeze-chute installments, the squeeze chute should be anchored to a concrete pad to prevent forward movement. When portable facilities are used, the squeeze chute should always be chained near the ground to a post on both sides of the working chute to prevent forward movement.

2. Adjust the squeeze chute to the size of cattle being worked. The bottom of the chute is most important. The inside width at the floor for 400- to 600-pound cattle should be 6 inches; for cattle of 600 to 800 pounds, 8 inches; and for finished cattle or cows, 12 inches.

3. Open the gate to the squeeze chute and begin urging the first animal in the working chute to enter. You may have to do some tail-twisting or rump-slapping to encourage the animal to enter. Opening the headgate on the squeeze chute will often entice the animal to enter, thinking it has found an escape route.

4. If the management technique requires complete head immobilization, secure the head bar over the top of the animal's neck first and then the nose bar over the bridge of the nose.

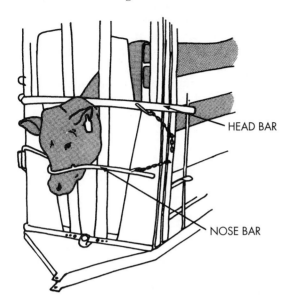

5. After the management technique is performed, release the nose bar, the head bar, and the squeeze mechanism, in that order, and finally open the headgate to allow the animal to exit.

> **CAUTION:** On squeeze chutes equipped with a V- or guillotine-type headgate, be sure to release the headgate before opening the exit gate to prevent choking.

Tilt Tables

Tilt tables offer total restraint, control, and maximum comfort to the animal and to the handler. There are many commercial types of tables, as well as workable homemade models.

With tilt-table restraint, the front end of the animal is restrained against a vertical platform with either a headgate or rope halter. In addition, the animal is secured to the same vertical platform with a wide canvas belt around its heart girth and flank. The table is tilted, and when the animal is positioned at the proper angle, its feet and legs are secured with ropes through holes in the table platform.

Step-by-Step Procedure

1. Halter the animal and lead it into position next to the vertical tilt table. If the animal is not halter-broken, portable gates and panels must be used to form a working chute leading up to the table.

2. When the animal is standing parallel to the table (it may be necessary to hold it in position by using a gate or panel as a crush), restrain its head by tying the rope halter into position. If the table has a headgate, secure the animal's head in it instead of using the rope halter.

3. Restrain the animal's midsection and rear end, using the wide canvas belts attached to the table for this purpose. Be certain that each of the three areas of restraint is securely fastened.

> **CAUTION:** The animal will probably resist being strapped to the table. Take care that a quick escape attempt does not injure you or the animal.

4. Tilt the table by operating the appropriate levers or wheels. Some tables can be stopped and safely held at any position from vertical to horizontal; others must either be in the full-vertical or full-horizontal position to be safely stopped with an animal aboard. Read the operating instructions carefully.

> **CAUTION:** Make certain that the table is securely anchored to the ground to prevent it from tipping over as the animal is tilted. If this occurs, serious injury to you or the animal is almost certain.

5. After the table is positioned as described, restrain each leg individually using ropes and the holes provided in the table. The legs should be positioned to allow you to perform the management technique most efficiently.

> **CAUTION:** The animal may try to resist restraint and move its legs vigorously in a paddling motion. To avoid injury, allow the animal to settle down before applying ropes.

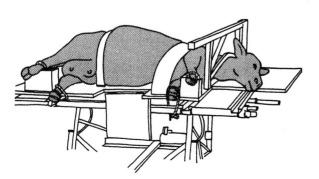

Foot-Trimming Chute

Foot-trimming chutes can be used for a variety of management techniques when cattle are gentle and easy to handle. The chute obviously works best to trim feet, but it can be used to restrain animals during artificial insemination (AI), implanting, dehorning, and castrating when a squeeze chute is not available. In small operations where cattle are trained to lead, this homemade chute design may be the only restraint that is required to perform most cattle management techniques.

The chute should be designed with a headgate, canvas belts on a rachet mechanism to support the animal when needed, removable 4 × 6s (inches) on both sides to support the animal's feet while trimming, and removable 2 × 10s on both sides for the animal to lean against when trimming.

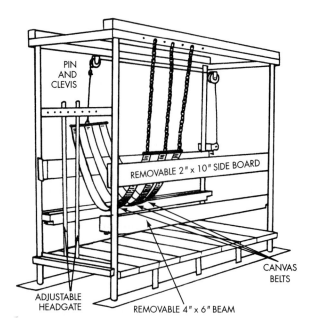

Grooming Chute

The grooming chute is used only when cattle are gentle and trained to lead. It is not constructed to replace the squeeze chute or foot-trimming chute for restraint. A grooming chute is used to clip and prepare an animal for show or sale. Many commercial and homemade designs are available for use. The best of them have a metal frame made of pipe or square tubing, rough wood floor, simple headgate mechanism, and swingable or removable side bars that allow easy access to the side of the animal.

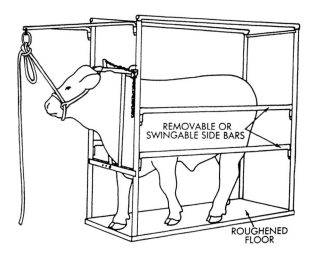

Step-by-Step Procedure

1. Lead the animal into the grooming chute.
2. Secure the animal's head in the headgate.
3. Remove or swing the side bars open on one side at a time to groom the animal's side or to "block" and clip. Replace the side bars to "bone" the legs.
4. Release the headgate and back the animal out of the chute when finished.

Casting—Half-Hitch Method

Casting refers to causing the animal to lie down on its side due to pressure exerted on its muscles and nerves by a series of carefully placed and tightened ropes. This restraint should be used only when near-total immobilization is required and no chutes or tables are available.

The half-hitch method that follows is only one of many, but it is the standard method for animals of any size. To minimize the risk of injuring the penis of a bull, wrap the rope with a soft material such as a cotton bag.

Before any animal is cast, the lay-down area should be checked for rocks, dirt clods, and any

Step-by-Step Procedure

1. Lead the animal into the trimming chute.
2. Secure the animal's head in the headgate.
3. For foot trimming, secure the canvas belts to support the animal, remove the 2 × 10 from the side of the animal that you are going to trim, lift one foot, place the lower part of the leg (below the knee) on the 4" × 6" beam for support, and begin trimming. Refer to Section 3.22 of Chapter 3 for details.
4. For management techniques other than foot trimming, all boards should be in position. The canvas belts are usually not needed.
5. Release the headgate and lead the animal out of the chute.

other debris that could cause cuts or bruises. If possible, bedding should be used to cushion the fall.

Step-by-Step Procedure

1. Halter the animal and lead it to the lay-down area. Secure the lead rope to a post or ring approximately 12" to 18" from the ground.

2. Tie the end of a 35- to 40-foot, $5/8$- to $3/4$-inch, three-strand, loose-weave cotton rope loosely around the animal's neck, using a bowline knot to form the loop. Position the knot approximately 6 inches below the ridge of the spine to one side or the other.

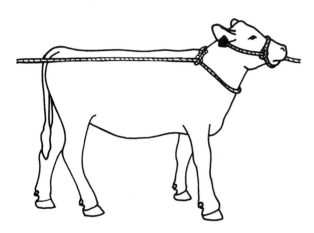

3. Place a wrap of rope immediately behind the animal's forelegs. Form this wrap by dropping the whole coil of the rope over the animal's back to the side opposite the bowline knot. Reach under the animal (carefully, to avoid being kicked), grasp the rope, and bring it out from under its body. *Note:* the use of a show stick here is helpful. Run the working end of the rope under the standing part connected to the bowline knot. This will form a half hitch about 12 inches behind the bowline knot.

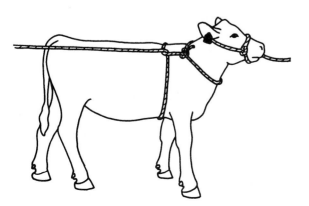

4. Place a second wrap of rope in the flank area, just in front of the udder if it is a cow, by repeating the processes in step 3.

5. Begin pulling on the working end of the casting rope. As pressure is exerted on the nerves and muscles, the animal will lie down. Keep pressure on the rope and have an assistant tie the animal's legs before beginning the management practice. A constant, steady pull should be maintained on the working end of the casting rope.

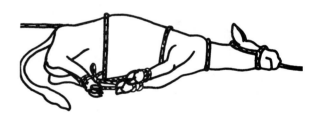

6. After the management technique has been completed, release the legs, loosen the casting rope, and allow the animal to regain its feet.

The halter rope may need to be released from the post or tie-ring so that the animal can rock itself to gain the necessary momentum to rise.

Casting—Crisscross Method

Step-by-Step Procedure

1. Halter the animal and lead it to the lay-down area. Secure the lead rope to a post or ring approximately 12 to 18 inches from the ground.

2. Place the center of a 40-foot rope over the neck/withers of the animal. Half of the rope will be on either side of the animal.

3. Pass each half of the rope between the front legs and then up and across to the opposite side of the animal.

4. Pass the ends of the crossed ropes between the hind legs and extend them straight back to the rear.

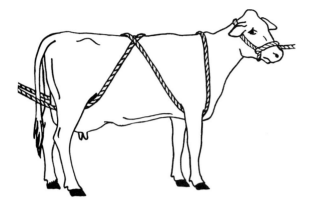

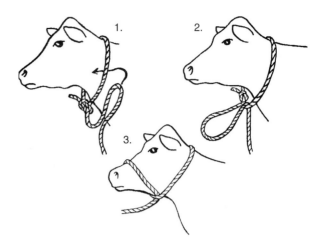

5. Begin pulling straight back. Pressure will cause the animal to lie down. Keep tension on the ropes as an assistant secures the legs of the animal.

> **CAUTION:** If the animal is a kicker, the handlers are in danger of being kicked during steps 3 and 4. To avoid this, toss the ropes under the legs instead of hand-placing them.

6. An advantage to the crisscross method of casting is that the udder, milk veins, or penis is not in danger of being damaged by the ropes crossing them as in the half-hitch method.

7. After the management technique has been completed, release the legs, loosen the casting rope, and allow the animal to regain its feet. The halter rope may need to be released from the post or tie-ring so that the animal can rock itself to gain the necessary momentum to rise.

> **CAUTION:** Use these techniques sparingly. They can cause pregnancy complications and displaced abomasums from the twisting, or bloat and pneumonia from having the animal lie on its side too long.

Halters

The halter is the least harsh of all restraint methods for the animal. It is also the most commonly used restraint method if all of its types are considered. The purpose of this section is to illustrate two of the more useful makeshift haltering techniques. For a detailed discussion of haltering with the standard manila adjustable rope halter, see Section 3.2 of Chapter 3.

Temporary Rope Halter

This makeshift halter is perhaps the most versatile, and works equally well on cattle, horses, sheep, and goats. Its versatility lies in the fact that it can be fashioned from any type of rope, from hard-lay manila or nylon to polypropylene. It is most commonly used as a temporary horse-restraint method. Directions for fashioning it are given in Chapter 1, Section 1.6.

> **CAUTION:** Do not tether (tie) not halter-broken animals unless you remain near them. Left unattended, they could fall and strangle themselves or become otherwise injured.

Slip Halter or Rope and Ring Halter

This halter can be made from a length (10 feet) of any type of rope, even extremely limp rope, and a 1- to 2-inch round ring. Use an eye splice or knot of your choice to fasten the ring to one end of the rope. To fashion a halter from this, form a bight in the rope, approximately 2 to 2 1/2 feet from the ring, and insert this bight into the ring, thus forming two loops. The last loop formed should be large enough to fit over the nose of the animal while the first goes over the head.

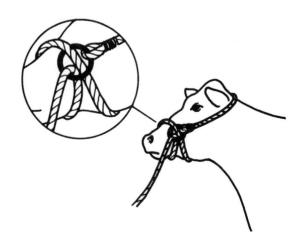

Nose Lead

The nose lead is an invaluable tool for restraining cattle, but it can be rather severe, especially if used improperly. With it and a headgate or stanchion, the animal can be made to stand still for many management techniques, such as an intravenous injection and foot-rot treatment. The nose lead can cause sufficient discomfort to discourage escape activity but does not injure the animal.

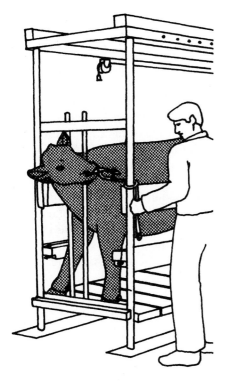

To place the nose lead, move the animal into a headgate, or otherwise restrict its movement, and back up to one side of the animal's head, facing the same direction it is facing. Grasp the nose by tightly clamping your index finger in one nostril and your thumb in the other nostril. This may be sufficient restraint without the use of a nose lead. (However, do not underestimate how strong a mature animal actually is. Even a strong human will be able to hold the head of a struggling animal for only a few seconds.) If not, quickly place the nose lead into position with your other hand and squeeze it shut. This will stop the animal from swinging its head. Do not expect the animal to stick its nose out and hold it still for you. It will be moving from side to side and up and down quite rapidly. The proper way to insert the nose tongs into the nostrils is to use a rotating motion. Place one side of the tongs into one side of the nose, then rotate the tongs across the end of the nose and place it into the other side. Do not attempt to push the tongs into the nose in a straight-in-motion from the front.

The nose lead should have a short length of rope attached to it. After the tongs have been inserted, pull the head to the side, take a double wrap around a pipe or post with the nose lead rope, and then have an assistant hold the end of the rope. Do not tie it. The animal may go down and tear its nose before you can undo even a quick-release knot.

Flanking the Calf

Flanking can be used very efficiently by a man of average size on baby calves up to 300 pounds. The calf can be approached, cornered if necessary, and caught, or it can be roped. Many beef and dairy management techniques, such as treating the navel with iodine, ear-tagging, tattooing, castration, and vitamin injections, can be performed when calves are flanked and restrained on the ground.

Step-by-Step Procedure

1. Approach, corner if necessary, and catch or rope the calf. Certainly, it is good common sense and good animal management practice to minimize stress on any animal, especially calves, whenever possible. Catch the calf as quietly and as efficiently as you are able. Shepherds have used a crook for centuries to snare the hind leg of a sheep they have singled out for handling. This method is efficient, quiet, and not stressful, as you can usually snare the animal before it realizes it is being captured. Very recently, cattle producers have rediscovered the shepherd's crook. It is being marketed as a "Calf Catcher" (Safe-T-Katch Calf Catcher is one brand name). It works well, and with a little practice you will understand the usefulness of the shepherd's crook in your cattle operation.

2. Two methods can be used when flanking the calf; the choice of method is dependent upon the approach used.

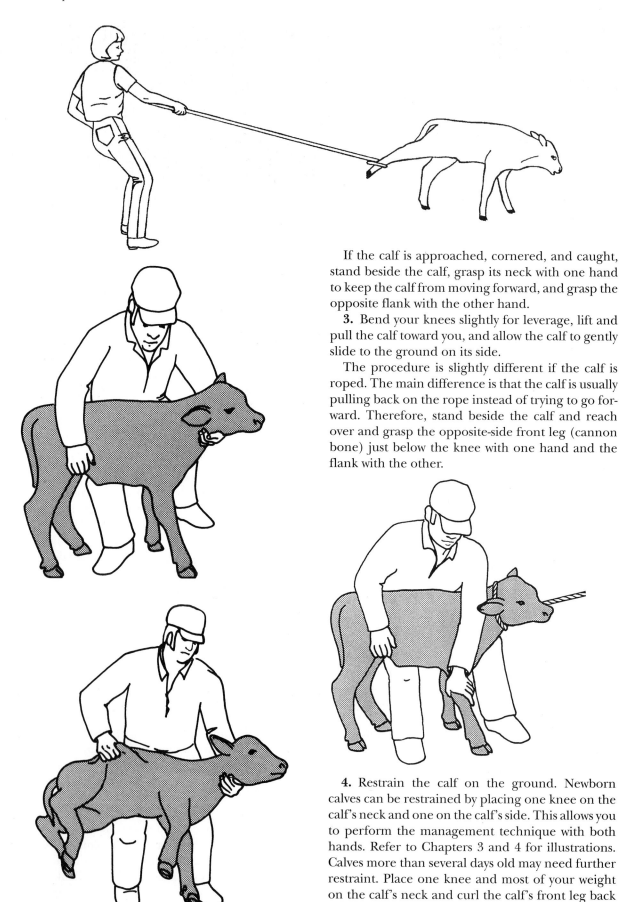

If the calf is approached, cornered, and caught, stand beside the calf, grasp its neck with one hand to keep the calf from moving forward, and grasp the opposite flank with the other hand.

3. Bend your knees slightly for leverage, lift and pull the calf toward you, and allow the calf to gently slide to the ground on its side.

The procedure is slightly different if the calf is roped. The main difference is that the calf is usually pulling back on the rope instead of trying to go forward. Therefore, stand beside the calf and reach over and grasp the opposite-side front leg (cannon bone) just below the knee with one hand and the flank with the other.

4. Restrain the calf on the ground. Newborn calves can be restrained by placing one knee on the calf's neck and one on the calf's side. This allows you to perform the management technique with both hands. Refer to Chapters 3 and 4 for illustrations. Calves more than several days old may need further restraint. Place one knee and most of your weight on the calf's neck and curl the calf's front leg back into a position that allows you to hold the calf down.

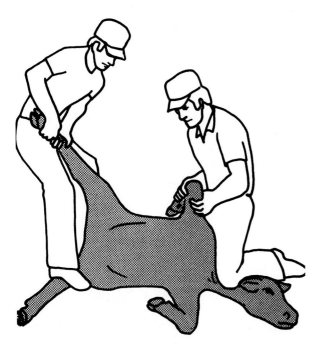

Calves approaching 300 or more pounds may require two people to restrain them on the ground. The first person places a knee on the calf's neck and curls the front leg back. The second person places his foot on the hock of the bottom rear leg and pulls back on the top rear leg to extend it.

5. Release pressure on the calf when the management technique is finished to allow the calf to get up.

Miscellaneous Restraint Techniques

Tail Twist. When an animal doesn't want to move through a chute or doorway, its tail can be twisted. Care should be taken that the tail is not twisted so hard that it breaks. Two methods can be used successfully, depending on the animal. The first method involves curling the tail into a loop. A second method involves grasping the tail and pushing it up to form a "lazy S" curve. When either method is used, stand to the side of the animal to minimize the risk of injury.

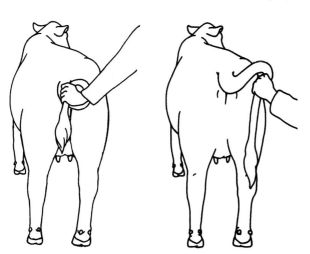

Tail Hold. Sometimes it is necessary to restrain an animal by distraction in addition to equipment restraint. The tail hold works well in a chute to prevent an animal from backing up. It also works well when the animal is tied, but is in need of an additional temporary distraction restraint. Stand beside the animal, grasp the tail near its base, and pull it straight up and over the animal's back. Generally speaking, the animal will not kick when the tail is in this position.

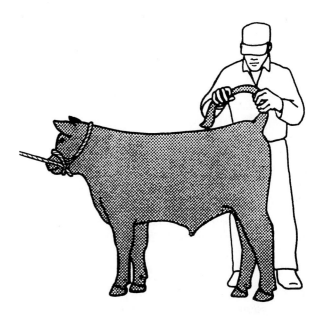

Nose Hold. Cattle often become contrary and lie down in a squeeze chute, foot-trimming chute, or grooming chute. When this happens, grasp the animal's nose by placing the palm of your hand over one nostril with your fingers tightly clamped over the other nostril to block breathing. When space and facilities permit, place the other arm around the animal's head to help restrain it. As the animal tries to

breathe, it may throw its head from side to side or up and down in an attempt to escape the nose hold. Continue holding the nose until the animal, in an effort to breathe, gets up. Some cattle will figure this out and begin to breathe through their mouths. When this happens, attempt to clamp the mouth shut in addition to holding the nose. The animal will get up before there is any danger of suffocation.

1.5 SWINE RESTRAINT

Restraining swine is a technique that is used routinely in day-to-day management. The manager or herdsman must know the appropriate techniques so that the restraint can be implemented in a safe and effective manner for both the animal and the handler.

When swine are taken away from their natural surroundings they become nervous and sometimes belligerent. Thus, it is essential that you work quickly and quietly, being gentle at all times. Make a determined effort to not mistreat the animals. Proper precautions should be taken to ensure that escape is impossible for the swine and that the equipment is in readiness for whatever management tasks are to be performed.

Restraining Young, Lightweight Pigs

For swine weighing less than 75 pounds, restraint is best accomplished by hand. The actual size of the pig that can be handled will depend upon the strength of the individual doing the holding and upon the task to be performed on the restrained animal. A pig can be restrained by hand in three ways: on its side, by its front legs, and by its rear legs.

Restraining the Pig on Its Side

Step-by-Step Procedure

1. Place the pig into the pen or alley where it is to be restrained. A small pen will make cornering and catching the pig much easier. In some instances, a hog hurdle or portable partition can be used to corral the animal.

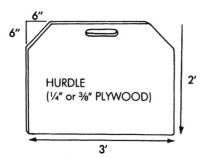

2. After cornering the pig, grasp it firmly by the hind leg with one or both hands. The hind leg is grasped and the entire rear end of the pig lifted off the ground so that the pig is deprived of its means and strength for movement and escape. The pig will continue to pull away from you, using its front legs, and sometimes squeal loudly, but you will be able to hang onto the rear legs and prevent the escape. With the small pig, grasp it by the rear legs and lift it completely off the floor with its head down. This will usually silence the squealing.

3. While holding a rear leg with one of your hands, use the other hand to grasp the front leg on the same side of the pig.

4. Using your hold on the front and rear legs, lift the pig completely off the floor and then gently put the pig back onto the floor, but on its side.

5. Use your knee to put just enough pressure on the side of the pig to help maintain control. Be careful when exerting pressure with your knee so that the pig's ribs are not cracked.

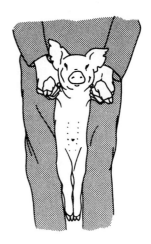

CAUTION: Be alert and watch for attempts by the pig to swing its head and mouth toward the hand restraining the front leg. Pigs can inflict a serious bite.

Restraining the Pig by Holding Its Front Legs

Step-by-Step Procedure

1. Place the pig into a small pen or alley as before and use a hurdle or a portable partition to maneuver it into a corner where you can grasp it.

2. After cornering the pig, grasp it by either of its rear legs with one or both of your hands. Hang on tightly—the pig's initial escape activities are quite strenuous.

3. After gaining control of the pig by grasping one of its rear legs, maneuver your hands and body so that you can grasp the front legs, one at a time. Your left hand should control the pig's left foreleg, and your right hand its right foreleg.

4. While controlling the pig in this manner, set it on its tail, with its back toward your legs. Its head should be up and its body squeezed between your legs.

CAUTION: When holding the pig by its front legs, be alert to the possibility of being bitten by the pig in its attempt to escape. Pulling back on the front legs and controlling the pig's head to some degree with your arms will help to prevent this. Although cumbersome, a durable pair of leather-type gloves should be worn to reduce the possibility of injury to the holder. Be careful when exerting pressure with your knees so that the pig's ribs are not cracked.

Restraining the Pig by Holding Its Rear Legs

Step-by-Step Procedure

1. Maneuver and corner the pig as in the first two restraint techniques.

2. Catch the pig by grasping a rear leg with one or both hands. Quickly adjust your grip so that your right hand is holding the pig's left rear leg and your left hand its right rear leg.

3. Lift the pig free of the ground by bringing both of its rear legs to about the height of your waist. This will position the pig's back against the front of your legs with its nose toward the ground. The pig's body can be squeezed between your legs for more control, if required.

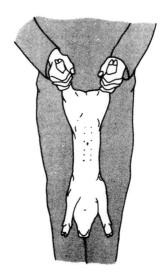

Restraining Older, Heavier Pigs

For restraining heavy pigs, several methods of restraint can be used, depending on the management task to be performed. These include restraint using a snare, laying the animal on its side, and using a headgate.

Using a Snare

Step-by-Step Procedure

1. To catch it with a hog snare, the pig must be confined to a very small pen, or crowded into a corner with a hurdle or partition. The crowding must be tight enough to prevent the pig from turning about and being able to avoid the snare. To facilitate snaring, the hurdle or panel should not be over 30" to 36" high.

2. Prepare the snare, making a loop at least 4" to 6" in diameter. Be certain that the snare, if wire cable, is of sufficient diameter (about $1/8$") that it will not twist into unmanageable shapes after being used a few times. The cable of the snare should retain its shape after being used. Other types of snares may be constructed from twine, rope with slip knots, or very small wire cable. The strength of the material must be proportionate to the animal's size.

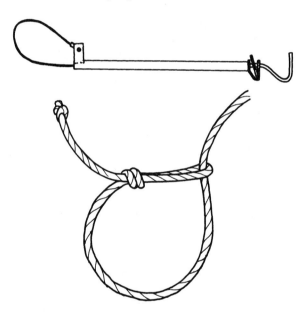

3. If you are using the panel to position the pig, have it placed between yourself and the pig and approach the animal from its right side if possible. Left-handed herdsmen may prefer the left side. Snaring can be attempted, and is often successful, without the aid of the hurdle or panel when pigs are confined to a very small pen.

4. With the handle of the snare in one hand, guide the loop of the snare into the mouth and over the nose or upper jaw. If a cable loop is used, it should be gently positioned as far back as possible into the mouth of the animal. The pig will naturally move its nose away from the snare as attempts are made to insert it into the mouth. Place your other hand upon the barrel of the snare to guide it as necessary. Make sure the snare is above the tongue and pulled back into the mouth. Make sure the loop is

not around the lower jaw. Placing the loop around both the upper and lower jaws is not satisfactory because it will slip off when the animal pulls back.

5. When the snare is properly positioned in the pig's mouth, pull the loop tightly around the upper jaw. The snare should be closed, as tightly as possible, by pulling upward on the snare handle with one hand and pushing downward on the barrel of the snare with the other. These pushing-pulling motions must occur simultaneously.

6. After the snare is in place, move to the front of the animal and pull forward. The pig will pull back into a rigid stance in its attempt to get away. With the animal thus restrained, maintain firm control until the required task is performed on the animal.

7. When swine are large (usually 400 pounds or more), they cannot be restrained with only a hand-held snare. For animals of this size, a rope snare should be used and the end secured to a sturdy object. When the rope is tied off, it is best to limit the animal to a foot or less of rope so it will not move forward, loosen the rope, and escape.

> **CAUTION:** Proper equipment that is strong and properly secured is important when preparing to snare and tie large swine. Animals that have been snared several times become wise to the snare and more difficult to catch. Therefore, they must be confined to an area from which they cannot escape until the snare is in place.

Laying the Animal on Its Side

Once an animal is snared, more complete means of restraint are often necessary to accommodate certain types of surgery, foot trimming, or other management procedures. There are two methods of laying the animal on its side.

Method One—For Animals Weighing Over 300 Pounds

1. Snare the animal as in the previous technique. Although the present technique is not primarily based upon snaring, firm control with the snare throughout the technique is essential to its success.

2. Form a loop in the end of a 15' length of $1/2$" to $5/8$" rope and place it around the neck of the pig.

> **CAUTION:** The noose that goes around the neck must be secured in place with a bowline knot so that it will not tighten and strangle the animal.

3. Position the loop so that the bowline knot is slightly to the right or left of the top of the pig's neck; then bring the rope back and place a half hitch around the body immediately behind the front legs.

4. Run the rope farther back along the topline and place a second half hitch just in front of the rear legs.

5. The pig can now be cast or laid over upon its side by pulling on the end of the rope that extends to the rear beyond the second half hitch. When pulled on with substantial force, the half hitches tighten, and the pressure exerted by the tightening

hitches will cause the animal to lie down. Keep the rope taut until the management technique has been completed and the animal is released.

6. As the animal begins to go down in response to the tightening hitches, it can be guided with the rope and snare to lie on one side or the other.

7. When the management task is completed, loosen the half hitches first and then the noose.

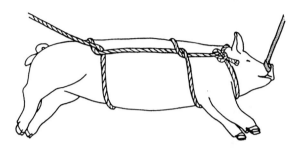

8. Remove the snare and observe the pig for a few seconds to see that he is recovering properly.

Method Two—For Animals Weighing Less Than 300 Pounds

1. Snare the animal and maintain control with the snare until the management technique is finished and the animal is released.

2. Using a *quick-release* knot, tie one end of a 4' to 5' length of $1/2$" to $5/8$" rope tightly around a front leg of the pig just above the knee.

3. Extend the rope to the rear leg on the same side and, while the pig is standing in a natural position, pull the rope taut and tie it with a quick-release knot to the rear leg above the hock. Move the rear leg if required.

4. Stand facing the same side of the pig where the rope has been attached to the legs.

5. Take a 1" × 6" board, 4' to 5' long, and hold it upright on the opposite side of the pig. Place the end of the board under the pig and behind the rope that connects the legs on your side, 3" to 4" from the end.

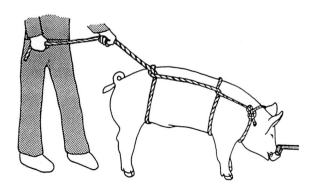

6. Pull toward yourself firmly and steadily on the top end of the board. Do not jerk sharply on the board as this could injure the pig's legs. As the front and rear legs of the pig are pulled toward each other, the animal will go down and roll toward you with both its front and rear legs off the ground.

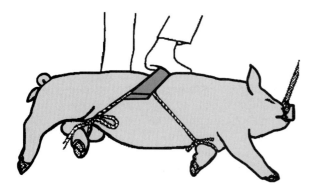

7. By pushing down and maintaining pressure on the board, you keep the pig firmly restrained. The snare must remain taut on the animal.

8. Perform the required management task.

9. Remove the board and rope from the pig's legs.

10. Release the snare and observe the pig for a few seconds.

Using the Headgate

The headgate is a useful, efficient means of restraint. It catches and restrains the pig by the neck, between the jaws and shoulders. It enables you to carry out techniques that need to be done on the head of the pig, such as ringing, ear-tagging, and obtaining blood samples from the ear, with a minimum of hassle.

1. Have all the necessary equipment, partitions, and the headgate in place and well anchored. The approach alley to the headgate should be narrow, usually about 18", so the animal cannot turn around.

2. The system works best if one individual drives the pig into the headgate while another operates the handle to the headgate. The individual operating the headgate should have his hand on the handle of the open headgate ready to catch the pig. The headgate is left in the open position to entice the pig to enter.

3. Drive the pig toward the open headgate. The headgate operator should not move, and if possible, be out of the line of sight of the pig so that it will move into the headgate more easily.

4. As the pig's ears pass the headgate opening, close it quickly, catching the pig at the neck. If the pig is not quite far enough forward, slightly reduce the pressure on the headgate handle while the person driving the pig pushes the animal forward into place.

CAUTION: With some headgates, it is possible to catch a pig directly on the cheek or jawbones. If too much pressure is applied, these bones may be broken. One can usually feel, with experience, whether or not the animal is caught properly by the neck. Fortunately, if the pig is caught by the jaw or cheek bones, it usually will slip back and get loose.

Some pigs move quickly enough and are strong enough to dart or smash through the opening of the headgate. The catcher is going to miss catching a pig on occasion. Provisions should be made for an exit pen where the pigs can be corralled and again guided through the chute.

5. After the pig is caught, latch the headgate to secure the pig while the management technique is performed. Once caught, the pig will fight the restraint by pulling backward with its rear legs. The reason for anchoring the headgate and chute is that it can be shifted out of position by this fighting. After a short time, the pig will usually quit fighting the restraint.

6. Release the pig. The person acting as the driver should make sure that the pig exits through the headgate.

1.6 HORSE RESTRAINT

The horse is a large, strong, quick-moving animal, and is potentially dangerous to handle. If the management technique you must perform is one that the horse is frightened by or one that will cause it some discomfort, remember that the horse can and will call upon its defensive behavioral characteristics: kicking (forward, backward, and sideways), striking out with one or both forelegs, rearing up to flail about with its front legs, biting, bolting ahead to run over the handlers, and leaning against the handler to smash him against a partition.

> **CAUTION:** Horses have a blind area immediately in front of them and to their rear. They literally cannot see anything in these areas, even something as large as a person. Couple this blindness to the horse's ability to sleep standing up and you have the potential for some dangerous moments for the nonthinking manager. A person can totally surprise a horse by approaching it directly from the front or directly from the rear if he does so without speaking to the horse. Approach the horse from a 45° angle and speak to it or rattle a feed bucket as you do so. Should you not follow these suggestions, you face a very real danger of being injured by a completely surprised horse.

The preceding is not to imply that every horse is mean or spoiled, or that they are all unsafe to handle; it simply is meant to encourage the handler to put safety foremost in his mind when handling a horse.

Because of the potential dangers involved, many horse management techniques must be preceded by appropriate restraint. However, do not use more restraint than is necessary to get the job done. For example, the horse that resists clipping does not need to be cast and tied if a nose twitch will distract it enough to cause it to stand still for the clipping. Horse restraint will vary from a chain lead over the nose to complete sedation. The horse's temperament, the task to be performed, and your handling ability must be evaluated before selecting the restraint method.

Chain Shank

The "lead rope" for a horse is frequently a 6' to 8' leather strap with a 12" to 18" chain attached to the end of it. This leather–chain lead should never be used to tie a horse, but it does lend itself to several mild, in-hand restraint uses.

Under the Jaw. Pass the snap of the chain lead through the nearside cheek ring of the halter, run it under the jaws or chin of the horse, and attach it to the cheek ring of the halter's offside. The bottom halter ring is skipped by the chain. The proper technique is to give quick, sharp jerks on the lead if the horse does not handle as you wish. This technique can be counterproductive; that is, it can produce exactly the wrong results! The horse will naturally jerk its head upward to escape the pain when the chain tightens under its lower jaw.

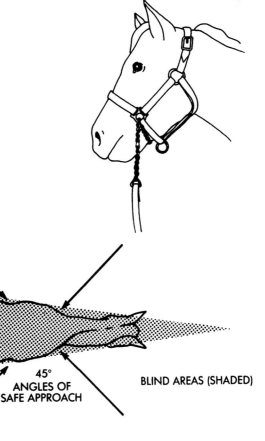

45°
ANGLES OF
SAFE APPROACH

BLIND AREAS (SHADED)

Through the Mouth. As before, pass the snap of the chain lead through the nearside cheek ring of the halter, run it from there into the horse's mouth—exactly as a bit would be used—and attach it to the cheek ring of the offside of the halter. The technique is to give quick, sharp tugs on the chain to make the horse respond to your control attempts. This can be a worthwhile restraint method, but care must be taken not to damage or "toughen" the horse's mouth. The most likely use of this restraint method is stallion handling during teasing or breeding.

Over the Nose. Pass the snap of the chain lead through the nearside halter cheek ring, over the top of the horse's nose, and attach it to the offside halter cheek ring. The technique is to tug with sharp, quick pulls upon the lead. The horse's attention may be diverted and the pressure on the nose may cause it to lower its nose as you desire, but the technique may backfire and cause the horse to fret, toss its head, and fidget more than usual! The primary value of this technique appears to be in leading high-spirited or ill-mannered horses.

Nose Twitch

The nose twitch is the basic tool of horse restraint. Twitches come in a wide variety of shapes, sizes, and configurations, homemade and commercially manufactured, good and bad. Whatever its shape and origin, a nose twitch is meant to apply pressure to the nerves of the horse's lip ("nose") and impart some degree of pain. This pain diverts the horse's attention toward the nose and away from the less painful work you are performing on it elsewhere. The directions that follow are for the use of the wooden-handle, chain-loop, or rope-loop twitch.

CAUTION: If the nose twitch is applied too tightly or for too long a time, serious damage can be done to the horse's lip.

The nose twitch must never be applied to the horse's ears. It may become totally ear-shy and lop-eared due to damage to the cartilage or to the nerves innervating the ear muscle.

The twitch must not be considered a cure-all. It should not be a substitute for proper training. Horses should be *taught* to lift their feet for foot work, to submit for clipping and ear trimming, and to allow washing of the sheath. The twitch should be considered primarily a crisis tool—a tool to be used in an emergency situation—not as a crutch for the poor horseman to lean upon!

Many horses will vigorously resist twitching. Placing the twitch can become a seemingly impossible and quite dangerous task. The frightened or determined horse may very well strike, rear, and bolt to escape its application.

Step-by-Step Procedure

1. Place a well-made halter and lead rope upon the horse. Have a knowledgeable assistant control the horse's offside.

2. Grasp the twitch handle in your right hand and insert three or four of the fingers of your left hand through the loop in the twitch end. Your thumb should remain outside the loop.

3. Approach the horse from its nearside in the safety zone off the shoulder. Use your left hand to quiet the horse by placing it upon the bridge of its nose. Slide your hand down the face to its lip or nose and grasp it securely in your left hand.

CAUTION: Do not lunge at the nose or otherwise frighten the horse by your movements. The horse may attempt to escape or defend itself at any point in this procedure—take care to remain alert and in a safe position.

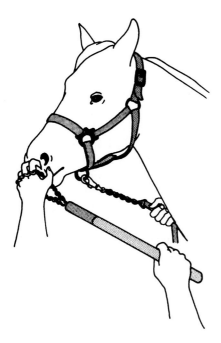

4. While holding the nose in your left hand, manipulate the twitch handle with your right hand so that the chain loop is positioned around the horse's nose. When it is in position, twist the twitch handle clockwise, causing the loop to tighten about the nose. At this point, release the left-hand grip from the nose. The twitch should be tightened only enough to keep the horse's attention; however, care

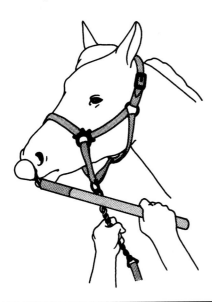

CAUTION: *Twist* on the twitch handle only. It is not a handle with which to pull on the horse's head. The assistant can control both the lead rope and twitch while the handler performs the management technique. It is not unusual to use three people in this situation—the lead-rope handler, the twitch handler, and the person performing the management procedure.

must be taken to avoid allowing the twitch to slip off during the management procedure.

5. Move the twitch handle back and forth in small arcs during the time the horse is restrained. This allows for slight relief from the numbing effect of long-term twitching. Keep the horse's attention by jiggling the lead rope or tapping upon the handle of the twitch.

6. To remove the twitch, place your left hand upon the end of the horse's nose and firmly grasp it. Carefully untwist and remove the twitch while at the same time beginning to rub and knead the nose. Continue rubbing the nose for a few seconds to stimulate blood flow.

CAUTION: The release of the twitch may be the signal for the horse to initiate escape activity. Take care to remain in a safety zone and to hold the twitch handle firmly in your right hand so that you are not injured by its swinging about.

Foreleg Restraint

Because the basis for a horse's escape activities involves its four strong legs placed upon sure, true footing, an excellent yet mild form of restraint involves "taking away" one of the horse's legs. With the horse now having only three legs to stand upon, it is less likely to dance around or to kick out with a rear leg, thereby leaving only two for support. Usually the front leg, on the same side as the rear leg you wish to work upon, is lifted.

Step-by-Step Procedure

1. Catch and halter the horse.

2. Place a web or leather hobble (strap with buckle and ring) around the pastern of the leg to be lifted. Fasten it securely so that it does not slip up above the fetlock.

3. Have a handler restrain the horse with a lead rope and halter while you tie one end of a $5/8$" or $3/8$" three-strand cotton rope into the ring of the hobble.

4. With the horse's four feet still upon the ground, pass the rope over the horse's topline, just to the rear of the withers. Bring it around the front of the horse's neck and run the working end of it between its standing part and the horse's shoulder. Bend it away from the horse and bring it back over the standing part toward the handler. The handler should grasp the end of the rope and snug it about the horse's neck. Lift the horse's leg and at the same time have the handler take up on the new slack created. With only moderate tension, the handler can keep the foreleg in a lifted position.

CAUTION: The horse may rear, strike bolt, or lunge just as it feels its leg being taken away. Be prepared for this and remain in a safety zone.

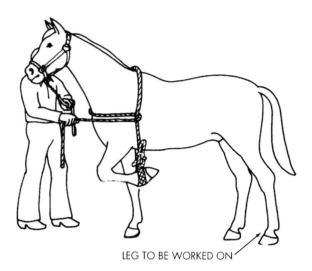

LEG TO BE WORKED ON

5. It is possible to tie the leg in an up position, either around the opposite-side elbow or with a quick-release knot to the rope segment that comes around the neck. This could make the procedure a one-man type, but this is not recommended for your or the horse's safety.

6. If a third set of hands is available, the rope is not a necessary part of this restraint technique. One person can control the lead rope and head, and another can lift and hold the leg in the up position while the third performs the management procedure.

Rear-Leg Restraint

The rear leg or legs of a horse are restrained for a variety of management procedures, including the prevention of kicking during rectal examination and breeding, hoof work (upon the lifted foot), minor surgical procedures, and as a training aid.

CAUTION: The procedures that follow are simple and quick to perform, but as with any techniques that threaten to take away a horse's footing, they may provoke a violent reaction. There is constant danger to you and to the horse (by falling) while it is being restrained.

Part of the need for rear-leg restraint is the tendency of the horse to kick. In fact, one reason for restraining the rear leg is to *teach* the horse not to kick. Take care that you do not get kicked while preparing to teach the horse not to kick you!

Sideline or Scotch Hobble

Step-by-Step Procedure

1. Have a capable assistant control the horse's head with a high-quality halter and lead rope.

2. Select a 20', $^3/_4$", three-strand cotton rope and, using a bowline knot, tie a large loop around the horse's neck. Position the knot over the shoulder on the same side as the rear leg to be lifted.

3. Run the working end of the rope from inside to outside, around the pastern of the rear leg to be lifted. A hobble, hobble and ring, or burlap loop can be used to encircle the pastern and prevent the possibility of rope burn.

4. Return the working end to the neck loop and tie it off there with a quick-release slipknot. If no hobble and ring is used, wrap the working end of the rope around the standing part twice to prevent the horse from stepping out of the loop.

5. The degree of rear-leg restraint will depend upon what management technique must be done. Rectal exams require only that the horse not kick. Training the horse not to kick requires slightly more lift so that the horse is jerked off balance when it attempts to kick.

Tail Tie and Leg Lift

Step-by-Step Procedure

1. Have a capable assistant control the horse's head with a high-quality halter and lead rope.

2. Place a hobble and ring or a burlap loop around the pastern of the horse's rear leg that must be lifted.

3. Attach one end of a 20', ³/₄" three-strand cotton rope to the tail just below the end of the caudal vertebrae. The knot used to join the tail hairs and rope is the *tail-tie knot;* it is tied in the following manner: (1) form a bight of 18" to 24" in one end of the rope (Figure A); (2) lay this across the horse's tail

rope bight, wrap it around the back of the tail, and bring it out and over the top of the standing part of the rope bight (Figure B); (5) wrap the working end of the rope bight over the tail-hair bight and insert the working end of the rope bight into the loop of rope formed on the left of the horse's tail (Figure C); (6) pull on the standing part of the rope to shape and secure the tail tie (Figure C).

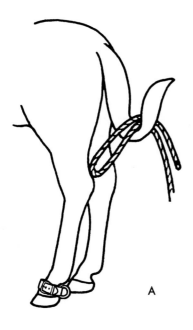

A

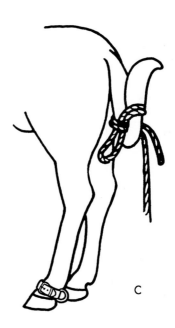

C

from right to left just below the end of the tail vertebrae (Figure A); (3) form a bight in the horse's tail by bending it backward and upward around the rope (Figure A); (4) take the working end of the

4. Insert the working end of the rope into the hobble ring from inside to outside, position the rope to the inside of the hock, run it up one side of the tail root, across the midline of the croup, and into the hands of the assistant, who is standing near the shoulder of the horse and who will hold the end of the rope and determine the height at which the hoof will be held.

CAUTION: This manner of restraint will not completely stop a horse from kicking, although some of the force will be taken away from the blow. Its primary use is to lift and hold a rear leg for prolonged work. Be careful and stay in the safety zone.

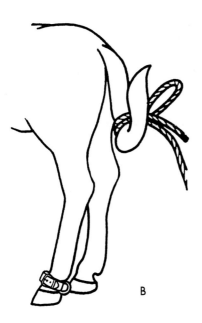

B

5. The initial lifting of the horse's rear leg should be done in the same manner as you would lift it for trimming. Do not use the rope to lift the leg; it should be used only to hold it.

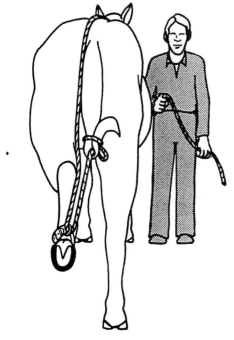

Breeding Hobbles

Step-by-Step Procedure

1. Have a capable assistant control the horse's head with a high-quality halter and lead.

2. Place a set of figure-eight hobbles with a ring or a burlap loop onto each hock of the horse. As an

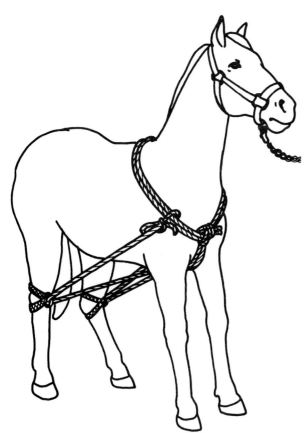

alternative, a hobble and ring or a burlap loop can be fastened around the pastern of both rear legs of the horse.

3. Tie a large doubled-bowline loop or a large bowline-on-a-bight loop into the center of 25' to 30', $^3/_4$", three-strand cotton rope.

4. Place this double loop over the mare's head and position the knot on the center of her chest. Run both ends of the rope between her front legs and place one end into each of the rings of the hock or pastern hobble. Depending on the length, the ropes can be tied at the ring with quick-release knots or run back up to the neck loop and tied off with quick-release knots.

CAUTION: It is possible for the stallion to become entangled in the ropes during the breeding mount or dismount. Be certain the quick-release knots are properly tied.

Miscellaneous Restraints

Neck Cradle

The neck cradle is a collar of sticks constructed so that it is impossible for the horse to lower its head or bend its neck to either side. Its primary use is to keep a horse from chewing at bandages or wounds on its body or legs. The sticks are 2" to $2^1/_2$" apart at the throat-latch end and 3" to 4" apart at the shoulder end. It is impossible for the horse to eat from a low manger with the cradle in place.

Side Stick

The side stick is a wooden dowel or light piece of pipe running from the halter lead ring to a side ring on a surcingle. Its purpose is to prevent the horse from reaching a wound or bandage on the rear half of its body. It has the advantage of being less restrictive than the neck cradle (the horse can eat with this one), but cannot be used for front-leg injuries because it would be possible for the horse to reach them.

Muzzles and Bibs

Muzzles and bibs are other contrivances used by man to prevent horses from chewing upon wounds and bandages. With a muzzle and certain types of bibs, the horse cannot eat, although drinking is possible. When the muzzles and bibs are removed

MUZZLE

BIB

for feeding, the horse must be observed closely or it will undo all the good that has been accomplished with only a few quick bites.

Earing Down

Earing down is an overstated title that refers to the technique of grasping a horse's ear in your hand and applying pressure to it by squeezing. If this technique is performed properly, the horse will not become ear-shy. The value of this temporary restraint is to distract the horse long enough to put another restraint upon it—for example, earing the horse for a few seconds until a nose twitch can be put into place.

Step-by-Step Procedure

1. Have a capable assistant control the horse's head with a high-quality halter and lead. He should stand on the horse's offside.

2. Approach the horse from the nearside and stand next to its shoulder. Place your left hand onto the cheekpiece of the halter and your right hand onto the upper neck and mane.

3. Speak to the horse, apply give-and-take pressure on the halter cheekpiece, and begin rubbing the neck, mane, and poll with the palm of your right hand.

4. Gradually work your right hand to the horse's near ear. The base of the ear should be

positioned into the web between your thumb and index finger.

5. Grasp the ear near its base and firmly close your grip. At the same time, pull the horse's head toward you with your left hand.

6. When you release the ear, massage it gently. It is good insurance to massage the ears daily as part of

CAUTION: Do not twist the ear or fold it over before squeezing. This is totally unnecessary and will make the horse ear-shy.

Do not lunge at the ear and try to grab it. Work up to it as directed. Expect the horse to attempt to pull away from your grasp. It may even attempt to rear. Be prepared for this and stay in a safety zone.

the grooming process, in preparation for the time when it will be necessary to grasp an ear for real.

Skin Roll (Skin Twitch)

The skin roll is another type of momentary restraint. Its primary use is to discourage escape activity by diverting attention until a more traditional restraint can be employed.

To use the skin roll, grasp the skin over the scapula in your hand or hands. "Roll" this grasp of skin forward and maybe slightly upward. This will serve to distract the horse only for a few seconds if the technique being performed is severe, but serves for a longer time if the management technique is mild. Your grasp will have to be firm, because the skin is tight and does not roll easily. Massage the skin after releasing it.

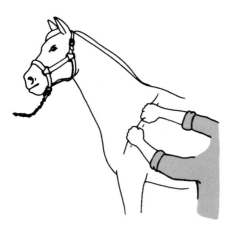

Temporary Rope Halter

A temporary rope halter can be made from any lead rope or lariat that is at hand. This technique has saved the day innumerable times when a horse without a halter had to be caught and led.

Step-by-Step Procedure

1. Place the loop of the lariat around the horse's neck. If the rope available has no honda, tie a loose loop with a bowline. (See "A")

2. Stand to the horse's nearside and form a bight in the standing part of the rope. (See "A")

3. Run this bight through the neck loop, bottom to top (see "B"), and as you do so enlarge the bight

enough so that it can be slipped over the horse's nose. (See "C")

4. With the temporary halter constructed in this manner, lead from the horse's nearside.

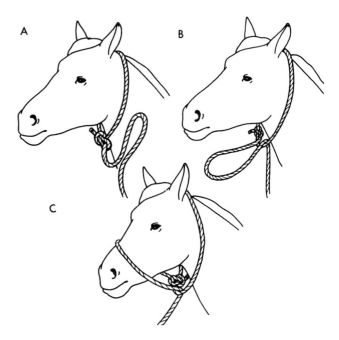

War Bridle

The war bridle, as a miscellaneous restraint technique, has been saved until last, and that is exactly where it belongs and how it should be used—as a last resort! It is a more severe form of restraint that should be used only on hard-case, mean horses, when no other method of restraint is available or when it is not possible to place any other method of restraint upon the horse. Admittedly, some people are expert in the use of the war bridle and cause the horse no more pain with it than with any other restraint technique. Regrettably, people with this knowledge are rare.

The war bridle brings pressure to bear upon the horse's poll, and upon the gums of the upper teeth or corners of the mouth. Like any other handheld restraint method properly applied, the war bridle works by causing discomfort, thereby distracting the horse from the discomfort of your management technique.

Step-by-Step Procedure

1. Form a loop in a 10' to 15' piece of $^3/_8$" rope by running the working end of the rope through a honda or eye splice installed in its standing end.

2. Place this loop over the horse's head and position it so that the honda is coming from over the top (versus up from the bottom) and lies on the horse's nearside. The rope should lie just behind the ears on the poll and pass through the horse's mouth like the mouthpiece of a bit. A harsher variation is to pass the mouth rope over the gums of the upper teeth.

3. The proper technique is to pull steadily or to use short, light tugs on the rope end. Heavy, vicious yanks will not accomplish more and run the risk of damaging the horse's mouth.

> **CAUTION:** The horse must never be tied with the war bridle. If the horse needs the war bridle, it is ill-mannered and likely unpredictable. Take care when approaching and handling such a horse.

Casting the Horse

There are times when it is necessary to have the horse restrained on the ground in a totally controlled situation. Three general methods can be used to accomplish this task: (1) general anesthesia, (2) casting and tying without anesthesia or sedation, and (3) the combination of tying, sedating, and casting. The following is the procedure for the third alternative, which is the most often used.

Step-by-Step Procedure

1. Have a capable assistant control the horse's head with a high-quality halter and lead.

2. Tie a doubled bowline knot or a bowline-on-a-bight into the center of a 60' to 70', $^3/_4$", three-strand cotton rope.

3. Place a hobble and ring onto the pastern of each rear leg.

4. Place the doubled bowline loop over the horse's head and position the bowline knot over the withers. One working end of the rope should extend to each side of the horse.

> **CAUTION:** It is important that the bowline loop be sized so that it is large enough to fall beyond (rearward of) the point of the horse's shoulder. This is to avoid damage to the nerves and muscles running across it when the horse goes down and the legs are flexed and tied to it.

5. Run the working end of each rope into the appropriate hobble rings, from inside to outside. After exiting the rings, the working end of each rope should then be inserted into the neck loop on its respective side, in a direction coming up from the rear leg, under the neck loop, and out to the front.

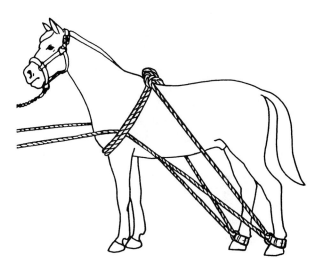

6. At this point, administer the sedative and decide to which side the horse will be cast. For our purposes, assume that we have chosen the right, or offside.

7. Take the nearside sideline back around the horse's hip so that the handler can apply tension at a 90° angle to the horse's body. This will assist in pulling the horse over at the appropriate time.

8. The offside rear leg is lifted by an assistant and the slack taken away by the offside sideline handler so that the leg is held in an up position—6" to 8" off the ground.

9. When the horse is becoming wobbly from the sedative, it should be pulled down onto its offside. The handler of the nearside rope puts tension upon the left hip, the lead-rope handler pulls the entire head to the left, and the offside sideline handler keeps the offside rear leg up.

10. As soon as the horse goes down, the lead-rope handler should place a double saddle blanket or other appropriate padding beneath and over its head (to help protect and quiet the horse).

> **CAUTION:** Take care to avoid the flailing legs of the cast horse. The horse will very likely struggle until its legs are secured.

11. All four legs should immediately be flexed and secured tightly against the body of the horse. This is for the manager's safety as well as for the horse's. The legs, fore and rear, are secured to the doubled neck rope with a series of half hitches and figure-eight loops.

> **CAUTION:** It is critical that the horse's legs remain flexed and securely fastened. You have plenty of rope remaining from the sidelines, so there is no excuse for insecurity.
>
> When releasing the horse, do so carefully, because it may be recovering from the sedative and partially regaining control of its limbs.

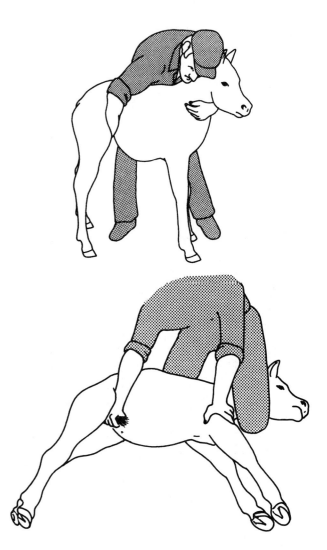

Foal Restraint

There are many times when a foal must be restrained, and because they usually are unschooled in the ways of the halter and lead rope, they must literally be held in position. The surest way to accomplish this is to corner and grasp the foal about the neck. Quickly maneuver yourself so that one of your arms is wrapping the chest of the foal, the other wrapping its hips just below the tailhead. Your knees can be placed in front of and behind the foal for added security.

If the foal must be cast, and you are of above-average strength and size, simply pick up the foal

from this position and lay it over upon its side. It is unnecessarily harsh for the foal and hard on your back to attempt this technique if it is beyond your strength. If such is the case, reach across the back of the foal with the arm that would normally encircle its hips, go up between its legs, and grasp its tail. Pull downward and toward the foal's abdomen until it gradually begins to relax and finally slumps to the ground. When it is on the ground, maintain your grip on the tail and control the foal's head and neck.

1.7 SHEEP RESTRAINT

Sheep that are managed as a flock, that is, not treated as individuals, as in a "pet lamb" situation, develop a wild, nervous temperament and are easily frightened. If you as the shepherd wish to handle sheep efficiently, you must recognize these behavioral traits and learn to work with them.

Because the sheep is wild, nervous, and timid, you must be steady, calm, and sure in your actions and voice. You must realize that one panicked animal can transmit its fright to the entire flock, and what was once a quiet, standing group can become a bleating, circling, jumping, senseless mass of hysterical animals. The single mistake that frightens the first animal can cause the flock to bolt and become unmanageable.

An unmanageable, frightened group of horses, cattle, or hogs is a very real danger to the handler. The superior size and weight of these animals, along with their well-developed defense mechanisms of kicking, striking, butting, goring, and biting, make them truly dangerous. Even in tightly confined quarters, however, the only hazard involved in handling a group of sheep that is frightened and panicked is that they will circle, mass, and pile up upon one another until they collapse and die of exhaustion.

To be certain, some mature rams and ewes are much bolder and will not panic or bolt as easily as the rest of the flock. This type of animal can deliver a quick, painful charge and butt which can knock a younger, lighter person from his or her feet. Serious injury is possible in these situations, and care must be taken to avoid it.

For the most part, however, sheep must be handled more with an eye toward minimizing the excitement, chasing, and wrestling that could harm them instead of with a strong concern for your own safety.

Catching Sheep

If the sheep are accustomed (conditioned) to coming to grain, minerals, or water at a given time and place each day, it is possible to capitalize upon this behavior to catch the animals. The water tank, feed bunk, or mineral feeder can be enclosed in some sort of pen. This can be a permanent corral adjoining the barn or loafing shed, a strongly built, rather elaborate "catch pen" out in the pasture, or a temporary, movable arrangement that can be shifted from pasture to pasture.

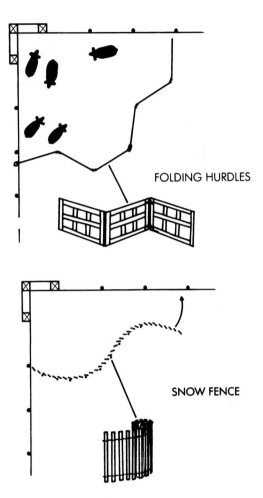

Temporary confinement fences for sheep usually consist of lightweight, easily handled, welded wire, wooden panels, folding hurdles, or rolls of snow fencing or woven wire. Steel posts provide the support, and an extension hurdle (expandable gate) acts as the gate.

> **CAUTION:** Sheep can easily jump over a 32" to 36" panel. For certain confinement, sheep partitions should be at least 40" to 42" high.

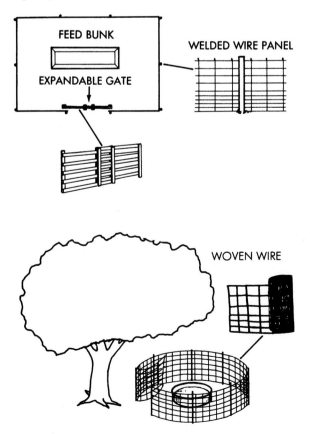

The sheep probably will have to be conditioned for several days before they will enter the catch pen freely, especially when you are present.

After the sheep have been lured into and confined in the temporary pens, they can be approached and caught and the necessary management techniques performed. After the technique is complete, each sheep can be released to the outside of the pen or "marked" with drover's chalk or a livestock grease stick and left inside until all members of the flock have been handled. The latter alternative is preferable, because with the entire flock confined and the sheep crowding one another, it is easier to catch the individuals. If each sheep is removed after the operations performed on it are completed, escape room is made more available for each remaining sheep, and some of the last few individuals can certainly tax a shepherd's patience. By the same rationale, the temporary pen should be no larger than is necessary to accommodate the group to be fed or watered in it.

Holding Sheep

When the sheep have been confined to an enclosure of the proper size (which is usually to say "small enough"), grabbing and holding on to them is more a matter of knowing *how* to do it than being strong enough to do it. To be sure, there are sheep that have learned that they can escape a jaw hold by lowering their heads and "bulling" forward, but those same sheep can easily be controlled by grabbing and elevating a rear leg.

Jaw or Chin Hold

Step-by-Step Procedures

1. Approach the sheep from whichever side is open to you. Keep in mind that even the meekest sheep isn't going to just stand there and allow you to walk right up and grab it. It will try to escape. Anticipate this and move quickly into the escape lane, then immediately continue walking toward the sheep. Keep your arms and hands extended out from your sides.

2. As soon as you are close enough, place one of your hands under the jaw or chin of the sheep and

lift upward immediately. Your intent is not to lift the sheep from the ground, but merely to gain control of its head. Most sheep stop struggling to escape when their jaws are grasped and canted slightly upward. An occasional sheep will try to jump over your jaw hold. If you are ready for it, this can be countered by flowing with the jump and by allowing the jump to swing your body and other hand into the sheep for further control.

CAUTION: Do not grasp the sheep by its wool. It could be torn away from its body, leaving a painful, unsightly injury, or at the very least, blemish the fleece and carcass.

3. As soon as possible, place your other hand behind the sheep's head. This will put your hands into a choke position around the sheep's neck, although this is *absolutely not* the intended purpose of this hold. The hand under the jaw is there to stop forward movement; the hand behind the neck is there to stop rearward movement. An alternative to this position is one hand under the chin and the other behind the dock or rear legs.

> **CAUTION:** Keep your center of balance well under yourself while holding a sheep in this manner. If you allow yourself to relax, you can easily be pulled off balance, fall to the ground, and lose the sheep if it should try to escape.

Leg or Flank Hold

When you are presented with the rear end of the sheep, you have two options to choose from in grabbing it. You can try to reach over the back of the animal and place a hand under its chin, or you can grab a rear leg just above the hock with your hand or with the shepherd's crook. If you reach over the back and take a chin hold, you may lose it if the sheep has any size at all and lunges forward, pulling you off balance. The recommended procedure is to grab a rear leg and pull backward and upward for control.

1. Approach the sheep as directly from the rear as possible, in a quiet, sure manner. Avoid an approach with enough angle for the sheep to see you, become alarmed, and attempt to escape.

2. With your strongest hand, grasp the sheep's leg on the same side as the hand you are using, anywhere from just above the hock to well up into the flank. Keep a firm grip and lift upward enough to clear the leg from the ground.

3. As quickly as you are able, maneuver your body so that you can grasp the jaw or chin with your other hand. When the jaw or chin hand is in position, release the leg or flank and control the sheep as in the jaw or chin hold.

Setting the Sheep on Its Rump

The whole purpose of catching and holding on to sheep is to enable you to perform the management techniques necessary for their well-being. Nearly all of these techniques require access to the sheep's underside. For all but the lightest lambs, this requires that you be able to "set up" the sheep, that is, catch it, hold it, "throw" or roll it to the ground, and reposition it so that it is sitting upright upon its rump with its backbone against your legs.

Step-by-Step Procedure

1. Approach the sheep from whichever side is open to you and catch it with either the jaw hold or flank hold.

2. Position yourself onto the sheep's left side. Hold the jaw or chin in your left hand and the dock in your right. The sheep's side should be against your legs.

3. Place the thumb of your left hand either over the sheep's muzzle or into its mouth just behind the incisor teeth. At the same time, move your right hand onto the sheep's right hip, just above and behind the stifle.

4. While controlling the sheep's lower jaw, bend its head sharply back over its right shoulder so that it is looking at its own rump. At the same time, press downward toward your legs with your right hand. This will cause the sheep to lose its footing and fall against your legs.

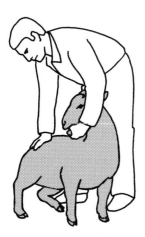

5. Step backward a half step or so, so that the rear of the sheep's body will slide off your legs and onto the ground. At this point, release the jaw and grasp the sheep's forelegs.

Straddle Restraint

Not all management techniques require that the sheep be set up. Drenching, bolusing or balling, eye or mouth examinations, and temperature taking are much more easily performed when the sheep is restrained in the normal standing position. If two people are available for the technique, one can hold the sheep in a jaw hold or a jaw and dock hold while the other examines, drenches, or whatever.

Sometimes two people are not available, or it is better management to use the second person in some other manner than helping with restraint. For such instances, the straddle restraint method is appropriate.

Step-by-Step Procedure

1. Catch, grab, and hold the sheep as in the jaw-hold restraint.

2. When the sheep is firmly under control, swing one of your legs over its back so that you are astride the sheep, as if you were going to ride it. Maintain control of the sheep's head with the jaw or chin hold.

6. Pull upward on the front legs to straighten the sheep. At the same time, step in behind the sheep and allow it to lie back against your legs. If the sheep should struggle and maneuver enough to begin to regain its footing, hold on to its front legs and step back a half step, causing the sheep to be rocked off balance and onto its dock and rump.

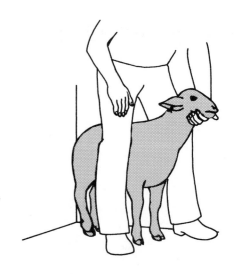

3. Clamp the sheep's shoulders or forward rib cage between your knees. Keep your feet solidly on the ground and use your hands to restrain the sheep from forward movement and your legs to restrain it from rearward movement.

CAUTION: Do not attempt to "ride" the sheep if it begins to escape and causes you to lose balance. You probably cannot, which means the sheep will escape anyway, and you run the risk of damaging the sheep's spine or legs. If the sheep tries to escape, stand fully upright and hold its jaw in your hand.

Halter Restraint

Admittedly, an adjustable rope halter is of absolutely no value to the shepherd as an aid to corraling, catching, and grabbing a sheep. Once the sheep is caught, however, the halter becomes a valuable tool in controlling it. With the halter properly applied and fitted, it is possible to tie a sheep for an extended period of time or to enable a smaller shepherd to control a very large sheep. Some of the larger breeds of sheep are handled in the sale and show rings in this manner. The halter is applied so that the lead rope comes out from under the chin on the left side. While it is possible to tie and restrain the movement of a sheep with the halter, do not expect the sheep to lead in the halter without proper conditioning or training.

Sorting (Cutting) Chute

If a large number of sheep must be handled in a relatively short period of time, a sorting or cutting chute is invaluable. Examples of its use include weighing (if a scale is set into the traffic flow), routine foot care (if a tilting squeeze is placed into the flow), weaning, sorting ewe lambs from ram lambs, and routine health care. If the sorting facilities are not available,

the alternative is to trap the animals in makeshift or permanent catch pens and catch and handle each individual as described earlier in this section.

Sorting chutes should be constructed with a curved crowding pen or holding area with pull-up entry and exit gates. The sides should be constructed of either plywood panels or plank lumber. Good working dimensions are 32" to 34" high and 14" to 15" inside width. Three-quarter-inch rods or steel pipes placed from side to side through the chute, 9" from the bottom and 4' apart, prevent the sheep from backing up. Attached immediately beyond the exit gate is the cutting gate, and depending upon your needs, it can be set up to allow the flock to be sorted into two or three separate pens. The lumber should be treated for maximum durability.

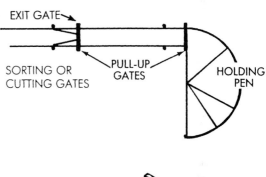

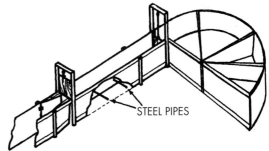

Step-by-Step Procedure

1. Move the animals into the curved holding pen connected to the sorting chute.

2. Open the pull-up gate, allowing the sheep entry to the chute. If the sheep are slow to enter the chute, have an assistant open the exit gate from the chute also. The assistant should remain near the gate so that he can drop it down once the sheep begin to enter the chute. If the assistant moves too far away, the sheep may bolt through the gate before he can recover.

3. After one of the sheep has entered the chute, the others will follow. Be patient and keep moving the sheep about in the holding pen until that first sheep commits. After that, rely on the $^3/_4$" pipes to prevent the sheep from backing out of the chute. A bellwether or "Judas goat" that has been trained to enter the chute freely is often kept with the flock. Because he is usually the oldest and boldest animal, the flock allows him to become their leader.

4. As sheep are released from the chute exit gate and sorted as needed, the open exit gate may serve as an inducement for the sheep next in line to move ahead in order to get out. It takes a practiced hand to allow the gate to remain open for a time long enough to entice the sheep to try to escape, but short enough to prevent it.

5. If the sheep do not move into and through the chute of their own choice, the shepherd must force them ahead. As explained previously, the sheep can be milled about until one of them blunders into it and the rest follow, or a trained sheepdog can be sent into the pen to start the sheep. If you have neither the dog nor the time to mill the sheep, movable partitions can be used as a crush to crowd the sheep toward the start of the working chute.

Tilting Squeeze Chute

The tilting squeeze chute (or sheep squeeze or tilting squeeze) is convenient for restraining sheep for examinations of one sort or another, for foot trimming or foot-rot treatment, or for checking the fertility of the semen of rams. It is particularly useful when several techniques are to be performed on one animal, or when help is at a minimum.

Step-by-Step Procedure

1. Place the squeeze chute into the flow of traffic through or just beyond the exit of the sorting chute.

2. Allow the sheep to enter the squeeze chute until its nose is even with the restraining panel at its exit end. If the sheep should hesitate to enter, or not enter all the way, it must be pushed into position.

3. When the sheep is properly positioned, grasp the handle of the squeeze and pull it toward you until the sheep is securely squeezed and thereby held in position. Depending upon the model of squeeze you are using, the handle is held in position with springs and catches or a drop-in pin system.

4. The chute can now be tilted into whatever position is necessary to perform the required management techniques. Most chutes allow a full-upright, on-the-side, and an upside-down positioning.

5. To release the animal, rotate the chute back to the starting, full-upright position, release the squeeze handle, see that the animal has regained its feet, and open the restraining panel.

Lamb Restraint

The restraint of a 10- to 20-lb, 15"-tall lamb is an entirely different matter than that of restraining adult sheep. Depending upon the management technique to be performed, the number of assistants available, and the actual size of the lambs, one of the following restraint positions will be appropriate.

In the first method, the handler stands, holding the right fore- and rear legs of the lamb in his right hand and the two left legs in his left hand. The lamb's head is upward and its back is held against

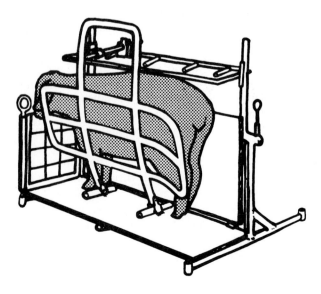

the handler's body. In the second method, the lamb is held in the same manner, but instead of standing, the handler sits and places the back of the lamb between his thighs.

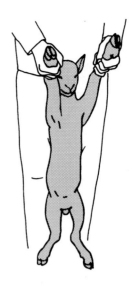

The next method, that of placing the lamb in an upright position (standing) and holding the lamb's rib cage or neck between your lower legs, is useful for drenching, or, if reversed, for tail docking. For the docking technique, it offers no advantage over holding the legs in your hand if an assistant is available.

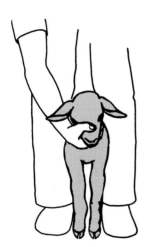

A modification of setting the lamb upon its rump can be used for injections in the axilla (foreflank or armpit) region. It is not particularly useful for rear-flank injections or other techniques because of the lack of firm rear-leg restraint. In this method, the forelegs are held in the hands of the handler, about waist high, with the rear legs dangling (or with the lamb sitting upon its rump). The lamb's back is against the legs of the handler. For added restraint, the lamb's hips can be squeezed between the legs of the handler.

1.8 GOAT RESTRAINT

Goats are extremely curious and have a strong need to give and receive affection. Effective restraint of the goat begins with an understanding of and capitalization upon this curiosity and the goat's need for affection. While the goat can be willful, it quickly conditions and is tenacious in maintaining habits. Unless you have an appreciation for both the physical and psychological aspects of the goat, your attempts at restraining them are likely to be just that—attempts. Success will come only when you realize that although the goat looks much like other animals, it must be handled quite differently. The handler is encouraged to refer to Sections 8.1 and 8.2 of Chapter 8 before reading this section.

The temptation is great, because the sheep and goat are nearly the same size and weight, to try to force sheep-handling and -restraint techniques upon the goat. This will work, if done properly, for kids and perhaps even yearlings, but as they mature, the adult goats must not be forced to endure sheep psychology.

Catching Goats

For a detailed discussion and supportive illustrations, refer to Section 8.2 of Chapter 8. Note that there is a decided difference between the approach to be used for a herd of intensively managed milk does and that to be used for a free-roaming herd of chevon producers or brush clearers.

Handling Goats

For a detailed discussion, refer to Section 8.2 of Chapter 8. Basically, the goat can be handled or held in the following ways:

1. Collar or identification chain.

2. Adjustable rope halter.

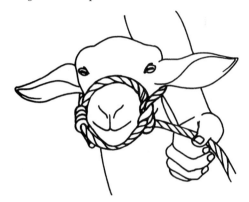

3. Jaw or chin hold.

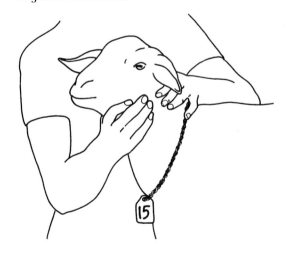

4. Halter, collar and ring, and bull staff.

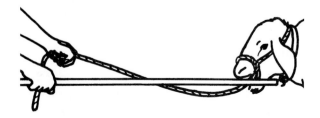

These can be used to hold and control a truly mean buck, although one really wonders if the buck could ever contribute enough to warrant keeping him under such circumstances.

Stanchion Restraint

Three main management techniques give cause for us to teach goats to climb upon, enter, and accept the stanchion as a restraint method—milking, grooming, and foot trimming. Each of these can be done on the ground, and thousands are done each day without a stanchion. The goat is either held by an assistant, tied to a post or tree, or, as in milking, conditioned to stand freely without restraint.

It is much easier on your back and patience if you properly restrain the goat for each of these techniques, especially foot trimming and milking (not to mention the increased hygiene that accompanies stanchion milking).

Most of the goats kept in the United States are kept as milk goats, which means that twice a day, every day, they are moved into and out of stanchions. It takes only a week or so of relieving the pressure of a milk-filled udder before the goats are waiting in line to hop on the stanchion! You need only concern yourself that the first milkings are gentle, so that the goat does not come to fear the milking operation itself.

In the case of mature animals that have never been taught to accept the stanchion, your task is quite different. You must physically pick up the animal and place it upon the platform, or push and pull it up the ramp, and then place its head into the squeeze bars of the stanchion. An assistant is necessary, and it is probably easiest if one person pushes the goat while the other pulls on a halter lead rope that has been run through the stanchion bars. Take care that the goat does not fall off the platform while its head is held in the stanchion bars.

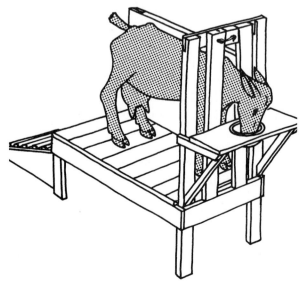

Setting the Goat on Its Rump

Some goat owners disdain this technique—probably because it is the most popular way of restraining a sheep! It is, however, an excellent method of restraining the goat for a variety of management techniques.

Step-by-Step Procedure

1. Place your left hand beneath the lower jaw of the goat and your right hand on its right side about midway between the hooks and stifle.

2. With your left hand, bend the goat's head back sharply over its right shoulder so that it is looking at its tail. At the same time, push downward and inward on the goat, toward your legs, with your right hand.

4. Release your grip with the left hand and grasp the front legs of the goat. Lift upward, straightening the goat, and at the same time step into the goat, positioning your legs against its back.

Laying the Goat on Its Side

Situations arise when the management technique requires that the goat be restrained on the ground on its side. Fortunately, this is not a difficult task to perform with the goat.

Step-by-Step Procedure

1. Proceed with steps 1, 2, and 3 of the technique for setting the goat on its rump (see above). At the conclusion of step 3, you have just stepped back about one-half step, and allowed or caused the goat's body to slide off your legs and onto the ground. Your left hand is still controlling the goat's head.

3. Step back about one-half step and allow the goat's body to slide off your legs onto the ground.

2. Instead of releasing the head, grabbing the forelegs, and stepping into the goat to support it, this time keep control of the head by keeping it bent backward and step back another step or so, allowing the goat's front quarters to fall to the ground.

3. Immediately have an assistant control the rear half of the animal by partially kneeling or leaning upon it.

4. Transferring control of the head to another assistant will free you to perform the management task at hand.

5. The goat's head can be allowed to lie flat upon the ground, unless it is struggling to escape. In that event, the assistant should use his left hand to lift the jaw and bend it toward the goat's tail only as much as is necessary to stop the struggling.

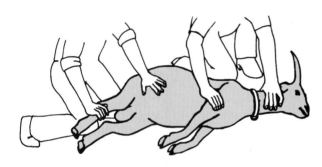

6. The goat's legs can be tied in any arrangement necessary to allow for the management technique.

Flanking

Flanking is another method of "throwing" or laying a goat upon its side. If it is done properly, there is no reason to avoid using this restraint method. It can be done gently and without physical or psychological trauma to the goat. It does require greater than average strength; and being about twice as tall as the goat helps, because it provides the necessary leverage.

A detailed step-by-step procedure for flanking calves (and other small ruminants) can be found in Section 1.4. It can and should be followed exactly for flanking goats. Basically, it involves standing on the goat's left side and placing your left hand over (some prefer under) its chin, near its neck, and grasping the loose skin in the right rear flank with your right hand. When you are so positioned, your right thigh and knee lift and push the goat away and upward, while your arms lift the goat away and upward. For a brief moment, the goat is flat on its side—in midair—just before you allow it to settle gently to the straw. At this point, steps 3, 4, 5, and 6 of the procedure for laying the goat on its side are repeated.

Kid Restraint

Kids can be restrained in a variety of ways, and the method used depends upon the management technique to be performed.

Holding Front and Rear Legs

This is an excellent method of restraint for castration and vulval inspections. The assistant can stand or sit and use his inner thighs to help restrain the kid.

Front-Leg Hold

This is a preferred method of restraining the kid for axillar injections and navel dipping. The rear legs can be left to dangle, rest on the ground, or be held clamped between your lower legs.

Held between Lower Legs

This method of restraint allows an unassisted manager with no holding box to securely hold the kid in position for disbudding and deodorizing.

Cradled in Arms or Lap

It is doubtful whether it's true (certainly it will never be proven), but some people feel that the kid feels

less pain if they hold it in their arms or lap while it is being disbudded and deodorized. You must remember that you do the kid no kindness if you allow it to jerk free during the hot-iron treatment. Hold it securely.

Holding Box

This is an excellent way to restrain the kid during disbudding and deodorizing. It securely, safely, and comfortably restrains the kid. Construct the box to the following inside dimensions: length, 24"; width, 5"; height, 16".

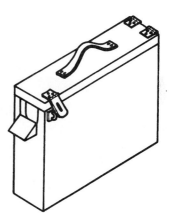

Miscellaneous Restraint

The side stick and Elizabethan collar, anti-self-sucking restraints, are covered in Section 8.7 of Chapter 8. Drenching, bolusing, and worming restraints are covered in Section 8.17 of Chapter 8, and tethering and controlling with fence are discussed in Section 8.2.

Consult these sections for detailed discussions and supportive illustrations.

CHAROLAIS BULL

Courtesy of American International Charolais Association

LIMOUSIN BULL

Photo courtesy of The Limousin World Magazine. Used with permission.

SIMMENTAL BULL

Courtesy of American Simmental Association

SHORTHORN BULL

Courtesy of American Shorthorn Association

BEEFMASTER BULL

Courtesy of Lasater Ranch

ANGUS BULL

Courtesy of Department of Animal Science, Oklahoma State University

HEREFORD BULL

Courtesy of Department of Animal Science, Oklahoma State University

BRAHMAN BULL

Courtesy of Department of Animal Science, Oklahoma State University

chapter TWO

Beef Cattle Management Techniques

2.1 INTRODUCTION

From the earliest days, the word *cattle* has had an all-embracing connotation in man's affairs. Originally the word meant property of all kinds, animate and inanimate. In this sense it was identical with *chattel* and *capital*. Today the term is not used in a general sense but refers to beef and dairy animals. With the development of specialized science came the classification of animals. Today's domesticated cattle belong to the order *Artiodactyla* (even-toed, hoofed animals) and the family *Bovidae* (hollow-horned). Cattle, buffalo, bison, antelopes, chamois goats, sheep, and musk oxen are classified as close relatives. Beef cattle belong to the genus *Bos* and the species *taurus* or *indicus*. *Bos taurus* are the descendants of European cattle and comprise a majority of the cattle found in the United States. *Bos indicus* are the humped Zebu cattle of India and Africa and the Brahman breed in America.

Cattle have a total of 32 teeth. They have eight incisors (cutting teeth) in the front of the lower jaw and none on top. Cattle do not have tusks (canine teeth), but they do have six molars (grinding teeth) on each side of the upper and lower jaws.

The ruminant digestive system has four compartments: the rumen, reticulum, omasum, and abomasum. The reticulorumen acts as a large fermentation vat for food and fiber particles. Inside these two compartments are billions of bacteria and protozoa.

The omasum does not seem to have a well-defined function. The abomasum or "true stomach" is similar to the stomach in man or the pig.

By the time man had learned to write, cattle had already been domesticated. Pictures in Egyptian tombs show various stages of cattle management. Engravings made before 2500 B.C. show children playing with hobbled calves, the delivery of a calf, men training a calf to drink from a vessel, the flanking or throwing of an animal, and cattle being branded with numbers.

The cattle industry in this country is strongly rooted in history and folklore. Without a doubt, the American cowboy is the greatest "hero" of all time. He has held that honor for over a century with good reason. The cattle industry has an attractive, romantic lifestyle in an outdoor setting.

Beef is tasty, rich in such nutrients as protein, iron, and B vitamins, and can be produced without competing against humans for cereal grains. Only 10% of the world's surface area is suitable for growing crops that produce food for human consumption. Another 19% of the world's surface area produces vegetation that has no food value for humans. Beef cattle have the ability to harvest this rough vegetation without the expenditure of fossil fuels and still produce large quantities of high-quality meat.

Beef tastes good and is extremely popular with Americans. In 2004, 11.3 billion beef servings were consumed in restaurants. Burgers, at 8.2 billion servings, lead the list, followed by steak entrees. The most popular steak cuts (in order) were Kansas City/New York Strips, Filet Mignon, Top Sirloin, T-Bone/Porterhouse, and Rib Eye.

One the home front, 88% of households will eat beef (at home) in the next two weeks. That amounts to 251 million people! Ground beef pounds lead the list of selections, but steak eaten "as is" is the most popular beef dish. Americans love to celebrate and entertain family and friends. When they do so, beef is the meat of choice. The 4th of July, Memorial Day, and Labor Day are perfect examples, and they rank in that order of popularity, measured by the pounds of beef consumed—totaling more than 150 million pounds—for those three holidays, alone.

The U.S. Beef Industry is made up of more than 1 million businesses, farms, and ranches, operating in all 50 states. To meet the demand for all of these consumer beef products, producers must breed, grow, finish, and market more than 35 million head of cattle. Cattle production is the largest provider of jobs in agriculture, and one out of every five jobs in corporate America is related to agriculture. Beef is big business!

Good management practices have been and will be essential to the fast, efficient, and economical production of beef. Because beef is not sold on a cost-plus basis, producers will have to pay close attention to all production costs if they are to make a profit. Historically, this has been done by increasing numbers to increase production. Today, however, we need to use better marketing and management practices to increase production and profit per beef unit.

Many producers overlook the need to know the basic concepts and costs of production—they have "some idea" of what it takes to produce a live calf or a pound of grain. In the years ahead, we need to have more than "some idea" of production costs and how to reduce them so that profit is increased. This means that it will be necessary to employ a good record-keeping system and the latest management techniques.

A knowledge of cattle behavior or "cow sense" is a must if one is to handle and manage cattle successfully. Most cattle are reasonably gentle, or can be made gentle by handling. Cattle become trustful of their owners and come to depend on them. This gentle trust can be destroyed, however, and cattle can become fierce when mistreated or when their safety is in jeopardy. Cattle have been known to charge such dangerous beasts as lions and bears and come away victorious. This innate trait can work for us as well as against us. For example, one of the outstanding characteristics of beef cattle is the strong maternal care exercised by the cow over the calf.

This inherited instinct helps to lighten the labor of cattlemen in caring for the young, but it often makes performing management techniques on the calf difficult and dangerous.

Understanding why cattle behave as they do can help reduce stress and injury to the animal and the handler. An understanding of cow psychology allows cattle to be handled more efficiently, and generally results in higher overall productivity.

Cattle have panoramic vision that allows them to see almost 360° without turning their heads. This type of vision is somewhat like looking through a fish-eye or wide-angle lens. Images are distorted, especially around the outer edges. Fences and other straight-line objects appear curved instead of straight. Cattle are easily spooked by moving objects because of this panoramic view. For this reason, solid sides on working chutes can be used advantageously to reduce distraction and allow cattle to be moved more easily.

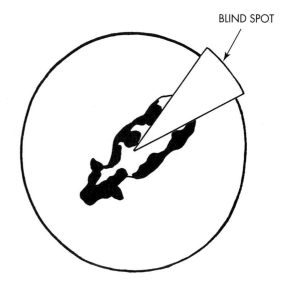

BLIND SPOT

Cattle are color-blind and have poor depth perception. This can create problems because cattle detect mostly movement and contrast between black and white. For example, in bright sunlight a shadow on the ground may appear to be a deep hole and scare cattle, or a jacket flapping on a fence post can spook them. Cattle also have a fear of stepping across anything with a grid pattern or design. This is one reason that cattle guards are so effective in keeping cattle confined. When facilities are designed, it is important to keep cattle vision in mind. Corrals and working chutes can hinder cattle movement if shadows create a grid pattern or if drains are placed in an alleyway. Diffuse lighting helps to minimize shadows, and the use of color or paint helps to minimize color contrast. Cattle react, like most animals, to loud, abrupt noises, but they can't stand high-pitched sounds. Cattle seem to react to a whip because of its high-pitched cracking noise more than because of the fear of being hit.

Cattle have a strong sense of smell and can react to odors over a considerable distance. The scent of new spring grass can take its toll on fences. Cows can smell their calves, and separating cows from calves works best when cows are moved upwind from their calves. If the cows detect that they are being separated from their calves, they can become reluctant to move. Odors are also a dominant means of sexual communication. A bull can detect sexual signals from a cow in heat that is located a considerable distance away. This accounts in part for sudden periods of aggressive behavior in bulls, and can result in broken fences.

Cattle tend to be gregarious, and they thrive best in groups. The separation of individual cattle from the group results in stress to those separated. A knowledge of herd behavior can lessen this stress. When separating cattle, do it quickly and get the groups as far apart as possible. Cattle should never be penned individually for an extended period of time. It is best to pen two or more animals together to minimize stress and maintain a high level of production. When new cattle are introduced, or when cattle are mixed, it is best to do it at dusk; there will be less fighting during the night.

Cattle like to follow the leader. When handling cattle, it is usually a sound practice to start one or two animals through the gate, doorway, or alley and allow the rest to follow. When the whole group is driven to an opening, they often turn around and come back toward you instead of going through the opening.

Mistreated cattle become trained to react in a wild manner. The overuse of electric prods or hitting with a stick or staff results in wild cattle with poor dispositions. A good cattleman will work cattle in as calm a manner as possible. Cattle should be allowed to follow the leader and move at their own pace. Pushing cattle too hard will excite them and make them unpredictable and hard to handle. This can occur when you are in a hurry or when inexperienced help is involved.

Chapters 2 and 3 describe 40 techniques that can help to increase beef production and profit. The techniques, which are set forth in a step-by-step manner, are written to provide the producer or the student of animal science a quick and comprehensive how-to reference on the handling and working of beef cattle. Cautionary notes are inserted at proven danger points and sources of error. You must be ready to react to dangerous situations and violent cattle behavior at all times, but especially when they are listed as a CAUTION.

Safe and responsible animal treatment should always be uppermost in a cattleman's thinking. Consider safety not only for yourself, but also the safety and well-being of the critter. Always move and handle cattle slowly and easily. Follow the Mexican saying, *poco a poco se andan lejos,* which means, "little by little they travel far."

2.2 IDENTIFICATION

Animal identification (ID) is a must in a beef herd. It is an important factor in managing herd health, artificial insemination, and performance testing, as well as routine observations and processing. Small-herd handlers can get by with identifying cows by color, cow name, or size, but this system begins to break down when more cattle are added to the herd. A numbering system must be developed to give each and every animal in the herd some type of identification.

There are three practical ways to identify an animal: *tattooing, ear-tagging,* and *branding.* Each of these methods serves a particular purpose. *Tattooing* is a permanent means of identification, but it cannot be read from a distance. Purebred organizations require that animals be tattooed in one or both ears before registration. Ear tags are an economical means of temporary identification that can be read at a distance. *Ear-tagging* has more flexibility than the other methods of identification; for example, the individual animal's ID number can be placed on the bottom of the tag, and sire and/or dam numbers can be placed on the top of the tag. The major disadvantage is breakage and loss of the tag. For this reason, ear tags should be used in combination with a permanent means of identification. *Branding* offers both readability at a distance and permanent means of identification. When several methods of identification are used in combination, it is important that all of them use the same numbering system. This eliminates the need of fumbling through record books to cross-reference a number.

Herd size will dictate how many numbers are used in the numbering system. Three or four numbers usually will be sufficient to identify all animals in a given herd. Each operation can tailor-make a numbering system that fits a particular need or objective. Several ideas should be considered for a cow herd: a numbering system that can span a time frame of 10 years without duplicating numbers, a system that includes the age of the animal, and a system that can tell something about the breed combination in a crossbred herd. One possible system uses the last digit of the year as the first number. The second number is optional and may use a coding system of 1 to 9 for sire breed or sire within a breed. The (second) third and fourth numbers can be used to identify the order of birth in a given year. The system can be used for a total of 20 years when heifers receive even numbers for 10 years and odd numbers for the next 10 years, and vice versa for bull calves.

An alternative to this is to use a code of 0–4 for sire breed or sire within breed for 10 years, and 5–9 for the next 10 years. This alternative does not require distinguishing sex by odd or even numbers.

Regardless of the system developed, it should be easy to use and understand. A good identification system can tell a herdsman how old a cow is, what the sire or breed of sire is, and, in operations where the breeding program involves a two-breed rotation, which breeding pasture the cow goes into. All of this can be determined by reading the ear tag or brand.

Feedlot and stocker cattle can utilize a much simpler system of identification, because in most cases they will be sold before they are two years of age. A simple system is 1, 2, 3, . . . for as many calves or yearlings as are owned, or some combination of the cow system just described.

Tattooing

Tattooing is a method of permanently identifying an animal. The tattooing instrument consists of a pliers-type device that has dies in the form of letters and/or numbers. The dies are made of sharp, pointed, needlelike projections that pierce the ear when the handles of the tattooing instrument are squeezed together. An indelible ink or paste is then rubbed into the small punctures. After healing, the tattoo is permanent.

Most of the purebred cattle breed associations require that animals be tattooed for registration. The advantages of an ear tattoo are that it is permanent and it does not disfigure the animal in any way. Two major disadvantages of tattooing are that the animal must be restrained so that the ear can be closely examined to read the identification, and, in a dark pigmented ear, the tattoo is nearly unreadable. A bright flashlight placed behind the ear will help read an indistinct tattoo. For this reason, most producers apply an ear tag or brand, in addition to the tattoo, so the animal can easily be identified from a distance. The permanent tattoo can then be used to identify animals when ear tags are lost or brands are illegible.

Animals can be tattooed at any age, but it is most convenient to tattoo baby calves. Tattooing, eartagging, and giving an injection of vitamins A, D, and E to a calf at the same time you treat the navel is a good, sound practice.

Equipment Necessary

- Squeeze chute or headgate
- Tattooing instrument
- Tattooing numbers and/or letters
- Tattoo ink or paste
- Alcohol
- Clean cloth
- Canvas (6' × 6' on which to lay calf and equipment)

Restraint Required

Newborn calves can be flanked and restrained on the ground. Calves over two months of age should be restrained in a squeeze chute or headgate. Large animals may require a head and nose bar in addition to the simple headgate to more completely immobilize the head.

Step-by-Step Procedure

1. Assemble the necessary equipment. It is important that the numbers be placed into the tattooing instrument in the proper order. As you look at them in the tattooing instrument, they should appear backward. Most tattooing instruments have space for four or five numbers. Always check the numbers (dies) on a piece of paper or cardboard before you begin, to make sure they are correctly placed. Some tattooing instruments have the numbers on a small drum which can be rotated to the desired number sequence. There are two advantages to the rotating-drum type of tattoo instrument: there are fewer losable parts and it is easier to keep up with the numbering system during calving and marking season.

2. Restrain the animal.

3. Locate the area of the ear that you wish to tattoo. Two ribs of cartilage divide the ear into top, middle, and lower thirds. The tattoo should be placed in the top third of the ear just above the cartilage rib and approximately equidistant from the base and the tip of the ear. Avoid tattooing in the edges of the ear or in the hairy portion of the ear because the tattoo numbers cannot be easily read later. Do not tattoo between the two cartilage ribs; this area is reserved for some types of ear tags or for a brucellosis vaccination tattoo in the right ear of heifers.

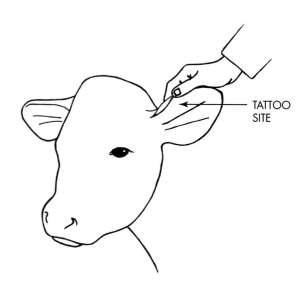

TATTOO SITE

4. Clean the inside of the ear where the tattoo will be placed with a cloth soaked in alcohol.

CAUTION: Infection or warts can result if a tattoo is placed in a dirty ear.

---

5. Position the tattooing instrument inside the ear so that the needlepoint dies are above the ribs as described in step 3. Squeeze the handles of the tattooing instrument together completely and quickly and then release them fully. Be sure that the handles are squeezed tightly together so all of the sharp, pointed, needlelike dies pierce the ear. It may be necessary to pull the needlelike dies gently out of the ear if they do not come out when you release the handles of the tattooing instrument.

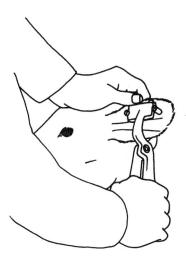

6. Rub tattoo paste or ink into all of the needle marks. The ink is applied with a roll-on applicator, but the paste should be rubbed on with the thumb. In either case, be sure to work the ink or paste well into the needle marks. Place the heel of the hand behind the ear for support. If you fail to do this, the tattoo may not be legible when you refer to it for identification at some later date.

7. Release the animal.

8. Clean the tattooing equipment with alcohol after each day of use.

Postprocedural Management. If the ear was clean before tattooing, no infection or swelling should result.

Ear-Tagging

Ear-tagging is one of the most popular methods of animal identification. This method of identification is not permanent because tags can break, pull, or rip out of an animal's ear. Ear tags should always be used in combination with a permanent method of identification such as a tattoo or brand.

Ear tags can be applied any time during an animal's life. In cow–calf operations, tags should be applied to calves at birth for identification and record purposes. In stocker and feedlot operations, cattle can be tagged upon arrival after purchase.

Ear tags can be purchased in a variety of styles, sizes, and colors. Basically, there are three styles of ear tags: one-piece plastic, two-piece plastic, and metal. The larger plastic tags are most popular because they are economical, easy to install, easy to read from a distance, stay pliable in cold weather, and stay in the ear longer than metal tags. The metal tags are difficult to read from any distance and tear out of the ear fairly easily. The plastic and metal tags can be purchased prenumbered or blank.

Equipment Necessary

- Squeeze chute or headgate (equipped with a head and nose bar for larger animals)
- Ear tag and applicator
- Marking fluid (for plastic tags)
- Metal stamp set (for stamping numbers in metal tags)
- Antiseptic (wound dressing)
- Canvas (6' × 6' on which to lay the calf and equipment)

Restraint Required

Newborn calves can be flanked and restrained on the ground. Calves over two months of age should

be restrained in a squeeze chute or headgate. For large animals, it may be necessary to use a head and nose bar in addition to the simple headgate to immobilize the head more completely.

Step-by-Step Procedure

1. Select a tag style. Both single and double plastic tags can be used successfully. The double tag is more versatile because one piece is used on each side of the ear, and the tag can be seen easily from the front and rear of the animal.

2. Select the tag size to be used. Usually, the largest tag available is best. The reason for this is that these large tags are easy to read at a distance. Don't be too concerned about placing a large tag in a small baby calf's ear and causing the ear to droop. The calf will grow and the ear will be able to support the weight.

3. Select contrasting ink and tag colors. The objective of the tag is identification at a distance, and the color combination is important. When only one tag color is used, consider a yellow tag with black marking fluid. This color combination fades less than most other colors and can be seen as well as or better than any other combination at a distance. Additional color combinations can be used to identify sire breed, individual sire, or sex in some operations that require more identification than animal number per se.

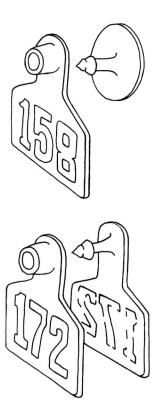

4. Select a numbering system for the ear tags. A three- or four-number system should be used, depending upon the number of cattle owned.

5. The next decision will be whether to purchase prenumbered or blank tags. Prenumbered are more convenient, but not as adaptable to your "system" as the blank tags can be. Make this decision based upon the unique needs of your operation.

If you choose the blank tags, number the plastic tags with marking fluid recommended by the tag manufacturer. Plastic tags should be numbered the day before they are inserted into the ear. You may want to number the tag on both sides. This allows the tag to be read from both front and rear in the working facilities. Several applications or coats of marking fluid will provide a longer-lasting and more visible identification.

For plastic tags, use the marking fluid or paint that "melts" and bonds to the ear tag. The magic marker–type inks should be used only to make temporary tags after an animal loses a tag. The magic marker–type inks fade and rub off more quickly.

Number the tags with large numbers along their bottoms so that they can be seen from a distance when hair grows in the ear.

6. Insert the ear tag into the appropriate applicator. Each tag manufacturer has an applicator designed specifically for its type of tag. Two-piece tags require that the male portion of the tag be slid over a pin and the female portion inserted into a clip. A metal tag should be inserted into the applicator until it snaps securely in place. Be sure to follow the manufacturer's directions when inserting the tag into the applicator.

When two-part tags are used, make sure that the male portion of the tag lines up with the hole in the female portion of the tag.

7. Select the ear to be tagged. This decision is based somewhat on your handling facilities. If cattle are worked from the right side, the ear tag should be inserted on the right ear, and vice versa, so that animals can readily be identified as they move through the facility.

If cattle are not identified by a tattoo or brand in addition to the ear tag, it may be advisable to tag both ears. This system will allow you to identify the animal even if one ear tag breaks or pulls or rips out of the ear. All cattle should be tagged in the same manner to minimize confusion.

8. Select the tagging site on the ear. The site selected will vary with the style of tag. Two-piece tags should be placed between the cartilage ribs, or below the ribs approximately halfway between the base and tip of the ear.

The two-part tags can be applied with the large tag in front or in back of the ear. It is easiest to have the male part of the tag in front of the ear and the female part of the tag behind the ear. The reason for this is that the male part of the tag pierces the ear, and you can locate it exactly where you want it. If the male portion enters from the back of the ear, you don't know where it will come through the ear in relation to the cartilage ribs.

When the large tag is used in the back, with the small button inside the ear, the large tag should be numbered on both sides and placed just below the cartilage ribs. The advantage of placing the largest tag in the back is that it improves readability. Hair will often cover the tag when it is placed on the front, but it can always be read from the back of the ear.

Metal tags, primarily used for brucellosis tags, should be placed into the top of the ear near the ear's base.

9. Hold the ear with one hand while using the other hand to insert the ear tag. Pay particular attention to the proper ear-tag site. The two-part tag is applied with a pliers-type applicator by squeezing the handles until the ear tag snaps together. The metal ear tag is applied in the same manner as the two-part tag.

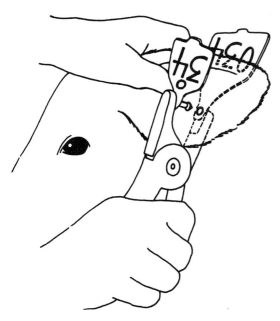

10. Treat the pierced ear around the tag with an antiseptic (wound spray) or iodine to prevent infection and fly irritation.

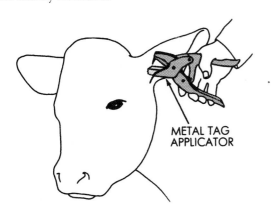

METAL TAG APPLICATOR

> **CAUTION:** Be sure to restrain the heads of large animals in a headgate equipped with a head and nose bar when applying the ear tags. The animal will often throw its head from side to side as the ear tag pierces the ear. Small calves can be restrained on the ground with a knee placed on their neck.

11. Release the animal. It can be returned to its pen, lot, or pasture immediately after tagging.

Postprocedural Management

Minor bleeding may occur at the tagging site. Bleeding should stop within several minutes after tagging unless a blood vessel was severed. When a blood vessel is severed, it may take 5 to 10 minutes to stop bleeding. The puncture wound will scab within 48 to 72 hours, and recovery (healing) will be completed in 7 to 10 days. If the scab is bumped or loosened during the recovery, the ear may bleed, and healing will be delayed.

In a limited number of cases, infection may occur at the tagging site. The infected site will swell, and a puslike discharge may be seen. If the tagging equipment is clean and an antiseptic solution is applied immediately after tagging, this infection should be minimized. If an infection does occur, the wound should be treated with antiseptic, and an injection of broad-spectrum antibiotic may be required.

Occasionally an animal will lose a tag and will need retagging. When this happens, a tattoo or brand can be used to help identify the animal. If animals are maintained in muddy environmental conditions, the tag may become covered with mud and need cleaning. During the winter, hair growth in the ear may cover the tag and make animal identification difficult. When this happens, the hair should be clipped in the manner described in Section 2.12.

Insecticide-Impregnated Ear Tags

Insecticide-impregnated ear tags for beef and dairy cattle are an asset to the livestock producer. Applied like an ordinary ear tag used for identification, insecticide-impregnated ear tags, commonly referred to as *fly tags*, provide a significant level of control against insect and arachnid pests. Insects controlled include horn flies, face flies, stable flies, and houseflies. Arachnid populations controlled include spinose ear ticks, Gulf Coast ticks, and lice.

There are several brands on the market. They vary as to their active ingredient(s), amount of insecticide contained in the tag, and longevity of tag effectiveness. Consult with your dealers, veterinarian, other producers, or your livestock extension agent/specialist as to which particular product to use in your area.

As with all insecticide, dewormers, and perhaps many other animal health products, it is good management practice to rotate products used from one season, or one year, to the next. This will help reduce the incidence of resistance development by the insects, internal parasites, or microbes.

The step-by-step procedure for applying insecticide-impregnated ear tags is the same as for applying ordinary ear tags.

CAUTION: Follow manufacturer's label directions, guidelines, and cautions. These ear tags are impregnated with organic insecticides. Handle them according to directions. Review label information for approval of use in lactating dairy cattle. Not all products are approved for use in dairy cattle.

Electronic Livestock Identification

Electronic livestock identification, commonly called *Electronic ID* or *EID,* is at the core of the now-mandated U.S. Animal Identification Plan (USAIP). USAIP defines the framework for a national animal identification system. This system is essential to the well-being of animal agriculture in the United States. The goal of USAIP is to achieve a traceback system that can identify all animals and premises potentially exposed to an animal with a foreign disease, within 48 hours after discovery. This "within 48-hour traceback system" will enhance the efficiency and effectiveness of our current animal health regulatory programs.

Additional benefits of the USAIP will include:

- Enhanced disease control
- Enhanced eradication capabilities
- Rapid containment of foreign animal disease outbreaks
- Enhanced ability to respond to bioterrorism threats
- Enable the establishment of source-verified marketing strategies
- Enable production trait "rewards" from packer to producer
- Enable identification of superior performance animals for producer decision making

Each of these benefits can be simultaneously achieved through the implementation of the single system developed for the USAIP.

To make the system work, two types of animal identification are necessary: individual animal and group/lot/premises identification. It is the goal of being able to traceback within 48 hours that requires the use of Radio Frequency Identification technology (RFID). Without RFID, it would be impossible to automate the recording of the millions of annual animal movements, much less do it in 48 hours!

RFID devices must be ISO (International Standardization Organization) compliant. There are various methods of "attaching" the RFIDs to the animal, including ear tags, ear tag attachments, implants, and boluses for the rumen or reticulum. Ear tags are currently the most often used method of attaching the RFID. There are strict standards and performance requirements for RFIDs. Companies that sell these devices are knowledgeable of these standards.

From a production point of view, all of these standards and performance requirements are beyond a cattle producer's control. They are in the domain of the manufacturer. Marketplace success will depend upon the manufacturer producing a high-quality, durable product at a reasonable cost. Producers should remember the following about an RFID system, regardless of the manufacturer:

RFID is only one form of animal identification. This chapter section discussed others.

RFID is simply an animal identification system. It is not a record-keeping system.

To maximize the potential benefit of RFID systems, producer effort will be necessary.

There are four component parts of an RFID system:

Transponder—the "tag." The transponder is either encased in a donut-shaped button, which acts as the female portion of a two-piece tag, or it is built into a traditional dangle tag.

Reader. This is a actually a receiver (or transceiver) that picks up the signal being emitted by the transponder. Readers are currently handheld wandlike devices, "paddles," or archlike portals through which animals walk.

Data accumulator. This is the collection point to which data (signals) flows from the transponder through the reader. It is actually a computer, like the laptop on your desk.

Software. Software manipulates the data (signals) and converts it into information you can use to trace animals or base management decisions upon.

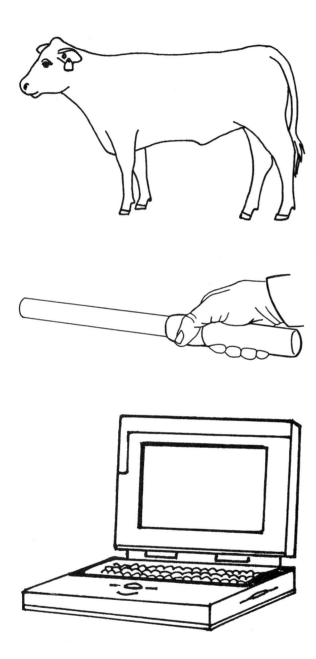

Equipment Necessary, Restraint Required, and the *Step-by-Step Procedure* are similar to the placement of the ear tags, discussed immediately following. The "tag," the pliers (applicator), and all of the procedures are exactly the same. However, there is an essential difference in the placement location of the transponder button. The transponder is located nearer the head and higher in the ear than the typical ear tag. This maximizes retention.

2.3 ORPHAN CALF CARE

Occasionally a beef cow will die or become paralyzed during or shortly after parturition and leave an orphan calf. This situation requires that the calf be fostered onto a lactating cow, fed fresh milk or a milk replacer from a nipple bucket or bottle until it is old enough to eat grain, or that a lactating cow be milked and the calf stomach-tubed. This discussion deals only with fostering an orphan calf.

Equipment Necessary

- Maternity pen, maternity chute, or rope halter
- Knife
- Heavy-duty string or twine
- Nipple bucket
- Milk replacer
- Water

Restraint Required

Ideally, a homemade maternity pen and maternity chute should be used to provide safety for both you and the calf. Some cows do not want a foster calf and will kick at you and the calf as the calf tries to nurse. The maternity chute allows the calf to nurse, restrains the cow, and minimizes injury to all involved. This type of restraint allows one person to perform the task quietly and safely.

A second type of restraint that is not nearly as safe as the maternity chute involves tying the animal with a rope halter and a quick-release knot. This second method usually requires an assistant, who holds the cow against a fence or wall to minimize movement when it is restrained with a halter. When the cow is extremely gentle and broken to lead, a second person may not be required.

Step-by-Step Procedure

1. Select a cow to be a foster dam. The cow that is selected must be in the proper physiological state to nurse a foster calf. The most logical candidate is a cow that just calved and lost her calf or calves soon after the foster calf was orphaned. This cow will have colostrum for the orphaned calf, and her instinct to nurse a calf is relatively high.

A second possibility for a foster dam is a cow with good milk production that calves the same day as the orphan calf was born. This arrangement allows the orphaned calf to receive colostrum, but the cow's natural calf often nurses more and gains faster than the foster calf.

A third candidate is a cow that loses her calf after it is 3 days of age. This alternative does not give the foster calf colostrum; therefore colostrum should be obtained from a cow that has just calved and given to the calf by a nipple bucket (or bottle) or by stomach tube.

CAUTION: Do not use a cow that lost her calf more than six or seven days before the calf was orphaned. A cow that has lost her calf two to seven days before a calf is orphaned should be restrained and checked to find out if she has milk. If she has some milk, her milk production should come back to near normal over the next several weeks as the calf nurses.

Orphaned, weak, or injured calves need milk or colostrum regularly. A newborn calf needs 6 to 8% of its body weight in colostrum during the first several hours of life. If you do not have colostrum on hand, try to obtain some frozen colostrum from a neighboring dairyman. Orphaned or injured calves must receive by stomach tube at least 2 quarts of milk two or three times per day until you can get them cross-fostered to another cow or trained to nurse a bucket or bottle.

2. Prepare the necessary equipment.

3. Move the cow into a clean, bedded, maternity pen, box stall, or small pen.

4. Give the orphaned calf a scent to help confuse the foster dam so that she will accept the calf more readily. There are several approaches to accomplishing this task. The first approach involves the placenta (afterbirth) of the foster dam if she has just calved and lost her calf. With a pair of plastic or rubber gloves, rub the placenta over the foster calf. If the placenta is not available, or if it has been more than several hours since the foster dam has calved, try a second alternative. This involves cutting the hide off the foster dam's natural calf that has just died. The hide can be rubbed on the orphaned calf and, if necessary, tied around the orphaned calf with string or twine. When tying the hide on the orphaned calf, be sure to cover the neck and back with the hide and leave it on for several hours before removing it.

5. After the calf is given a scent from the placenta or hide, or the hide is tied onto the calf, move the calf into the pen or stall with the foster dam.

6. Allow the calf to nurse the foster dam. Remain in or near the pen and observe the situation closely. If the cow accepts the calf and allows it to nurse, the calf is successfully fostered. Some cows, however, do not want to nurse a foster calf; they will kick the calf and may try to kill it as the calf tries to nurse. Likewise, some calves do not want to nurse.

When the calf does not nurse or gets kicked, proceed to step 7.

7. Restrain the cow in a maternity chute or use a rope halter and a second person to hold the cow against a wall or fence.

8. Move the calf to the cow-restraint area.

9. Hold the calf by placing one of your legs behind it to keep it from moving backwards. The other leg should be placed along the calf's shoulder to help support the calf and to keep the calf from moving sideways. One hand should be placed on the calf's head to gently push it down toward the udder, while the other hand directs the cow's teat into the calf's mouth. If the calf does not want to nurse, it may not be hungry, or perhaps it doesn't know what to do. If the calf isn't hungry, wait about four hours and try again. If the calf doesn't know what to do, allow the calf to nurse your thumb or first two fingers while holding the calf in the position just described. Usually, the calf will learn to suck, and your thumb can be replaced with the teat. If this doesn't work, try squirting a little milk into the calf's mouth. Be patient. Keep trying until the calf nurses.

10. Turn the cow and calf loose together in the maternity pen or box stall. Observe immediately and periodically for the next several days to make sure that the cow accepts the calf as her own and does not try to butt it or kick it. After 2 or 3 days, the cow can be released from the maternity pen and returned to the rest of the herd.

CAUTION: Cows may try to kick you during this step, especially if restrained with only a rope halter. The udder of a cow that has not been nursed for several days can become sensitive, and the calf's nursing can cause pain, which can also make some cows kick.

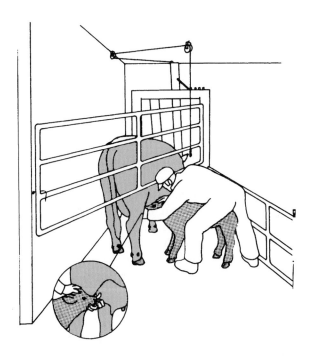

CAUTION: If the cow does not accept the calf in the first several hours, you should try another cow or start feeding the calf milk replacer.

2.4 MILKING

Milking is not practiced extensively in beef operations, but occasionally a situation arises in which beef cows need to be milked. Some of the situations that call for milking result when a cow produces more milk than the calf can consume and the possibility of mastitis exists, when a calf is orphaned early in life and requires colostrum from another cow, or when an injury occurs to the cow or calf that prevents the calf from nursing.

Milking a beef cow is accomplished by hand-milking. This technique is discussed in detail in Chapter 4 (Section 4.8, "Milking"), but there are several points at which the procedures for milking beef and dairy cows differ.

Step-by-Step Procedure

1. The beef cow should be restrained in a maternity chute or squeeze chute to minimize the danger of personal injury. A halter can be used to restrain the cow when a second person holds the cow against a fence or wall to minimize its movement. Stress should be minimized whenever possible.

2. Select a cow to be milked. This cow may be a cow that has a large, swollen, sore udder, or it may be a cow that has just calved and has colostrum. A cow will generally have colostrum two to three days after calving. If a calf is orphaned after it is three days old, the gentlest lactating cow in the herd can be selected for milking.

3. If the cow is nursing a calf in addition to being milked, only two-quarters should be milked out, saving the remaining two-quarters for her own calf.

4. Repeat the milking process twice daily for as long as needed.

5. If calves are to be fed this milk, refer to Section 4.14 of Chapter 4.

6. The teats need not be dipped after milking.

2.5 IMPLANTING

The use of growth-stimulating compounds, in the form of pellets implanted subcutaneously in the ear, is common practice for cow–calf producers and cattle feeders. By far the most common reason for implanting is to increase the efficiency of gains via increased rate of gain and decreased pounds of feed per pound of gain.

Implants contain either a naturally occurring hormone or a synthetic hormone that acts like a natural hormone. These natural or synthetic hormones are released into the bloodstream in very small, but constant, amounts. The net effect is that they increase the deposition of protein, in the form of muscle, in the animal's body.

Implanting, properly planned and conducted, can provide the cattle producer with a return of "seven-to-one" on the investment. Research over the years has documented an improvement in average daily gain of as high as 30% and an increase in feed efficiency of up to 18%. Both benefits contribute directly to more pounds of beef for sale at a reduced cost per pound to produce it. This translates to increased profit for the cattle producer. There is no management tool that shows a greater return for dollar invested than implanting. Each type of implant compound has an effective lifetime that is defined by the manufacturer. As one implant lifetime is expiring, the producer usually has the option of reimplanting with the same product or with an alternative. Recommendations and animal health regulations change from time to time, so follow directions carefully.

Implanting cattle is a solid recommendation. However, there are several factors to be considered: (1) There is a small reduction in the quality grade of slaughter cattle that have been implanted, to which the market responds with a slightly reduced price to the feeder. (2) Heifer calves and bull calves that will be returned to the breeding herd need to be managed differently than calves headed for market—follow the manufacturer's recommendations. (3) There are drug-withholding times for animals that will be taken to slaughter—these must be planned for and adhered to very carefully. (4) There is some resistance from the public regarding animals produced with the aid of added "hormones or chemicals"—even though there is no evidence that eating meat produced with the aid of implants increases the rate of any adverse health conditions. (5) Implanting does take a bit of time and does add to the initial expense of production. (6) Implanting is easy to accomplish, but manufacturer's guidelines and directions must be carefully followed.

Some of the most common implanting errors are as follows:

1. Depositing the implants at the wrong site in the ear. Follow directions carefully.

2. Depositing the implants between the layers of the skin, instead of below the skin and on top of the cartilage. Take care to follow the directions regarding how to insert the needle.

3. Depositing the implants into the cartilage, instead of just on top of it, immediately below the skin layers. Take care to follow the directions regarding how to insert the needle.

4. Sticking the needle through the skin on the back of the ear. This will result in the implants being deposited onto the ground when the trigger is squeezed.

5. Severing a blood vessel. Hemorrhage will occur and the absorption will be too rapid. Be careful; this can be avoided.

6. Crushing the implants. To avoid this, insert the needle to its full length, then withdraw it a distance equal to the length of the implants before squeezing the trigger. This can be avoided by some of the newer implant guns, which have needles that automatically withdraw when the trigger is squeezed, thus allowing for implant space.

7. Infections can be set up if your procedures do not adhere to some fundamental sanitation guidelines. Follow the guidelines from the manufacturer.

> **CAUTION:** Be sure to follow the manufacturer's recommended withdrawal time before slaughter to minimize potential carcass residues. *Withdrawal time refers to the length of time required between the last implant date and time of slaughter.*

Equipment Necessary

- Squeeze chute, headgate (head and nose bar desirable for large feedlot cattle)
- Implants and implant gun
- Cotton
- Alcohol
- Pipe cleaner

Restraint Required

The age of the calf to be implanted will determine the amount of restraint required. The best way to handle calves of less than one month of age is to approach, corner, catch, flank, and restrain the calf on the ground. A headgate, squeeze chute, or calf squeeze chute (tilting) should be used on steer calves from 3 to 7 months of age. Feedlot cattle will require restraint in a headgate or squeeze chute. In the case of large feedlot cattle, it may be necessary to further restrain the animal's head with a nose lead or head and nose bar for your safety. Without working facilities for older calves, implanting becomes impractical if not impossible.

Step-by-Step Procedure

1. Place the implant cartridge into the implant gun according to the manufacturer's directions.

> **CAUTION:** Make sure that the implant cartridge is inserted correctly, because crushed pellets can result if pellets are forced out the wrong side of the cartridge. Crushed pellets are absorbed too rapidly, which may cause side effects and reduce the growth response.

2. Insert the needle into the implant gun. Make sure that the needle is sharp.

3. Determine which ear you want to implant and adjust the needle and cartridge so that the needle can be positioned next to and parallel to the ear, with the slant side of the needle facing upward. Implant all calves in the same ear to minimize confusion. When calves are reimplanted, the opposite ear should be used.

4. Restrain the animal.

5. Select the implant site on the back side of the ear. Most types of implants, such as those from Synovex, should be placed in the middle third of the ear. The skin in this area is relatively loose, and insertion of the needle is easy. The implant site should be located on the back of the ear between the cartilage ribs. Ralgro implants should be placed

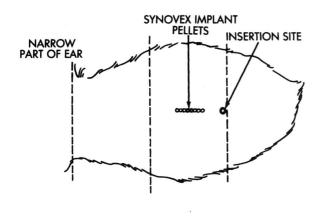

NARROW PART OF EAR

SYNOVEX IMPLANT PELLETS   INSERTION SITE

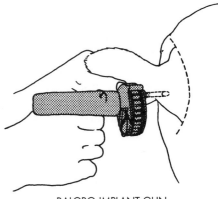

RALGRO IMPLANT GUN

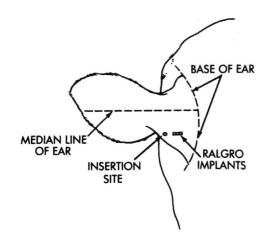

BASE OF EAR

MEDIAN LINE OF EAR

INSERTION SITE

RALGRO IMPLANTS

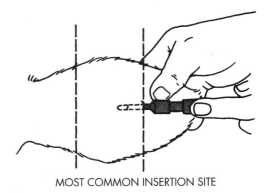

MOST COMMON INSERTION SITE

1 inch from the baseline of the ear, below its median line. The skin is relatively loose there, and there are no major blood vessels in the area.

6. Clean the needle and implant site with cotton or a sponge soaked in alcohol to reduce contamination of the needle wound. This type of contamination can cause the implant to be "walled off," thus reducing the growth response.

7. Grasp and manipulate the ear with one hand while the other hand positions the implant gun parallel to and nearly flush with the ear. The point of the needle should be against the ear, with the beveled part facing you.

As you begin to insert the needle into the ear, you may find that the tip of the ear is partially in your way. If this happens, manipulate the ear by folding or bending it gently so that the needle can be inserted easily.

8. Use the needle tip to lift the skin slightly in the implant area. Insert the needle between the skin and the cartilage of the ear. If you are not certain where the needle tip is positioned, you can feel it below the skin.

9. Insert the needle fully, up to the hub, where it joins the implant gun.

CAUTION: When the ear is grasped and the needle inserted into the ear, the animal may throw its head and injure you. This can be prevented by the use of a nose lead or a headgate equipped with a head and nose bar. If help is available, your assistant can turn the animal's head to the side and restrain it against the headgate with his leg or hip.

Avoid piercing or cutting ear veins with the needle, whenever possible. Observation and experience will help to minimize this problem. When the vein is cut and minor hemorrhage results, the implant pellets are immersed in blood, which causes softening and rapid absorption of the implant compound. This can result in a condition known as *bullers*. Bullers are animals that exhibit increased mounting activity, thus reducing the expected growth response.

Do not allow the needle to gouge or pierce through the cartilage. If you feel resistance as you insert the needle, it is quite probable that the cartilage has been gouged, and the pellets may be crushed. Crushed pellets can cause side effects in the animal as well as clog the needle. In addition, implants deposited in gouged cartilage may be covered with scar tissue and "walled off." This will result in very poor absorption of the drug.

Intradermal (between layers of skin) or intramuscular (in the muscle) implants result in poor absorption of the implant compound and very little growth response.

10. For the Ralgro implant gun, pull the gun and needle back about ⅜" to create a space or cavity for the implant pellets to prevent crushing.

11. Squeeze the trigger of the implant gun to insert the prescribed number of pellets according to the manufacturer's directions. Keep the trigger depressed and withdraw the needle.

12. When using the Synovex gun, squeeze the trigger fully, but do not draw the needle back to create space for the pellets or withdraw the needle manually. The Synovex implant gun needle automatically retracts when the trigger is squeezed. At this point, the pellets will be in the proper spot and the needle will be withdrawn.

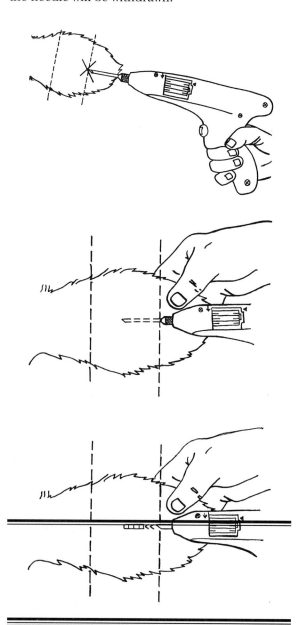

CAUTION: Proper dosage is important. Don't take the attitude, "If a little bit is good, a lot is better."

13. Feel the ear for the implants under the skin. If the implants are not felt, examine the ear more closely to see if the implants were inserted properly. Reimplant only if pellets are not in the ear.

14. Press your thumb onto the implant site opening. This will seal the opening and help to seal out dirt and bacteria that may cause abscesses.

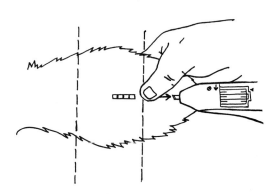

15. Release the animal.

16. Clean the needle, if necessary, with cotton dipped in alcohol. If the inside of the needle is filled with blood or a crushed implant, it should be cleaned by inserting a pipe cleaner.

17. Heavy hair and thick skin will dull needles. In addition, bumping the needle on the headgate or other object can cause a burr on the tip. The burr can be removed with a whetstone.

2.6 PASTURE BREEDING

The natural mating or breeding of cows on pasture is an important aspect of most beef operations. Pasture (natural) mating is the process of allowing bulls and cows to be together during the breeding season, and mating occurs without the herdsman's supervision. Natural mating may be the only method of breeding that is used in an operation, or it may be used in combination with an artificial insemination (AI) program. Some beef operations use a limited AI season and then turn in a "cleanup" bull for an additional time to breed those cows that did not conceive after two AI services.

The herd bull plays an important role in the cow herd. Herd bulls are responsible for more than 75% of the genetic improvement of the herd when their high-performing heifers are selected as replacements. Bull selection is very important, therefore, and it should receive serious consideration.

Several things must be considered when you are deciding which bull to buy or which bull to cull. A bull represents a sizable investment, and his influence on the performance of the herd will have an effect on the profit potential of the cow herd in future years. A bull must be capable of (1) producing fertile sperm, (2) mounting, (3) erection, (4) ejaculation, (5) settling a cow, and (6) siring efficient, fast-growing calves.

Equipment Necessary

- Record book
- Chin-ball marker

Restraint Required

The herd bull usually need not be restrained during the breeding season, but bulls should be restrained in a squeeze chute or headgate for evaluation of semen and breeding soundness before the breeding season. Bulls should be restrained in a squeeze chute or headgate when chin-ball markers are put on and refilled with ink.

Step-by-Step Procedure

1. Purchase a herd bull at least 30 to 60 days before the breeding season. This will allow time for the bull to become accustomed to his new surroundings. New bulls should be placed in a small pasture or trap to improve their general physical condition. Be sure that the lot is free of junk, wire, and boards that could cause injury or lameness.

2. Cattle producers should select a breeding bull for their commercial or purebred herd on the basis of the animal's estimated *breeding value* and predetermined *breeding soundness*. *Breeding value* is defines as the ability of the bull to transmit the selected trait to his offspring. *Breeding soundness* addresses the entire reproductive function including libido (sex drive, desire to mate), ability to produce live, fully functional sperm in a healthy semen environment, and the structural (anatomical) ability to pursue and mate with heifers and cows.

One measure of the breeding value of a bull (or cow) is *expected progeny difference* or EPD. EPDs provide us with a means to compare breeding animals. They enable us to predict the performance of future progeny from one bull to the predicted performance of future progeny from another bull. Note the key word in these sentences is *compare*. EPDs give us the tool to compare two or more animals for the trait being selected for.

There are EPDs for several traits, but the most important to the cattle producer are birth weight, weaning weight, yearling weight, and milk production. These are important because they directly affect the bottom line, the profit and loss statement, of the cattle operation.

While the actual calculation of EPDs involves extensive knowledge of animal breeding theory and significant computer hardware and software, the day-to-day employment of EPDs in a breeding operation is a matter of knowing what you wish to select for coupled with a good dose of common sense.

For example, if you are a commercial producer, and all of your calves, bull and heifer, are going to be for sale at weaning or as yearlings, you would select a breeding bull with a low birth-weight EPD (because you do not want to be awake all night "pulling" more calves than absolutely necessary), a high weaning-weight EPD (because you want the calves to grow rapidly and accumulate as many pounds as possible to sell at weaning), a high yearling-weight EPD (more growth potential, in case you retain ownership), and you would not pay very much attention to milk EPD (because you are not keeping the heifer offspring).

Naturally, it is never quite as simple to put this into practice in the producing herd as it is to explain an example of how it should work. This is because there are different accuracies of EPDs based upon number of offspring, age of the bull, age of the cow, crossbreeding programs, and the laws of chance. However, EPDs are the best tool we have today for the cattle producer to use in making selection decisions for the economically important production traits.

Bull buyers should insist on EPDs accompanying each bull they purchase or use in the future. AI sire summaries provide this information, and purebred producers can have these figures calculated for them by their respective breed associations. So whether you select from an AI catalog or from a sale catalog, EPD figures should be available. Insist upon them, learn how to interpret them, and then use them.

3. Confirm your pedigree and performance-based selection of a breeding bull with a *breeding soundness examination* (BSE). Breeding value is the single most important factor when selecting a bull, but if the bull is *not* sound he will not be able to perform.

First, consider skeletal and structural soundness. A bull with excellent records cannot transmit that superiority unless he has good feet and legs and is willing and able to seek out cows in heat and breed them. Avoid bulls that are extremely sickle-hocked, cow-hocked, or post-legged.

Second, select a bull with ample bone, adequate skeletal size as indicated by shoulder or hip height and body length, slightly above average muscling in

CAUTION: Avoid massive shoulders or abnormally heavy muscling that could result in calving difficulties. Avoid bulls that are excessively fat. These bulls are often less fertile and lack the sex drive necessary to breed cows. Conversely, thin, half-starved bulls will not have the stamina to mate with and settle a large number of cows during a short breeding season.

the forearm, rump, round, and stifle, and a trim brisket and underline.

4. Examine the bull externally and internally for abnormalities and indications of disease. Testicle size or scrotal circumference is an important consideration. Both testicles should be large, similar in size, and firm. Small testicles indicate low sperm production. Don't buy bulls with cryptorchidism (one or both testicles retained in the body cavity) or scrotal hernias, because both conditions are heritable and they can impair reproductive performance.

The penis should be checked for adhesions, tumors, "broken penis," and other injuries and deviations that will reduce the bull's breeding ability. Check the scrotum and testicles for lesions, scabs, and frostbite that may reduce fertility. In addition, a rectal palpation of the seminal vesicles for abnormalities is recommended.

5. If it is provided, evaluate the circumference measurement of the bull's scrotum. If it is not provided, arrange to have it measured. Progressive breeders are selecting bulls that meet or exceed the threshold of acceptable scrotal circumference measurement. Today, the commonly accepted threshold scrotal size in yearling bulls is 30 to 32 mm.

Research has demonstrated positive impacts, as scrotal size increased from 30 to 40 mm, on semen quality. The two most important correlations demonstrated are greater sperm reserves and higher concentrations of active motile sperm in bulls with larger scrotal measurements. These two factors in semen quality assure that bulls with larger testicles, as measured by scrotal circumference, will get more cows and heifers pregnant in a shorter period of time than small-testicled bulls.

In addition, heifers sired by large-scrotal-circumference bulls will cycle and breed at an earlier age than those sired by bulls with smaller scrotal measurements. Younger-age cycling translates to earlier-in-the-season cycling, which translates to older, potentially heavier calves at sale time.

Measure the circumference of the scrotum in the following manner: (1) Restrain the bull in a chute. (2) Carefully position yourself to the rear and side of the chute. (3) Grasp the bull's scrotum as shown in the figure at the right. (4) Position the scrotal tape around the greatest diameter of the scrotum. (5) Note the reading.

Take care to grasp the scrotum as shown in the figure, and be certain to pull the testes firmly downward into the lower part of the scrotum before measurement. Be certain the testes are forced downward and toward one another before you take the measurement. If the testes are separated during measurement, inaccurate measurement will result. Any tape measure will do, but it helps if it is graduated

in millimeters and if it is somewhat rigid and will form a loop that will stand by itself as it is being slipped over the scrotum.

If care is taken to properly restrain the bull and to position the measurer so as to minimize the danger of being kicked or stepped on, this procedure is quick, easy, painless to the bull, and not especially dangerous to the handler. However, it will seem as if you do not have enough hands to do all of the things that need doing. The self-looping tape certainly is an aid.

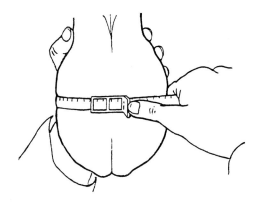

6. Check semen quality 30 to 60 days before the breeding season. If the semen is of poor quality, this will give you a little time to get the semen reevaluated or to find another bull without delaying the breeding season. Yearling bulls that are bought from bull test stations in the spring or fall often have had a semen evaluation. Semen quality is determined by collecting a sample by either electroejaculation or with the aid of an artificial vagina. The sample is then examined with the naked eye for a milky white color, and under a microscope for sperm concentration, motility, and morphology. Make sure that a qualified technician evaluates the semen sample. Have the semen evaluated by an animal reproduction specialist, an AI representative, or a veterinarian. If semen quality is poor and cows return to estrus, much time and money are lost.

7. Check and trim the bull's feet, if necessary, four to six weeks before the breeding season. This will allow ample time to correct temporary soreness and lameness that may result from hoof trimming. The bull's feet and legs should also be inspected for corns, arthritis, and abscesses that may impair mobility and reproductive performance.

8. Check the bull's mating desire, or libido. There is no practical way to measure the potential mating desire except to turn the bull in with cows in heat and determine whether he has the desire and ability to mount and serve a cow.

9. If all cows are to be mated naturally, place the bull into the breeding pasture for a 60-day breeding season. If AI is used for the first 30 to 45 days of the breeding season, use a cleanup bull for an additional 15 to 45 days. Be sure to observe the bulls early in the breeding season for sexual activity and mounting behavior. Some producers use a chin-ball marking device on bulls so that breeding dates and sexual activity can be recorded. These records can be helpful during the calving season when cows that are expected to calve near a given date are being observed. A good rule of thumb for estimating the number of cows that the average bull can service during a 60-day continuous breeding season is one yearling bull for 10 to 15 cows, one two-year-old bull for 20 to 25 cows, and one mature bull for 20 to 35 cows. The number of cows can then be adjusted to meet the limitations imposed on bulls by the environment (type of terrain and amount of heat), management (pasture size), and individual ability.

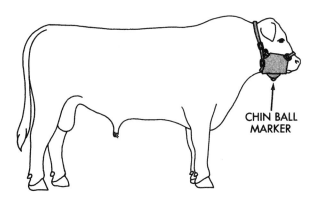

CHIN BALL MARKER

When several bulls are used in the same pasture—a multibull breeding pasture—care should be taken to select and group bulls of equal size and age.

10. Young bulls are still growing and need additional energy during the breeding season. Yearling bulls may have just lost their baby teeth, and they are expending a lot of energy. Yearling bulls should be fed approximately ½ pound of high-concentrate ration per 100 pounds of body weight daily during the first 3 to 4 weeks of the breeding season to help maintain breeding performance. This type of feeding regimen should not get young bulls overly fat. If a bull is getting too fat, reduce the amount of grain slightly.

11. Observe the bulls and cows daily during the breeding season. Pay close attention to the bull to see if he is following and mounting cows in heat. Watch for signs of swelling, injury, lameness, infection, or severe infestations of flies that may affect the bull's activity and mating behavior. If one of these problems occurs, restrain the bull for closer examination and treatment. If you have any questions or don't know how to proceed with treatment, consult your veterinarian. Entire breeding seasons have been lost because of undetected injuries early in the breeding season.

12. Remove the bulls from the breeding pastures after a 60-day breeding season. If your present breeding program has cows calving over an extended period of time (more than 90 days), gradually reduce the breeding season to 60 days over a 2- to 3-year period. Do not try to reduce the season to 60 days in one year, because late-calving cows will not cycle and conceive during this shortened breeding season and will turn up open when you test for pregnancy. The 60-day breeding season makes your calves more uniform at weaning and allows the routine management techniques to be performed on all calves at the same time. After the 60-day season is established, a cow that is determined to be open should be culled; such a cow often has reproductive problems. In addition, it will be 24 months before this cow weans her next calf in a herd with one 60-day calving season per year.

CAUTION: Remember that a bull is potentially dangerous. Before entering the bull pen, be sure to find and establish an escape route so the bull cannot get you down or pinned against a wall or fence. It is highly recommended that you carry a club with you when working around bulls. Never turn your back on a bull, and have a second person close by who can distract the bull if necessary.

When bulls start to become mean, send them to market. Many beef producers have been maimed or killed by so-called tame bulls. Be careful! No bull is worth the risk of getting seriously injured or killed. Be especially careful that young children, less capable of escaping, are not allowed to play in or walk into the bull pasture or lot.

2.7 SORTING

Sorting is the process of dividing a large group of cattle into two or more smaller groups. Cattle can be sorted by age, sex, stage of production, slaughter grade, state of health, or breed, depending upon the management technique that is to be performed. An example would be sorting cows from calves at weaning or steers from heifers at castration time.

This technique is difficult for the individual who has not had experience working with livestock. Sorting cattle requires second-guessing the animals and anticipating their next moves. These maneuvers are difficult to describe in a step-by-step manner and require firsthand experience. To gain experience, observe and work with an experienced cattleman before attempting to sort by yourself.

Equipment Necessary

- Corral
- Sorting alley, working chute, or large pen

Restraint Required

The restraint of individual animals is not necessary for this technique, but corral pens are needed when large groups are to be worked. Portable or permanent corrals work well to sort cattle when such corrals are equipped with a sorting alley or working chute. A sorting alley approximately 12 to 14 feet wide allows one person to sort cattle into two groups, or several people to sort cattle into three or more groups. A working chute can be used to sort cattle, but bruising, stress, and shrink are greater than when a sorting alley is used. Large pens can also be used on small operations, or when small numbers of animals are to be sorted from a group. The disadvantage of large pens is that labor requirements and animal stress are greater and animal control is more difficult than in the other two methods.

Step-by-Step Procedure

1. Decide which cattle will be sorted from the group and how they will be identified.

When separating cows and calves, always sort (move) the cows away from the calves, because the cows are more apt to move than are the calves. To reduce stress, avoid working animals more than once when they are being sorted. This can be accomplished by sorting out several groups of cattle into two or more pens at the same time.

2. Prepare a sorting area to accommodate the group to be sorted. You may need to set up a portable corral or check the permanent corral for protruding objects.

3. Be sure that the group of cattle that you are moving is of manageable size, and move the group into the corral slowly and quietly to avoid stress. When only one animal is to be sorted from the group, several people are often needed to move slowly and quietly through the group, isolate the desired animal, and move it along the fence to the gate.

4. Move the cattle into the sorting alley or working chute. Cattle prefer to go back the same way they came. This behavior is used to best advantage by moving cattle from point *A* to point *B* in a sorting alley. Then animals can be sorted into pens as they go back the same way they came (to point *A*). Notice how the gates open when cattle are at point *B*.

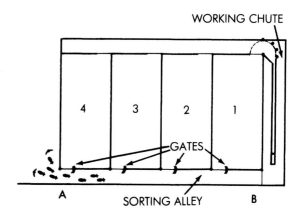

CAUTION: The susceptible zone for bruises on cattle is 28 to 58 inches from the ground. Check the sorting alley and pens for nails and bolts that stick out in this zone, and remove or cover them with a piece of worn-out tire. Be sure to eliminate those nails and bolts that are found with tufts of hair on them. These precautions will minimize bruising and injury.

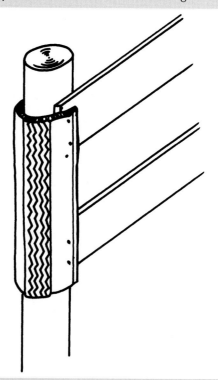

When gates sag, drag on the ground, and remain partially open, they create a serious bruise hazard to cattle and increase the chance of injury to man when nervous or excited cattle are worked.

Good footing is important in the sorting alley to help prevent the animals and the handlers from slipping and injuring themselves. Do not sort cattle when the alley is coated with ice. If you have to work cattle on ice, it is best to spread sand or crushed limestone on the icy surface to give the animals some footing.

When a working chute is used, a cutting gate should be included so that cattle can be sorted in three ways (*A, B, C*).

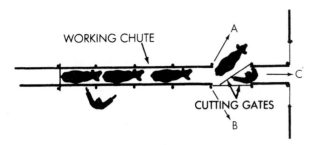

WORKING CHUTE

CUTTING GATES

5. The following steps are for the sorting alley. Identify the pens. Designate the last pen along the sorting alley (pen 4) for the largest group (example: cows). This allows the gate to be tied open (across the alley), and cattle destined to be in this group to be moved down the alley and into a pen without opening and closing gates. Designate pen 1 or pen 2 for the smallest group. This allows more time to identify the animal and open the proper pen gate. The person identifying the animal can refer to pens by number, or as "steer," "heifer," or "bye" to minimize confusion.

6. Position one or more people at the pen gates along the alley and one or two people at point *B* to hold and sort the cattle.

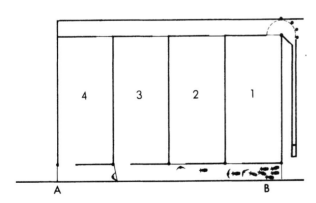

7. One person at point *B* should use a canvas slapper, broom, or whip lightly to sort and move one animal down the alley at a time.

CAUTION: Cattle like to follow the leader, and you should be ready for this. After one animal is sorted down the alley, both people at point *B* should face the cattle and, if needed, use the canvas slapper or whip on a few animals' noses. Each situation is different, but you may have to back up a few steps, wave your arms, and holler a little bit to stop the whole group from going down the alley.

When sorting cows and calves, always sort the cows away from the calves, because cows will move individually down the alley easier than will calves.

8. Identify the animal and call out the pen it goes to as the animal is sorted down the alley. Say it loud enough that the people operating the gates understand which gate to open. Repeat if necessary.

9. The person operating the identified pen opens the gate and allows the animal to enter.

10. Repeat steps 7 to 9 until sorting is completed.

CAUTION: Be patient if the animal does not sort easily. If the animal has never before entered the gate through which you are trying to move it, this task can be difficult. The herding instinct of cattle is strong, and the animal will try to return to its penmates.

Do not use an electric prod in the pen, corral, or sorting alley. A canvas paddle about 4 inches wide by 2 to 3 feet long, attached to a wooden handle, is very effective in moving cattle. The paddle should be dry (not frozen) to prevent bruising. An ordinary broom also works well when used as an extension of one's arm.

2.8 CASTRATION

Castration is the removal of the testicles of a bull by either surgical or nonsurgical methods. Bulls are castrated because of consumer preference at the marketplace, for economic reasons, and for better animal disposition. Consumers tend to prefer meat from steers because they believe castration improves the color, texture, tenderness, and juiciness of the meat. Cow–calf and feeder producers benefit from castrating bulls because the market pays a premium for steers, partly because of consumer preference and partly because of the quieter dispositions of steers and the ease with which they are handled while in the feedlot. If market discrimination against bulls is reduced in the future, more of them may be fed; they have better growth rates, feed efficiencies, and carcass cutability than steers.

Bulls can be castrated at any time during their lives, but it is recommended that it be done when the bulls are less than 3 months old to minimize stress. Calves can be castrated at birth or when they are ear-tagged. In some beef operations, it is not practical to castrate 1- to 3-month-old bulls because potential herd bull prospects will not be selected until performance records are calculated after weaning. Bulls that weigh more than 500 pounds, however, generally bleed more and tend to be affected more by stress and infection. The best seasons to castrate are the spring and fall, so that flies and maggots do not increase irritation and infection of the wound. Avoid castrating during screwworm season in the South, during periods of intense heat, and on cold, wet days whenever possible.

CAUTION: Do not attempt to castrate animals without someone experienced in the castration techniques to guide and teach you.

Several methods of surgical castration have been used with varying degrees of success. All of the methods can accomplish successful removal of the testicles, but not all techniques allow the scrotum to drain properly. Good drainage is one of the most important aspects to consider. One method of surgical castration that gives excellent drainage is completely removing the lower third of the scrotum.

Another thing that needs to be considered when bulls are surgically castrated is the removal of the tunica vaginalis (referred to as *tunics* in this discussion). The tunics are the thin membranes that surround each testicle individually. Some castration techniques open the scrotum, cut the tunics, remove the testicle, and leave the tunics in the scrotum. If the tunics are not removed, they can swell to ½ inch or more in thickness within the first day after castration. This swelling can aggravate and enhance the possibility of infection, edema, and the tunics' building up into a large mass of scar tissue.

The last thing that must be considered in surgical castration is hemorrhage control. This is especially important in bulls that weigh over 500 pounds and are not allowed to immediately lie down and rest, allowing blood to clot normally, after castration. When no hemorrhage control is used, a large blood clot often forms inside the animal in the flank area. This large blood clot provides an ideal medium for bacterial growth and can act as a wick for bacterial invasion of an area of the body that can become seriously infected. The presence of both the blood clot and swelling is largely responsible for postcastration complications.

Two nonsurgical castration methods can be used on bulls. The first method involves the use of an emasculatome (Burdizzo®). Sometimes the emasculatome is referred to as a clamp or pincher. The emasculatome method is a bloodless castration that can crush the cord and blood vessels associated with the spermatic cord without disturbing the central septum of the scrotum. The advantage of this method is that it can be done during the fly season as long as the weather is favorable. The disadvantage of the method is that the inexperienced operator may miss one of the cords completely, so the animal will remain a bull and develop a staggy appearance at a later date.

Another nonsurgical method involves the use of an elastrator and a rubber ring. This procedure works best on animals of less than 1 to 2 months of age. The band is placed on the scrotum between the animal's body and testicles. The rubber rings cut off blood circulation to the testicles and lower scrotum, which will atrophy and slough off in 3 to 4 weeks. The advantage of the elastrator is that it is bloodless, but the disadvantages outweigh the advantages. The disadvantages are that tetanus and infection can be real problems as the scrotum atrophies and sloughs off. When this method is used, tetanus antitoxin should be administered at the time of castration.

Restraint Required

Bull calves younger than 1 month of age can be approached, cornered if necessary, caught, flanked, and restrained on the ground. Bulls younger than 6 or 7 months can be restrained in a calf (tilting) squeeze chute, headgate, or regular squeeze chute. Bulls over 7 months of age should be restrained in a headgate or squeeze chute. When bulls are castrated standing up, the tail should always be pulled straight up over the back and held firmly (see Chapter 1). This will help to keep the animal from moving during the castration, as well as minimizing the danger of your being kicked.

Surgical Castration

Equipment Necessary

- Squeeze chute, headgate, or tilting calf squeeze chute
- Nose lead
- Scalpel or sharp knife
- 8- or 10-inch curved or straight hemostats (alternative: dull knife or emasculator)
- Pan
- Disinfectant
- Aerosol wound dressing (Furacin base)
- Tetanus antitoxin
- Syringe (1½", 18-gauge needle)
- Canvas (6' × 6', on which to lay a baby calf and equipment)

This technique will describe how to castrate bulls with a scalpel or sharp knife. This method maximizes drainage and hemorrhage control, and should be considered the method of choice when bulls of more than 500 pounds are castrated. The use of a dull knife and an emasculatome will also be discussed as alternative methods of testicle removal.

Step-by-Step Procedure

1. Assemble, clean, and sanitize the necessary equipment.

> **CAUTION:** Sterilization can be accomplished by boiling the instruments in water for 30 minutes. This should be done before beginning to castrate, and the instruments placed in a pan of disinfectant between castrations. The disinfectant solution should be changed periodically to maintain its antibacterial activity. The number of calves that can be castrated before changing the disinfectant will depend on how clean the cattle are and the type of disinfectant used. A good rule of thumb is to change the disinfectant every 15 calves.
>
> When bulls are castrated by surgical means on farms or ranches that have had a previous history of tetanus, vaccinate the bulls with tetanus antitoxin at the same time they are castrated (see Chapter 10).

2. Restrain the bull. If you are castrating a bull that is standing, be sure to have an assistant pull the tail straight up and over the back to maximize your safety. The tail should be held firmly in this position until the castration technique is completed.

3. Palpate the scrotum to make sure that both testicles are present.

Sometimes the bull is a cryptorchid, or has a scrotal hernia that requires more complicated surgery and the services of a veterinarian. Cryptorchidism is a condition in which one or both testicles have not descended into the scrotum. A scrotal hernia is the result of a weakness in the inguinal canal, which allows the viscera (intestines) to protrude into the scrotum.

4. Wash the scrotum and your hands with an antiseptic soap. Keep your hands, the scrotum, and the instruments as clean as possible.

Do not handle the chute or animal more than absolutely necessary so that contamination of the scrotum is minimized.

5. Grasp the lower third of the scrotum and push the testicles up toward the body with one hand while holding the scalpel or knife in the other hand. Be certain that the testicles are pushed upward as far as possible, and maintain this grasp while proceeding with step 6.

> **CAUTION:** Make sure that the testicles are above the point where you intend to insert the blade.

6. Position the scalpel or knife with the other hand, above your fingers, and make several controlled strokes to completely remove the lower ⅓ to ½ of the scrotum.

> **CAUTION:** Be sure that you cut the scrotum below the testicles to reduce pain to the animal and minimize injury to yourself.

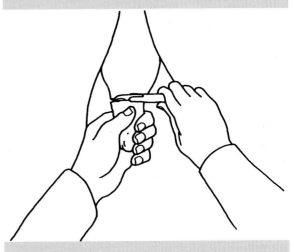

> Be careful. If the animal moves, or if a large bull with tough skin is being castrated, this can be dangerous to both the bull and you. A large vein (saphenous) runs down the inside of the bull's back leg. In mature bulls this vein is ¾ to 1 inch in diameter. If the blade slips or the bull moves as you cut this vein, the bull can bleed to death before you can get the bleeding stopped. In addition, the blade is in very close proximity to the hand holding the scrotum. Be extremely careful that you don't slip and cut your hand severely.

7. Place the scalpel or knife in a pan of disinfectant.

8. Grasp one testicle and apply slight downward tension to the spermatic cord with one hand while the thumb and index finger of the other hand separate the testicle and cord from the surrounding connective tissue. The thumb and index finger should be placed on the cord and moved up and down the cord to separate and tear the connective tissue.

9. Once the connective tissue is separated, separate the cremaster muscle so that hemorrhage-control measures can be used. Use your index finger and thumb to grasp the cord as high as possible inside the scrotum, and pull downward on the testicle with a slow, steady pull. The cremaster muscle will break between the top of the testicle and your fingers. This will require some judgment on your part.

> **CAUTION:** The spermatic cord consists of an outer membrane, the cremaster muscle, the vas deferens, an artery, and a vein. As the muscle begins to separate and break, you will be able to feel the tension of the cord relax. When this happens, *stop.* You *do not* want to separate blood vessels and vas deferens at this time. Breaking the blood vessels before measures are taken for hemorrhage control can cause serious complications in cattle over 500 pounds, cattle with blood-clotting deficiencies, or cattle that are not allowed to lie down and rest in a clean area immediately after the operation.

10. Crush and clamp the cord with a hemostat to control hemorrhage. Apply slight downward tension on the testicle with one hand and apply an 8- or 10-inch curved or straight hemostat to the cord (blood vessels and vas deferens surrounded by a membrane) with the other hand. The cord should be crushed several times along the length inside

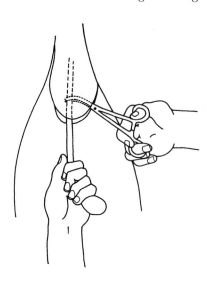

the scrotum. Crushing the cord in this manner damages the epithelial cells, stimulates the blood-clotting mechanism, and will reduce the amount of bleeding that will occur when the testicle is removed. After the cord has been crushed several times, lock the hemostats tightly onto the cord just below where the cord has been crushed in several places.

> **CAUTION:** Make sure that the hemostat is positioned inside the scrotum so that when the testicle is removed, the remaining cord will not hang below the scrotum and prevent scrotal healing.
> The locked hemostats will serve as a handle later in the technique and will aid in hemorrhage control. Without the hemostats locked on the cord, the cord will disappear up into the scrotum and it will be impossible to observe the cord for hemorrhage control.

11. Remove the testicle. Squeeze the cord tightly just below the hemostat with the index finger and thumb of one hand, and pull downward on the testicle with the other hand until the cord breaks and the testicle (including the tunic) is completely removed.

Alternative Method for Removing the Testicle: Dull Knife. This method replaces steps 8 through 11. Begin by grasping the testicle and applying downward tension on the cord with one hand. Use the other hand to position the dull knife on the cord inside the scrotum, and scrape up and down on the cord until it is abraded completely and the testicle (including the tunics) is removed.

Alternative Method for Removing the Testicle: Emasculator. This method replaces steps 8 through 11. Begin by grasping the testicle and applying downward tension on the spermatic cord with one hand. The other hand should position the emasculator as high as possible on the cord. The crushing part of the emasculator should be positioned toward the body and the blade toward the testicle. Squeeze the handles of the emasculator together and sever the cord to remove the testicle

> **CAUTION:** Abrading the cord with a dull knife does provide some hemorrhage control by damaging the epithelial cells, but it is not as effective as the hemostat mentioned in step 10. This method is not recommended for use on bulls weighing over 500 pounds or bulls that are not allowed to lie down and rest in a clean, dry area immediately after castration.

(including the tunics). Keep squeezing the handles for 10 to 15 seconds to reduce hemorrhage.

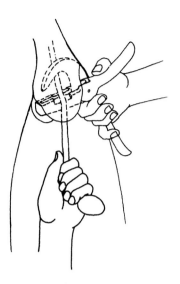

> **CAUTION:** The emasculator is not totally effective in hemorrhage control, because it does not damage the epithelial cells to the same extent as does the hemostat or dull knife. This method, therefore, is not recommended for use on calves over 500 pounds or calves that are not allowed to lie down and rest in a clean, dry area immediately after castration.

12. Remove the second testicle in the same manner as the first (steps 8 through 11).

13. Apply an antiseptic powder to the scrotum. A powder that contains Furacin works well for the prevention of bacterial infection.

> **CAUTION:** Be sure to place some powder well up into the scrotum so that any contamination that may have occurred during the operation or that may occur after the operation is over does not have an opportunity to produce a serious infection.

14. Whenever possible, release the animal into a clean, dry area immediately after castration. Provide a pen, lot, or pasture where animals can lie down and rest while the blood clots normally. The area should also be big enough to allow the animal to walk around at its own discretion to help minimize edema and swelling in the scrotal area. Calves that are nursing their dams should be returned to them, provided that the area is clean. Weaned calves that have been castrated should be penned together for at least a day before they are returned to a larger group containing noncastrated calves.

Nonsurgical Castration—Emasculatome (Burdizzo®)

Equipment Necessary

- Squeeze chute, headgate, or tilting calf table
- Nose lead
- Emasculatome (Burdizzo®)

The emasculatome is the preferred method of nonsurgical castration. This method can be used at any time of year without concern about open-wound infection and parasites.

Step-by-Step Procedure

1. Assemble the necessary equipment.

2. Restrain the animal. If you are castrating an animal that is standing, be sure to have an assistant pull the animal's tail straight up and over its back to maximize your safety. The tail should be held firmly in this position until the technique is completed.

3. Palpate the scrotum to make sure that both testicles are present.

> **CAUTION:** Sometimes the bull is a cryptorchid or has a scrotal hernia that requires special techniques and the services of a veterinarian. Cryptorchidism is a condition in which one or both of the testicles are retained inside the body cavity. A scrotal hernia is the result of a weakness in the inguinal canal; the viscera (intestines) protrude into the scrotum.

4. Work one testicle down to the bottom of the scrotum and use one hand to position the spermatic cord to the outer edge of the scrotum. The cord will feel somewhat like the seam on an old pair of blue jeans. With the other hand, position the open emasculatome approximately a third of the way across the scrotum and 2 inches above the top of the testicle, being certain that the cord that was worked to the edge of the scrotum is within the jaws of the emasculatome.

> **CAUTION:** Do not position the emasculatome across the center septum of the scrotum. If you crush across the center septum, the bottom of the scrotum may slough off and expose the testicles. The reason for this is that the blood supply to the bottom of the scrotum runs through the center septum.
>
> You may not be able to get the emasculatome positioned 2 inches above the testicle on bull calves less than 1 month old. If you cannot get it 2 inches above the testicle, place it as far above the testicle as possible.
>
> Before you close the emasculatome to clamp the cord, make sure the cord is located between the jaws of the instrument. Be careful that the jaws do not clamp the bull's rudimentary teats or the skin above the scrotum.

5. Squeeze the handles of the emasculatome together and hold them in position for approximately 30 seconds. Some emasculatomes have a leg strap on the end of one handle that enables you to

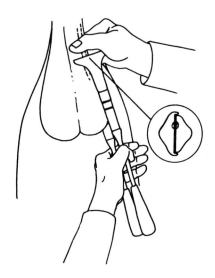

squeeze the handles together with one hand and your leg while the other hand holds the cord between the jaws. If the emasculatome does not have a leg strap, both hands must be used to squeeze the handles shut. Check the cord to be sure that it is between the jaws. If you have missed it (it does have a tendency to slip), release the emasculatome and repeat steps 4 and 5.

6. Release the emasculatome and repeat steps 4 and 5 with the emasculatome positioned ¼″ to ½″ below the original clamping site.

7. Repeat steps 4 through 6 for the second testicle, operating from the other side.

8. Release the animal.

Nonsurgical Castration—Elastrator

Equipment Necessary
- Elastrator
- Elastrator bands
- Tetanus antitoxin
- Syringe (1½″ 18-gauge needle)

This method is the least desirable of the castration methods because tetanus can be a serious problem when the bottom of the scrotum atrophies and sloughs off. This method, if used at all, should be used only on calves less than 1 month old.

Step-by-Step Procedure

1. Assemble the elastrator and the elastrator bands. Place one rubber elastrator band over the four prongs of the elastrator.

> **CAUTION:** When bulls are castrated with elastrator bands on ranches or farms that have had problems with tetanus, vaccinate them with tetanus antitoxin at the time of castration.

2. Restrain the calf.

3. Palpate the scrotum to make sure that both testicles are present.

> **CAUTION:** Sometimes the bull is a cryptorchid or has a scrotal hernia that requires special techniques and the services of a veterinarian. Cryptorchidism is a condition in which one or both of the testicles are retained inside the body cavity. A scrotal hernia is the result of a weakness in the inguinal canal; the viscera (intestines) protrude into the scrotum.

4. Prime the band with several squeezes of the elastrator handle, at the same time cupping your hand over the band to avoid injury to your eyes if the band should break or slip.

5. Squeeze the elastrator handles and position the elastrator near the scrotum, with the prongs toward the calf's body.

6. Grasp the scrotum and both testicles and pull them through the stretched elastrator band.

7. Position the elastrator band between the testicles and the body. Release tension on the band and remove the elastrator, leaving the ring on the scrotum above the testicles. Check to make sure that both testicles are below the elastrator band. If both testicles are not palpated below the band, cut the band with a sharp knife and repeat steps 1 through 6.

After the band is placed on the calf's scrotum, the calf usually will lie down, kick, and wring its tail. This may continue for about 30 minutes, until the scrotal area becomes numb.

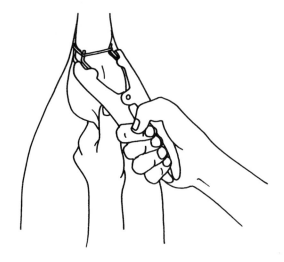

8. Release the animal.

9. Some producers recommend reworking the calves in 7 to 10 days, removing (cutting off with a knife) the necrotic portion of the scrotum (and testes) below the band.

10. Treat with antiseptic powder or aerosol wound dressing.

Postprocedural Management

Surgical. Hemorrhage and infection are always potential problems after castration. To minimize these problems, place the steers in a clean, dry area where they will not be disturbed. The area should be large enough for the animal to walk around at its own discretion. This will minimize swelling and edema. Generally, the calves are still nursing, and they should be placed on a clean pasture or in a clean, freshly bedded pen with their dams. If the steers have been weaned, they should be penned as a group for at least one day, until the effect of castration subsides, before they are returned to a larger group that contains noncastrated animals.

Castrated animals should be observed closely for several days after surgical castration. Some swelling and stiffness will occur within a day after castration, and then gradually will go away by the end of the first week. The scrotum should be healed approximately two weeks postcastration. If superficial infection occurs around the edges of the scrotum, a broad-spectrum antibiotic should be administered. Any sign of pus, severe swelling, or abscess formation should prompt you to call a veterinarian.

Nonsurgical. The main concern after nonsurgical castration is swelling. Usually the animal will have some swelling within 1 to 2 days after castration; the swelling will gradually go away by the end of the first week. When the emasculatome is used, the scrotum will not slough off if the central septum was not clamped. The testicles should show some signs of atrophy within 2 to 3 weeks after castration. When the elastrator is used, the bottom of the scrotum and testicles will atrophy and slough off in 2 to 3 weeks. If preventative measures have not been taken, tetanus can be a real problem when the testicles and scrotum slough off. If swelling appears excessive after either method, call a veterinarian for assistance.

2.9 WEIGHING

Animal weights can be used in cow–calf, stocker, and cattle feeding operations. Scales are used routinely in performance-tested cow herds to identify potential herd replacements and for culling cows and calves with low performance. Periodic animal weights can be used to identify needed ration changes and possible herd health problems. In addition, animal weights after purchase or before marketing can be used to calculate percentage shrink and aid in management decisions.

Animals can be weighed at any time during their lives. When calves are weighed at birth and weaning, a 205-day adjusted weight and ratio can be calculated to assist in selection decisions. If calves are also weighed when they are one year old, a 365-day adjusted weight and ratio can be calculated. Animal weights before, during, and after the growing and finishing periods in stocker and feedlot operations can be used in calculating gain, feed efficiency, and cost of gain.

Equipment Necessary

- Canvas sling
- Calving chains
- Handheld scale
- Working chute
- Livestock scale
- Record sheets

Restraint Required

Newborn calves can be flanked and restrained on the ground so that the canvas sling can be applied and the calf can be weighed. Older cattle should be restrained in a working chute with a portable livestock scale positioned directly in front of or in the place of the squeeze chute. When permanent scales are used, the working chute should be designed so that the cattle do not need to go through the squeeze chute to reach the scale. Cattle that are trained (broken) to lead do not require restraint in a working chute because they can be approached, haltered, and led to the scale.

The step-by-step procedures are divided into two age classifications: newborn calves and older cattle. The older-cattle classification includes all cattle that are more than several days old.

Newborn Calves

Step-by-Step Procedure

1. Assemble the necessary equipment. This will include a canvas sling and a handheld scale capable of weighing up to at least 150 pounds.

2. Catch and restrain the newborn calf before it is 24 hours old.

CAUTION: A cow may be very protective of her calf and attack with her head, causing injury to you. This is especially true if the calf bawls. If the cow appears to have this tendency, you must plan an escape route before performing the technique. If the cow and calf are in a maternity pen or lot, you may want to let the cow out of the pen before trying to weigh the calf. If the cow and calf are in a pasture, a pickup truck can be positioned in such a way that the weigher can escape around or into the back of the truck.

6. Hook the canvas onto the scale, lift the scale high enough that the calf does not touch the ground, and read the calf's weight. Hold the scale so that you can read the calf's weight. This step may require two people if the calf is extremely heavy. An alternative to holding the scale to read the calf's weight is to use a tripod or bracket that hooks on the fence or over the side of the pickup truck.

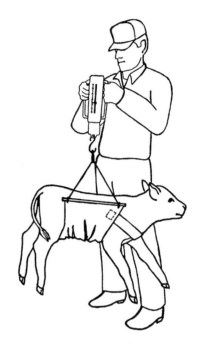

3. Move the calf over to where the handheld scales are located. Grasping the skin under the chin with one hand and the rump with the other will allow you to move the calf around.

4. Flank and restrain the calf on the ground. Depending upon your size, availability of help, and where you are for the weighing, it might be easier to position the sling while the calf is standing.

5. Position the canvas sling, with the four corners of it attached to some sort of rope, wire, or cable, under the midsection of the calf. Gather the two loops of this rope together so that they can be placed onto the scale hook. Before attempting to hook the sling to the scale, be certain that the calf's legs are properly inserted into the opening between the wide body band and narrower neck band of the sling.

7. Record the calf's weight and the identification numbers of the cow and calf. It may also be convenient to castrate, ear-tag, and tattoo the calf at this time.

8. Release the calf and return it to its dam.

An alternative to using the canvas sling, although the sling is much more desirable, is to use a calving chain. To make use of the chain to weigh calves, restrain the calf as you did in the sling technique, but instead of placing the sling below the midsection, make loops in both ends of the chain, slip them over the joined rear and forelegs of the calf, and then place the chain over the scale hook. This method is just as accurate as using the sling, but the calf is more comfortable and will struggle less with the sling.

Older Cattle

Step-by-Step Procedure

1. Assemble and prepare the necessary equipment. Place a portable or permanent scale at the end (or near the end) of the working chute. Check

the scales when they are in position to be sure the platform is free to move. Sometimes dirt, manure, leaves, and rodent nests become lodged around or under the platform and affect the scale's accuracy. Clean and repair the platform if necessary. Prepare a weigh sheet with animal numbers.

2. Move the cattle into the restraint area.

3. Tare (balance) the scale to zero. This should be done initially, and then again after the first or second animal. If the scales do not balance after several animals have been weighed, this step should be repeated after every one or two animals until it balances on its own.

Cattle often track mud and manure on the scales, which can affect the scale. Check the scale after each lot is weighed, or after each 15 to 20 animals (in a dry environment) or each 5 to 10 animals (in a wet environment).

4. Open the scale door near the entering animal's head and allow the animal to walk in. Sometimes cattle hesitate and do not want to enter the scale. When this happens you can twist the animal's tail, gently prod it in the rump with a dull nail, or open the front scale door to entice the animal to enter.

CAUTION: When you open the scale door, be ready. Once the animal sees the opening, it may lunge toward the open door and escape before you can close it.

---

5. Shut the scale door and weigh the animal. Each scale has its own peculiarities, and the weigher must become familiar with these. Follow the manufacturer's recommendations.

Some animals will rock back and forth when they get on the scale. This makes weighing more difficult because the scale can fluctuate by 10 or more pounds. Depending on the accuracy of the weights desired, you can wait until the animal stops moving before you read the weight, or you can estimate the weight as the scale fluctuates. Sometimes an animal will stop moving long enough that the scale can be read if an assistant scratches the animal on the back or places a hand by or on its head.

6. Record the weight and animal identification number.

7. Open the scale door and allow the animal to exit.

8. Close the scale door and repeat steps 2 through 7 until all cattle are weighed.

2.10 WEANING

Weaning involves removing the calf from the cow and terminating lactation. Usually this management technique is performed when the calf is approximately 7 to 8 months of age. When the recommended 60-day breeding season is used, all calves in the herd can be weaned at the same time. Extended calving seasons (longer than 3 months) or split calving seasons (spring and fall) will require weaning in at least two groups.

Special circumstances may require early weaning (weaning after 2 to 3 months and before 7 to 8 months). When grass gets short in the summer and milk production goes down, it may be economically advantageous to wean the calves early. Calculate the costs involved in such a decision at a given point in time. Usually it is more economical and efficient to feed the calf directly than to feed the cow to produce milk. A second situation where early weaning may be advantageous is when a young cow begins to lose weight and condition while nursing a calf. Again, an economic evaluation will dictate how to proceed, but the cow will gain more weight and condition in a shorter period of time if she is not lactating. A third situation is when cows calve in the fall and cows are fed harvested feed. It is possible to feed lower-quality forages to nonlactating cows than to lactating cows because lactating cows require more nutrients for milk production. Consider also the environmental conditions surrounding the cow herd. If cows are in confinement with extremely muddy, unsanitary conditions and calves are scouring, it may be beneficial to wean calves early and place them in clean, dry pens. Anytime you decide to wean early, a good calf ration should be formulated and fed.

Equipment Necessary

- Sorting alley, corral, or small lot
- Working chute
- Squeeze chute
- Pen or pasture (for weaned calf)
- Record sheets
- Scale
- Multiple-dose syringe
- Syringe needles (1", 14-gauge)
- Vaccines

Restraint Required

Special restraint is not required for this technique. The important thing is that cows and calves need to be confined in an area that allows sorting with a minimum amount of stress on the animals. After the cows and calves are separated, the calves should be placed in a pen or pasture equipped with a feed bunk or hayrack. Ideally, this pen or pasture should be out of sight and sound of the cows.

Step-by-Step Procedure

1. Preparation for weaning comes several months before the cows and calves are actually separated. This program may or may not be followed completely, but it is a recommendation. Replacement heifer calves should be vaccinated between 2 and 4 months of age for brucellosis, and never after 8 months of age. Heifers will need to be worked before weaning to meet these age limitations.

CAUTION: Heifers vaccinated between the ages of 2 and 4 months do not show persistent antibody titers to brucellosis. Approximately 10% of heifers vaccinated at 6 to 8 months, 25% of those vaccinated at 8 to 10 months, and 40% of those vaccinated at 10 to 16 months show persistent antibody titers when tested later. A persistent titer indicates that the animal has been exposed to the disease by vaccination or actual disease. The law requires that these animals be disposed of. When a positive reactor (persistent titer) is discovered in a herd, that herd will be placed under quarantine until the herd tests clean.

It is also a recommended practice in spring calving herds to vaccinate for blackleg when cows and calves are placed on spring pasture, and a blackleg booster should be given two to three months after the initial vaccination. Calves should be vaccinated for infectious bovine rhinotracheitis (IBR, or red nose), parainfluenza (PI_3), and bovine viral diarrhea (BVD) at least two weeks before weaning or at weaning. In addition, calves should be wormed and treated for grubs at weaning. For labor efficiency, such operations as castration, dehorning, branding, and some of the above vaccinations can all be done in one fall roundup.

2. Assemble all materials, supplies, and record forms.

3. Move animals into a sorting alley, corral, or small lot for sorting.

4. Sort cows from calves. Do not try to sort calves from cows.

5. Move cattle quietly and gently to a work chute and squeeze chute if they are going to be vaccinated, implanted, wormed, or treated for grubs.

6. Return the cows to their pasture or lot.

7. Weigh calves so that accurate production records can be kept.

8. Place calves in a lot or pasture equipped with sturdy fences and gates, a feed bunk, and a hayrack. If cows and calves are not separated far enough so

that they cannot hear and see each other, cows will often go through and over fences while attempting to get to their calves.

9. Whenever possible sort calves by weight and pen them into separate groups. If you do not, the larger calves will deprive their smaller, less-aggressive herdmates of adequate feed and water. Sorting into two groups should be the minimum, with three groups being more desirable. If sorting and penning by size is not an option, be certain there is adequate feed bunk and waterer space for each calf. Placement of the bunks and waterers along the fenceline (where the calves will be walking) will help the nervous calves locate the feed and water as rapidly as possible.

10. Offer calves hay immediately after they are weaned. Free-choice access to a good-quality grass–legume hay—is important the first several days after weaning. No moldy or spoiled hay should be used. Start feeding hay by scattering some on the ground and leave a trail that eventually leads to the feed bunk. This way they can't miss it. Water and a salt–mineral mix should be available at all times.

11. On the evening of the second day you can start feeding some mixed ration on top of the hay in the bunk. This ration may include grain, silage, medication, or a complete starter ration.

CAUTION: Do not overfeed the grain-containing ration (concentrate). Start at 1 to 2 pounds per head per day and gradually increase the amount of the ration over the next 10 to 14 days until they are on full feed. Creep-fed calves can be started on 2 to 3 pounds of grain the first day after weaning and the grain ration increased over a period of 7 to 10 days.

If you feed an antibiotic or medicated feed, make sure the manufacturer's recommended dosage on the label is followed. Don't overfeed, but don't underfeed either. Underfeeding can be ineffective against the respiratory diseases that often result after cattle are weaned and especially after they are transported. Medicated feeds are not quite as important after creep-fed calves have been weaned.

Postprocedural Management

Observe animals three to six times daily after weaning. Check for droopy ears, runny noses, and coughing. The stress of weaning often precipitates a respiratory infection. Part of this is caused by throat and lung irritation during the bawling period. If any signs of sickness, respiratory infection, or pneumonia are observed, sort the sick calves out of the group and take their temperatures. The animal's temperature is a measure of how sick it is. Record the temperature, animal identification number, type of treatment given, and calf's response. Consult your veterinarian for treatment and medication needed for respiratory infections.

The first night after you wean calves, be ready to listen to cows and calves bawling. Bawling will begin to subside after the first day, and the animals will be back to normal in two to five days.

Cows may show some udder swelling the week after weaning. This is normal. Beef cows generally do not milk very heavily after a lactation period of 7 to 8 months, and they will dry up in 7 to 14 days.

2.11 BRANDING

Branding is a permanent means of identifying cattle. Two methods of hide branding are available: hot and freeze branding.

Brands are used to establish ownership or to identify individual animals. The brand consists of letters of the alphabet, numbers, designs, or combinations of these. Some states require registration of the ownership brand and designation of its location on the animal. The ownership brand helps to discourage cattle rustling, and also serves as a cattleman's trademark. When ownership brands are applied, it is best to use hot brands. Success or failure can then be determined shortly after branding and animal-to-animal variation does not affect the results as greatly as it does with freeze branding.

The individual animal identification brand permits quick identification of an animal from a distance. This brand usually consists of three or four numbers in the same sequence as the tattoo and ear tag. Either hot or freeze brands can be used satisfactorily for identifying individual animals.

Identification and ownership brands can be applied at any time during the year, but this procedure is usually performed in combination with one or more other techniques such as weaning, castration, and vaccination. When ownership brands are used, it is usually done before or during weaning because the probability of a calf going astray is greater after weaning than it is before.

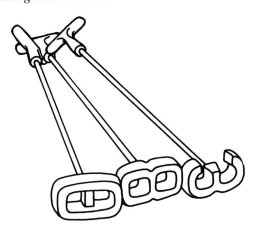

The size of the branding iron is important. Each character should be 4 inches high by 3 inches wide (outside measurements) for calves less than one year old and 6 inches high by 3½ inches wide for older cattle. The face of the branding surface should be about ⅜ of an inch in width, with the surface edges slightly rounded.

Brands can be applied to several different areas of the animal, but the most common locations are the hip, rib, thigh, and shoulder. The area you brand may be determined by the branding office in your state. One reason that these locations are popular is that the brand can be applied easily to these areas in a squeeze chute, and the brand is seen easily from a distance.

Keep in mind that while we have spoken of the value of branding, the process is not without its cost. Hot branding damages the hide, making it less valuable for leather goods. This costs the beef industry nearly $15 per head, for a total annual loss of $120 million.

Equipment Necessary: Hot Branding

- Branding irons (steel or iron) or electric irons
- Small propane tank with burner or wood fire
- Squeeze chute
- 30-gallon drum

Equipment Necessary: Freeze Branding

- Copper or copper alloy branders
- Liquid nitrogen or dry ice + 99% isopropyl alcohol
- Styrofoam cooler (inside a wooden box)
- One-quart squeeze bottle
- Funnel
- Electric clipper
- Stiff-bristled brush
- Clock (with second hand)

Restraint Required

In the past it was a general practice to rope calves and drag them to the place of branding. The calves were then flanked and restrained on the ground for branding. In recent years, however, flanking has been largely replaced by moving cattle to a squeeze chute. This method is easier on both man and animal. Animals should be placed into a squeeze chute equipped with a headgate and tailgate.

Hot Branding

A good hot brand is recognizable because it destroys hair follicles located under several layers of skin and leaves a permanent bald scar on the hide of the animal.

Step-by-Step Procedure

1. Assemble and prepare the necessary equipment. The irons used in hot branding should be iron or steel, and should be free of dirt and hair. If two or more pieces of metal join at the face of an iron, a notch should be filed or cut away at the junction to allow some heat to escape. The notch should be ¼" deep by ¼" wide. This notch will prevent too much heat from being applied at the junction, which could cause blotching.

2. Heat the branding irons. The lowest-cost method of heating branding irons is to use the hot coals of a wood fire (coal should not be used). To obtain a good bed of hot coals, a small pit measuring approximately 3 feet long, 1½ feet wide, and 10 inches deep should be dug. The fire can be built over the pit by using wood more than 3 feet long placed over the length of the pit. Allow time for the fire to burn down to the point that plenty of red-hot coals appear in the bottom of the pit. The branding irons should be placed on top of the coals, not in the flames of the fire.

A second and more convenient way to heat irons is to use a small propane tank and burner. The heat produced by the burner should be contained within a 30-gallon drum or comparable structure with one end open. The irons are placed into the open end of the drum, with the branding surface facing the burner. This type of heating apparatus is easily moved and stored, and it is clean-burning.

A third method is to use electric branding irons. This method has gained in popularity recently, but electric irons are expensive. A second disadvantage is that very often two or three characters are clustered on one handle to cut cost. When the numbers are clustered, they are generally too small and too close together, which precludes rocking the iron and obtaining a good application of each brand character. Time lapses between cattle being branded are lengthened because of the additional time required to reheat the iron.

3. Move the animals to the corral area and into the working chute or squeeze chute. Restrain the animal in a squeeze chute and drop the tailgate so the animal cannot back out. The animal's head should be in the headgate, and the sides of the chute squeezed against the calf. Most chutes are designed with hinged sidebars that allow access to the hip and shoulder regions of the animal. One or two of these should be lowered to allow access.

4. Put on a pair of leather gloves to prevent burning your hands when handling hot irons. When using electric irons, proceed to step 7.

5. Take the branding iron out of the fire or drum and check the number or character to be used to be sure it is the right one.

CAUTION: The amount of squeeze put on the animal by some chutes is excessive and may cause hip damage, especially when hydraulic chutes are used. It is best to have a chute in which the amount of squeeze tension can be regulated manually.

CAUTION: The branding iron handles can be very hot, and leather gloves should always be worn by the person who is doing the branding.

6. Check the irons for temperature. The amount of heat required for a good brand is difficult to describe. Some experience is needed to establish the correct iron temperature. The color of the hot iron is a good indicator of the temperature. A *black iron* is too cold. This color is caused by carbon accumulation and usually results shortly after a steel iron is placed into a fire. The black iron is probably hot enough to burn the hair, but not hot enough to form a permanent brand. A *red-hot iron* is too hot. Using this type of iron is cruel to the animal and causes a large sore, which results in an indistinct or blotched brand. An iron that is the color of *gray ashes* is at the proper temperature to do a good job of branding.

If the iron is too hot, allow it to cool until the ash-gray color appears. If it is too cold, put the iron back into the heat source and check the amount of red-hot coals (or the burner) for the amount of heat given off. It may be necessary to increase the amount of wood in the fire (or propane pressure) to increase heat output.

7. Firmly press the ash-gray-colored branding iron against the hide, and rock the handle slightly to vary the pressure and obtain uniform application of the entire character. The color of the branded hide should be light tan, or the color of new saddle leather. If the iron is the proper temperature and the cattle have a light (summer) hair coat, the time required to brand should only be 3 to 5 seconds. Cattle with a heavy (winter) hair coat should be *clipped* before branding; otherwise a longer (5 to 10 seconds) application is needed.

8. Apply one iron at a time. If two irons are applied by the same person, the chance of slipping and blotching the brand is increased greatly. Place brand numbers or characters at least 1 inch apart. Repeat steps 5 through 8 until the animal is branded with all of the desired characters.

9. Place the iron back into the heat source as in step 2. Make sure the branding iron is clean and free of hair before it is used again. Usually, the hair will burn off during the reheating process, but check to be sure.

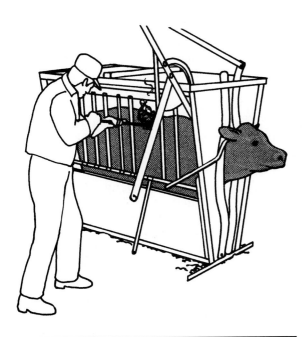

1. Prepare the branders. The branders should have the dimensions discussed in the introduction and be free of dirt and hair. The freeze branders should be made of copper or copper alloy.

2. Cool the irons in a refrigerant. One method of doing this is to place the branders in liquid nitrogen ($-320°$F). Place 3 to 4 inches of liquid nitrogen into a styrofoam cooler or insulated bucket before the irons are added. It will take about 5 liters (quarts) of liquid nitrogen for 20 head of cattle.

The second method of cooling branders involves placing them in a mixture of 99% isopropyl alcohol and dry ice. The recipe for 20 head of cattle is 1 gallon of 99% isopropyl alcohol plus 20 pounds of dry ice placed into a styrofoam cooler. The ratio of alcohol to dry ice is not critical, but 1 pound of dry ice per animal branded is a good rule of thumb.

Both methods require more refrigerant to cool the branders initially than to rechill between animals. Additional refrigerant (liquid nitrogen or alcohol and dry ice) should be added as needed to ensure that the branders are covered by refrigerant.

3. Fill the quart squeeze bottle with 99% isopropyl alcohol.

4. Restrain the animal in a squeeze chute equipped with a headgate and tailgate as in step 3 of the procedure for hot branding.

5. The irons are ready for use when the refrigerant stops boiling. Initially, this will take about 20 minutes, and it will depend on how many branders you are trying to cool at one time. The cessation of boiling indicates that the brander has reached the temperature of the surrounding refrigerant.

6. Clip the area to be branded as closely as possible. This can best be done by using a No. 40 surgical clipper to remove the hair and "underfur" which acts as insulation and increases the time required for proper branding. If necessary, a stiff-bristled grooming brush can be used after clipping to remove dirt, hair, and dandruff.

CAUTION: Only one application of the brand is necessary when it is done correctly. If a second application is necessary to touch up a spot or to obtain more burn, it should be done with extreme care. The iron must be placed into the exact position of the first application so that the character is not blotched.

Holding the iron on too long causes unnecessary pain and a large wound that has difficulty healing.

Do not permit the iron to slip or slide during application. When this happens, a blotch will usually result.

Do not brand wet animals. An iron applied to a wet animal loses temperature rapidly and tends to scald rather than burn the hide. The result is a serious scar that is slow-healing and hard to read.

10. Release the tension on the squeeze mechanism of the chute.

11. Make sure the tailgate is dropped behind the animal. This prevents the animal from backing out of the squeeze chute and the next animal from slipping through the squeeze chute before you are ready.

12. Release the headgate and allow the animal to exit.

13. Repeat steps 2 through 12 until all animals are branded.

Freeze Branding

Freeze branding involves the application of very cold branders to the hide. When applied properly, the melanocytes (color-producing cells) are destroyed and the hair grows out white. On white cattle, deliberate overbranding will kill the hair follicles in the same way that a hot brand does.

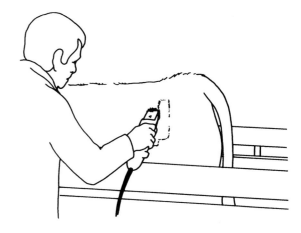

7. Liberally apply 95% isopropyl alcohol from the squeeze bottle over the branding site. Soak the area but don't waste alcohol. It need not be rubbed in.

8. Put on a pair of leather gloves, take the brander out of the refrigerant, and check the character to be used to be sure it is the right one.

> **CAUTION:** The handles of the brander can be very cold and may freeze-burn your hands if you aren't wearing leather gloves.

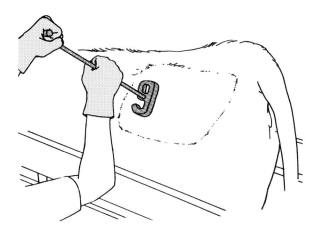

9. Check the clock to ensure the proper brand application time.

10. Apply the brander to the clipped, alcohol-soaked area, and apply pressure to the brander by leaning on it.

The refrigerant used, color of the cattle, hide texture, and stage of hair growth will determine the length of time required to apply a good freeze brand. When either liquid nitrogen or alcohol–dry ice is used as a refrigerant, the minimum time of application is 30 seconds. When liquid nitrogen is used in the winter, hold the iron on black calves for 45 seconds and on red calves for 60 seconds. White calves can be overbranded for $2\frac{1}{2}$ minutes to kill the hair follicles and produce a bald brand similar to a hot brand. When dry ice and alcohol are used during the winter, the brander should be held in place for at least 60 seconds.

Overbranding white calves with alcohol and dry ice does not produce consistent results.

Underbranding may produce a few white hairs, but the resulting brand is usually not legible. In some cases, the underbranded hair will remain colored but will grow faster than unbranded hair, while other cases result in a marked darkening of the branded hair.

The calf usually will jump and squirm for the first 10 seconds after the brander is applied to the hide. The reason for this is that the extreme cold activates the nerve endings. After about 10 seconds, the nerve endings are frozen and inactivated and the animal usually stops moving. You should be ready for this and keep the brander in the same position for the entire time to ensure a good, clear freeze brand.

11. Apply one brander at a time. Two people can each apply one brander, but two branders should not be applied by the same person at one time because the chances of slipping and ending up with a poor brand are greatly increased.

12. Place the brander back into the refrigerant and make sure that the refrigerant covers the branding iron. If it does not cover the irons, add more liquid nitrogen or alcohol. If the alcohol–dry ice refrigerant does not seem cold enough, add more dry ice. Make sure that the brander is clean before placing it into the refrigerant.

13. Release the tension on the squeeze mechanism of the chute.

14. Make sure the tailgate is dropped behind the animal. This prevents the animal from backing out of the squeeze chute and the next animal from slipping through the squeeze chute before you are ready.

15. Release the headgate and allow the animal to exit.

16. Repeat steps 2 through 15 until all animals are branded.

Postprocedural Management

Hot Branding. The brand will be indented into the hide and saddle brown in color immediately after branding. The brand area will scab within the first week, and the scab will slough off in seven to nine weeks. When a brand is blotched, it should be treated with a topical wound spray or powder to prevent fly irritation. The brand will scab over and the skin will slough off, taking a longer period of time to heal than when the brand is not blotched. If the brand becomes infected or does not show signs of healing, it should be treated with a topical wound dressing.

Freeze Branding. The hide will be indented where the branders were in contact with the skin. This is normal because you have placed pressure on the branders and the surrounding skin has frozen. As the skin begins to thaw, a marked reddening of the skin occurs around the branded area. The skin may also swell around the area. Both of these occurrences are normal. Within 1 to 2 weeks after branding, some skin will slough off. The white hair will not grow until the next cycle of hair growth. For example, if the cattle are branded in late fall, the white hair may not appear in any quantity until the spring. When overbranding is used, the skin as well as hair may slough off in 1 to 2 weeks after branding.

2.12 CLIPPING

Animals can be clipped to aid in animal identification or for exhibition and sale purposes. The most common areas clipped by commercial cattlemen are the long hairs inside the ear so that ear tags can be seen, and the area over a hot brand so that brands can be read. Freeze brands on dark cattle will not need clipping, but light-colored cattle that have been "overbranded" to produce a bald brand should be clipped. These areas are clipped during the winter when the hair is long, or just before calving. Clipping for exhibition and sale requires more skill and practice as well as a mental picture of how the ideal animal should look. The objectives of this type of clipping are to make the animal's weaknesses less obvious and to accentuate its strong points.

Equipment Necessary

- Currycomb
- Electric clippers
- Lightweight oil
- Scissors
- Scotch comb or plastic miracle brush
- Working chute, squeeze chute, grooming chute, or rope halter

Restraint Required

The restraint method used will be determined by the type of clipping to be done. Ear and brand clipping can be done at the same time some other management techniques are performed. The brand clip requires the least restraint. Cattle can be clipped in a working chute, headgate, or squeeze chute. The ear clip requires some head restraint so the points of the scissors do not accidentally poke the animal as it moves its head about. This restraint should be in a headgate or squeeze chute. Clipping for exhibition and sale is more time-consuming and requires some animal restraint for your convenience as well as safety. Restraint should be in a blocking chute or a hoof-trimming chute. All of the clipping methods can be accomplished with only a halter tied with a quick-release knot if the animal is halter-broken. Extreme caution should be used when only a halter is used on any animal. Most animals tend to get nervous when people work around their heads.

Ear Clip

Step-by-Step Procedure

1. Obtain a sharp pair of scissors with rounded blade tips or an electric clipper. Roaching scissors that are used to trim manes on horses work well to trim the long hairs in ears; however, they are somewhat expensive. Lower-cost scissors can be purchased, but keep in mind animal safety and the restraint method used when sharp, pointed scissors are used.

2. Restrain the animal.

3. Grasp the ear that has the ear tag with your left hand and trim the hair inside the ear with your right hand so the ear tag is visible. Position the scissors on the inside of the ear near the tip, and trim the hair as you move toward the head. It may be necessary to cut additional hair out of the center of the ear before the ear tag is readily visible from a distance. Cut the hair as short as possible so the process does not have to be repeated in the near future.

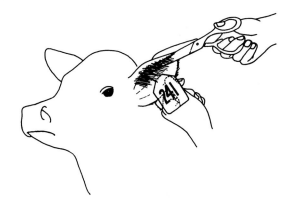

CAUTION: Animal head movement can be a problem in this step. Most cattle do not like to have people hold and clip their ears. Be careful that you do not poke or cut the animal's eye or ear with the point of the scissors during the clipping process.

4. Release the animal.

Brand Clip

Step-by-Step Procedure

1. Assemble the electric clippers, lightweight oil, and currycomb.

2. Apply lightweight oil to the electric clipper head while it is running. The oil can be purchased in an aerosol spray can, or sewing machine oil can be used.

Another method is to place the cutting edge of the clipper head in a 1-pound coffee can filled half full or less with diesel fuel. Turn the clipper head on for a few seconds while the cutting edge is immersed in oil. Turn off the clipper and wipe off excess oil with a clean rag or cloth.

3. Clean the brand area, if needed, with a currycomb. It is especially important to do this if the brand area is full of mud or manure.

CAUTION: Some hot brands have a raised scar when they heal after the branding process. Be careful not to make the scarred area bleed by scraping too hard with the currycomb or by getting too close with the clippers.

4. Turn the clipper on and adjust the tension on the clipper blades. This is accomplished by turning the wing-nut screw on top of the head of some clipper models. Tighten the screw until the motor of the clipper begins to drag, and then loosen the screw until the motor runs freely again. It may be necessary to make this adjustment periodically.

5. Clip over and around the brand area so the entire brand is visible. Trimming around the brand will prevent uncut hair from covering the numbers of the brand.

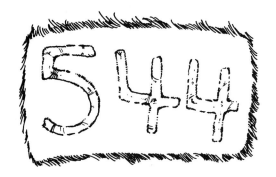

Oil the head of the clippers periodically to keep the blades sharp and clean. If animals are especially dirty, you may need to oil the head every 3 to 4 animals. If the animals are relatively clean, you may be able to clip 10 to 20 animals without oiling the head.

The clipper blades may become dull after several animals have been clipped, especially if animals are extremely dirty. This calls for blade sharpening or replacement. Blades are expensive to buy, and it is practical and economical to learn how to sharpen blades from an experienced cattleman. Blade sharpening is more of an art than a science.

6. Release the animal.

Exhibition and Sale Clip

1. Assemble the electric clippers, lightweight oil, and scotch comb or miracle brush. Determine which type of clipper head will be needed. To trim the underline, brisket, head, and tail, use a clipper head with a fine comb and blade (teeth short and close together). Blocking can also be done with a hair head (fine comb and blade), or it can be done with a goat-shearing head (intermediate comb and blade) or sheep-shearing head (coarse comb and blade), depending on personal preference.

2. Attach the head to the motor assembly and lubricate the head with a lightweight oil. The oil can be purchased as an aerosol spray or a liquid. Sewing machine oil or diesel fuel can also be used. If diesel fuel is used, it should be placed into a 1-pound coffee can filled half full or less. Turn the clipper head on for several seconds while the cutting edge is immersed in the fuel. Turn the clipper off and wipe the excess oil off with a clean rag or cloth.

3. Wash the animal.

4. Restrain the animal for clipping.

5. Thoroughly dry the animal with an electric blow-dryer or by continuous brushing of the hair in an upward and forward motion.

6. Start the clipper and allow the animal to get used to the sound before starting to cut. If this is the first time the animal has been clipped, it may be desirable to position the clipper near the animal's head so the animal can see where the noise is coming from. This step should take only 1 or 2 minutes.

7. Decide where to clip hair on the animal.

8. With the clipper turned on, approach and stand alongside the animal. Scratch the animal as you approach so that it knows where you are. Place the clippers on the underline.

CAUTION: Stand facing the same direction as the animal, because if the animal gets scared and tries to kick, it is less likely to hit you in the head.

Pull the clippers backward over the hair on the underline several times before starting to cut hair. This will allow the animal to become accustomed to the vibration of the clippers before you start to clip.

9. Clip the animal's underline after the animal becomes accustomed to the clippers. Most of the underline should be clipped as short as possible to give the animal a clean and trim appearance, but not all of the hair on the underline is clipped as short as possible. Hair on the underline in the flank

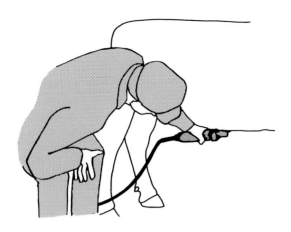

CAUTION: When you start cutting hair on the underline, the animal may kick because of the vibration and the slight pulling of the hair as it is clipped. Position yourself an arm's length away from the animal as you begin to clip, and keep your head out of the path of a potential kick of the rear legs.

and cod or udder area is only trimmed. The objective in this area is to trim the tips of long hair and make the underline appear fairly level instead of appearing shallow in the flank area.

Clip all of the hair in front of and around the navel or sheath as close as possible. Continue this short clip forward between the front legs of the animal. The skin between the front legs has several folds which will require manipulation with the free hand to get the hair clipped short. You may need to go in several directions on the underline to accomplish a short clip. You need to remember that hair is cut shorter when the clippers go against the grain (lay of the hair) than when they go with the grain.

Visualize an imaginary line that separates the underline from the side of the animal. This line runs from the foreflank to the inside rear flank. All hair from this line toward the ventral midline should be clipped short. Make sure you cannot see where the hair was trimmed when you stand near the animal. Hair along this area needs to be blended and tapered so that an unnatural clipper line is not seen. This is easily accomplished with a hair head, goat head, or sheep head by placing the fingers of the free hand in a "C" on the clipper head. The fingers should be along the base of the bottom blade to act as a guide, the palm along the side of the clipper head, and the thumb on top of the clipper head near the blade-adjustment screw. This position of the clipper head will allow you to taper and blend the clipped hair of the underline into the unclipped hair on the sides as you slowly bring the clippers toward you.

The fingers under the clipper head not only give the clipper head stability, but also allow you to regulate depth of cut and prevent cutting too close as you start the blending. You may need to repeat the process several times to get the desired effect. Hair can also be blended by turning the clippers over and using the clipper head as a guide.

10. Clip the loose skin of the brisket (dewlap) as close as possible to give the animal a trim appearance. This should be done from where you left off between the front legs to the front edge of the bottom jaw. Clip the sides of the brisket to give the animal a clean, trim appearance, but don't go past an imaginary line that separates the brisket from the neck. Hair in this area should be blended together so that an unnatural clipper line is not seen. To blend the hair in this area on cattle with a short, summer hair coat, take the clippers and go in a downward motion (with the lay of the hair) on the neck. Generally, the clip will not be too close if you keep the clippers on a slight angle as you go with the lay of the hair. To blend hair in the brisket–neck area on cattle with a fairly long or winter hair coat, place your fingers under the clipper head as in step 9, and move the clippers in an upward and outward motion. By holding the clippers at a slight angle, you can make the final blending of hair.

11. Clip the animal's head as close as possible when the breed or sex is exhibited with a clipped head. Do not clip the ears, eyelashes, or hair on the muzzle. Clip the head from the muzzle to an imaginary line straight down from the base of the ear. Visualize the line by running an imaginary string from the base of the ear, around under the jaw, and over to the base of the other ear. Hair along this line should be tapered in the same manner as in step 10. Clip around the poll area and between the ears. Hair at the base of the ears should be tapered as before. Behind the poll of the animal, do not clip more than about 1 inch past the crease in the skin where the head and neck join.

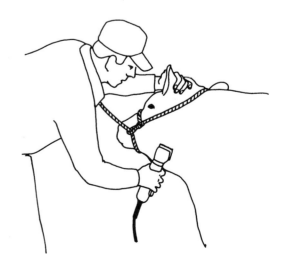

12. Clip the animal's tail. To emphasize bone, length, and weight of the animal, the tail is clipped only along the face and not on the sides. Start with the clippers near or just above the twist area and clip straight up the tail. Do not clip the top of the tail where it curves up over to the tailhead. Trim the tips of the long hairs along the side of the tail with a scissors or clippers to give the animal a neat appearance.

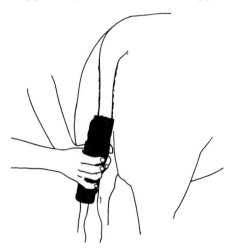

13. Plan how you are going to block the animal by developing a mental picture of the ideal animal and evaluating the animal to be clipped for strong and weak points of conformation and muscling. Areas to evaluate are topline, neck, shoulders, sides of the paunch (belly), hooks (hip bone), rump, tailhead, rear quarter, and legs. Leave as much hair as possible. Every inch of hair makes an animal appear bigger, fleshier, and to have more bloom.

14. Begin to block the animal according to the plan developed in step 13. The objective of blocking is to make the animal appear as smooth and trim as possible, and involves trimming and blending the tips of hair until this is accomplished. Be sure to use a scotch comb or plastic miracle brush to pull the hair upward and forward like you will have it during the exhibition or sale. Do this before you start blocking, and do it frequently during the blocking process. After the hair is groomed, begin to trim the tips of long hair along the underline and gradually work your way up the side of the animal. The amount of "bulge" an animal has in the area of the paunch, shoulder, stifle, and quarter areas will dictate how much of the hair should be trimmed and how to blend it together to present the animal to best advantage. An animal with long, wild hair may need more hair cut off than will an animal with short, curly hair.

Go slowly and don't take off a lot of hair in any one clipper pass. Speed will come with experience. If you go fast, the chances of nicking or gouging the hair increase. Be patient and keep at it. If you are unsure of yourself, or if you have an unsteady hand,

place the fingers of your free hand under the head of the clippers to act as a depth gauge.

Practice on some calves during the off season. This will increase your ability and confidence. Likewise, you should take the opportunity to watch established exhibitors prepare their animals before and during a show or sale whenever possible.

15. Trim the topline. Here is where you can bring out the length, height, straightness, and thickness of the animal. Begin near the rear of the animal, leaving the tailhead until last. Locate the lowest spot on the animal's back and make the rest of the hair along the topline even with it. Clip forward to the crest of the animal's neck (in front of the hair swirl). Be careful not to clip too much hair and so make the animal appear low-fronted.

16. Trim the tailhead. Trim the hair on the back side of the tailhead so that the animal's rump looks long and straight. Finish the top of the tailhead by cutting only wild, unmanageable hair and hair ends to give the full advantage of height and smoothness.

17. Last-minute clipping and trimming can be accomplished after the animal is groomed and the legs "boned" with glycerine soap. Boning is a process that makes the hair on the legs stand up. The hair on the legs is not all the same length and will require some trimming and blending. This trimming can be done with a pair of scissors or the clipper. Hair can be trimmed above the hoof head, behind the hock, and inside the leg so the leg does not appear crooked. Long hair can be trimmed around the knee and joints so that the animal does not appear coarse-jointed.

2.13 BODY CONDITION SCORING

Feed cost is the number one expense facing most cattle producers and certainly is the top economic concern of the commercial producer. Much effort and thought are expended attempting to control this area of cost . . . and are often in conflict with the attempts to maximize production. The dilemma revolves about the problem of determining how much feed is enough. There are excellent guidelines for feeding cattle, but they are general guidelines that pertain to the average cow under average conditions. Cattle producers often have a difficult time determining whether they are in the average spectrum.

A better way of determining whether your cattle are being fed adequately is to feed them to a body condition that has been proven to be correlated with optimum performance. In essence, body condition scoring will allow you to periodically evaluate the "flesh" your cows are carrying and adjust their feed intake so that they are receiving neither too much nor too little. In addition to feed costs, the body condition of beef cows also has a strong influence on the reproductive performance of the cow herd. On the average, as measured in multiple herds across the United States, cows with a *body condition score* (BCS) of 5 average 85 to 95% pregnancy rates, cows with a BCS 4 average 55 to 65%, and those with a BCS 3 average less than 50%. These percentages represent rebreeding pregnancy rates following calving. They illustrate the fact that cows must have adequate body condition—BCS 5 or higher—in order to withstand the rigors of nursing a calf and still have enough energy (body energy reserves in the form of fat) for normal physiological functioning and rebreeding.

There are nine body conditions, and therefore nine *body condition scores,* ranging from emaciated to extremely fat. They are BCS 1—severely emaciated; BCS 2—poor or very thin; BCS 3—thin; BCS 4—borderline or slightly thin; BCS 5—moderate; BCS 6—high moderate or slightly fleshy; BCS 7—good or fleshy; BCS 8—fat or obese; and BCS 9—extremely fat or very obese. Each of these scores describes the amount of body reserves in the animal. They are subjective, visual appraisals based upon apparent external fat cover, appearance of muscle, and obvious skeletal features.

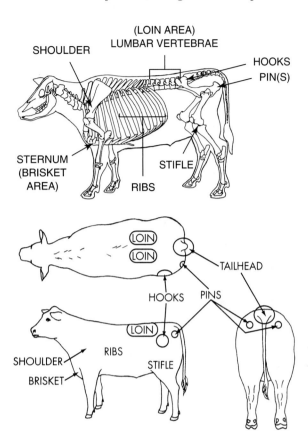

Equipment Necessary

- Working corral
- Body condition scoring guideline pictures and verbal descriptions
- Cow record book

Restraint Required

No restraint, other than a small working corral, is normally required. Cattle can be worked through a chute system, especially if you wish to manually validate your visual appraisal with some hands-on palpation of the cattle.

Step-by-Step Procedure

1. Study the following anatomical models with particular emphasis upon the key reference points indicated. While studying the models, visualize where these anatomical points will be exhibited on the live, hide-covered cow.

2. Approach the cattle to be scored (it is best to start with only a couple of cows in the pen), and carefully watch them move. Watch particularly for hipbones, ribs, spine, tailhead, and obvious soft fatty areas as they move. Make some initial notes.

3. Move around the pen slowly, and look for features to verify your initial scoring notes. Don't be concerned if you cannot remember all of the points in each BCS description. No one can, at first. It will come with practice.

4. Assign one of the following scores to each animal.

5. Record the BCS information for each animal. Be certain to verify animal identification.

6. Study and analyze the condition scores. If they are above 5–6, reduce the feed intake. If they are below 4–5, increase the feed level. A one-level change in condition score represents approximately an 80-pound rise or fall in body weight.

Postprocedural Management

Cows are kept for reproduction. Calving performance can be optimized when cows are fed to the proper condition, usually between BCS 5–6. This allows for intrauterine growth, uncomplicated calving, postcalving milk production, and rapid rebreeding. BCS lower than 5–6 will negatively affect each of these performance traits, and overconditioning, above BCS 6, will cost more money in feed bills than is necessary.

It makes good management sense to keep your cows as near the optimum BCS 5 as possible. A good system would have the cows at BCS 5 for breeding (June). If normal summer and fall moisture is good, adequate pasture will be available and winter hay supplies will not be short. After weaning, push the cows to BCS 6–7 by calving time (Feb/Mar). After calving, as they meet the increasing demands of the nursing calf, they will lose 1–2 BCS by good-grass

BCS 1: Severely Emaciated. No fat detectable, by sight or touch, over the spinous processes of the backbone, edge of the loin (transverse processes), edges of the hipbones, or ribs. There is visible and palpable space between the spinous process of the vertebra of the back. Tailhead is quite prominent. All ribs and entire bone structure are easily visible. There is severe muscle loss (atrophy) in the shoulders, loins, and hindquarters. All ribs are highly visible and hard (lacking any fat cover) to the touch. Animal appears lethargic and weak.

BCS 2: Poor or Very Thin. Slightly more tissue is palpable over the spinous processes of the backbone, but the processes still appear to have space between them, as in the BCS 1. Tailhead is slightly less prominent to the eye than the BCS 1. Ribs are slightly less visible to the eye, but still lacking in fat cover. There is still muscle loss (atrophy) in the shoulders, loins, and hindquarters. Nearly as emaciated as the BCS 1, but animal does not appear as weak.

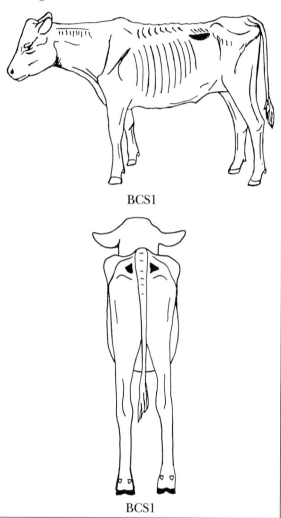

BCS1

BCS1

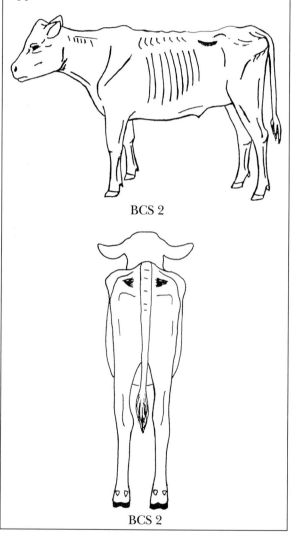

BCS 2

BCS 2

time, in the late spring. If summer grass is good, mature cows will be able to provide for the calf and maintain near the desired BCS 5. If grass is poor, all lactating females and certainly all first- and second-calf heifers will need additional feed.

A cow needs to gain 80 to 100 pounds, in her own body mass (fat and muscle), to increase her BCS by 1. In other words, to feed a cow from BCS 4 to BCS 5, she must consume enough feed to gain 80 pounds. If she is pregnant or lactating, she

BCS 3: Thin. There is a slight fat cover beginning to appear over the spinous processes of the backbone. Spinous processes of the backbone are still highly visible, but the spaces between them are less pronounced. Each rib is still visible, but upon palpation, they are less hard (have some fat cover) than the BCS 2. Slight fat cover beginning to appear over the tailhead. There is still some slight muscle loss (atrophy) in the shoulders, loins, and hindquarters.

BCS 4: Borderline or Slightly Thin. Individual spinous processes are no longer visible, nor are the spaces between them. The individual spinous processes can still be palpated. Only the rear 2–4 ribs can be seen, often only ribs 12 and 13. Hipbones are still obvious to the eye. There is fat cover over the edges of the loins and shoulders. There is no visible muscle atrophy, and the muscling appears full, but flat (not bulging) in the hindquarter.

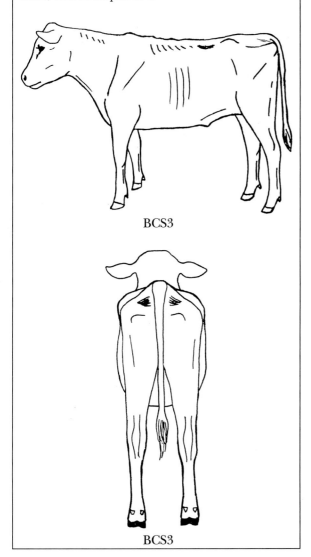

BCS3

BCS3

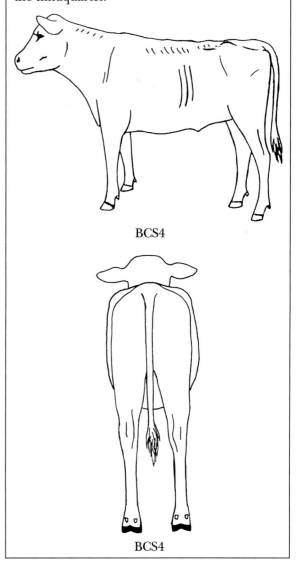

BCS4

BCS4

must consume enough feed to gain the necessary 80 pounds and she must consume enough feed to provide for the growth of the calf or for the amount of milk produced. If a cow drops several BCS, rather than the desired 1–2, it can be almost im-

possible to feed her enough to "catch up" or "keep up" with the normal biological cycle of the rest of the herd. *Evaluate BCS several times per year, at least quarterly. Make adjustments to the feeding program when necessary.*

BCS 5: Moderate. The cow looks good, healthy, and finished "just right." The last two ribs are obvious to the eye unless the animal has been "shrunk." There is fat cover over the shoulders, foreribs, and loins. Fat cover is "springy" over the ribs. The tailhead has fat cover on each side, but the fat is not mounded on each side. Spinous processes are palpable only with pressure. There is no fat in the brisket and very little fat over the hooks and pins.

BCS 6: High Moderate or Slightly Fleshy (Sometimes Termed "Good"). All ribs are fat covered, and none are visible to the eye. Spinous processes palpable only with firm pressure. Hindquarters are becoming plump and full and slightly rounded in profile. There is considerable fat cover around the tailhead, and it is spongy when palpated. There is some fat in the brisket. Obvious fat is covering the loins, shoulders, and foreribs.

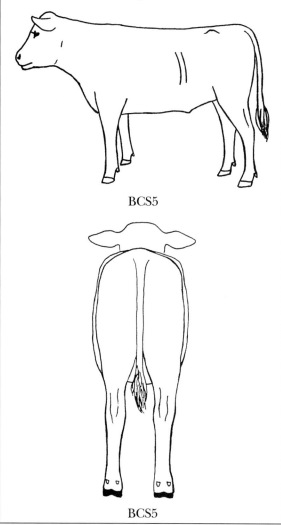

BCS5

BCS5

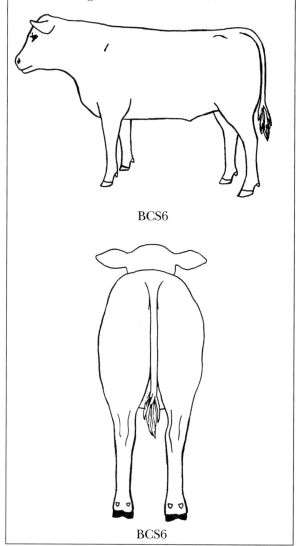

BCS6

BCS6

BCS 7: *Good or Fleshy.* Cow appears smooth due to fat processes. Tailhead has "pones" of fat on each side. Fat is filling the brisket and flanks. There is considerable fat covering the shoulder, loins, and foreribs. It is still possible to feel the ends of the spinous processes, but only with a good deal of pressure.

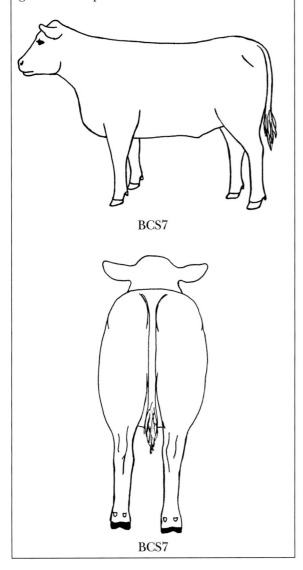

BCS7

BCS7

BCS 8: *Fat or Obese.* Cow is carrying too much condition. She will appear square in profile when viewed from behind. Tailhead is lost in pones of fat. Fat deposits will be seen below vulva. Flanks appear deep due to fat fill, and brisket will be full of fat. Spinous processes impossible to palpate.

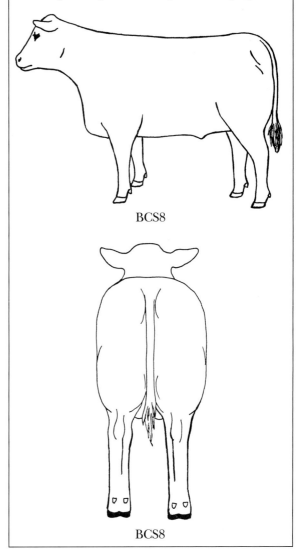

BCS8

BCS8

BCS 9: Extremely Fat or Very Obese. Bone structure is no longer visible from any aspect. All definition is lost. Brisket entirely distended with fat and neck appears short due to this. Pones of fat are obvious at many places on body. Loin, hip, and tailhead take on a "rippled" look due to excess fat.

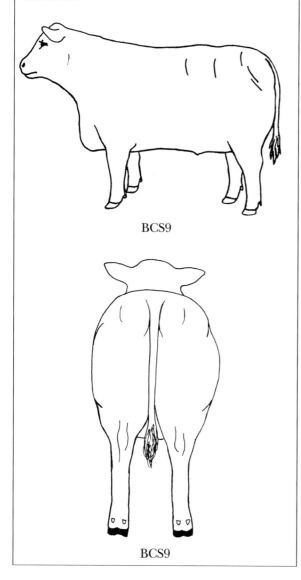

BCS9

BCS9

2.14 BEEF QUALITY ASSURANCE

"Beef Quality Assurance," a program custom-designed for your operation with the goal of producing the highest-quality beef possible, should be the goal of every cattle producer who cares about the industry we have today and the one we would like to have tomorrow.

To understand this concept better, cattle producers must realize that we, as producers of beef, are part of the finest food industry in the world. We also need to realize that the food industry that pays us for our product is consumer-driven. If the beef industry loses the confidence of the consumer, economic disaster and the loss of our livelihoods will be the end result. Beef, properly raised, processed, marketed, and prepared, is a premium product, full of taste, predictable, and wholesome. Consumers have come to expect this top quality. If we deliver less, we will lose their confidence.

The following discussion of ways to assure beef quality (beef quality assurance) is directed toward helping cattlemen put management procedures into place that will assure quality, maintain profitability, and retain consumer confidence.

Think in terms of total quality management. Quality management is not a sometime thing. You either manage an operation to be top quality from start to finish or you are going to sacrifice some degree of product quality. How much quality you sacrifice depends upon how much "slippage" there is in the total operation.

Keep the animals' environment clean. Clean will not always equate to healthy and safe, but if you care enough to be certain that manure, mud, standing water, and any other debris are removed, you are well started on your way to total quality management.

Keep the facilities and working equipment in good repair. Nail heads, broken boards, corners of tin, rusted and weakened hinges, rotted floor boards, and one-too-many patch-ups on something that should have been fixed properly and permanently the first time can damage animal hides, tear or bruise muscle, or provide a site for infection to develop. You are in this business to stay. Take a few minutes and fix things that need fixing.

Have the type and size of working equipment necessary for your operation. There is nothing more frustrating than trying to work 200 head of 1,500-lb cows through a facility put together to handle 50 head of 1,000-lb cows. All corrals, chutes, and holding pens should be large enough for the operation using them. If you have some remodeling to do, consider making the sides of your chutes solid and curved. Solid eliminates some of the distractions for the animals, and curved takes advantage of a cow's natural tendency to circle.

Take the time to do the job correctly. Everyone is in a hurry to get to the next job, because time is money. This should not be an excuse for hurrying too fast to do the job at hand correctly. If you do not take the time to do it right the first time, the end result is going to be doubly or triply bad . . . you have totally wasted the first effort because the job still needs doing; now you have to take the time to do it over, and your total quality management program has very likely suffered. To a large degree, time *is* money and you should learn to do each job as rapidly as you are able . . . as long as the level of speed you use allows you to do the task exactly right!

Treat your animals with care and respect. Think about it. These animals are your bread and butter. Taking care of them will put money into your pocket. These animals are also sentient, feeling critters and we should respect that fact. To needlessly cause them unnecessary stress and pain is costly, senseless, and inhumane. If your animals are used to being handled gently, they will resist less and the whole process will take you less time.

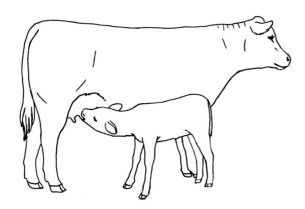

Properly restrain your animals for the task at hand. If the animals are restrained in a chute or pen suited to the animals and type of management techniques you are planning to administer, there will be less opportunity for the animal to resist and less opportunity for the animal to be injured. The NCA's (now NCBA's) Beef Quality Audit shows that bruising costs the industry $25 million annually, and that is only part of the story. Stressed animals do not "cut" or cook as consumers have come to expect. They will lose confidence and reduce demand if we are not careful.

Properly train your work crews. Each worker should know exactly how to use the restraint equipment, handle the animals, administer the animal health products, or perform whatever management task you ask to be done. If there is any doubt, train, retrain, or reassign until you have people doing tasks that they are capable of doing totally correctly. Money will be wasted, animals unnecessarily stressed, people may get hurt, and your frustration level will rise rapidly if you omit this training advice.

Pay the strictest attention to drug-withdrawal times. We cannot sell animals that will test positive for drug residues. This includes young animals destined for steaks as well as cull bulls and cows destined for the hamburger trade. Sometimes it takes a great deal of willpower and memory to keep an unexpected cull from being sold before it has

cleared the drug-withdrawal time. Again, we must do this because we are dealing with product safety and wholesomeness and the related consumer confidence.

Keep good records. Good records will keep you in compliance with drug-withdrawal regulations, allow you to time your vaccination boosters correctly, assure the best possible breeding performance . . . and, in general, be the cornerstone of your total quality management program.

Set up a beef team network. Establish a relationship with your veterinarian, extension worker, animal health supply house, and feed supplier. These are all resources that you should integrate into your management program. They have as much at stake in your success as you yourself do. Use this team to help you think total quality management.

The foregoing suggestions should help provide the proper mind-set and highlight some broad, general areas where we, as producers, can watch carefully to see that we are doing all we can to maximize quality. They are suggestions that we can follow every day of our beef-producing year. The following are specific opportunities to put your total quality management approach into play.

Calving Time. To assure top-quality beef at market time, the first days of the calf's life must be as stress-free as possible. First of all it needs the opportunity to nurse a properly fed and healthy mother cow to assure it of a full dose of disease-fighting colostrum. Nothing you can do as a manager will do more for the calf than to assure that it gets full doses of colostrum in its early days.

More and more managers are finding out that iodining navels at birth does pay a dividend, and that it makes economic sense to identify their calves at the same time. If you are going to castrate and you have to dehorn, consider doing it at this time also.

Branding. The next specific opportunity to put your beef quality assurance program into effect is at branding, and there are several things you can do at this time to assure top-quality, top-dollar, top-consumer-confidence-building beef. For certain, if you need to dehorn, and didn't do it at birth, do it

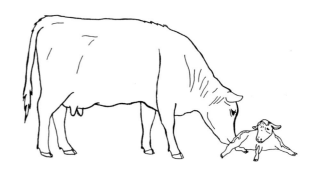

now. Waiting only provides more opportunity for bruising and hide damage.

When branding, be certain to place the brand into a clean site to minimize infection. Use as small a brand as is possible for your needs. Consider placing the brand on the back of the hip instead of on the ribs or on the quarter, if your local regulations allow. Use a branding iron that is not glowing cherry red, and do not rock it back and forth. Try freeze-branding instead of hot-branding if this will work for your breed of cattle.

Now is the time to start a vaccination and implanting program.

Weaning. Now is the time to separate cows from the calves. This is a naturally stressful time for the calves. This is also the time when calves are usually implanted for the second time and when the second round of vaccinations is given. There is a lot of opportunity for things to go wrong at this time, and hence the opportunity for you to have a real impact on beef quality by doing things correctly and minimizing the stress.

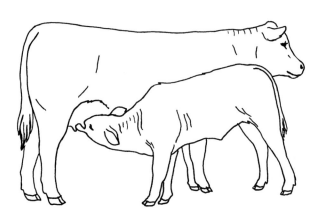

Shipping. This is the last opportunity for you as the owner to have an impact upon the quality of the beef that will come from your cattle. Rather than being thankful to get rid of them in a day or two and rushing to finish any remaining management items, stop and consider that this is the last time for you to assure top-quality beef and a satisfied customer.

Vaccinate the calves for the last time 2 to 3 weeks before shipping. Bunk-train the calves so they will hit the ground running at the lot. When you load them for your last time, make it as stress-free as possible. Your calves are getting ready for a long, stressful ride, and they'll benefit from easy loading.

Use animal health products properly. Throughout the entire management cycle of your beef operation, there is hardly a time when some animals in your herd are not being administered some sort of animal health product. There is plenty of opportunity to have an impact on the improvement of beef quality by being certain that these products are handled

and administered properly. Not only will the consumers benefit, the calves will be more stress free, and your bottom line will be improved.

1. Read and follow label directions.

2. Use transfer needles to reconstitute vaccines. These are double-ended needles that are not used to vaccinate the animals. Insert one end into the sterile diluent and the other end into the dessicated (dried) vaccine. The vacuum in the dried-vaccine bottle will suck the correct amount of diluent into the vaccine bottle, where shaking thoroughly will properly mix the two.

3. Don't mix more vaccine than you will use at one time. After an hour or so in the hot sun, a lot of vaccines begin to rapidly degrade.

4. Keep vaccines out of the sun and in a cooler to prevent their degradation.

5. Do not mix up your own concoctions so you can give a homemade "two-in-one." They may wind up canceling one another out. Do not combine vaccines.

6. Use separate syringes for each vaccine being given. An example of why this is important: bacterin residue in a syringe will destroy a modified live vaccine that is loaded into the same syringe.

7. Use proper syringe technique; i.e., fill syringes properly by voiding air from them, and be certain the air is out of the syringe and needles before you inject products into the animals.

8. Use the injection route recommended. Mostly, these will be subcutaneous and intramuscular (at the proper injection site).

9. Use the injection site that will damage the least amount of high-priced muscle. Basically, the neck muscle is the first choice.

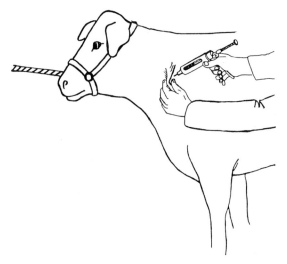

10. Keep syringes, vaccine bottles, and needles clean.

11. Use proper-size needles for the injection route, site, and product being given.

12. Keep the animal needles out of vaccine bottles.

13. Discard the needles when they are bent or dull.

14. Always restrain the animal properly when administering animal health products.

At times, the total quality approach to beef quality assurance seems like a lot of fuss and bother. Each of us may know some cattlemen who have been "getting by" for years, taking all sorts of shortcuts and doing hardly any of these extra steps to assure top-quality beef. Why should we? Because things have changed and are continuing to change drastically. The consumer has become much more demanding and quality conscious. The mass media people have caused a much-increased accountability level. Packers are not going to absorb the losses any longer. Times have changed . . . if we wish to stay in the cattle business, then we must change the way we do business, as well.

These steps will cost a few dollars more, but when you compare that extra cost to a greatly reduced demand for and price for beef, it makes those beef quality assurance dollars appear to be the very good investment that they are.

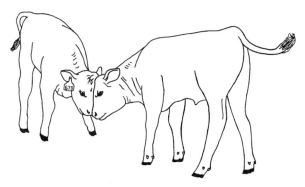

Following is the checklist of Beef Quality Assurance management practices that members of the Idaho Beef Quality Assurance Program agree to implement:

- Provide adequate nutrition to facilitate growth. Include vitamins and trace minerals when necessary.
- Use only feed additives that are approved for cattle. Do not feed animal by-products.
- Administer medications and vaccinations according to label directions, unless directed by your veterinarian to do otherwise.
- Be certain that the injection site is free of manure and dirt before administering medications or vaccinations.
- Inject no more than 10 cc of medication per injection site. Adjoining injection site should not be closer than 4 inches. Give injections only in the approved "injection triangle" in the neck and foreshoulder area.
- Use subcutaneous injections instead of intramuscular injections, whenever possible.

- Keep all syringes as clean as possible. Truly sterile needles and syringes may not be possible in farm/ranch conditions.
- Change needles every 10 to 20 head processed.
- Vaccinate and/or precondition cattle according to one of the following programs:
  - Vaccinate calves at 3 to 4 weeks before weaning. Ship calves at weaning.
  - Vaccinate calves preweaning or at weaning. Revaccinate (booster) calves in 2 to 4 weeks. Wean calves 45 days before shipping.
- Keep detailed records on individual animal treatments and on total herd vaccinations.
- Use implants according to manufacturer's directions.
- Dispose of medication and vaccine containers according to USDA guidelines.

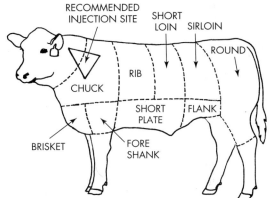

1200# STEER → 750# CARCASS → 500# RETAIL CUTS
RETAIL CUTS
   110# STEAKS
   110# ROASTS
   130# HAMBURGER & STEW MEAT
   150# FAT & BONE & SHRINK

- Check facilities for corners, nails, and other objects that can cause hide damage and bruising.
- Maintain facilities to minimize slick or frozen floors.
- Handle cattle quietly and calmly, with only as much force as necessary.
- Minimize stress during branding. If you must brand, use as small a brand as possible. Place brands in front of shoulder or behind hook bones, in order to minimize hide damage.
- Castrate and dehorn all calves as early as possible.
- Develop a training program for all employees. Retrain all employees yearly. Make certain that all new employees receive training before they work with cattle.
- Establish a "non-fed" cattle marketing program that adheres to the same quality assurance standards as your weaner, stocker, and fed-cattle marketing program does.

BLACK ANGUS

Courtesy of Department of Animal Science, Oklahoma State University

BEEFMASTER–LASATER

Courtesy of Lasater Ranch

SALERS

Courtesy of Department of Animal Science, Oklahoma State University

CHAROLAIS

Courtesy of American International Charolais Association

LIMOUSIN

Photo courtesy of The Limousin World Magazine. Used with permission.

SHORTHORN

Courtesy of Department of Animal Science, Oklahoma State University

SIMMENTAL

Courtesy of American Simmental Association

RED ANGUS

Courtesy of Department of Animal Science, Oklahoma State University

chapter THREE

Cattle Management Techniques

3.1 INTRODUCTION

There are four types of cattle: beef, dairy, dual-purpose, and draft. *Type* is defined as an ideal or standard of perfection combining all the qualities that contribute to an animal's usefulness for a specific purpose. This definition does not include breed specifics. The specific breed characteristics have value as trademarks of purebred or "blue blood" origin for promotion and sale, but they do not always contribute to an animal's usefulness. The two most common types in the United States are beef and dairy. Most of the dual-purpose and draft-type animals in the United States are used primarily in beef production systems; however, the larger draft breeds are used as a source of power in many parts of the world.

The beef-type animal was selectively developed for meat production. Its purpose is to convert grain, forage, by-product feeds, and nonprotein nitrogen (NPN) to a high-quality meat product for human consumption. Beef animals are characterized by muscle, substance of bone, rapid growth rates, and trimness.

The dairy-type animal was selectively developed for milk production. Its purpose is to convert grain and forage into a high-quality milk product that is easily digested. Dairy animals are characterized by a lean, angular appearance and a well-developed mammary system.

Most cattle are naturally shy. Their instincts tell them to beware of new surroundings or changes in daily routines. Cattle tend to react slowly when you try to teach them to do things they have never done before. Attempts to force an animal to do something it does not want to do often result in failure and can cause the animal to become confused, frightened, or angry. Handling livestock requires that they be outsmarted rather than outfought and that they be outwaited rather than hurried.

Most of the work involved in working cattle can be done without heavy physical exertion. This is especially true if the pastures, corrals, and handling facilities are planned and organized properly.

Regardless of the type of cattle, many management techniques are performed the same way. This chapter contains those techniques that are common to both beef and dairy cattle. The same step-by-step format used in the preceding chapters is followed. Cautions are included at proven danger and error points to maximize the safety of both the cattleman and animal; however, these should *not* be considered the only dangers involved or the only times you should practice caution around cattle. Differences between beef and dairy management considerations are discussed individually.

3.2 APPROACHING, CATCHING, AND HALTERING

Haltering is a method of restraint that can be used by all producers. For the small operator lacking facilities, for animals that are trained (broken) to lead, or for very tame, docile animals, this method can be used to restrain the animal during simple management techniques. The rope halter, however, is not a substitute for working facilities such as the headgate, squeeze chute, or stanchion used in many cattle techniques. In some techniques, the halter does not completely restrain the animal, which increases the possibility of injury to the handler.

Approaching, catching, and haltering is the first step in training (breaking) an animal to lead. On small operations with a minimum of facilities, animals that are trained (broken) to lead can be handled more easily.

Two basic types of halters are used in this technique. A manila rope halter will give you a better grip and allow less rope to slip through your hands during the handling of the haltered animal. In addition, when a manila rope halter is used, a quick-release knot tends to hold better. The second type of halter is made of nylon and will not swell when it gets wet. For this reason, nylon halters should be used when animals are tied in the rain or when they are being washed.

Equipment Necessary
- Rope halter (manila or nylon)
- Small box stall or pen

Restraint Required
Animals confined in large, open lots or pastures should be moved to a small box stall or pen unless they are extremely gentle and halter-broken. In the latter case, these animals can be approached and haltered almost anywhere.

> **CAUTION:** To prevent foot injuries, always wear protective leather shoes or boots when working cattle. This is especially critical in working with haltered cattle and when breaking to lead.

Step-by-Step Procedure

1. Move the animal to be haltered into a small stall or pen. In confinement operations, this can be done by one or two persons gently and quietly sorting and moving one animal into a small area. In pasture or range operations, an entire group of animals might have to be moved into a corral or pen area before the desired animal can be sorted.

Sorting one animal out of a group on pasture doesn't work very well; the whole group of cattle will get excited. The sorted animal usually tries to circle as it attempts to return to the rest of the group. Occasionally the sorted animal will move pretty well until you get to the gate area, and then it circles around and escapes. In this situation, it is usually best to move all or part of the group to the working area before sorting.

2. Prepare the halter by loosening the chin rope so that this section measures 2 to 3 feet from the adjustment loop to the eye loop. This makes a loop large enough to flip over the nose and under the jaw in step 7 (see Section 1.3 of Chapter 1).

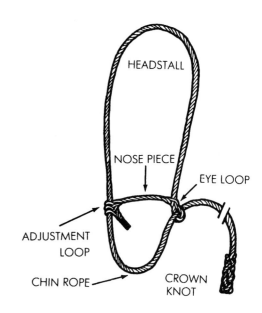

ADJUSTABLE EYE-LOOP HALTER

3. Adjust the nose piece to the size of the animal's head by moving the adjustment loop. You will need to estimate how long the headstall (rope that goes over the ears) needs to be so that the nose piece will fit just below the eyes of the animal.

4. Hold the halter at the top of the headstall with the right hand (hand nearest the animal upon approach); hold the lead rope with the left hand.

One common mistake is grabbing the chin rope instead of the headstall. Be careful.

Be sure that the eye loop is on the left side so that after the animal is haltered, it will lead from its left side.

5. Approach the animal at a 45° angle from the side. The approach should be smooth and at a normal walking speed. As you approach the animal, scratch its back and talk to the animal in a quiet voice. This should help quiet the animal and make it aware that you are there.

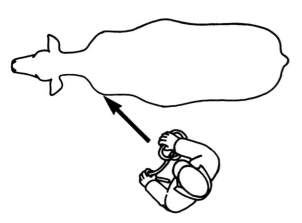

CAUTION: The animal can be approached from other angles, but you should be aware that the animal can kick as you approach. Each animal is different. There are no hard-and-fast rules about how to halter an animal; determine the best and safest approach with each particular animal. Remember that animals can sense fear. Proceed with confidence, but don't be careless.

6. As you approach the point of the animal's shoulder, use the hand nearest its neck to raise the halter by the headstall over the neck. This may or may not be the right hand, because by this time you may have gone around the pen several times and the animal may have changed directions several times.

CAUTION: Steps 5 and 6 can be very frustrating at times if the animal does not want to cooperate, but be patient and proceed with confidence. Eventually the animal will give up, and you will be able to proceed to step 7.

Be sure that the lead rope will come off the left side of the animal's head, even if you change hands or directions.

7. This step is done in one smooth motion and requires some skill. Extend the hand holding the halter, as described in step 6, over the animal's head, flip the halter over the animal's nose, and pull the headstall back over the poll and at least one ear of the animal. If you are lucky and the animal is extremely cooperative, you may be able to get the headstall rope over both ears before the animal moves. As you place the headstall over one or both ears, pull the lead rope so that it will tighten under the chin. Once the animal is caught, place the headstall over the second ear.

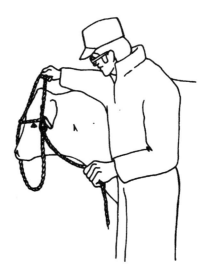

CAUTION: If this step is not completed in one smooth, almost simultaneous movement, the animal will move to escape. If the animal escapes, it is generally harder to approach it and halter it the second time, but be patient and try again.

If the animal will not cooperate or you fail to halter the animal the first time, place the animal in a squeeze chute or stanchion to apply the halter. Well-broken animals can sometimes be tricked into thinking they are caught before the halter is applied. To do this, place approximately 3 feet of the lead strap over the animal's neck and grasp the dangling end to complete a loop around the neck. Once this movement is completed, the halter can be placed on the animal.

8. There are some potential problems associated with haltering an animal. One common problem is that some animals remember being haltered at other times when they were given a shot, had a blood sample taken, or had other techniques performed on them. These animals often acquire the ability to escape the halter approach just described by dropping their heads as you begin step 7. One way to get around this is to place the headstall of the halter over the ears first. Leave enough slack in the lead rope that you don't inadvertently pull it and tighten the chin rope before the animal raises its head. Be alert and move with the animal until it picks up its head and the chin rope drops under the jaw. As soon as this happens, jerk the lead rope and you should have the animal haltered.

A second bad habit that makes haltering difficult is the action of the animal as it throws its head and moves just before you pull the lead rope to tighten the chin rope. When this happens, the animal will probably escape haltering. If the animal throws the halter off its head, try again. If the halter remains draped over one or both ears, it is possible to follow the animal around the pen without pulling on the lead rope; readjust the halter, and continue as before.

3.3 TRAINING (BREAKING) TO LEAD

Animals that have been trained to lead are much easier to handle in the day-to-day management of the cattle operation. They are usually more gentle and tame, can be moved from one place to another for treatment, and are easier to show or exhibit.

Training to lead is best done as early in the animal's life as possible. A young animal is much easier to handle physically and will be much easier to train to lead than older and heavier animals. One person can usually accomplish the training of the young animal, while training older animals may require two or more people.

Training animals to lead is a must in most herds with registered or show cattle and in small herds that lack working facilities. Showing and exhibiting animals is an important part of the registered cattle business. Shows and exhibits are an important way in which these herds advertise their breeding programs, type of cattle, and herd performance. Small operations that lack facilities are more efficient when cattle are broken to lead. Many techniques can be performed without elaborate facilities on cattle that are trained to lead. In all operations, the animal that is trained to lead is a joy to work with compared to the animal that is wild and difficult to approach or handle.

The training-to-lead technique requires patience. The animal should always be handled as gently as possible. Excessive force, use of whips, clubs, and prods, and excessive noise and yelling will only add to the confusion for the animal and make the job more difficult. Care should be taken not to injure the animal by jerking, dragging, or excessive pulling on the lead strap. Patience, repeated attempts, and a reward system can make the job considerably easier to accomplish.

Equipment Necessary

- Tie halter
- Rope halter (manila or nylon)
- Leather neck strap (with metal ring)
- Short rope or strap (buckle at each end)
- Headgate
- Leading gate (mounted on a two-wheeled cart)

Restraint Required

Restraint required is a rope or tie halter. Young animals can be restrained with a rope halter with one person on the lead rope. Older, larger animals often require two rope halters and two people. Stubborn animals may have to be tied to a leading gate or tied to a second animal.

Step-by-Step Procedure

1. Assemble the necessary equipment.

2. Approach, catch, and halter the animal.

3. Tie the animal to a manger, wall, or fence with a quick-release knot approximately 2 feet from the ground. This step will help the animal to learn that it cannot get away, and will help the animal become accustomed to the halter.

Keep the animal tied for 1 to 2 hours the first day and repeat this during the next 1 to 2 days depending on animal cooperation. Animals that have been halter-tied and remain high-spirited can sometimes be gentled by tying the rope very short and high so that the animal has to stretch to be comfortable. After a while the animal begins to tire, and it is usually easier to handle.

CAUTION: Do not leave the animal unattended while it is tied initially. The first time that an animal is tied, it usually pulls against the wall fence or manger and fights the rope. If the animal should slip and fall, it could choke itself.

Make sure that the rope is fairly short (12 to 18 inches between the halter and fence) so that the animal does not get its legs tangled in the rope.

Be sure to use a quick-release knot so that the animal can be untied quickly if it falls down.

Do not tie an animal to a post without a fence. If animals circle the post, they choke themselves before you can get the rope untied and unwound.

Always carry a sharp pocketknife when animals are tied. This will allow you to cut the rope if you can't get the knot untied in emergency situations.

4. While the animal is tied in step 3, scratch, brush, or curry it. This step is optional, but it usually helps gentle the animal and reassures it that you aren't going to hurt it.

5. Prepare to lead the animal in a small area the first time.

CAUTION: Do not lead the animal in a large open lot or in an area that is not enclosed by fences the first time or two you try to lead it. If it should get away from you, the training process will be much more difficult the next time.

Use your right hand to grasp the lead rope approximately 12 inches from the animal's head. Position the left hand on the rope a comfortable distance from the right hand (approximately 18 inches).

CAUTION: The lead strap should never be wrapped around the arm or wrist and should not be tied around the body. This is a very dangerous practice and can result in serious injury to the handler. If the animal escapes and the lead strap is wrapped or tied around you, you can be dragged on the ground and seriously injured.

The first session of training the animal to lead should be relatively short, perhaps 30 minutes. This should be followed by a rest period of at least half a day before another attempt is made; however, too much time between attempts will make your job more difficult. When training is started, stay at it until the animal is trained to lead to your satisfaction.

6. Attempt to lead the animal. Young, tame animals (less than 400 pounds) are fairly easy to handle in most cases. Begin by leading the animal slowly and gently for a short distance and then return to the housing area. If you are alone, try leading the animal in a counterclockwise circle with yourself on the inside. This helps to keep the animal's head pointed toward the inside of the circle and helps to discourage it from escaping.

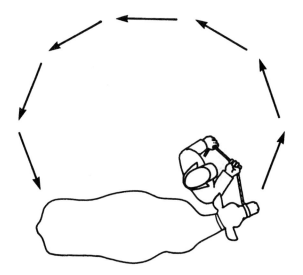

Some animals will plant all four feet and refuse to move, while others will try to escape by running. Several tricks can be used to prevent the animal from running. First, the person on the lead rope needs to be sufficiently strong in comparison to the size of animal. Second, keep the animal's head up and don't let the animal lower it between its front legs. When the animal lowers its head, you lose a leverage advantage. Third, keep the animal's head angled slightly toward you instead of away from you. The last two tricks allow you to pull the animal's head around to its shoulder to prevent a runaway or to help you make the animal move in a small circle.

Don't let the animal escape by running away during the training process. Once an animal escapes, it will try to do it again, and training is more difficult. If you have any doubts about whether or not the animal will escape, see the training aids in step 7.

The counterclockwise circle is used only when training the animal to lead. After the animal is trained, begin leading the animal in a clockwise manner so you are on the outside of the circle. This will prevent you from blocking the judge's view of the animal in the show ring.

It is sometimes helpful to lead a slightly thirsty animal to water. This gives the animal a reason to lead, and gives it a reward for moving in the desired direction.

If the animal plants all four feet and refuses to move, obtain the assistance of a second person. This person can sometimes be of assistance just by his or her presence. If this doesn't work, the assistant should pat the animal's rump or gently twist the animal's tail until the animal moves. Do not abuse the animal by using a club, whip, or prod (hot shot) to get it moving. This type of abuse will make the training process more difficult, if not impossible. If the animal absolutely refuses to move, one or two assistants may have to push the animal physically.

Large or high-spirited animals may require a training aid (next step). If an animal ever gets away from you during the training process, it will try to escape every time you try to train it.

7. Several training aids can be used for large, high-spirited, or stubborn animals.

Double Halter. The animal that wants to run is perhaps the hardest to train to lead. It is wise to place a second halter on the animal by reversing it so the lead strap comes off the right side of the animal's head. In this manner, you can have one person on each side of the animal, and you will have more physical-restraint capabilities. This animal may alternately refuse to move, then be convinced to move, then try to escape by running. The people involved must be aware at all times that the animal may do this and be ready to stop it.

Leading Gate. This training aid utilizes a homemade, two-wheeled cart and gate pulled by a powered vehicle (tractor or pickup). The gate is made of pipe welded to the cart frame. The gate measures approximately 8 feet wide by 4½ feet high. The purpose of the gate is to prevent the animal from becoming entangled in the bumper of a pickup or the wheel of a tractor.

CAUTION: Do not use a pickup or tractor without this training gate. Animals can be injured or choked if they bolt forward when they are tied to a pickup hitch or tractor drawbar, or when the rope becomes tangled in the tractor tire tread.

Two persons are necessary when this training aid is used—one person to drive the vehicle and a second person to watch the animal closely. The lead rope should be approximately 18 inches long (halter to gate) and 18 inches from the ground, and tied in the center of the gate. Tying the animal in this manner will prevent animal injury, and the head will be at a comfortable height.

Drive the vehicle very slowly and stop frequently to prevent injury to the animal. The person watching the animal should pat the animal on the rump or twist the animal's tail gently when the animal plants all four feet and refuses to move. Do not drag the animal. If the animal does not walk on its own or after some coaxing, stop the vehicle and scratch

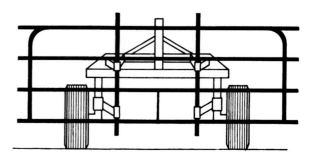

the animal. After several minutes, proceed forward for a short distance (maybe 5 to 10 feet) and repeat the process until the animal begins to follow the cart on its own.

Limit the training process to sessions of approximately 30 minutes once or twice a day until the animal is trained to lead.

3.4 WASHING

When preparing for an exhibition (show) or sale, the ability to properly wash an animal becomes important. Frequent washing stimulates hair growth, keeps the hair loose and fluffy, and keeps the animal clean. The technique involves wetting the animal, working the soap into a good lather, and rinsing thoroughly.

Animals can be washed at any time during the year, but a heated wash area should be used on extremely cold days. Animals should be washed on a weekly basis during the 6- to 8-week period before a show or sale to stimulate hair growth and keep the hair loose and fluffy. Cattle often get dirty in transit and they should be washed shortly after arrival at the show or sale. Finally, the cattle should be washed on the day before the show, and shortly after it to remove any excess hair preparations.

Equipment Necessary

- Nylon rope halter
- Wash chain
- Rice root brush or plastic miracle brush
- Currycomb (steel)
- Pail or bucket
- High-sudsing soap (Orvus® or Castile®)
- Milk oil dip, vinegar, or creosote dip
- Water hydrant
- Garden hose (at least 10 feet long)

Restraint Required

The animal should be trained (broken) to lead before washing. After the animal is trained to lead, the restraint should be a rope halter or wash chain tied with a quick-release knot or fastened to a fence or wall.

Step-by-Step Procedure

1. Assemble the necessary equipment and prepare the wash area. The wash area should have a roughened floor surface to keep the animal from slipping.

> **CAUTION:** Do not wash cattle on polished concrete or wooden floors, because these surfaces become very slippery when wet.
>
> In cold weather, a heated wash area should be used. This may be a hospital area, a veterinary room in your barn, or a separate heated building. A wet animal can lose a tremendous amount of body heat to the environment. If the environmental temperature is too low, the animal will be stressed and could get sick. Do not wash cattle when building temperatures are below about 50° F. Remember also that wind and drafts increase the chill factor.
>
> You may want to wear a rubber wash apron and rubber boots during the washing process to keep dry.

2. Approach, catch, and halter the animal and lead it to the wash area. A halter made of nylon rope should be used so that the rope under the jaw can be loosened easily when wet.

> **CAUTION:** Do not use a manila rope halter during the washing process because the rope will swell when it gets wet. This restricts breathing and could cause the death of the animal if the animal pulls back and applies tension to the lead rope.

3. Tie the animal to a sturdy fence post or wall with a quick-release knot.

> **CAUTION:** Do not tie the animal to a post in the middle of a pen or lot. The animal has too much freedom of movement and can injure itself or you. If the animal lunges, it can injure a shoulder; if it keeps going in circles around the post, it can choke itself before you can untie and untangle the rope.

4. Use a currycomb to remove caked-on mud and manure from the animal's hair. When large knobs or clumps of manure are encountered, it may be necessary to use rapid downward and sideways motions with the currycomb to work them loose. One problem with this method is that some hair may also be removed in the process.

> **CAUTION:** Pulling on these knobs of hair can be irritating to the animal. Be careful that you do not curry the hide so hard that the teeth of the currycomb cut or lacerate the hide. Moistening the knobs with water may help.
>
> The animal may try to kick, step on your toes, or pin you against the wall or fence. Keep one hand on the animal's shoulder, hip, or loin so that any movement the animal makes will be telegraphed to you by muscle tension.

5. Turn the water on to a slow stream (low pressure) and stand at the animal's shoulder to minimize danger from kicking.

6. Begin washing the animal by placing one hand on the animal's shoulder and directing a slow stream of water on the animal's feet and legs with the other hand. This allows the animal to adjust gradually to the water temperature. When high-pressure washers are used, begin the washing process by directing a stream of water at the feet and legs without having the pump motor turned on. After the animal has adjusted to the water, turn on the pump and wait a few seconds before directing the high-pressure spray at the feet. After a few seconds, the animal should adjust to the noise, and the high-pressure stream can be applied to the front feet, then the hind feet, and finally the legs. Be careful! Don't place the tip of the spray nozzle too close to the animal's hide; some units produce enough pressure to lacerate the hide.

7. Increase the water flow and thoroughly wet the animal after it has adjusted to the water temperature. Thorough wetting is more quickly accomplished by cupping the hand over the end of the hose rather than shooting the water onto the hide from a distance. When washing the head, be careful. Most cattle do not like to have their heads washed. To keep water and soap from entering the ear, as well as to detect any impending movement by the animal, grasp the ear like a banana and hold it down firmly. If the animal does not cooperate well, wash only one side of the head at a time while firmly holding the ear on the side being washed.

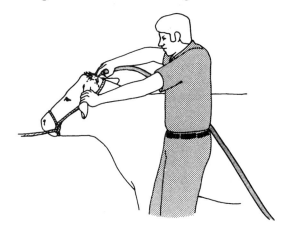

8. Apply soap or soap solution to the entire animal. This can be done by directly rubbing a handful of Orvus or a bar of Castile soap on the topline and side of the animal or by dissolving Orvus, Castile, or coconut oil shampoo in warm water before applying it.

> **CAUTION:** Try not to get soap in the eyes or ears, because this can cause irritation. Use a coarse towel to clean the inside of the ears.

9. Use a rice root brush or a plastic miracle brush to obtain a good lather and get deep cleaning action. It may be necessary to apply more soap (and water if the animal dries out fast) from time to time to obtain a good lather. The animal usually enjoys this step, but it is still a good idea to keep one hand on the animal's shoulder, loin, or hip at all times. Make sure that all parts of the animal are thoroughly scrubbed and clean. Parts of the animal that are frequently overlooked in washing are the head, tail, brisket, belly, legs, and hindquarters.

10. Wash the switch by placing the tail in a pail of soapy water or by applying a handful of soap directly. A brush or manual manipulation of the switch can be used.

11. Rinse soap from the animal several times. Be thorough! Don't forget those hard-to-get areas. Dried soap left on the hide can cause dandruff and itching.

12. Extremely dirty animals, animals with dandruff, and animals with white areas should be washed a second time. Repeat steps 8 to 11.

13. After thoroughly rinsing it, you may want to rinse it again with a dilute hair conditioner to make the hair more manageable and show more "bloom." One to two ounces of milk oil dip or a quart of vinegar dissolved in 3 gallons of water can be used. After the rinse solution is applied over the top of the animal (be careful of the ears, nostrils, and mouth), use your hand to rub the solution down the sides, under the belly, down the legs, over the head, and down the brisket. Do not rinse the animal with water after the rinse solution is applied.

14. Scrape excess water off with a brush, the backside of a scotch comb, or a specially designed scraper.

15. Brush or blow dry the animal until the hair is completely dry. This is especially important in cold weather. For beef animals, it is appropriate to pull the hair upward. For dairy, it is appropriate to brush the hair down. Do not take wet animals out of the warm wash area on a cold day. Animals can be placed in a clean, dry, draft-free, bedded area after they are washed.

Postprocedural Management

Observe the animal for several days after washing. Ears may droop, but the condition is usually temporary and will rectify itself within a week. Respiratory ailments and death may result if water is forced into the nostrils or mouth during the washing process. If the animal develops a cough, a runny nose, or a temperature, treatment may be required.

3.5 MOVING CATTLE

In the day-to-day management of both beef and dairy operations, it is necessary to move animals from one place to another. Most of the techniques discussed in the cattle chapters involve moving animals at some point. The move may be to a holding pen, breeding chute, or stanchion, to another lot or pasture, into the squeeze chute for treatment, or to a loading area. Moving cattle is more an art than a science, and requires some planning and knowledge to accomplish it with the least amount of time, effort, and stress to the animal. Animal and handler safety should always be considered.

Equipment Necessary
- Halter
- Whip
- Web or canvas slapper

Restraint Required

Because the technique is one of movement, no restraint equipment is necessary other than the pasture, lot, pen, or alleyway in which the animals are kept or to which they will be moved.

Step-by-Step Procedure

1. Plan the move. Decide where the animals should go and what path will be followed. Have sufficient help. Plan a course of action to use should one animal escape. A farm truck or livestock trailer can facilitate the moving of cattle if the move is to be across an open area or over a long distance.

> **CAUTION:** The area through which the animals will be moving should be clear of obstructions—broken gates, old fencing, extraneous equipment, or junk. Obstructions in the area can result in injuries to the cattle or to the operator, can furnish escape routes for the cattle, or can separate the group as they dodge obstructions.
>
> Be sure to check alleys, pens, and working areas for protruding objects. The susceptible zone for bruises on cattle is 28 to 58 inches from the ground. Bolts and

nails that stick out at this height should be removed or covered with a piece of worn-out tire tread. These precautions will minimize animal bruising and injury.

When moving cattle, good footing is important to prevent leg and pelvis injuries (spreader injuries). Do not try to move cattle when the ground or alleys are covered with ice.

2. Identify the animals to be moved. The move will be easier if the animals are sorted into one group before movement of the group is attempted.

3. Check your gate and door openings. Open securely all gates and doors in the planned path so that once the cattle move in the desired direction they will not be stopped by a gate that has swung shut or a door that is only partially open. Close securely gates and doors *not* in the planned path to prevent cattle from going in the wrong direction.

4. Move the cattle.

Driving. The move can be an upsetting and stressful experience unless procedures are followed. It is best to move cattle quietly, without rushing or exciting them. Do not holler and yell unless absolutely necessary. Be patient! The move should be relatively slow to help prevent the animals from falling, especially if they are moving across a slippery area. Do not move cattle by running them, because you will lose control. If the cattle do run, let them proceed as long as it appears that the running will not interfere with the general movement.

Some cattle won't move. A light tap on the rump with the whip or canvas slapper will usually move them along.

CAUTION: Do not use a cattle prod (hot shot) when moving cattle in this manner. The prod will excite cattle into running and darting.

A group of cattle housed in a small pen may run in all directions when first turned loose. It is best to turn them into a larger area, preferably a dry lot or corral, before moving them any distance, especially if they are to be moved across a slippery area.

Leading. Leading can be very simple for cattle that are broken to lead. The halter is applied and the animal led wherever you wish. Many beef operations and some dairy operations can effectively use a feed sack, bucket of grain, or bale of hay to "lead" cattle from one pasture to another. To make this work, cattle must be accustomed to eating grain or harvested forage. As you feed or supplement the animals, call them. A variety of calls such as ŚĂ B̌OOOOOSS, COME B̌OOOOOSS or ẂOOOOOW

can be used. The call should be repeated several times in a deep, forceful voice. If cattle have previously been fed from a pickup, sounding the horn on the pickup can work. Once the cattle learn that the call means they will be fed, they will usually come running. This is true even after a week, month, or season without the call to feed. The feed or supplement that is used to lead the cattle must be preferred to (more palatable than) the pasture or other roughage they are grazing.

Do not call the cattle unless you are going to feed or move them. If you call them needlessly, they will soon learn that you are going to trick them. The cattle should be hungry. If you know that you are going to move cattle on a certain day, don't feed them until you are ready to move them.

To lead cattle by enticing with feed, you may be able to stand in the pasture, by the barn, or in an open gate and get them to come to you by calling. If this doesn't work, rattle a feed sack, clang the feed bucket, or spread a small amount of feed on the ground in one location. Move in the desired direction and repeat the call until all cows begin to follow. This can be done on foot, on horseback, or in the back of a pickup with the tailgate down.

CAUTION: If cattle are very hungry and you are on foot, be careful. Cattle will crowd, push, and shove to get feed; they can injure or even trample you.

5. Anticipate what the animal is going to do. The experienced person is able to anticipate the animal's movement to prevent its escape. Watch very closely for head movement. The animal may turn its head ever so slightly in the direction it intends to move just before moving. By anticipating the animal's move and moving to prevent it, you can usually keep the animal going in the desired direction.

CAUTION: If one of the animals breaks away, not all the helpers should leave the group to round up the escaped animal. Moving the remaining group to its destination is usually more important at this time than rounding up one animal. If all the animals succeed in starting back toward the original pen, it will be very difficult to stop them. It will be more difficult the second time to start moving in the proper direction.

It can be dangerous to jump immediately in front of animals that are determined to leave the group and turn back. They can knock you down and seriously injure you. You can try to turn them back if you are far enough away to escape if they refuse to stop.

Watch for any slight movement of the feet in one direction just before the animal wheels around to move in that direction. An animal that slows down for any reason may be planning an escape. Keep the group moving together to discourage escape.

6. Be positive in your movements. Make your moves smoothly and quickly to keep the cattle going in the direction you want them to move. Hesitation and indecision on the part of the handler is a major cause of cattle escaping or separating from the group.

7. A well-trained dog or horse that responds to directions from its handler can make moving cattle much easier. A poorly trained or "stray" dog can cause cattle to run and scatter in all directions. A horse that lacks training or is not responsive to its rider is of little help in moving cattle.

3.6 LOADING AND TRANSPORTING

Cattle operations all require that cattle be moved from one location to another. This may be movement to and from a pasture, a farm or ranch, a livestock show, a terminal market, or an auction.

Transportation of livestock is very expensive, and usually the producer pays this cost either directly or indirectly. In addition to the actual shipping costs, there may be other hidden costs from shrinkage, bruising, crippling, disease (especially shipping fever), and the possible death of an animal. Much of this loss can be eliminated or minimized if proper skills are used in preparing, handling, working, and transporting cattle.

Equipment Necessary

- Holding pen
- Working chute (optional)
- Loading chute
- Livestock trailer, truck, or railroad car

Restraint Required

The only restraint that is necessary in this procedure is a pen, lot, or pasture equipped with a holding pen or corral and a loading chute.

Step-by-Step Procedure

1. Obtain health certificates and permits when needed. The need for these and the time required to obtain them will be determined by the location to which cattle are to be transported, state regulations, and sex of the animal. Consult your veterinarian for the latest information at least one month before shipment to avoid costly delays and frustrations. If cattle are to be exported to another country, get the necessary information about 6 months before you plan to ship. The regulations require many blood tests and records, especially in breeding stock.

2. Check weather forecasts. Plan to transport cattle during the most desirable time of the day in extreme climates. Cattle should arrive at their destination in the morning hours so that ample time is allowed for them to find feed and water before dark.

3. Select the mode of transportation that best fits your situation. Trucks and railroad cars require a loading chute at both the loading and unloading points. Gooseneck and bumper-hitch livestock trailers do not require a loading chute.

4. Select a reputable firm to haul your cattle. The trucks and trailers should be clean and disinfected. The drivers should be reliable. The lowest-cost firm is not always a bargain. If bruised, sick, or dead animals arrive at the destination because of poor transportation, it is your reputation that usually suffers, not the trucker's. On extremely long hauls with beef cattle, some provision must be made to unload the cattle and allow them to rest. With dairy cattle, it is very important that cows be milked on schedule and be allowed to rest to maintain satisfactory milk production.

5. Prepare cattle for several days in advance when transporting them long distances. Beef shrinkage is a major hidden cost that can be minimized if cattle are fed properly before they are transported. Several feeding suggestions are: (1) Several days before shipping, reduce high-shrinkage feeds. These include green feeds, such as pasture and silage, and high-concentrate feeds like grain. (2) Increase the amount of good-quality sun-cured grass or grass–legume hay. Cattle should have access to this hay until loaded. Do not feed a straight legume hay because it has a laxative effect. (3) Cattle should be held off high concentrates (grain rations) and succulent green feeds for about 12 hours, and held off water for about 2 hours before loading for shipment. These feeds cause scouring and excessive urination, resulting in more shrink. In addition, the truck floor becomes slippery and animals can be bruised or injured. (4) Avoid abrupt ration changes of any kind before shipment. Do not feed excess quantities of salt before shipment to obtain maximum water consumption at the marketplace. This is cruel to the animals, and good buyers will not be deceived.

6. Prepare the loading chute. This may be accomplished by simply moving the portable loading chute into position at the end of the working chute. If a permanent loading chute is used, it should have a fairly flat angle with stair steps instead of cleats, and solid sides so that animals cannot see out. It is best to have stairstepped loading chutes with a minimum of a 4-inch rise and 12-inch run. Ideally, the run should be 18 inches.

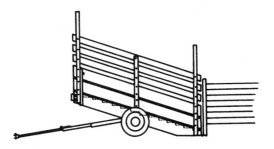

7. Check the alleyways, working chute, and crowding pens for protruding objects. The zone in which cattle are susceptible to bruising is 28 to 58 inches from the ground. Nails, bolts, and corners of boards that stick out in this zone should be removed or covered with a piece of worn-out tire. Bolts and nails that are found to have tufts of hair on them should be eliminated. These precautions will help to minimize bruising and injury to the animal.

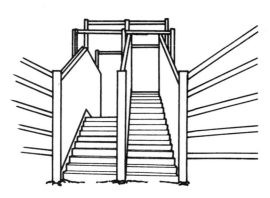

8. Back the transport vehicle up to the loading chute. The entry door should be centered within the loading chute, and the vehicle should fit squarely and snugly along the front edge of the chute.

CAUTION: If a gap exists between the transport vehicle and the chute, an animal can get a foot caught and break a leg. It may try to escape if the gap is large enough.

9. Check the transport vehicle for cleanliness and bedding.

CAUTION: If the inside of the vehicle is not clean and does not have some clean sand, straw, or limestone bedding to prevent animal slippage in transit, *do not* load cattle. If an animal slips and falls, it can be trampled and injured by other cattle in the truck.

If there is a lot of black soot on a semitrailer behind the exhaust stack, do not load cattle. Cattle breathing this type of exhaust usually develop respiratory problems and shipping fever. Make sure that the exhaust stack is tall enough to clear the trailer.

10. Check the transport vehicle for ventilation. In extremely cold weather, ventilation should be reduced by blocking some of the open slots in the trailer. Extremely hot weather requires maximum ventilation.

11. Make necessary adjustments in the transport vehicle as described in steps 8 and 9, or reject the hauler. Make sure that latches and protruding objects are not sticking out in the doorway to cause severe loin bruising and injury to the animals. Remember, the cattle buyer has the right to reject your cattle upon arrival if they are not healthy because of poor management and planning; you have the right to reject the truck before loading.

12. Move a manageable number of cattle, depending on labor and facilities available, quietly to a holding pen near the loading area. The holding pen can be the same pen that is used with the working chute or a specially designed pen adjacent to the loading chute.

CAUTION: Avoid striking the animals, using prods (hot shots), and loud shouting. Hot, excited animals are more susceptible to excessive shrink, disease, and injury.

Don't rush or crowd cattle through gates and chutes. This causes unnecessary stress, excitement, bruising, and injury to the animal. Avoid prolonged processing, working, and sorting before loading.

Do not mix cattle from different pens. Mixed cattle tend to fight before and after transit, which causes bruising and dark cutting carcasses. Fighting during transit is usually fairly minor because of the close animal contact and the motion of the truck.

If animals are moved long distances to the loading area, they should have time to rest, eat, and drink moderately before loading. Do not ship cattle that have not been fed or watered recently.

13. Move the group into the loading chute. Avoid loud shouting, hurrying, and striking. Loading can be frustrating at times, but be patient and don't lose your temper. Never beat an animal with such objects as forks, canes, pipes, sticks, or frozen canvas slappers. A whip, broom, or soft, dry canvas slapper used sparingly can minimize animal stress and injury.

14. Load cattle into the transport vehicle with the same patience and cautions as in steps 11 to 12.

CAUTION: Do not overcrowd or underload cattle. Overcrowding causes injuries and high shrinkage; underloading causes cattle to shift when the truck starts, stops, or turns corners. When cattle shift, some may slip and fall, causing injury. Excess space should be partitioned off.

Do not partition bulls and cows, cows and calves, or cattle from different pens together if it can be prevented. This will help minimize bruising and injury.

---

15. Make sure that the driver of the transport vehicle knows how you want the cattle penned upon arrival at the destination. Cattle should be penned with members of their original group whenever possible. It may be necessary to give the driver a piece of paper indicating which pen of cattle is in each compartment.

16. Transport the cattle.

CAUTION: If you are the driver, slow down on sharp turns and *do not* make sudden starts and stops. Inspect the load periodically to prevent trampling of any animals that may fall in transport.

Do not make lengthy stops while transporting cattle. Get to your destination as quickly and safely as possible. On long hauls (over 18 hours), take advantage of good rest stops and allow beef cattle to eat and drink.

When dairy cattle are carried on a transport vehicle for more than 12 to 14 hours, they should be unloaded, milked, and fed. Make arrangements ahead of time to have facilities and feed available at the rest stop. Failure to milk the cows after 12 to 14 hours can result in a serious decline in milk production and may cause mastitis.

---

17. Back the transport vehicle slowly and squarely against the loading dock.

18. Slowly unload one compartment of cattle at a time. Avoid overcrowding the loading chute and alleyways. Crowding causes cattle to bump against gates and posts and bruise themselves.

19. Check cattle for runny noses, raspy coughs, and droopy ears. Cattle showing these symptoms should be treated promptly.

20. Lactating dairy cows should be milked immediately after unloading. The udders should be thoroughly examined for cuts, bruises, and signs of mastitis. If any of these conditions are found, they should be treated.

Postprocedural Management

All animals will lose weight (shrink) in transport. There are two kinds of shrinkage. *Excretory shrinkage* is the loss in animal weight due to excretion of feces and urine. This type of shrinkage can be regained rapidly in 3 to 7 days when cattle are fed and watered regularly. *Tissue shrinkage* results in loss of flesh and body water and is regained more slowly over 15 to 30 days. Several factors affect the amount and type of shrinkage; some of these factors are age, temperature, previous feeding, and length of the trip. When heavy cattle are transported after feeding over a short distance on a hot day, the weight loss results mainly from excretory shrinkage. Milk-fed calves shipped on a cold day over a long distance will incur heavy tissue shrinkage. It is dangerous to list "normal" shrinkage for cattle because each situation is different. It is not uncommon, however, for calves to shrink 10% or more, while heavier, more mature cattle may not shrink more than 4 to 5%.

Sick, bruised, and crippled animals may be a problem when cattle are unloaded. Do not handle this type of situation unless you consult your veterinarian. If cattle are not treated promptly, the situation becomes worse.

Recovery Sequence

The time required for animals to recuperate from shrinkage losses varies considerably. The recovery time may vary from less than a week for older, heavier cattle to as much as a month for calves or stale cattle (cattle bought from several locations and put together at one location over a 1- to 2-week period before transport). Any outbreak of disease during the recovery period (for example, shipping fever) can drastically increase the time required for cattle to return to pretransport weight.

Lactating dairy cows usually show a marked decline in milk production after transport. This lowered production can be temporary for cows early in lactation; for cows late in lactation (over 6 months), it can last until they have turned dry.

After hauling, provide free access to good-quality hay first. After a long haul, one of the most important considerations for the animal and for the rumen microbes is energy. The hay should be available 4 to 6 hours before cattle have access to water. Don't hide the hay; spread some hay on the ground leading up to the bunks. When water is available to cattle immediately after a long trip, they engorge themselves and won't eat hay, and they may then suffer from rumen dysfunction. When lightweight or weaned beef calves are received, water should be made available after 4 to 6 hours in stock tanks, not in automatic waterers. Many young calves, especially those from the southeast, have not drunk from an automatic waterer; the hissing noise frightens them, and they refuse to drink.

Observe cattle several times daily for the first three weeks after arrival. Upon entering a pen, observe the general appearance of the cattle. Look for differences in the cattle's attitude or the way they react to a person walking or riding through the pen or herd. An animal that doesn't feel well may move slower than the rest and may hold its head at a lower angle. The sick animal may move a short distance and stop. Any of these actions may be taken as clues that the animal is getting sick and may need closer observation. Make a closer inspection of such individuals and determine whether they need to be removed for special treatment.

The primary purpose of checking pens is to detect early symptoms. Making an exact diagnosis of an animal in a pen is not nearly as important as deciding that the animal is not competing and needs individual attention.

3.7 HEAT DETECTION

A number of both beef and dairy producers use artificial insemination (AI) and hand mating each year. One of the major problems with these programs is heat (estrus) detection. An experienced observer with a well-trained eye can do a good job by observing the herd for ½ to 1 hour early in the morning and ½ to 1 hour in the evening. Heat detection is more an art than a science, and not everyone can do a good job of it. The observer must be taught to recognize the signs of estrus and must be aware that some cows show only abbreviated signs of estrus. If heat periods are missed or incorrectly identified, increased calving intervals will result, with longer dry periods and reduced profits. It is extremely important that cows be inseminated at the proper time in relationship to ovulation. Estrus synchronization can be used to help reduce the time spent on heat detection.

Estrus (heat) is a fairly well-defined period that occurs in nonpregnant cows once each 19 to 23 days. This period is characterized by increased sexual activity and acceptance of the bull by the postpubertal heifer or cow. The period basically begins with the first acceptance of the bull and ends with last acceptance. European cattle have a 12- to 18-hour heat period, while zebu (Brahman) cattle may exhibit estrus only for 3 to 6 hours. Approximately 5% of the herd will exhibit estrus each day during the breeding season.

Heat-detection aids can be used to good advantage in beef and dairy operations. Dairy operations prefer the use of heat patches and grease marks on the rump. These aids are especially useful when cows are confined in a small area or on concrete. One of the most popular aids in beef operations is "teaser" or "marker" animals equipped with a chin-ball marker. These animals are usually altered in some way so that they cannot breed a cow in estrus by natural means.

Heat-detection aids are designed only to aid or assist the person doing the detection and should not be relied on totally. It is still mandatory that the herd be observed often enough to determine which cows have been mounted and when the mounting has occurred. When teaser animals or gomer bulls are used, a good rule of thumb is to use 1 active teaser per 50 cows in a small heat-detection pasture. This number can be adjusted up or down depending on herd size and type of pasture.

There are a number of factors that affect the expression of heat. Adverse weather and sudden changes in the weather can suppress the display of estrus. Hunger and thirst also play a role. If nutritional levels are too low over an extended period of time, cows will not cycle. Many types of stress, such as rough handling, noise, hauling, and excess movement, will suppress or inhibit estrus and ovulation.

Equipment Necessary

- Pocket record book
- Heat patch
- Sale-barn glue
- Wax marking stick
- Chin-ball marker
- Hormonally prepared cow (usually a cull cow)
- "Gomer bull" (penectomized, vasectomized, Pen-O-Block, deviated penis)

Restraint Required

The technique of heat detection itself does not require physical restraint of the animal. If heat-detection aids are used, however, the animal will have to be restrained in a squeeze chute, headgate, or stanchion. All cycling cows must be restrained if heat-detection patches are to be applied to the rump. "Gomer" bulls must be restrained to fill the chin-ball marking device with a pint of dye every several days during the breeding season. Hormonally treated cows used as marker animals to detect heat are restrained in a squeeze chute for administration of hormones before and during the breeding season on a prescribed injection schedule. Such cows must also be restrained in a headgate or squeeze chute to fill the chin-ball marking device.

Visual Heat Detection. In most cases, animals can be observed during normal activity or as they move to and from a housing, feeding, or pasture area. In dairy operations, cows can also be observed as they are moved to and from the milking parlor. Heat detection in beef operations often requires observation in the pasture or feeding area. This requires more time because beef cows are confined to larger areas and are not always easily visible.

Timing is extremely important when observing for estrous activity. Behavioral heat activity is usually seen in the early morning between 2:00 and 6:00. Since many producers are not doing chores or milking that early, it is extremely important that an observation be made the first thing in the morning. Behavioral estrous activity is reduced from mid-morning to mid-afternoon but increases toward evening. Evening observations are essential to a good visual heat-detection program.

Do not disturb the cows when making your observations. Any noise or movement attracts their attention and causes them to stop showing visual signs of heat activity.

Observation for heat in a stanchion or tie-stall barn is similar to the procedure just discussed. The main difference is that cows are not running loose and cannot ride or mount each another. When the vulva is swollen and a clear discharge is dripping from the vulva, turn the cow loose with a few other cows and watch for signs of standing heat.

A good set of records will help you to know which animals to observe more closely. These records should include previous heats and breedings. The normal heat cycle is 19 to 23 days. Before you make your observations, look at your records and decide which cows are most likely to be in heat. Your record should include dates of a bloody discharge on the vulva or the tail of the cow. This condition indicates which cows were probably in heat a couple of days earlier. These cows are likely to be in heat 18 to 20 days after the blood is observed.

Cows that are not caught in heat should be examined by a veterinarian who can give an estimate of when they should be in heat again. If they are not cycling, they should be treated by the veterinarian.

Don't guess at which cows are in heat. Estimates are that as many as 10% of the animals inseminated are actually not in heat at the time of insemination. This is a total waste of time and, in many cases, of expensive semen.

The following steps describe the visual signs of estrous activity. Become familiar with each and every sign. Some cows display all the signs, while other cows exhibit only one or two.

Step-by-Step Procedure

1. The onset of behavioral estrous activity is usually gradual and may occur over a period of 4 to 24 hours. Animals can be characterized as restless and nervous. Some females tend to isolate themselves from the rest of the herd and pace the fences. It is not uncommon for these cows to twitch their tails nervously, walk around with their tails elevated, and show a clear, thick mucous discharge from the vulva. The discharge may be observed first on the tail, but after close inspection it will be clear that it originates from the vulva.

> **CAUTION:** A cow with a cloudy, milky-white, or pussy discharge may have a vaginal or uterine infection. She may be in heat, but she has a poor chance of conception. Consult your veterinarian for advice and treatment.

2. The most obvious behavioral estrous activity is mounting. When a cow in heat tries to mount another not in heat, the mounted cow will not stand quietly and allow the other cow to mount. A cow in heat may also follow others, stand beside them, and put her head on their rumps or backs. She may throw her head as if to mount, wrinkle her nose, and snort. If you are unsure which cow is in heat, be patient and continue to observe the herd. If the suspected cow continues in this activity, more than likely she is in heat and should be watched until she allows other cows to mount and ride her.

The time interval between a cow's attempts to mount may be as long as 20 minutes and can depend somewhat on the lot surface. If the lot is concrete, it can become slick and reduce visual activity, and a casual look at the herd will not reveal the number of cows in heat.

Two cows may be in heat at the same time. When the cow being mounted does not attempt to escape, she usually is in standing heat.

3. Cows in standing heat or estrus generally exhibit restlessness and often remain standing throughout the day. The cow may bellow frequently and exhibit signs of a reduced appetite.

4. Some cows become more friendly to other animals or man during the heat period. The cow may smell and nose other cows or stand head to head with another cow. As the cows stand head to head, they may or may not butt heads.

CAUTION: When cattle are mixed or new cattle are introduced into a group, they will fight by butting their heads together as they attempt to restructure the pecking order. Don't confuse this situation with standing heat.

5. A cow in estrus often has her tail raised, and she may have a long string of clear mucus hanging from the vulva or on her tail or rear quarters. Upon closer observation, the cow generally shows a somewhat reddened, loose, slightly swollen, and relaxed vulva.

6. A roughened tailhead or mud on the back or sides of the cow indicates that the cow has been ridden. The hair on the tailhead normally lies down, with the tips of the hair pointing toward the tail. When the cow has been ridden, the hair stands up and the tips of the hair point forward, the hair between the hooks (hipbones) and tailhead is rubbed off, or the hair has a roughened, disorganized appearance.

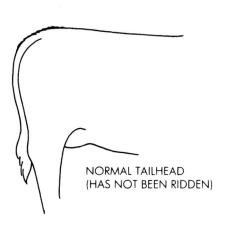

NORMAL TAILHEAD
(HAS NOT BEEN RIDDEN)

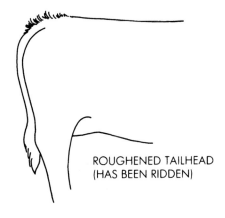

ROUGHENED TAILHEAD
(HAS BEEN RIDDEN)

7. In dairy cattle, a decrease in milk production for no apparent reason may indicate that the cow is coming into heat or is in heat. Beef cows may also reduce milk production during estrus. Hungry, bawling calves could be an indication of heat, though this should not be the only sign you use. When you observe this, look carefully for other signs of estrous activity.

8. Blood on the vulva, tail, or rear quarter of a cow or heifer usually indicates that she was in heat 2 to 3 days previously. It is generally too late to breed this type of cow during this cycle. This should be recorded in your record book, however, because it is helpful in estimating the next heat period. Bleeding that occurs after a natural or AI service does not mean that conception has not occurred.

9. Record the identification numbers of cows that show estrous activity in your record book, breeding chart, or, for cows not bred, on an expected-heat date list.

10. If the cow shows estrous activity in the morning, inseminate her in the evening. If the cow shows estrous activity in the evening, inseminate her on the next morning.

11. There is data showing that about 25% of cows exhibit signs of estrus between 6 P.M. and midnight and another 40% from midnight to 6 A.M. These are some of the least workable times for the cattle producer to observe his or her cows. These 65% of the cattle are the main reason that visual heat-detection efficiency is low. The best recommendation to accommodate for this is to observe the cows as late as the light will allow in the evening and as early as possible in the morning.

12. There are some secondary signs of heat that can be observed after the primary (or standing) heat signs have subsided. They include a roughened tailhead or mud on the rump; pacing, licking, sniffing, or resting of the chin upon the tailhead of another cow; the presence of clear, stringy mucus discharge hanging from the vulva or smeared about the rump; mounting of another cow or allowing another cow to mount—but not willing to stand for riding. When you notice a bloody mucus being discharged from the vulva, estrus has occurred 2 to 3 days prior.

13. An abrupt change in weather, particularly an extreme change in temperature, can alter the normal estrous activity. When this occurs, the normal signs of estrus will likely be diminished and heat-detection efficiency will drop sharply.

Heat-Detection Aids

These aids are designed to help detect cows in heat, but problems can result with each detection aid used. Heat patches can fall off or be activated unintentionally (for example, when a cow walks under a tree limb).

The following are some of the common and popular ways in which estrous activity is detected.

Heat Patch. The plastic heat-detecting device (patch) is glued to the top of the sacrum (back) between the tailhead and the hooks (hipbones) with a branding-tag glue used in sale barns. The heat patch contains a small vial of fluid with a smaller vial of color that breaks and discolors the total fluid when pressure is applied by mounting activity. Look for additional signs of estrus to be sure that the cow hasn't accidentally ruptured the color vial. This can occur if the cow rubs against a low-hanging tree limb.

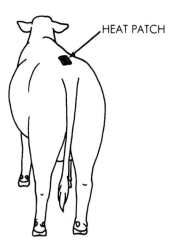
HEAT PATCH

The well-known Kamar patch and the newer Bovine Beacon patch are two examples of heat patches. The Kamar patch turns from white to red. The Bovine Beacon turns from white to bright pink. Cattle that are wearing heat patches should be observed at 12-hour intervals.

There is a new patch heat-detection system on the market that is based upon the "scratch and sniff" or "scratch-and-reveal-your-prize" technology commonly used in advertising, fast-food games, and lottery programs. It is called the *Stampe's* <u>ESTRU$ ALERT</u>® heat-detection system. This system utilizes a 2" × 4" self-adhesive patch that comes in a variety of base colors covered by a "scratchable" silver coating. It is placed just above the tailhead, like other heat-detection patches. As the cow or heifer in heat is ridden by another animal, some of the silver coating is scratched away and the base color is revealed. As successive ridings are allowed, more and more of the base is revealed. Different colors can be used for successive heat cycles.

Electronic Heat-Detection Systems. The *Heat Watch System* is based upon the measurement of the pres-

sure applied, by a "riding" cow, to a small transmitter embedded in a "tag" that is affixed to the tailhead region of the cycling cow. The transmitters are approximately ¾" × 2" × 3". The signal transmitted by the tag containing the transmitter is picked up by a receiver, which is direct-wired into an on-farm computer where the data is accumulated for later analysis.

The transmitters are coded so that each mounted cow sends a unique animal-identifying signal. This unique code and the clock built into the computer enable the manager to know exactly when standing heat(s) began and which cow(s) is (are) ready to be bred . . . without having to be in the barnyard to learn all of this.

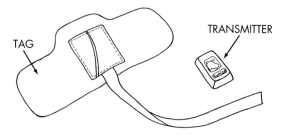
TAG TRANSMITTER

A combination of glue and a 12" tail tie is used to keep the pressure transmitter/tailhead tag in place. System directions are explicit regarding attachment methods.

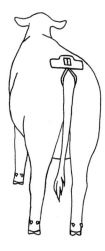

The *Afimilk system* is a computerized, multifaceted system designed for managing the complex and sophisticated dairy operations we see today. It is an information gathering system that automatically records each individual cow's performance (such as milk yield) and activity level. Activity level is measured by a pedometer, which is strapped to the ankle of the cow. A pedometer counts leg movements (steps), which are a direct function of cow activity. The Afimilk pedometer can be used as a quite accurate heat-detection device because as the cow comes into heat, and should be inseminated, there is a sharp increase in her activity level.

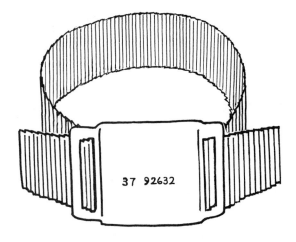

The pedometer is worn about the ankle in conjunction with an electronic identification tag, which in the case of the Afimilk system is called the Afi-Act ID tag. As the cow comes into the milking parlor, information stored in the Afimilk pedometer and her Afi-Act ID number are automatically "dumped" into an Afimilk recording and computer interface device. These are graphed and reviewed by the cow manager. The cow depicted in the following graph should be inseminated (note the very sharp rise in her activity level).

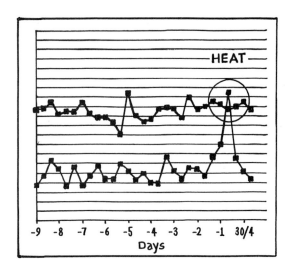

The *De Laval ALPRO Activity Meter* functions similarly to the Afimilk pedometer. It too interprets increased cow activity as a rather precise indicator of the onset of estrus. The ALPRO Activity Meter is worn, along with a transponder, on a collar suspended from the neck of the animal. When the cow enters the milking parlor, data is "dumped" into the processor (computer interface). This information is graphed and interpreted by the cow manager.

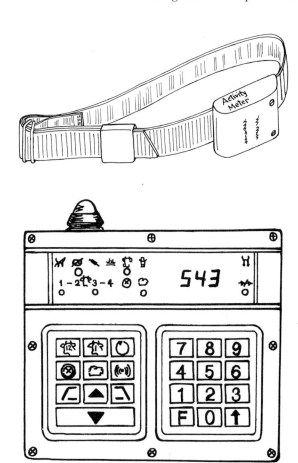

Colored Wax Marker (Tailhead Marking). Apply a thick deposit of marking wax to the very top of the tailhead of the cow. When a cow is mounted, this wax is spread down the sides of the tailhead onto the rump. This indicates that the cow has been ridden by another cow. Be sure to look for additional signs of estrus.

Most large herds have self-locking stanchions at the feed bunks. Tailhead marking is a popular

**GREASE MARKS
(HAS BEEN RIDDEN)**

method to aid in heat detection in these operations. Tailheads are usually marked daily while the cows are in the lockups. Cows can also be bred via AI in these locking stanchions. Most herd reproductive health checks are conducted here as well.

Chin-Ball Marker. The chin-ball marker is simply a metal paint reservoir with a large ball bearing held by a leather harness under the chin of the marker animal. This marker operates somewhat like a large ballpoint pen. As the marker animal mounts and dismounts the cow in heat, a characteristic mark is left on the cow's back and rump. The paint reservoir contains enough latex-type paint for about 25 cows, depending on mating behavior. The paint well must be checked and refilled every several days during the breeding season. This marking device is used on all "gomer" bulls and hormone-treated cows.

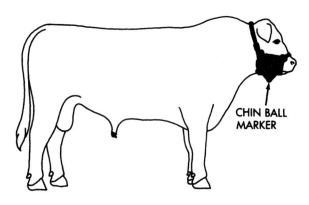

CHIN BALL MARKER

"Gomer" bulls are those that have been surgically or nonsurgically altered so that they cannot breed a cow naturally. Sometimes these bulls "fall in love" with one cow and won't leave her side, and therefore miss other cows in heat. In addition, some gomer bulls may lose their sex drive and interest when they discover that they cannot breed a cow when they mount. The hormone-treated cow seems to have more to offer to the cattleman.

Vasectomized Bulls. The vasectomized bull has been used in the United States for many years as a marker bull. The operation involves removing part of the vas deferens from the spermatic cord. The blood and nerve supply of the spermatic cord are left intact. The testes and penis function normally, but transport of spermatozoa to the urethra is blocked. The sexual activity of the bull remains unaltered, and he will detect heat as well as an intact bull. The vasectomy should be performed by a veterinarian. The disadvantage of this operation is that the bull continues to enter cows and can be the

transmitter of venereal diseases from one cow to another.

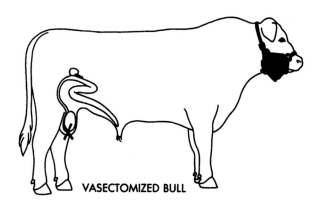

VASECTOMIZED BULL

Deviated Penis and Prepuce (Sheath). This type of gomer bull is altered by a surgical technique that transplants the penis and sheath from their natural position to the fold of the flank. This method permits normal erection, but does not allow the bull to enter the cow. The disadvantage of this system is that the bull can become frustrated and lose his sex drive. The advantage is that venereal disease cannot be transmitted with this type of gomer bull.

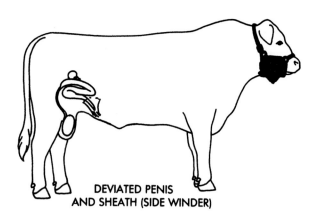

DEVIATED PENIS AND SHEATH (SIDE WINDER)

Pen-O-Block. This method involves the insertion of a commercially available plastic tube inside the sheath. The device blocks the preputial opening, preventing erection and penetration. The Pen-O-Block is secured by making a small opening through the skin on each side of the sheath, inserting the plastic tube, placing a pin through the openings in the sheath and tube, and securing the pin with washers and cotter pins. The disadvantage of this system is that bulls can lose their sex drive. The advantage is that venereal disease cannot be transmitted and the Pen-O-Block can be removed after the AI season, which would allow the bull to be used as a cleanup bull.

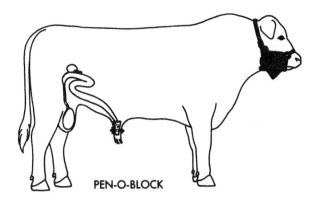

PEN-O-BLOCK

Penectomy. This type of gomer bull has undergone surgical removal of part of his penis. Some techniques remove approximately two-thirds of the penis and exteriorize the remaining end between the anus and scrotum. The disadvantage is that sex drive can be reduced. The advantages are that sexual contact is impossible and veneral diseases cannot be transmitted.

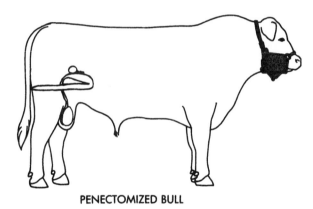

PENECTOMIZED BULL

Hormonally Treated Cows. This system utilizes a cull cow, a hormone preparation available from your veterinarian, and a systematic injection schedule. The cost of preparing marker animals of this type is relatively low, and almost everyone has a cull cow. Animals injected with a testosterone-based solution are usually sexually aggressive and mount other cows in heat within a few days after the last injection. The injections are then given periodically during the breeding season to maintain sexual activity. The system has many advantages: the hormonally treated cow can be prepared cheaply, the cow can't spread venereal diseases, maintenance cost of a marker animal is reduced, and the cow can be sold for slaughter after the breeding season.

If you plan to use a hormonally treated cow for heat detection, select a cull animal with sound feet and legs—one large enough and aggressive enough to compete with the herd. The number of treated cows needed to detect heat in a herd situation varies with the cycling activity of the cow herd and the size and terrain of the breeding pasture or lot. In herds with a large breeding pasture or high cycling activity, 1 treated cow is needed for each 20 to 30 breeding females. If cycling activity is low or the breeding area is small, 1 treated cow can be used to detect heat on 30 to 40 cycling cows.

Do not select a cow that is pregnant, nursing, or in the milking herd. Do not expect to rebreed the cow after the treatment, because she will be subfertile.

In beef operations, the hormonally treated cow should be a mature cow that has lost her calf at birth or one discovered open during the calving season. This type of cow is the most economical to use and can be slaughtered after the breeding season.

Crossbreds are generally more aggressive than purebreds. Purebred Brahmans have shown the poorest response to hormone treatment and are not recommended for heat detection. Brahman crossbreds seem to work very well as heat detectors.

Cows can be treated successfully with several different hormone combinations and injection schedules. Two methods that have been effective and easy to administer are presented here. The hormones must be ordered and purchased from a veterinarian, but the herdsman can give the injections. Be sure to read the label and follow the directions.

> **CAUTION:** Not all products contain the same concentration of testosterone. Do not use the hormone treatment on cows that are producing milk, because it will contaminate the milk.

The first hormone method requires administration of 500 mg of testosterone enanthate (in oil) intramuscularly for a fast-acting effect, and 1,500 mg of testosterone enanthate (in oil) subcutaneously for an initially slower, but longer-lasting, response. This should be done at one time with a maximum of 10 cc (mL) injected at any one location. Most of the cows will become active and ready to use in 3 to 7 days after treatment.

To maintain an active cow, administer 500 to 1,000 mg of testosterone enanthate subcutaneously (maximum of 10 cc per location), depending on cow activity, 3 to 4 weeks after the initial treatment. Administer the same dosage for as long as the cow is used as a detector.

> **CAUTION:** Discontinue hormone use at least 14 days before treated animals are slaughtered for human consumption.

The second hormone method requires administration of 200 mg of testosterone (aqueous suspension) and 500 mg of testosterone propionate (repositol) intramuscularly on day 1. The injections should be made with a maximum of 10 cc (mL) per location. On the 8th and 22nd days, inject 500 mg of testosterone propionate (repositol) intramuscularly as before. After the second injection (8th day) the cow should be active and ready to use. The cow may require additional injections every 14 days, depending on her activity. These additional injections should consist of 500 mg of testosterone propionate (repositol) and be administered as before. After the breeding season, allow at least 14 days before marketing the animal to reduce potential residue problems.

3.8 ARTIFICIAL INSEMINATION

Artificial insemination (AI) is a technique that deposits semen into the cow's reproductive tract with an inseminating rod or instrument. The use of AI allows the producer to utilize a larger variety of genetically superior bulls at a reasonable cost without the danger and expense of owning a herd bull. Other advantages of AI include improved reproductive health of the herd, the use of several different breeds in a crossbreeding program, and improved recording of cows' breeding and calving.

A good AI program requires a well-trained and experienced inseminator, good herd management, suitable handling facilities, and a well-trained, interested herdsman. Without these, AI becomes impractical if not impossible.

Equipment Necessary

- Working chute or rope halter
- Liquid nitrogen semen storage tank
- High-quality semen
- Inseminating instrument
- Shoulder-length disposable plastic glove
- Lubricant (mineral oil, liquid soap)
- Widemouth thermos (for a thaw box) or electrical thaw unit
- Thermometer
- Tweezers
- Scissors
- Paper towels
- Record book

Restraint Required

Cows in estrus should be quietly sorted from the herd and moved to the restraint area. The cow to be inseminated should be restrained in a working chute or squeeze chute. If a working chute is used

for restraint, it should be equipped with a gate that goes completely across the chute. This gate will prevent the cow from going forward during the insemination. A bar should be placed across the alley above the cow's hocks to prevent her from backing up. If the cow is broken to lead, it is possible to restrain her with a rope halter tied with a quick-release knot. When a halter is used, a second person should hold the cow against a wall or fence while the cow is inseminated.

Step-by-Step Procedure

The AI industry utilizes semen stored in "straws."

1. Identify the cow in estrus (heat) by observation when estrus synchronization is not used. Cows have a 3-week (19 to 23 days) estrous cycle. This means that once every 19 to 23 days, a cow sheds an egg (ovulates). At the end of this estrous cycle, the cow will show signs of estrus and normally will accept the bull. Observe the herd for estrus at least twice daily. Recommended times for heat detection are early morning and late evening. The signs of estrus include some combination of the following: (1) swollen vulva; (2) nervousness, bawling, and excessive walking; (3) mounting other animals; (4) standing and allowing other animals to mount; and (5) paint, chalk, or grease marks on the back of a cow when a marker animal has been used.

2. Identify and record the identification number of the cow in estrus.

3. Cows observed in standing estrus in the morning should be restrained and inseminated in the evening. Cows observed in the evening should be restrained and inseminated in the morning. This schedule is important because it ensures that the sperm reaches the egg at the most optimum time for fertilization.

4. Restrain the cow in a working chute. The cow(s) in estrus should be moved and worked quietly for high conception rates.

5. Prepare a thaw box for the semen. A warm-water thaw box is used for straws. The AI technician or semen supplier will give you specific instructions on water temperature and thaw time. These instructions should be followed for best results. The thaw box should be a widemouthed thermos with a thermometer or a commercial electric thaw unit. The temperature should be checked before the straw is dropped into the water, and a thermos of hot water should be prepared to adjust thaw-box water temperature as needed. The straws should not be thawed in a shirt pocket. This can reduce semen quality.

6. Open the liquid nitrogen tank and remove the circular plug.

7. Locate the canister that contains the semen to be used for the insemination. Keep a record of semen that identifies the location of each sire's semen in the tank. The canisters and canes of semen should be organized in the nitrogen tank by breed and/or sire so that a given ampule or straw can be located quickly and easily.

8. Lift the wire that holds the canister with the semen to be used for insemination. Pull the canister out only far enough to identify the cane containing the semen to be used.

9. Lift the cane and remove the semen.

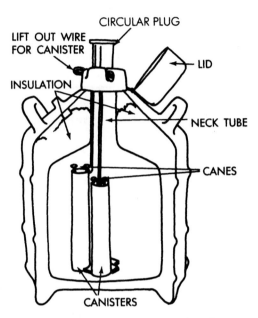

Straws. Lift the cane with one hand and use a pair of tweezers to remove a straw from the goblet. This should be done in the neck of the tank. After the straw is removed, quickly replace the cane in the canister, check the identification on the straw, and place it into the thaw-box unit.

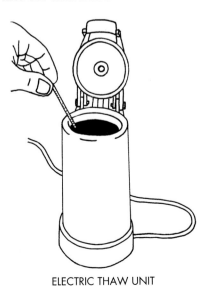

ELECTRIC THAW UNIT

CAUTION: Do not lift the semen on the cane out of the neck of the tank. Removal of the straw should take place in the neck to minimize damage to the remaining semen. Do not replace a straw once it has been removed from the cane and lifted out of the tank. If a straw is removed from the tank, discard the exposed semen, or expect poor chances for conception. Quick replacement of the cane in the canister is very important to maintain the quality of the semen remaining on the canes.

10. Replace the circular plug in the nitrogen tank and close the lid.

11. Thaw the semen. Follow the semen supplier's thawing recommendations. In extremely cold weather, the semen may start to refreeze when removed from the thaw box. You will have to warm the inseminating instrument before loading it. Do this by rubbing a paper towel briskly along the insemination instrument. Refrozen semen is of low quality and should be discarded. Use semen within 15 minutes of thawing.

12. Prepare the inseminating instrument. Assemble the breeding gun (tube and plunger) and place it between your teeth. Keep the breeding gun clean to reduce the possibility of uterine infection.

13. Remove the ampule or straw from the thaw box and dry it with a paper towel; water can kill sperm.

14. Prepare the semen for loading. Cut off the sealed end of the straw (opposite the cotton wad). This cut has to be either a square cut or a diagonal cut, depending upon the style of French gun being used; otherwise the straw will not fit snugly (seat) against the sheath. The cut can be made with a scissors or with a commercially available snipper.

15. Load the inseminating instrument. Insert the cotton-wad end of the straw into the breeding gun. The size of the breeding gun must be compatible with the size of the straw being used. The cut end of the straw will stick out of the breeding gun and the plunger will be behind the cotton wad of the straw when properly inserted. Slide the plastic sheath over the breeding gun and secure it with retainer rings. Make sure that the straw fits snugly and squarely into the sheath to prevent semen from leaking back into the sheath.

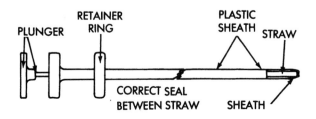

16. In warm weather, place the loaded inseminating instrument between your teeth to free both hands. In cold weather, the inseminating instrument should be wrapped inside several paper towels and perhaps even placed inside the front of the inseminator's coveralls to prevent cold shock to the semen.

17. Put on a disposable glove and lubricate the hands and fingers. It is most important to lubricate the fingers, back of the hand, and wrist area because they receive the most contact. Liquid soap or mineral oil can be used as a lubricant.

18. Stand perpendicular to the cow and grasp the tail with the ungloved hand. With the gloved hand, wipe lubricant across the anus.

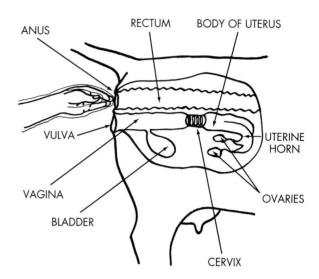

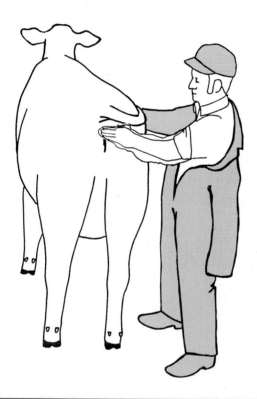

CAUTION: If this step is done suddenly, the cow may kick. Standing sideways (perpendicular to the cow) can minimize the danger.

Do not get lubricant on the lips of the vulva; it may come in contact with the inseminating instrument, be carried into the reproductive tract, and kill the sperm.

19. Form the gloved hand into a wedge and insert it with a thrust into the anus as far as possible. As the hand enters the anus, it should be folded to form a loose fist. This will present a blunt surface, instead of pointed fingers, and will straighten the rectum without damage. The cow will exert a fair amount of pressure against the hand and arms, and she will try to expel them.

CAUTION: Moving the hand into the rectum gradually will irritate the cow, and rectal contractions will be greater. Once the hand is inside the rectum, the cow will gradually arch her back and danger from kicking generally is minimized.

20. Manually remove feces from the rectum so that the feces are not deposited on the inseminating instrument during insemination. Then wipe the vulva area with a clean, dry paper towel. Always wipe the vulva from top to bottom. Never make two wipes with the same towel.

21. Move the tail to the outside of the palpating arm so that the ungloved hand is free to guide the inseminating instrument without fear of fecal contamination.

22. Keep your fingers close together and gently work forward and side to side in the tract without digging with the fingers. If rectal contractions are encountered, it may be necessary to relax the contracting

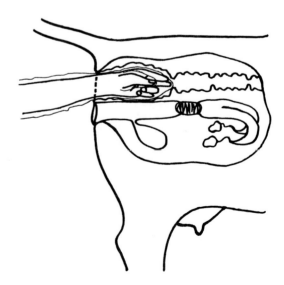

CAUTION: Some bleeding may occur during the palpation process. This results from scraping the mucosa and rupturing small blood vessels at the surface of the rectal wall. This is not serious, but extreme care should be taken that the inseminator does not rupture the rectum wall. The rectum wall is fairly thick, and it is unlikely that a rupture will result if the tract is carefully palpated. If a rupture does occur, it usually leads to peritonitis and the death of the animal. When a rupture is identified, it is usually best to send the cow to slaughter as soon as possible.

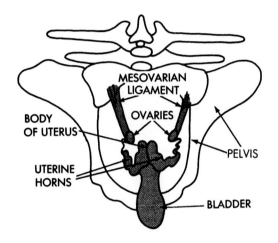

ring by inserting two fingers through the center of the ring and gently massaging back and forth.

23. Feel for landmarks. The forward (anterior) portion of the pelvis (*pelvic rim*) is the best landmark because it is a bone and does not move. The pelvis forms a cradle for the nonpregnant reproductive tract. The *cervix* is the second landmark to be located. The cervix is about 1 inch in diameter, 4 inches long, and has a gristlelike feel resembling a turkey neck. Locating the cervix is not as easy as locating the pelvic rim because it does not remain in one area. With the palm of the hand down, the fingers extended, and the thumb pointing forward, the hand can be pressed downward. With downward pressure, the hand can be moved from side to side to locate the thick-walled cervix. This process may have to be repeated several times if the cervix is not located the first time. Beginning inseminators often insert the hand too deeply into the cow; in such a case the arm may have to be pulled back a few inches.

24. Once the cervix is located, it must be grasped and controlled so that the inseminating instrument can be inserted. Encircle the cervix with the thumb and fingers in such a way that the thumb is on top of the cervix and the fingers are under the cervix. The thumb should stay on top without rotating the

wrist. This can be difficult and tiring, because it requires effort to stretch the rectal wall.

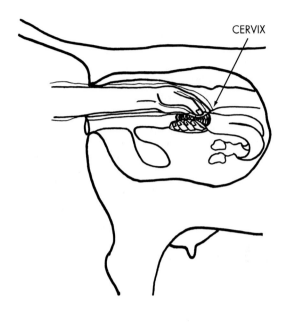

CAUTION: The gloved arm and hand can get very tired during this technique because of continued contractions of the rectum. Be careful not to fear the rectal wall.

25. Once the cervix is grasped and controlled, spread the lips of the vulva by applying downward pressure in the region of the vagina with the heel of the gloved hand.

26. Insert the tip of the inseminating instrument through the vulva without touching the lips. The tip of the instrument should be pointed upward at a 30 to 40° angle to avoid entering the suburethral diverticulum (urethral opening) on the floor of the vagina. After the tip of the instrument is inserted 3 to 4 inches into the vagina, it can be straightened so that it is approximately parallel to the plane of the rectum.

27. As the inseminating instrument is slid forward, the cervix must be pushed forward to straighten out the vaginal folds. Remember, the cow is trying to force the cervix toward the vulva and this creates more folds.

CAUTION: If the cervix is not pushed forward, the tip of the instrument can get caught in a fold and stretch or puncture the vagina. If the tip should become caught in a fold, withdraw the instrument a short distance and point it in a slightly different direction. Be patient and keep trying.

28. To enter the cervix, encircle the end of the cervix nearest the vagina with the thumb and first two fingers. Again, the thumb will be on top with the first two fingers wrapped around the cervix. This allows the base of the hand and last two fingers to form a funnel to guide the tip of the instrument into the cervix. The opening of the cervix (*cervical os*) protrudes into the vagina and toward the vulva, creating a blind pouch called a *fornix* all around the cervical os. It usually helps to push the cervix forward to reduce the size of the fornix.

29. Locate the opening of the cervix with the tip of the breeding instrument. The opening of the cervix is generally, but not always, in the center of the cervix. When the tip of the instrument comes in contact with the cervix, a gristlelike texture will be felt. It may be necessary to probe lightly with the tip of the instrument to locate the opening of the cervix.

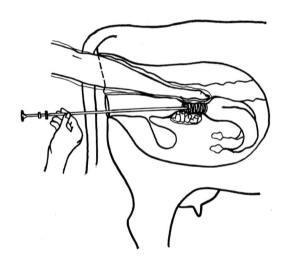

CAUTION: If the opening of the cervix is not located with the tip, do not force the instrument. Try placing the first two fingers on top of the cervix and pinning the cervix to the pelvic floor. Try to locate the opening with your thumb. Maneuver the tip of the instrument until it strikes your thumb, remove the thumb, and insert the tip of the instrument into the cervix.

30. Once the rod or gun is inside the opening of the cervix, apply slight forward pressure on the instrument and regrasp the cervix as in step 25. The thumb and first two fingers should always be in front of the instrument tip to manipulate the cervix.

31. Manipulate the cervix over the tip of the rod as the inseminating instrument is slid forward. The cervical rings often hinder the passage of the

instrument, and this will require turning and twisting the cervix. It may be necessary to bend or turn the cervix at a right angle to move past some cervical rings.

CAUTION: On first service, the semen will be deposited as in steps 33 to 35. On second service, it is recommended that the length of the cervix be estimated and the semen deposited at the midpoint of the cervix. The mid-cervical deposition is done to prevent interruption of a pregnancy in cows that show estrus after conception.

32. After it has been inserted through the cervical rings, the inseminating instrument will slide without resistance. This indicates that the tip is in the uterus or uterine horn.

33. It has been shown through multiple research trials that "horn breeding," as compared to "uterine body breeding," produces equal or better conception rates and non-return-to-heat rates. No trials have shown a statistical disadvantage for horn breeding.

Therefore, after you have passed the inseminating rod tip through the cervix, continue to insert it through the uterine body (which is very small—about the size of a quarter) and into one of the uterine horns. As long as the passage is easy, continue. When passage becomes difficult, you are probably approaching the greater curvature of the uterine horn. Stop passing the rod, pull the rod back slightly (perhaps a half inch), and deposit the semen.

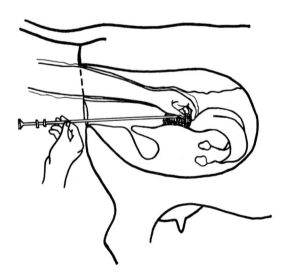

34. Deposit the semen slowly over at least 5 seconds. Make sure that the thumb or index finger is not over the tip when semen is deposited, or some of the semen will be deposited in the cervix and vagina.

35. Remove the arm and inseminating instrument from the cow.

36. Record the cow ID number, bull ID number, and date in a record book.

37. Release the cow and dispose of the rod or sheath, ampule or straw, paper towels, and disposable glove.

Postprocedural Management

Watch the cow for return to estrus. Normally this will occur in 3 weeks (19 to 23 days) if conception has not occurred.

3.9 ESTRUS SYNCHRONIZATION

One of the major drawbacks of artificial insemination is the huge amount of time required to detect estrus (heat) in the females. This problem can be partially overcome through the use of estrus-synchronizing compounds.

In addition, there are several other advantages that can be claimed for an estrus-synchronization program: it makes embryo transfer programs possible; it allows the scheduling of the calving season to coincide with your management scheme at the farm or ranch; it allows the grouping of the calving heifers so that the extra time necessary to calve them properly will not be superimposed upon the normal demands and minicrises related to calving the rest of the herd; and finally, more calves can be dropped in a shorter period of time with the result being a more uniform group that has the potential of demanding top dollar at sale time.

Estrus synchronization is not designed or recommended to be a substitute for good herd management practices. It will not cure reproductive problems or be a cure-all management practice.

Equipment Necessary

• Working chute
• Squeeze chute or stanchion
• Syringe with needle
• Implant jar
• Estrus-synchronizing compound
• Two pens (to separate cows and calves)

Restraint Required

Cows and heifers to be synchronized should be restrained in a working chute, squeeze chute, headgate, or stanchion.

Step-by-Step Procedure

The step-by-step procedure includes general areas and precautions that must be evaluated when considering an estrus-synchronization program. The step-by-step procedure does not list specific procedures in the process of synchronization; these procedures should be discussed with an extension specialist, company technical representative, or veterinarian.

1. Evaluate your operation and determine whether or not an estrus-synchronizing compound could be used successfully. A successful program requires a critical analysis of the entire operation. Correct problem areas before you start.

2. Cows will cycle after calving, and therefore be able to respond to the synchronizing program, if enough time has elapsed since calving (50 to 60 days is a good average) and if the nutritional program of the herd has been high enough to enable the cows to come through calving in moderate to good condition (body condition score 5–6).

3. The cow herd must be free of such diseases as leptospirosis, infectious bovine rhinotracheitis (IBR), vibriosis, and trichomoniasis, because these diseases affect reproductive performance. A good vaccination program should be followed to prevent these diseases.

4. Good handling facilities are a must because cattle may have to be worked two or three times in a 15-day period, depending upon which synchronization program you select. The facilities should be designed so that cattle can be worked quickly and quietly. The breeding area should be covered so that insemination can be done on schedule regardless of the weather.

5. The AI technicians should be experienced, with a history of good conception records. Each technician should artificially inseminate only 10 to 12 cows in one work period; poor conception can result when the technician becomes fatigued.

6. Cows must be cycling before treatment. Cows that are fewer than 50 to 60 days postpartum (after calving) should not be placed in a synchronizing program.

7. High-quality semen must be used to ensure good conception.

8. Heifers must be 14 to 15 months of age and must be cycling. In English breeds this occurs at a weight near 650 pounds; in the larger exotic breeds, it occurs at 750 pounds.

9. Obtain the estrus-synchronizing compound.

10. Follow product directions. Several treatment schedules are available. Treatment regimens are being developed at an increasing rate.

CAUTION: Estrus-synchronization compounds can cause pregnant cows to abort. Make sure that cows are not pregnant before treating them.

Following are the most common estrus-synchronizing protocols being used.

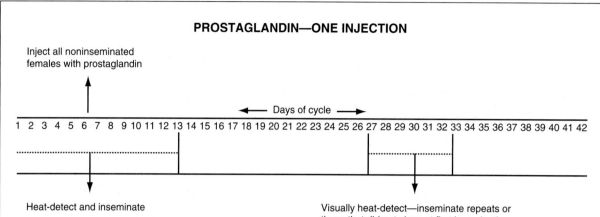

Prostaglandins. Prostaglandins are available only on the prescription of a veterinarian. The drugs are administered either intramuscularly or subcutaneously, depending on the product being used. Prostaglandins currently on the market include Lutalyse (Phormacia/Pfizer), Estrumate (Schering Plough), and PROSTA-MATE (Phoenix Scientific). There are two systems for prostaglandin synchronization: the one-injection method and the two-injection method.

Step-by-Step Procedure: One-Injection Method

1. For the first 6 days, day 1 to day 6, follow a conventional heat-detection and insemination program.

2. On day 6, inject all females not already detected in heat with the prostaglandin of choice.

3. From day 6 through day 12 or 13, heat-detect and breed as females come into heat. This should bunch most females and catch the bulk of the inseminations. Stop breeding activity after day 13.

4. From day 27 to day 33, heat-detect and inseminate. These cattle represent females that synchronized from the injection on day 6 but did not settle.

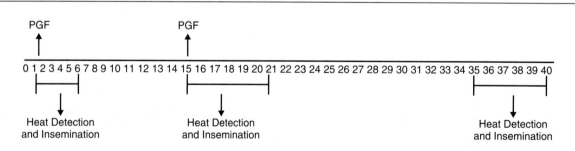

Step-by-Step Procedure: Two-Injection Method

1. On day 1, inject all cows with the prostaglandin of choice.

2. From day 1 to day 6, follow a conventional heat-detection and insemination program.

3. Fourteen days after the first injection, inject all animals not detected in estrus after the first injection with a second prostaglandin doses.

4. Heat-detect and breed for the next 6 days.

5. From day 35 to day 40, heat-detect and inseminate. These cattle represent females that synchronized from the injections on day 1 and day 14, but did not settle.

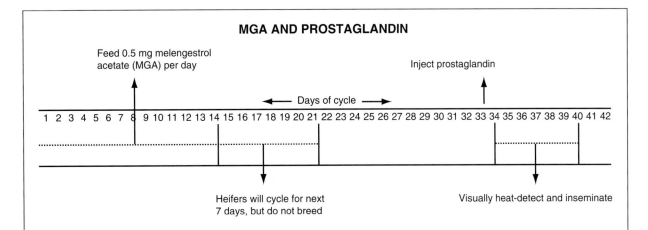

MGA AND PROSTAGLANDIN

MGA and Prostaglandin. Melengestrol acetate (MGA) fed in conjunction with a prostaglandin injection is an inexpensive, easily administered system for synchronization of estrus in cattle. The most commonly used MGA/prostaglandin system is the 14-day system.

Step-by-Step Procedure

1. From day 1 to day 14, feed 0.5 mg of MGA per head. MGA can be fed as a top-dressing, or it can be mixed into the ration.

2. On day 15, stop feeding the MGA. The heifers will come into heat during the next 3 to 5 days. Do not breed during this heat as the fertility is very poor.

3. On day 33, inject the prostaglandin of choice. Females will respond to the prostaglandin injection and come into heat during the next 5 days.

4. From day 34 to day 39/40, heat-detect and breed the females as they come into estrus. Some animals will respond quickly to the injection on day 33, but most will take from 3 to 5 days to exhibit signs of estrus.

5. Heifer often will not initiate an estrous cycle, so timed inscminations are not recommended.

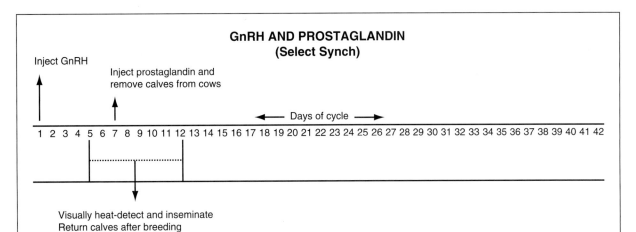

GnRH AND PROSTAGLANDIN
(Select Synch)

GnRH and Prostaglandin. GnRH, gonadotropin releasing hormone, is available commercially from your veterinarian as Factrel, Fertagyl, or Cystorelin. With this synchronization regimen, the cows receive both gonadotropin and prostaglandin injections.

Step-by-Step Procedure

1. On day 1, inject GnRH.

2. Start heat-detection and insemination 24 to 48 hours before the prostaglandin injection and continue for 7 days.

3. On day 7, inject prostaglandin.

4. Remove the nursing calves from the cows, and observe cows for heat.

5. Observe for signs of heat (estrus), and inseminate as the cows come in.

6. After the cows are bred, the calves can be returned.

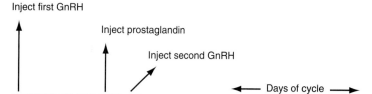

**GnRH (TWO INJECTIONS) AND PROSTAGLANDIN
[Ov-SYNCH]**

Inject first GnRH

Inject prostaglandin

Inject second GnRH

⟵ Days of cycle ⟶

1 2 3 4 5 6 7 8 9 10 11 12 13 14 15 16 17 18 19 20 21 22 23 24 25 26 27 28 29 30 31 32 33 34 35 36 37 38 39 40 41 42

Fixed-time inseminate all females

Ov-Synch. The Ov-Synch system is used in both the dairy and beef industries. It is quite extensively used by dairy producers. It involves the use of GnRH, prostaglandin, and a timed insemination.

Step-by-Step Procedure

1. Inject GnRH on day 1.

2. Seven days after the GnRH injection, on day 8, inject prostaglandin.

3. Two days later, on day 10, inject GnRH for the second time.

4. Inseminate all females 12 to 18 hours later.

5. All cows can be time-inseminated at the second GnRH injection. No heat detection takes place. This is the Co-synch protocol.

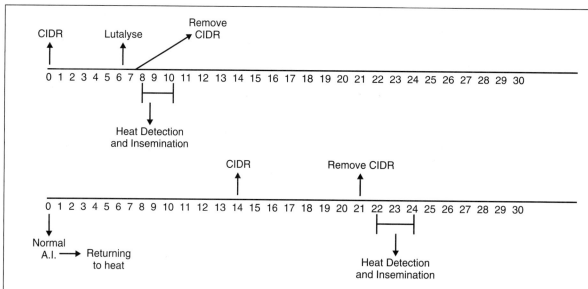

CIDR   Lutalyse   Remove CIDR

0 1 2 3 4 5 6 7 8 9 10 11 12 13 14 15 16 17 18 19 20 21 22 23 24 25 26 27 28 29 30

Heat Detection and Insemination

CIDR   Remove CIDR

0 1 2 3 4 5 6 7 8 9 10 11 12 13 14 15 16 17 18 19 20 21 22 23 24 25 26 27 28 29 30

Normal A.I. ⟶ Returning to heat

Heat Detection and Insemination

CIDRs. CIDR, controlled internal drug releasing devices, are T-shaped flexible silicone and nylon devices containing 1.38 grams of progesterone. When the CIDR is placed into the vaginal tract, it releases progesterone at a constant rate. This mimics the corpus luteum activity and stops cycling until the Lutalyse is injected and the CIDRs removed. It is approved for beef cattle, dairy heifers, and dairy cows returning to heat following AI.

Step-by-Step Procedure

1. Insert the CIDR on day "0."
2. Inject Lutalyse on day 6.

3. Remove the CIDR on day 7.
4. Heat-detect and inseminate on days 8, 9, and 10.

For lactating dairy cows, use the following protocol:

Step-by-Step Procedure

1. Insert the CIDR 14 days following normal artificial insemination.
2. Remove the CIDR 7 days after insertion (on day 21).
3. Heat-detect and inseminate on days 22, 23, and 24.

Inserting the Eazi-Breed CIDR

Equipment Necessary

- CIDR inserts
- CIDR insertion gun
- Disposable latex or nitrile gloves
- Lubricant for insemination
- Wash pail and disinfectant
- Paper towels

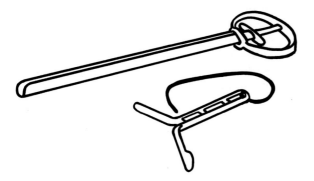

Restraint Required

Normal restraint procedures for insemination or palpation are appropriate for CIDR insertion. A workable setup consists of a breeding chute or squeeze chute that limits movement forward and backward, as well as side to side.

Step-by-Step Procedure

1. Wear a pair of the disposable gloves. A bit of talcum powder on the inside of the gloves will help prevent sweating and hard-to-remove gloves after the procedure is finished.

2. Prepare a pail of disinfectant solution. This is used to wash the insertion gun, and your hands, after placing the insert into each cow.

3. Place the body of the CIDR into the insertion gun, with the tail of the insert along the slot of the insertion gun.

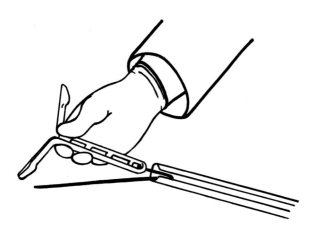

4. Lubricate the tip of the insert.

5. Clean the vulva of the cow, using paper towels.

6. Open the lips of the vulva and insert the insertion gun at a slight upward angle. Move up and over the pelvic bone until resistance is met. Stop at that point.

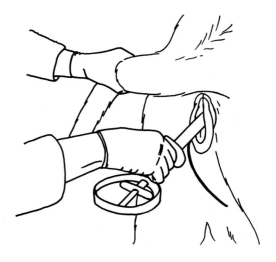

CAUTION: Be certain the tail of the insert (blue cord) is on the underside of the insertion gun and that it curls downward. If it does not, it is too obvious to curious penmates and will very likely be the target of much licking and mouthing, until it is removed. A good recommendation is to clip the tail of the insert to a length of 2 to 3 inches. This makes it much less noticeable to penmates

7. Deposit the insert by depressing the plunger on the insertion gun. Slowly withdraw the insertion gun.

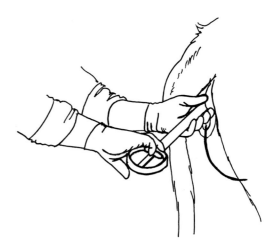

8. Clip the tail of the insert so that no more than 2 to 3 inches protrudes from the vulva. This will keep inquisitive herdmates from pulling at it and possibly removing it.

9. Refer to the synchronization protocol sheet. Day 6 following insertion is the time to give heifers a dose of prostaglandin.

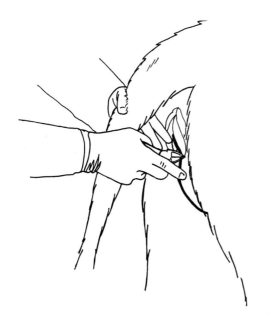

10. After 7 days, remove the CIDR by pulling on the "tail" of the insert. Use a gentle but firm pull.

CAUTION: Be certain to recover a CIDR from every animal that received one. Do not assume that a penmate removed the insert if you do not see a tail protruding. Glove up—go in and determine what the situation is. The tail could have been "sucked" into the vulva. For the cow or heifer to settle, it must be removed.

---

11. Refer to the synchronization protocol sheet. After removal of the CIDR, heat watch should be established and cows inseminated when detected in estrus.

3.10 PREGNANCY EXAMINATION

The purpose of the pregnancy examination is to identify "open" or nonpregnant cows early in the gestation period. Open cows in beef operations result in increased feed cost and reduced profit potential. Open dairy cows result in longer lactation and dry (nonlactating) periods, lower total milk production for the herd, and increased costs.

Pregnancy examinations should be performed when calves are weaned in beef operations and 35 to 45 days after breeding in dairy operations. The difference between the two cattle types dictates different management objectives. In beef cattle, the main objective is to get a cow bred during a 60-day breeding season and to wean a live calf at 7 months. When she fails to do either, she should be culled. The two most logical times to cull a beef cow are when she does not produce a live calf at calving time

and when she turns up open after weaning a calf. A pregnancy examination in most beef operations is appropriate at weaning time.

In dairy operations, the main objective is to get a cow bred as soon as possible after calving to produce a calf and to initiate lactation. If a cow is not pregnant 35 to 45 days after breeding, the cow should be treated and rebred as soon as possible to maximize herd milk production.

The pregnancy examination involves a rectal palpation of the reproductive tract for signs of pregnancy. These signs may include enlargement or displacement of the uterine horn, uterine artery changes in early pregnancy, and presence of the fetus and cotyledons in mid-to-late pregnancy. The procedure requires a delicate feel and a knowledge of anatomy. It is not a technique that can be performed by the average producer without some training.

Many beef and dairy operations use a veterinarian for their pregnancy checks. The pregnancy examination is part of the normal herd health program.

CAUTION: Do not attempt palpation unless a trained and experienced palpator is present to assist and talk you through the procedures. Unless you know what you are doing, you may cause the cow to abort.

Equipment Necessary

- Working chute or squeeze chute
- Bar or pipe (5 feet long)
- Disposable plastic glove (shoulder length)
- Rubber band or alligator clip
- Lubricant (liquid soap or mineral oil)
- Record sheet

Restraint Required

Most cows, regardless of temperament, should be placed into a working chute or a squeeze chute. The chute should permit the animal to stand in a normal way on solid footing. Place a bar or pipe behind the cow just above the hocks to eliminate the animal's kicking and to protect the palpator during examination.

The chute should be designed to include a small gate at the rear of the animal to allow entrance and exit for the palpator. Provide a palpation cage or a gate that will swing across the working chute in front of other animals coming behind the palpator.

Step-by-Step Procedure

1. The palpator must become familiar with the anatomy of the abdominal organs and reproductive tract. The uterus is supported by the broad ligament

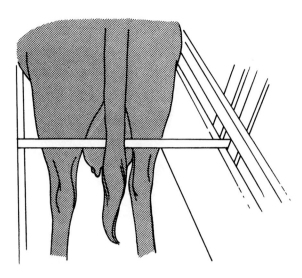

and may be forced to the right, forward, and down as it enlarges during pregnancy.

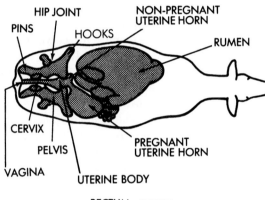

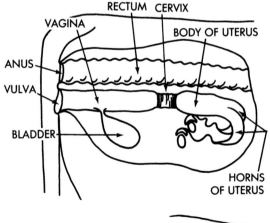

2. Restrain the cow.

CAUTION: If a headgate is used, the bar behind the cow is still needed to protect the palpator from being kicked.

3. Put on the disposable glove; either hand may be used. A rubber band or alligator clip can be placed at the top of the glove to keep the glove from sliding down.

4. Lubricate the glove with liquid soap, mineral oil, or other suitable lubricant. Lubricate the hand area of the glove carefully, because the hand enters the anus first.

5. Stand perpendicular to the cow and grasp the tail with the ungloved hand. With the gloved hand, wipe lubricant across the anus. See steps 19 to 25 in Section 3.8 for a description of the next several steps.

CAUTION: If this step is done suddenly, the cow may kick. Standing sideways (perpendicular to the cow) can minimize the danger.

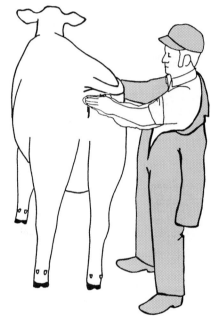

6. Form the gloved hand into a wedge and insert it with a thrust into the anus as far as possible. Fold the fingers into a loose fist as the hand is inserted. This will give a blunt surface instead of pointed fingers, and will straighten the rectum without damage. The cow will exert a fair amount of pressure against the hand and arm as she tries to expel them.

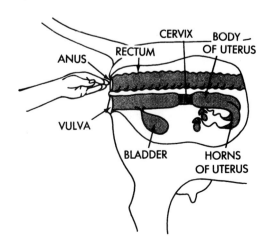

7. Remove feces from the rectum if they seem to be a problem. It is often possible to "submarine" under the feces.

8. Keep fingers close together and gently work forward and side to side in the tract without digging with the fingers. If rectal contractions are encountered, it may be necessary to relax contracting rectal rings by inserting two fingers through the center of the ring and gently massaging back and forth.

> **CAUTION:** Moving the hand into the rectum gradually can irritate the rectal lining and cause the cow to increase rectal contractions. Once the hand is inside the rectum, the cow will usually arch her back, and danger from kicking is usually minimized.

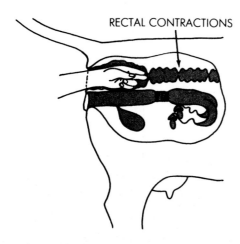

RECTAL CONTRACTIONS

> **CAUTION:** Some bleeding may occur during the palpation process. This results from scraping the mucosa and rupturing small blood vessels at the surface of the rectal wall. Extreme care should be taken not to rupture the rectal wall. If the rectal wall begins to feel rough like sandpaper, it means that the mucosal lining has been rubbed off and the animal should be released and checked at a later time. The rectal wall is fairly thick, and it is unlikely that a rupture will result if the tract is carefully palpated. If a rupture does occur, it usually leads to peritonitis and the death of the animal. When a rupture is identified, it is best to send the cow to slaughter as soon as possible.

9. Feel for landmarks. The forward (anterior) portion of the pelvis (pelvic rim) is the best landmark because it is a bone and does not move. The pelvis forms a cradle for the reproductive tract. In the nonpregnant cow and in early stages of pregnancy, the reproductive tract is usually located within the pelvic cradle and easily felt with downward pressure. As pregnancy advances, the uterus and cervix move over the anterior portion of the pelvis and into the body cavity.

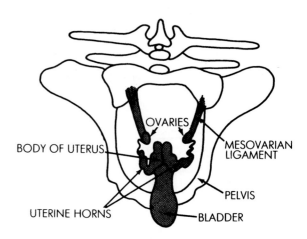

BODY OF UTERUS

OVARIES

MESOVARIAN LIGAMENT

PELVIS

UTERINE HORNS

BLADDER

The cervix is the second landmark. The cervix is approximately 1 inch in diameter, 4 inches long, and feels like gristle, resembling a turkey neck. Locating the cervix is not as easy as locating the pelvic rim because the cervix does not remain in one area and moves with advancing pregnancy. In the open cow and early stages of pregnancy, the cervix is on the floor of the pelvis. As pregnancy advances, the cervix drops over the pelvic rim as the uterus pulls down.

10. Locate the uterus. After the cervix is located, move your hand forward to the body of the uterus. The uterus is soft and pliable in comparison to the gristlelike cervix. The horns of the uterus are normally coiled at the front edge of the pelvic rim in the nonpregnant heifer and in early stages of pregnancy. In the older cow, the uterine horns may hang over the anterior portion of the pelvis into the abdominal cavity.

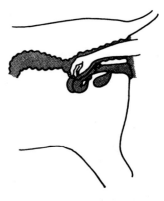

11. Use your middle finger to separate the uterine horns. The junction of the two uterine horns (bifurcation) can be located fairly easily by separating the two horns with slight downward pressure of the middle finger.

> **CAUTION:** Do not try to force the middle finger downward. Although the rectal wall is fairly thick, it can be ruptured.

12. *Determining if the cow is open (nonpregnant).* Remember that the reproductive tract of the heifer is usually located within the pelvic cradle. The reproductive tract of the older cow may have slipped over the anterior portion of the pelvis. After the horns of the uterus are separated, the thumb and index finger can be slipped around one horn of the uterus. Gentle milking of that horn should indicate the presence or absence of fluid or membranes within the uterus proper. If no bulge, fluid, or membranes are present, the cow is either open or in a very early stage of pregnancy.

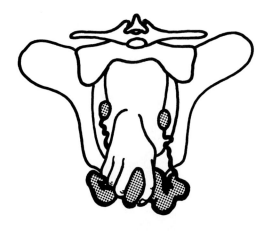

CAUTION: Do not squeeze or pinch the membranes of the uterus or uterine horns. Doing so can cause the cow to abort.

13. *Determining pregnancy at 1 month.* An experienced palpator can detect pregnancy in a beef female as early as 30 days after breeding. If palpation is to be done at this time, good records of the cow's breeding date are helpful.

The cervix and the uterus in a pregnant cow are located in about the same positions as in an open cow. The uterus will feel slightly thinner in a pregnant cow than in an open cow, and some fluid will fill it. When the horns of the uterus are palpated, one horn will feel slightly larger than the other.

CAUTION: To the inexperienced palpator, an infected uterus may feel the same as a pregnant uterus at 1 month to 6 weeks of pregnancy.

The sack of fetal fluids can be felt at 30 days by gently milking the horn as in step 12. The fluid-filled sack is about the size of a small cherry tomato (¾ inch) and is firm to the touch. The embryo in-

side the sack is approximately ½ inch long, but not yet palpable.

CAUTION: An inexperienced palpator should not try to palpate the fluid-filled fetal sack because rupture of these membranes means death for the embryo. The experienced palpator sometimes will gently allow the membranes of the fetal sack to slip between his fingers as a method of diagnosis. A corpus luteum will be present on the pregnant horn side; however, the corpus luteum of a cycling cow and a pregnant cow are indistinguishable. Palpation of the ovary is of little value.

14. *Determining pregnancy at 45 days.* Most experienced palpators prefer to palpate cows a minimum of 45 days after termination of the breeding season.

The locations of the cervix and uterus are similar to those of the open cow, but the pregnant horn is noticeably larger than the nonpregnant horn and appears thinner-walled. This difference can be felt by cupping the hand under the uterus.

The fluid-filled sack surrounding the fetus has enlarged to the point that it is of the size and shape of a pullet egg (1 to 1½ inches). The fetus at this point is about 1 inch long, but not yet palpable. The fetal sack contains considerable fluid and may be felt through the uterine wall.

CAUTION: Membrane attachment has just taken place (38 to 40 days), and the fetus should not be moved around in the uterus because the "buttons" can be detached, causing the death of the fetus.

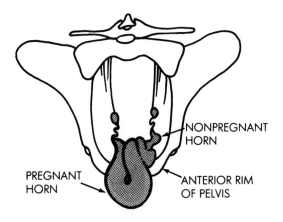

15. *Determining pregnancy at 2 months.* At this point, the cervix will probably be near the anterior rim of the pelvis. The uterus begins to migrate forward over the anterior portion of the pelvis as its contents increase in weight. The pregnant horn has enlarged to the point that it feels somewhat like a large banana. The nonpregnant horn is much shorter and smaller. By cupping the hand around the uterus, the mass and weight can easily be felt without feeling the fragile fetus. The fetus is now about 2½ inches long. The fetal sack filled with fluid can be felt without actually feeling the fetus. Presence of the fetus can be detected by gently tapping the uterus, causing the fetus to swing like a pendulum and bump against the uterine wall.

16. *Determining pregnancy at 3 months.* This is the first time that the inexperienced palpator has a chance to be relatively sure of a diagnosis. The cervix and uterus are located over the anterior rim of the pelvis in the abdominal cavity. They are sometimes difficult to feel, and the palpator should reach down because the uterus lies deeper in the abdominal cavity than at earlier stages of pregnancy. The pregnant horn is considerably larger than the nonpregnant horn. In most cases at this stage, the fetus can be palpated. The fetus is about 6 inches long; the head is about the size of a golf ball. The fetus and surrounding fluids feel like an average head of lettuce.

At 90 days of pregnancy, the cotyledons or "buttons" and uterine artery can also be palpated. The cotyledons are soft and difficult to palpate, even though they are about 1 inch in diameter. The uterine artery enlarges as pregnancy advances. At

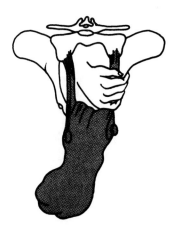

90 days the artery is about ⅛ inch in diameter and is located by inserting the hand past the anterior rim of the pelvis, cupping the hand to one side, and pulling the hand back to entrap the broad ligament against the front edge of the pelvis. By using the index finger and thumb, the front edge of the broad ligament can be palpated and the uterine artery located. As the blood passes through the artery, a buzzing sensation can be felt. It is possible to palpate mistakenly the femoral artery that lies to the side of the uterine artery. The femoral artery, how-

ever, does not have the buzzing sensation and does not move when palpated.

17. *Determining pregnancy at 4 months.* The structures palpated at 90 days can be palpated at 120 days; the main difference is size. The uterus is usually located far down in the abdominal cavity, and the palpator must reach down deeply. This displacement of the uterus is a key to determining pregnancy. The fetus at this time is about 10 inches long and has a head about the size of a tennis ball. The cotyledons are about 1½ inches in diameter, very firm, and fairly easy to palpate in the body of the uterus. Another key to determining pregnancy is the size of the uterine artery; at this stage it is about ¼ inch in diameter.

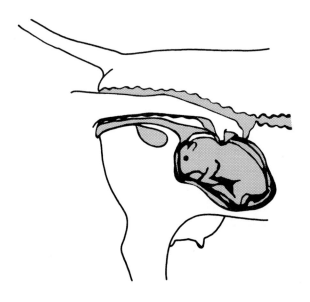

18. *Determining pregnancy at 5 months.* The major change that occurs during the last 4 to 5 months of gestation is simply an increase in size. Palpation of the fetus becomes easier as pregnancy advances.

Postprocedural Management

Beef. Cows diagnosed as open should be recorded in a record book and sold (culled) as soon as possible. If the cow is kept and rebred with the herd during the next breeding season, it will be approximately 24 months before she will wean a calf, assuming that there is one 60-day calving season per year. With two 60-day calving seasons (spring and fall), it will be approximately 18 months before the cow weans a calf.

Dairy. In dairy operations, high-producing cows not diagnosed as pregnant should be examined by a veterinarian for uterine infection. After treatment, the cow should be rebred as soon as possible. If the cow does not conceive after treatment, consider culling her after lactation ceases.

3.11 NORMAL CALVING

The most important and critical time in the life of a calf is calving time. A little extra attention at this time will pay big dividends. More than 6% of all beef and dairy calves born in American cow herds die at or soon after birth. Most of the calves lost at birth are anatomically normal but die of injury or suffocation resulting from difficult or prolonged parturition (dystocia). Calving difficulty is more prevalent in beef herds when large European breeds of bulls are mated to small breeds of cows. In both beef and dairy herds, bulls should be selected for calving ease. The factors affecting calving difficulty fall into three main categories—calf effects (heavy birth weights), dam effects (age and pelvic area), and fetal position.

Normal calving can be divided into three stages—preparatory, fetal expulsion (delivery), and expulsion of the placenta or afterbirth (cleaning). A general understanding of the calving process is important to proper calving assistance.

Equipment Necessary

- Clean, lighted maternity pen (approximately 12' × 12') or clean-sod calving pasture
- Flashlight
- Record book

Restraint Required

Parturition in most cows occurs while the cow is lying on her side. If the cow appears to be having difficulty while she is standing or gets up when assistance is given, it is desirable to move the cow carefully to a clean, dry, well-lighted maternity pen if she is not already there. If restraint is required, the cow can be approached, haltered, and tied securely with a quick-release knot to a gate or manger so that assistance can be given. The cow can be placed into a maternity chute when one is available.

Last Trimester of Gestation

1. In the last trimester (3-month period) of gestation, the cow shows some distension of the lower right abdomen as the calf increases in size and lies on the floor of the abdominal cavity.

2. Up to several weeks before calving, some cows have increased distension of the udder. This occurs because the cow is initiating the process of milk production. The fluid from the udder changes from a watery substance several weeks before parturition to a thick, milky colostrum at the time of calving. This milk should not be removed from the udder because opening the teat ends at this time can allow bacteria into the teat canal and cause mastitis. Distension of the abdomen and the presence of colostrum are general considerations and not good indicators of when a cow is going to calve. Some cows may not readily display these two signs before calving.

3. Place the cows due to calve into a well-lighted, clean, dry maternity pen or a small, clean-sod pasture so that cows can be observed frequently (every several hours) for signs of calving. The heavy-springing cows (cows about to calve) should be moved to these areas several days before anticipated calving. At this time breeding dates and records can be very useful.

4. Several days before calving, the ligaments around the tailhead and in the pelvic area relax and sink to increase the size of the birth canal. In addition, the vulva becomes swollen and begins to sag as fluid accumulates in the tissue.

Stage 1—Preparatory

This stage includes a time span of from 2 to 6 hours before the expulsion of the fetus. The normal sequence of events in the process is described next; the steps include management considerations where appropriate.

1. Just before the onset of labor, the calf is rotated into a position of least resistance.

2. Within a few hours of calving, most cows become nervous and uneasy. As contractions increase, cows will often separate from the rest of the herd.

3. Labor begins shortly before calving. The cervix begins to dilate and rhythmic contractions of the uterus begin. Initially, contractions occur at approximately 15-minute intervals. As labor progresses, the contractions become more frequent and more intense. These contractions, which resemble swallowing, begin at the back of the uterine horn and continue toward the cervix, tending to force the fetus outward.

> **CAUTION:** Any unusual disturbance or stress during this period, such as excitement or movement, may delay calving by inhibiting contractions. A good rule of thumb during this period is to be nearby but not in sight.

4. At the end of the preparatory stage, the cervix expands to create a continuous canal from the uterus to the vagina. A portion of the fetal membrane (water sac) is forced into the cervical canal and pelvis. The pressure caused by this force ruptures the membranes. Fluid normally escapes to lubricate the birth canal. This portion of the water sac usually hangs from the vulva and can be seen easily.

Stage 2—Delivery

This stage begins when the fetus enters the birth canal. The cow usually will lie down shortly before or during this stage.

1. Uterine contractions increase in frequency (about 2 minutes apart), duration, and intensity once the water sac is broken. They are accompanied by voluntary contractions of the diaphragm and abdominal muscles. Several layers of membrane line the birth canal for protection; fluid lubricates it.

2. In a normal delivery, the calf's forelegs and head, surrounded by membranes, are forced through the canal and protrude from the vulva. Steps 1 and 2 occur in approximately 15 to 30 minutes. If delivery does not appear normal, assistance should be given.

3. Before hurrying to summon professional assistance, the careful attendant can help hasten a hard, but not abnormal delivery, by helping the cow to dilate her vulva. Even if the assisted dilation does not result in the cow or heifer being able to deliver unassisted, the dilation will help with the pull that will be necessary.

If the front feet (pads down) and head are visible, lubricate your gloved hands, then place your hands, one on each side of the feet and nose of the calf, and work your hands and arms between the calf and

the vulva of the cow for 5 or more minutes. This will actually expand the vulvar tissue and make the delivery easier. If the feet and nose are not visible, use your arms and hands to expand and help dilate the vulva for the eventual passing of the calf. If other than the front feet (pads down) and nose are present in the canal, proceed to Section 3.12, which addresses the abnormal deliveries you may be encountering.

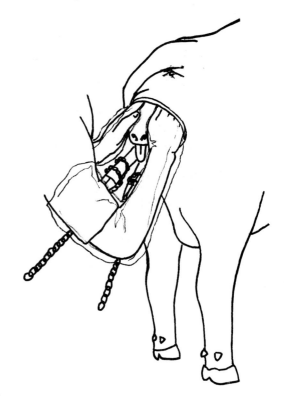

> **CAUTION:** Do not use soap as part of your lubricating products. It will indeed "lubricate," but it also dissolves and removes the cow's or heifer's natural birth canal lubricants. This situation will actually make the following delivery more difficult. Use a commercial product designed for obstetrics. This is the harvest time for your operation, not the time to save a few pennies.

4. With the calf's head exposed, the dam exerts maximum straining to push the shoulders and chest of the calf through the pelvic girdle. After the shoulders have passed the pelvis, the abdominal muscles of the calf relax and its hips and hind legs extend back to permit easier passage of the hip region.

5. Even though this is a normal delivery, if you are nearby and the calf appears to be mildly "hung-up" on its hips (chest and rib cage exposed), step to the cow and rotate the calf's hips 45 degrees. This will often enable the cow to deliver the calf on its own.

In fact, after this rotation of the hips, the weight of the calf and membranes, plus the effects of gravity, will often pull the calf from the birth canal.

6. The delivery stage is normally completed in less than 1 hour. If delivery is normal, proceed to stage 3. If delivery is not completed in 1½ to 2 hours after the water sac ruptures, if the cow is straining unproductively, or if any combination of calf part is visible except two feet (pads down) and the muzzle, special assistance is needed.

7. The calf is normally born free of fetal membranes because these remain attached by cotyledons or "buttons" to the uterus until after the calf is born. This membrane attachment during the calving process provides an oxygen supply for the calf during birth. When the umbilical cord of the calf passes through the pelvis and vulva, it usually breaks, and the lungs become functional.

CAUTION: If the umbilical cord does not break and the uterine cotyledons release from the placental caruncles, the calf may be born while inside the placenta. If this happens, the calf will drown in its own placental fluid. If membranes are covering the nose, remove them immediately and wipe the nostrils clean of mucus.

Check immediately to see if the calf begins breathing. If breathing does not start within 10 to 60 seconds after expulsion, there are several ways to initiate it: (1) *Tickle the calf's nose* and inside the nostril with a piece of straw. This can help cause a reflex action and start the calf breathing. (2) *The calf can be rubbed vigorously* with an old gunnysack or coarse towel. (3) It may be necessary to "slap" the sides of the calf with the flat part of the hand. (4) A 6-inch piece of ¾-inch *garden hose can be inserted 2 inches into one nostril* of the calf. Hold the mouth and nostrils shut so that air enters and leaves only through the hose. Alternate blowing into the hose and allowing the calf to exhale. Repeat blowing at 5- to 7-second intervals until the calf begins to breathe. (5) The calf can be supplied with *supplemental oxygen.* Commercial respirators are available and may be a wise investment for large herds. If a commercial respirator is not available, the oxygen tank from an oxyacetylene welding outfit can be adapted for use during the calving season. Attach a hose of desired length to the oxygen tank regulator. The other end of the hose can be attached to a 6-inch-diameter plastic butter dish. The butter dish *should not* clamp tightly over the calf's nose, because too much pure oxygen can be harmful to the calf. The calf should receive approximately a 50:50 ratio of atmospheric air to oxygen.

Stage 3—Cleaning

This stage consists of expulsion of the placenta. This is accomplished by continued uterine contractions.

1. The cow normally will lick the calf immediately after birth and remove any remaining membranes covering it. The licking motion of the cow serves to massage and dry the calf as well as to give the calf security.

2. Treat the calf's umbilical (navel) cord with a strong tincture of iodine (7% solution) as soon as possible after birth to prevent navel infection.

3. The calf generally will try to stand within the first 30 minutes after calving and start searching for the cow's udder. The legs are usually very wobbly at first, but they become more stable in the next several hours.

4. The calf should receive colostrum (cow's first milk) as soon as possible after birth to provide a mild laxative action and protective antibodies that protect the calf from respiratory and gastrointestinal infections early in life.

5. The placenta (afterbirth) is usually expelled by continued uterine contractions during the next 2 to 12 hours.

6. The placenta should be removed from the calving area immediately after expulsion.

CAUTION: Occasionally the placenta fails to separate from the uterus and remains after 12 hours. Retained placentas are more frequent in cows that have a premature calf, twins, induced parturition, vitamin A deficiency, or uterine infection. If any of these conditions occurs, contact a veterinarian.

Postprocedural Management

Beef. The cow and her calf should be left alone as soon as possible after birth. This allows the cow to lick the calf and claim it as her own. Ideally, it is best to leave the cow and calf in the calving pasture or maternity pen for at least 1 to 3 days after calving. After that time they can be moved back with the rest of the herd. Sunshine, exercise, and a warm, dry place to lie down are important to a healthy, rapid-growing calf.

Dairy. Observe the cow for signs of milk fever frequently (every 1 to 2 hours) for 12 hours after calving. This condition may occur from several hours before calving to 12 or more hours after calving. The condition is serious, and it must be treated immediately or the cow will die.

Normal Recovery Sequence

The period from parturition to conception is a very important time in the life of a cow. During the first several days after calving, the cow may show some swelling around the vulva. This is normal, and a noticeable reduction in the swelling should be evident by the end of the first week after calving. During the next several weeks, the uterus of the cow will recover from the pregnancy and prepare for the next pregnancy. On the average, cows will return to estrus 40 to 60 days after calving. When a cow has not shown signs of estrous activity 60 days postpartum, the producer should be concerned, and professional help contacted. Cows and heifers should be bred on the second, not the first estrous period. This is usually a normal cycle; estrus with conception is acceptable. If the cow is to calve at 12-month intervals, she must conceive within 82 days of parturition.

3.12 ABNORMAL (ASSISTED) CALVING

This technique deals with the problems that may arise during the process of calving. Unless cervical dilation is completed, assistance can be given too early. This can result in injury to the cow, the calf, or both. Final dilation is, however, quite rapid, and assistance can be given too late. Timing is important to providing proper assistance, and frequent observations are a must. Calving assistance should not be attempted without proper preparation of facilities and equipment.

Equipment Necessary

- Clean, lighted maternity pen
- Obstetrical chains and two handles
- Calf-puller
- Disinfectant
- Pail and scrub brush
- Lubricant (mineral oil or obstetrical lubricant)
- Flashlight
- Rope halter
- Paper towels
- Boots
- Coveralls
- Plastic gloves

Restraint Required

Parturition in most cows occurs while the cow is lying on her side. When assistance is given during the calving process, the cow should be restrained in a maternity chute or maternity pen. If the cow is restrained in a maternity pen, she should be haltered and tied with a quick-release knot. When the cow is out in the pasture and can't be moved, try to approach, catch, and halter the cow and tie her to the bumper of a pickup, tree, fence, or post. This last method is the most dangerous and should be used only as a last resort.

Examination

An examination should be performed just as soon as you determine that the delivery is going to be abnormal. Following are some hallmarks of an abnormal delivery:

- You see something other than the two front feet (pads down) followed closely by the nose of the calf, coming from the birth canal.
- You have observed the water sac protruding or hanging from the vulva of the cow for a period of 2 or more hours.
- The cow has strained *constantly* (no breaks) for 30 to 45 minutes with no progress.
- The cow quits working at the delivery for more than 20 minutes, after the delivery has truly started.

These time estimates are not meant to be precise. They are good guidelines and if followed will allow time for the cow or heifer to do what nature equipped them to do (if it is possible). Also, if the birth is going to require assistance, you have not waited so long that the mother and offspring are exhausted and likely beyond the skills of even the best professional help.

1. The calving attendant must be familiar with the normal calving process and calf presentation of single and twin calves.

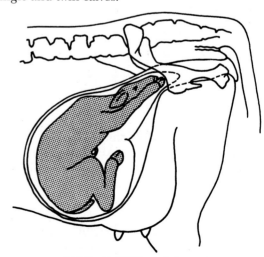

NORMAL PRESENTATION

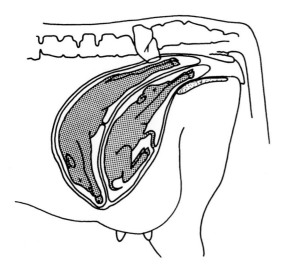

NORMAL PRESENTATION

2. Restrain the cow.

3. Wipe all fecal material from around the anus and vulva, using paper towels soaked in a pail of disinfectant.

4. Place a disposable sleeve or glove over the hand and arm.

5. Lubricate the glove liberally with obstetrical lubricant or mineral oil. Try to keep the disposable glove as clean as possible.

6. Insert the hand into the vagina while holding the tail to the side with the ungloved hand. Form a cone with the fingers and thumb to make insertion easier and to cause less damage.

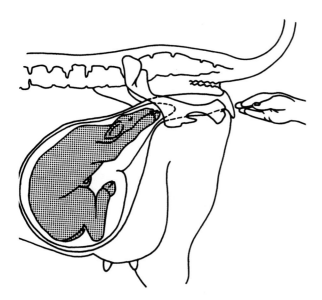

Steps 7 to 12 should be conducted with the gloved hand to minimize infection in the uterus and to prevent transmission of disease from the cow to the palpator.

> **CAUTION:** Proceed in the reproductive tract gently so that ruptures do not occur and cause peritonitis.

7. Determine the extent of cervical dilation by palpation. If the hand cannot pass through the cervix easily, the hormones and muscular contractions involved with parturition are not responding normally, and a veterinarian should be called immediately for assistance.

8. Determine the size of the calf relative to the birth canal. The forcing of a large calf through a small pelvic opening may result in the death of the calf and injury or paralysis of the cow. If this examination can be done before the feet and head are in the pelvis, the opportunity for a successful cesarean section still exists. If it appears that the cervix is too small, seek professional help immediately.

9. Determine whether the calf is alive or dead. Either way, the calf must be removed as soon as possible. If the calf is alive, it will respond, by moving, to pinching between the hooves, sticking a finger in the nostril or eye, or pinching the anus if the calf is in a backward presentation.

10. Examine the birth canal and fetus. If they are dry, you must provide additional lubrication. A commercial obstetrical lubricant, Ivory soap dissolved in warm water, and mineral oil make suitable lubricants.

> **CAUTION:** The lubricant should be kept clean. If it becomes contaminated, the chances of uterine infection are increased.

11. Determine the position of the fetus by feeling the various parts of the calf. The orientation of the calf can easily be determined. When the calf is presented in the normal position, the hoof pads (bottom of the feet) are facing downward and the head (nose and nostrils) can be felt between the front legs. The hoof pads face in the same direction that the pastern bends when it is manually flexed. For example, if the pastern bends at an angle in the downward position, then the hoof pads face down and the calf is in the upright position. The opposite is true when the calf is upside down. Another way to determine the position of the calf is to feel for dewclaws. The dewclaws face in the same direction as the hoof pads. If the dewclaws are palpated in the downward position, then the hoof pads face downward, and vice versa.

When the calf is palpated and no head is felt, then the calf has one of several abnormal presentations. To determine if the calf is backward (rear end first), feel for the tail. Palpation of the tail is a sure indication that the calf is backward. If no tail is felt, or only one foot is felt, some searching must be conducted to identify where the head and the other foot are located.

12. Determining how far the labor has progressed will indicate what needs to be done first when the calf is abnormally presented. If the calf has not entered the birth canal, proceed to the following description that best describes the situation. If the calf's feet and head are exposed, there is no chance of returning the calf to the uterus or performing a cesarean section. Assistance should be given immediately.

If the calf has entered the birth canal, but is not exposed, it can be pushed back manually into the uterus. This is very difficult to do because of the counteracting force of the labor contractions and because of space limitations. Pressure should be applied to the fetus between contractions of the cow. During contractions, apply enough preasure to maintain any progress that was made in getting the fetus back into the uterus. This procedure must be repeated until the calf is back into the uterus and out of the pelvic girdle. It may be necessary to rotate the calf's head slightly if progress is not made. Drugs or hormones can sometimes be given to cause the cow to reduce the intensity of contractions and allow the corrections to be made. Once the calf is back in the uterus, apply the corrective measures appropriate to the problem.

> **CAUTION:** If the fetus is abnormally presented and cannot be pushed back into the uterus, call a veterinarian immediately. The loss of both the cow and the calf at this point is a definite possibility without professional help.

Abnormal Presentations

1. *Head first with one or both legs bent backward.* In this case, the calf's head is easily palpated, but the front legs are bent at the knee with the hooves tucked close to the body.

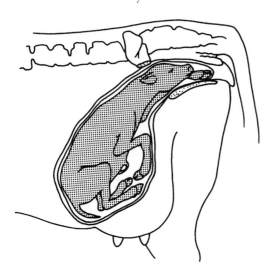

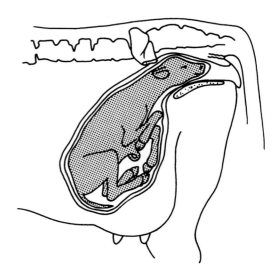

Corrective Measures. Push the calf back into the uterus. Grasp the calf's hoof if it can be reached and pull upward and toward the birth canal. If the hoof cannot be reached, grasp the cannon bone of the leg near the pastern and pull toward the birth canal. Gently position the front legs and the head in the birth canal. When moving the feet into position, be careful that the hooves or your fingers do not lacerate or puncture the uterus. Chains should be applied to each foot as it is lifted into the birth canal to prevent the feet from slipping back down in the uterus. Proceed to "Assisting the Difficult Delivery" (p. 150).

2. *Head and foot first with one leg crossed over the neck.*

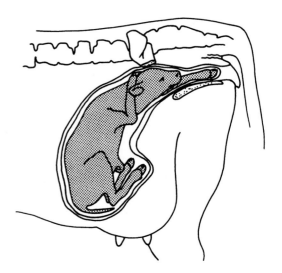

Corrective Measures. Push the calf back into the uterus. Grasp the leg over the neck by the cannon bone above the pastern. Raise the leg over the head and pull it into the birth canal. Chains should be applied to the legs. Proceed to "Assisting the Difficult Delivery" (p. 150).

3. *Front feet first with the head twisted upward and backward.* The head is palpated along one side of the calf.

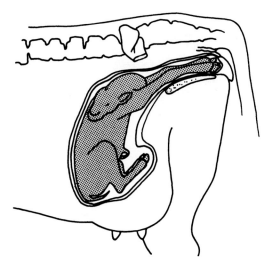

Corrective Measures. This presentation is often corrected when the fetus is pushed back into the uterus. If the head position is not corrected, the calf's head must be turned. Grasp the nose of the calf and pull the head to the side, downward, and then toward the birth canal. Care should be taken that the calf's neck is not broken. Proceed to "Assisting the Difficult Delivery" (p. 150).

4. *Front feet first with the head turned down between the front legs.* The head is palpated between and below the front legs. The head is unable to move to the proper position without manual assistance because the legs act as a clamp around the neck.

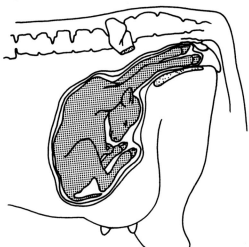

Corrective Measures. Apply calf chains to the front legs so that they can be easily pulled back into the birth canal later. Push the calf back into the uterus. Raise one leg up and to the side to allow room to move the head. Grasp the head of the calf at the muzzle and pull upward and toward the birth canal. Proceed to "Assisting the Difficult Delivery" (p. 150).

5. *Front feet first, but the calf is upside down.* The calf's dewclaws can be palpated on top of the leg, and the eyes can be felt near the floor of the pelvis instead of near the top.

Corrective Measures. Push the calf back into the uterus. Rotate the fetus 180° to the normal upright position. This will be very difficult to do, but be patient and keep trying. If one or both legs are bent back after rotation, see step 1 and then proceed to "Assisting the Difficult Delivery" (p. 150).

> **CAUTION:** A fetus may be upside down because of a rotation of either the fetus or the uterus. If it is a twisted uterus, a veterinarian's help is needed.

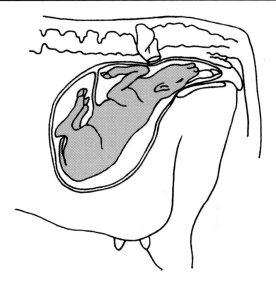

6. *Backward presentation with rear feet first.* Both the tail and the feet can be palpated.

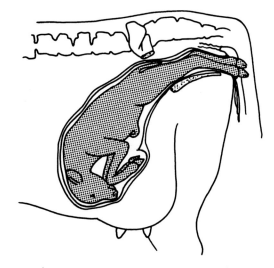

Corrective Measures. Delivery is usually uncomplicated, but the calf may suffocate during prolonged labor. The reason for this is that the umbilical cord will usually rupture before the calf's nose is exteriorized and the calf is able to breathe. In this case, it is best to apply calf chains and pull the calf. Proceed to "Assisting the Difficult Delivery."

7. *Breech presentation—calf backward with rear legs tucked under the body.* The calf's tail can easily be palpated, but the rear legs are positioned near the uterine floor.

Corrective Measures. Push the fetus forward. Grasp the calf's hooves if they can be reached and pull upward and toward the birth canal. If the hooves cannot be reached, grasp the legs at the cannon bones just below the hock and pull upward and toward the birth canal. Whenever possible, grasp the hooves so that the pastern will bend and your hand will protect the uterus from lacerations and punctures from the hooves. Proceed to "Assisting the Difficult Delivery."

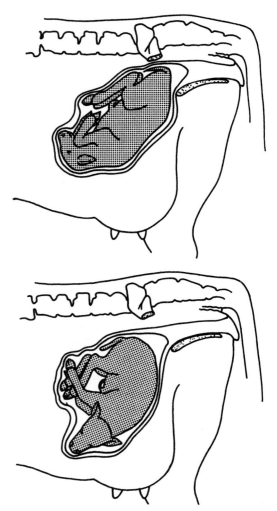

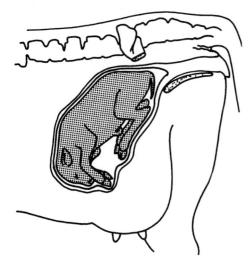

8. *Calf upside down and backward with rear legs tucked up to the body.* The tail can easily be palpated, but the hocks are located near the spinal column of the cow instead of near the uterine floor.

Corrective Measures. Push the fetus forward. The calf must be rotated 180° to the upright position. Be patient and persistent, because this will be a difficult task. After the calf is rotated, it is still backward, and the feet are most likely still tucked close to the body. Refer to step 7 to properly locate the rear legs in the birth canal. Proceed to "Assisting the Difficult Delivery."

> **CAUTION:** A fetus may be upside down because of a rotation of either the fetus or the uterus. If it is a twisted uterus, a veterinarian's help is needed.

9. *Back of the calf is presented first.* The head, tail, feet, or hocks are not easily palpated. The calf's head can be either in the down position or in the upright position.

Corrective Measures. The fetus must be rotated so that either the head and front legs or else the rear legs can be started through the birth canal. Whenever possible, try to position the head and front legs first. Grasp the neck, head, or front legs (in that order of preference) and try to pull it toward the birth canal. If this is unsuccessful, grasp the tail, hock, or rear leg (in that order of preference) and try to pull it toward the birth canal. It is likely that the calf will still remain in an abnormal position when one end or the other is presented. Refer to those positions of steps 1 to 8 that best describe the new situation.

> **CAUTION:** This cow is a good candidate for a cesarean section. Call for veterinary help.

Assisting the Difficult Delivery with Obstetrical Chains

The following steps are presented in two parts when appropriate. The term *frontward* indicates how to proceed if the calf is positioned in the birth canal with front feet and head first. The term *backward* is used to indicate the procedure to follow if the calf is presented with the rear legs first. Those steps that do not indicate calf position apply to both presentations.

> **CAUTION:** All backward presentations should be considered emergencies because the umbilical cord is cramped between the fetus and pelvis early in delivery. This means that blood and nutrient flow to the fetus is slowed, and the fetus will die or sustain brain damage unless delivery is rapid once it has entered the birth canal.

1. Make a loop in the obstetrical chain by passing the chain through the oblong ring at the end of the calving chain.

2. Slip this loop over the gloved and well-lubricated hand. This will allow for easy application and maneuverability in the birth canal or uterus.

3. Attach the loop of the chain to one leg of the calf and slide it up on the cannon bone 2 to 3 inches above the dewclaws. It may be necessary to maintain a slight tension on the chain so that it does not slip off the leg.

It is a good idea to attach the chains to any visible leg or legs, just as soon as you seen them. They represent progress made and you do not want to "lose" them while you are examining the malpresentation.

It may be your decision, after the examination, that the calf has to be pushed back into the uterus and the entire process restarted; but release the chains after you have made that decision. The malpresentation may be something as simple as a single leg bent backward. You'll be glad you have the one leg secure as you push the calf back into the uterus to clear room for the other leg.

4. Half-hitch the chain between the dewclaw and hoof head. The half hitch can be made on your hand outside of the cow and then applied to the leg. The half hitch in a hard pull helps to distribute the stress imposed on the bones over two locations instead of one, reducing the possibility of bone fracture.

> **CAUTION:** Do not apply a single loop of the calving chain around both legs at the same time. The danger of breaking one or both legs is very high, especially if the birth is difficult and a hard pull is used.

5. Repeat steps 1 to 4 for the second leg.

6. Before applying the calving-chain handles to the calving chains, make sure that the chains will pull from the bottom of the leg (dewclaw side). This will ensure that the legs will be pulled straight and not at an angle or with a bend.

7. Attach two calving-chain handles to the chains and pull gently, making sure that the loop and half hitch of the chain have not slipped from the desired position on the calf's leg.

8. Some calves can be delivered by pulling both legs evenly; however, it is best to pull alternately on one leg and then the other a few inches at a time. When the legs are "walked out" in this manner, the shoulders or hips are allowed to pass through the pelvic girdle one at a time.

9. Once the calf's legs are exposed, the calf should be pulled downward (toward the cow's hocks) at a 45° angle.

10. One or two individuals using chains with manual strength should be able to pull a calf. If

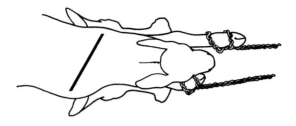

CAUTION:

CALF FORWARD. If the shoulders should happen to become lodged in the pelvis, apply traction to the calf's head. This will reduce the compaction of the head against the sacrum (top) of the birth canal and reduce the dimensions of the shoulder and chest region. The chain can be applied by making a loop as in step 1 and applying it to the nose and muzzle of the calf between the muzzle and the eyes. Care should be taken when pulling on the nose so that you do not break the nose or jaw.

CALF BACKWARD. In a breech delivery, the extraction of the fetus is against the normal direction of hair growth, and the birth canal should be liberally lubricated with an obstetrical lubricant. It may be necessary to rotate the fetus about one-eighth of a turn to take full advantage of the greatest diameter of the cow's pelvis. If delivery proves extremely difficult a cesarean section is probably necessary and should not be delayed.

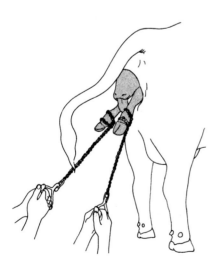

the birth is extremely difficult, however, it may be necessary to use a mechanical calf-puller. When this situation arises, it is best to seek experienced help. If a calf-puller is used incorrectly, it may cause permanent damage to both cow and calf.

CAUTION: As the calf's head and shoulders come through the birth canal, the chance of uterine or cervical lacerations is greatest. This damage may lead to infection and future reproductive problems. Because pressure dilates the cervix and birth canal, traction that is applied gradually can usually prevent damage. If assistance should happen to be given to a cow too early, the slow application of traction would not interfere with normal cervical dilation and would again minimize the potential of damaging the birth canal.

Assisting the Difficult Delivery with a Calf-Puller

CAUTION: A good rule of thumb is . . . the less you use a calf-puller, the better. However, sometimes there is no other way. For instance, if the calf is backwards, you are out at the remote calving barn 2 hours from any help, you are alone, and the calf-puller is the only way you have of pulling the calf quickly enough to save its, and maybe the cow's, life, it is a very good option for you! Follow the directions carefully. With as much power as the puller can deliver, it is easy to damage the calf and/or the cow.

Step-by-Step Procedure

1. Restrain the cow. It is cumbersome enough to use the puller when the cow's ability to jump around is limited. It is almost impossible, at times, to use the puller properly when the cow has freedom to run up and down the alley or around the calving pen. Remember to tie her about 15 to 18 inches from the ground, so she will not choke if she goes down.

2. Position the calf-puller, with the spanner against the cow, just below the vulva. The support loop or strap will be over her back, between her hooks and pins.

3. Position the obstetrical chains onto the calf's legs. Be certain they are double-looped and that one loop is above and one loop below the fetlock

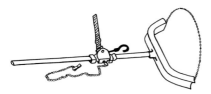

CAUTION: As the frightened cow jumps around, the long shaft or "handle" of the puller will be swinging about, and you or a helper may receive a sharp and painful blow. Be patient. Allow the cow to settle down. As the next contractions set in, they will take her mind off escaping.

joint. For the correct procedure to place the chains, see steps 1 to 8 immediately preceding.

CAUTION: The double loop of chain on the leg is essential. Do not forget it! Second, the half-hitches must be on the bottom of the fetlock joint. This will allow for a straight pull on the leg and minimize the danger of breaking the leg by twisting it.

4. Tie the ends of the two chains together, a foot or less from where they are attached to the calf's legs. This loop of chain is where you will attach the hook of the cable, which is attached to the ratchet on the calf-puller. It is very important that you use short chains coming from the calf. You will need a lot of pulling length to get the calf out. The way to ensure this is to attach the hook from the ratchet as close as possible to the calf. And be certain that all the slack cable is wound onto the ratchet. Once you start, you will have to finish the task without delay, especially if the calf is backward. You will not have a chance to re-hook the ratchet if the calf's head and thorax are stuck inside the cow and you find that you are out of "pulling room."

5. When you are ready to start pulling the calf, lift the shaft of the puller upward, to a position parallel to the ground, and take up all the slack you can with the ratchet. Then, lower the shaft *slowly* to the ground. As the spanner pushes against the cow, the cable will tighten and the calf will be pried outward. Now lift the shaft upward again, and take up all the slack again. Now lower the shaft and again pry the calf outward. Repeat this raising and lowering, or pumping motion, until the head and shoulders or hips, if the calf is backwards, are free of the cow.

6. Once the head and shoulders or the hips are free, you can stop the pumping and use the ratchet directly to pull the calf from the cow.

CAUTION: Uterine prolapse is an inversion of the uterus that can be caused by pulling the calf too rapidly. In this situation, always contact a veterinarian for recommendations and drugs needed for treatment. If the cow is not treated promptly and correctly, she may die.

Postprocedural Management

Check the cow to be sure that she can get up after calving. Occasionally a cow can become temporarily or permanently paralyzed during parturition. When this occurs, see Section 4.11 of Chapter 4 and call a veterinarian if necessary. Refer to Section 3.11 for more information on the normal recovery process.

3.13 NEONATAL CARE

One of the most critical times in the life of an animal is shortly after birth. A little extra effort and attention at this time can result in more calves available as replacements or for sale.

It is estimated that about 6% of all calves born in the United States die at or shortly after birth. A large portion of that loss is a direct result of navel infections or navel ill, respiratory infections, and scours. This section describes those management procedures that can be performed from the time the calf is born until it reaches several days to several months of age. In each case, some recommendations will be made about the time at which to perform the various techniques.

Equipment Necessary

- Piece of canvas (6' × 6')
- Scissors or pocketknife
- Strong tincture of iodine (7%)
- Navel clamp
- 2-cc syringe
- Injectable vitamins A, D, and E
- BoSe (injectable selenium)
- Nasal vaccine
- 10-cc syringe (or larger)
- Record book

Restraint Required

The best way to restrain a calf is to approach, corner, and flank it. See Chapter 1 for details of calf restraint. Calves younger than 2 months of age can be restrained easily on the ground by one or two people, or straddled and restrained against a wall or fence. When conditions are wet and muddy and the calf is restrained on the ground, it is advisable to use a piece of canvas measuring approximately 6' × 6' to keep you, the calf, and your equipment clean. This piece of canvas should be kept with your calving equipment.

CAUTION: Anytime the calf's dam is in the same pasture, lot, or pen, be careful. Most cows are very protective of their calves, especially during the first few days after birth. If the calf bawls or the cow thinks you are going to hurt the calf, she may attack and injure you with her head. If you have a choice, restrain the cow or separate the cow from the calf before performing the following techniques. If this is not possible, allow yourself an escape. This escape may be a pickup truck that you use as a barrier. In this situation, it is best to have a second person who can distract the cow and keep her away from the person performing the techniques.

Step-by-Step Procedure

1. Treat the navel as soon as possible after birth. This technique is extremely important in operations that have a history of navel infections, when environmental conditions are wet and muddy, or when cows calve in confinement. To treat the navel, obtain a squeeze bottle with a pointed nozzle and fill it with a 7% iodine solution. If the umbilical (navel) cord is extremely long, it should be cut off with a clean pocketknife or scissors to a length of 2 to 4 inches.

CAUTION: Do not jerk or pull on the navel. If the navel is pulled off the calf near the body, bacteria can enter the calf's body and cause serious infection. Under normal situations the 2- to 4-inch navel will dry, shrivel, and seal this open avenue to the calf's body in 1 to 2 days. The purpose of the iodine is to help prevent bacterial invasion before the navel is sealed.

Insert the point of the squeeze bottle into the end of the navel cord and squeeze a generous amount of iodine solution into the cord.

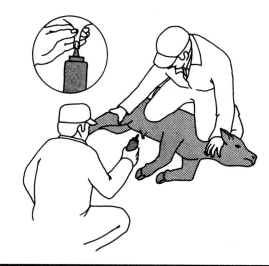

CAUTION: Be careful not to *force* iodine excessively into the cord. This will result in iodine being forced into the body cavity.

Coat the outside of the navel cord and the area surrounding the cord with iodine solution.

2. In operations where navel ill is a real problem, it may be desirable to place a plastic clamp on the navel cord after treating it with iodine so that infectious agents do not move up the cord. This can be accomplished by opening the clamp, applying it to the cord ¼ to ½ inch from the body, and snapping it shut.

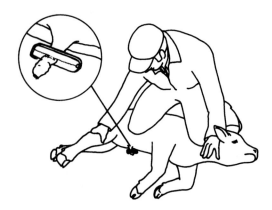

CAUTION: Timing is critical. If the cow has not licked or dried the calf before the clamp is applied, she may pull both the clamp and the navel cord off the calf. If the calf is dry, this usually is not a problem. The value of using the clamp after the calf is 1 to 2 days old is questionable.

3. A simple, slightly modified (more oval than round and more shallow than deep), aluminum soup ladle is the basis for a good alternative method to the squirt bottle for applying iodine to the navel. It provides excellent saturation of the navel area, while virtually preventing the strong-odored iodine from saturating hands and clothes. And as an added bonus, you can see what you are doing without crawling under the calf of sitting or kneeling on the ground.

4. The calf ideally should receive colostrum (cow's first milk) within 15 to 30 minutes after birth to provide a mild laxative and the antibodies that help protect the calf from respiratory and gastrointestinal infections early in life. Usually, the calf will find the cow's udder on its own or with the help of the dam within 1 hour after birth. Some calves try to nurse the brisket, behind the front leg, or other places on the cow and may not have found the udder after 3 to 4 hours. One way to tell if a calf has nursed is to look at the cow's udder. Hair around the teat will be matted and stuck together if the teat has been nursed, and straight and relatively fluffy if it

has not been nursed. The front teats are usually the first to be nursed, in which case they will be shiny.

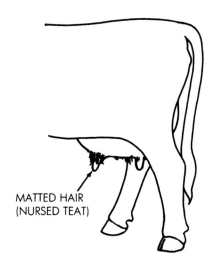

MATTED HAIR
(NURSED TEAT)

5. If the calf has not nursed within 1 to 2 hours after birth, assistance should be given. The ability of the calf to absorb antibodies from the colostrum directly through the intestine and into the bloodstream declines as time after birth increases; it is important that the calf get colostrum as soon as possible after birth for resistance to infection. Three approaches are possible.

The first approach to getting colostrum into the calf is to milk the cow and give the colostrum to the calf with a nipple bottle. This approach is probably more popular among producers that keep constant watch over calving cows, and it ensures that the calf receives colostrum immediately after birth. If the cow does not cooperate and you can't get her milked, use some frozen colostrum that you may have stored in the freezer or obtained from a dairyman. Three to four pints of colostrum will be sufficient.

The second method is to place the cow in a maternity chute, or to approach, catch, and halter the cow and tie her with a quick-release knot to a fence, wall, or manger. When a halter is used to restrain a cow, a second person should be there to hold the side of the cow against the structure to which she is tied. Move the calf toward the cow's udder. Place one knee between the rear legs of the calf to help hold the calf in the proper position. By applying pressure to the calf's head with one hand, it can be positioned near the teat. With the other hand, place the cow's teat into the calf's mouth. The calf should begin to nurse if it has the nursing instinct.

You may need to be patient and continue to place the teat into the calf's mouth several times before the calf leans what to do. If results are not satisfactory, maintain the same position and calf restraint just described and see if the calf will nurse your thumb. If the calf nurses your thumb with good

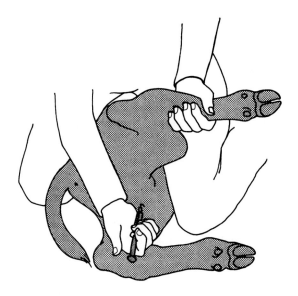

suction, it has the nursing instinct. At this point you can substitute the cow's teat for your thumb. You may have to squirt some milk from the cow's teat into the calf's mouth while holding the head in position with the other hand. Once the calf tastes the milk, it usually begins to nurse on its own. If it doesn't, keep trying.

Fresh 2-year-old heifers may have plugs in the ends of their teats. This is nature's way of helping prevent bacterial invasion of the udder through the teat canal. The plug is easily removed by applying pressure to the teat as would occur in hand milking.

The third method is esophageal feeding—or feeding with a stomach tube. It is the least desirable method of giving colostrum because it is the most dangerous to the calf, and research has shown that the absorption of antibodies is the least efficient. If this is the only method available to you, carefully follow the step-by-step procedures detailed in Section 3.14, "Stomach Tubing."

6. Within 12 to 24 hours after birth, the calf can be given an injection of vitamins A, D, and E. The vitamin shot is especially important in cases when cows may have been fed marginal or low-quality feeds without supplemental vitamin A or when calves are confined indoors without access to sunlight and therefore lack the opportunity for natural synthesis of vitamin D. The injection should be given intramuscularly with a ½- to 1-inch, 16-gauge needle.

In cold weather the vitamin solution becomes rather thick and hard to inject. You can either keep the bottle fairly warm so that it flows easily, or use a 14-gauge needle. Check the label to find out which temperatures are best for storing the vitamins.

Keep the vitamin bottle out of direct sunlight. It should be placed in a box with other calf equipment instead of on the dash of the pickup.

7. The use of injectable selenium (BoSe) depends on geographic location and must be considered in areas where white muscle disease is a potential problem. The injection is given intramuscularly in the same manner as the vitamin shot.

8. Consider immunizing the calf against respiratory infections and viral ("hot") scours within the first 12 to 24 hours after birth. This is especially necessary if there is a history of respiratory infections and scours on the premises, if the cows produce large amounts of milk, which can cause scouring and lower calf resistance to both types of infections, or if the animals are housed inside with poor ventilation. Calving in confinement during the muddy, wet season also predisposes calves to infection. This immunization may consist of an intramuscular injection or a nasal spray. Either way, the immunization material should be obtained from a veterinarian or veterinary supply house to ensure freshness and quality. Some of these immunizations, packaged in 1, 10, or 25 animal doses, must be mixed with the sterile diluent provided and administered immediately after mixing. Wait until there are enough calves born to utilize most or all of the dosage prepared. Follow the directions given on the label for dosage and administration site.

To give a nasal injection, use a disposable plastic round-tipped tube designed to fit a syringe. Fill the syringe with vaccine to the proper dosage level. Replace the needle with the plastic tube, insert the tube into the calf's nostril, and depress the plunger of the syringe. If a multidose syringe is used, be sure to set the syringe for the correct dosage as recommended by the manufacturer. Make sure the calf's head is tilted upward and not downward so that the entire dose is administered into the calf's nostril and, ultimately, the lungs.

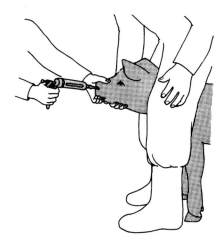

9. Other techniques can be performed at the same time as the navel is treated and shots and immunizations are given. These include tattooing, ear-tagging, castration, and dehorning. These techniques are easier to perform when the calf is less than 24 hours old than when the calf is older and stronger.

10. Calf scours can be a serious problem and a common killer of young calves several days after calving. This is especially true in herds where cows produce large amounts of milk, or the premises and herd have a history of scours. The scours can be of two basic types: viral and bacterial. Viral scours can be controlled by immunizing the beef cow before parturition or the calf for this type of scours. Bacterial scours can be caused and aggravated by a high level of milk production, overfeeding of milk replacer in bottle-fed calves, or dirty, poorly ventilated housing.

The milk or milk replacer intake should be limited to about 9% of the calf's body weight; thus, a 100-pound calf would receive about 9 pounds of milk or milk replacer. This amount is usually fed twice a day and should be uniform from feeding to feeding in the amount, concentration (if milk replacer), and interval between feedings.

In either type of scours, the calf has a gaunt appearance, severe diarrhea, a rough hair coat, and a messy tail and rear quarter. Scours must be treated in several ways. The basic principle in treating scours is to control the bacterial infection with antibiotics and to administer a source of electrolytes to counteract loss of fluid and dehydration (see Section 3.14 for the technique and electrolyte recipes). Products are also available commercially to treat calf dehydration.

11. Another calfhood disease that can cause reduced performance and possible death is pneumonia. Conditions that can cause pneumonia include overcrowding, drafts, poor ventilation, and poor sanitation. A calf with acute pneumonia usually has a raspy cough, runny nose, rapid, shallow breathing rate, and fever. Some calves can have subacute pneumonia without showing any sign of a runny nose or cough, but die suddenly after several weeks. Detection of a subacute pneumonia is difficult. These calves may have droopy ears, high fever, rapid breathing, reduced intake of milk or grain, and a slow growth rate. In cases where you suspect pneumonia, it is best to obtain a rectal temperature and use a stethoscope to check for lung congestion. Normal rectal temperature is 101.5±1°F.

> **CAUTION:** Be certain that you choose a type of housing that is either warm or cold. The type that is halfway between will usually be drafty and act as a moisture trap (so much so that at times it will literally rain inside). Drafts and the excess moisture will result in pneumonia and scours problems in the calves.

3.14 STOMACH TUBING

This job involves passing a flexible tube into the mouth, down the esophagus, and into the reticulum or rumen. Stomach tubing can be used to relieve gas pressure in bloated animals or to administer fluids containing electrolytes and nutrients to scouring-dehydrated calves, colostrum to orphaned calves, and milk to calves with mouth injuries.

Equipment Necessary

- Squeeze chute, headgate, stanchion, or halter
- Plastic or rubber stomach tube ($\frac{1}{4}$" × 2' for newborn calves or $\frac{1}{2}$" × 4' for older animals)
- Fluid (milk, electrolyte solution, colostrum, mineral oil)
- Funnel
- Stomach tube kit—commercially available

Restraint Required

The age of the animal to be stomach-tubed will determine the extent of restraint required. A newborn calf can be approached, cornered if necessary, and restrained by backing it against a fence and standing astride its neck. As the calf gets older (200 lbs or more), it will have to be moved into a squeeze chute, headgate, or stanchion, then haltered and tied, or roped with a lariat. Most mature cattle will have to be restrained in a squeeze chute, headgate, or stanchion to provide some safety to the handler. Without these facilities for older calves (400 lbs or more) and mature cattle, stomach tubing becomes dangerous and almost impossible. If restraint equipment is not available and cattle are broken to lead, it is possible to restrain the animal with a rope halter, but this is dangerous and not recommended.

Step-by-Step Procedure

1. Assemble equipment and necessary treatment fluids. The cow must be milked to obtain milk or colostrum.

2. Restrain the animal.

3. Prepare the tube by rounding its inside and outside edges. This will minimize damage to the lining of mouth, esophagus, and reticulorumen.

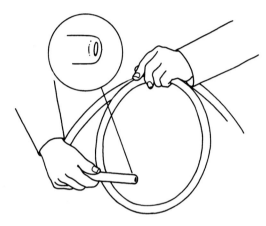

4. Open the animal's mouth. The method of restraint will dictate where you stand and how you proceed.

Newborn Calves. The calf should be in the standing position to prevent fluids from going into the lungs. Back the calf up against something solid,

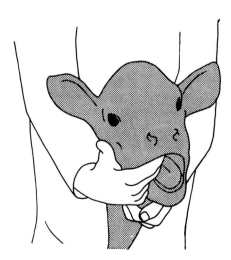

CAUTION: Calves have sharp teeth on the top and bottom in the back of the mouth. You may get small cuts and bruises on your fingers when performing steps 4 to 5.

such as a fence, wall, or tree. A second person can assist. Stand astride the calf with the back of your legs pressed against the calf's shoulders. This will prevent the calf from going forward, while the solid object behind keeps it from going backward. To open the calf's mouth, grasp its head with one hand and insert several fingers of the other hand into its mouth. Apply pressure to the calf's palate (roof of the mouth) with the fingers to open its mouth.

CAUTION: If possible, the stomach tube should be used with the calf in a standing position. Also, two people should perform the task whenever possible. Calves kick sharply, and they can move very quickly. If one person is trying to control the calf and pass the tube, a couple of problems can arise pretty quickly: you can jab the tube into the pharyngeal area (back of the throat) and injure it, or the calf will aspirate (breathe in or suck in) colostrum into the trachea and lungs when it starts jumping around. With one person holding the calf and one person managing the tube, these risks will be minimized.

Animals Restrained in a Squeeze Chute, Headgate, or Stanchion. When animals have their heads restrained in a headgate or stanchion, face the same direction as the animal and position the arm nearest the animal over and around the animal's head. Open the animal's mouth with the hand of the arm around the animal's neck. To do this, insert your thumb into the corner of the animal's mouth and apply pressure to the tongue as you squeeze the animal's jaw between your thumb and fingers.

CAUTION: When your arm is wrapped over and around the animal's head, the animal will telegraph its movements to you. The animal may throw its head up and down or sideways, so be alert. The animal may also cut and bruise your thumb with its teeth.

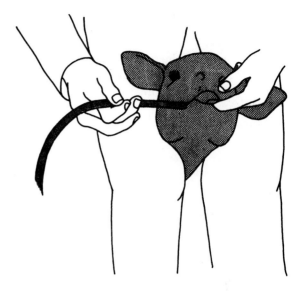

CAUTION: The haltered animal's head has considerable freedom of movement, and you must be alert when using this restraint to stomach-tube the animal. The animal may fight, and it may pin you against the fence or wall, step on your feet, or throw its head from side to side and up and down.

If the animal pulls back quickly on the lead rope while you are attempting to open its mouth, your thumb can be seriously crushed, cut, and bruised.

Mature, Haltered Animals. The approach to opening the haltered animal's mouth is much the same as opening the mouths of animals restrained in a headgate or stanchion. The biggest difference is that the animal's mouth can be clamped shut by the halter if the animal pulls back and applies tension to the lead rope. When this happens, the animal must be brought forward so that tension is relieved on the lead rope, allowing the mouth to be opened. The technique used to open the animal's mouth is the same as the one used to open the calf's mouth.

5. Pass the stomach tube inside the mouth, along the side of the tongue, and down the esophagus. Do this slowly to minimize damage to the lining of the mouth and esophagus. Do not force the tube. Take advantage of the animal's swallowing after the tube reaches the base of the tongue.

In newborn calves, about 18 inches of the tube should be inserted into the mouth and down the esophagus so that the end of the tube will enter and stay in the reticulum and not the rumen. The reason for this is that the rumen of newborn calves is not yet functional. The fluids, especially milk and colostrum, will remain in the rumen and sour before being passed to the lower tract. In older animals, the tube should be placed into the rumen. Determine if you have done this by listening at the exposed end of the tube for gurgling noises or smelling the tube for escaping rumen gases.

CAUTION: Make sure that the tube goes down the esophagus and not down the trachea (windpipe). To check this, feel the underside of the neck for the tube. It is easily felt when inside the esophagus; but if the tube is inside of the rigid ribbed structure of the trachea, it cannot be felt easily.

6. Apply the funnel to the end of the tube and pour the desired fluid down the tube. You cannot administer fluids too fast, because tube size will sufficiently regulate flow rate. The quantity and type of fluid administered will be determined by the size of the animal and the type of treatment.

CAUTION: Make sure that the animal is standing with the head angled upward before administering any fluids. This step usually requires holding the head up while a second person pours the fluids down the tube.

Scouring Calves. In addition to medication, scouring calves need electrolytes and nutrients to counteract loss of fluids and dehydration. Commercial preparations can be purchased; label directions

should be followed. If commercial preparations are not available, a homemade preparation can be made as follows.

Be sure to mix the electrolyte solution thoroughly before administration to make sure all ingredients are in solution.

Electrolyte Formula 1*	
White corn syrup (dextrose)	8 tbsp
Salt (sodium chloride)	2 tsp
Baking soda (sodium bicarbonate)	1 tsp
Warm water (to give a total of 1.25 gal)	
Total	1.25 gal

*Feed 2.5 pints to a 90-pound calf four times per day.

Electrolyte Formula 2*	
Beef consommé (e.g., Campbell's)	1 can
Jam and jelly pectin (MCP)[†]	1 pkg.
Low-sodium salt (Morton's)	2 tsp
Baking soda (sodium bicarbonate)	2 tsp
Warm water (to give a total of 2 qts)	
Total	2 qts

*Feed twice daily.
[†]Can be replaced with 100 cc (ml) of white corn syrup.

After 1 to 2 days on electrolyte solution, the calf should show signs of improvement. If scours appear to be serious, the feces show blood, or the animal does not respond to treatment, get a veterinarian's help immediately.

Orphaned, Weak, or Injured Calves. Calves that are orphaned or weak need milk or colostrum, and may need to be fed by a stomach tube. A newborn calf will require about 6 to 8% of its body weight in colostrum during its first 6 to 10 hours of life. If you do not have colostrum on hand, obtain frozen colostrum from a neighboring dairyman. Orphaned and injured calves will need to be fed at least 2 quarts of milk two to three times per day until you can get them fostered or trained to nurse a bucket or bottle (see Chapter 4).

Bloated Animals. Bloated animals often require stomach tubing to relieve gas pressure in the rumen and for dosing with mineral oil, poloxalene, or other commercial bloat-treatment products (see Section 3.15). The animal should be stomach-tubed with 1 pint to 1 quart of mineral oil, 2 ounces of poloxalene, or commercial product according to directions. This may have to be repeated if the animal bloats again.

7. Slowly remove the tube from the esophagus to prevent excess irritation.

8. Release the animal from restraint.

9. Wash and clean the equipment after treatment.

CAUTION: Some fluids, especially milk, make excellent media for bacteria and infectious agents inside the stomach tube and funnel if these are not thoroughly cleaned.

Postprocedural Management

Watch animals closely for continued scouring, dehydration, listlessness, coughing, and bloat. If the animal does not respond favorably after administration of the fluids just described, seek professional help.

3.15 TREATMENT OF BLOAT

Bloat is a noninfectious disease of ruminants that is caused by high-concentrate diets, grazing on legume pastures, peritonitis, and genetic abnormalities such as dwarfism. The exact cause of bloat still remains unclear, but the problem always leads to a buildup of gases in the rumen. The gases that are produced during fermentation of feeds normally are expelled from the rumen by eructation (belching). In the case of bloat, the animal's ability to eructate is reduced, and the continuous production of gases increases pressure inside the rumen. The first indication of bloat is a slight distension of the paunch on the left side of the animal in front of the hipbone (hook). If the animal does not appear to have any discomfort, it may relieve the problem on its own as it walks around. This type of animal should be observed closely and frequently because the condition may develop very rapidly into acute bloat.

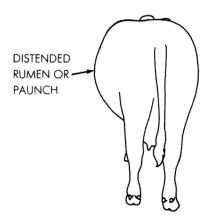

DISTENDED RUMEN OR PAUNCH

In acute (severe) bloat, there is severe distension of the left side of the animal. This can be accompanied by some distension of the right side, protrusion of the anus, respiratory distress, severe pain and discomfort, and death from suffocation if not treated immediately.

People often refer to two types of bloat. The first type is *feedlot* or "free gas" *bloat.* This type of bloat can occur at any time during the year, but it is more common during hot, humid weather or when barley and alfalfa are fed in combination. The second type is *pasture bloat,* or "frothy bloat." It generally occurs on lush legume or wheat pastures, and especially on clover and alfalfa pastures. Legume pastures are particularly hazardous when hungry animals are allowed to graze on them, when pastures are moist after a rain or heavy dew, and when pastures have been recently frosted.

Equipment Necessary

- Squeeze chute, headgate, stanchion, or rope halter
- Loading chute or blocks of wood (2 feet long)
- Stomach tube (½–⅝" diameter × 6' long)
- Poloxalene
- Surfactant
- Rubber hose (⅜–⅝" diameter × 12" long)
- Rope (⅜–⅝" diameter × 3' long)
- Trocar or pocketknife

Restraint Required

The age of the animal and the stage or amount of bloat will determine the restraint required. Calves that are nursing their dams seldom have problems with bloat. Cattle on pasture, in drylot, or in the feedlot should be moved into a squeeze chute, headgate, or stanchion if the animals can stand and walk. Without these facilities, many of the steps listed next become dangerous and almost impossible to perform. If restraint equipment is not available and cattle are broken to lead, it is possible to restrain the animal with a rope halter, but this can be dangerous and is not recommended. In serious cases of bloat where the animal is lying on the ground and struggling to breathe, the technique can be performed without restraint.

Step-by-Step Procedure

The procedural steps are presented starting with initial stages of bloat and progressing to severe cases of bloat. In each case, several alternatives for treatment exist, and they are discussed individually. Acute bloat must be treated immediately because the animal will die before professional help arrives.

1. If the conditions that cause bloat (mentioned earlier) are present, observe cattle frequently during the day by walking or driving through the herd. If an animal is down or off by itself, check it for bloat.

2. Assemble necessary equipment when you observe an animal beginning to show distension of the paunch on the left side. If you don't prepare during the early stages, you may be caught with a severe case of bloat without the proper equipment on hand when needed.

3. In mild cases of bloat, try to keep the animal on its feet and moving. Forcing the animal to walk can help the animal to eructate and relieve gas buildup. If this does not work, or it cannot be done, proceed to step 4.

4. Restrain the animal when it begins to show signs of discomfort or when it appears that the animal's paunch is continuing to distend instead of going down. Proceed to one of the suggestions contained in steps 5 to 9 depending on facilities, severity of the paunch distension, and disposition of the animal.

5. Elevate the front feet of the animal. The purpose of elevating the front feet is to elevate the cardia (place where esophagus enters the rumen) so that it is not covered by rumen fluid or stable foam. This will enable the animal to eructate in most cases. This can be done with gentle animals that are broken to lead by placing them in a loading chute.

For animals that are not broken to lead, place some wooden blocks where the animal's front feet will be before it is moved into the headgate, squeeze chute, or stanchion. Elevate the front feet at least 6 inches. This step can relieve pressure in the rumen by itself, but it works best in conjunction with step 6.

6. Run the end of a 3-foot rope through a 12-inch piece of rubber hose. The rope can vary in size as long as it fits inside the hose, which can be a piece of rubber garden hose. Place the hose into the animal's mouth from side to side above the tongue and to the back of the mouth. Tie the rope snugly, but not so tightly that the animal is gagged, over the top

of the animal's head. This will allow the animal to chew on something fairly soft and help initiate eructation. Keep the animal standing to aid in eructation. If the animal does not begin to eructate and relieve some pressure in 10 to 15 minutes, proceed to step 7.

7. Stomach-tube the animal as in Section 3.14. Make sure the animal is standing so that gases can be released easily. If the tube is positioned correctly in the reticulorumen, gases will be released immediately and the paunch will become smaller. While the tube is in position, proceed to step 8.

8. Administer a surfactant or antifoaming agent into the stomach tube or by drenching to reduce surface tension of the otherwise stable foam. Some of the products being used for this purpose are vegetable oils, mineral oils, turpentine, and poloxalene. One pint to 1 quart of the oil (preferably mineral oil), 2 oz of poloxalene, or 2 oz of turpentine diluted with 1 pint of cold water should be administered. These can be administered down the stomach tube by using a funnel or by drenching. Keep the animal on its feet so that it can void the gases as soon as it can eructate. In most cases, steps 7 and 8 will be effective and eliminate the problem.

9. A trocar and cannula or pocketknife can be used as a *last* resort. This should *not* be used until all other possibilities are exhausted. Consider using the trocar or pocketknife only in extreme cases when the animal is down on its side, legs extended, and in severe respiratory distress. Prepare the trocar by placing the cannula over the trocar. Make sure that the beveled edge of the cannula is near the pointed end of the trocar. Use your left hand to place the point of the trocar on the animal's left side equidistant from the last rib, hipbone (hook), and transverse processes of the lumbar vertebrae (loin). Strike the trocar sharply with the palm of the free hand in the direction of the animal's right

front knee. Hold the cannula in place and withdraw the trocar to permit gas to escape slowly from the rumen. After the gas stops escaping, withdraw the cannula.

A *pocketknife* can be used in lieu of a trocar. Be sure that you follow the same procedures as noted, but be careful not to let the pocketknife close on your hand as you make the puncture. It may be necessary to keep the knife blade in the puncture and turn it at a 90° angle to allow space for gas to escape.

When a trocar or pocketknife is used, the gases that escape may be somewhat offensive to you. In addition, some rumen fluid and contents may be propelled out of the stab wound. Stand back so you don't get it in your eyes; it may burn.

CAUTION: Animals treated for bloat with a trocar should be treated for peritonitis. Veterinary assistance will be needed after using the trocar because additional medication is required. Follow the veterinarian's recommendations for postprocedural management.

Postprocedural Management

Move the bloated animal to a small pen or stall after treatment. Check the animal at 30-minute intervals until the bloat symptoms subside. Change the animal's ration to include high-fiber feed such as grass hay (not legume hay) for 2 to 3 days and gradually raise the amounts of the regular ration until the animal is back on full feed. This may take 7 to 10 days from the time the animal is treated until it is back with the rest of the herd.

No effective drugs are available to prevent feedlot-type bloat. Increasing the roughage in the ration will help to control bloat, but almost any combination of alfalfa hay and barley can cause bloat. Therefore, use grass-type forages instead of legumes in combination with barley.

Pasture or frothy bloat can be prevented or controlled by following some recommended practices. (1) Don't place hungry cattle on legume pastures. Fill cattle on grass hay before turning them on legume pastures. (2) Leave cattle on legume pasture 24 hours a day. Don't bring them in and pen them at night. (3) Dry forage, hay, or grass pastures should be available to cattle grazing legume pastures. (4) Use pasture seed mixtures that are at least 50% grass. (5) After a frost, avoid using pastures that contain a large percentage of legume. After the legume turns brown and goes dormant for the winter, it usually is safe to graze again. (6) Keep salt and water conveniently accessible at all times.

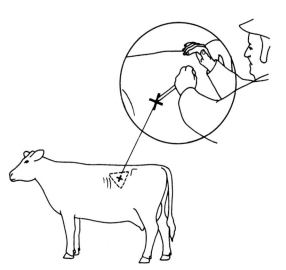

(7) Mix ground corn cobs or oats with grains fed on legume pastures. (8) Provide poloxalene as a feed supplement. Feed according to manufacturers' directions.

3.16 DEHORNING

The objective of dehorning is to reduce the possibility of injury and bruising of animals as well as the herdsman. Any animal with horns can severely cut or bruise other animals in the herd and reduce carcass value in beef and milk production in dairy. Animals with horns require more shed and feeding space, and they are harder and more dangerous to handle during routine management practices.

> **CAUTION:** Dehorning should be performed as early as possible in the animal's life (preferably at less than 2 months of age) to minimize animal stress and setbacks. As animals increase in age, the job becomes more difficult. The technique causes more trauma to the animal by exposing the sinus cavity within the head, and there is increased bleeding.

Do not dehorn cattle by any of the cutting methods during the fly season or during extremely cold weather. Maggots can be a problem during hot weather. The exposed sinus cavity and blood that appear after dehorning provide an ideal medium for parasite infestation. The open sinuses can lead to respiratory complications during extremely cold weather. The ideal times of year to dehorn by any of the cutting techniques are in the spring before flies appear and in the fall after flies disappear. The caustic chemical or hot, bell-shaped dehorners can be used throughout the year.

Equipment Necessary

- Squeeze chute or headgate (equipped with a head bar and nose bar)
- Stanchion, rope halter, and nose lead
- Scissors
- Petroleum jelly or Vaseline
- Caustic stick or paste
- Bell-shaped dehorner (electric or hot iron)
- Spoon (gouge), tube, or knife
- Barnes-type dehorners
- Dehorning clippers or saw
- Antiseptic
- Forceps or tweezers
- Cotton
- String
- Anesthetic

Restraint Required

Young calves (less than 4 weeks of age) can be approached, cornered if necessary, flanked, and restrained on the ground. Older cattle must be restrained in a squeeze chute or headgate equipped with a head bar or nose bar to immobilize the head. A stanchion can be used, but the head must be secured with a rope halter and nose lead to prevent operator injury. See Chapter 1 for each of the restraint methods.

Age is an important factor in dehorning, and often determines the method to be used. There are several methods of dehorning cattle; each is discussed separately below.

Caustic Stick or Paste. This system is designed to dehorn very young calves (younger than 1 to 2 weeks) in which the horn has grown very little. The technique involves applying a caustic paste (or stick) to the horn button.

1. Assemble the petroleum jelly and dehorning solution, paste, or stick.

2. Restrain the calf.

3. Clip the hair around the base of the horn button as close as possible with a scissors or electric clippers.

4. Clip off the end of the horn button with a sharp pocketknife so that the dehorning chemical can penetrate the horn and destroy the modified skin tissues that produce the horn. When the horn has not penetrated the skin, as on very young calves, nick the skin over the horn button with a pocketknife so that the chemical can penetrate.

5. Apply a ring of petroleum jelly or Vaseline around the base of the horn button.

> **CAUTION:** The dehorning chemical can cause severe burns on unprotected skin. The ring of petroleum jelly will also prevent any excess caustic chemical from running down the side of the head and into the eye, where it can cause discomfort and injury.

6. Apply the dehorning chemical. The caustic stick or paste must be applied exactly according to label directions. If too much solution is used, or if the chemical flows out of the horn button site, it can cause a serious scar. If the solution is not applied in large enough quantities, a stub horn can result.

The caustic stick should be dampened with a moist paper towel or cotton to increase its effectiveness.

Apply the caustic stick in a circular motion on top of the horn button and the area *immediately* around the horn. Care should be taken to minimize skin contact. The caustic stick should be rubbed on the

CAUTION: Too much water will cause the caustic to become a liquid, which can flow out of the intended site and cause skin burns; dampen the caustic lightly. Be sure to wear a pair of leather gloves or wrap the caustic stick in a paper towel so that you do not get chemical burns on your hands.

horn until blood appears. The degree of destruction of the horn growth area depends on the pressure used, area covered, amount of moisture present, and the number of applications.

CHEMICAL DEHORNING

If paste is used, apply it to the horn until it is approximately the thickness of a dime over the top of the horn button.

CAUTION: Keep the calf away from the dam for several hours to allow the caustic compound to dry and prevent burning the cow's flank.

7. Release the animal.

Bell-Shaped Dehorners. Use of the bell-shaped electric dehorner or hot iron is an excellent method of removing horns from the calf at any age when the horn button is less than ¾ inch in length. This is best done when the calf is under 4 weeks of age, since at that time one person can restrain the calf on the ground while the second person applies the electric dehorner or hot iron. Calves can be dehorned with a hot dehorner at any time before they are 4 months old.

1. Assemble the electric dehorner or hot irons. If hot irons are used, select the size of iron that fits the

horn to be removed. These irons are available in sizes to accommodate different horn sizes.

2. Heat the dehorner. Electric dehorners are not difficult to heat, but the manufacturer's recommendations should be followed. Hot irons should be heated in the same manner as branding irons (see Chapter 2).

3. Restrain the animal and secure its head in a squeeze chute with head and nose bar. Small animals can be flanked and restrained on the ground.

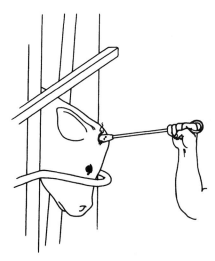

DEHORNING WITH BELL-SHAPED HOT IRON

4. Apply the heated dehorner over the horn button and horn base until a ring of copper-colored hide appears around the horn base. This will require 10 to 20 seconds, depending on horn size. The time involved may seem longer than this because of the offensive odor of burned hair and skin.

5. Repeat steps 3 and 4 for the second horn.

6. Release the animal.

Spoon (Gouge), Tube, or Knife. This system of dehorning involves the actual cutting out and removing of the small horn button. The method works best on young calves (less than 60 days old, or having horns less than $1\frac{1}{2}$ inches long). Very little pain or bleeding results. This method can also be used on calves 2 to 4 months old. Before 4 months, the horn buttons on most calves can be well developed, but they remain soft and loosely attached to the skin. This type of horn can be easily removed with a knife-like dehorning tool that separates the horn button from the adjoining tissue with very little blood loss.

1. Assemble the necessary equipment. When tubes are used, select the correct one to fit over the base of the horn and about $\frac{1}{8}$ inch of skin around the horn. Tubes of different sizes can be purchased in sets of four. The equipment should be sterilized whenever possible before dehorning is started, and then placed in a pan of disinfectant.

2. Restrain the animal and secure its head in a squeeze chute or headgate.

3. Clean the horn and area around the horn with soapy water or a disinfectant to minimize gross contamination during the dehorning process.

4. (Optional.) Apply local anesthetic around the base of the horn to deaden the pain. On small calves, pain is not a problem and anesthetics are not necessary.

5. Place the cutting edge of the instrument on the skin around the base of the horn.

Tube. Place the tube over the base of the horn so that approximately $\frac{1}{8}$ inch of skin around the base of the horn is included within the tube. Push and twist the tube each way until the skin has been cut through. A cut $\frac{1}{8}$ to $\frac{3}{8}$ of an inch deep is required. With the tube in this position, rapidly twist inward and downward (toward the jaw) on the tube so that the cutting edge of the tube cuts under the area of the horn button to spoon it out.

Spoon (Gouge) or Knife. Position the spoon or knife along the skin at the base of the horn and cut around and under the horn. You should take about $\frac{1}{8}$ inch of skin around the horn and go about $\frac{1}{4}$ to $\frac{1}{2}$ inch below the surface of the skin as you spoon or cut beneath the horn.

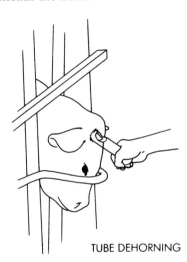

TUBE DEHORNING

6. Apply an antiseptic solution with a Furacin base (or suitable substitute) to the horn area to prevent infection. The area where the horn was removed should not bleed severely if the procedure is performed when calves are small.

7. Repeat steps 3 to 5 for the remaining horn.

8. Release the animal.

CAUTION: After the horn is removed, bleeding can occur. If bleeding appears excessive, it is advis-

able to pick up the artery with forceps and pull it straight out so that it breaks in the soft tissue of the head. Bleeding should stop because the epithelial cells of the artery are damaged and the blood-clotting mechanism is activated.

Barnes-Type Dehorners. In large beef herds where cows and calves are dispersed over a wide area, dehorning is often postponed until weaning, when calves are approximately 7 months old. The Barnes-type dehorner is designed to dehorn animals that are 4 to 12 months old. It lifts the horn out by the roots and crushes the blood vessels so that only a small amount of bleeding occurs.

1. Assemble, clean, and sterilize the necessary equipment, which includes forceps (or tweezers), Barnes-type dehorner, cotton, and a pan of disinfectant in which to place the equipment.

2. Restrain the animal and its head securely.

3. Place the dehorners, with the handles together, over the horn and down against the skull of the animal. They should be placed close enough to the animal's head that a ring of hair and skin $\frac{1}{4}$ to $\frac{1}{2}$ inch wide is removed with the horn.

4. To remove the horn, make sure the knives of the instrument are correctly placed and spread the handles apart quickly. This will close the knives and remove the horn.

BARNES-TYPE DEHORNING

CAUTION: Be sure that the horny tissue of the horn is removed completely. If it is not removed, a horn stump or scar can grow back.

5. After the horn is removed, bleeding will occur because the artery is exposed. The bleeding can be reduced by using forceps (or tweezers) to pick up the main artery on the ventral side (underside) of the cut. Twist and pull the artery out of its position until it breaks off within the soft tissue. The broken artery will retract into the softer tissue, and bleeding will be stopped by the clotting mechanism.

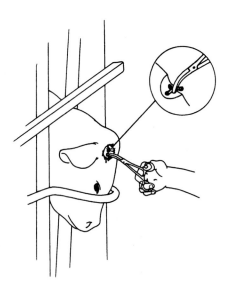

6. Treat the wound with an antiseptic spray or solution to prevent infection. The antiseptic should contain Furacin or a suitable substitute.

7. Place a thin layer of cotton over the exposed sinus cavity. The cotton will stick to the wound, help prevent foreign particles from entering the sinus, and help reduce infection.

8. Repeat steps 3 to 7 for the second horn.

9. Release the animal.

Dehorning Clippers and Saws. The dehorning clipper is the most efficient instrument for dehorning older cattle (1 to 2 years). Dehorning saws can be used to tip (cut ends off) horns or to remove the entire horn. The use of saws is necessary when the horn base is too large for clippers or when abnormal horn growth prevents the use of clippers.

> **CAUTION:** Hard, brittle horns of mature cattle should be removed with a saw. Clippers are likely to sliver or crack the bone that forms the horn core.

1. Assemble the clippers or saw, forceps (or tweezers), and cotton, and place them into a pan of disinfectant. If a saw is used, it should be a fine-toothed saw similar to a miter-box saw.

2. Restrain the animal and tightly secure its head.

3. Local anesthesia should be administered around the base of the horn.

4. Apply the dehorning clippers over the horn or place the saw at the base of the horn.

5. Make sure that the tool is positioned to cut deep enough to remove a ring of skin with the horn. The deep cut destroys the modified skin cells from which the horn grows.

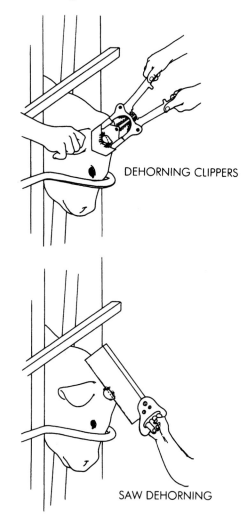

DEHORNING CLIPPERS

SAW DEHORNING

6. Mature cattle tend to bleed more than younger cattle. Be prepared to act rather quickly. Take the forceps, grasp the artery, twist it, and pull it straight out of its position in the ventral (under-) side of the cut so that it breaks off deep in the base of the horn. The remaining artery will retract into the soft tissue and bleeding will be stopped by the clotting mechanism. Use of the saw results in less bleeding because the action of the saw blade lacerates the blood vessel instead of making a clean cut, but the wound produced by the saw will take longer to heal.

7. Apply an antiseptic solution or spray to the wound to prevent infection. The antiseptic should contain Furacin or a suitable substitute.

8. Place a thin layer of cotton over the exposed sinus cavity. The cotton will stick to the wound, help prevent foreign particles from entering the sinus, and help reduce infection.

9. Repeat steps 3 to 7 for the other horn.

10. Release the animal.

Postprocedural Management

Caustic Stick (or Paste) Method. Calves dehorned in this manner should be protected from rain for several days after treatment so that the caustic chemicals do not dissolve in water and run down the animal's head and into its eyes. A few days after chemical dehorning, a scab will appear over each horn button. This will drop off in approximately 1 to 2 weeks and leave a smooth spot of skin about the size of a dime devoid of hair. These spots will eventually be hidden by the growth of surrounding hair.

Bell-Shaped Dehorners. Unless an extremely deep burn is made, these calves will not require any additional attention. The horn or button will usually slough off in 4 to 6 weeks, leaving a relatively smooth area devoid of hair.

Surgical or Cutting Method. Calves less than 4 months old normally will not bleed severely. Owners of recently dehorned cattle should observe them closely for squirting blood or excess blood running down the side of the head for 12 to 24 hours. If blood begins to squirt out of the horn at any time after dehorning, restrain the animal as before, pull arteries that haven't already been pulled, tie heavy string around the area just below the base of the horn, or use bandages to apply pressure to the wounds to stop bleeding. If string is used, be sure to remove it within 12 hours to allow restoration of circulation.

Place animals in an area that is easy to observe and as free as possible from falling dust and chaff. Do not, for example, feed freshly dehorned animals in a slant-bar hay feeder that can cause particles to drop into the open sinus cavity. The sinus cavity usually swells closed in 2 to 3 days after dehorning.

If any signs of infection, such as excessive swelling, maggots, or pus, are observed, they should be treated as soon as possible.

3.17 BOLUSING AND DRENCHING

Bolusing refers to placing a large tablet, capsule, or magnet into the mouth and behind the base of the tongue so the animal can swallow it. Medication can be given this way for scours, digestive disturbances, and internal parasites; a magnet can be introduced into the reticulorumen to prevent "hardware" disease. *Drenching*, on the other hand, refers to the administration of liquids and medications into the mouth near the base of the tongue. These liquids could be administered for bloat, internal parasites, or other sicknesses.

Restraint Required

Animals should be restrained in the standing position. A newborn calf can be approached, cornered, and restrained by backing it into a corner and standing astride its neck. Most cattle that are more than 3 to 4 months of age will have to be restrained in a headgate, squeeze chute, or stanchion to minimize animal movement and maximize the handler's safety. In cases where the animal is gentle and broken to lead, it can be approached, haltered, and tied with a quick-release knot.

Bolusing

Equipment Necessary

- Squeeze chute, headgate, stanchion, or halter
- Bolus (balling) gun
- Bolus (tablet, capsule, or magnet)

Step-by-Step Procedure

1. Assemble the necessary equipment. Select the bolus gun and bolus gun head of the size that best fits the bolus to be administered. Also, take into account the size of the animal when selecting the size of the bolus. Don't administer a large bolus to a small calf because the bolus gun can cause unnecessary throat irritations and the calf may not be able to swallow the bolus.

2. Restrain the animal.

3. Load the bolus gun by inserting the bolus into the head of the bolus gun. Hold the gun upright so the bolus does not fall from the gun and onto the ground or into the manure.

4. Open the animal's mouth. Stand near the animal's head and face the same direction it is facing. Place the arm nearest the animal over and around the animal's head and insert your thumb into the animal's mouth. Apply downward pressure on the animal's tongue as you squeeze the jaw between your thumb and four fingers.

To open the calf's mouth when you are standing astride its neck, insert four fingers into the calf's mouth and apply pressure to the palate (roof of mouth). See Section 3.14 for illustrations.

> **CAUTION:** The animal may pinch and lacerate your fingers in this step and steps 5 and 6. This is one of the hazards of the trade and hard to eliminate completely because leather gloves are cumbersome to work with. Remember that cattle do not have top teeth in front, but there are teeth on both the top and bottom in the back of the mouth. Try not to place your fingers between the upper and lower teeth in the back part of the mouth.
>
> The animal may throw its head from side to side or up and down during steps 4 to 6. Be careful that this does not injure you or cause you to injure the animal with the bolus gun.

5. Insert the bolus gun into the animal's open mouth, along the side of the mouth and tongue to the base of the tongue. Gently push the gun back into the animal's mouth and allow the animal to swallow the head of the bolus gun.

> **CAUTION:** Do not force or jam the bolus gun down the animal's throat. This will cause irritation and may make the animal go off feed.

6. After the animal has swallowed the head of the bolus gun, depress the plunger to dispense the bolus and pull the gun out of the mouth slowly to prevent irritation.

7. Observe the animal for a few seconds before releasing it. The animal may spit the bolus out if it was not placed far enough down the esophagus.

When this happens, pick up the pieces and determine how much of the bolus was consumed. You may have to bolus the animal again if too little medication was consumed.

8. Release the animal.

Drenching

Equipment Necessary

- Squeeze chute, headgate, stanchion, or halter
- Drenching syringe or long-necked bottle
- Drench

Step-by-Step Procedure

1. Assemble the necessary equipment. The gaskets of the drenching syringe usually must be soaked in water if they are leather, and the syringe must be assembled. The long-necked bottle can be a wine bottle or soft drink bottle.

2. If the medication to be administered comes in a powder form, prepare the liquid according to label directions.

3. Restrain the animal.

4. Load the drenching syringe or bottle. Insert the drenching syringe into the liquid to be administered. Pull back on the plunger to fill the syringe. Hold the syringe upright so the liquid does not run out of the end. When a long-necked bottle is used, place a funnel in the bottle and pour the liquid into it. Hold the bottle upright.

5. Open the animal's mouth. Stand near the animal's head and face the same direction it is facing. Move the animal back in the headgate so that the backs of its ears are touching the headgate. This will help to minimize animal movement during the drenching process. Place your leg (the one nearest the animal) under the animal's head. Then place the arm nearest the animal over and around the animal's head and insert your thumb into the animal's mouth. Apply downward pressure on the animal's tongue as you squeeze the jaw between your thumb and four fingers.

> **CAUTION:** As with bolusing, avoid putting your fingers between the animal's upper and lower teeth in the back of the mouth. Watch that rapid movements of the head do not injure you or the animal.

6. Raise the animal's head and insert the drenching syringe or bottle into the animal's mouth. The syringe or bottle should be placed along the side of the animal's mouth and tongue. Do not force the syringe or bottle past the base of the tongue.

> **CAUTION:** When a long-necked glass bottle is used *do not* place the bottle where there are both top and bottom teeth. The animal could crush the bottle in its mouth. It is very important that you allow the animal time to swallow *at its own pace. Do not* try to force it. If you force it, the animal may get fluid in the lungs, which can result in pneumonia. After the animal has finished, slowly pull the syringe or bottle out of its mouth.

7. Observe the animal for a few seconds before releasing it. The animal may cough and choke if the fluid was placed in the lungs. When this happens, place the animal in a small pen and observe it closely for at least 24 hours. Watch for continued coughing, fever, and droopy ears.

8. Release the animal.

Postprocedural Management

Each drug or material that is used can cause side effects, such as pneumonia when liquids are drenched, and scours after administration of worming agents.

Read the label on the medication for any abnormal conditions that might occur. Observe animals for scouring, listlessness, and coughing. If the condition persists for more than a day, seek professional help.

3.18 CONTROL OF INTERNAL PARASITES

Internal parasites have the same effects and are subject to the same methods of diagnosis and control in feedlot, cow–calf, stocker–feeder, and dairy operations. The major difference among these enterprises is the way the effect of parasitism is measured: decreased feed efficiency, lower rate of gain, decreased calf weaning weight, lower milk production, or failure to come into estrus (heat) and conceive.

There are several internal parasites that infest cattle and reduce profits. These include the liver fluke (occurs in low-lying, wet areas infested with snails), tapeworms (primarily in the southwest United States), gastrointestinal roundworms (widespread and a common problem), lungworms (limited and generally not a problem), and coccidiosis (a common problem). Based on surveys, there is a high probability that most herds have a worm problem, primarily intestinal roundworms. Many cattle operations have some type of parasite control program, and symptoms of severe parasitism such as bottle jaw, severe diarrhea, and death are not common. Any parasitism, however, can cause slower gains, reduced milk production, and digestive disturbances that reduce efficiency and lower weaning weights.

Generally, moderate to heavily infested cattle have a rough, dull hair coat. Positive diagnosis can be made for internal parasites by microscopic examination of a fecal sample for egg counts or by postmortem examination of an animal as soon as possible after death.

Egg counts can be misleading in some cases because: (1) only mature worms lay eggs; (2) high-roughage feeds seem to increase worm egg production while high-concentrate rations tend to decrease worm egg production; (3) worms do not lay eggs at a uniform rate; (4) some worms lay eggs only at certain times of the year; (5) some heavily infested animals have depressed worm egg production because of nutrient competition among worms; (6) worm egg production varies with time of the day; and (7) some highly pathogenic worms lay few eggs while some worms of low pathogenicity lay large numbers of eggs. Low egg counts can be misleading because worms may still be present and causing damage.

Intestinal parasites affect cattle and cause damage in different ways. Worms, especially hookworms, can suck blood, secrete a substance that prevents blood clotting, and cause a loss of blood serum from the animal's digestive tract. Worms can produce small nodules in the lining of the intestinal tract that affect secretion and digestion and reduce the availability of nutrients for production. Finally, worms can produce scars and small lesions in the intestine that can be invaded by bacteria and toxins. The end result of worm damage is increased susceptibility to infection and reduced performance.

Internal parasites can be controlled by administering worming agents with a bolus, drench, injection, paste, pour-on, or feed additive, and by good pasture management practices. The best time to worm cattle varies with the type of enterprise. Feeder calves should be wormed in a preconditioning program or as soon as possible after they arrive at the feedlot. In cow–calf operations, cattle should be wormed at the end of the grazing season as they enter the winter supplemental feeding period, just before the breeding season when cows are moved from early spring pasture, or when cattle are worked and the worm infestation is expected to be high. Dairy operations should consider worming cows during the dry period, when the cows are moved from the drylot to the calving area, or immediately after calving. Young cattle such as heifer and bull replacements should be wormed at least yearly and sometimes more often depending on need and grazing intensity.

CAUTION: Make sure that you use only approved worming agents on dairy cattle. Check the approved list available from your milk sanitarian, plant fieldman, or county extension office. Failure to use approved products can result in contaminated milk that will be rejected by the milk processor as unfit for human consumption. A mistake of this kind can result in substantial losses. For a more complete CAUTION regarding animal health products, see Chapter 10, Section 10.4, "Drugs."

Equipment Necessary

- Balling (bolusing) gun
- Bolus
- Drenching syringe
- Drench
- Syringe
- Injectable wormer
- Caulking gun
- Tube of paste wormer
- Bag of feed-type wormer—crumbles or block
- Pour-on wormer

Restraint Required

The amount of animal restraint will vary with the type of administration selected. If the worming agent is fed, the animals will not require any restraint other than the pen, lot, or pasture where they are normally fed. When injectable wormers are used, animals should be restrained in a working chute. Bolusing, drenching, pour-ons, and pasting require animal restraint in a squeeze chute. If restraint equipment is not available and cattle are broken to lead, it is possible to restrain the animal with a rope halter. This can be dangerous, however, and is not recommended.

Step-by-Step Procedure

1. Determine whether you need to worm. If cattle have a rough, dull hair coat, worming should be seriously considered. If pastures were overgrazed, overstocked, and very short during the grazing season, then intestinal parasites could be a problem. Most herds would benefit from a routine worming program. If there is any doubt, a microscopic examination of several fecal samples by a veterinarian is recommended.

2. Decide on a method of worming. Several factors need to be considered, namely, labor, facilities, preference, and effectiveness.

Feeding a Worming Agent. This method requires the least amount of time, labor, and facilities. Cattle must be on a feeding program where a grain-type ration or supplement is being fed before the worming agent is added to the feed. This type of wormer is less effective than other types of wormers because some animals will tend to consume more than their allotted dosage while other animals (usually the meek, timid, and heavily worm-infested) do not get enough. This is the reason that these types of worming products have a wide safety margin to prevent overdoses. To ensure the proper dosage for each animal, mix the worming agent thoroughly in a complete mixed ration.

Injectable Wormer. This method requires less restraint than blousing, drenching, pour-ons, or pasting, but more restraint than feeding. The major advantages of injectable wormers are that each and every animal gets the required dosage based on its weight, and they are effective and easy to administer.

Bolus, Drench, Pour-on, and Paste Wormers. These methods require the greatest amount of restraint unless the animal is very tame. The disadvantage in bolusing is that some animals will spit out the bolus if it is not administered properly. Drenching and pasting have the disadvantage that they are somewhat messy, but the animal cannot spit them out. The advantage of all three methods is that each and every animal gets the required dosage based on its weight.

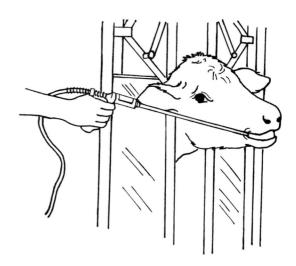

CAUTION: Do not administer any combination of two insecticides, wormers, or drugs that contain the class of compounds known as organophosphates (or cholinesterase inhibitors) at the same time. Read the label to determine the active chemical ingredients. If the name of one of the active ingredients in either compound includes the term *phosphate* or *phospho*, the compound is an organophosphate. If the label of the second compound being considered says it contains cholinesterase inhibitors, don't use the two compounds together, even if the two insecticides, wormers, or drugs are designed to perform different functions. If two such compounds are used together, the animal's nervous system can react violently, causing respiratory failure and death. The same result may occur if a double dose of one such compound is given.

See Chapter 10, Section 10.4, "Drugs," for a more complete discussion.

3. Purchase the worming agent and assemble the necessary equipment.

4. Restrain the animal as necessary.

5. Weigh the animal or estimate its weight so that the correct dosage will be administered.

6. Administer the worming agent. Be sure to follow the manufacturer's recommended dosage based on the animal's weight and age. Don't take the attitude that "if a little bit is good, a lot is better" or "they don't need that much, I'll cut it in half."

Bolusing and Drenching. Drenching, as a method to administer oral dewormers, has increased in popularity. This is especially true when a large number of cattle are being worked through a chute. . . and if you have one of the hooked applicator tubes to use with an automatic-fill drench gun. This type of applicator tube allows you to stand to the side of the restrained animal (in a chute) and insert the hook of the tube into the corner of the animal's mouth. With this process, the person administering the drench does not need to touch the animal. For a complete discussion of drenching, see Section 3.17.

Pasting. Stand beside the animal's head and face the same direction the animal is facing. Back the animal up in the chute so that the backs of its ears are against the headgate. Using the arm and leg nearest the animal, place the leg under the animal's head, the arm over and around the animal's mouth, and your fingers under the jaw. Squeeze the tongue and lower jaw between your thumb and fingers to open the animal's mouth. Slide, don't jam, the caulking gun loaded with a tube of paste into the mouth and toward the base of the tongue. Elevate the animal's head as much as possible. Squeeze the trigger of the caulk gun the recommended number of clicks based on the animal's weight. Pull the caulk gun and tube out of the animal's mouth. Continue holding the head up for several seconds.

Most of the time, a few seconds spent holding the animal's head in an elevated position will assure that the paste works its way to the rear of the tongue and will be swallowed by the animal. However, if the paste does not arrive at the base of the tongue, the animal will slobber some of it out and a margin of dewormer efficacy will be lost.

The hook adapter for the tube end of the paste applicators (caulking gun) takes a large part of the guesswork out of how long to hold the head up, and nearly eliminates the question of whether the animal slobbered away too much. To properly use this hook-tube applicator, restrain the animal in a chute, hook the tube into the corner of the animal's mouth, and squeeze the trigger. The bend, or hook,

in the applicator tube ensures that the paste is deposited at the rear of the mouth and base of the tongue.

Injection. Load the syringe, remove the air bubble in the syringe by depressing the plunger slightly, and administer the correct dosage by an intramuscular injection.

In meat-producing animals, injections should be administered into an area of the body involving the lowest-quality meat cuts. . . providing the injection site selected is satisfactory for the product to be administered. The muscles and skin of the neck are the first choice.

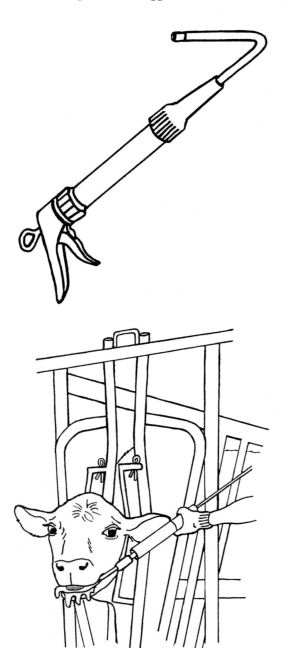

For a more complete discussion of injections and injection sites, see Chapter 10, Section 10.5, "Administering Pharmaceuticals and Biologicals."

7. Release the animal.

8. Repeat steps 4 to 7 on each animal until finished.

9. Clean and sterilize equipment after worming all animals.

Pour-on. Follow procedures for pour-on external parasite control—page 173.

Postprocedural Management

Cattle severely infested with worms may tend to scour within several days after treatment. Unless diarrhea becomes serious, this should not be a problem. Close examination may show some worms present in the feces of cattle recently wormed. Ideally, cattle should be wormed again after 14 days to eliminate any worms that hatched from eggs. This practice is not routinely followed in the industry unless severe worm infestations are diagnosed.

Evaluate your pasture grazing program. (1) Don't overgraze or overstock your pastures. When cattle graze forage very close after many cattle have been on a pasture, they can consume large quantities of parasite eggs and compound the problem. (2) In the spring of the year before the cattle are placed on pasture, harrow the pasture to scatter manure piles that have accumulated. This allows the sun to penetrate, dry out, and kill parasite eggs and larvae. (3) Rotate pastures to allow regrowth (if grass is continually grazed close, parasite problems increase).

3.19 CONTROL OF EXTERNAL PARASITES

External parasites cause substantial losses in the cattle industry each year. These losses are in the form of increased feed costs, reduced weight gains, lower milk production, impaired value of products, and even the loss of animals. Close observation and good production records are necessary to minimize the losses.

External parasites are divided into two main categories: (1) insects, which includes flies, lice, and mosquitoes, and (2) arachnids, which includes ticks and mites. These parasites significantly affect animal performance by feeding upon tissue, blood, or secretions, and by annoying the animal, causing nervousness, reduced feed consumption, reduced milk production, and reduced weight gains. Some of these parasites are directly or indirectly responsible for the transmission of pinkeye, mastitis, anaplasmosis, blue tongue, and other infectious diseases. In addition, rubbing caused by irritation can leave bare patches of raw skin and increase the cost of maintaining fences against which infested animals scratch themselves.

There are several methods for controlling external parasites. These include pour-on systemics, insecticide-impregnated ear tags, spraying, dipping, fogging, dusting, cattle oilers, dust bags, feed additives, sanitation, and the genetic makeup of the animal itself. The following discussion describes pour-on systemics, spraying, fogging, dusting, and insecticide-impregnated ear tags.

Dipping is an effective method of external parasite control, but most producers do not have a dipping vat at their disposal. Cattle oilers and dust bags are easy to install. Generally, however, they are not as effective as some other applications when used as the only method of control because the application is not uniform over the animal. Feed additives can be used to prevent maturation of larvae in the feces (manure); but if one's neighbors do not control parasites, the additives' effectiveness is limited because many parasites can migrate over the fence. Sanitation, including manure removal, is a must around buildings to control flies.

Follow label instructions on all insecticides. In the case of beef cattle, be sure to allow the recommended time between insecticide application and slaughter. In dairy cattle, be sure to check the approved drug and chemical list before insecticide application so that milk will not be contaminated.

CAUTION: Prior to the administration of any combination of insecticides, wormers, or drugs, consult with your veterinarian to ensure compatibility. This will prevent any negative interactions plus result in optimum benefits to the animal from the treatment.

Try not to get the insecticide on your skin or clothes. This is especially true of the systemic pour-ons because these compounds will be absorbed by your skin and can cause dizziness and nausea. If you should come in contact with the insecticide, wash it off immediately with soap and water and change your clothes. Always wear rubber gloves when handling these materials.

Use only approved insecticides, chemicals, or drugs on dairy cattle. Check the approved list available from your milk sanitarian, plant fieldman, or the extension office in your country or state.

Equipment Necessary

- Working chute, small pen, or halter
- Insecticide
- Insecticide-impregnated ear tags
- Sprayer (hand or high-pressure)
- Long-handled measuring cup
- Bottle (equipped with specially equipped graduated applicator)
- Dusting container (with holes in the top)
- Fogger
- Rubber gloves
- Insecticide jug, hose, pistol-like applicator

Restraint Required

The restraint used will depend upon the number of animals to be treated and the type of application selected. When only one or two animals are treated for external parasites, they can be approached and haltered or they can be confined in a small pen. When large numbers are to be treated, restraint in a working chute is best. In this situation, application is relatively easy; the chance of animals being treated twice or overdosed is minimized, and safety to both man and animal is maximized.

Pour-ons

Step-by-Step Procedures

1. Purchase an insecticide designed to prevent or treat a particular external parasite problem. Read and follow the label directions. Be sure to follow the organophosphate caution mentioned earlier.

2. Assemble the pour-on insecticide, graduated applicator, rubber gloves, pour-on bottle, or pistol-like applicator.

3. Determine how many animals you will treat that day, put on a pair of rubber gloves, and mix or prepare the insecticide mixture. Make sure that you are in a well-ventilated area and that you thoroughly mix the solution. You may need to stir or agitate the solution periodically. Some pour-ons are ready to use and need not be mixed with water.

When treating for grubs with systemic insecticides, timing is critical and cutoff dates in your area should be observed. The cutoff dates reflect the latest time in the year that cattle can be treated safely. This date varies from area to area and corresponds to the stage of grub migration in the body. Grubs migrate to tissue along the esophagus or to the spinal column before they reach the back. Grubs killed while they are along the esophagus can cause a host–parasite reaction which results in the swelling of that area. This causes difficulty in swallowing, eructation (belching), and breathing, which may lead to death of the animal. Grubs killed along the spinal column can cause paralysis. Before treating for cattle grubs, check with your local extension agent for the cutoff dates that pertain to your cattle and their source of origin, if they were obtained outside your area.

4. Place the ready-to-use or mixed insecticide into a suitable container. Some of these compounds can be purchased in a plastic bottle with a specially designed pouring applicator that screws on the top or with a pistol-like applicator that is attached, via a hose, to a premixed container. Other insecticides must be placed into a bucket or a widemouthed container. Containers must be unbreakable, portable, and designed to minimize spilling.

5. Restrain the animal.

6. Estimate the age and weight of each animal to be treated. Previous records will be helpful in determining this. The necessity for weight estimation can be minimized when uniform weight groups are treated.

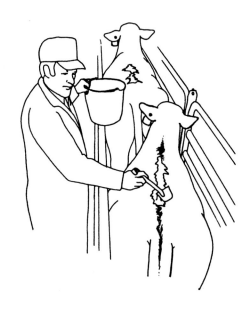

7. Fill the applicator only to the level recommended by the manufacturer for that age, weight, and type of animal.

8. Apply the insecticide to the animal. Some manufacturers recommend applying the insecticide along the entire length of the back, while others recommend applying it only to the withers.

9. Release the animal.

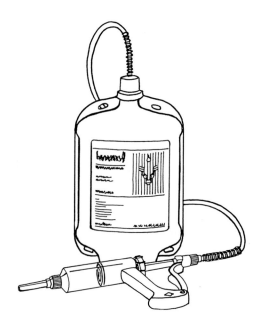

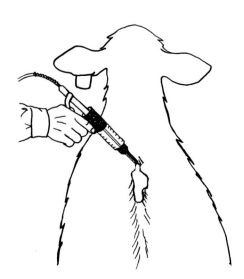

Spraying and Fogging

1. Purchase an insecticide designed to prevent or treat a particular problem. Make sure that you read the label and follow the directions. Be sure to follow the organophosphate caution mentioned earlier.

2. Assemble the sprayer, insecticide, and rubber gloves.

3. Determine how many animals will be treated and prepare the insecticide. Be sure to wear rubber gloves to prevent the concentrated insecticide from coming in contact with your skin. Mix the solution thoroughly. You may have to restir or agitate the solution periodically. Be sure to follow the caution concerning skin contact mentioned earlier.

4. The insecticide should be placed into the tank of a hand sprayer or large high-pressure sprayer depending on the number of animals to be treated. Foggers can also be obtained in various sizes.

5. Restrain the animals in a small pen or lot when a high-pressure sprayer or fogger is used. If a hand sprayer is used, animals should be restrained with a rope halter for complete animal coverage.

6. Apply the insecticide to the animal. Apply enough spray to wet the animal thoroughly to the skin. The size of the animal and the length of its hair will determine the amount of spray needed. Complete coverage may require 2 to 4 gallons for mature animals in winter coat. Two to four quarts is usually sufficient to treat short-haired cattle. When fogging or misting, apply the insecticide mixture at a lower application rate than when spraying. This requires that the mixture have a higher concentration of insecticide. The mist or fog is allowed to drift over the animals while they are contained in a small pen or lot. This method is simple for one man but requires very calm wind conditions. It is difficult to get uniform treatment and hard to determine whether a particular animal has been treated adequately.

> **CAUTION:** Do not spray or fog animals on extremely cold days to avoid animal stress that can cause pneumonia.

7. Release the animal.

Dusting

1. Purchase an insecticide designed to prevent or treat a particular external parasite problem. Read and follow the label directions.

2. Assemble the insecticide, applicator, and rubber gloves.

3. The powdered insecticide is often purchased in a suitable container. Punch holes in the top of the container with a pointed object. If the material does not come in this type of container, you can use a coffee can with a plastic lid; in that case, punch holes in the bottom of the can.

4. Restrain the animal in a suitable manner.

5. Powdered materials designed for direct animal application are applied in the same way as pour-on materials. Apply approximately 6 oz. of dust along the back of the animal unless otherwise specified on the label. Rub the animal with a rubber-gloved hand, if necessary, to disperse the powder through thick hair. An alternative to this is to place the powder into a self-applicator called a *dust bag*. These bags are hung where the cattle can rub against and under them.

6. Release the animal.

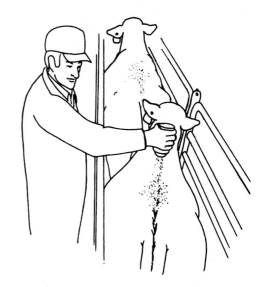

> **CAUTION:** Avoid getting the insecticide powder in your eyes and on your skin. If you should get insecticide in your eyes, flush them with water immediately to minimize eye irritation.
>
> Do not dust cattle when it is raining. Rain reduces the effectiveness of the insecticide.

Insecticide-Impregnated Ear Tags. Insecticide-impregnated ear tags for beef and dairy cattle are an asset to the livestock producer. Applied like an ordinary ear tag used for identification, insecticide-impregnated ear tags, commonly referred to as *fly tags*, provide a significant level of control against insect and arachnid pests. Insects controlled include horn flies, face flies, stable flies, and houseflies. Arachnid populations controlled include spinose ear ticks, Gulf Coast ticks, and lice.

There are several brands on the market. They vary as to their active ingredient(s), amount of insecticide contained in the tag, and longevity of tag effectiveness. Consult with your dealers, veterinarian, other producers, or your livestock extension agent/specialist as to which particular product to use in your area.

As with all insecticides, dewormers, and perhaps many other animal health products, it is good management practice to rotate products used from one season, or one year, to the next. This will help reduce the incidence of resistance development by the insects, internal parasites, or microbes.

The step-by-step procedure for applying insecticide-impregnated ear tags is the same as for applying ordinary ear tags. (See Chapter 2, Section 2.2, "Identification.")

CAUTION: Follow manufacturer's label directions, guidelines, and cautions. These ear tags are impregnated with organic insecticides. Handle them according to directions. Review label information for approval of use in lactating dairy cattle. Not all products are approved for use in dairy cattle.

Postprocedural Management

Store unused insecticide in the original container and dispose of empty containers. Do not store insecticides in areas where feed or food are stored. *Make sure* insecticides are kept out of the reach of children, pets, and unauthorized people. Empty glass insecticide containers should be broken and buried. Metal containers should be punctured, crushed, and buried. Bury insecticide containers at least 18 inches deep in sanitary landfills or isolated places where water supplies will not be contaminated.

Pour-ons. The pour-ons have good residual effect if all animals in the herd are treated at the same time. Generally, it is not necessary to repeat the treatment during that season for a given parasite problem. It may be necessary, however, to use a second pour-on for another parasite problem the following season. You may, for example, treat for grubs in midsummer to early fall, depending on cattle origin, and then treat lice during the winter.

Spraying and Fogging. Sprays can give very effective control of flies when applied on a routine basis. Generally, it is necessary to spray every 2 to 3 weeks during the peak fly season to obtain good control. Fogging is less effective and may have to be repeated more often than spraying because animal coverage is not uniform. With both methods, two applications are needed about 2 weeks apart for lice control.

Dusting. Dusting is less effective than pour-ons and sprays when all are used to control the same parasite problem. Powders do not penetrate long, dense hair, but they are useful in cold weather when wetting animals may be detrimental. Treatment frequency for effective control is similar to that for spraying and fogging.

Insecticide-Impregnated Ear Tags. Follow manufacturer's directions regarding tag replacement and posttagging management. Refer to "Postprocedural

Management" explanation for ear-tagging beef cattle. (See Chapter 2, Section 2.2, "Identification.")

3.20 TREATMENT OF RINGWORM

Ringworm or barn itch is a widespread disease in the United States. It is a contagious disease of the outer layers of the skin caused by a fungus. All domestic animals and man are susceptible to it. With cattle this skin condition is mainly a problem during the winter when animals are closely confined and exposure to sunlight is minimal.

The disease is easily identified by a circular patch of roughened, scaly skin devoid of hair. These areas may become gray and crusty in appearance and increase in size if not treated. Ringworm can occur anywhere on the animal, but it mainly occurs around the head, neck, and root of the tail. A mild itch often accompanies ringworm, and the disease can be spread from animal to animal through stratching devices, fence posts, gates, currycombs, and brushes. The disease should be treated promptly to prevent its spread and to relieve animal discomfort.

> **CAUTION:** Ringworm is easily spread from animal to man. Use rubber gloves whenever possible when handling cattle with ringworm.

Equipment Necessary

- Knife or corncob
- Rubber gloves
- Strong tincture of iodine (7%)
- Squeeze chute, stanchion, or rope halter

Restraint Required

Most infected cattle should be moved to a squeeze chute or stanchion and restrained. If cattle are gentle and easily handled, they can be approached, haltered, and tied with a quick-release knot.

Step-by-Step Procedure

1. Identify the infected animals.

2. Assemble, clean, and disinfect the necessary equipment.

3. Restrain the animal.

4. Put on a pair of rubber gloves.

5. Use a clean pocketknife or corncob to scrape or rub the crusty, scaly skin off the infected areas. Continue scraping until the skin appears pink or red. This will allow good penetration of the fungicide into the infection.

> **CAUTION:** Remember, ringworm is contagious and man is susceptible. One area that may become infected is under the fingernails, where it is painful and hard to treat. It can also occur on the head, causing temporary loss of hair.

When working around the head, be careful not to poke the eyes with the knife or corncob. If the ringworm is located on the head, a squeeze chute equipped with a head and nose bar may be necessary to immobilize the head. This will maximize your safety and prevent the animal from throwing its head around while you scrape the infected area. A rope halter and nose lead can sometimes be used to restrain the animal when you are working around its head.

If a knife is used, it should be dipped in a solution of disinfectant between each use and thoroughly disinfected after use to prevent infecting yourself or someone else with the ringworm fungus. In addition, all other equipment should be disinfected between each use whenever possible. Equipment can transmit the fungus from animal to animal.

Corncobs should not be used on more than one ringworm patch. This will help prevent spreading the fungus to yourself and other animals.

When large numbers of infected animals are worked, it may become impractical to perform this step. In this case, proceed to step 6 without taking off the crust. When the crust is not removed, however, treatment is not nearly as effective.

6. Treat the scraped area with a strong tincture of iodine or another prepared medication. Usually this is done with a squirt bottle or can. Be sure to apply a liberal dose to the scraped area and the area around the infection.

7. Repeat the treatment every 2 to 4 days until the infection is cleared up. If you do not repeat the treatment in 2 to 4 days, the fungus will tend to spread more readily between animals. The ring of infection will continue to increase in diameter. Be sure to remove the crust each time you treat the animal.

Postprocedural Management

Sunlight (ultraviolet radiation) helps to reduce the incidence of ringworm. Place cattle in an open lot or on pasture with exposure to sunlight after treatment.

Normal Recovery Sequence

After two or three treatments, the ringworm should be killed. The circular patch should begin to reduce in size, and short hairs will begin to grow in 2 to 3 weeks.

3.21 TREATMENT OF PINKEYE

Pinkeye is the common name for *infectious bovine keratoconjunctivitis* (IBK), an infectious disease of cattle characterized by inflammation and watering of the eye. The disease is widespread in the United States and often becomes more severe in certain years and locations. The infection can affect cattle of all ages, but it is more common in young cattle. The exact cause of pinkeye is complex and not completely understood. Pinkeye is most commonly associated with the bacteria *Moraxella bovis.*

Bacterial pinkeye can occur at any time during the year, but it is most common during warm weather when flies, bright sunlight, dust, wind, pollen, and unclipped pastures can contribute to eye irritation. This form of pinkeye is transmitted primarily by flies and other parasites that feed on eye secretions, but it can be spread by direct animal-to-animal contact. The bacterial organism causing the problem produces a toxin that actually erodes the outside covering of the eye. In early stages, the eye begins to water, and the animal tends to close its eye. As the infection progresses, the eye and eyelid become red and swollen. Finally, an ulcer develops on the eye that can cause cloudy spots on the cornea or cause the eye to rupture. If the disease is not treated in the early stages, permanent eye damage and blindness can result.

Complete prevention of pinkeye is almost impossible. Bacterial pinkeye can be controlled to some extent by controlling face flies and other external parasites, by good nutrition (including adequate vitamin A), and by isolating infected animals.

In addition to being an aggravating, time-consuming condition to treat, and the fact that cattle appear to be distressed with the condition, pinkeye can have serious economic effects on the herd. Calf performance suffers in direct proportion to the amount of time they are affected, the severity of the outbreak, and whether one or two eyes are involved.

Equipment Necessary

- Squeeze chute, headgate, or stanchion and halter
- Ophthalmic ointment, powder, or spray
- Eye patch
- Eye patch adhesive (similar to adhesive used on cattle tags in sale barns)
- Syringe (1-inch, 18-gauge needle)
- Injectable antibiotic (with or without cortisone)

Restraint Required

The amount of restraint required will be determined by the treatment method and size of the animal. Young calves (less than 300 pounds) can often be approached, cornered, flanked, and restrained on the ground when they are housed in a small area. Most cattle, however, should be placed in a squeeze chute or headgate equipped with a head and nose bar to minimize head movement and provide maximum safety to the handler. It is possible to use a stanchion and halter on tame animals, but the animal will throw its head when the eye is treated with an aerosol or injectable antibiotic.

Step-by-Step Procedure

1. Observe all animals in the herd, especially cattle under 3 years old, at least once daily during the fly season. Look for animals that have watery eyes, one or both eyes closed, white scum on the eyeball, a pink or red ring around the cornea, or ulcers on the cornea. Early detection and treatment of pinkeye can be difficult when cattle are on pasture or range.

2. Identify those animals showing the symptoms of pinkeye and move them to the restraint area.

3. Restrain the animal.

4. Examine the eye. Make sure it's free of foreign material. If it is not, clean the eye gently with a piece of soft tissue paper dipped in clean water. If the eye shows signs of serious ulceration, pus, or rupture, it may be necessary to inject some combination of corticosteroids, antihistamines, antibiotics, and foreign protein into the eyelid or systemically. If the eye has a deep ulcer, it may require a surgical procedure.

5. Injection of the eyelid with a long-acting antibiotic for infection and cortisone for inflammation is optional but highly recommended to help save the eye and reduce animal stress. This is done by inverting the top eyelid and injecting a combination of antibiotic and cortisone under the pink

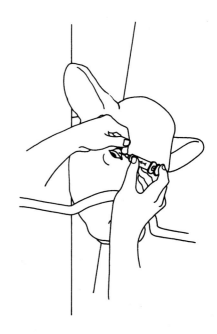

lining of the eyelid with a 5-cc syringe (1-inch, 18-gauge needle).

6. In most cases of pinkeye, the eye should be treated with an ophthalmic ointment, spray, or powder. These products are available commercially. Apply the ophthalmic spray, powder, or ointment liberally to the eye so that the medication gets on the eyeball and under the eyelid.

7. Apply a liberal bead of special patch adhesive approximately ½ inch wide around the eye patch and ¾ inch from the edge. The eye patch can be purchased commercially, or it can be made from old blue jeans or a clean, new burlap bag. The object of the patch is to keep sunlight, dust, dirt, pollen, and flies out of the eye, keep medication in the eye, and reduce the need for isolation of the animal.

> **CAUTION:** Do not go all the way around the patch with adhesive. Leave approximately 2 inches near the bottom of the patch (the side that goes below the eye) to allow for drainage.

8. Apply the eye patch over the eye. Do not get adhesive in the eye. Position the patch over the eye and determine where the adhesive is located in relation to the eye before applying the patch. Be sure that the 2-inch break in the adhesive bead is positioned below the eye.

9. An alternate treatment plan involves the injection (intramuscular) of a long-acting oxytetracycline, such as LA-200.

10. R.elease the animal.

Postprocedural Management

The eye patch will drop off in about 7 to 14 days after application. If the pinkeye is detected early, the eye should be healed and cleared up by this time. If the eye was seriously ulcerated, it may take from 4 to 8 weeks to heal, and a second patch may be necessary to minimize irritation from flies, dirt, and pollen. In the latter case, small, dense, white scars often remain at the site of the healed ulcers and the animal may be partially or totally blind in that eye.

3.22 HOOF TRIMMING

Foot trimming is a technique that can be used to extend the useful life of individual animals in the herd, prepare animals for shows and sales, and correct the hoof problems of young splayfooted animals. The ability of the producer to determine which animals need their feet trimmed and to trim the foot properly is an important management skill.

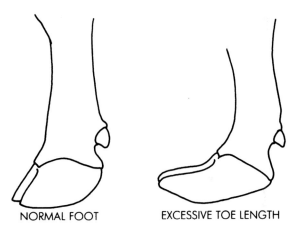

NORMAL FOOT   EXCESSIVE TOE LENGTH

The rate at which the foot grows and the speed at which it wears are influenced by the environment. An animal's foot will replace itself about once a year. With this much growth, adequate wear must take place to maintain a normal foot conformation.

The type of housing influences how much wear takes place on the foot or hoof. Animals housed in free-stall barns, wet or damp lots, or barn stalls tend to have softer-textured hooves than do animals that have access to pasture or drylot. This increased hoof softness and growth result from the soft, moist lot conditions.

The most common abnormality is excessive toe length. In severe cases, the toes will curl up and perhaps be crossed. When this occurs, the sole is bearing more weight than the wall of the hoof. If this condition exists over an extended period of time, severe internal hoof damage can result. Excessive length of the toes can also cause foot problems. The rolling effect can trap foreign material inside the hoof, and bacterial infection may result. Feet of these types are susceptible to bruises, which can develop into foot abscesses.

Make a complete and thorough observation of all the animals in the herd every month or two.

Prepare a list of those animals that need immediate attention. A clean, flat surface will help you to identify animals that need their feet trimmed. In dairy herds, the milking parlor is a good place to make observations because the cows are on a raised surface and you can easily see the length of the toes. In beef herds, observations are not as easy and time should be taken to check animals while they are standing on concrete at the water tank, mineral feeder, or feed bunk, or while working them. The animal should also be observed for abnormal walking patterns. A buildup or growth of the hoof center causes the foot to rock from side to side or from front to rear as the animal stands or moves.

Animals housed continuously in dry tie stalls or small pens with little or no exercise seem to develop the most severe hoof problems. The lack of moisture makes the hoof hard and brittle, while lack of exercise eliminates normal wear and the toes become extremely long. Trimming in this situation is more difficult because the hoof is very hard and brittle.

Young animals are often neglected when it comes to foot care. Corrective trimming can be accomplished when animals are young and growing rapidly. If animals are kept on a soft manure pack, the feet often become extremely overgrown. This can affect development of the total leg structure during growth if it is not corrected. The hooves of young animals that are growing rapidly can be trimmed to correct not only excessive growth but also pigeon-toed, splayfooted (toe-in or toe-out) conditions.

Beef. Don't neglect checking the herd bull. Bulls should be checked 1 to 2 months before the breeding season for excessive hoof growth. A well-planned program can result in improved reproductive performance. The end result will be an increase in conception rate, the number of calves weaned, and profit.

Dairy. Most dairymen should have a well-planned foot-trimming program. It may be necessary to identify cows that need foot trimming every 6 months. Try to stay away from trimming gestating cows that are within 6 to 8 weeks of calving. The best time to perform this technique is several weeks before the end of lactation so that the feet can become toughened before the cows are turned out to drylot or pasture. This program can result in cows that will stay in the herd longer and can boost milk production.

Commercial foot trimmers, with portable equipment, are used on a regular basis on large dairy operations. Cows to be trimmed are selected by the herd manager.

Equipment Necessary

- Hoof-trimming chute, tilt table, rope halter
- Hoof trimmers
- Wood chisel
- Hoof nippers
- Search knives (farrier's knives)
- Foot-trimming knife
- Mallet or hammer
- Rasp
- Electric sander
- $4' \times 8' \times {}^{3}\!/_{8}''$ plywood (or wood door)
- Block or box ($24'' \times 16''$)
- Manila rope ($75' \times {}^{1}\!/_{2}''$)
- Manila rope ($6' \times {}^{3}\!/_{8}''$)
- Iodine (7% solution)

Restraint Required

Animal restraint is required while trimming the feet or hooves of cattle. This restraint may be complete, as in the use of a tilt table and hoof-trimming chute, or partial when manually lifting individual feet for trimming. A rope halter to hold the animal's head and a second person to assist are required regardless of the restraint method used.

Restraint of the animal and its legs is rather unique to this technique. For the sake of discussion, foot-trimming restraint is divided into four common systems.

Manually Lifting. This is the safest method of restraint for the animal and requires two people, one person to lift the leg manually, and the other person to trim the foot. It causes the least amount of stress and injury to the animal. It is, however, very difficult for the person holding the animal's leg.

The manual lifting method of restraint is used more in dairy operations than in beef operations. Extremely high-strung, nervous, wild, or mean animals are difficult, if not impossible, to restrain in this manner.

When lifting the front foot, stand in front of the front leg, facing toward the rear of the animal. Clasp the hands around the front leg at the fetlock joint, with the thumbs placed under the dewclaws. Pull upward on the leg and dewclaws to pick up the foot. Flex the leg at the elbow and the shoulder, positioning the foot so that the trimmer can proceed. If only one person is available, the trimmer can pick up the foot in this manner and place it on a block or box for support. This is sometimes difficult, especially if the animal does not cooperate.

The hind foot can also be lifted manually by one person. The procedure is to stand alongside the leg, facing the rear of the animal. Place your thigh just below the stifle, grasp the foot below the dewclaws at the fetlock joint, and pull upward in the same manner as with the front foot. With one motion, pull the leg up while flexing the fetlock. If the animal cooperates, the foot can then rest on the block or box for trimming. Lifting the hind leg almost always requires two people. Once the foot is placed

of chute can be homemade and could be used for techniques other than hoof trimming. This method works best with animals that are tame and broken to lead, but it can be used on any animal. Animals should be moved or led into the hoof-trimming chute and the head secured in the headgate. Tie the animal's head down with the halter and a quick-release knot to minimize animal movement; three 6- to 8-inch belts should be placed under the animal to help support it. The belts should be positioned behind the front legs, just in front of the navel or sheath, and near the rear flank (in front of the udder on cows). Apply only slight tension to the belts—just enough that you *do not* lift the animal off its feet.

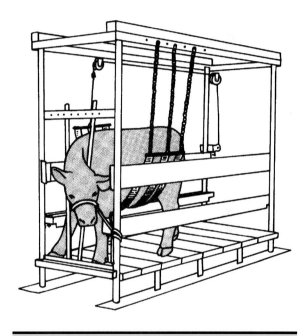

CAUTION: If too much pressure is applied to the belts, or if the belts are not used, the animal can get twisted in the chute. If the legs become caught between the floor of the chute and the bottom of the 4 × 6 (the 4 × 6 supports the foot and leg while trimming), the animal can break a leg as it struggles to regain its feet. The chute should be designed with the 4 × 6 secured on each end with a bolt and wing nut so that it can be removed quickly.

on the box, one person applies downward pressure on the leg just above the hock. This helps to keep the leg immobilized while the second person trims the foot.

This system allows you to place the foot on the ground periodically to make sure that the bottom of the foot lies flat on the ground at the proper angle.

Hoof-Trimming Chute. The use of the hoof-trimming chute can require two people. This type

A 6- to 8-foot piece of ³⁄₈-inch rope equipped with a honda loop (not a quick-release honda) should be placed around the animal's back leg and slid down between the dewclaws and hoof head. One person should pull backward and upward on the rope to pull the animal's leg up. Place the front part of the animal's leg on a 4- by 6-inch board (top approximately 18 inches from the floor of the chute) while the second person applies downward pressure to the leg just above the hock. This will help to immobilize the leg.

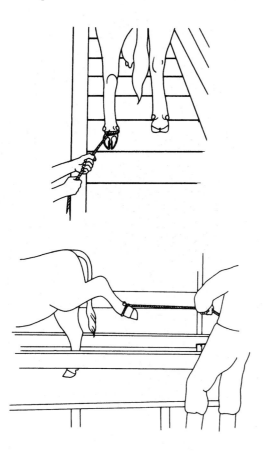

As you pull the animal's leg upward, it may begin to kick with the raised leg. In most cases the animal is not trying to kick you; it is simply experiencing a new situation and attempting to avoid the restraint. Once the animal realizes that it cannot escape the rope, and the second person applies downward pressure to the leg just above the hock, it will usually stop kicking. Continue to restrain the leg with downward pressure on the hock until trimming is completed.

Raise only one foot off the floor at a time so the animal can remain standing.

Wrap the rope around the 4 × 6 several times, once between the dewclaws and hoof head and once or twice above the dewclaws on the cannon bone. The person applying presure to the hock should then hold the rope tightly.

CAUTION: The animal may try to move its leg back and forth on the board, and may rub some hair off the leg if it is not secured tightly with the rope. It may be necessary to apply additional downward pressure on the hock as you anticipate animal movement.

If the animal goes down in the chute, release the rope securing the foot and relieve tension on the

hock immediately to prevent injury to the animal. Make the animal stand up by placing your hand over its nose, completely cutting off its air intake. In 15 to 30 seconds the animal will move its head from side to side as it tries to get air. When the animal finds that it cannot get air this way, it usually stands up. This method is not cruel, because the animal can breathe through its mouth.

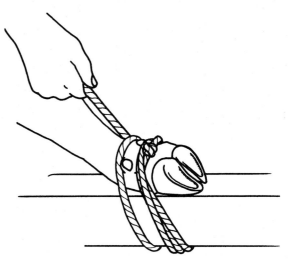

The front foot is lifted and restrained in the same way as the back foot, with one difference. Instead of applying downward pressure on the hock, hold the front leg with a hip and one hand. Position your hip in front of the leg on the 4 × 6 while holding the leg against your hip with one hand and the rope in the other.

This restraint method allows the foot to be placed on the chute floor for checking without a great deal of difficulty. If more trimming on the bottom is necessary, then the leg can be lifted and restrained as before. The experienced hoof trimmer generally does not need to put the foot down for checking until he has finished trimming the bottom of the foot.

Tilt Tables. Tilt tables, which offer total restraint, control, and comfort to the animal and the handler, are becoming increasingly popular. There are many types of tilt tables, from the modern hydraulically controlled types to homemade models that work well.

The animal is restrained against a vertical platform at the head with a headgate or a rope halter. The animal is also secured with a wide belt around the heart girth and around the flank. The table is then tilted either by a hydraulic cylinder or with a ratchet-type mechanism. When the animal is lying flat, each foot is secured separately to allow accessibility to the entire foot surface and to prevent the animal from kicking and thrashing its feet.

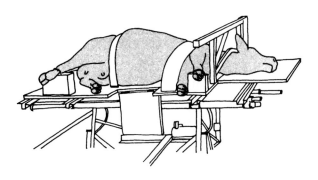

CAUTION: Make sure that the tilt table is secured or anchored to the concrete or ground to prevent the table from tipping over as the animal is tilted.

This method, again, requires experience in trimming the foot in an unnatural horizontal position, and you must be able to visualize what the foot will look like when the animal is standing. This restraint is probably the easiest for the trimmer and the quickest way to trim feet.

CAUTION: The animal should not remain on the tilt table for an extended period of time. Thirty minutes is the maximum.

Casting. This method requires that the animal be haltered with a rope 50 to 75 feet long and ½" in diameter. The rope is tied around the animal's neck with a bowline knot and brought back to the withers, where it is passed around the heart girth in a half hitch. After the half hitch is made at the withers, run the rope straight along the back and make a second loop under the body just in front of the udder and the hook bones. Secure the rope hitch in that position. The loose end of the rope is then pulled by a person standing at the rear of the animal. In pulling the rope, you are applying pressure to vital nerve areas, which causes the animal to fall.

CAUTION: Keep tension on the rope during the entire hoof-trimming process to keep the animal immobilized. If you don't, the animal will begin to thrash its feet or stand up.

This type of restraint should be applied in a well-bedded pen, a pasture, or a dirt lot. It is not advisable to use this method for cows that are within 6 to 8 weeks of calving.

Casting does not allow the trimmer to check the hoof on a flat surface before releasing the animal. If the animal is allowed to stand up for checking, and more trimming is found to be necessary, the animal must be recast. This would cause more stress to both the animal and the handler. This method of animal restraint should be used only as a last resort when other equipment is not available.

Regardless of the restraint method that you choose, remember that the animal's safety as well as your own must be foremost in your mind.

Step-by-Step Procedure

1. After the animals have been identified and assembled, assemble all of the necessary equipment. Be sure that all the cutting equipment is sharp.

2. Restrain an animal.

3. Look closely at the foot that you are about to trim. Try to visualize what you want to trim.

4. Start trimming by taking some of the excess growth off the front of the toes. Hoof trimming usually employs the use of the hoof trimmers or the wood chisel (or hoof knife) and hammer when the foot is manually held or when a trimming chute is used. The hoof trimmer is fast and, with its long handles, safer for the person doing the trimming. It can, however, easily cut too deeply and injure the foot. It is best to use one cutting edge at a time by moving only one handle. In this manner the toe can be shortened and rounded at the same time.

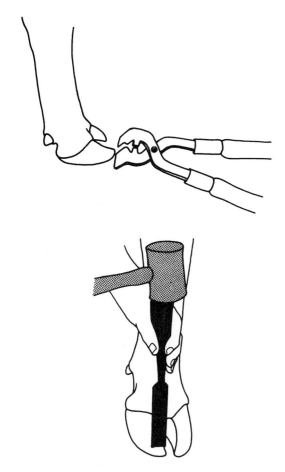

If the chisel (or hoof knife) and hammer are to be used, the wood floor of a hoof-trimming chute can be used, or the foot can be placed on a 4 × 8 sheet of ³/₈" plywood or an old wooden door. Place the chisel on the toe at the location you wish to cut. Hit the chisel (or hoof knife) a sharp blow with a hammer or mallet. When animals are restrained on a tilt table or by casting, the hoof nippers are usually used to shorten the toes. The main principle is to make the toe length in proper proportion to the animal and hoof size. Each situation is different and there are no hard-and-fast rules to follow. This takes experience.

CAUTION: Cut or trim in small bites. If you try to cut too much at one time, you run the chance of cutting too deeply.

The old saying, "You haven't gone far enough unless you draw blood," is absolutely incorrect. Trimming too short will cause lameness and possibly permanent hoof damage. "When in doubt you've gone far enough."

If you want to correct a splayfooted condition in a young animal, it will be necessary to take more off the front of one toe than the other toe. If the animal toes out, trim the front of the inside toe slightly shorter than the outside toe. If the animal toes in, trim the front of the outside toe slightly shorter than the inside toe.

5. Lift the foot and clean the sole with a search knife. A normal hoof that needs trimming has some degree of concavity to the sole (raised horny rim around each individual toe), which causes the horny outer wall or rim to bear most of the animal's weight.

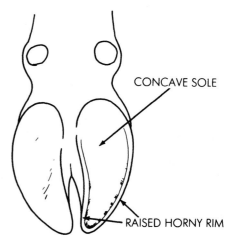

6. Begin trimming the bottom of the hoof at the heel with the hoof nippers, hoof chisel, or hoof knife and work at an incline toward the toe.

CAUTION: Usually, very little trimming is required at the heel. Heels are often shallow or worn, and they can be soft and tender. Trimming the heel in this case can cause lameness and defeat the purpose of hoof trimming. By trimming more off the tip of the toe, you make the animal put more weight on the toe and allow it to stand in a more nearly correct manner.

As you trim the bottom of the hoof, trim the rim on both sides (inside and outside) of each toe so that the bottom of the hoof is smooth. This will help to distribute the weight of the animal over the entire (bottom) surface of each toe and prevent accumulation of stones and manure in the center of the toe.

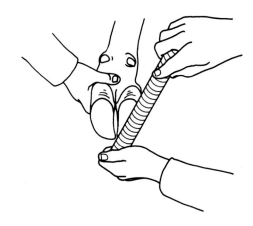

CAUTION: Be sure to trim with small bites. Any deep cut can go too deep and draw blood. If you trim slowly, you can watch the color of the hoof; as it gradually turns pink, it is time to quit. Some animals do not show the pink color, and it is necessary to take your thumb and press it against the center of the toe from time to time. When the center of the toe begins to feel soft, it is time to quit regardless of color.

If you should go too deep and draw blood on the bottom of the hoof, apply a strong tincture of iodine (7% solution) or another prepared medication. This will help reduce infection, but the animal will almost always limp for at least 1 to 2 weeks.

CAUTION: Extreme care must be taken in the use of the electric sander, because it rapidly produces excessive heat while sanding. This heat can cause internal damage to the foot and defeat the purpose of foot trimming. Apply the sander to the hoof for a few seconds and then let it cool. Repeat this process until the hoof is smooth.

7. If the tips of the toes come together after the top and bottom of the hoof have been trimmed, it will be necessary to trim the inside of the toe tips. Place the hoof knife or chisel between the toes and slice off the curved portion of the tips so that the toe tips are straight and some space exists between the tips.

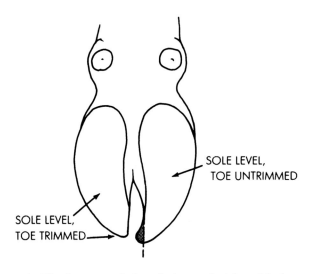

SOLE LEVEL, TOE UNTRIMMED

SOLE LEVEL, TOE TRIMMED

8. The bottom of a hoof trimmed with a chisel or hoof knife remains fairly smooth. When much of the trimming on the bottom of the hoof is done with the hoof nippers, however, the edges of the hoof and some of the bottom of the hoof may be rough. A rasp is useful in smoothing these surfaces. A power sander often is used for smoothing the sides and bottoms of the hooves.

9. Take another look at the finished foot and try to visualize the normal or ideal foot that you are trying to form. By doing this, you should be able to avoid retrimming or redoing the job immediately. One way to check the bottom of the foot of an animal on a tilt table or in a foot-trimming chute is to use a piece of plywood (6" × 6") and push it against the bottom of the foot. This will give you some idea of what this foot will look like when the animal is standing. If the plywood rocks on the foot, you will be able to determine where the high spots are and where you should concentrate your trimming.

10. If any soft spots or abscesses have been found, they should be thoroughly cleaned and treated. A strong tincture of iodine (7% solution) is commonly used for this treatment. Severe abscesses may require the attention of a veterinarian.

Postprocedural Management

Be sure that the animal has proper care after having had its feet trimmed. This is especially important when the animal's feet were very long or abnormal before trimming. Trimming these types of feet exposes sensitive tissue that is vulnerable to bruises and infection. In severe cases where much trimming has been done, the animal will often limp for several days to 2 weeks because of a change in the angle of the foot. To minimize problems after hoof trimming, the animals should be kept in a clean area for approximately 1 week whenever it is possible to do so. Most important, do not expose animals to abrasive surfaces, gravel lanes or pastures, poorly bedded stalls, or rough, frozen ground. These types of surfaces can cause bruising and possibly infection.

If you cut too deeply and the hoof bleeds during the trimming process, thoroughly clean the infected area and apply a strong tincture of iodine (7% solution) or another prepared medication daily to prevent infection. The animal should be placed in a clean, dry area without abrasive flooring. If any infection appears during the healing process, a broad-spectrum antibiotic should be injected intramuscularly.

3.23 CONTROL OF FOOT ROT

Foot rot is a bacterial or fungal infection which causes swelling and cracking of the soft tissue between the toes and in the hoof head. This condition usually results in lameness. Lameness can cause reduced milk production and performance in animals. Foot rot occurs most frequently in cattle housed or fed in muddy or wet surroundings, but can occur in any housing or feeding area. Exposure to frozen ground, extremely rough, hard soil, and large crushed rock also contributes to infection of the foot. Winter thaws and wet spring conditions contribute to the problem. The ability of the producer to detect the problem early and to treat it immediately will minimize the harmful effects of foot rot.

Equipment Necessary

- Topical dressing
- Search knife
- Squeeze chute, headgate, stanchion, or halter
- Syringe
- Antibiotic
- Footbath
- Copper sulfate (blue vitriol)

Restraint Required

To treat cattle for foot rot, restrain them in a squeeze chute, headgate, or stanchion. For cattle that are broken to lead and are tame, a halter restraint may be sufficient.

Step-by-Step Procedure

1. Observe and identify animals showing signs of foot rot. These animals are usually lame, the area above the foot is swollen or red, and the foot is inflamed between the toes.

2. Move the suspected animals to a handling facility.

3. Assemble the necessary equipment. Be sure that the equipment is cleaned and sanitized.

4. Restrain the animal.

5. Manually lift the foot (see Section 3.22).

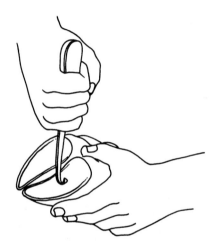

6. Clean the underside of the hoof with a search knife. This is particularly important because limping or lameness could be caused by conditions

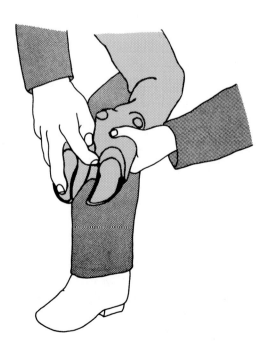

other than foot rot—for example, puncture wounds or abscesses in the hoof or heel of the foot. Foot rot is best identified by the smell of the infection and exudate (fluid oozing from the infected area). It has a pungent odor different from the odor of manure. Passing a finger or paper towel between the toes will help you to identify this odor.

7. Apply the topical foot-rot dressing to the infected area or areas. The topical dressing is applied by pouring it directly from the container onto the infected area. Use enough to cover the infected area thoroughly.

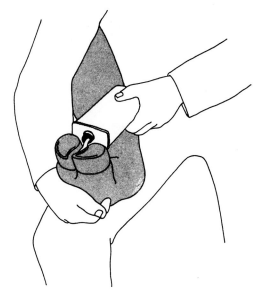

8. In persistent or severe cases of foot rot, it may be necessary to inject a broad-spectrum antibiotic intramuscularly.

9. Release the animal from the restraint and place it in clean, dry quarters.

10. In dairy herds where the problem is prevalent or an outbreak of foot rot appears, a footbath containing copper sulfate should be used. The footbath should be emptied, cleaned, and a new solution added once a day for a week or 10 days. Dairy herds can use the footbath easily when it is located in the exit of the milking parlor.

A footbath is usually 3 or 4 inches deep and long and wide enough that the animals have to walk through it and cannot jump or walk around it. It can be constructed of metal, plastic, or concrete. The footbath is so constructed that it can be easily cleaned.

Some animals may resist walking through the footbath. Locating a footbath in a narrow entrance to a feeding area or in the exit of the milking parlor in a dairy reduces the reluctance of animals to walk through it.

CAUTION: The animal may jump or move when you touch the infected areas. Be prepared to move away or to restrain the animal if necessary.

CAUTION: Milk-withholding times must be followed on lactating dairy cows treated with antibiotics. In addition, only drugs approved by the FDA for use in lactating cows can be used without a prescription from a veterinarian. Label treatment guidelines must be followed.

Postprocedural Management

After treatment, the animals should be observed twice a day. If swelling and lameness are not reduced by the second day after treatment, treat the animal again. If the condition persists beyond the fourth day after the initial treatment, you may be treating for the wrong problem, and a veterinarian should be consulted.

DAIRY CATTLE BREEDS

JERSEY COW

Reprinted with permission from Hoard's Dairyman.

HOLSTEIN COW

Reprinted with permission from Hoard's Dairyman.

GUERNSEY COW

Reprinted with permission from Hoard's Dairyman.

BROWN SWISS COW

Reprinted with permission from Hoard's Dairyman.

AYRSHIRE COW

Reprinted with permission from Hoard's Dairyman.

RED AND WHITE COW

Courtesy of Red & White Dairy Cattle Association

MILKING SHORTHORN COW

Courtesy of Department of Animal Science, Oklahoma State University

MILKING DEVON COW

Courtesy of Department of Animal Science, Oklahoma State University

chapter |FOUR|

Dairy Cattle Management Techniques

4.1 INTRODUCTION

Dairy animals are large, ranging in size from 900 to 1,800 pounds or larger for females, with males weighing as much as 3,000 pounds. Birth weights vary from 60 to 125 pounds. The female reaches puberty at about 10 months of age, but breeding is usually delayed so that the first calving occurs at from 22 to 26 months of age.

Dairy animals are *polyestrous*, meaning that they come into estrus (heat) every 21 days throughout the year. Thus, it is possible to have calves born at any time of the year. Because the gestation period is from 270 to 280 days, the cow will usually have just one calf each year. At times, the cow may have twins or triplets, but multiple births are not common.

Dairy cows were first introduced into the United States in the early 16th century. From this first importation, the dairy cow population of the United States cows grew to 9 million by 2004. The United States exports breeding stock to many parts of the world.

The dairy cow is important to animal agriculture in the United States. She is an efficient producer of human food—milk. The average milk production of the U.S. dairy cow is 18,957 pounds of milk in 10 months. The top milk production is 76,064 pounds of milk, 2,434 pounds butterfat, and 2,358 pounds protein in 365 days. This great production was

achieved in 2004, by a Holstein cow owned by Hartje Meyer Dairy. With these production figures, it is easy to see how the dairy cow has earned the title "foster mother of the human race."

The genetic potential of the dairy cow has been increased through the use of artificial insemination using production-tested bulls. Artificial insemination of dairy cattle became an accepted practice around 1946.

The average age of the dairy cow in most herds is nearly 5 years. Unprofitable cows are removed from the herd annually. In large commercial dairy operations, cull rates will exceed 30% each year. Some cows will remain in the herd for over 10 years; however, most cows in the herd will be younger. During each year of her life, a good dairy cow should produce a calf and be in milk production for 300+ days. The female calves she produces are used for replacement animals, while the male calves are used for the production of veal or finished out as beef-type steers. When the cow reaches the end of her useful life as a milk producer, she becomes a source of beef for human food.

There are six major breeds of dairy animals in the United States. Listed in alphabetical order, they are Ayrshire, Brown Swiss, Guernsey, Holstein, Jersey, and Milking Shorthorn. The Holstein and Brown Swiss are the largest of the dairy breeds in size and produce large quantities of milk containing a low percentage of butterfat. The Jersey is the smallest of the dairy breeds and will produce smaller amounts of milk of a higher butterfat and higher protein content. The Ayrshire, Guernsey, and Milking Shorthorn breeds are the medium-size animals that produce milk intermediate in volume and butterfat content. The Holstein is the most common breed in the United States.

Dairy animals are easily trained so that day-to-day management is easy. They can be trained to lead with a halter, to enter a milking parlor, to open swinging exit doors, to return to their housing area, or in the case of comfort or tie stalls, to know which stall is theirs and to go to this stall when returned to the barn. With gentle and easy handling, most dairy animals become tame and can easily be approached in the feedlot or pasture.

Dairy cows are *homeotherms* like most farm animals, which means that they maintain a constant body temperature regardless of the environmental temperature. They are *ruminants*, which means that they are cud-chewing mammals, and they have a stomach consisting of four divisions or chambers—the rumen, reticulum, omasum, and abomasum. Being ruminants, they have the ability to eat large amounts of forages (roughages) such as grasses and legumes and, by the action of the rumen and regurgitating and chewing of the cud, are able to digest these long-fibered materials and produce a nutritious and wholesome human food: milk. They need a balanced ration each day that contains the right proportions of energy, proteins, vitamins, and minerals. The cow must have large amounts of fresh water available at all times. A cow usually consumes between 3 and 4 gallons of water for each gallon of milk produced.

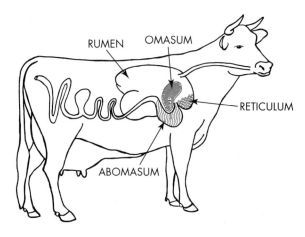

Dairy cattle are adaptable to many types of housing. They are capable of efficient production within a wide range of temperatures, but the optimum temperature is between 45 and 75°F. Extremely warm temperatures (above 85°F) and high humidities (above 75%) and temperatures below 10°F tend to reduce the efficiency of production. The effects of high temperature can be reduced by furnishing shade or by housing in areas where air movement can be increased by the use of fans. In hot, semiarid climates, corral misting, exit-lane cooling, and holding-pen cooling are used to reduce heat stress in high-producing dairy cows. Fans and misters are used in addition to shades. The effect of low temperature can be reduced by housing that will protect the animal from the extreme cold, particularly the direct wind.

To increase the efficiency of the cow, cleanliness is a must, not only for the comfort of the cow but for the production of high-quality milk. Dryness is important because it helps to keep the cow comfortable and will help to reduce the incidence of respiratory problems. Freedom from drafts helps to reduce stress on the animal and controls respiratory problems. A comfortable cow is easier to handle and produces more efficiently than a cow that is stressed.

The type of housing that is suitable varies with geographic location. Open lots with shade are used in warmer climates and covered barns are usually used in colder, moister areas.

Dairy barns are usually one of four different types.

1. *Free-stall housing.* This is a system where the cow can pick a free stall that is 4 feet wide and 8 feet long. The stall should be well bedded with clean, dry material in a well-ventilated barn. The walkway or alleyway behind the free stalls serves as a manure collection area and should be scraped free of manure at least once each day. Flush systems are used in many new facilities to clean alleyways in free-stall barns. These systems reduce labor associated with cleaning facilities. The cow should be fed in another area of the barn or outside in a feedlot.

2. *Loose housing.* In the loose housing system, the cow is allowed to rest in any area of the barn she chooses. The whole barn is bedded to make it clean and comfortable for the cow. Feeding is usually done in a separate area of the barn or in an outside feedlot.

3. *Comfort stall–tie stall–stanchion barns.* In this type of barn the cow has her own stall in which she is tied, and all of her feed is brought to her. The stall is bedded to make it more comfortable, and the manure collects in a gutter behind the cow. The manure is removed mechanically at least once each day. The main consideration in all types of housing is to keep the cow clean, dry, and free from drafts.

4. *Open lots.* Large dairy operations often use open lots to separate milking cows into production groups. These facilities are constructed to hold 80 to 200 cows per pen. These dirt lots are spacious, with concrete feeding and lane areas. Straw-bedded areas near windbreaks are used during winter months to provide comfort. Feed areas are scraped or flushed daily. Manure is removed from the dirt areas on a regular basis. Open lots provide low-cost facilities in large operations.

The efficiency of the cow is influenced by many things—housing, management, genetic potential, and above all, a balanced ration. The herdsman's ability to combine all of these factors will determine the success or failure of the dairy enterprise.

4.2 IDENTIFICATION

Identification of dairy animals is a requirement in the everyday management of the herd. It is essential in recording parentage, in registration of offspring, for heat detection, milking, and feeding, for the health program, and at sales and exhibits. An ideal identification system should be: (1) permanent, (2) visible and readable at a distance, (3) visible in the milking parlor, particularly in the milking parlor pit, (4) acceptable to the dairyman, and (5) reasonably priced. Unfortunately, none of the present systems of identification meet all of these requirements.

The identification methods presently in use include: (1) ear-tagging; (2) neck chains or rope and number; (3) ankle band; (4) brisket tag; (5) tail tag; (6) temporary identification such as the back tag, chalk or grease marker, paint, and tape; (7) picture or sketch; (8) tattooing the ear; (9) udder tattoo, (10) electronic scanner; and (11) branding (freeze or hot). The characteristics of each of these systems are given in Table 4.1.

TABLE 4.1   Advantages and Disadvantages of Dairy Herd Identification Systems

System	Advantages	Disadvantages
1. Ear-tagging (plastic tag)	Easily applied, easily read	Easily lost (plastic tags seem to last about 2 years, then break off). Not permanent. Can't be changed from one animal to another. Can't be seen from parlor pit.
Ear-tagging (metal tag)	Easily applied	May last longer than plastic but not permanent. Hard to read from a distance or in parlor.
2. Neck chain or rope	Easy to apply, painless to the animal, fairly easy to see	Not permanent. Can be dangerous to the animal—chain can catch on a nail or bolt or fence and choke the animal. New rope types with slipknots prevent this. Hard to see in parlor or when animals are closely bunched.
3. Ankle band	Easily seen in milking parlor, easy to apply	Not permanent; may be hard to read because usually dirty. May cause irritation to the ankle.
4. Brisket tag	Usually more permanent than ear tags	Hard to see in milking parlor or when animals are in close groups.
5. Tail tag	Easily seen in parlor, easy to apply	Not permanent. Many times they are dirty and hard to read. Not easily seen in groups of animals.
6. Temporary methods; usually used to identify cows treated with antibiotics when milk must be discarded, sale animals, cows in heat, dry cows, for judging contests	Easily used	Not permanent.
7. Picture	Permanent (for spotted animals)	Difficult to carry around; not easily used on a day-to-day basis, especially in large herds.
Sketches	Permanent (when drawn accurately)	Difficult to carry around; not easily used on a day-to-day basis, especially in large herds. Difficult to sketch accurately.
8. Tattoo (ear)	Permanent, easily applied	Hard to read from a distance; usually have to catch the animal. Can't be used to identify animals in a group or in the milking parlor.
9. Tattoo (udder)	Can be seen in the parlor	Not permanent; hard (and dangerous) to apply. Hard to see when cows are in a group.
10. Electronic	Quick, easily used, usable in milking parlor	Hard to use in feedlot or pasture on day-to-day basis. Often, electronic identification is used in addition to another, visible identification system.
11. Branding	Permanent, easily read	Not accepted by dairymen with registered herds, hard to get a good brand (an open 8 can look like a 3), may be obscured by dirt when animal is ridden by others when in heat, can't be seen from the parlor pit unless applied on lower thigh.

Ear tags are the most commonly used method of identification of dairy cattle. They come in many sizes and shapes, can be one- or two-piece, preprinted (numbered or lettered) or blank, and metal or plastic. The larger plastic tags are the most popular. Two-piece plastic tags are gaining in popularity because they can't be changed from one animal to another and seem to last longer.

Blank tags should be numbered or lettered the day before you plan on using them because it takes time for the ink to dry properly. For plastic tags, use marking fluid or paint; these seem to "melt" into the plastic and bond securely to the tag. Magic Marker–type inks may work well for temporary identification, but they fade and rub off too quickly. Blank metal tags should also be numbered before you start the step-by-step procedures.

The neck chain or rope and number is the second most accepted identification system. The brisket tag, ankle band, and tail tag are the least used for the identification required for day-to-day management of the herd. For permanent identification, required for registration papers or records in the herd, a picture or sketch of the color markings is the method most often used. Tattooing is also used, particularly in the Brown Swiss and Jersey breeds.

Many herds are identified with two methods. One is usually permanent (and often very difficult to read easily), while the second is visible and easy to read. The two methods are cross-referenced in the herd records. For example, a herd might be using a nonvisible permanent tattoo and a visible number on a neck chain.

The temporary identifications such as back tags, chalk or grease marker, paint, and tape are used to identify animals that have been specially treated in some manner for short-term identification of new animals entering the herd. In some herds the dairyman recognizes each animal by sight. Identification can still be a problem in these herds when it is necessary for someone else to perform the management of the herd.

Restraint Required

Animals that are identified soon after birth need very little restraint. Older animals require restraint with the halter, headgate, stanchion, or squeeze chute. A squeeze chute equipped with head and nose bars makes the job considerably easier.

Because each system of identification requires a somewhat different method of application, a step-by-step procedure is given for each.

Ear Tags

Ear tags are the most commonly used method of identification in both dairy and beef herds. For a complete discussion of ear tags refer to Section 2.2 of Chapter 2.

Neck Chain or Rope and Number

Equipment Necessary

- Neck chain or rope
- Number (for the neck chain or rope)
- Halter

Step-by-Step Procedure

1. Assemble the necessary equipment.

2. Restrain the animal. Usually a stanchion or halter is sufficient. Some animals require the restraint offered by the nose tongs.

3. Position the chain around the neck, making sure that the tag has first been slipped on the chain.

4. The chain or rope should be tight enough that it will not slip over the head, but loose enough that it will not interfere with breathing.

CAUTION: Neck chains or ropes put on young animals must be checked periodically because as the animals grow, the chains will become tight. If neglected, they can interfere with breathing and may choke the animal.

Be on guard against the animal's making a quick move with its head and hitting you while you are applying the neck chain.

5. Release the animal and allow it to return to the housing area.

6. If the number on the neck chain is different from the herd number of the animal, write down the identification number applied to the animal so that it can be recorded in the permanent records.

Ankle Bands

Equipment Necessary

- Ankle band
- Marking paint or fluid
- Kow-Kant-Kick

Step-by-Step Procedure

1. Assemble the necessary equipment.

2. Restrain the animal. The stanchion or headgate is usually sufficient.

3. Apply the ankle band around the rear leg. Follow the recommendations of the manufacturer.

> **CAUTION:** The animal may kick while you are trying to apply the ankle band. If she persists in moving about or kicking, the tail hold may be necessary to restrain her; in extreme cases, the Kant-Kick should be used.
>
> The ankle band should be loose enough to permit free blood circulation but tight enough to keep the band from slipping or rubbing.

4. Release the animal and return it to the housing area.

5. Enter the number of the ankle band on the permanent record of the animal.

Postprocedural Management

Observation of the animal for 3 or 4 days is required to make sure that the ankle band is tight enough and that no problems are developing.

Brisket Tag

Equipment Necessary

- Brisket tag
- Brisket tag applicator
- Marking paint or fluid
- Antiseptic
- Squeeze chute
- Halter

Step-by-Step Procedure

1. Assemble the necessary equipment.

2. Restrain the animal. Restraint must be complete because the brisket tag must go through the skin of the dewlap and the animal will resist the piercing of its skin.

3. Apply the brisket tag as recommended by the manufacturer.

> **CAUTION:** The animal will react to the application of the brisket tag. She may strike at you with her front feet or jump and move about rather rapidly. Be prepared to protect yourself, particularly your hands, by more complete restraint of the animal or by moving away.

4. Apply an antiseptic to the pierced area.

5. Release the animal and allow it to return to the housing area.

6. Record the brisket tag number on the permanent record of the animal.

Postprocedural Management

Make daily observations for a week to check for infections and treat them as necessary.

Tail Tag

Equipment Necessary

• Tail tag

Step-by-Step Procedure

1. Assemble the necessary equipment.

2. Little if any restraint beyond a halter or stanchion is necessary.

3. Apply the tail tag as recommended by the manufacturer. The tail tag is usually applied in the switch (long hair) and is not likely to cause any damage or discomfort to the cow.

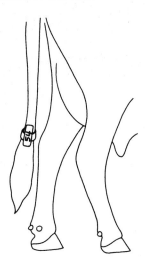

4. Release the animal and allow it to return to the housing area.

5. Record the number on the permanent record of the animal.

Temporary Identification

Equipment Necessary

• Back tag
• Chalk or grease marker
• Paint (spray can)
• Tape (masking or adhesive)
• Halter

Step-by-Step Procedure

1. Assemble the necessary equipment.

2. Restrain the animal. This normally is done with a stanchion or halter.

3. Apply the temporary identification to any part of the animal's body. The location chosen depends on the reason for the temporary identification. If it is for identification in the milking parlor, apply tape around the rear leg so that it can easily be seen from the parlor pit. If it is for sale purposes, apply a back tag on each side of the rump. Chalk markers, grease markers, or spray paint applied to the back or side of the animal provide a good temporary means of identifying it.

4. Release the animal and return it to the housing area.

5. Record the temporary identification on the permanent record of the animal.

Picture or Sketch

Equipment Necessary

• Camera
• Film
• Blank sketch form
• Halter

Step-by-Step Procedure

1. Assemble the necessary equipment.

2. Move the animal to a satisfactory location for taking the picture. This is usually a smooth, dull background that will allow the markings of the animal to be easily seen in the picture. Animals that are trained to lead are easily moved and posed for a picture.

3. Brush or clean the animal's coat of all manure or dirt. Manure or dirt on the animal will look like spots or will cover the marking on the animal and cause faulty identification.

4. Attempt to pose the animal so that you can see the insides of the legs on the opposite side of the animal, the forehead, and the tail.

5. Take the picture.

6. Turn the animal.

7. Take a picture of the other side.

8. Move the animal back to the housing area.

9. The picture is mounted on the permanent record of the animal. If pictures are used for the registration application to the breed registry office, mount the pictures on the application certificate and send it to the registry office.

Before taking the photo, it is a good idea to have the animal's herd number or her name mounted on a card above the animal so that it will show in the photograph. You are then less likely to make an incorrect identification of the animal.

The sketch is drawn on the outline of the animal in much the same manner as the picture is taken.

Both sides are sketched to make identification easier. Be sure that the markings on one side of the animal match up on the opposite side as they come across the top line.

Ear Tattoo

The ear tattoo method of identification is used in both dairy and beef herds. It is required when registering Brown Swiss and Jerseys. For a complete discussion of the method, refer to Section 2.2 of Chapter 2.

Electronic Scanner

Some dairy herds are now using electronic identification (ID) systems. These systems make use of an ID implant and an electronic scanner that reads the ID and initiates a readout on a computer. These systems show promise, especially in larger operations.

Electronic identification can be used in conjunction with milking parlor computer software programs. Cows can be auto-identified during entry into the milking parlor. Information, such as milk yield and milking times, can be recorded for each cow during the milking process. Electronic identification is usually supplemented with a second, visible method of cow identification.

Branding

Branding is probably the least used of the identification systems for dairy herds. While branding is the most permanent of the identification systems, it has not been accepted by very many dairymen because it detracts from the general appearance of the animal and because it is difficult to apply. If it is used, it should be applied on the lower part of the thigh so that it will be visible from the milking parlor pit. A detailed discussion of branding appears in Section 2.11 of Chapter 2.

With all of the preceding identification systems, be gentle and patient and do not mistreat the animal.

4.3 NATURAL BREEDING

Most dairy herds are now using artificial insemination (AI) in their breeding programs. With AI, heat detection in cows can be a major problem, especially in larger herds. The proper use of a "cleanup" bull (natural breeding) can assist the dairyman in getting cows bred that are difficult to detect in standing heat or are "hard to settle," thus keeping them in the herd longer. Cows that have been in milk for 150 days since calving and have not been detected in standing heat, or cows that have been bred by AI three times or more and are still not pregnant, are prospects for the "bull lot."

The use of the cleanup bull usually creates problems in herd management. If the dairyman uses a cleanup bull, it is necessary to have a separate lot for the cows that are to be serviced by the bull. This means two groups of cows in milk and, in many operations, the housing facility makes this difficult.

Running a bull with the entire herd is not recommended; it allows cows to be bred back too soon after calving. Short lactations and short dry periods may result because accurate breeding records are not usually kept. In addition, some cows may be bred naturally when they should have been bred by AI.

In herds where a separate lot cannot be established, it is best to keep the bull in a bull pen and hand-mate the cows by moving them to the bull pen as they are observed in standing heat. Obviously, if the primary problem is in detecting (catching) cows in standing heat, one of the benefits of keeping the cleanup bull is defeated by this method.

> **CAUTION:** Make sure the bull is fertile. Have a qualified technician such as an animal reproduction specialist, AI representative, or a veterinarian evaluate the semen sample for sperm concentration, motility, and morphology.

Another disadvantage of using the cleanup bull is the possible poor genetic potential of the bull. Cleanup bulls are not production-proven sires as are most AI sires. Many dairymen raise their own cleanup bulls, usually out of one of their best cows, sired by the best AI bull being used. This, however, does not guarantee that this cleanup bull will sire heifers that will eventually improve the milk production of the herd; they might actually reduce the herd average.

Cleanup bulls will be placed with the cows in most large dairy operations. Usually, cleanup bulls will be added to selected groups of cows based upon reproductive parameters. Most cleanup bulls remain on the dairy for only a short period. Larger, older cleanup bulls create a greater danger to dairy employees. Larger bulls also increase the risk of injury to the cows during natural service.

Producing the Cleanup Bull

The selection of the cleanup bull, any bull to be used in natural mating, begins long before the animal reaches sexual maturity or puberty. Selection should begin while the animals are still calves. At this point, selection must be based on the prospect's pedigree. Milk production is the primary trait to be selected for in the dairy business.

Dairy producers should evaluate the potential of a cleanup bull prospect through use of an index called the "parent average." To calculate the parent average index, the producer needs access to the Predicted Transmitting Ability figures maintained by the USDA. If the sire of the calf prospect is a high-ranking bull (80th to 90th percentile) and if the dam of the calf is at or very close to "Elite Cow" status, then the prospect is a worthy one. In the day and age of flushing donor cows and embryo transfers, it is possible for a fortunate dairy producer to have several calves of this quality being considered for cleanup bull. When such is the case, other traits such as structural correctness become deciding factors.

Health and Breeding Soundness Exams

Before a cleanup bull is turned in with the cows, he must undergo a thorough health exam. He must be free of any diseases that are capable of being transmitted to the cows. In addition, he should undergo a thorough examination of his reproductive system. Both testes must have descended into the scrotum. If one of the testes has been retained in the abdominal cavity, the calf fails the breeding soundness exam and must be rejected. A sperm sample must be taken and evaluated. Bull calves reach puberty and start producing sperm at 7 to 9 months of age, some as early as 6 months. The sperm count should be high, but it must be remembered that the small testes of the bull calf will produce only about one-half the sperm of the fully mature bull testes. The average ejaculate of a mature bull contains about 1,000 million (one billion) sperm per milliliter of ejaculate, with no more than 20% "abnormals."

Using the Cleanup Bull

> **CAUTION:** Remember that any bull is potentially dangerous. He may seem to be a tame animal, but on any given day he may turn and severely injure or perhaps kill you. This is especially true when a cow is in heat and you try to remove her from "his" group or move the group to the holding pen for milking. Never handle the bull by yourself, and never turn your back on the bull.

Equipment Necessary

- Bull pen
- Breeding chute
- Bullring
- Staff
- Halter
- Headgate

Restraint Required

Young bulls less than 18 months old will usually require very little restraint; older bulls may require considerable restraint. The use of a leading staff snapped into his bullring will keep the bull away from you.

The cow usually requires no restraint. She may require restraint if you move her to the bull pen for breeding. This restraint would involve nothing more than a halter or a headgate in the breeding chute.

Step-by-Step Procedures

Before deciding to keep and use a cleanup bull, plan ahead about how you are going to house him and what system of mating is to be used—pen mating or hand mating. If you keep the bull with the cows in a pen, develop a plan to house the animals that have breeding problems separately from the other cows in the herd.

If you choose to hand-mate, use a bull pen that includes a breeding chute that allows the cow to be moved in and out safely. Plans for the construction of bull pens and breeding chutes are available from many universities and commercial plan services.

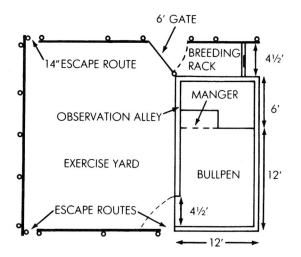

> **CAUTION:** The bull pen lot fence must provide escape routes and be built with heavyweight materials capable of restraining the bull. Heavy 2-inch-thick lumber bolted to 6" × 6" posts set every 4 feet and set 3 feet into the ground is recommended. Four-inch boiler pipe bolted or welded to steel beams set in concrete is also very satisfactory. Heavy cable run through steel beams or boiler pipe set in concrete also makes a good fence.

Regardless of the system that you use, as the bull approaches 12 months of age, put a bullring in his nose. The bullring can be put in by a dairyman or veterinarian using proper animal restraint. The use of a squeeze chute equipped with a head and nose bar makes the job easy and safe. Restraint using only a halter and tying the head securely can be used, but the job is more difficult and more dangerous because the head can still move. Use a self-piercing ring and place it about 1 inch from the tip of the nose through the soft, thin tissue that separates the two nostrils.

Hang a lightweight chain from the ring as an additional safety measure. The chain should be just

> **CAUTION:** Keep your hand and fingers away from the nose as you pierce the soft tissue with the bullring to prevent puncturing or cutting your hand.

long enough to reach the ground (3 or 4 feet); the bull will slow down to avoid stepping on it.

> **CAUTION:** The chain should not be hung from the ring if the bull will run in a wooded area or where he will not be observed at least once a day. The chain can become entangled in underbrush or in a fence and trap the bull away from feed and water.

Use of the Breeding Chute with Elevated Floor for the Bull

1. If the bull is running with the problem cows, the cows should be observed and breedings recorded. If the actual service has not been observed, a roughened tailhead and mucus around the vulva usually indicate that the animal has been bred.

2. If the bull is rather young and not quite large enough to reach and penetrate the cow for the mating, it may be necessary to move both the cow and the bull to a breeding chute. Some breeding chutes are constructed so that the bull will have a slight advantage in height.

3. The cow is moved into the breeding chute and the bull put into the chute behind her. If she is in standing heat and the bull is young and active, there should be no problem in his servicing the cow.

4. If the bull is unable to mate properly, he is too young or too small.

Bull Pen System

1. When using the bull pen, move the cow that has been observed to be in standing heat to the breeding chute near the bull pen.

2. The gate from the breeding chute to the bull pen is then opened, and the bull is given access to the cow.

3. After the bull has bred the cow, the bull dismounts and is still in his pen. It is therefore not necessary to enter the bull pen.

4. The cow is then returned to the housing area.

> **CAUTION:** A cow in heat may require two or more persons to move her from her group.
>
> Remember that the bull is dangerous. Do not enter the bull pen. Even younger bulls can become dangerous near a cow in heat. If you must enter the pen for any reason, be sure that another person is present. Before entering the pen, find and establish your escape route so that if the bull does start to move toward you, he will not trap you. The bull pen should always provide escape routes. At the first signs of consistent meanness in the bull, it is best to send him to market and secure another. Many dairymen have been killed by bulls that were "tame."

4.4 DRY-COW CARE

Care of the dry cow is one of the important tasks involved in managing a dairy herd. Proper care during this time can influence milk production during the next lactation, the health of the calf she is carrying, and the general well-being of the cow. Cows need a dry period to get ready for the next lactation. Cows should have a dry period of around 55 to 60 days. Cows with dry periods greater than 70 days or less than 45 days reduce herd profits. Neglect during the dry period can result in mastitis, feet and leg problems, weak calves, and obese cows that will have health problems after calving. Dry cows that get too fat may develop "fat cow syndrome," a term applied to general health problems that develop after calving. These health problems may include reduced appetite, ketosis, displaced abomasum, and reduced kidney and liver function.

Equipment Necessary

- Dry-cow mastitis treatment tubes
- Alcohol pads
- Collodian

Restraint Required

Restraint in a stanchion or headgate will be required if dry-cow treatment for mastitis is necessary.

Step-by-Step Procedure

1. Prepare a list of cows due to dry off (terminate lactation). Making this list in advance and keeping it current will allow the dairyman to plan ahead for their care. This list is furnished to the dairymen on the Dairy Herd Improvement (DHI) Production Testing Program.

2. Plan to provide a separate housing and feeding area for the dry cows. Several dry-cow pens may be required in large dairy operations. Groups might include cows recently dried, intermediate dry cows, and close-up dry cows. The close-up group would include cows 2 to 3 weeks prior to calving. This arrangement allows for special feeding for this group of cows prior to calving. A separate area will make it possible to feed a "dry-cow ration" to the dry cows and will keep them out of the milking parlor, thus making the milking operation more efficient. Dry-cow housing should be clean, dry, and free from drafts. In the spring, summer, and fall, and when available, a pasture or woodlot is an excellent place to keep the dry cows.

3. Mastitis treatment at dry-off is the most effective time to eliminate subclinical mastitis. Dry treatment can also help to prevent new infections during the dry period. Make a decision as to whether you are going to use selective or all-cow treatment. Selective treatment would be based upon a mastitis evaluation of each cow. Cows with mastitis problems during the current lactation would be likely candidates for treatment. Selective treatment has the dual advantages of reducing the total cost of treatment and the risk of antibiotic residues in the milk tank. However, a complete program of treating all cows in all quarters will eliminate existing infections and lower the incidence of new infections.

4. Secure a supply of dry-cow mastitis treatment tubes.

5. Dry off the cow. Dry-off occurs as the milk pressure in the udder equals the blood pressure, thus stopping the production of milk. There are two ways to dry off the cow. In both systems, it is helpful to reduce the feed and water intake during the drying-off process (usually a few weeks).

The first way to dry off the cow is simply to stop milking her. This is one method that is often overlooked by many dairymen. Some feel that the udder will fill with too much milk and thus cause mastitis. Most cows will not have this problem.

The second way to dry off the cow is to skip every other milking (milk once a day) for about a week, and then to stop milking completely. After three more days, milk out the cow and discard the milk. If the udder pressure builds up again, milk out the cow again. The problem with this method is that as you remove milk from the udder, the cow will continue to produce milk. This procedure involves more labor and prolongs the time required to dry off a cow.

Treat for mastitis immediately after the last milking (only for cows scheduled to be treated). See Section 4.9 for the proper procedure. Teat dipping (see Section 4.8) twice a day for a week after drying off is also a good practice.

CAUTION: Dry-treated cows should be moved immediately from the milking herd to the dry pen or clearly identified as "treated." Leaving treated dry cows in the milking herd creates a major antibiotic risk to the herd milk supply. Record all dry treatments. Dry-cow antibiotics have extended milk-withholding times. Cows calving well in advance of expected dates may produce milk with drug residues after calving. Good records are required to verify these withholding times.

6. Move the cow to the dry-cow housing area.

7. Feed a balanced ration available from your feed supplier or as recommended by a dairy nutritionist. The amount of ration to be fed depends on the body conditions of the dry cows. Most cows need to gain in body condition during the dry period, but they should not be allowed to get fat. Clean, fresh water must also be available at all times. Where the herd health program indicates the need, inject the dry cows with vitamins A, D, and E and a selenium preparation as recommended by the manufacturer.

8. Observe the dry cows daily. This observation is a must to spot cows that develop mastitis, which is indicated by an udder quarter that is swollen or "hot" (warm) to the touch. Observe the herd for cows that are limping, breathing hard, coughing, or not eating. If these conditions are observed, they must be checked and treated immediately.

9. As calving time approaches, the cow should be clipped and moved to a maternity pen, stall, or area. Move her into the maternity area 3 or 4 days before calving. With experience, it will become easier to estimate when the cow is nearing calving time. The first signs to look for are swelling of the udder, filling of the teats with milk, and a relaxation of the muscles around the tailhead.

In the spring, summer, and fall when the weather is fairly warm and reasonably dry, a maternity stall is not necessary if the cows are on pasture or have access to a clean, dry area where shade is available. A woodlot is excellent for this.

4.5 CARE OF THE YOUNG DAIRY CALF

In a beef operation, the brood cows raise their own calves. In fact, one of the "cull/keep" decision points for any rancher, relative to each of his cows, is whether she successfully raises a healthy calf each year. The fewer number of times management has to intervene to keep the calf healthy or growing rapidly, the better. Beef calves are tagged at birth, or shortly after, and they might be given some other management attention prior to weaning, but it is usually minimal. If the mother beef cow does her job, little attention is necessary until the calves are "worked" at weaning.

The management of the young dairy calf is nearly the opposite in every respect. It is not that the mother dairy cow could not care for her young heifer or bull calf; rather, the intense human management is necessary because the calves are taken away from the cows immediately after birth because the cows are turned into the milking string, where they will spend the next 300 days of their lives being milked twice a day. Any attention the dairy calves receive must be given by human attendants.

Bull calves and heifer calves, born in a dairy herd, take decidedly different routes after birth. The largest percentage of the bull calves ultimately become steers, are fed in a feed lot, and enter the human food chain as some form of beef product. The genetically superior bull calves may be raised to maturity, used in a bull stud, and have a tremendous impact on the future of the dairy industry. Heifer calves are either raised as replacements for the milking herd or fed in a feed lot and enter the food chain like the bull calves. In times of expanding dairies, most heifer calves with strong genetic backgrounds will be raised for replacements.

Today, replacement heifers are either raised on the farms where they were born and where they will enter the milking string as 2-year-olds or "farmed out" to contract replacement heifer growers. These replacement growers will take the heifers directly from the farm when they are 1 or 2 days old and return them 2 years later—healthy, ready to calve and "freshen," and then enter the milking string.

Step-by-Step Procedure

1. The first step in raising the dairy calf is to successfully calve out the cow. Detailed discussions of normal and abnormal calving procedures can be found in Chapter 3, Sections 3.11 and 3.12. It is important to review and thoroughly understand these calving procedures.

Dairy cows, because of the way they are managed on most dairies, will normally calve in a dry-lot group setting. This leaves much to be desired in terms of cow comfort, hygiene, protection from the elements, and visibility. Many dairies are recognizing these shortcomings of group calving and are attempting to establish some type of calving barn where cows can deliver their calves in a stalled area, protected from the elements, well-lighted for those times when assistance is needed, and attended by a calving manager. These calving barns are not elaborate, but they are entirely functional. Most are open-sided, but some are enclosed, intensively managed facilities.

2. After the calf is delivered, it must be cared for immediately. The first thing to do is to be certain that the calf is breathing. If it is not, there is no time to waste. Action must be taken. The following are procedures that the calving barn manager must be capable of performing.

- Be certain that the mouth and nostrils are free of the placenta. Sometimes the placental membranes are tough and do not tear during delivery. They could be covering the nose and mouth and prevent the calf from breathing. Clearing the airway is a simple process. Simply tear the membranes away.
- Be certain the mouth and throat of the calf are clear of mucus. A big breath, taken just as the calf is passing through the birth canal, can cause mucus to be swallowed. This can block the breathing process. Clear the throat with a small hand or with a few fingers inserted into the mouth, to the back of the throat.
- If the calf is still not breathing, alternately press and release the rib cage of the calf. Sometimes the calf just isn't vigorous enough to get it all started without the stimulation of the rib cage "pumping." The calf's ribs can take some pretty vigorous pumping, but use common sense. A 200-pound man can break the ribs of the calf if he exerts full weight upon them.
- Use a twig of straw, insert it into the nose of the calf, and "tickle" the inside of the nostril. Sometimes this will stimulate breathing or a "sneeze."

- Use an artificial resuscitator made especially for calves. Small dairy farms may not have one of these on hand, but if many calves are being born each day, it will pay for itself many times over. Consider this: A 5,000-cow dairy will have a calf born every 2.5 hours, around the clock, day in and day out. In that type of operation, a few pieces of specialized equipment will more than pay for themselves in peace of mind alone!

3. Iodine the calf's navel. This ought to be the first thing done, immediately after you have made certain the calf is breathing properly. The iodine can be sprayed on the navel, but it is more effective if the navel can be dipped. See Chapter 3, Section 3.13, "Neonatal Care." Whatever method you choose, thoroughly saturate the navel.

4. Check the calf for umbilical hernia, over- or undershot jaws, cleft palate, supernumerary teats, and leg/fetlock soundness. Make note of any problems.

5. Identify the calf. Even if you have only a handful of cows, it will be impossible to remember which calf belongs to which cow and in what order the calves were born if you do not have a system established for identifying the calves. See Chapter 2, Section 2.2, "Identification," for step-by-step procedures. Whatever method you use, do it early in the calf's life. The best time to identify the calf is at the first contact following birth. Some producers will put off identification until they are certain whether they wish to retain the heifer or sell it. This is not sound logic. Identify all calves with a preliminary system and replace it later with the main system when appropriate.

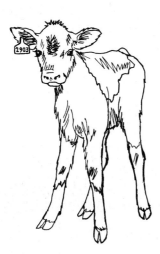

6. Make certain the calf receives colostrum. A calf should receive 1 quart of colostrum for every 40 pounds of body weight. Thus, an 80-pound calf needs 2 quarts, a 100-pounder needs 2.5 quarts, and so on. Many producers give every Holstein calf a full

3 quarts of colostrum, regardless of weight. The calf should receive the colostrum in the first hour after birth, for optimum antibody absorption from the colostrum into its bloodstream. The way to make certain of this, even if the calf is with the cow for a day, is to milk out the colostrum (or take some frozen colostrum that you have on hand) and feed it from a bottle to the calf. If the calf will not nurse the bottle, use a stomach tube. See Chapter 3, Section 3.14, "Stomach Tubing," for step-by-step procedures.

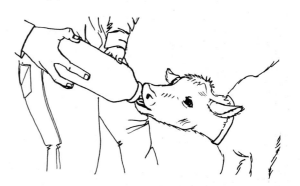

7. Separate the calf from the cow and place it in appropriate housing. There are several types of housing arrangements that will serve to protect the dairy calf from the elements. Most managers agree that it is best to house the dairy calf singly until it is weaned from a liquid diet and is eating dry feed. The sucking instinct is strong, and if the young calves are group-housed, they will suck on any and all appendages they can reach. This is not hygienic, can cause infections, and can result in a loss of ears, tails, and the like.

Individual calf hutches are a good option. These can be homemade or purchased from a variety of suppliers. Calves can be tethered to the hutch, or wire panels can be used to fashion a small enclosure for each hutch. Hutches are movable, and disease

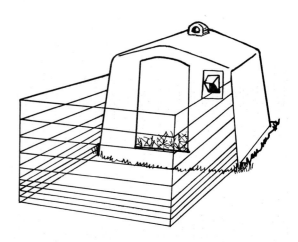

buildup can be prevented by periodically moving the hutches to unused ground. Provide lots of bedding and keep it clean and fresh.

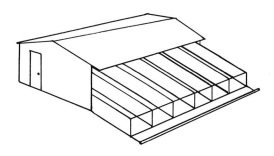

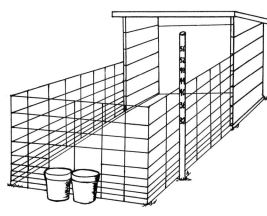

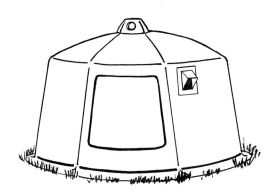

The "cold" calf barn with outside runs is another popular and entirely serviceable option.

After the calves are weaned and have outgrown their individual hutches, they should be moved to a type of group housing. An open, cold barn is the best option for most producers.

Closed barns can and do work for some producers, but careful attention must be paid to ventilation, and a more careful attention to preventive health measures must be strictly adhered to.

8. Feed a quality milk replacer. Research has shown that a 22% milk replacer is the best for rate of growth and return for dollar invested in liquid supplements. Follow the manufacturer's label directions for feeding guidelines. Keep the feeding equipment clean.

9. Train the calves to eat dry feeds. Calves can be weaned successfully as early as 5 to 6 weeks. From the first day, tempt the calf to eat a mouthful or two of a high-quality calf starter (18 to 20% protein). This is most easily accomplished if the grain is introduced immediately following the milk replacer. See Section 4.14, "Weaning and Training to Drink and Eat." When the calf is eating a pound of starter each day, begin to cut back on the milk replacer. Reduce the amount of milk replacer to one-half the normal daily amount for 1 week, keeping the starter in front of the calves at all times. At the end of the week, stop providing the milk replacer altogether.

10. *Miscellaneous management procedures.* At the appropriate times, the calves need to be dehorned, castrated, retagged, and vaccinated. See Chapters 2, 3, 4, and 10 for the timing and step-by-step procedures for these techniques.

11. *Heifer health programs.* The first weeks and months of a calf's life are critical if the calf is to turn into a healthy dairy cow. If the calf is crowded into a poorly ventilated, inadequately sanitized housing setup, and is maintained on a marginal nutrition program, respiratory and digestive distress will be frequent occurrences in its life. The way to minimize the stressors and avoid the extreme health consequences is to establish a preventive health program. A strong health program involves deworming (beginning as early as 3 weeks of age), treatment for external parasites, and a comprehensive vaccination program.

Any strong heifer health program should be devised with the consultation of the attending herd veterinarian. Comprehensive health programs will vary from region to region depending on specific pathogens common to the area. However, most programs will focus on preventing the major viral respiratory diseases—Bovine Viral Disease (BVD), Infectious Bovine Rhinotracheitis (IBR), Para-Influenza 3 (PI3), and Bovine Respiratory Syncytial Virus (BRSV)—the several serotypes of leptospirosis, the clostridial diseases, and brucellosis. Johne's disease is an emerging crisis that cannot be overlooked. It is caused by *Mycobacterium paratuberculosis* and can be transmitted to the calf immediately after birth when the calf nurses its dam's dirty udder.

See Chapter 10 for the timing and step-by-step procedures for a herd health program.

12. *Micronutrient nutrition.* Micronutrient nutrition has a huge effect on the functioning of the animal's immune response. Today, we know that copper, zinc, iron, selenium, vitamin A, and vitamin E must be present in proper amounts to enable the animal's immune machinery to respond strongly to the vaccination challenge. The herd nutritionist and the attending veterinarian should work together in building proper micronutrient nutrition into the herd health program.

13. *Breeding heifers.* A well-fed Holstein or Brown Swiss dairy heifer will reach about 850 pounds when it is between 12 and 15 months of age. Consider 850 pounds as the target and have 12 months of age as the absolute minimum. Ayrshires will be approximately 75 pounds lighter, Guernseys approximately 100 pounds lighter, and Jerseys about 150 pounds lighter than the Holsteins. If these targets are met, mature size of the adult will not be affected by calving as a 2-year-old. Everything being equal, heifers that calve as 2-year-olds will have a higher lifetime production total than heifers that calve as 3-year-olds.

Birth weight is inherited. Sires vary considerably in their genetic potential to sire large, heavy calves. Avoid sires that produce heavy-birth-weight calves when breeding dairy heifers. There are Holstein sires with proven milk improvement records that will sire light-birth-weight calves. Select these sires for freshening heifers.

Use estrus synchronization for heifers. Heifers can be difficult to "catch" in heat using visual techniques. Synchronizing the heifers, and using artificial insemination at a predetermined date, is an excellent management option. See Chapter 3, Sections 3.8 and 3.9, for step-by-step procedures for estrus synchronization and artificial insemination.

4.6 CLIPPING

Management and milking of cows can be made easier if the animals are clipped twice a year. The udder should be clipped to make it easier to keep clean and easier to wash and dry in preparation for milking. Milk marketing regulations include cleanliness of the cows, and clipping the long hair off the udder makes it easier to meet this requirement. The flanks and underline (belly and brisket) should also be clipped to help keep the cow clean. Keeping the hair on these areas short will help to prevent manure and mud from sticking to them. Cutting off the end of the switch on the tail about 1 foot above the floor will keep manure from caking in it and help the cow's general appearance. Often, all the hair is clipped from the tail of cows milked in parallel parlors. Some dairymen clip down the topline (vertebrae) and over the neck. Shortening this hair reduces problems with lice.

A good time to clip the udder, underline, and flank of the cow is just before she is due to freshen. This will ensure that her udder has been clipped when she enters the milk line.

Dairy animals are also clipped for shows, exhibits, and sales. The purpose of clipping these animals is to improve their general appearance and increase their acceptance by the show judge, classifier, prospective buyers, and the general public. Clipping for show, exhibit, or sale requires that different areas of the animal be clipped than are clipped for milk production. This clipping requires more skill and practice as well as a mental picture of what the ideal animal should look like. The object is to present the animal to its best advantage.

If you lack experience in clipping for show, exhibit, or sale, practice on some other calves first. This will increase your ability and confidence. Take the opportunity to watch experienced exhibitors prepare their animals before and during a show or sale.

Equipment Necessary

- Halter
- Clippers
- Extension cord
- Currycomb
- Brush
- Fuel oil
- Can or plastic jug with large mouth
- Antiseptic
- Squeeze chute
- Kow-Kant-Kick
- Nose tongs (bull leader)

Restraint Required

Cows that have been handled regularly and with gentleness may require only a halter, stanchion, or headgate for clipping. Some cows will need an additional restraint such as the tail hold or Kow-Kant-Kick.

Clipping for show, exhibit, or sale is more time-consuming and requires more animal restraint for your convenience as well as for your safety. Restraint in a headgate, stanchion, or squeeze chute may be necessary, and a halter must be used to help restrain the head. A set of nose tongs can also be helpful.

Step-by-Step Procedures

General

1. Assemble all the necessary equipment.
2. Restrain the cow as necessary.
3. The clippers must be sharp and clean. To help keep them that way, the head (clipper blades only) should be dipped in fuel oil (in the can or jug) while the clipper is running. Shut off the clipper and wipe off excess oil. This should be done three or four times while clipping each cow.

Dull clippers make the job slow and "pull" as you clip, which may make the animal resist more than is normal. Sharpen dull clipper blades or keep an extra pair of sharp blades on hand to use when necessary.

CAUTION: Use only approved, properly grounded extension cords. Clipping should be done in a dry area to help prevent electrical shock.

---

4. Remove dirt, manure, and bedding from the areas to be clipped. This is best done by using the currycomb and brush.

For extremely dirty animals, it may be necessary to wash the areas to be clipped. If the area is washed, it should be allowed to dry before you start to clip. Removing the dirt and manure makes clipping easier and keeps the clipper blades sharp much longer. It also allows you to do a smoother and closer job of clipping.

5. Start the clippers, holding them some distance from the cow. Hold the clippers so that she can see and hear them. This will take only a few minutes and will save time as you proceed with the clipping. It is done to get the cow used to the noise. Starting the clippers close to her udder may scare her, and she may kick in self-defense.

6. When you start clipping, clip against the lay of the hair. Move the clippers slowly and be as gentle as possible. Hold the clippers tightly so you won't drop them if the cow moves or kicks.

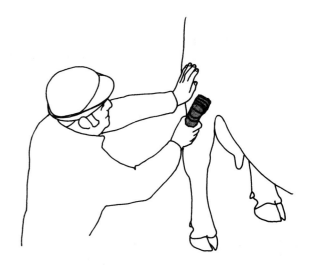

Clipping Milk Cows

1. Clip the udder, flanks, and underline.

The udder is clipped by starting at the bottom of the udder and moving the clippers up toward the body of the cow. In this manner, you are clipping against the lay of the hair, and a closer clipping job will result. The whole udder is clipped as closely and smoothly as possible.

The flanks are clipped starting at the front udder attachment and moving up and on an angle toward the pinbones. Everything below this line on the flank and thigh is clipped as closely as possible by clipping against the lay of the hair.

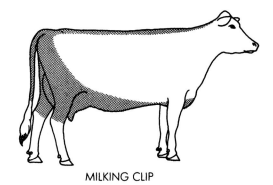

MILKING CLIP

The underline is clipped in the same manner, starting at the front udder attachment and moving toward the front legs. The brisket is also clipped. Clipping is usually limited to the underline. The body above the underline is not clipped.

2. Clip off the switch about 1 foot above the floor.

3. The topline is also clipped by many dairymen. Clip against the lay of the hair, starting at the tailhead and moving forward to the neck. About two clipper widths on each side of the backbone are clipped.

> **CAUTION:** The cow may kick during clipping, especially while you are clipping the udder. Care must be taken to prevent her from hitting you or the clippers. Put your forearm against her rear leg, so that you can feel her slightest muscle movement or leg motion. Push against the leg to keep her from hitting you. If this doesn't work, apply restraint by using the tail hold. The Kant-Kick may be required for some cows.
>
> Hold the clippers flat against the skin. Be cautious around skin folds that you do not cut the skin. Stretch skin out where necessary to avoid cutting it.

Clipping Show, Exhibit, or Sale Animals

1. Clip the entire head. Start at the poll, clipping as closely as possible. The forehead and face are clipped next. Because the hair often lies in a swirl on the face, the clippers must be moved in several directions so that the hair is clipped as closely as possible. The sides of the head, jaw, and nose are also clipped against the lay of the hair. Do not clip the eyebrows, eyelashes, or hair on the muzzle.

SALE OR SHOW CLIP

The ears are clipped both inside and out. Clip against the lay of the hair, starting at the tip of the ear and moving the clipper toward the base of the ear.

> **CAUTION:** The animal may resist clipping of the head. Start the clippers at a distance from the head, where the cow can see them. Be as gentle as possible and do not use restraint by squeeze chute with nose and head bar unless absolutely necessary.
>
> Be prepared to protect yourself if the animal throws her head from side to side or kicks. Keep one hand on the area being clipped to anticipate movements by the animal. Keep a firm grip on the clippers at all times so that you won't drop them if the animal hits them.

2. Clip the neck against the lay of the hair to make it look longer and thinner, thus making the animal look more "dairy." Blend the clipped area with the unclipped area as you approach the shoulder; turn the clipper over and comb the long hair. Clip the dewlap and the brisket.

3. Clip the tail, against the lay of the hair, starting about 2 inches above the switch. Clip up the tail until you reach the tailhead near the pinbones, then blend by clipping with the lay of the hair.

4. Always clip the udder of a cow that is being prepared for show or exhibit. Clip the udder against the lay of the hair as smoothly as possible. The udder of a heifer is seldom clipped, although it may be clipped if the hair is very long. The underline of the heifer is seldom clipped. The underline of the cow may be clipped to make the milk veins look more prominent.

5. The insides of the thighs may be clipped to make them look thinner, denoting more "dairy" character. Keeping the animal clean helps her general appearance. Knees and hocks are usually stained, and cleaning them is difficult. Clipping the longer hair off the knees and hocks makes it easier to clean them.

6. Clipping the body of the animal is not necessary if the animal has been brushed regularly to remove the long and dead hair. If the hair is long and you attempt to clip it, it is very difficult to blend the clipped and unclipped areas. If you must clip on the body (for instance, over the shoulders and withers to make them look sharper), clip with the lay of the hair to make blending easier.

CAUTION: Go slowly and don't take too much hair in any one pass. If you go too fast, the chances of nicking or gouging the hair increase.

---

7. Check the clipped area for any nicks or cuts that you may have caused with the clippers. Treat nicks and cuts with an antiseptic.

8. Release the animal and return her to the housing area.

4.7 TRAINING TO MILK

The technique for training (breaking) the first-calf heifer to milk can be trying and sometimes dangerous. It is, however, one of the most important management skills on the dairy farm. The handling of the heifer during the first few days after her first calf is born can determine her attitude toward the milking procedure for the rest of her life. She can become a calm and easily handled cow or an "outlaw" that is nearly impossible to milk.

Study this section and Section 4.8 before attempting to break the just-fresh, first-calf heifer to milk. The steps involved should be followed as closely as possible to prepare the heifer properly for milking.

Equipment Needed

- Milking machine
- Parlor or stanchion
- Water hose or bucket of water
- Paper towels
- Strip cup
- Halter
- Kow-Kant-Kick, Cattle Controller, or 30 feet of ½-inch rope
- Teat dip

Restraint Required

The heifer should be handled as easily and calmly as possible. If you are milking in a parlor, getting her to enter the parlor can sometimes be difficult. If you move her with a group of experienced cows, she may enter without difficulty.

If the heifer is broken to lead with a halter, she can be led into the parlor or stanchion. For some heifers, two or more people may be needed to force her into the parlor. This should be avoided, if

possible, because it causes more stress on the animal. Sometimes, however, this is the only method that will work.

If she is going to be milked in a stanchion, the ease with which you put her into the stanchion and handle her can affect how she reacts to the milking procedure.

Some dairymen prefer to have the springing heifers enter and go through the parlor or stanchion area (without being milked) a few days before calving. The heifer will become accustomed to the surroundings and procedures and will be easier to handle after calving.

Many fresh heifers need very little restraint while actually being milked. This is especially true if the dairyman is gentle and easy while handling the heifer. Some heifers, however, need a considerable amount of restraint. The type of restraint required depends on the reaction of the heifer to the milking procedure. The heifer that continues to kick or jump up and down during the milking process needs the restraint afforded by the Kow-Kant-Kick, Cattle Controller, rope, or tail hold.

Step-by-Step Procedure

1. Assemble all the necessary equipment. The equipment must be clean and sanitized; dirty milking equipment can cause mastitis.

2. Move the fresh heifer into the parlor or stanchion. Avoid loud shouting and the use of whips, clubs, or any other such instruments when attempting to move the heifer. Heifers that have gone through the parlor or have been handled before this time will be much easier to move to the parlor and the stanchion.

3. Prepare the heifer for milking as in steps 2 and 3 of Section 4.8. When washing the udder, be careful that the heifer does not kick you. Any quick or fast movement is likely to cause her to kick. The fact that you are squirting water on her udder may scare her and cause her to become very excited and kick in self-defense. Some dairymen prefer to massage the udder gently without the use of water for the first one or two milkings, thus preventing the possible reaction to the spraying of water on the udder. This practice is acceptable now because the milk is not salable until three days after calving. The gentle massage with a dry towel will also serve to remove any dirt that might be on the udder. If the heifer reacts to this procedure and tries to kick, use the hand and arm nearest to her back legs to reach across the front of the back leg nearest you. Grasp the back of the hock of the opposite leg with your hand, thus pushing your arm against the leg nearest you. In this manner, you can keep that leg back and use the other hand to clean and massage the udder.

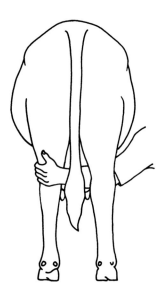

> **CAUTION:** Some heifers may react violently at this time. Be alert and ready to move away or pull your arms and hands away if she kicks. Try again, being as gentle as possible. If the heifer still reacts violently, it may be necessary to apply a restraint to keep her from kicking. Refrain from using restraints as long as possible. The Kant-Kick, Cattle Controller, or a rope tied tightly around the heifer's body just in front of the udder over the back and in front of the hooks (hips) will assist in restraining her.

4. Use the strip cup to determine if the milk is normal. "Normal" milk for any just-fresh cow is colostrum (first milk after calving), but it should not contain flakes, clots, serum, or blood. Colostrum may appear yellow and be thick and sticky.

KOW-KANT-KICK

5. Apply the milker. Start with the farside of the udder, putting on the two teat cups, and then apply the teat cups to the nearside. Applying the teat cups to the farside first makes it likely that if the heifer kicks, she will do so on the side away from you.

> **CAUTION:** Take care that you are not hit by the animal as you attempt to apply the milking machine. Many animals will not react violently to this procedure; they soon find out that you are relieving pressure in the udder and that milking will make them more comfortable.

6. Milking should be completed in 3 to 5 minutes. Many fresh animals, particularly first-calf heifers, will not have a milk letdown during the first milking or two after calving. The milking machine should, therefore, be removed after 3 to 5 minutes and the procedure tried again at the next milking. Remove the machine in the manner described in Section 4.8.

7. Remove any restraint immediately after removing the milking machine. Continue to be calm and gentle. Move slowly and easily with the least amount of disturbance to the animal.

8. Follow the procedures given in Section 4.8 to dip teats and clean equipment.

> **CAUTION:** The udder of the fresh heifer may swell considerably, which is normal. This may cause reduced blood circulation in the udder, and in cold weather can result in frostbite or frozen teats. The heifer should not, therefore, be exposed to extreme cold (10°F or lower) until after the swelling has decreased. She should be housed in clean and dry surroundings.

It is again important to remember that the cow can kick the milker and cause severe injury. Being alert and anticipating the kick will help to prevent the milker from being injured. If, after a week or 10 days, the animal is still an "outlaw," it is best to cull her from the herd and not risk injury to the most important person on the dairy farm—the milker.

4.8 MILKING

Milking the cow carries more responsibility than any other job on the dairy farm. The regular use of a proper routine can result in more milk, less mastitis, less milking time, and longer cow life in the herd. The result is more profit per cow.

CAUTION: Milk-withholding times must be followed on lactating dairy cows treated with antibiotics. In addition, only drugs approved by the FDA for use in lactating cows can be used without a prescription from a veterinarian. Label treatment guidelines must be followed.

All milkers should be familiar with the mechanisms of milk secretion and the makeup (anatomy) of the udder of the cow. They should have an understanding of how the cow produces milk and how a milking machine works.

Information on milk secretion and milking machine operation is available in many books and from milking machine manufacturing companies.

As the cow is prepared for milking by washing and drying the udder and teats, the nerves in the teats send a message to the pituitary gland located at the base of the brain. The milk letdown hormone, *oxytocin*, is released into the bloodstream— reaching the udder in about 1 minute. This hormone causes the muscle fibers surrounding the

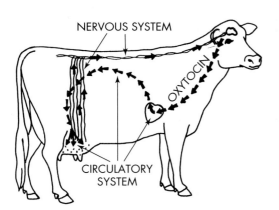

alveoli to contract and force the milk out into the ducts and udder cistern, making milking possible. Alveoli are the very tiny milk-secreting cells in the udder. The effect of oxytocin is diminished in about 7 minutes, so rapid milking is necessary if one is to obtain a complete emptying of the udder.

When cows are mistreated or unduly disturbed or excited just before or during milking, adrenaline may be released into the bloodstream by the adrenal glands. This hormone counteracts the effects of oxytocin and prevents or stops milk letdown. This reaction can cause incomplete milking and lead to udder problems and reduced milk yield.

Because milking is the most important task on the dairy farm, it should be performed by the most capable person on the farm. To milk properly, the cow and the milker must both be comfortable. The provision of a clean, well-lighted building (parlor) with temperature and humidity control will help the milker to do a good job.

Equipment Necessary

- Milking machine
- Parlor or stanchion
- Water hose or bucket of water
- Paper towels
- Strip cup
- Teat dip
- Milk bucket
- Sanitizer

Restraint Required

A properly trained cow will not require restraint during the milking operation. For the procedure used to handle a first-calf heifer, see Section 4.7.

Step-by-Step Procedures

Machine Milking

1. Provide a stress-free environment for the cow. If the cows are forced or abused in any way, they will not cooperate, and milk production will be reduced.

2. When the cows enter the parlor, their udders, teats, and especially the teat ends should be thoroughly washed with warm sanitizer solution. A small hose should be at hand with which to spray a limited amount of warm water (about 110°F) containing the sanitizing solution on the udder. The udder should then be thoroughly massaged and dried with an individual towel for each cow. The udder of the cow must be thoroughly dried. This should, of course, include the teat and particularly the teat end. If the udder is not completely dried, the remaining water on the udder can drain down the udder to the top of the teat and into the teat cups, thus contaminating the milk. This must be done before the milking units are applied. This step usually requires 30 to 45 seconds.

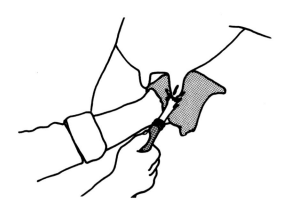

An alternative to water wash is the use of a commercial pre-dip. Pre-dipping replaces the water wash in the milking procedure. Each teat is dipped and the dip allowed to remain on the teat for the manufacturer's recommended contact time prior to removal by drying. Pre-dip can also be applied by handheld sprayer or hose sprayer. Massaging dipped teats helps to loosen dried materials, plus it acts as a stimulus for milk letdown. Wearing smooth rubber or disposable gloves helps to prevent the spread of microorganisms and prevents human skin contact with the pre-dip materials. Be cautioned and take care to remove all of the dip prior to the attachment of the milking unit to prevent milk contamination. Pre-dipping with a germicidal dip reduces environmental mastitis. Cows must be clean for pre-dipping to be successful.

3. The foremilk (first milk) from each quarter should be stripped into a strip cup and discarded. The strip cup is used to determine if the milk in each quarter is normal. Strip cups are seldom used in large dairy operations. A common procedure in most parlors is to strip directly onto the floor, followed by hosing the floor with water. Milk should not be stripped onto the hand. Hand contamination can lead to cow-to-cow transfer of mastitis organisms. If the milk appears abnormal (contains clots, flakes, blood, or serum), it should be milked into a separate container and discarded. Stripping also stimulates milk letdown.

4. Thoroughly clean and dry each teat, using one paper towel per cow. Pay particular attention to the teat end, as you wash and dry. Do not use one towel for more than one cow. Multiple cows per towel will spread bacteria and hasten the spread of mastitis organisms.

5. The milker should be applied within 30 seconds to 1 minute after the stimulating massage takes place. Because there is a variation in the amount of time required for milk letdown, the milker should not be attached until letdown has occurred. This can be determined by observing the

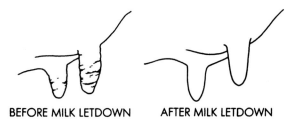

BEFORE MILK LETDOWN   AFTER MILK LETDOWN

teats and attaching the milker when the teat becomes full of milk. Applying the machine too early can cause injury to the very delicate teat membranes. Waiting too long after stimulation can cause loss of the letdown effect of stimulation and reduce milk production.

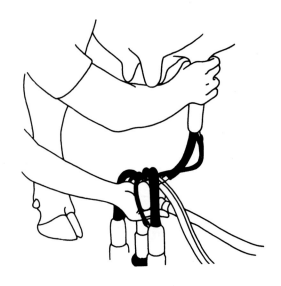

Some of the new milking machines have electronically controlled or air-controlled vacuum levels and pulsation rate. The vacuum level and pulsation rate are automatically increased or decreased with changes in the rate of the milk flow. This will help to reduce the amount of damage done before milk letdown and at the end of milking when the milk flow is reduced.

6. The milker should be adjusted under the cow so that it is at right angles to the floor of the udder. Some milking machines are easy to adjust and others are difficult. A downward pull on the milker helps to straighten out the milk ducts in the udder and allows for a more complete milkout. Milking is completed when the milk flow through the milk hose or milker claw is reduced to a trickle. A few cows may require slight downward tension on the milker toward the end of milking to assist in complete milkout. This action is called *machine stripping*.

7. Remove the milking machine upon completion of milking. The vacuum should be turned off and the machine removed gently.

Some newer milking machines have a detaching mechanism. At the completion of milking, the detacher shuts off the vacuum, removes the milker unit, and pulls it out from under the cow.

> **CAUTION:** Grabbing the machine and pulling it off the udder without proper release of the vacuum can cause damage to the udder and the teat. The machine should be removed immediately after milkout of the udder or when the milk ceases to flow. Overmilking is thought to be one of the major causes of mastitis.

8. Immediately after the milkers are removed, the teats should be dipped in an effective teat-dip solution. This practice removes the last drop of milk from the end of the teat and helps to reduce new infections of mastitis. Some teat dips also help to keep the teats pliable and guard against chapping and sunburn.

> **CAUTION:** Immediately after milking, when the cow is returned to the housing area, take care that she is not exposed to extreme cold or wind; the udder, and particularly the teat ends, might freeze. To prevent freezing, the cow's udder and teats should be thoroughly dry before she leaves the milking parlor.
>
> Cows that have recently calved (freshened) have swollen udders and should not be exposed to extreme cold or direct wind. A fresh cow, especially a 2-year-old heifer, is especially susceptible to frostbite because circulation is reduced in the udder and the teat due to swelling. Problems with frostbite are usually not a concern in the tie-stall or stanchion-type barn.

9. All milking equipment must be thoroughly cleaned, sanitized, and properly stored immediately after milking. Follow the approved procedures for the equipment used and satisfy the requirements of the milk marketing regulations.

Hand Milking

The process of hand milking starts with the same procedures as does machine milking. All of the important steps leading up to the actual attaching of the machine are exactly the same. For differences between beef and dairy hand-milking procedures, refer to Section 2.4 of Chapter 2.

1. With a hand-milking system, you do not apply the milker. Instead, place a milk bucket under the cow's udder; a milk stool will make the milker more comfortable.

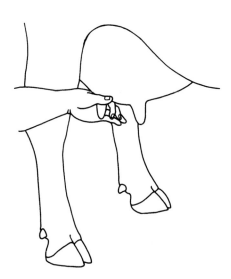

2. Cows are milked by hand pressure applied at the upper portion of the teat with the thumb and first finger. This pressure traps the milk in the teat.

3. The teat is squeezed against the palm of the hand by the remaining three fingers, with pressure first applied by the middle finger, then the third, and then the fourth finger. This causes milk to squirt from the canal.

4. When the milk is squeezed out of the teat, the pressure applied by the fingers is released and the milk is free to flow from the cistern of the udder into the teat. You can again apply pressure to the teat using the above system to remove the milk. Continue this procedure until the cow is milked out.

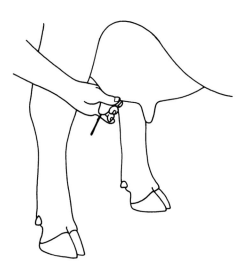

When you first start to milk, it may seem complicated. Learning to coordinate the hand motions rapidly takes some practice. With practice, however, the cow can be milked rather rapidly, usually within 10 minutes.

5. Some cows have very short teats and may be difficult to milk by hand. Use only the thumb and first two fingers; with extremely short teats, stripping may be necessary.

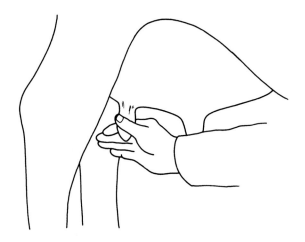

Stripping is milking by applying pressure with the thumb and first finger at the top of the teat as in step 2. Because there is no room on the short teat for the rest of the fingers, the milking is accomplished by pulling down on the teat, letting it slide between the thumb and first finger. When the end of the teat is reached, the pressure is released and the procedure started over in the same manner. Some hand milkers prefer this system and milk almost as rapidly as with the full-hand system.

6. Some cows are hard milkers. This is usually caused by a small opening in the end of the teat. The opening can be surgically enlarged by a veterinarian, but the probability of infection and resulting mastitis is great.

Some cows may also have a side opening on the teat detrimental to hand milking, but not to machine milking. This condition can also be corrected surgically, but this is seldom done.

CAUTION: The milker should be prepared to protect himself; a cow may kick because of an injured teat, a sore on the teat, rough handling, or meanness. Some hand milkers protect themselves by leaning forward on the cow's flank against the cow's leg. The milker can feel the muscle of the leg move before the leg actually moves. He can use his forearm to push against her leg and keep it from hitting him. If the cow persists in kicking, a previously mentioned restraint should be used (refer to Section 4.7).

---

7. Immediately after milking, the teats should be dipped in an effective teat-dip solution. This practice removes the last drop of milk from the end of the teat and helps to reduce new infections of mastitis. Some teat dips also help in keeping the teats pliable and guard against chapping and sunburn.

CAUTION: Observe the same cautions as for machine milking.

---

8. All milking equipment must be thoroughly cleaned, sanitized, and properly stored after milking. Follow the methods approved for the equipment used and satisfy the requirements of the Grade A milk ordinance and code.

4.9 MASTITIS TREATMENT

Mastitis, an inflammation or swelling in the udder caused by an infection, is a costly disease. This infection can be caused by several different types of bacteria. Mastitis can infect one-quarter or the whole udder.

Mastitis has been called a "man-made" disease because man may cause mastitis by improper care and treatment of the cow. Poor milking procedures, overmilking, improperly functioning milking machines, poor housing conditions, and injuries caused by poor surroundings can all lead to mastitis.

It is estimated that at any one time, 40 to 50% of all dairy cows in U.S. dairy herds are infected with mastitis in one or more quarters of their udders. This can reduce milk production by as much as 20% and shorten the productive lives of the cows.

A check for mastitis is performed before each milking by checking the foremilk. If this is not conclusive, the California Mastitis Test (CMT) can be used. CMT test materials are available from livestock supply companies. The CMT is an easy-to-use cowside test that can identify cows with high levels of somatic cells. Elevated somatic cell levels usually indicate mastitis infections.

Monthly somatic cell counts (leucocyte counts) are helpful in identifying problem cows. This test is available through private testing laboratories, some dairy herd–improvement testing labs, and some dairy plant testing labs.

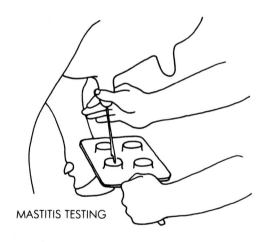

MASTITIS TESTING

Fresh (recently calved) 2-year-old heifers with their first calves may show some blood in their milk for a few days. This should not by itself be considered a result of mastitis. Rather, it is normal and usually clears up in 2 or 3 days.

Even in well-managed herds with proper milking procedures and good housing and feeding practices, cases of mastitis do occur. The proper and prompt treatment of these cases can reduce both the severity of the infections and the damage that they cause.

Equipment Necessary

- Treatment drug
- Alcohol packet
- Bucket
- California Mastitis Test kit
- Strip cup

Restraint Required

Little restraint is necessary in the treatment of mild mastitis infections. It may be necessary to restrain cows with more severe mastitis infections in stanchions or squeeze chutes and to apply the tail hold when handling them.

Step-by-Step Procedure

When a case of mastitis appears, immediate steps should be taken to treat the cow; the longer one waits, the greater the damage. When a mild case of mastitis occurs (flakes or clots in the milk), the dairyman can usually treat the cow. If the cow has a high temperature, poor appetite, or a sharp reduction in milk yield, or if the quarter is extremely hard or cold to the touch or contains serum instead of milk, a veterinarian should be called to treat her immediately. If the dairyman is treating the cow, he should employ the following procedure.

1. The cow may be treated in the milking parlor or in another treatment area. Milk out the udder completely into a bucket and discard the milk. The whole udder should be milked out; milk letdown has occurred, and incomplete milking could cause problems in the other quarters of the udder. Milk from cows with slight infections can be used to feed calves, but it cannot be sold; it is considered to be abnormal milk. Because an antibiotic is inserted into the teat and cistern of the udder, milk should be removed first so that there is less dilution of the drug.

2. Wash your hands. Open the packet containing the alcohol pad and thoroughly clean the end of the teat, particularly the indentation at the tip. (This packet is included with the antibiotic.)

CAUTION: The cow may kick during the cleaning of the teat with alcohol. Be prepared to protect yourself with your forearm against her leg, restrain the cow in a stanchion or squeeze chute, or apply the tail hold.

3. Grasp the teat to be treated with the thumb and forefinger of one hand. Use your other hand to insert the antibiotic treatment tube into the teat canal. When inserting this tube, be sure that the teat is held straight down. Do not turn or bend the teat sideways while inserting the tube. Full insertion of a conventional mastitis syringe cannula can lead to new mastitis infections as a result of the treatment. Partial insertion of the syringe cannula is recommended. With this method, the cannula is inserted only 2 to 3 mm into the teat canal. This reduces the likelihood that mastitis organisms will be forced into the teat cistern during treatment. Many commercial mastitis tubes are designed for partial insertion.

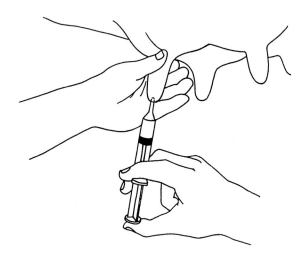

CAUTION: Be sure to use only approved mastitis treatment drugs.

Because the udder or the quarter of the udder may be very sore at this time, a normally gentle cow may react by trying to kick while she is being treated. Care should be taken to protect yourself.

---

4. Once the tube is inserted, inject the drug slowly into the teat. Each tube contains one treatment for one quarter. Dip the teat with a postdip after the treatment is complete. Discard the used tube where it is out of the reach of children and away from the milk supply.

5. The udder immediately above the teat and the cistern should be massaged to help distribute the antibiotic throughout the udder.

6. Immediately identify the cow as treated. Also, make a permanent record of the treatment date and type and milk-withholding information for future reference.

CAUTION: Milk from cows treated with antibiotics must be discarded for the length of time indicated on the label of the antibiotic container. This time varies with the drug used. The milk discarded as a result of treatment for slight mastitis infections may be used to feed calves, but only if the calves are housed in individual pens where it is impossible for them to suck each other.

---

7. Sometimes an intramuscular injection of an antibiotic must be used to treat severe mastitis. The proper procedure for giving an intramuscular injection should be followed. Severe mastitis usually results in loss of the milk-producing ability of the infected quarter or quarters of the udder. Some types of mastitis can be so severe that they can cause the death of the cow in a very few days or even hours.

8. Wash your hands and all equipment used. Discard milk and treatment tubes. Washing your hands and the equipment will help to prevent the spread of mastitis to other cows.

Postprocedural Management

Treated cows should be checked for improvement at the next milking, usually 10 to 12 hours after treatment. The mastitis check described in the introduction to this section should be used.

In mild cases, the cows can remain in the normal housing unit. In more severe cases, they should be separated from the rest of the herd and housed in the hospital area or a box stall.

Is is a good practice to milk out the infected quarter(s) often. Removing the infected milk tends to lessen the severity of the mastitis and allows the antibiotic to become effective more rapidly.

4.10 MASTITIS PREVENTION

Mastitis can result from environmental or contagious sources (*Staphylococcus aureus, Staphylococcus agalactiae,* and *Mycoplasma* spp.) and present a complete range of symptoms, from the severe acute case requiring immediate and comprehensive treatment to the subclinical case, which you can be totally unaware of. Treatment of mastitis is best done under the care and direction of the herd's attending veterinarian. A standard protocol for treatment of acute mastitis is presented in Section 4.9. It is the prevention of the insidious subclinical mastitis that we will address in this section.

It is estimated that fully 20% of all lactating cows are suffering from subclinical mastitis. Largely unnoticed, subclinical mastitis nevertheless has a profound impact on the dairy herd, causing a reduction in production approximating 10%. A little arithmetic: In a herd of 1,000 cows, there are likely 200 subclinically mastitic cows (20%). Their production is reduced by 10%. Assume a rolling herd average of 25,000 pounds. Each cow will produce about 82 pounds per day. A 10% reduction in this level of milk for the subclinical mastitis-affected cows amounts to about 8 pounds per day. Over a 305-day lactation this amounts to 2,440 pounds lost per each of the 200 cows. A loss of

2,440 pounds of milk is about 24 cwt of milk at $12/cwt. This represents an annual loss of $360 per cow. If 200 cows are affected, $72,000 in potential income is lost.

The example just calculated may not be accurate or appropriate for your situation. But, whatever the actual numbers, no dairy producer can afford to operate without attempting to control the prevalence of all mastitis occurrences—acute and subclinical.

The following are management principles that will reduce the prevalence of environmental and contagious mastitis in your herd. Each recommended practice can be implemented with a minimum of expense and change of management style. All should be implemented.

1. Prepare teats and teat ends prior to milking. One method is to use warm water to wash and single-use towels (paper of cloth) to thoroughly dry the udder and teats. Pay particular attention to the teat ends. Couple this practice with step 2.

2. Treat each teat with a premilking sanitizing dip solution (pre-dipping), and treat each teat with an iodophor or chlorhexidine sanitizer after milking (post-dipping).

3. Instead of steps 1 and 2, use a waterless udder, teat, and teat end preparation by covering most of the udder and all of the teats with a sanitizing solution, such as an iodophor or chlorhexidine compound. Use a single-use towel to dry the teats.

4. Have milkers wear latex or nitrile gloves for all of the milking. We know that milkers transmit bacteria from one cow to the next. One way to prevent this spread is to wash hands often. However, sanitizers can be harsh after repeated use. Wearing gloves and washing them between cows, while on your hands, will reduce the spread of bacteria and prevent sensitive hands.

5. Be gentle with the teats. Any injury that occurs to the teats will in all likelihood end up causing a case of mastitis. Sources of injury include leaving the milking machines on for too long, worn-out inflations, poorly adjusted pulsators, poor bedding choice, poor stall design, and frostbite.

6. Keep your cows clean at all times. This is a complex issue involving bedding, stall type, animal density, type of waste removal system, and use of pasture. Whatever the exact mix on your farm, keep your cows as clean as possible. Clean cows have clean udders. Clean udders help prevent environmental mastitis and speed up the entire milking process.

Following is a cow hygiene scoring system. As you study it, keep in mind the overarching principle that clean cows, with clean udders, have fewer problems with environmental mastitis than dirty cows with dirty udders. In the hygiene scoring system, pay particular attention to the udder, the lower leg, and the upper leg and flank.

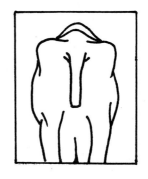 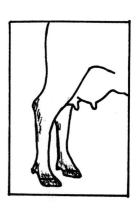

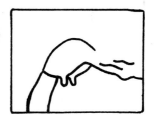

HYGIENE SCORE - 1

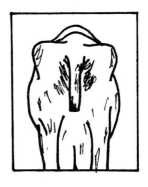 

HYGIENE SCORE - 2

HYGIENE SCORE - 3

HYGIENE SCORE - 4

7. Monitor cows for new cases of mastitis. This includes cases of subclinical mastitis, as well. There are several tools to assist with this. They include forestripping, the California Mastitis Test (CMT), and somatic cell counts.

8. Institute a dry-cow mastitis treatment program. Consider treating all quarters at the onset of the dry period and treating each freshening cow with lactating cow antibiotics at the end of the dry period.

9. Work with your attending veterinarian to assure the proper selection, administration, and withdrawal of mastitis treatment drugs.

10. Treat infected cows aggressively. An acute case of mastitis can easily become a chronic case. It is much harder to control the spread of mastitis organisms in a herd where there are chronically affected cows.

11. Keep meticulous written records on all mastitis management procedures—subclinical and acute. Review them frequently to assure that you are gaining ground.

12. Be certain that your home-raised replacements are mastitis free. Management practices should include feeding mastitis-free milk to the calves, preventing teat suckling, controlling flies and rodents, providing clean bedding and housing, and pasteurizing hospital pen milk before feeding it to the calves.

13. Treat all purchased replacement calves, heifers, and cows as infected. Ask the seller for mastitis records, somatic cell count histories, and bulk tank culture reports.

14. Assure proper mineral nutrition. The immune response mechanisms can afford resistance to mastitis only if the animals are well nourished. Pay particular attention to selenium, copper, zinc, and vitamins A and E.

15. Control the fly, rodent, and bird populations. Keep the environment clean, apply insecticide ear tags, and establish a control program for larger "pests."

16. Consider milking more than twice daily. The third and fourth milkings help to maintain udder and teat health by keeping the teat canal flushed out, thereby preventing the colonization of bacteria in the canal.

17. If you have a backflush system in your milking equipment, use it. These systems help assure that the spread of pathogens from one cow to the next does not occur.

18. Train and retrain your milkers and herdspeople.

4.11 DOWNER COWS

The "downer cow" is a problem that occurs periodically on most cattle farms. Down cows cannot get up and stand by themselves. The problem results from injuries caused by a cow slipping on slick floors or being bumped by other cows, by paralysis as a result of calving, or from such diseases as milk fever. Because she is not able to get up and move around, the down cow must have some special care if she is to recover. The type of care necessary will depend on the size, weight, and condition of the animal involved. It is no longer a legal option to haul the downer cow to market and have her become part of the human food chain.

Equipment Necessary

- Cow lift
- Wide belts
- Skid
- Tractor with front-end loader
- Overhead beam with pulleys
- Feed bucket or box
- Water tub
- Halter

Restraint Required

Because the cow is down, little if any restraint is required. A rope halter is helpful in holding her head while she is being treated or moved.

Step-by-Step Procedure

Downers may occur at any location in the operation; it is usually necessary to move them to an area where they can be more comfortable and more easily treated. Proper care and procedures can make this job relatively safe for the animal and the operator and make recovery faster. Neglect, rough treatment, or improper treatment at this time can increase the amount of injury and slow the recovery rate.

1. Assemble all the necessary equipment.

2. Be as gentle as possible so that you do not frighten the cow. Approach her and place the halter upon her head.

3. Move the skid as close to the animal's back as possible. Because she is not able to get up or do much moving, it is necessary to move her onto the skid. Usually, the easiest method of accomplishing this is to roll her onto it. To facilitate this, the skid should be low, no higher than the width of 2-inch lumber. By grabbing the cow's feet, two or three people can roll her over and onto the skid.

CAUTION: If the animal starts to kick or thrash her feet, it is best to tie a rope around them. Tie the two back feet together and, if necessary, the two front feet together. Most downer cows will not fight much with the two front feet. Roll the cow over by

pulling on the rope(s), staying a safe distance from her feet. One person should be positioned on the halter to hold the animal's head. If this person holds the head down, usually with his knee over the neck just behind the jawbone, the animal will resist treatment less.

If the animal has an obvious injury, moving should be done only with the advice of a veterinarian.

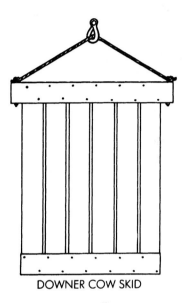

DOWNER COW SKID

4. After the animal has been rolled onto the skid, pull it to the area where you can keep the animal comfortable and treat her easily. A box stall is fine; in good weather, a shaded area in a drylot or pasture will suffice.

5. Remove the animal from the skid by the same procedure you used to move her onto the skid; namely, roll her over by grabbing her feet, and then pull her up and over, or use the ropes if she reacts too much.

6. When the cow has been moved or rolled off the skid, she should not be left lying flat on her side. She should be rolled onto her sternum with her back higher than her feet. If she lies flat on her side or with her feet higher than her back, rumen fluids can move into her throat or lungs and choke her. If she cannot maintain this position on her own, she should be propped up or supported by bales of hay or straw.

7. Allow the cow to lie on one side for part of a day and on the other side for the rest of the day. Turning her is accomplished in one of two ways: (1) roll her over using the same method used to roll her on and off the skid, or (2) tuck the cow's feet under her and, with two or three other people, push toward her feet so that she rolls over and lies on her other side.

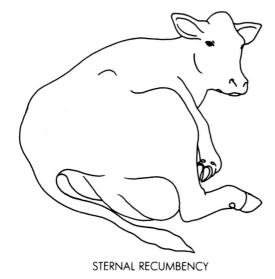

STERNAL RECUMBENCY

8. Obviously, the animal that cannot get up and move around must have feed and water brought to her in shallow containers. Many people prefer to feed a good-quality legume or legume–grass hay at this time because it is easier to handle and feed. The water should be fresh. In hot weather, make sure that the cow is given all she wants to drink. Downer cows must be milked twice a day; hand milking is required.

9. Some cows will try to get up on their own and may need only a little help from the dairyman to succeed. This help can be given by grasping the tail just below the tailhead and lifting as the animal tries to get up. If the cow does not get up on her own or with assistance in two or three days, a "lift" should be used to get her to stand on her feet. Once standing, many animals will make an effort to remain upright. Lifting the cow can be accomplished by several different methods, two of which are described below.

Cow Lift. A cow lift is clamped over the hooks and loin of the animal. Lifting is accomplished by using a front-end loader or overhead pulleys hooked to a support beam.

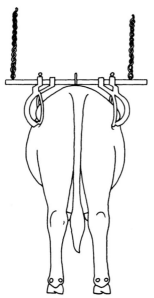

Wide Belts. One belt is passed just behind the front legs, and the second belt is passed under the cow just in front of the udder. If a third belt is used, it is passed under the cow between these two belts. The belts are then hooked to the tractor or the overhead pulleys.

The cow is lifted until her legs are straight and her feet touch the ground. She should not be lifted to where the feet just barely touch the ground; rather, they should be so positioned that if she wants to stand, she can do so. Keep the support reasonably snug until you can determine whether she will stand on her own. The cow should not be supported by the sling for more than an hour at any one time. If she has recovered sufficiently to stand and does so for several minutes, the lift can be gradually lowered. If she continues to stand, the support can be removed. The supports should be removed very quietly and easily to prevent scaring the animal and causing her to move too rapidly. She may fall and injure herself again.

Cows that cannot or will not get up or stay up after the lift has been used may have permanent damage. A decision about the destruction and disposition of permanently disabled animals should be made in consultation with the herd health adviser.

Postprocedural Management

If you are successful in getting the downer cow to stand and move around, she should be left in comfortable surroundings for several days until she is totally recovered. Moving her back to the rest of the herd too quickly will result in her being bumped or pushed around and again slipping or falling.

Careful observation for mastitis during this period is necessary because the cow has been lying on her udder and you have been moving her around; she could have a bruised udder. Cows should also be observed for any cuts, bruises, or other lacerations, and if any of these are present, they should be treated.

4.12 EUTHANASIA

Anyone who produces livestock, poultry, reptiles, or fish understands that animals are born and animals will die. It is the way of nature. Our responsibility, as caretakers of animals, is to provide a humane environment for the animals to live in while they are under our charge and when the time is right, to provide for a respectful and compassionate death. This is a harsh thought for most people. We become attached to our animal charges and ending their life is something we are willing to deal with only reluctantly. But, eventually, we or someone else must.

Downer cows are no longer allowed to enter the food chain, since a single cow was found to have Bovine Spongiform Encephalopathy (BSE) in December 2003, in the state of Washington. Prior to that time, if a cow became a downer she could be loaded onto a trailer and transported to the market, where she would promptly be slaughtered and entered into the food chain. That is no longer an option.

Today, your choices include ministering to the cow until she shows improvement, regains her feet, and reenters the productive string, or making her as comfortable as possible and allowing her to die a slow death, or you can humanely and respectfully euthanize her. At that point, the renderer must be called to haul away the carcass, or you can make arrangements to compost her carcass on your farm.

As harsh as it may seem, at times there is little choice but to euthanize the animal. This can be done with a lethal injection administered by your veterinarian, a captive bolt-stunning device, or a gunshot to the head. The latter is detailed below.

Equipment Necessary

- Halter
- Revolver (.38 Spl +P, 9 mm, .357)
- Solid-point bullets
- Front-end loader or other means to move carcass
- Compost pile for carcass disposal or arrangements for pickup by renderer

---

CAUTION: Be certain you know how to handle the firearm safely. This is a time of stress and little things can be forgotten in the heat of the moment. Review safe firearm handling procedures before you go to the pen or corral. This is something that is best done in the company of another experienced animal handler. This is not the place for young or excitable personalities.

CAUTION: Study the following figures carefully. Become familiar with the location of the brain and with the external landmarks on the cow's forehead used to locate it. Think through the process— correct placement top to bottom and right to left . . .

and correct angle of the gun barrel so that the bullet will travel through the brain front to back. Do not do this if you are uncertain. Call your veterinarian for assistance or contact a neighbor who has done it before. Ending the downer animal's life is a necessary process and somebody has to do it. However, it does not need to be you.

Step-by-Step Procedure

1. Approach the animal calmly and quietly. Since you have been treating her, trying to get her on her feet again, she should not struggle and attempt to rise and escape. If she does, back off a bit and allow her to relax.

2. Place the end of the revolver barrel approximately 6 inches from the forehead. Be certain to align the barrel as indicated in the figures—both top to bottom and right to left.

3. Squeeze the trigger slowly. Constantly reaffirm that the barrel is aligned correctly. You must control the placement of the bullet with certainty. Do not jerk the trigger, or perhaps even look away, at the last moment. This must be done correctly, if it is to be a humane and respectful end to the animal's life. Done properly, there is immediate and total loss of consciousness and massive brain destruction.

4. Return the revolver and ammunition to their respective places of safekeeping.

Postprocedural Management

At this point, the animal's carcass must be properly disposed of. Check local and state regulations to be certain you are in compliance.

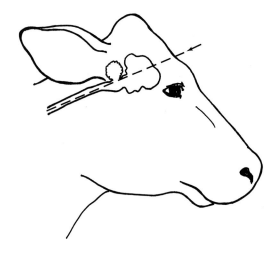

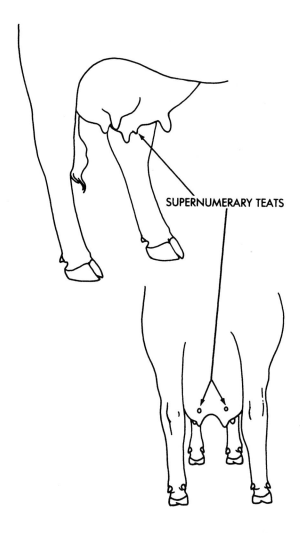

SUPERNUMERARY TEATS

4.13 REMOVAL OF SUPERNUMERARY TEATS

Many dairy calves are born with more than the usual four teats. These supernumerary (extra) teats can develop and grow just like a normal teat. They cause problems because mastitis infections may develop in them, they may interfere with either hand or machine milking, and they detract from the general appearance of the animal. For these reasons, it is a good practice to remove these extra teats as early as possible in the calf's life.

Bred heifers should be reexamined for the presence of extra teats about 2 months before they are due to freshen (calve). Extra teats on cows in milk that are objectionable from a management standpoint or that detract from the cow's general appearance can be removed. If they are to be removed, it should be done during the cow's dry period. During this time the cow's udder is usually smaller and not distended with milk, and healing is more rapid.

Equipment Necessary

- Sharp scissors or knife
- Antiseptic
- Disinfectant

Restraint Required

Because extra teats are best removed soon after birth, little restraint is necessary. If removal is delayed until animals are older, a headgate or stanchion is required. The use of a halter to help secure the heifer is also recommended. With older heifers, help from one or two additional persons will be required.

Newborn Calves

These extra teats can be removed immediately after birth at the same time the navel is treated. If it is done at this time, the calf is easy to handle, and one person can accomplish the job.

Step-by-Step Procedure

1. Assemble your equipment—a sharp scissors or knife and an antiseptic material for treatment of the wound. The scissors or knife should be clean and disinfected. A clean cloth on which to lay the equipment will prevent contamination.

2. Approach, catch, flank, and restrain the calf on its side.

3. Raise the back leg and expose the udder so that you can determine which of the teats is extra. The extra teat is usually located higher up on the rear of the udder or between the normal and larger front and rear teats. If there is any doubt whatsoever in your mind, do not proceed beyond this point. An experienced dairyman or your veterinarian should be consulted before you go any further. The extra teats can be removed at the same time the calf is vaccinated for brucellosis.

4. Grab the end of the extra teat and use the scissors or knife to cut it off close to the udder. Because the calf is very young, the cut bleeds only slightly.

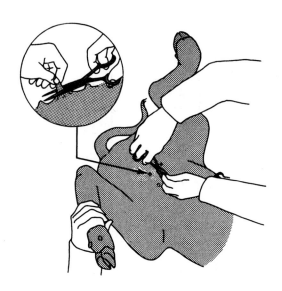

CAUTION: The heifer may try to kick the person doing the cutting. Plan on preventing the heifer from hitting you by firmly applying the proper restraint.

Care must be taken to prevent the knife or scissors from slipping so that you don't cut either yourself or the heifer.

5. After the teat is removed, an antiseptic should be applied to reduce the chance of infection.

6. The calf is now ready to be returned to its pen.

7. The equipment should be cleaned thoroughly and stored so that it will be ready for use next time.

Older Calves or Heifers

With older heifers, a good time to remove extra teats is when the animal is vaccinated against brucellosis. This vaccination is done during the third or fourth month of life. If you choose to do it yourself at that time or at a later date, the older calf or heifer must be restrained in a stanchion, headgate, or squeeze chute.

1. Assemble the necessary equipment.

2. The equipment should be thoroughly cleaned and disinfected.

3. The calf or heifer should be moved to the headgate, stanchion, or squeeze chute and secured.

4. At least one other person will be necessary to help restrain the animal. His duties will include applying the tail hold, nose lead, or other restraint. He may also pass the equipment and antiseptic to the person removing the extra teats.

5. Examine the udder for extra teats and identify the ones to be removed. Be very sure that you have made proper identification of the extra teats.

6. Grasp the end of the extra teat. Cut it off with the scissors or knife close to the udder.

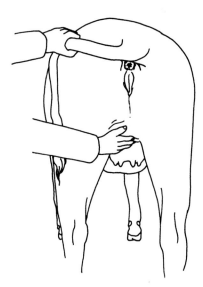

7. After the teat has been removed, apply the antiseptic to reduce the possibility of infection.

8. The animal can now be released from the restraint and returned to her housing area.

Postprocedural Management

It is a good idea, regardless of when the extra teats are removed, to observe the animal for several days to see that no infection develops. If infection does develop, treatment will be necessary.

4.14 WEANING AND TRAINING TO DRINK AND EAT

Most dairy calves are weaned from the cow soon (immediately or within 3 days) after birth. This means that the calf must be trained to drink milk from a bucket, pail, or nipple immediately and to eat grain as soon as possible after weaning. With the proper approach and gentle handling, this job can be accomplished easily with little stress to the animal.

Equipment Necessary

- Calf bucket
- Grain box or bucket
- Nipple bucket
- Milk or milk replacer
- Nipple bottle
- Headgate, stanchion, or halter
- Sanitizing solution

Restraint Required

Little restraint is necessary to train the animal to drink from the nipple bucket or bottle. Some restraint is necessary when training the calf to drink directly out of a bucket. An overprotective cow should be restrained by a stanchion, headgate, or halter.

Weaning the Calf

Because most dairy calves are weaned shortly after birth, there is not much of a problem in completing this task. The calf is simply removed from the cow and moved to the calf housing area.

Calves should be housed in clean, dry, well-ventilated buildings. This type of environment can be provided in either cold or warm housing. Cold housing must provide protection from drafts and be clean and dry.

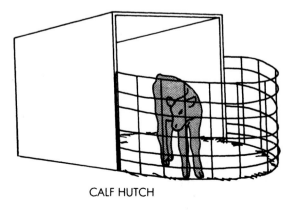

CALF HUTCH

The calf hutch is an example of this type of housing. Warm housing must also provide protection from drafts and be clean and dry. Properly ventilated, insulated, and heated buildings are used.

The cow is then moved to the cow housing area where she will join the rest of the herd in milk. Many dairy calves are removed from the cow after 12 to 24 hours. The calf should receive colostrum (the cow's first milk after calving) for the first 3 days of life.

The longer the calf is left with the cow, the more difficult it will be to train the calf to drink from a bucket. In addition, the longer the calf remains with the cow, the more the cow will fuss for the calf by bawling and walking the fence.

Training the Calf to Drink

Training a calf to drink from a bucket or a nipple is a rather simple task if a few suggestions are followed.

First, because the calf has been with its dam before weaning, it probably has nursed and is not hungry. It is much easier to break a calf to the bucket or nipple if it is hungry at the start. This means that you should wait for 8 to 12 hours after the calf has been removed from its dam before you try to feed it.

Second, after deciding whether you will use an open-top bucket, a nipple bucket, or a bottle, put the milk in the chosen container and offer it to the calf. The calf should receive about 9% of its body weight of milk or milk replacer per day. More milk than this can cause dietary scours. The calf should be fed twice a day. No extra water is necessary.

Because the calf has a natural instinct to nurse, the nipple system is the easiest to use. For this reason, many people use the nipple bucket or bottle. One problem with the nipple is that it is difficult to clean.

1. Enter the calf pen while holding the bucket or bottle firmly in one hand. It may help to back the calf up against a partition or into a corner. Attempt to get the calf to nurse by placing the nipple near its mouth. If the calf will not nurse, it may be necessary to open the calf's mouth with your other hand. This is done by inserting your fingers into the side of the mouth between the front and back teeth. Then apply pressure on the roof of the mouth and the calf will open its mouth wide enough for you to insert the nipple. If the calf still refuses to nurse, a little pressure on the nipple to let some of the milk flow into the calf's mouth will start the calf nursing.

the calf by straddling it and putting its head between your legs. This will let you restrain the calf so that you can handle the bucket with one hand and use the fingers-in-the-mouth technique with the other.

CAUTION: Take care to use no excessive force in pushing the calf's head into the milk. Doing this forces the calf to draw its breath and actually draw milk into its lungs. This can cause pneumonia, and the calf may die. Keep the calf's nostrils out of the milk by pulling up on the nose.

Remember to be gentle with the animal. Mistreatment at this time can scare the calf enough that it will, by sheer fright, refuse to drink at all.

If the calf absolutely refuses to drink, the best policy is to wait 2 or 3 hours until he becomes hungry again. Try once more, being as gentle as possible.

Use one bucket or bottle per calf to prevent the spread of bacteria from calf to calf.

---

2. If you have decided to use the open-top bucket, you will find that it is more difficult to get the calf to drink. Some calves will be hungry enough to start drinking without much training. Others will need to be caught in the corner of the pen and held so that you can put your fingers in their mouths and try to get them to suck the fingers. If the calf will do this, lower its head into the milk while he is still sucking on your fingers, and he will pull the milk into his mouth.

If you separate your fingers and leave space between them so the milk will flow more easily, the calf will very soon learn to drink on its own. This procedure may have to be repeated for one or two feedings. Some calves take as long as 2 or 3 days to learn to drink from the open bucket.

3. Holding the calf can be a problem, especially if it is a rather large calf. Most of the time, you can hold

4. Dismantle the nipple bucket or bottle. Wash all calf-feeding equipment thoroughly, sanitize it, and store it on a rack to dry.

Training the Calf to Eat Grain

Most successful calf-raising programs follow the practice of getting the calf to eat a good calf starter (grain) as quickly as possible. A good home-mixed calf starter is ⅓ each by weight of whole shelled corn, whole oats, and 26% natural protein pellets. Adding a little dried molasses to the mix will make the starter more palatable. The grains may be rolled or cracked but should not be finely ground. With only a little training, calves can be taught to eat grain as early as the first week of life. The easiest way to get the calf to start eating grain is described next.

1. Immediately after the calf has finished its milk, it will still have the urge to nurse. Place a small amount of grain in the palm of your hand, let the calf nurse your finger, and slide the grain into its mouth as it is nursing. Most calves will start chewing on the grain, find it tasty, and start to eat.

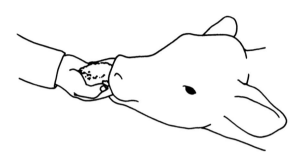

2. After two or three hand feedings of grain, put a grain box or bucket with some grain in it into the calf stall. Place some grain into your hand and encourage the calf to nurse as before. Slowly move your hand toward the bucket or box that contains grain. Remove your fingers from the calf's mouth and hold the grain in the palm of your hand. Put some of this grain on the calf's muzzle. Lower your hand into the grain box or bucket. You may have to push the calf's head into the grain box or bucket to help it find the grain. With any luck, the calf should be eating grain within a week or 10 days of birth.

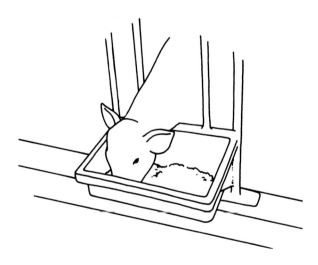

CAUTION: Do not try to force the animal to eat grain by forcing the grain into its mouth or down its throat. This can cause choking and irritation or laceration of the throat.

4.15 BODY CONDITION SCORING OF DAIRY COWS

Dairy cows are kept to produce milk, which in turn is processed and sold as fluid milk, cheese, ice cream, butter, and a whole host of other high-demand and healthful food products. In order for the dairy cow to lactate, or produce milk, she must cycle and re-breed, have a calf each year, and her managers must keep her in a general state of good health. All of this production, the calf, and the ability to stay healthy are based upon adequate levels of energy in the diet.

Body condition scoring is a tool that the dairyman can use to estimate the adequacy of the feeding program for the cow herd. Because of the tremendous demands on the cow during early through peak lactation, as compared to the relatively easy time of early lactation and very late lactation, it is impossible to maintain a constant BCS year-round.

Follow the cycle of a cow for a year and you will quickly see that the demands on her system fluctuate a great deal. A dairy cow is usually allowed a 60-day dry period, a time when she is carrying the calf in her uterus but is not being milked. Certainly, the calf is in the last trimester of growth, but this is a time when the dairy cow's condition can be raised. Following calving, the demands on the cow change from those of providing for the growth of the calf to providing for the increasing milk flow of early lactation. Following early lactation, the cow moves into her peak lactation period, a time when it is essentially impossible to maintain her body condition. A good-producing dairy cow will lose condition during this time. Somewhere about three months into her lactation, she will be rebred, thus adding to the demands upon her body. She will carry through with the 300-day lactation, which tapers off toward the end. During late lactation, she will start gaining condition again and finally will enter the dry period of a year earlier.

Since the high-producing dairy cow cannot consume enough energy to meet the very large energy demands of peak lactation, the dairy manager works hard at having her enter this period with a high BCS, indicating a high level of stored energy—energy she will use to make up the difference between what she takes in from the diet and what she needs for lactation. To optimize the cow's performance during each period, the manager must understand this phenomenom and take advantage of it.

The wise dairyman adds energy to the diet and builds BCS any time possible so the cow's genetic potential is not limited by the manager's nutritional limits.

There are five body conditions, and therefore five body condition scores (BCS). They are BCS 1, BCS 2,

BCS 3, BCS 4, and BCS 5. Each of these scores describes the amount of body reserves in the animal. They are subjective appraisals based upon both visual and touch estimates of apparent external fat cover and obvious skeletal features.

Equipment Necessary

- Halter, stanchion, or other minimal restraint
- Body condition scoring guideline pictures and verbal descriptions
- Cow record book

Restraint Required

Minimal. Dairy cattle are handled and observed so frequently that the opportunities for appraisal can be worked into the normal routine of the farm. Halter and lead rope, stanchion, crowding area, or holding pen will be satisfactory.

Step-by-Step Procedure

1. Study the anatomical model below with particular emphasis upon the key reference points indicated. While studying the model, visualize where these anatomical points will be exhibited on the live, hide-covered cow.

2. Approach the cow to be scored (it is best to start with only a couple of cows if they are in a pen setting), and carefully watch it move. Watch particularly the hipbones, ribs, spine, tailhead, and obvious soft fatty areas as they move. Make some initial notes.

3. Move around the pen slowly, and look for features to verify your initial scoring notes. Don't be concerned if you cannot remember all of the points in each BCS description. No one can, at first. It will come with practice.

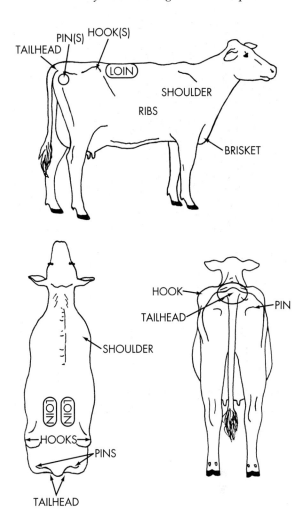

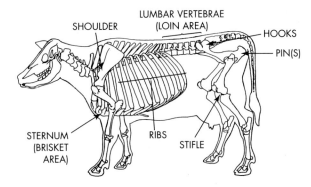

4. For the dairy cow, BCS is primarily determined by the amount of fat covering the tailhead and rump regions. Most BCS evaluation is done from directly behind the dairy animal, so begin your scoring process by standing there. Check the degree of depression about the tailhead. Then move to the rump area, where you will palpate the pelvic bones ("hooks" and "pins"), short ribs (transverse processes of lumbar vertebrae), and loin. Some find it helpful to step to the side and evaluate the degree of depression in the loin area.

5. Assign one of the following scores to each animal.

6. Record the BCS information for each animal. Be certain to verify animal identification.

7. Study and analyze the condition scores.

Postprocedural Management

Cows are kept for reproduction. Calving performance can be optimized when cows are fed to the proper condition, usually about BCS 3 to 4. This allows for intrauterine growth, uncomplicated calving, postcalving milk production, and rapid rebreeding. BCS lower than this will affect each of these performance traits, and overconditioning will cost more money in feed bills than necessary.

BCS 1   *Tailhead:*   Deep cavity around tailhead. No fatty tissue exists between the pins.

          Rump area: Bones of pelvis are sharp and easily felt. No fatty tissue is present in the pelvis area.

          Loin area: Ends and upper surfaces of short ribs can be felt. Deep depression is visible in the loin area. No fatty tissue is present in the loin area.

BCS1

BCS1

BCS 2   *Tailhead:*   Shallow cavity around tailhead, lined with fatty tissue. Some fat can be felt below pinbones.

          Rump area: Pelvis easily felt.

          Loin area: Ends of short ribs can be felt, but they are rounded and upper surfaces can be felt only with slight pressure. There is depression visible in the loin area.

BCS2

BCS2

BCS 3   *Tailhead:*   No visible cavity around tailhead. Some fatty tissue is present over the entire tailhead area.
  Rump area: Pelvis can be felt, but slight pressure is necessary. Skin appears smooth.
  Loin area: Ends of short ribs can be felt only with firm pressure. Thick layer of fatty tissue covers the loin. Very slight depression is visible in the loin area.

 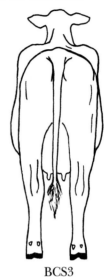

BCS3   BCS3

BCS 4   *Tailhead:*   No cavity visible and folds of fatty tissue surround tailhead.
  Rump area: Pelvis can be felt, but firm pressure is necessary. Pinbones are covered with and surrounded by patchy fat.
  Loin area: Short ribs cannot be felt, even with firm pressure. No depression is visible between backbone (spine) and hipbone.

 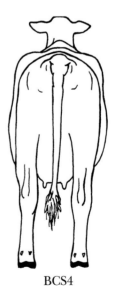

BCS4   BCS4

BCS 5 *Tailhead:* Tailhead is buried in thick, fatty tissue layer.

Rump area: Pelvis cannot be felt, even with firm pressure.

Loin area: Thick layer of fatty tissue covers the short ribs. Bone structure cannot be felt, even with firm pressure.

BCS5

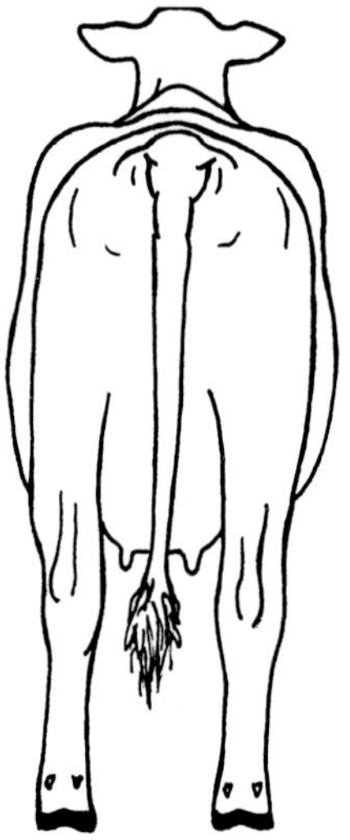

BCS5

CHESTER WHITE BOAR

Courtesy of Department of Animal Science, Oklahoma State University

DUROC BOAR

Courtesy of National Swine Registry

HAMPSHIRE BOAR

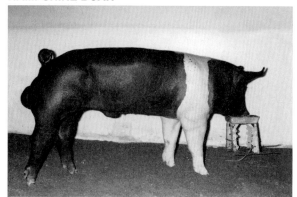

Courtesy of National Swine Registry

LANDRACE BOAR

Courtesy of National Swine Registry

POLAND CHINA BOAR

Courtesy of Department of Animal Science, Oklahoma State University

SPOTS BOAR

Courtesy of Department of Animal Science, Oklahoma State University

YORKSHIRE BOAR

Courtesy of National Swine Registry

TAMWORTH BOAR

Courtesy of Department of Animal Science, Oklahoma State University

chapter |FIVE|

Swine Management Techniques

5.1 CHARACTERISTICS OF SWINE

Pigs are among the most intelligent of domesticated farm animals. They grow rapidly from a few pounds at birth to 250 pounds market weight in about 6 months. The generation interval is short. The male and female reach puberty at $5\frac{1}{2}$ to 6 months of age, and they may mate at 6 to 8 months of age. It is possible for females to farrow when they are about 1 year old.

Swine are *polyestrous*, which means that they come into estrus, or heat, at 21-day intervals throughout the year. It is possible therefore to farrow pigs at any time of the year. Pigs are also very prolific. At each estrus they ovulate about 16 to 18 ova and implant large numbers of fertilized eggs, delivering 10 to 12 live pigs at the completion of each pregnancy. A pig's average birth weight is about 3 pounds, much smaller and lighter than other farm animal species. By 3 or 4 weeks after farrowing, the pig usually will have increased its birth weight by a factor of 6 to 8. Pigs are weaned at between 2 and 4 weeks of age.

The highly productive female can give birth to and nurse 2 to 2.4 litters of pigs per year. With average litter size, this would be 18 to 22 pigs annually. If her offspring are marketed at 6 months of age, when they weigh 250 lbs, the annual production of a sow may well be from 4,500 to 5,500 lbs of live pork. The females—sows or gilts—are excellent mothers. They are very protective of their young during farrowing and lactation.

Longevity in swine is good. Sow productivity (number and weight of pigs weaned) peaks at about the fourth or fifth litter and continues to be acceptable until the sows reach an age of 5 to 6 years (about 10 litters). Sires, if not permitted to get too heavy, also retain good productivity until they reach an age of about 5 to 6 years.

The pig can smell, see, and hear very well, and is an animal that will respond promptly to training. For example, many routines such as leaving or entering farrowing crates, moving to and from mating areas, and the use of feeding and watering devices are promptly learned. This learning potential improves the overall efficiency of the enterprise and challenges the herdsman or manager to make optimum usage of this trait.

Pigs are *homeotherms,* like most farm animals, which means that they maintain a constant body temperature regardless of environmental temperatures. Normal body temperature is $102 \pm 1°F$. They are *monogastric,* which means that they have one stomach. Like humans, swine need a balanced diet each day, or a ration that contains the right proportions of energy (corn or other grains), protein, vitamins, and minerals. Swine never overeat and founder, so they can be self-fed their rations. They are omnivores and use their snouts to seek out grubs, earthworms, and other material from the soil. This searching or hunting results in considerable damage to pastures or turf and necessitates the practice of ringing. Pigs can easily become infested with internal parasites when given access to hog lots or pastures contaminated with worm eggs or larvae.

Pigs have coarse, bristly hair which is sparse. There are four toes on each foot, two of which are functional. The pig's body is well insulated with a layer of fat. The exception to this is the very young pig, which has little or no fat, and thus requires a better environment to keep warm. A baby pig's eyes are open at birth. Pigs can shiver. They tend to pile up in a corner when they are cold. Pigs have very few sweat glands, and so protection from the sun and hot weather may be required during the summer months.

One of the biggest mistakes made when working with pigs is that of trying to hurry them. The pig thinks through many situations and responds favorably if given the time. If rushed, the pig is prone to panic, and then attempts escape by going over, under, or through equipment to return to more familiar surroundings. This return reaction is normal and very strong in swine. Herdsmen moving animals must anticipate and be prepared for this.

The pig is very strong, and the fact that the center of gravity of its body is low makes it difficult to stop the animal without specialized equipment, especially if it is frightened and attempts to run.

The pig is industrious, and is given to rustling, nosing, chewing, and exploring the premises in which it is confined. This trait makes the design and arrangement of equipment very important if major repairs and constant maintenance are to be avoided.

5.2 HOUSING OF SWINE

Optimum environmental temperatures for the pig range between 55 and 85°F, depending on the age or stage of development in the life cycle. Temperatures that are greatly different from these may stress the animal and detract from its maximum performance. This fact is responsible for the use of shelter or buildings in swine rearing. A shelter lessens the pigs' exposure to extremely hot or cold climactic conditions. It provides protection for both the animals and their handlers.

All or part of an existing older building or a small individual house will suffice as a hog shelter. The important thing is that it keep out drafts, snow, and rain, provide shade in hot weather, and have a dry floor.

The simplest housing is probably an A-frame type—a watertight roof that forms the sides of the building, and a rear wall. The front is usually open, but it can be fitted with a door. If it is movable, locate the A-frame at a spot where water won't puddle, and face the front away from the prevailing wind.

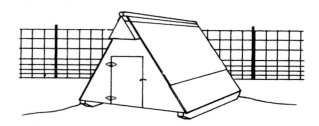

Keep the inside of the house dry, clean, and well bedded with straw, peanut hulls, or wood shavings. Remove the bedding when it gets wet and dirty, and spread it on a field or pasture.

Hog lots must be fenced "hog-tight." For lots several acres in size, use 32-inch woven-wire fencing, with a strand of barbed wire at the bottom just above the ground. Pigs are more apt to go under fencing than over it.

For smaller lots, a temporary or permanent fence of 1" × 6" boards or wire panels about 35 inches high is easy to construct. Attach the boards or panels to steel or wood posts.

Pigs can be fed daily from a hog trough or pans. Provide enough trough space that all the pigs can eat at one time. If a self-feeder is used, provide a feeder hole for each four or five animals, and keep it in adjustment to prevent waste of feed.

Hogs should have access to plenty of clean water at all times. A 35-pound pig drinks about ½ gallon per day; a 225-pounder drinks about 1 to 1½ gallons; and a brood sow suckling a litter drinks about 5 gallons.

For dispensing water, you can use either a heavy trough or pan that the pigs can't upset, a home-made waterer, or a nipple unit connected to a water line. Check daily (twice a day during hot weather) to make sure that water is always available. Protect waterers from freezing in the winter.

The type and overall design of more complex housing depend upon climate and geography. Two types of buildings are most commonly used to house swine. One of these is the open-fronted building, which possesses natural ventilation, minimum insulation, and provides minimal protection during severe cold weather. These buildings generally have a low initial cost, but require more labor for their operation. The second type is an environmentally regulated building, which has a higher initial cost but low labor requirements. Pig performance in the open-front building is excellent in the summer, but somewhat lower in the winter. The totally enclosed buildings provide excellent wintertime performance, but tend to be hot during the summer months. Pig performance in the two houses balances out on the basis of a full year's usage.

OPEN-FRONT BUILDING

Most producers use a totally enclosed building for the farrowing house and nursery; other types of buildings are used for growing and finishing and for the pregnant sow.

A good swine house includes provisions for temperature control, proper ventilation or controlled air change, good sanitation, and physical equipment for partitioning, feeding, watering, and material handling. A substantial tonnage of feed must be moved into buildings when they are in use, and a large amount of manure must be removed. Design and planning to incorporate these types of features into a building will add to its success. No one building design best suits all farm-production situations. The herdsman's managerial ability will determine the success or failure of any one kind of building in the total swine-production program.

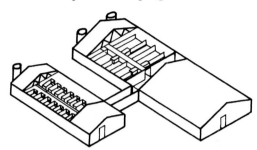

ENVIRONMENTALLY REGULATED BUILDINGS

5.3 HEAT DETECTION: TEST MATING

A profitable farrow-to-finish swine operation is dependent on good reproductive efficiency. This includes high conception rates and large litters of strong pigs farrowed, weaned, and marketed.

Heat detection is the process of determining when the gilt or sow is showing estrus or is in standing heat, ready for mating. To achieve high conception rates and big litters, the sperm must be deposited in the female's reproductive tract a few hours before ovulation occurs. Female swine will ovulate 8 to 12 hours before the end of standing heat, or 27 to 40 hours after the start of standing heat. When mating occurs too early or too late, the conception rate is reduced.

Test mating involves using a new, young boar in order to check his fertility, libido (sex drive), and his ability to mount and breed a female. This practice of test mating can improve reproductive efficiency by identifying a problem boar before the breeding season starts.

Mating is usually done by pen mating, hand mating, or artificial insemination. Pen mating allows the boar to be with a group of females at all times. The hand-mating system controls the matings; the actual breeding date is known and recorded and the number of services is controlled. Boars and females are maintained in separate pens, and matings are controlled by checking heat, mating, and then penning separately after mating.

Artificial insemination requires special training. It involves collecting semen from a boar and then inseminating the sow. Four to ten sows can be bred with the semen collected from one boar; the exact number is dependent upon how the semen is handled and processed.

Facts about Female Swine

Gilts will reach puberty or start coming into heat (cycling or showing estrus) between 5 and 8 months of age, when they weigh between 200 to 280 pounds. The length of the estrus, or heat period, averages about 2 days and occurs about every 20 to 21 days if the animal does not become pregnant.

Time of Puberty, Estrus, and Ovulation in Gilts	
Age at puberty	5–8 months
Weight at puberty	200–280 pounds
Length of estrus cycle	18–24 days (21 average)
Duration of estrus (heat)	2 days (average)
Time of ovulation	40 hours before onset of estrus
Weaning to estrus	3–8 days (5 average)

Gilts are usually not bred on their first heat because ovulation rate (and litter size) increases in the second and third heat. To help synchronize and initiate the first heat (that is, to bring all gilts into heat about the same time), confinement-reared gilts can be moved, mixed, and given fence-line contact with boars.

Sows will not normally come into heat while they are lactating. Sows whose pigs have been weaned will come into heat 3 to 8 days after weaning.

Facts about Boars

Boars should be purchased 45 to 60 days before they are to be used for breeding to check them for health, get them conditioned to your farm, and evaluate them. Newly purchased boars should be isolated for at least 30 days. They should be housed in clean and disinfected quarters.

Boars reach sexual maturity at about 7 months of age. They should not be used for breeding before this. Test-mate boars when they are about 7½ to 8 months old, after the isolation period. Test mating is the process of allowing a new boar to mate in order to check the boar's breeding abilities.

1. Take a gilt in heat (estrus) to the boar, and observe the boar for libido (sex drive), aggressiveness, and desire to mate.

Heat is the time the female accepts the male for mating. Females that are in standing heat will allow a person to sit on their back, and at the same time will stand rigidly and attempt to stiffen their ears in an erect position. The latter action is called "popping their ears." If females do not stand solidly and pop their ears, they are not in heat. A boar should be present for best success in heat detection. The vulva may be swollen, especially in gilts, and nervousness may be noticed before and after standing heat.

Sometimes it is impossible to move a gilt or sow that is in standing heat, because when touched they stand rigidly and resist being pushed or moved. To facilitate heat detection, a pen system utilizing a "hot lane" can be used. With this arrangement, females are penned next to a boar pen, and the females in heat will find the boar quicker than boars find the females. Females in heat tend to walk or stand along the fence next to the boar pen, making it easy to identify them.

The mating behavior of swine starts with the boar approaching and sniffing the vulva of the female. As standing heat time approaches, the boar will nudge or nose the female's flanks, and may attempt to mount her. When the gilt is finally in standing heat, the boar (or even other females) will mount her.

2. When test-mating a boar, give assistance during the first service or two. For example, sometimes young boars will mount a gilt from the front. If this happens, gently move the boar to the proper

SNIFFING

NUDGING

MOUNTING
ATTEMPTS

MOUNTING AND MATING

position by pushing his rear legs around toward the rear of the gilt, or gently push him off the front end and guide him toward the rear of the gilt. Do not hit or injure the young boar. Most young boars will soon learn to mount correctly.

3. Inspect the young boar's ability to enter the vulva of the gilt. Anatomical abnormalities of the penis may prevent copulation. These include a limp, infantile, or tied penis. A "tied" penis is an anatomical defect and is technically an attached frenulum. An attached frenulum will prevent the penis from extending normally and entering the female reproductive tract. This defect is easily remedied by clipping the restrictive tissue, freeing the penis. Mate several gilts, and then observe them 18 to 22 days later to determine if they return to heat or become pregnant. The ejaculation process (copulation) takes about 5 to 7 minutes. It is important that the boar be allowed to complete the entire ejaculation process.

4. If possible, have a qualified technician collect a semen sample and check it for quality. The volume of semen and the motility, concentration, and morphology of the sperm should be checked.

For high conception rates, adequate boar power should be provided for the groups of females to be bred. A young boar (8½ to 12 months) can pen-breed 8 to 10 gilts during a 4-week breeding period, and a mature boar (over 12 months) can breed 10 to 12 gilts.

When hand mating, a young boar should service no more than once a day, five times a week. A mature boar can be used for two services per day, not exceeding seven per week.

Do not turn a young untried boar in with a group of sows that are just weaned and coming into heat. The sows will come into heat 4 to 7 days after weaning, and the boar could die from exertion caused by the agitation from the sows in heat.

Breeding females two or three times during the heat cycle (at intervals of from 12 to 24 hours) will increase conception rate and litter size.

If several pens of females are being bred at the same time, consider alternating or rotating boars among pens each 12 or 24 hours. This helps to prevent a group of sows not becoming bred because of a sterile boar.

Breeding pens should be free of any junk or materials which can cause injury. Pens or floors that minimize slippery conditions will maximize breeding performance and success.

Pen boars separately, or pen a group of boars together if they are to be turned in together with a group of sows.

Provide dry, comfortable housing.

5.4 DIAGNOSIS OF PREGNANCY

The gestation period in swine is about 114 days. Maintaining throughout gestation sows that are bred but do not conceive increases feed and overhead costs for the swine operation. Pregnancy diagnosis is a management practice for improving reproductive efficiency; females that do not become pregnant are culled.

There are two practical methods for pregnancy diagnosis. One method is to observe the herd each day for signs of estrus (see Section 5.3). Any female returning to estrus about 21 days after mating is nonpregnant and can then be culled or returned to the breeding pen.

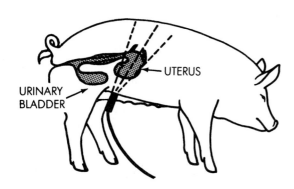

The second method of pregnancy diagnosis is to use an ultrasonic machine on sows or gilts. Ultrasonic waves from the machine are reflected off a fluid-filled uterus, and diagnosis with these devices should be done at least 30 days after mating for best results. Depending on the machine, diagnosis can be done up to 60 to 100 days after mating. With proper technique, these machines are 90 to 95% accurate in predicting pregnancy. The cost, indicator (screens, sounds, light), and weight of these instruments vary. Some also measure backfat and loin-eye area.

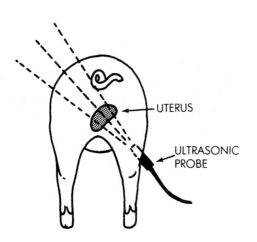

Equipment Necessary

- Ultrasound machine
- Oil (corn oil works well)
- Chalk marker

Restraint Required

No physical equipment is required, but a small pen is required so that pigs cannot escape.

Step-by-Step Procedure

One individual can check for pregnancy, but two people make the job easier. Results are indicated by a beep or are visible on the screen of the ultrasonic device.

1. Read and follow the directions for the machine you are using. Make sure that batteries are charged and the machine is working properly.

2. Females to be tested should be put into a small pen, or they can be driven through a chute designed for performing this skill. Animals should be relaxed and not crowded. The pen should not be so large that animals can escape or the operator has to do a lot of running around. Do not mix animals from different pens, as they may fight, making the pregnancy diagnosis dangerous and difficult.

3. With a chalk marker, oil, and pregnancy-checking machine in hand, enter the pen where females are to be checked. Place the oil and marker in the left hand and the probe in the right hand, or vice versa. The machine is strapped around the neck.

4. The female to be checked should be in a normal standing position. Other sow positions, such as

lying down, may give false results. Stoop over, take a squatting-type position, or kneel down on one knee beside the female to be checked. Your position will depend on the position of the animal. The animal should not be moving when the check is made. Place a few drops of oil on the tip of the probe. Touch the probe to the lower flank of the female, about 2 inches behind the navel and just outside the nipple line. The probe should point toward about one o'clock at a 45° angle from the floor. Check the indicator for pregnancy.

A false-positive reading can be obtained by improper probe placement or aim. When this occurs, the probe will detect the urinary bladder instead of the fluid in the uterus. Overconditioned (fat) animals may give a distorted signal resulting in a false diagnosis.

5. Mark all the females as either pregnant or not pregnant.

6. If one side of the animal gives an unclear diagnosis, try the other side.

7. Cull nonpregnant animals.

5.5 FARROWING

The act of farrowing, or parturition, is the process of giving birth to pigs. It occurs about 114 days (plus or minus 1 to 3 days) after conception. It is a time that is labor-intensive in that it demands the application of certain skills by the manager or herdsman if a high percentage of the pigs farrowed alive are to survive. On the average, swine producers lose about 12 to 15% of live pigs before they are weaned. With certain diseases, losses may reach 100%. Greatest losses result from stillborn pigs, crushing or injury, and starvation. Pig survival is extremely important to the profitability of the swine herd.

A large percentage of sows or gilts are farrowed in a central maternity area or building called a *farrowing unit* or house. Other farrowing operations include the use of A-frame houses that are portable and are occupied by one female.

Equipment Necessary

- Disposable plastic gloves, shoulder and wrist lengths
- Paper towels
- Can of spray paint or marking chalk (orange preferred)
- Oxytocin
- Routine medicines that may be needed
- Supplemental heat sources, heat lamps, pads, or hovers
- Bedding
- Milk replacer

- Bottle with nipple, or 50-ml plastic syringe fitted with 2 inches of plastic tubing
- Shallow pan
- Antimicrobial lubrication

Restraint Required

No physical restraint equipment is required. However, each sow or gilt must have a farrowing crate, pen, or individual house to provide comfort and safety for the litter.

Step-by-Step Procedure

Three logical steps requiring management skills are: (1) preparing for farrowing, (2) watching for signs of approaching farrowing, and (3) the farrowing or birth process.

Preparing for Farrowing

1. Sows and gilts become restless and irritable immediately before farrowing, and some will try to get out of the pen or crate. Make sure that all materials used for farrowing pens or stalls are well built, sturdy, and in good repair. Listen to weather forecasts and be prepared to protect sows and their litters from severe weather.

2. Farrowing crates are used to protect the baby pigs from being lain on and crushed by the sow. Farrowing crates are usually 5' × 7'. The sow is confined to a 2' × 7' area in the middle of the crate. One and one-half feet remain on each side of the sow for the newborn pig creep area or sleeping area. The sow cannot turn around, and thus the risk of crushing the newborn pigs is reduced. The bottom rail of the partition confining the sow to the 2' × 7' area is 8" to 10" above the floor, making it impossible for the sow or gilt to crush the newborn. There are design variations in farrowing crates that allow more freedom of movement for the sow. Most crates are equipped with a feeder and watering device, and some provision is made for allowing baby pigs access to water.

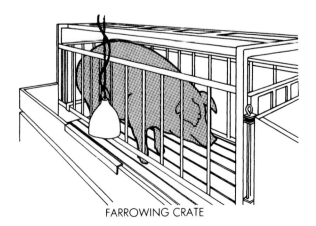

FARROWING CRATE

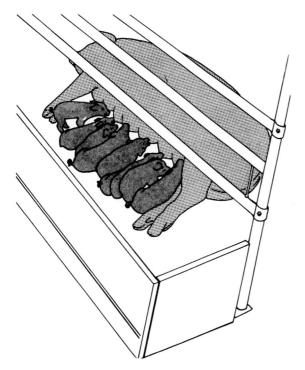

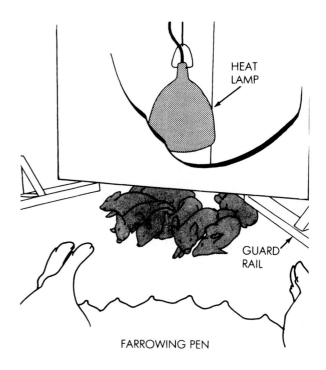

FARROWING PEN

If a pen is used for farrowing, guardrails located about 6" out from the wall and 8" up from the floor on each of the four walls will reduce the number of pigs crushed by the sow. Pens should be no smaller than 6' × 8'. The guardrails give the pig an area for escape when the mother lies down. A creep area should be provided in a corner of the pen.

All protective equipment is extremely important during the first 72 hours after birth. After this time, the baby pigs usually are strong enough to escape crushing. Fifty percent of the newborn pig loss will occur in the first 48 to 72 hours after farrowing.

3. The farrowing house must provide two environments. The sow is comfortable at a temperature of about 55 to 75°F, and newborn pigs at 85 to 90°F, depending on the floor type. To provide a room temperature that is comfortable for the sow, use space heaters in the winter and wall fans or snout cooling in the summer.

Use a supplemental heat source and a generous supply of dry bedding in the creep area to provide comfort for the baby pigs. Some swine producers have reported that wood shavings used for bedding may irritate the navel, resulting in excessive bleeding. Coarse bedding should not be used in farrowing units that have fully slotted floors with manure pits because it will cause removal problems. A piece of plywood, a heating pad, or other commercially available flooring materials placed in the creep areas of slotted-floor units can help provide for the baby pigs' comfort, particularly during the cold winter months.

4. Have the farrowing pens and equipment cleaned (all organic matter removed) and disinfected and left unused for 5 to 7 days before the animals are placed in the unit.

5. Wash the sow before putting her into the farrowing pen or crate, if weather and washing facilities permit.

6. Prepare a lactation ration that is laxative to prevent constipation. Constipation can be prevented or corrected by feeding a bland or bulky diet from 3 to 4 days before farrowing until about 1 week after farrowing; or add 20 lbs of epsom salts or 15 lbs of potassium chloride per ton of ration and include this in the diet throughout lactation. Use of linseed meal as part of the protein provides a laxative effect. Oats or wheat bran added to the grain in a 1:4 ratio can be used to provide bulk. Other fibrous feeds such as alfalfa meal or beet pulp (80 lbs/ton) may also be used. Seek help if necessary from an animal nutritionist in the formulation of rations for swine.

Signs of Approaching Farrowing

If the exact breeding date is known for each female, on the 109th day of gestation move her into the farrowing house or pen after washing. If they are to be turned out for feeding, use spray paint (when the animal is dry) to mark each sow on her back with the number of her crate or pen to help in returning her to the correct pen or crate. Animals should be fed and watered at least twice a day, in the morning and evening. In hot weather, provide water more often. If the exact breeding date is unknown, the manager must watch for signs that farrowing is near.

If animals farrow outside the farrowing unit they may be interfered with by other animals. Newborn pigs may be scattered over a wide area, and pigs may be farrowed in a totally unsatisfactory location, thereby jeopardizing the welfare of the litter. The skilled manager can, through self-training and keen observation, learn to ascertain when farrowing is imminent.

1. Females within 24 to 48 hours of farrowing will have enlarged vulvas, and in many cases will be secreting a clear mucus from the vulva.

2. Females within 24 to 48 hours of farrowing will have enlarged udders that may have a tight, glossy appearance. While the sow or gilt is eating or resting, carefully squeeze and pull down on the nipple with the thumb and forefinger to ascertain the amount of milk secretion. Presence of milk in the form of a large droplet indicates that farrowing may occur within 24 hours. The milk may be grayish in its earliest stage and become whiter as farrowing time approaches. Move the sow to the farrowing facility immediately if a full squirt or spray of milk is present. Do not wait. In a few instances, some sows will have a small quantity of milk several days before they farrow.

3. A few hours before farrowing, the female will gather sticks, straw, cornstalks, and other bedding material in her mouth and carry them to an isolated location where a nestlike arrangement is fashioned. Before this happens, the manager should predict the approach of farrowing and confine the animal to a farrowing crate or a farrowing pen if these are used. Once the female has settled herself in her nest and has lain down, farrowing time is near; she will be extremely difficult to move because she will fight and insist on remaining near the nest.

If the animal is housed in a crate or pen, she will try to build a nest if bedding is available, or she may attempt to make a nest by scraping or pawing the floor, simulating the gathering of materials and the arrangement of the nest. If a farrowing crate is used, make sure that it is constructed so that animals cannot jump out the top, turn themselves around in the crate, or injure themselves.

4. Within 3 to 4 hours of farrowing, the sow will be lying down most of the time, and her respiration rate will become higher, even if the environment is cold. She may be restless and nervous, changing position often, rising and lying down frequently. She may chew or nudge the equipment that is accessible to her. It is important that the farrowing unit be comfortable (55 to 75°F) and not so hot that the sow is heat-stressed. If possible, check sows every

couple of hours if they are about to farrow to observe signs that farrowing is proceeding normally.

The Farrowing (Birth) Process

1. Most sows and gilts farrow without help or trouble. One of the first signs that a sow is starting to farrow occurs when the sow goes into labor and starts straining. Fluid, probably blood-tinged, will flow from the vulva, especially while the sow is straining (labor) and pushing the pig through the birth canal. Some milk may drip from the nipples and can be readily milked from the teats.

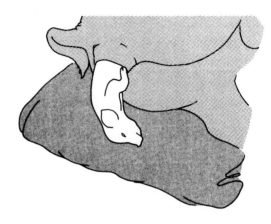

2. After labor starts, the first pig should arrive within 30 minutes to 1 hour. Pigs are normally born either head first or breech (rear legs first).

When the pig is born, the umbilical cord will still be intact in the birth canal. Don't worry about this. It will eventually be broken by the pig's moving toward the nipple. The average birth interval between pigs is about 15 to 20 minutes, but it can vary from simultaneous to several hours in rare cases. If the interval between pigs is longer than 15 to 20 minutes, the incidence of stillborn pigs is greatly increased.

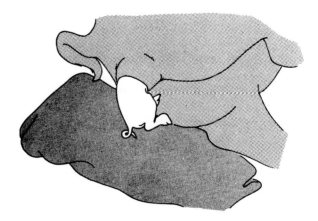

3. Most newborn pigs will immediately start breathing and moving toward the underline to obtain colostrum, which is very important to the baby pig;

CAUTION: The entire birth process for a litter will average about 4 hours. Shorter time for farrowing is of no concern, but if the process takes over 4 hours, a check may be required to determine whether the sow needs assistance.

If there has been an unusually long interval between pigs and the sow has been in labor or has quit straining for some time, she may need assistance. If available, put a shoulder-length plastic glove on your arm and hand and apply some water-soluble lubricant. Do not use Vaseline or other oily materials. With the gloved or clean hand, enter the reproductive tract of the sow or gilt, and search for a pig that may be caught. Grab the pig with your hand and try to maneuver it free. Gently and slowly pull the pig completely out.

Do not administer an oxytocin injection to any sow or gilt until at least one pig has been farrowed. Oxytocin is a hormone that causes the uterus to contract. It also causes immediate milk letdown. Most females should farrow without the aid of oxytocin. Seek professional help if the sow does not start farrowing when it appears she should, or when there is a long interval between pigs. As you gain experience in working with farrowing sows, it will become easier to know when assistance should be provided.

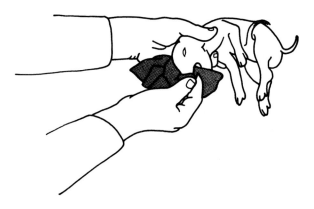

CAUTION: Remove any membrane that covers the head of a newborn pig to prevent suffocation. Wipe the pig off with a paper towel, if available. If one appears lifeless, breathing can sometimes be started by rubbing or slapping its side.

CAUTION: Make sure the sow or pigs cannot reach the lamp or electric cord, and that the lamp cannot touch bedding or other flammable material. Hang the heat source high enough that it does not scorch pigs. Do not hang the heat source by its cord; fasten it securely at least 24 inches from the bedding or flooring.

it should be ingested within 36 hours, and contains antibodies that give passive immunity and protect the newborn pig against certain diseases until the pig starts producing antibodies on its own (active immunity) at about 3 weeks of age. If you are present during farrowing, proceed quietly, and only if the sow does not get up or become excited, to move the newborn pigs to the sow's underline if the umbilical cords have broken. The young pig can be picked up by one of its hind legs. If the sow is disturbed, rubbing the underline may cause her to lie down again.

After being born, the pig knows which way to go to find the nipple, apparently because going the wrong way will cause him to rub up against the bristle of the mother's hair coat.

4. If newborn pigs are piling on top of one another or shivering, they are probably cold. In cool or cold weather, make sure that the newborn pig has access to supplemental heat, such as heat lamps, pads, or hovers in the creep area.

5. Toward the end of the farrowing process, usually after the last pig is born, a larger portion of the placenta (afterbirth) is shed. Be aware that many times the last few pigs are more likely to suffocate because they may be encased in these membranes. If you are

present, remove these membranes from the pigs immediately.

After all pigs have been farrowed, remove the placenta from the farrowing pen and bury it or burn it in an incinerator.

If the sow can move around, she may eat the afterbirth. If this occurs, do nothing about it; go ahead and remove any that may remain.

In slotted-floor farrowing units, lay a board, burlap bag, or similar material at the back of the sow so that the afterbirth does not fall through the floor into the manure pit. This tissue can clog manure-handling equipment.

6. It is best to let the sow lie still, suckle the litter, and rest a few hours after farrowing. Sows have a characteristic grunt or song when they nurse their pigs. After this she should be encouraged to get up and take a drink of water. Do this by slapping her gently on the back. If she doesn't drink, try again later. If she shows an interest in feed, provide 2 to 3 lbs of the lactation ration. Sometimes mixing the feed with water will stimulate eating. After the pigs are 2 to 3 days of age and the udder of the sow is functioning normally, the feed supply can be increased until the sow is on a full hand-feed by 5 days after farrowing. Some producers successfully provide a full feed immediately after farrowing, but sows must be observed more carefully for milking problems.

CAUTION: The mother sow can be very protective of her young, so do not enter the pen or stick your hands near her head—she may bite, and even attack. Use a hurdle. If sows are turned in and out of their crates or pens for feeding, be very careful; if a sow accidentally steps on a pig and the pig squeals, all the sows in that group may attack you. Handle one sow at a time, if possible.

Watch sows that have farrowed for signs of a hardened udder or an unusually hot udder, which indicates the presence of infection. If discharges from the uterus continue after 24 hours, this may also indicate infection. Use a rectal thermometer to check body temperature. Seek professional help if necessary. It is also important at this time to watch the sow or gilt for constipation. Should this condition occur, it should be alleviated at once by giving the sow exercise and adding a tablespoon of epsom salts to her morning and evening ration for a few days.

7. Some sows may become temporarily vicious or hysterical at farrowing and may trample or lie on their pigs, or kill them with their mouths. Some may try to eat their pigs. Try to eliminate this loss by removing all pigs to a warm place until farrowing is completed. The sow may accept her pigs in a few hours. Test the sow by placing only one pig with her and watching her reaction.

8. Some swine producers keep a radio on in the farrowing unit. Sows then become accustomed to the noise and are less easily disturbed by unusual noises caused by other animals or people working in the unit.

9. Remove any wet bedding and manure daily to keep the pen dry.

10. If the baby pigs are getting ample milk, are warm, dry, and free from chilling drafts, and the sow and pigs look like they are comfortable, the farrowing process has been successfully completed.

11. Very soon after birth, pigs develop a nursing order—which means that each pig nurses the same teat throughout lactation. Variation in pig growth may result from inequalities in milk yield of the sow's mammary glands. Pigs with heavier birth weights are more likely to claim the anterior (front) teats. About two to three pigs in a litter will consistently nurse two teats.

12. Although their eyes are open soon after birth, newborn pigs are unable visually to identify teats. The teat is found by feeling with the nose.

Soon after a pig finds a teat, a contact with a littermate will cause a defensive maneuver and some degree of fighting takes place. Most fighting will take place when the pig is 2 to 6 hours of age.

13. *Care of orphan pigs.* Baby pigs may be unable to nurse and are orphaned because of the death of the mother sow, a hostile mother, a too-large litter, failure of the sow to have milk, or the fact that the pigs are small or weak. When this occurs, pigs may be transferred to foster mothers, or they may be fed a sow-milk replacer.

If possible, all newborn pigs should be allowed to nurse to obtain the first milk of the sow (colostrum). If for some reason the pigs' mother does not have milk, colostrum might be provided by allowing newborn pigs to nurse another sow that is just starting to farrow or has recently farrowed; this is accomplished by removing the foster sow's litter for about an hour. Another source is cow's colostrum. It can be preserved by freezing until needed, or it can be used fresh. Pigs can be bottle-fed this colostrum and then permanently transferred to a foster dam.

After pigs have received colostrum, they should be transferred to a foster sow. Transfers should be made between sows that have farrowed within 72 hours of each other. Another method is to wean all pigs from a sow that has weaning-age pigs and then to use her as a foster mother. Give her one less pig than the number of functional udder sections.

Check the foster mother after transferring pigs to make sure that she accepts the new pigs. She may refuse them by biting at them or by refusing to nurse. If this happens, try another sow. Sometimes applying a disinfectant or other odorous compound to all the pigs will make her accept them.

Commercial sow-milk replacers are nutritionally adequate for newborn pigs, but they do not contain antibodies. They do contain antibiotics which help to control the growth of unfavorable bacteria. Effective use of sow-milk replacers requires that equipment be cleaned and disinfected often. This should ideally be done after every feeding to control bacterial growth.

Scouring or diarrhea is a problem with newborn pigs reared artificially. Wetness, chilling, and engorgement promote diarrhea.

Baby pigs can be fed in a shallow pan, bottle fed, or fed by allowing them to suckle a syringe (50 ml) with a 2-inch plastic tube attached.

Train baby pigs to drink from a shallow pan by letting them get hungry. Then push their mouths (not their noses) down into the milk. They should start drinking after getting a taste of the milk. Do not feed too much, as this will cause diarrhea. If diarrhea results, reduce the intake. Feed pigs individually, if possible, to avoid fighting and spilling. A baby pig will consume no more than about 15 to 20 ml (cc) of liquid at a feeding. It is best to feed a newborn pig every ½ to 2 hours. At 1 week of age, pigs can be allowed a measured quantity of liquid in the morning to consume during the day, and a similar amount in the evening. Increase the amount gradually as the pig grows.

Runt pigs in a litter can be given supplemental feedings of a milk replacer to increase survival rate. Feed 15 to 20 ml (cc) of milk replacer. A milk replacer can be made by mixing 1 qt of milk, ½ pt of half-and-half, and 1 raw egg. Another mixture is glucose (dextrose) and skim milk in equal proportions. Do not use table sugar for baby pigs because they will scour more rapidly. Antibiotics should be included in all mixtures.

Hand feeding milk replacers to baby pigs is time-consuming, and results are many times discouraging because of the high death rate. Pigs must be kept in a warm and dry environment to increase chances of success.

Automated feeding devices are available commercially. They work well if directions are followed.

An alternative to milk replacers for orphan pigs is dry feed formulated for early weaning. These feeds are formulated using high-quality, palatable ingredients such as whey, fish meals, blood plasma, and soy protein concentrate. Commercial starter feeds are available for pigs as light as 5 pounds and as young as 5 to 7 days. Mixing the starter feed with a small amount of water or milk replacer and feeding it in small amounts in shallow pans will help ease the pigs onto the dry feed.

14. Neonatal care of the baby pigs is done during their first day of life. This may include (in suggested order of performance) navel cord care, clipping of needle teeth, tail docking, iron injections, ear notching, and castration.

Store the cleaned equipment needed for these techniques in a plastic bag, toolbox, or similar container. A bushel basket, tub, or a specially built box on wheels can serve as a container in which to put the baby pigs while these tasks are being performed. Bed the container with fresh, clean straw or similar material if available.

5.6 NAVEL CORD CARE

Navel care in swine refers to the treatment and care of the navel or umbilical cord stump of newborn pigs. Ideally, this task is completed as soon after birth as possible. Through most of the 114-day pregnancy period, the fetus obtains nutrients and voids excreta through the umbilical cord (navel cord). When this cord is broken, as the pig being born leaves the birth canal, the passageway within the cord provides bacteria a potential passageway into the body of the newborn. Bacteria that do manage to ascend the cord sometimes cause infections, usually later in the pig's life, around the navel. The infection may also localize in the hock or knee joints and cause lameness. Navel-related infections can be reduced by using clean farrowing facilities and providing navel care at the time of birth.

In most cases navel care is provided when other tasks such as ear notching, tail docking, and teeth clipping are done. In many swine operations these tasks are performed as the first chores in the morning when all litters farrowed the previous day are cared for.

Equipment Necessary

- Heavy string
- Side cutters or scissors
- Disinfectant (iodine—USP 2% solution, Merthiolate, or equivalent)
- Optional: cotton swab, artery forceps

Restraint Required

No restraint equipment is required.

Step-by-Step Procedure

The state of the navel cord will vary from being very wet (in pigs less than 1 hour of age) to very dry (in pigs about 1 day old).

1. Restrain the pig gently, grasping the pig over its shoulder or back, forefinger on one side of the pig and thumb on the other side. The holding position will vary with method and the preferences of the individual.

CAUTION: Do not perform the navel care in the farrowing pen or crate because the sow may bite you in her effort to protect her pigs, especially if the pig squeals.

2. In many swine-farrowing operations, both wet and dried navel cords are cut off (leaving about 1 inch) without tying.

Sometimes newborn pigs will bleed excessively immediately after the umbilical cord breaks as the pig is born. This sometimes occurs in all pigs in a particular litter. Tie off the navel cords immediately when this is observed because pigs may die or become very weak from the loss of blood.

3. Thoroughly treat the navel cord by wetting (wet or dry navel cords) with tincture of iodine

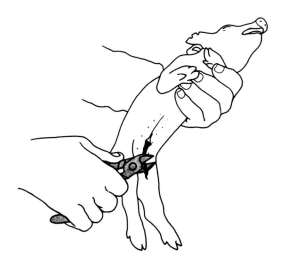

(USP, 2% solution), Merthiolate, or equivalent disinfectant that can safely be used on the skin. The treatment can be given by using a cotton swab on the end of an artery forceps or similar tool. Soak the cotton ball thoroughly in the iodine and then dab the navel, wetting it thoroughly.

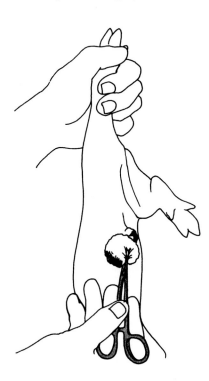

Another method is to use a bottle (preferably unbreakable) containing the iodine disinfectant. Insert the navel cord into the mouth of the bottle. Push the bottle up against the baby pig's body. Then shake the bottle and pig enough to wet the navel cord thoroughly.

The third method is to use commercially prepared navel disinfectants, which may be sold in various forms as liquids or aerosols.

When using tincture of iodine, be sure of its disinfecting power by using a fresh solution each day or after every 80 to 100 pigs.

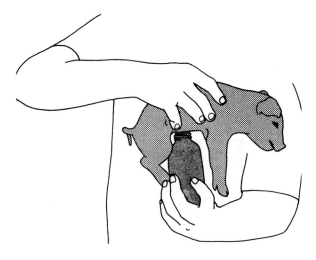

Postprocedural Management

After the navel cord is treated, observe it for any bleeding. Pigs may be placed in the farrowing pen or crate immediately after this treatment.

Recovery Sequence

1. The navel should shrivel and be dried up in 24 hours.

2. The dried cord should have dropped off by the time the baby pig is 3 to 4 days old.

5.7 CLIPPING NEEDLE TEETH

The pig is born with only eight teeth, called *needle teeth* or *wolf teeth*. Four of these teeth are incisors and four are canine teeth. They are fairly long and sharp. They are located on the sides of the upper and lower jaws, with two positioned on top and two on the bottom of each side of the head.

In many litters of pigs there is intense competition at nursing time, and fighting and biting very often result. These teeth are the pigs' only method of defense, and they use them quite promptly after birth in maintaining possession of or in gaining access to a nursing spot. This is especially true when the sow or gilt possesses, for example, only 12 teats on the underline and there are 13 or more pigs in the litter. In the process of maintaining or acquiring a nipple, the pig may fight by biting a littermate, making small cuts around the nose and face which create openings for infection. Even worse, very often a pig will mistakenly bite its mother, sometimes lacerating the mammary gland or nipple. In severe cases, the mother sow may become so irritated by the biting on her udder that she chooses to lie on her stomach and refuses to nurse the pigs. This stress causes abnormal milk letdown and interferes with normal lactation.

To prevent these problems, and to provide more comfort to both the mother sow and her offspring, the needle teeth usually are clipped within 24 hours after birth.

Equipment Necessary

- Pig teeth nippers (sometimes referred to as side cutters)
- Disinfectant

Restraint Required

No restraint equipment is required.

Step-by-Step Procedure

The clipping of the needle teeth is usually done at the same time such other tasks as ear notching, tail docking, and navel care are completed.

1. Restrain the pig by grasping the head with one hand. Place the fingers of the same hand near the back edges of the pig's mouth and firmly force its mouth open far enough that the points of the needle teeth are exposed. Be careful that you do not choke the pig. Keep your fingers away from the throat area of the pig. The pig may squeal and fight this restraint, but good firm control is essential to completing the task. The exact method for holding the pig may vary slightly from one person to another. With experience, each individual will find his or her preferred method.

CAUTION: Do not do the clipping of the needle teeth in the farrowing pen or crate because the sow may bite you in her effort to protect her pigs, especially if the pig squeals. When forcing the pig's mouth open with your fingers, do not put them in a position near the needle teeth where the pig could bite you.

2. Grasp the disinfected teeth nipper with the other hand. The open clipper should enter the mouth directly in front of and as perpendicular as possible to the teeth.

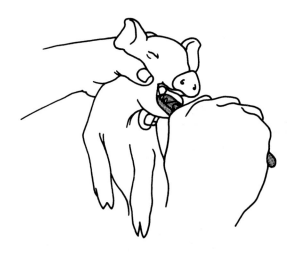

3. If pigs are less than 2 days of age, completely cut off both teeth at once, reasonably close to the gum. One tooth at a time may be cut if this is preferred. Do not cut the gum. If the clipping is done soon after birth, the tooth is still somewhat soft and usually will not shatter or chip, which would cause the gum or tooth to be exposed to infection.

If pigs are 2 days of age or older, clip off about $\frac{1}{3}$ to $\frac{1}{2}$ of the exposed teeth. Older pigs have teeth that are harder and will tend to shatter or chip more. Clipping off just the points of the teeth will tend to reduce both chipping and the chances of infection. Care must also be taken that the gums, lip, or tongue of the pig are not clipped. If a cut is made so short as to draw blood, do nothing about it, but be more careful as you proceed with the next tooth.

4. After clipping the teeth on one side, turn the pig to give access to teeth on the other side of the head. Clip the remaining teeth.

Postprocedural Management

After teeth clipping, pigs may be returned to the farrowing pen or crate immediately. There should be no adverse effects on the pig, and benefits to littermates and the mother's underline are substantial.

5.8 TAIL DOCKING

In swine confinement systems, swine are very often provided between 3 and 8 square feet of space per pig. This amount of space provides adequately for the pigs' comfort. On occasion, however, the pig succumbs to the temptation or tendency to chew or bite on portions of another pig's anatomy, sometimes because of irritability that may be caused by stress—for example, a chilling draft. The most accessible part of the anatomy for a pig to bite or chew on is the tail. This leads to injury and possibly infection. To prevent tail biting and the complications of infection, tails are docked or cut off the newborn pig. This procedure is the best method known for controlling tail biting (cannibalism of the tail), and thereby improving the comfort of pigs that will be raised in confinement facilities. If pigs are to be reared in nonconfining facilities where they have access to pasture or outside hog lots, it is not necessary to dock tails.

Tail docking should be done within 24 hours after birth for the following reasons: the pig is small and easy to hold; at this age littermates are less likely to investigate and nip or bite a newly docked tail; the pig and farrowing area are still clean; and the pig is

> **CAUTION:** Older pigs that have been weaned should not have their tails docked, because excessive bleeding may occur, there is a greater chance for infection, and it may stimulate other pigs in the pen to bite at the cut tails.

well protected with antibodies obtained from the colostrum milk of the sow.

Equipment Necessary

For Baby Pigs
- Pig teeth nippers (side cutters)
- Disinfectant

For Older Pigs
- Elastrator bands and applicator

Optional For Baby Pigs
- Electrical tail-docking and cauterizing instruments
- Rose pruners
- Poultry debeakers

Restraint Required

No restraint equipment is required for baby pigs. Restrain older pigs by catching and holding, or by using a hog snare (see Chapter 1).

Step-by-Step Procedure

Baby Pigs

1. The pig is held suspended by the rear legs with one hand. Little or no squealing by the pig will occur with this method of restraint.

> **CAUTION:** Do not perform the tail docking in the farrowing pen or crate because the sow may bite you in her effort to protect her pigs, especially if the pig squeals.

2. With your other hand dock or cut off the tail with a disinfected side cutters. Dock the tail no closer than $\frac{1}{2}$ inch from the place where the tail joins the body of the animal, and leave no more than half the tail.

> **CAUTION:** Cutting the tail too short could interfere with the muscle activity around the anus later in the pig's life. If too much tail is left, tail biting may still occur in a pen of pigs raised in confinement.
>
> It is best to use a side-cutter pliers with a relatively blunt cutting edge for tail docking because less bleeding results than when a very sharp instrument such as a scalpel is used.

Older Pigs

3. If it is necessary to dock the tails of older pigs, use the elastrator. This is a bloodless method of using a specialized pliers-type applicator to place a strong, heavy rubber band or ring on the tail. The pig is caught and restrained, and a rubber band or ring is placed over the tail at the point at which the tail is to be removed. The rubber band will cut off the blood circulation, and the tail will drop off in 7 to 10 days, leaving a healed stub. Be aware that penmates will try to bite, chew, and break the rubber band. Pen pigs with tail bands individually to prevent this.

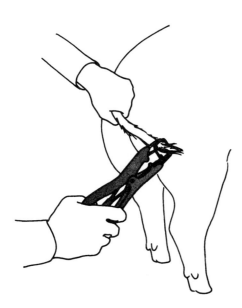

Postprocedural Management

If docking is done with clean tools within 24 hours after birth, no excessive bleeding or hemorrhaging should occur. If excessive bleeding does occur, tie off the tail just above the docking site with a piece of cord, using a square knot. After docking, pigs may immediately be placed in the farrowing pen or crate.

Recovery Sequence

1. There should be only slight bleeding immediately after the removal of the tail.

2. The cut surface will scab over within 48 to 72 hours.

3. Unless the scab is accidentally loosened, complete healing will be accomplished in 7 to 10 days.

4. In a few instances, infection at the docking site may occur. The tail (or stump) will swell and become discolored (very dark reddish color) and may have a liquid discharge from the cut. The pig may show some reduction in appetite. By careful observation after docking (at least daily observation until healing occurs), these infections can be found before they have advanced. An intramuscular injection of an appropriate antibiotic should remedy the infection and the cut should heal. Keeping equipment, animals, and facilities clean will minimize these infections.

5.9 IRON INJECTIONS

There are times in the life of the pig when injections of medications are given. These injections are described as routine because they do not involve the diagnosis, prevention, or cure of a disease. The best example of a routine injection is the administration of iron to the newborn pig.

Iron is an important part of the hemoglobin within the red blood cell. When there is a deficiency of iron the baby pig cannot synthesize an adequate amount of hemoglobin and becomes anemic. Iron-deficiency anemia develops rapidly in nursing pigs because of low storage of iron in the newborn pig, the low iron content of sow's colostrum and milk, the lack of contact with iron in the soil, and the rapid growth rate of the nursing pig.

In pigs that are farrowed in confinement with no access to soil, iron-deficiency anemia may result within 7 to 10 days after birth. Iron is administered soon after birth to prevent anemia and to maximize pig performance and health.

Equipment Necessary

- Disinfectant
- Iron solution (iron dextran and iron dextrin are two of the best forms to use)
- Syringes, clean plastic, 10 to 20 cc
- Sterilized needles, 20-gauge and 14- or 16-gauge, ½ inch long

Restraint Required

No restraint equipment is required.

Step-by-Step Procedure

Iron injections can be given at any time within the first 3 to 4 days of life. If pigs are to be weaned by 3 weeks of age, a single injection of 100 mg of iron is adequate.

If pigs are to be weaned at more than 3 weeks of age, then 150 to 200 mg of iron should be injected. Although a single injection is usually adequate, in cases where sows are heavy milkers with rapidly growing pigs that do not consume creep feed, a second iron injection may be necessary before weaning.

> **CAUTION:** Injections of some forms of iron, which may be quite safe orally, can cause iron toxicity and death. Use the injectable forms of iron (iron dextran and iron dextrin) that have a high margin of safety. Do not overdose. Use proper administration of the iron.
>
> Do not perform the iron injection within the farrowing crate or pen because the sow may bite you in her effort to protect her pigs, especially if the pig squeals.

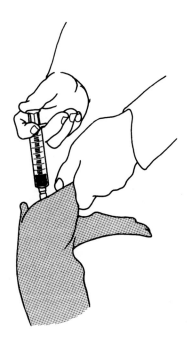

1. Read the label and instructions for the iron product you are using. Usually 1 ml (or 1 cc) contains enough iron.

2. Select a clean syringe and needle. Use a large-diameter needle (14- to 16-gauge) to speed up the filling of the syringe, because iron compounds are thick. Use the smaller-diameter (20-gauge) ½-inch needle for injecting the iron into the muscle. The gauge of the needle should be indicated on the needle. Disposable syringes and needles are preferred.

It is advisable to inject a small quantity of air into the bottle of iron before filling the syringe. Pull back on the syringe handle, drawing in several cc of air. Stick the needle into the rubber stopper of the bottle and turn the bottle upside down. Inject the air into the bottle and then gently pull the handle of the syringe back to fill the syringe with iron solution. To prevent contamination of the iron left in the bottle, remove the syringe from the needle, leaving the needle in the bottle for the next filling. Put a clean needle on the syringe for the injection. Point the needle and syringe upward, and push slowly on the plunger to remove any air in the syringe and needle. You are now ready to inject the iron.

Iron should be injected intramuscularly (within the muscle), either into the ham or neck muscle. The neck muscle is the most desirable because there is a possibility of a residual iron stain later in the carcass of market hogs. A stain in the ham muscle of a market hog would be more detrimental than one in the neck.

3. Inject the iron solution in the manner described next.

Injection into the Muscle of the Ham

Grasp and hold the pig by one of its rear legs, with the pig's head toward the floor. Preclean the injection site by rubbing it with a cotton swab containing disinfectant. To help prevent runback of iron, force the skin of the pig slightly to one side with the thumb of the hand that is holding the pig. With the syringe in your other hand, push the needle with a little jab through the skin at the cleaned site. With the thumb on the plunger of the syringe, inject the proper dosage slowly into the muscle. Care should be taken to inject into the thick muscle. Do not strike the bone. Withdraw the needle, and let the tightened skin move back over the injection site to prevent runback.

Injection into the Neck Muscle

Grasp and hold the pig by putting your hand under its throat without choking the pig, with the thumb and forefinger just behind the ears. To prevent runback of iron from the injection, turn the pig's head slightly away from the site of the injection. Inject the iron into the muscle on the side of the pig just off the topline. Be careful not to strike the spinal area. Use the same cleaning and injection procedures as those used for injection into the ham.

Another method of holding the pig for the iron injection into the neck is to put the pig between your knees. Stretch its head to one side and then follow the same cleaning and injecting procedures used for the iron injection in the ham.

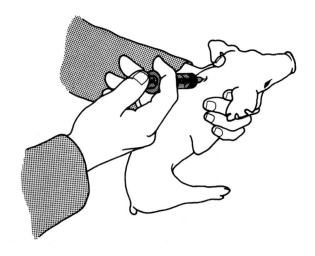

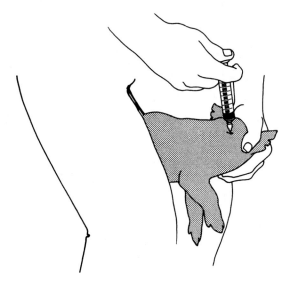

Postprocedural Management

Pigs may be returned to the farrowing crate or pen after the iron injection. Wash all syringes and needles and sanitize them.

5.10 EAR NOTCHING

The swine enterprise requires both financial and production records to identify strengths and weaknesses in the operation. Some production records may require identification of individual pigs. Ear notching is the most used method for baby pig identification.

Ear notching is the process of removing a portion of the ear. The notches or holes grow as the pig grows. This results in a permanent identification unless an ear or portion of an ear is accidentally torn off.

Ear notching should be done soon after birth so that there is immediate identification. Pig birth weights and subsequent treatments can then be recorded. At birth there is less bleeding than there is with older pigs. Older pigs are usually not ear-notched because the process could stimulate ear biting or cannibalism between pigs, causing infections.

Ear tagging should be considered as the best alternative for gilts, sows, and boars.

Each pig must have an individual ear notch in many seedstock herds because it is advantageous and a requirement for pedigree performance records for individual breeding stock. The information then is used in swine genetic-improvement programs.

In those swine operations where all hogs except replacement gilts are marketed for slaughter, it is not necessary that each pig have an individual number. Each litter, or all pigs in a farrowing group, or only the gilts to be considered for replacements might be ear-notched at birth with the same number. The system will vary with the preferences of the swine manager.

Ear-Notching Systems

The most common individual pig and litter ear-marking system is known as System A; it is the identification method required by the purebred swine associations in the United States. Study the key to learn which number each notch location represents.

The litter number is notched in the pig's right ear and the individual pig number in the pig's left ear. When animals are registered with the breed association, litter number and pig number and a sketch of the actual notches are recorded on the pedigree.

System B is also an individual pig and litter ear-marking system. Study pig No. 5972 in the illustration to learn the location of the notches and the number that each notch represents. For example, the first litter marked would be litter 0. Ear notches in the top of the right ear indicate thousands, top left hundreds, bottom left tens, and bottom right ones. Ear-notch all the pigs of one sex first in litter 0

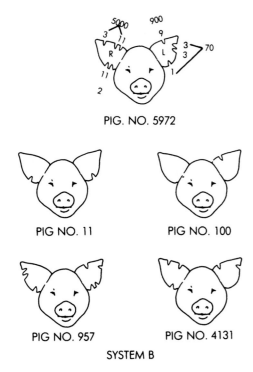

PIG. NO. 5972

PIG NO. 11

PIG NO. 100

PIG NO. 957

PIG NO. 4131

SYSTEM B

System C is a litter-marking system. All pigs in a litter may be marked with the same number. The commercial swine producer might use the system to mark only the potential gilt replacements. One mark might be used for a group of litters from one farrowing, or litters farrowed during a certain week or month.

Other systems can be developed to meet the needs of individual swine producers.

Equipment Necessary

- Ear notcher
- Disinfectant
- Farrowing record forms

Restraint Required

No physical restraint equipment is required when ear-notching newborn pigs. A snare or hog chute is required for older pigs.

CAUTION: Do not ear-notch pigs in the farrowing pen or crate because the sow may bite you in her effort to protect her pigs, especially if the pig squeals.

---

using numbers 0 (no notches) through 9 (one notch in the middle of the lower right ear). When there are more than 10 pigs in the litter, start over with pig number 0, which is likely to be of a different sex. This marking system uses both pig number and sex for identification. The next litter is litter 10, pigs 10 through 19; next litter 20, pigs 20 through 29; and so on up to litter 9990. Use the sequence 1,000 to get away from notches of some numbers on pigs of the same age that might be mistaken for another number. Avoid litter numbers of eight to reduce the number of notches in the ear.

Step-by-Step Procedure

1. Separate one litter of newborn pigs from the sow.

2. Separate by sex if your ear-notch numbering system or record-keeping system requires this. Notch gilts first. Replacement gilts then have low numbers that are easier to read later.

3. Count the pigs and record the pig numbers (if applicable) for this litter on farrowing record forms.

4. Of the management skills done on newborn pigs (navel care, teeth clipping, tail docking, iron injection), ear notching or castration should be done last because more bleeding occurs. Work on one pig at a time, completing all tasks for an individual pig before starting on another one.

5. Grasp the pig firmly but gently, taking care not to choke him. Put your thumb on one side of the head or face and the other four fingers on the opposite side. The pig may fight the operation slightly. The pig's right and left ears are on the *pig's* right and left side, respectively. The notching system used will determine the location of the litter number and the individual pig number. Use an ear notcher that is designed for newborn pigs. Notching too shallowly can result in errors in reading the numbers. If the notch is too deep, the pig may lose or more easily tear off part of the ear. Do not put notches too close together. Leave at least $\frac{1}{4}$ inch between notches.

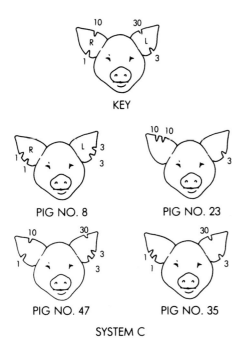

KEY

PIG NO. 8

PIG NO. 23

PIG NO. 47

PIG NO. 35

SYSTEM C

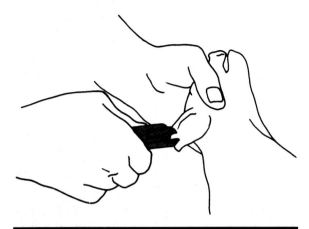

> **CAUTION:** Do not get notches too close to the head on the lower part of the ear, as these can be missed when reading a number in an older pig. Do not get the points of notches made toward the tip of the ear on both the top and bottom too close, as the pig is apt to get those notches caught in equipment and more easily rip off the tip of the ear.

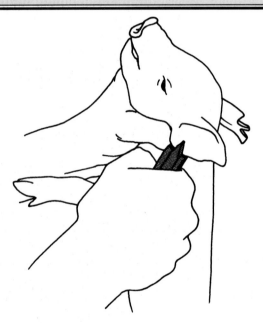

6. Grasp the disinfected ear notcher (from a shallow container of disinfectant). Check the record form (if applicable) for the number of the pig to be notched. Notch the ear with the litter number first. Proceed to notch the pig number in the other ear. After notching, double-check the notch to make sure it is correct. If it is incorrect, correct it if possible. If the notch cannot be corrected, record the mistake on the record form.

Develop a routine procedure for ear notching. For example, litter notch first, pig notch second, and then check notches. Stick to your system and concentrate on the job to be done. Errors may result if you are visiting with someone or not concentrating on the numbers and their location in the ear.

7. Some notchers are designed to punch a hole in the ear. These can be used in operations where only one or two marks are needed. Usually only one hole is punched in an ear. The hole punch or notch can also be used in combination with the V-notch system.

Older Pigs

8. Do not identify older pigs by ear notching because penmates may be stimulated by the sight of blood and start ear biting (ear cannibalism). It is best to use another method of identification, such as ear tagging. If ear notching is the only method available, then an ear notcher designed for older animals must be used. Pen pigs individually for several days until the ear has started to heal if ear biting does occur after notching.

Postprocedural Management

Baby pigs may be returned to the farrowing pen or crate after ear notching.

Recovery Sequence

Bleeding from the notches should stop within 1 to 2 minutes. If bleeding does not stop, sprinkle the area with ordinary household flour. There are also commercial preparations available to stop blood flow if excess bleeding occurs.

After 1 day, healing should be started and a scab will form on the incisions. The notches should be completely healed in about 1 week to 10 days.

Weighing Baby Pigs

Taking the birth weight of baby pigs immediately after farrowing can be a useful practice, especially in

a seedstock herd. If you do weight your baby pigs, do it immediately after the pigs are ear-notched, but before the rest of the processing routine. For a step-by-step procedure for weighing the baby pigs, see Section 5.21, "Weighing Pigs."

5.11 EAR TAGGING

The ear-notching system and ear tags are the most common methods of identifying individual animals in swine herds. Swine seedstock producers or pure-bred breeders are required to use the ear-notching system for all registered animals. For commercial swine producers, ear-notching pigs and reading the system of notches takes too much time.

Swine producers must be able to identify breeding animals quickly, easily, and without animal restraint. The seedstock producer can ear-notch pigs at birth and then use ear tags to facilitate identification. The commercial swine producer may wish to use a simple ear-notch system for baby pigs to identify potential herd replacements, and then insert ear tags at the time a gilt is selected and enters the breeding herd. Ear-tagging the breeding herd will greatly aid the keeping of accurate swine records. Ear tagging is a process of inserting a plastic or metal tag into the pig's ear with a knife or pliers-type applicator.

Ear tags, as compared to notches, offer the advantage of being visible and easily read from a distance. Ear tags should be inexpensive, permanent, easy to apply to the animal, and cause little pain when inserted into the animal's ear.

There are many kinds of ear tags available commercially in a variety of colors, sizes, and costs. Most tags can be purchased either prenumbered or blank. Blank tags are numbered by the swine producer to meet the needs of a particular herd.

Equipment Necessary

- Ear tags (consider selecting a tag that can be read from both the front and rear of the animal)
- Ear-tag applicator
- Marking ink for tags (if tags are not numbered)
- Disinfectant

Restraint Required

Options: Hog snare;
hog-catching chute;
narrow (18 to 22 inches) chute, pen, or feeding stall in which an animal can be penned for tagging.

Step-by-Step Procedure

1. Check and have ready all the ear tags that will be needed. Number ear tags with permanent ink if

they are not already numbered. The use of several coats of ink will provide a longer-lasting and more visible ink mark. A metal toolbox works nicely for storing and transporting materials to be used for ear tagging. Read and follow the directions of the manufacturer of the tags and applicator you are using.

2. Restrain the animal with either a snare or a head chute. A narrow chute may be the only restraint needed for applying some types of tags.

3. Insert the tag into the applicator.

> **CAUTION:** Even though infections caused by ear tagging are rare, have all equipment clean and disinfected. All knife edges or points that will penetrate the skin should be clean and disinfected before each tag is inserted.

4. Clean and disinfect the tagging site with a cotton swab.

5. Securely hold the ear to be tagged, stretching it slightly but firmly. Quickly, but accurately, insert the tag into an area of the ear that is free of blood vessels. Following are some tips in the use of two kinds of applicators.

Pliers-Type Applicators

With tags that have two parts and require a pliers-type applicator, make sure that the parts are inserted correctly (lined up) in the pliers. If it is not lined up, the tag may not be fastened securely and the animal will lose it. If one part of a double tag has a tip or edge, insert the tag so this edge is on the inside of the ear. When the incision heals, the animal may want to rub the ear and the tag to relieve itching. If the tip or edge is placed on the inside of the ear, the animal cannot use it to rub against a fence or other object, causing more injury and perhaps infection.

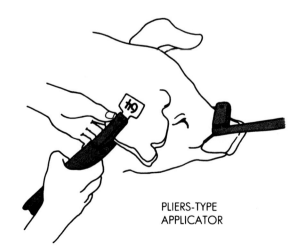

PLIERS-TYPE
APPLICATOR

Knife-Type Applicators

Avoid any obvious blood vessels (which can be more easily seen on white pigs) when inserting the tag. Be alert when using the knife-type applicators. Sometimes tags will slip loose from the applicator before the tag is completely pulled through the ear.

   6. Record the ear notch and tag number.

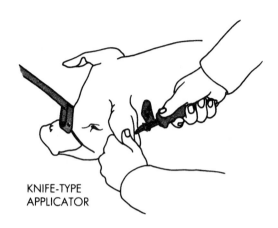

KNIFE-TYPE
APPLICATOR

Postprocedural Management

Animals may be returned to their pen immediately after tagging. An animal may sometimes lose a tag and must be retagged. If animals are maintained in lots or pastures, a tag may get covered with mud and require cleaning.

Recovery Sequence

Minor bleeding at the site of insertion will occur—the larger the puncture, the more the bleeding and the longer the irritation and healing time.

   Bleeding should stop within a couple of minutes. If a blood vessel was severed in the process, some bleeding may occur for several minutes.

   The cut or hole will scab over within 48 to 72 hours. Should the animal bump or loosen the scab, some additional bleeding may occur during the healing process. Unless the scab is accidentally loosened, complete healing will be accomplished in 7 to 10 days.

   In a very limited number of cases, infection may occur at the tagging site. The infected area usually will swell and may have a pus-like discharge. By careful observation after tagging, these infections can be found before they have advanced. Clean equipment and sanitizing procedures will minimize the occurrence of these infections. If an infection does occur, an injection of antibiotic will be required.

5.12 BRANDING

Paint branding for temporary identification is the most common type of branding used on hogs.

Although both hot and freeze branding are accepted methods of identification for cattle, these methods of identification for swine are seldom practiced and are not recommended.

   Paint branding is used for temporary identification that is easily visible from a distance. Irons in the shape of numerals are used to paint or imprint a number on the back of the animal. Sows may be paint-branded or spray-painted (aerosol) to identify them in relation to a specific farrowing crate or pen in swine operations where sows are turned out to a feeding and watering area each morning and evening. This method is useful for those workers who are unaccustomed to the sows and their farrowing crates or pens, especially when large groups of sows are turned out at one time. The sow's paint-brand number should correspond to the crate or pen number in which she is housed.

   Animals to be shown or sold in a sale may also be paint-branded and numbered. The paint brand very often is the same as the number that appears in the sale catalog for that individual animal. Prospective buyers then can look over the sale offering before the sale and easily obtain the number and evaluate any performance data that are provided in the sale catalog. Paint branding thus makes the job of evaluation, buying, and selling easier.

Equipment Necessary

- Paint brands—2½ to 4 inch, depending on pig size
- Paint (quick-drying implement enamel—yellow is a preferred color)
- Container for paint, drip pan for branding irons
- Burlap
- Supplies for cleaning brands and equipment and a supply of rags
- Table for holding branding equipment
- Gloves (optional)
- Protective clothing such as coveralls or apron (optional)

Restraint Required

A small pen or narrow alley in which pigs can be held while branding is necessary.

Step-by-Step Procedure

   1. If a large number of animals are to be paint-branded, try to arrange for the efficient movement of animals through a narrow alley where an individual pig can be held without being able to turn around. In some cases, pigs may have to be weighed and then paint-branded.

   2. Have sufficient help for weighing (if this is done), branding, recording data, driving pigs, and opening and closing gates.

3. Prepare branding materials. If possible, use a table, preferably an old one (or use a cover on the table to protect it from paint). Locate it near the area where pigs will be paint-branded. The individual doing the branding should stand outside of the pig alley or chute and, preferably, on the pig's left side. The table should have on it the paint brands, paint container, and rags. A pie tin lined with two or three layers of burlap makes an excellent paint container and pad for coating the irons.

4. Line up the brands in numerical order, and keep them in order as you use them to help avoid branding mistakes.

5. Pour enough paint into the pie tin that the burlap is saturated. Placing too much paint in the tin or using paint that is too thin will cause the paint to run when it is applied to the pig's back, thus making the number illegible.

6. Select the brand for the first digit. Cover the brand face with paint by pressing the brand into the burlap soaked with paint (in the pie tin container).

7. Paint-brand the first digit on the animal by pushing the brand down on the back of the animal behind the shoulders. The number should be branded perpendicular to the backbone unless it is one or two digits. It should be read from head to tail when reading on the pig's left side. Rotate the brand back and forth over the contour of the animal's back to obtain a complete and clear brand. The animal may move; be alert and do not smear the brand.

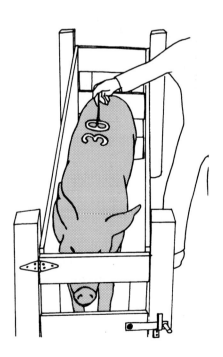

8. Return the branding iron to the drip pan in its proper location.

9. Proceed to paint-brand the remaining digits onto the back of the animal.

10. Use the 1 digit to underline the 6 and 9 digits with a paint line to prevent errors in reading the brand. Some sets of paint-branding irons have either a 6 or a 9, but not both. This digit is then reversed to provide the other digit.

11. Record paint-brand numbers, if this is required.

12. When finished, clean the equipment. Follow the cleaning instructions provided with the paint being used.

Postprocedural Management

The paint brand will smear for 20 to 30 minutes after branding. Crowding animals should be avoided during this time if the numbers are to remain easy to read.

Recovery Sequence

Most paint brands will not last for more than 30 days. They will be easiest to read for several days to a week, depending on the amount of washing, brushing, or similar grooming activities that are performed.

5.13 TATTOOING

Tattooing involves the placement of a mark (tattoo) into the skin of the animal. A tattooing device contains changeable letters or numerals. The letters or numerals are outlined in a series of sharp points (dies) which pierce the skin in the tattooing process. Indelible ink is put into the small skin punctures, and a permanent mark results. Tattooing is done in the ear of the pig for herd identification. It is usually done on the shoulder or ham for hog carcass identification.

Ear tattooing is limited in use because of the amount of time required to perform the actual tattooing and the difficulty of reading a tattoo mark. Tattoos are easier to read in ears of white pigs than the colored breeds. In many instances the pig must be restrained to obtain an accurate reading of the tattoo.

Tattooing can be a valuable option for identification in the swine breeding herd. The entire herd in seedstock or purebred operations can be tattooed in the ear, providing permanent identification. The more visible ear notch or ear tag then is used to supplement the tattoo system. The tattoo provides positive identification in cases where the ear notch may not be clear as a result of frostbite or if the ear is torn. If tattooing is properly done, it should last for life. In most herds, however, ear notching or ear tags will be sufficient.

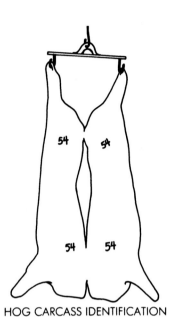

HOG CARCASS IDENTIFICATION

Tattooing of swine is also used for hog carcass identification during the slaughtering process. Pre-kill tattooing on the shoulder or ham will provide identification after the animal has been slaughtered and the hair and head removed. In most slaughter plants the skin remains on the carcass, as it must for tattoo identification. Identification by tattooing is done on slaughter animals to collect carcass information and diagnostic data relating to herd health.

Tattooing may be used to identify animals involved in quarantine associated with disease outbreaks and in disease-eradication programs.

Equipment Necessary

- Tattoo marker (pliers-type for ear tattooing)
- Hog carcass tattoo and sponge inking pad (5" × 5") or brush for slaughter animals
- Tattoo ink or paste (blue, purple, or green is best)

Restraint Required

Ear tattooing: No restraint equipment is required for young pigs; a hog snare or headgate is needed for older pigs.

Carcass tattoo: Small pen or alley.

Step-by-Step Procedure

Pliers-Type Tattoo Marker

1. Prepare for tattooing by arranging and inserting the numbers or letters (dies) in the tattoo marker. Test the mark by imprinting a piece of paper or cardboard. Prepare any record forms that will be needed.

2. Restrain the animal. For older, heavier animals use a snare or headgate. For young pigs, grasp the animal by the front legs with the pig's back next to the person holding it and with its ears readily exposed.

3. Wipe the inside of the ear clean in the area to be tattooed. Pick an area between the ear ribs (cartilage), avoiding blood vessels if possible. Only white pigs may be tattooed on either the inside or outside of the ear.

4. Thoroughly cover the points of the digits or letters with tattoo ink or paste.

5. Tattoo the ear by holding the ear with one hand, locating the site, and squeezing down on the handles of the tattoo instrument with the other hand.

Hog Carcass Tattoo

1. Prepare for tattooing by arranging and inserting numbers or letters in the tattoo marker. Test the mark by imprinting a piece of paper or cardboard. Have record forms and tattoo numbers ready.

2. Animals need not be restrained. They should be put in a small pen. Carry the tattoo and inking pad or brush with you as you tattoo a hog.

3. Use an inking pad or brush to thoroughly cover the points of the digits or letters with tattoo ink or paste.

4. Swing the tattoo hammer toward the ham, shoulder, or belly of the animal with sufficient force

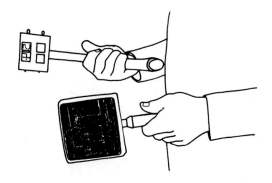

that the points penetrate the skin. Hit the hog squarely so that all tattoo numbers are imprinted clearly. If an animal is moving, run with it in order to hit it squarely.

5. For best results, ink the tattoo points after each tattoo is applied.

6. Record the tattoo and ear notch or pig number, if required.

7. Change and double-check the tattoo number for the next pig.

5.14 CASTRATION

Castration is a standard management practice in swine herds. In commercial herds all males are castrated before the growing–finishing phase of the life cycle. In seedstock or purebred herds, males are culled and castrated, and only the superior animals remain intact for use as breeding stock.

Castration is the process of removing the two testicles of the male pig. The testicle is an organ responsible for the production of sperm (the male germ cell) and the male hormone, testosterone. Although there may be some promise for chemical castration in the future, the current method is surgical removal.

Pigs are castrated if they are to be marketed for slaughter, because the meat from boars or uncastrated male pigs may have an odor when cooking

that is very offensive to many people. This odor is called a "boar odor" or a "tainted" odor. To eliminate this characteristic in the meat and to make pork a more acceptable product for consumers, pigs are castrated at a young age.

There are many techniques for castration. The position of the animal during surgery and the method and degree of restraint are dictated by the age and size of the animal. The best time to castrate the pig is when it is between 1 and 14 days old. At this time the young pig is easier to restrain; the sow, farrowing pen, and the pig itself should be cleanest; the animal bleeds less from the surgery; and the pig has antibody protection from the sow's colostrum. The younger the pig, however, the smaller the testicles, which makes castration more tedious in nature. However, pigs can be castrated successfully on day 1. One of the major disadvantages of castrating early is that scrotal hernias are difficult to detect.

Equipment Necessary

- Castration knife, scalpel, razor blade (one that has a long handle and uses blades that are easily replaced is preferred), or side cutters
- Disinfectant, container for disinfectant, and cotton swab
- Mechanical pig holder (optional)

Restraint Required

For young pigs, one person holds the pig by the rear legs while the other person does the castrating. When only one person is available for castration, a mechanical pig holder can be used. It is possible for one person to hold the baby pig with one hand or between the knees and also do the castration. For older, heavier pigs, use a hog snare or restrain the animal on its side (see Chapter 1).

Scalpel, Knife, or Razor Blade Method

Step-by-Step Procedure

1. If a litter is still nursing the sow, her male pigs should be placed in a box, container, or cart and taken to a location where their squealing will not upset or be disruptive to the mother or other sows in the farrowing house or room.

CAUTION: Do not perform the castration surgery in the farrowing crate or pen because the sow may bite you in her attempt to protect the pigs.

2. Restrain the pig. Mechanical holding devices may be used or, when available, one person may hold while a second performs the surgery. The preferred method for holding the pig is to grasp the rear legs, one in each hand, with the back of the pig next to the holder. The hands of the holder should be about waist high.

3. Once the pig is restrained, clean the scrotum and surrounding area with the cotton swab soaked in a mild disinfectant solution.

4. If you are right-handed, put your left hand around and in back of the pig's left leg and pull the skin tight above the scrotum just below the tail with thumb and forefinger. This will tighten the skin, thus outlining the testicle and the location for the incision.

> **CAUTION:** Do not put your hands or fingers below or in the path of the surgical tool. If a slip is made or the animal suddenly jumps or moves, you could cut your fingers or hand.

5. Before either incision is made, examine the testicles to determine that there are two and that they are the same size. If there is a scrotal enlargement, it could indicate a scrotal hernia. Do not castrate the pig unless you are trained to repair the hernia at the same time. The pig's intestines will be forced through the incision if the pig has a scrotal hernia. This condition is best left to a veterinarian because it requires a more complicated type of surgery. Sometimes the testicle is removed before a scrotal hernia is discovered. If this happens, the herniation must be repaired or sewed immediately. If you are unable to repair or sew the internal rupture (herniation), sew the castration incision (outside skin) with a household needle and thread until veterinary help arrives.

If one or both testicles are not found, the pig may be a cryptorchid, meaning that the testicle(s) failed to descend through the inguinal canal from the abdomen during the embryological development. This condition should be identified by a specific ear notch and the pig then referred to a veterinarian for a more complicated surgical procedure.

6. Use a disinfected castration knife. With the knife in your right hand, make two incisions about as long as the testicles over the center of each. Cut deeply enough to cut through the outside body skin and the white membrane that surrounds the testicle. The testicles should protrude through the incision. Pull the testicle through the incision. If it is difficult to get the testicle through the incision, lengthen the incision slightly at the end closest to the tail. The testicles can then be squeezed out and pulled through the incisions.

Twist the testicle for two revolutions, and then while pulling gently on the testicle, sever the attachment to the body (a cordlike structure) with a scraping movement on the cord with the castration knife.

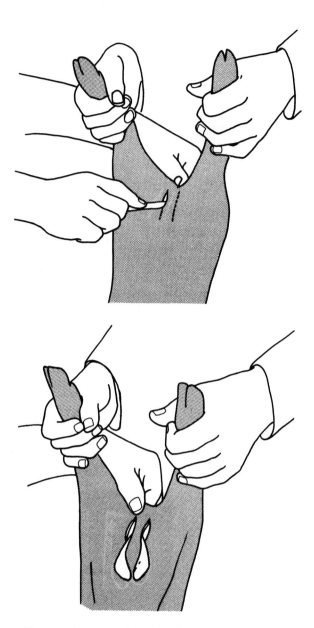

Repeat the procedure for the second testicle.

7. After castration is completed, antiseptic powder or spray may be applied to the incision.

8. Pigs may be returned to the farrowing crate or pen immediately after castration.

Postprocedural Management

All animals that have been castrated should be observed frequently for a few hours after surgery to make sure that excess bleeding or the development of a hernia has not occurred. Observe the pigs by entering the pen, or check each litter if they are still in a farrowing crate or pen.

Check for blood running down the rear legs or blood dripping from the incision. If a pig does appear to be bleeding excessively, separate it from the group. Spray the wound with a blood coagulant and keep the pig quiet.

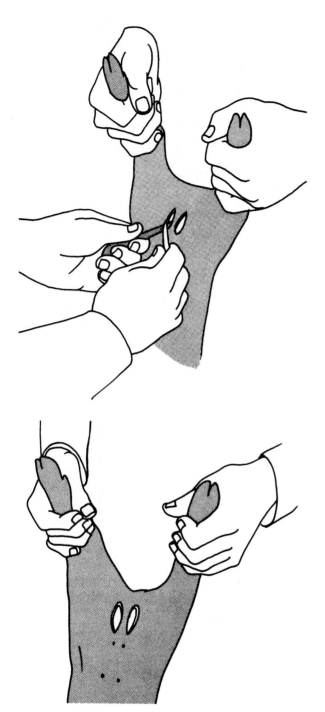

Check for any excess tissue or intestines which extend through the incision. Catch the pig, and if the tissue protruding is the cord from the testicle, cut it off. If part of the intestine is protruding, then the intestine must be put back and sutures put in to prevent this from recurring.

Minimize stress by providing a clean, draft-free environment for the pigs.

Recovery Sequence

Healing is rapid and should be completed in 5 to 7 days. In older animals this time may be extended

several days. The degree of swelling after surgery depends on the age of the pigs—the older the animal, the more the swelling.

Side Cutter Method

Step-by-Step Procedure

1. The side cutter method of castration should be performed when the baby pigs are between 4 and 10 days of age. Pigs younger than 4 days are more prone to hernia, and pigs older than 10 days are more easily castrated using a scalpel.

2. While a single person can restrain (hold) a 4- to 10-day-old pig and perform the side cutter castration procedure by himself, it is easier if two people are available. Hold the baby pig by the rear leg or legs, with the belly facing outward.

3. Disinfect the area around the testes using a cotton swab soaked in the disinfectant solution.

4. The person with the side cutters should use his fingers to roll, push, and fold the skin just below the testes of the baby pig. This will create a fold of skin below the testes, and the testes will be more pronounced above the fold of skin.

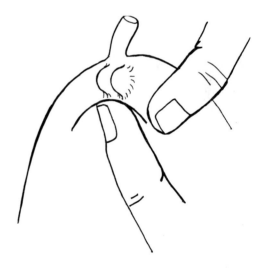

5. Using a disinfected side cutters, position them about two-thirds of the way into the fold of skin and

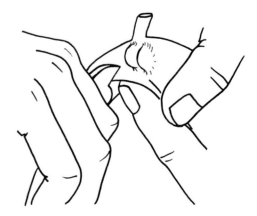

make a sharp, clean cut into the scrotal skin just off the midline of the baby pig. After this first cut is made, make another cut through the scrotal tissue on the other side of the midline.

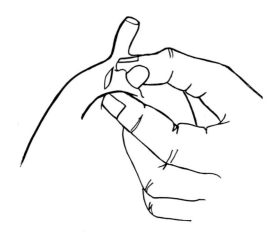

6. Pinch the testes, one at a time, between your thumb and forefinger. The purpose of this "pinching" is to cause the testes to be forced out through the incisions you just made with the side cutters.

7. Using the side cutters, more as a pliers than a cutter, pull the testes, one at a time.

CAUTION: Press your thumb firmly against the baby pig's body, where the testes are coming out, as you pull the testes. If you do not, there is a good risk of causing a hernia. There will be little bleeding with this method.

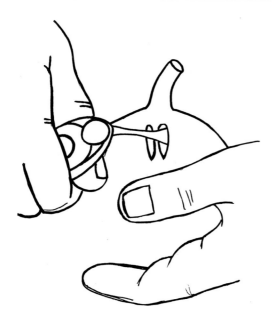

8. The cord will break where the thumb is being pressed against the body. Be certain to pull the cord

slowly and steadily. The intent is to "break" the testes and cord free or pull them from the baby pig, not to cut them away.

9. After the testes are removed, cut away, using the side cutters, any cord tissue left outside the incision.

10. After castration is completed, an antiseptic powder or spray may be applied to the incision.

11. Baby pigs can be returned to the farrowing crate or pen immediately after castration.

Recovery Sequence

Healing is rapid and should be completed in 5 to 7 days. In older pigs, the healing time may be extended several days. Older animals will also swell more following the castration surgery.

5.15 WASHING SWINE AND EQUIPMENT

Washing is the process of cleaning swine to remove any mud or dirt, and excrement, which may contain pathogenic (disease-causing) organisms or infective larvae of internal parasites. Pigs normally pick up these materials on their bodies and hair-coats as a result of being housed in hog lots and pens.

The washing of swine, if the weather and facilities permit, is most often done to sows just before moving them into a clean farrowing facility to remove these sources of potential infection from the newborn pig.

Other swine are sometimes washed just before a swine show or a swine seedstock or breeding stock sale or auction. Washing these animals at home before these events improves their appearance and may help to reduce the possibility of disease transmission.

Pigs marketed for slaughter are usually not washed before shipping.

Equipment Necessary
- Small pen or crate
- Source of water
- Brush, long bristles
- Household soap or animal shampoo
- Power sprayer (optional)
- Disinfectant
- Garden sprinkler can

Restraint Required

The restraint required for washing a pig is a pen about 5' × 5'. The pen should not be so large that the animal escapes or runs around the pen during the washing process. The pen should be constructed of spaced board fence or wire panels about 4 feet high. If solid fencing or walls are used in a wash pen it is more difficult to drive a pig or sow into it unless alleys or hallways are available for animal control. The pig is apparently very aware of being trapped and will resist efforts to drive it into a solidly walled area.

Optional methods of restraint include the use of a crate about 2' × 5' or 2' × 7'. This method makes access to the animal more difficult. Groups of animals can also be put into a larger pen where several can be washed at the same time, but do not put animals from different pens or crates into the same wash pen or area because the animals may fight each other.

Step-by-Step Procedure

1. Put the sow or pig into the wash pen or area.

CAUTION: Do not wash pigs outside when temperatures are below 55°F, as this may cause the pigs to become sick.

You should wear workshoes or overshoes for protection from water and from being stepped on. Also consider using a rubber apron or suit that will provide protection from the water.

If power spraying equipment is used, take precautions so that the high-pressure water does not frighten or irritate the animal. Direct the water slowly and from some distance away on the feet first. Be sure to heed any directions provided by the manufacturer of the power equipment.

Do not attempt to wash a sow that has started farrowing. Let her finish parturition if the facility will allow her separation from other hogs, or move her directly to a farrowing crate or pen.

---

2. Wet the animal's entire body first, starting with the feet, legs, head, and then the rest of the body. Lukewarm water is preferable to cold water, especially in the winter. Do not use hot water (above 80 to 85°F). Check the water temperature with your hand before putting it on the animal.

Be careful not to spray water directly into the ear of the animal. To prevent this from happening, gently grab the ear and squeeze it shut while wetting the portions around the ear. If water does get into the ear, the animal may shake its head from side to side trying to get rid of it. The animal with water in its ear may also hold its head to one side for a period of time. If you do accidentally get water in the ear, do

nothing about it and proceed with the washing process. The animal will eventually rid itself of the problem.

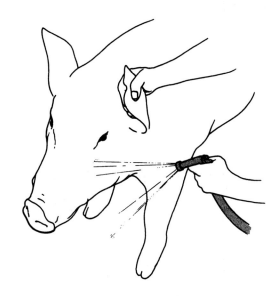

3. After the animal is thoroughly wet, apply generous amounts of household soap, liquid detergent, or animal shampoo. Then use the long-bristled brush to scrub the entire body while a rich lather is created during the washing and cleaning process.

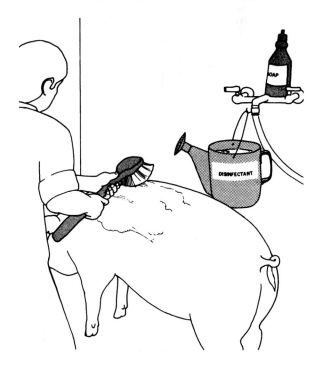

4. Be sure to clean the feet and legs thoroughly. Remove all dirt and manure from between the hooves or toes and around the dewclaws. Use a long-bristled brush to clean between the toes. The action of the foot in contact with the wet floor along

with the brush between the toes should be sufficient to clean the bottom of the foot.

5. When washing sows scheduled to be housed in the farrowing unit, clean the area around the vulva and the entire underline, with special emphasis on the concave surface on the end of the teat. Inspect the end of each teat and remove any dirt with your hand. These areas are the parts of the sow nuzzled by the newborn pig as it begins its search for milk. The teats must be clean to prevent the newborn pig from ingesting contaminated material.

6. Remove soap and dirt by rinsing the animal with lukewarm water. If the animal is extremely dirty, repeat steps 2 through 6.

7. (Optional.) After the sow has been washed, disinfectant should be applied. A garden sprinkler can work well for this. Sprinkle the disinfectant solution over the back, head, and sides of the animal, wetting it thoroughly. Use your hands to force the disinfectant down to the underline and teats. Follow the directions on the product container and be aware that many disinfectants cannot be used directly on the skin.

Washing Facilities and Equipment

8. It would be foolish to thoroughly clean the pregnant sow or gilt and then place it in a dirty farrowing crate or pen. If it is at all possible, therefore, excrement, dirt, and any material that might be a source of disease or parasitic contamination is cleaned from the farrowing and nursery facilities on modern swine farms. High-pressure sprayers using 150 to 1,500 lbs of pressure are used to loosen and wash off undesirable substances. If sprayers are not available, buckets of warm water and long-handled brushes will provide the same results but will require a much longer time to accomplish the task. Feeders, drinking fountains, the floor surface, and pen sides as high as the pig reaches are areas that must be cleaned.

Postprocedural Management

After washing the animal, move it to clean quarters free of all drafts. It is not necessary to dry the animal.

5.16 RINGING

The pig, when allowed access to dirt or pasture lots, will follow its natural instinct to root in the soil in search of grubs, worms, and some types of roots, or perhaps just to provide a moist hole to lie in and keep cool. This rooting generally occurs in the spring or late fall when the soil is moist and the worms and grubs are abundant. Unfortunately, the rooting process can be very destructive to pasture crops and creates holes and an unevenness on the ground that makes travel over it with farming equipment slow and tedious. Pigs can also root under fences and escape. Rooting may result in mud holes which collect water and create unsanitary conditions for hogs and a good breeding place for parasites and insects.

To remedy these problems, specially constructed wire rings are inserted into the pig's nose with a pliers-type applicator called a *hog ringer*. The nose ring creates a sensitivity in the tissue of the nose so that when rooting is started and pressure is put on the ring some discomfort occurs and the pig stops rooting. This results in conservation of pastures and hog lots and more sanitary conditions for the pigs.

Generally, only animals weighing 40 pounds or more cause serious rooting damage. These are animals that should be rung if kept in drylots or pastures. Ringing is not necessary when hogs are raised in confinement on concrete or slotted-floor systems.

Equipment Necessary

- Hog ringer
- Shoat rings for hogs weighing 40 to 125 lbs
- Hog rings for hogs weighing 125 to 225 lbs
- Hog rings for hogs weighing 225 to 500 lbs
- Pinchers or cutting pliers

Restraint Required

Headgate or hog snare

Step-by-Step Procedure

Ringing hogs is most easily done with two people assisting.

1. Check the hog ringer by clamping it shut as though you were putting a ring in the nose of the hog. Check to make sure the points of the ring come together and overlap slightly as you squeeze the handles of the hog ringer shut. If the points do not come together, adjust the thumb screw on the handle of the ringer until ring closure is complete. If there is a gap between the points of the ring, the pig is apt to work the ring off its nose and lose it in a very short time. Once the ringer is adjusted, it is not necessary to check it after each pig. Check the adjustment occasionally if a large number of pigs are to be rung.

2. Restrain the pig by using a headgate or a hog snare. It is possible to place lighter pigs (40 to 50 lbs) in a small pen and catch them by hand. After catching the pig by one of its rear feet, grasp the front legs one at a time and set the pig on its tail with its back toward the holder, with its head up and positioned between the holder's legs.

CAUTION: When holding the pig by its front legs, be alert to the possibility of being bitten by the pig in its attempt to escape. Pulling back on the front legs and controlling the pig's head with your arms will help keep you from being bitten. Although they make holding the pig a little more difficult and clumsy, a durable pair of leather-type gloves should be worn to reduce the possibility of injury to the holder.

---

3. *Option 1.* Rings are inserted in the rimlike tissue on the end of the nose. Care must be taken to insert the ring deeply enough, but not too deeply. If it is inserted too deeply, the ring may cause constant pain to the pig and may cause infection. If it is inserted too loosely, the ring catches on various objects and may be torn from the nose. When inserted properly, the ring should fit snugly on the rim of the nose.

With the hog ringer containing a ring, center the ring over the rim of the nose, but without touching the pig's nose. When the ring is in position, squeeze the pliers-type applicator quickly so that the ring is in place before the pig has a chance to jerk its head. Sometimes the pig will move just as you are in the process of closing the ring. If the ring appears to be inserted incorrectly, use a pinchers to cut the ring, remove it, and insert a new one.

For growing and finishing pigs, one ring in the nose is sufficient. If more than one ring is inserted, it may interfere with a pig's normal operation of lids on self-feeders and waterers.

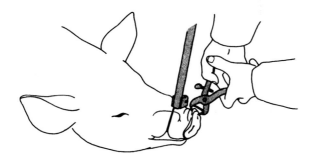

For sows, two to three rings spaced about ½ inch apart at the top of the nose will be required to prevent rooting.

Boars used as herd sires are not rung because the rings may interfere with the mating process. The male hog uses his nose to nuzzle and nudge the female in the foreplay preceding mounting and breeding. If a boar creates a major problem in rooting, however, try ringing as a remedy and then observe him to make sure that he mates.

Option 2—Humane Rings. The humane ring is of a design different from that of hog rings. This ring is inserted with a pliers-type applicator into the muscular bridge on the end of the nose between the nostrils. This leaves the rim of the nose free for manipulating feeders, but still tends to discourage deep or destructive rooting. Do not get the ring too deep or too shallow. Pigs with humane rings can have difficulty in operating paddle-type automatic waterers.

Postprocedural Management

There may be slight bleeding at the ringing site, but usually there is no infection. Observe pigs for several days after ringing for signs of discomfort, which may be expressed by a loss of appetite. If this happens, the pigs must be caught and the ring cut and removed from the nose. Ring the hog again, putting the ring at a more shallow location in the nose.

5.17 REMOVAL OF TUSKS FROM BOARS

One of the most obvious secondary sex characteristics of the mature male pig is the specialized type of tooth called a *tusk*. These tusks are the elongated canine teeth.

At birth the pig has four temporary incisors plus the four canine teeth. These eight teeth in the newborn pig are called the *needle teeth,* or *wolf teeth.* The four canine teeth grow and develop into tusks because of the stimulation of the hormone testosterone, which is produced in the testicles of the uncastrated male pig (boar). These teeth become quite large in the boar and project outside the mouth. They are positioned on each side of the upper and lower jaw and are about halfway between the end of the nose and the rearmost portion of the mouth. The action of the lower tusks rubbing against the upper ones causes the formation of sharp edges and points which extend outside of the mouth.

The tusk is a weapon for defense, and when two mature boars are penned together they may fight viciously to establish a pecking order. When tusks are left intact, the boars may suffer severe cuts from thrusting their heads against each other. In addition, sows or gilts may be severely injured by the boar's tusks in the foreplay which precedes mating. Even the most docile boars have been known to turn on the herdsman or other animals if they become angered or disturbed.

Cuts and wounds inflicted by the teeth of pigs are very painful and may be slow to heal because of the infective nature of such an injury. To help prevent injury to humans and other pigs, the boar's tusks should be removed periodically.

Equipment Necessary

- Hacksaw
- Broom handle or similar material (18" in length)
- Hog hurdle

Restraint Required

Rope or hog snare (The rope should be about 10' in length and $\frac{3}{8}$" to $\frac{1}{2}$" in diameter. One end of the rope should have a loop fastened with a slipknot.)
Strong portable gate
Pen with a well-anchored post
Headgate (optional)

Step-by-Step Procedure

The tusks to be removed are easy to identify. The tusks from the upper jaw tend to grow straight out of the side of the mouth and somewhat perpendicular to the jaw. The lower tusk grows almost straight upward from the lower jaw. The upper tooth is oval in cross section, the lower is triangular. The tusk on the 8- to 10-month-old pig will be about 1 inch in length, with a diameter of about $\frac{3}{16}$ inch. The older boar may have tusks 2 to 5 inches long with diameters of up to $\frac{1}{2}$ inch, depending upon

how long they are allowed to grow. Tusks should be removed every 4 to 6 months in herd boars. At least two people should assist with this task.

1. The first step in removing tusks from a boar is to put the boar into a well-constructed pen. There should be a well-anchored post in the pen for tying the boar.

> **CAUTION:** When handling or moving a boar, always place a hurdle, hog panel, or partition between you and the boar for protection should he try to attack or bite you.

2. Restrain the boar behind a strong portable gate by squeezing the animal up against a pen partition.

> **CAUTION:** The gates and pen partitions must be well secured at the boar's head, because once he realizes he is trapped he will attempt to escape. The animal may put its nose in the openings of the partition and throw it into the face or body of the herdsman performing this task. If the equipment is not strong enough to hold the animal, it will break and both animals and people can be injured. A sturdy headgate is an option for restraining the boar.

3. Snare the boar by inserting the loop at the end of the rope into the boar's mouth. The rope must be above the tongue, well back of the tusks to be removed. Pull the slipknot tight over the snout. The slipknot should be about 8 to 10 inches from the end of the rope. This 8- to 10-inch piece of rope can then be used to safely pull the slipknot loose after detusking.

4. Keep the rope taut. Tie the rope to a well-anchored post. Leave no more than 12 inches between the post and the boar's snout.

5. Remove the portable gate. If necessary, maintain additional control of the animal by having a second person secure a firm hold of the boar with a hog snare. After being tied to the post, the boar will generally pull back, keeping tension on the rope in his attempt to escape. Boars occasionally try to release the tension on the rope by walking forward toward the post. To maintain control, it is important to permit only 12 inches of rope or less between the post and the boar's snout.

6. Once the restraint has been proven adequate for holding the boar, place the 18-inch section of the broom handle in the boar's mouth on top of the tongue and toward the back of the mouth. With one hand, pry downward on one end of the broom handle, forcing the boar's mouth open. This exposes the tusks.

7. With the other hand, grasp the hacksaw and saw the tusks. Be certain that the hacksaw blade is tightened and sharp. Cut both upper and lower

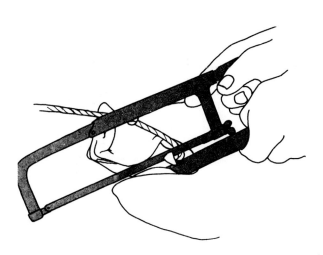

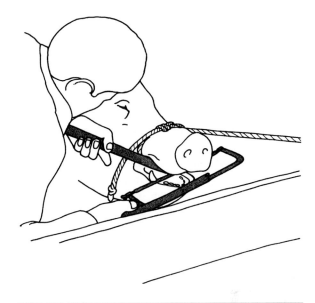

CAUTION: Be sure to watch carefully the entire length of the blade as you saw so that no part of it is damaging the gums or lips of the boar.

tusks as close to the gum as possible. Do not cut the gum. The blade of the hacksaw should point toward the back of the mouth. The boar may have to be pushed, shoved, or moved around to gain access to the tusks on the opposite side of the mouth.

8. When you are finished, remove the rope from the post. Loosen the slipknot on the boar's snout by pulling on the end of the rope next to the slipknot. Remove the rope from the boar's snout.

CAUTION: Some boars become upset or angered by the detusking process and should be watched carefully once they are released and moved to their quarters. Guard against an attack from them by using a hurdle.

Postprocedural Management

No special treatment is required. After removal, tusks will grow out again in about 4 to 6 months and should again be removed.

5.18 COOLING SWINE

The temperature best suited to the physiological needs of the pig ranges from 55°F for the older animal to 85°F for the small pig. The pig, like all farm animals, is homeothermic. (A *homeotherm* is an animal that maintains a constant internal body temperature, irrespective of the environmental temperature.) Normal body temperature in swine is 102 ± 1°F.

The pig responds to the ambient temperature. An increase in respiration rate is an excellent indicator of heat-stress discomfort. It can result from high temperatures and humidity, excitation, and movement or forced exercise.

Temperatures that depart much from the optimum result in stress on the animal. One of the pig's first responses to excess heat is a loss of appetite, resulting in a reduction in feed intake, which then reduces daily weight gains and affects lactation performance in sows. Hot weather also may reduce reproductive performance in the breeding herd. Both the male and female are affected. In the boar, environmental temperatures above 85°F cause lower fertility, resulting in lower conception rates and litter sizes. In the pregnant female, embryonic death and stillborn pigs could result.

The pig has a very limited capacity for dissipating excess body heat by sweating; it has almost no functional sweat glands. There are a few sweat glands on the nose, but these are not very effective in terms of total heat dissipation. The pig loses considerable heat through respiration, the normal rate of which is 20 to 40 breaths per minute. Swine must be cooled during periods of heat stress.

Equipment Necessary (Options)

- Shades
- Wallows
- Ventilating system
- Evaporative cooling system
- Source of water
- Spraying system
- Zone cooling system

Restraint Required

None.

Step-by-Step Procedures

The best indicator of heat stress is respiration rate. With sunny, calm weather conditions above 70°F and high humidity, the older, heavier hogs such as sows and boars will be affected first. Hogs will need protection from the sun. This can be provided by some type of shade. As it gets hotter, some type of cooling may be needed. Cooling devices include shades, wallows, spraying devices, ventilation fans, evaporative coolers, and air conditioning. Make sure that the animals have an adequate supply of drinking water. Check the automatic waterers at least twice a day in hot weather.

CAUTION: If a pig is overheated (very fast respiration rate) and must be cooled to prevent death, the cooling should be done slowly and carefully. Do not hose down or pour water over the entire body of the pig all at once. Begin by providing the pig with a wet or moist place to lie down by pouring water on the floor or ground. If the pig is so stressed that it cannot move or lie down, it may be necessary to gently throw or cast the pig in the moistened area. Do not fight the animal to do this. Let the animal stand, if necessary. Provide shade, if possible. After the pig is lying down, water should be slowly and continuously placed on the ham, feet, and legs. After a few minutes (usually 15 to 20), the respiration rate will decline and the pig may then be wet down slowly over its entire body surface. After the pig is cooled, be careful that the animal is not permitted to lie in the wet spot until it is chilled. Chilling can result in stress just as overheating does. After the animal has cooled, provide a well-ventilated, shady place. Check about every 30 minutes for a couple of hours to make sure that the animal remains cool and does not become overheated again. Water has an excellent thermal conductivity and is a valuable tool in the total cooling process.

Shades

1. In hot weather, pigs of all ages need protection from the sun and heat. Hog houses without mechanical ventilation should be constructed so that they can be opened, preferably in the front and back, for ventilation.

2. Keep pigs fenced out of airtight structures in hot weather.

3. Trees give good shade, but livestock should be fenced away from valuable trees because they could damage them by rooting, rubbing, or gnawing.

4. Another method of providing shade is to place at least four posts in the ground or on skids. The posts should stand 4 to 5 feet above the ground. This is high enough for a breeze and low enough to keep the sun out. The framework on top of the posts can be covered with woven wire or wooden snow fence. Then cover it with reflective or insulative materials such as straw. These shades can be made portable by mounting them on pressure-treated skids, but they

must be staked or anchored at the four corners to prevent their being overturned by the wind.

5. In emergencies, use a farm implement such as a grain or bale wagon in the hog lot to provide shade. Implements should not be used as permanent shades because animals can injure themselves on the undersides of the equipment or can become stuck beneath them.

6. Allow shade space of 15 to 20 sq ft per sow, 20 to 30 sq ft per sow and litter, 40 to 60 sq ft per boar, 4 sq ft per pig up to 100 lbs, and 6 sq ft per pig over 100 lbs.

Wallows

1. Wallows are shallow, watertight structures that usually have 40 to 75 square feet of floor space, are 10 to 15 inches deep, and are made of pressure-treated lumber, metal, or concrete in which hogs can wade or wallow to wet themselves.

Mudholes made by the pig are wallows for cooling; hence the pig's reputation of being dirty because of attempts to keep cool. Mudholes are messy and unsanitary, and hogs should be rung to prevent them from making mudholes.

2. Wallows should be constructed with a ramp on one end so that pigs can get in and out easily without injury.

3. The wallow should have a drain plug.
4. One 75-sq-ft wallow will hold about 8 to 12 market hogs.
5. Wallows should have skids so that they can be moved.
6. It is recommended that wallows remain in the sun so that pigs will wet themselves and leave to lie in the shade. This requires a smaller wallow than a shaded one. The wallow may be covered with some type of shade 4 to 5 feet high constructed from material such as picket fencing. To prevent fighting, it may be necessary to provide wallow space for every pig.

Spray Cooling

Spray cooling is a method of wetting down pigs with a spray nozzle system to improve their environment and performance. In extremely hot weather, relief from heat can be accomplished by simply wetting the animals down once an hour with water from a handheld hose. This requires more labor than do automatic sprayers.

1. Research indicates that pigs can be cooled effectively by spraying them for as little as 2 minutes out of every hour. This can be done with the aid of automatic timers. The water required per pig is 0.045 gallons per minute. The nozzle size can be selected by multiplying the number of pigs per pen by 0.045 gallons. *Example:* 20 pigs per pen × 0.045 gal/minute = nozzle size of 0.9 gal/minute.

Some systems release a fog or vapor continuously. The disadvantage of this system is that when there is a wind the amount of drift is greater than it is with a spray.

2. The optimum temperature at which to begin spraying is 85°F.

3. Use noncorrosive spray nozzles and in-line filters that can be cleaned easily.

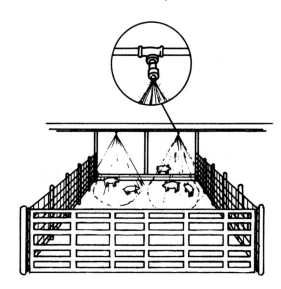

4. To obtain maximum wetting, nozzles should discharge large droplets in a solid cone spray, not a hollow cone.

5. Install an automatic timer, thermostat, and valve controls. Obtain help from the manufacturer of the equipment or a professional familiar with such devices.

Ventilation Fans, Evaporative Cooling, and Air-Conditioning

1. In environmentally regulated swine buildings, mechanical ventilation is one method used to keep hogs cool. The movement of air produced by the fans causes heat to be transferred away from the animals' bodies.

2. Reproductive performance is reduced in hot weather, so when the breeding herd is kept in confinement, evaporative cooling is sometimes incorporated into the structure. Evaporative cooling is most efficient in dry climates, but it is also used by swine producers in some of the humid areas of the midwest to provide some relief from hot weather to maximize reproductive efficiency.

Evaporative coolers have different designs, but in general they consist of pads that are moistened with cold water. A rigid plastic pipe with spaced holes allows the water to drip uniformly over the pads. With a fan system, air is pulled into the building through the pads, thus providing a cooling effect. A gutter beneath the pads collects excess water and returns it to a sump, from which it is recirculated. The system should be flushed with clean water periodically. A temperature drop of 8 to 10° can be expected with evaporative cooling.

3. Air-conditioning is used by a few swine producers to cool swine in completely confined breeding buildings. Before deciding to use this method of cooling, check on costs and maintenance requirements.

4. Zone cooling directed toward the animal's head can be effective when animals are housed in crates or individual pens. A supply of high-velocity air around the animal's head enables it to lose more heat and remain cooler. Zone cooling can use uncooled or cooled air. In farrowing houses, zone cooling helps the swine producer to maintain a cool environment for the sow while allowing higher temperatures for baby pigs in the creep area.

> **CAUTION:** Seek professional help when designing or selecting environmentally regulated swine buildings so that the heating, ventilating, and cooling aspects of the building match the requirements of the hogs to be housed.

5.19 INTERNAL AND EXTERNAL PARASITE CONTROL

External parasites are widespread in the swine industry. There must be constant checks by the swine producer if severe infestation of the herd is to be avoided. Lice and mange mites are the two external parasites of particular concern to swine producers. These parasites cause irritation, which leads to itching, restlessness, and decreased feed intake and growth rate. Hog lice may cause anemia in young pigs. These parasites can also transmit such diseases as swine pox and eperythrozoonosis, also referred to as "Epy."

The hog louse is a blood- and lymph-sucking parasite that feeds only on the pig. It is about ¼ inch long, bluish black in color, and clings to the hair of the animal. Hog louse infestation starts in the neck and around the head of the tail and expands to the tender parts of the body. Lice can be seen with the naked eye. They usually feed in groups, and can sometimes be seen moving and crawling. They are especially easy to see on white pigs. Hog lice do not burrow into the skin.

Mange is a skin condition in swine caused by a mite that burrows into the skin. Infested areas of the skin will be thickened with some scab formation and loss of hair. Mites first invade the ear, then the neck, head, and other parts of the body.

Both parasites are transmitted by pig-to-pig contact. The parasites may occasionally leave the pig and survive for a few days in bedding or on equipment until a pig comes in contact with them. It is therefore important to isolate newly purchased animals and carefully examine them before they are added to the herd.

Controlling external parasites requires a well-organized, timely treatment program. Maximum effort should be expended on the mature breeding herd. If eradication is accomplished there, there is much less chance of infestation occurring in newborn pigs. Most serious infestations occur during the winter when animals huddle together for warmth and comfort. A summer program of eradication is therefore likely to yield the best results.

Internal parasites include roundworms (*Ascaris suum*), nodular worms (*Oseophagostomum* spp.), whipworms (*Trichuris suis*), lungworms (*Metastrongylus* spp.), red stomach worms (*Hyostrongylus rubidus*), threadworms (*Strongyloides ransomi*), and kidney worms (*Stephanurus dentatus*). Internal parasites cause millions of dollars in damage annually to the swine industry in the United States. Loss estimates are as high as $5 per pig marketed, or $500 million dollars annually. Over 90% of the hog farms in the United States are infected with one or more internal parasites. Internal parasites or worms compete directly for food in the stomach and intestines and cause serious damage to other organs by migrating through the tissues. This damage can lead to other diseases such as pneumonia and diarrhea.

Roundworms

Roundworms are the most common internal parasite of swine. Pigs are infected by ingesting eggs which hatch into larvae in the small intestine and penetrate the intestinal wall, moving into the bloodstream. The larvae move to the liver and lungs of the pig and are then reswallowed and mature in the large intestine. Roundworms are the largest of the internal parasites, reaching a length of 12 inches at maturity. Mature worms continue the cycle by laying eggs in the large intestine. The eggs are then excreted and may infect other hogs.

Nodular Worms

Like roundworms, nodular worms are very common in the United States. They reside in the large intestine of an animal. Nodular worms are much smaller than roundworms, reaching a mature size of about ½ inch. Infective larvae hatch in feces and are ingested by the hog. Once in the large intestine they bore through the mucosal lining, causing small nodules that interfere with nutrient absorption. The larvae move through the bloodstream and return to the large intestine in about 1 week. There they mature and begin laying eggs.

Whipworms

Whipworms are whiplike in shape and reach a mature size of about 2 inches. Like roundworms, infection is initiated through ingestion of infective eggs, which hatch in the large intestine. The larvae then bore into the mucosal lining and remain for 2 weeks. The larvae then return to the large intestine to mature and begin laying eggs. Large numbers often cause inflammation, resulting in diarrhea, dehydration, anorexia, and death. Damage from whipworms is relatively extensive in the industry because whipworms are more difficult to control than other worms.

Threadworms

Threadworms are very small, reaching only ¼ inch in length, and are common only in warmer climates, particularly in the southeastern United States. Adult worms embed themselves in the lining of the small intestine and lay eggs, which are passed with the feces and do little damage in mature animals. Infection is initiated through penetration of the skin or oral mucosa by infective larvae, which migrate through the circulatory system to the lungs, where they move up the bronchial tree and are reswallowed. However, a considerable number of larvae move into the mammary tissue of the animal, where they remain dormant. The dormant larvae become active in the lactating sow and move into the baby pigs through the colostrum, where they mature in 4 days. Large numbers of threadworms in baby pigs can cause severe diarrhea and lead to extremely high mortality levels.

Stomach Worms

Stomach worms are small worms, ⅜ of an inch in length, that invade the stomach mucosa of the pig and feed on blood. Stomach worms are ingested as larvae and spend about 2 weeks feeding in the mucosa before reentering the stomach to mature and begin laying eggs. Large numbers of stomach worms can lead to anemia and stomach inflammation.

Lungworms

Lungworms are different from other internal parasites in that they require an intermediate host to become infective. Adult lungworms are approximately 2 inches long and lay in the bronchi and bronchioles of the lung. Eggs are coughed up and swallowed and passed out of the animal. For the eggs to mature and produce infective larvae, they must be eaten by an earthworm, which is then eaten by the pig. The larvae enter the lymphatics and migrate to the heart and lungs, where they mature. Severe infections can lead to hemorrhage and respiratory stress or infection.

Kidney Worms

Kidney worms are 1 to 1½ inches long and reside in the ureters between the kidneys and the bladder and in adjoining areas. Mature females lay eggs, which are passed in the urine, mature and hatch into infective larvae, and are reingested or penetrate the skin. Larvae then invade the lymph system and move to the liver, where they mature and develop for several months. The extended

Parasite	Dichlorvos	Febendazole	Levamisole	Lvermectin	Pyrantel tartrate	Piperazine	Thiabendazole
Roundworm		XXX(+)	XXX	XXX	XXX	XXX	
Nodular worm	XXX	XXX	XXX	XXX	XXX	XXX	
Whipworm	XXX	XXX(+)					
Lungworm		XXX	XXX	XXX			
Red stomach worm		XXX		XXX			
Threadworm			XXX	XXX			XXX
Kidney worm		XXX(+)	XXX	XXX(+)			

XXX = 90% parasite removal
(+) = effective against immature stages of parasite

development time spent in the liver can cause severe damage. Eventually the worms mature and move to the kidneys. However, they may also move into other organs and tissues, including the spinal column, creating posterior paralysis.

Controlling internal parasites requires a well-organized program of control and prevention. There are numerous products available that control one or more adult parasites, and many are effective against larvae. These products include oral and injectable products for various applications. Some products, such as ivermectin, are very effective in controlling most adult internal and external parasites, but may not completely eliminate all larvae or individual species such as whipworms. Follow label directions and rotate products to provide complete protection.

The table above is a listing of the commonly encountered internal parasites of swine and the currently approved drugs used for their removal. Each drug indicated with "XXX" removes 90% of all adult parasites under test conditions. Those indicated with "(+)" are also effective against immature stages of the parasites.

It should be the goal of every swine producer to eliminate problem parasites from the herd completely. Modern dewormers and organized plans for control make it possible to attain that goal.

Coccidia

Nursing pigs, especially those raised indoors, can be struck by the disease *coccidiosis*. It is an important disease caused by the internal protozoan parasite *Isospora suis*. Coccidiosis strikes nursing pigs between 5 and 15 days of age. Affected pigs develop a yellowish, very fluid diarrhea. The baby pigs do not lose their appetite and continue to nurse, but they become dehydrated and weak. Infected pigs do not respond to antibacterial treatments. Mortality rates will be moderate to high and will be increased if concurrent bacterial, viral, or parasitic infections are present.

There is no FDA-approved treatment for coccidiosis. Prevent the disease through strict sanitation of farrowing quarters and equipment.

Equipment Necessary

- Pesticide for lice and mange mites
- Sprayer
- Protective clothing such as rubber boots, apron, or waterproof suit
- Anthelmintics—oral and injectable

Restraint Required

No physical restraint equipment is required, but animals should be penned to allow easy access to the pigs with the handgun of the sprayer. Pigs should not be penned in large groups or be so crowded that it is impossible to treat every pig.

Step-by-Step Procedure

Spraying is the most common method of applying the pesticide. When spraying cannot be used, pesticide granules, dusts, and liquids can be used directly on the animal or in the bedding. Hand sprayers may be used, but for treating mange mites it is best to have equipment capable of 200 to 250 psi to force insecticide into the tunnels where the mites have burrowed.

A swine external-parasite control program should be initiated during the warm months when spraying can be done outside without stressing the pigs. There are numerous spray materials on the market. Seek out professional help (an entomologist or veterinarian), if needed, in selecting the correct pesticide. Not all products will control both lice and mange.

Spraying every 10 to 14 days for at least three consecutive treatments is the most effective procedure for control or eradication of swine parasites. Sows should be sprayed for mange control starting about 45 days before farrowing. Then spray pigs when they reach 8 weeks of age, with a follow-up spraying 10 to 14 days later. Additional applications may be necessary.

CAUTION: When operating a high-pressure sprayer, read and follow safety instructions of the manufacturer.

1. Set up a routine lice- and mange-control program for the swine operation by selecting dates when treatment will be given. Clean and repair equipment and make sure it is working before the spraying is to be done.

2. Select the product or products to be used on your farm.

3. Prepare the spray material. Read and follow label directions on the pesticide container. For complete wetting, at least 2 to 3 quarts of liquid will be needed for each mature animal. Make sure withdrawal directions are followed for market swine.

4. Animals should be herded into a small enclosure that leaves enough room for the spray operator to get within 3 to 4 feet of the animals.

5. Wear waterproof, protective clothing and boots.

6. Direct the spray onto every part of the animal until very thorough wetting has been accomplished. The pig's ears (inside and out), neck, legs, and belly are the most critical areas for complete wetting.

Swine will move away from the direction of the spray in their attempts to escape getting wet. The sprayer operator must be prepared to move the pigs around in the pen while spraying so that all parts of the pigs' bodies, especially around their heads, are thoroughly wet.

Postprocedural Management

1. Sprayed swine can be returned to a cleaned pen.

2. Some pigs may hold their head to one side after spraying because of the presence of spray within their ears. Do nothing about this, as pigs will return to normal in a short time.

5.20 TRAINING, GROOMING, AND EXHIBITING SWINE

Some producers present swine for show or sale at such events as fairs, expositions, and breed-type conferences. Seedstock producers often sell breeding stock at an auction, or utilize special observation-pen areas or rooms for private sales. It is important that animals look their best at these types of events so that they can be examined thoroughly by prospective buyers. A good job of selecting and feeding will make the tasks of training, grooming, showing, and selling easier and more enjoyable.

In selecting swine for show or sale, know the economically important physical traits of the animal. These include sound and structurally correct feet and legs; a well-developed underline including 12 to 14 well-spaced, prominent teats; good length, depth, and width to the body; obvious evidence of muscling and freedom from fat; a quiet temperament and disposition in the prospective brood sow; and a rugged and aggressive disposition in the boar.

Be informed about required show or sale information, which may include birth dates; performance data, such as daily gain, feed efficiency, and backfat measurements; weight classifications; health regulations; requirements for registering animals with a breed association; pedigrees; guarantee policies for breeding animals; animal transport arrangements; breeding dates; and arrival and release times for pigs. At seedstock sales, much of this information is presented to prospective buyers in a sale catalog.

Equipment Necessary

- Brush
- Short cane, whip, or other "show stick"
- Small clipper
- Soap and water

- White talcum powder or white cream shoe polish
- Hair-coat dressing

Restraint Required

Hog snare
Small pen

Step-by-Step Procedure

Handling and Training

1. When training the pig, practice gentleness and patience while the animal is becoming accustomed to being handled. Most pigs respond and learn quickly. Do not abuse or mistreat the animal at any time by hitting, beating, or causing injury. Never allow the animal to become tired or overheated when training. Pen the pig or pigs to be handled in a separate lot or pen at least 2 weeks before the show or sale so they can become accustomed to the training routine.

2. Remove any hog rings that may have been placed in the pig's nose. Use a hog snare or headgate to hold the pig, cut the ring in two with a pinchers, and then pull the ring out of the nose.

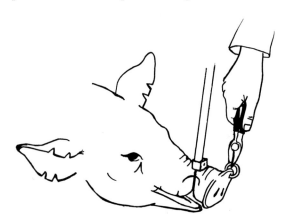

3. Brush the hair coat daily for the 3 or 4 weeks before the show, for the purpose of training, cleansing, and bringing out the luster in the hair. Do not apply enough pressure to cause discomfort to the pig. The pig will shy away from the brush if you are pressing too hard. Brush the hair in the direction that it grows, down and back from the middle of the topline. Brushing can be started when the pig is eating. Apply a soft oil, such as baby oil, to the skin two or three times before leaving for the show. Start doing this about 5 to 7 days before the show, and continue until the skin is no longer rough and scaly. The oil will soak into the hide and hair on your pig, making it soft and smooth. The oil will also help remove and hide flaking, dry skin, scratches, and other imperfections.

4. Pigs' feet seldom need trimming, but if it is necessary (about 2 weeks before the show), restrain the pig and trim as required. Do not cut too deeply, because doing so will cause sore feet.

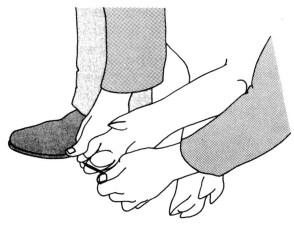

FOOT TRIMMING

5. About 1 week before the show, try driving the pig. Teach the animal to guide or turn by gently touching the side of its head, shoulder, or ear.

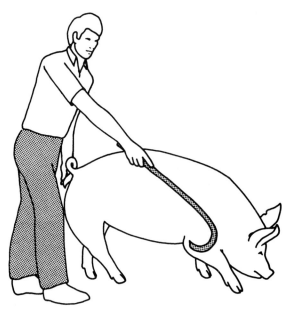

6. Before going to the show, fit your pig by clipping or blocking the hair. Each hog is an individual and should be clipped to accentuate its best features and hide its flaws. Evaluate your pig and fit it accordingly. For example, if your pig is light-boned and refined, trim only the long guard hairs on the belly, rim of the ears, and flanks to make the pig look cleaner. If your pig is extremely wide and massive, clipping hair on the ears and tail may make it look larger-framed and trimmer. If your pig has extremely long hair, use a blocking trimmer to shorten all the hair on the pig. This will make your pig look younger and more productive.

7. Keep the pig bedded in clean straw or similar material. This will keep the pig clean and will add luster to its hair coat.

At the Show

1. If weather permits, the pig should be washed. Wash it about 1 to 2 hours before showing, or long enough before the show that the pig will be dry, but not so long that the animal will get dirty again. In some instances, the pig will have to be washed on the day before the show.

2. Groom the pig. If the pig is black, red, or spotted, apply a light coat of oil evenly with a sprayer or a woolen rag. An oil mixture, if not purchased as a commercial preparation, can be made by mixing 2 parts of light oil (such as paraffin oil or mineral oil) and 1 part of kerosene by volume. Do not apply so much oil that the pig has a greasy appearance. Never put oil on a white pig, because it causes a darker, greasy appearance. If the pig is oiled and the weather is hot, wash the animal after showing it. If the weather is extremely hot, do not oil the pig. The pig can be kept cool by thoroughly wetting it with water and brushing its hair smooth just before entering the ring.

Use white talcum powder on all white breeds. Animals must be dry. Use a soft-bristled brush to spread the powder evenly over the entire body.

Talc or white cream shoe polish may be used to whiten the white points on the legs of dark-skinned hogs.

Exhibiting Swine

1. Practice with your pig at home and study it while you practice. Learn at which gait and in which position the animal looks best. Work hard at learning how to maneuver your pig into these positions and gaits. When you get into the show ring, you will want to move your pig at those gaits and into those positions as many times as possible.

2. Make eye contact and smile at the judge immediately upon entering the ring. The judge will remember you and will look for you later. Throughout the drive, make frequent eye contact with the judge . . . but keep close tabs on your pig and be alert for any directions the judge may give you.

3. Do not let your pig smell or rub noses with any other pig in the class. Most pigs will eventually fight if given the chance. *If your pig does fight, move away and let the ringmen separate the pigs.* Pigs that are fighting are very aggressive and are temporarily dangerous. Move away so that you are not bitten or stepped on. If you are the only one available to break up the fight, use a hand hurdle to place between the pigs. Do not use your hands to try to separate the pigs. It is best to be alert and prevent the fight before it begins.

4. Keep your pig moving. Do not stop your pig and set it up in a stationary position. A short pause at a flattering angle is fine, but the traditional stationary pose with the showman squatting behind the pig was developed to show fat, lard-type hogs. Modern, lean, muscular hogs look best at a slow, relaxed walk 15 to 20 feet in front of the judge. When a hog stops moving it often begins to dig and root, which makes it look shorter and higher-topped. Digging also makes a mess of the pig and the show ring.

5. Keep your pig in the proper position at all times. Watch the judge and maintain eye contact, but make sure your pig is in position first. Your pig should be moving at a slow, relaxed walk as much as possible. Avoid corners and traffic, and avoid moving behind the judge. Anticipate where your pig is heading and turn it before it gets out of position. Drive your pig toward open areas in the ring. If the judge is not looking at you, you are most likely in traffic or out of position.

6. Keep your show stick off of your pig unless you need to use it. Often, exhibitors distract their pig and the judge by using their show stick unnecessarily or by rubbing the pig's back or belly. These movements confuse the pig and distract the judge. Never hit your pig on the top or rump. It accomplishes nothing of value and you may bruise your pig. Never hit a hog on the hock or cannon bone (lower leg) with a cane to move your pig. This action is painful to the pig, distracting to the judge, and should be discouraged.

Use your show stick to turn your pig by shading the eyes or with pressure contact (not a sharp blow) to the side of the head or jowl. Use your show stick to move your pig with gentle contact to the flank or side. If your pig stops and begins digging, use your show stick to lift the head before attempting to move your pig.

7. Keep your pig between you and the judge. Remember that the judge needs an unobstructed view of your pig to do a proper evaluation. Stay in proper position to show your pig effectively without obstructing the judge's view.

8. Do not overshow or be too aggressive with your pig. Learn to move smoothly and efficiently in the ring. Excessive or jerky movement distracts the judge. Some showmen work too hard in the ring. Often, these showmen are noticeably excited and aggressive. This distracts the judge and the pig. If you are overshowing or being too aggressive your pig will become agitated and tire quickly. Let your pig be a pig. Top showmen always have a pig that is relaxed and comfortable.

9. Be courteous to the judge, ringmen, and other exhibitors. Don't be afraid to smile! Be polite and receptive to instructions. Show respect for the ringmen, other exhibitors, and the judge. Be proud of your pig and of your hard work. Accept criticism and defeat with grace and restraint. Accept praise and success with respect and humility.

10. If the judge so indicates, try to stop the pig for the judge's close observation. The judge may want to examine the pig's underline for soundness or to handle the pig for some other reason. Attempt to stop the pig by standing directly in front of it and putting your hands in front of its eyes. Do not fight with the pig. It may be that you will only be able to slow him down.

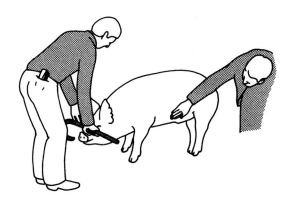

If the pig is hot when returned to the pen after the show, stay with it until it quiets down. A hot pig should have water to drink, but water should be placed only on its feet or nose until it is cool.

5.21 WEIGHING PIGS

Pigs are weighed to obtain information on animal performance, which is used for selection or research purposes, and to obtain weights for use in marketing.

For purposes of evaluating animal performance, weights may be obtained at various ages: on the newborn pig, when pigs are from 14 to 28 days of age, or when pigs reach market weight at 230 to 260 pounds. Pigs in research projects may be weighed weekly, biweekly, or monthly. These weights are obtained to measure sow productivity and to obtain data for calculating gains and feed efficiencies, which are important selection criteria when choosing replacement gilts or herd sires. These measures are also used to evaluate swine management practices and research treatment variables.

There are two times in the life cycle of the pig when it is most frequently weighed and marketed. The first occurs when pigs weigh between 20 and 60 pounds and they are sold as feeder pigs. The second is at a weight of 230 to 260 pounds, when pigs are sold for slaughter.

Equipment Necessary

- For newborn pigs, a scale that will weigh up to about 10 to 15 lbs.
- For pigs over 3 weeks of age, a portable or permanent scale
- Pens and alleys to move pigs on and off the scale
- Hurdle

Restraint Required

The restraint required to weigh older pigs is a well-constructed system of holding pens, a narrow alley or area leading to the scale, which is also enclosed, and a holding area for pigs that have been weighed. There are many arrangements for weighing, sorting, and loading hogs for shipment; these will vary with the needs of the swine operation and the personal preferences of the manager.

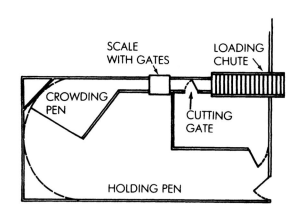

Newborn Pigs

Many swine producers, especially in commercial operations, do not weigh newborn pigs because of the time and labor involved and because the weights are not really needed for record keeping. If, however, it appears that birth weights are low, then obtaining these weights for a period of time may be useful in evaluating how extensive the problem may be. If swine producers need to keep more detailed records, then birth weights may be required. A good birth weight for a newborn pig is between 2¾ and 3¼ pounds.

For weighing newborn pigs, a scale that will weigh up to about 10 to 15 pounds will provide sufficient accuracy. The scale may be either the hanging type or a table model. Baby pigs should be weighed soon after birth so that a true birth weight is obtained. Weighing is usually done after the sow has completed farrowing and the baby pigs are dried. For labor efficiency, birth weight should be obtained when performing such other tasks as teeth clipping, tail docking, and ear notching.

Step-by-Step Procedures

1. The first step is to have record forms ready to record the litter information. If individual birth weights are obtained, then the pig must also be identified, usually by an ear notch.

2. Check the scale for zero adjustment and adjust it if necessary.

3. Pick up the pig with your hands by gently grasping it around the back and belly, and set it on the scale. The pig probably will move around at first, causing the pointer of the scale to move back and forth. Wait until the pointer stops to obtain an accurate weight.

CAUTION: The weighing of newborn pigs should not be done within the farrowing crate or pen. Weighing is usually done by putting the pigs in a portable cart in which they are transported, a litter at a time, to a service area where processing takes place.

---

4. Record the weight to the first decimal point (for example, 3.2 lbs). Record the sex of the pig. Ear-notch the pig. If a hanging scale is used, let the

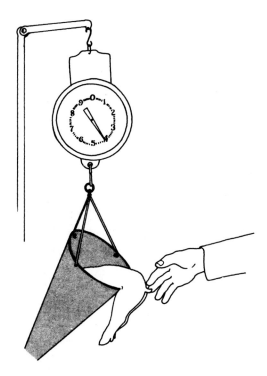

rear legs of the pig hang over the edge of the balance; the pig will then be less apt to move and fight to stand up. Different techniques, depending on the type of scale, may have to be tried to arrive at the easiest and most accurate method for obtaining birth weights.

Weighing Older Pigs

Pigs older than 3 weeks will require larger scales. In most cases, some type of portable scale is used when weighing pigs on the farm. Permanently installed scales may be used on some farms and at livestock markets. Portable scales are usually lightweight so that they can be lifted into place, or they may have wheels to facilitate movement. Scales usually have hinged gates with latches on each end or some type of pull gate. The passageway approach to the scale is usually identical in width to that of the scale, and narrow (about 18"), so that mature pigs cannot turn around. The pig is driven through an entry gate and both gates are latched shut. The pig is then weighed and leaves through the exit gate. Hinged gates are usually much easier and faster to operate than gates that slide up and down. Pigs usually are individually weighed on the farm or at swine research facilities. When pigs are sold at a market, they are weighed as a group and their individual weights determined by averaging. The price the producer receives is determined by weight, and in some instances is based on the quality and carcass potential.

Scales have a balance beam, a dial, or an electronic digital readout. Be aware that in cold weather, dial scales that use oil are sometimes slow in indicating the weight, thus lengthening the time required to weigh pigs. Make sure there is adequate lighting over the scale area to facilitate the reading of ear notches or identification. Scale construction that allows light to enter through the gates and sides will be helpful in providing necessary lighting for moving pigs over the scale more rapidly.

Be prepared for the weighing operation. Make the following preparations in advance so the actual weighing of thc pigs can take place without delay. Make sure the scale is in good repair—no broken or worn parts. Make sure the scale will adjust to zero. Test the scale by weighing yourself. Repeat the weighing once or twice. Have pens, alleys, and partitions used to move pigs through the scale in good repair and ready. If the scale is permanently installed and some of the mechanisms are under the floor, make a thorough check and perform any cleaning and maintenance that may be required. Rodents sometimes build nests in the mechanisms. Some scales (at markets or auctions) must be checked by employees of the scale company. Don't wait until weighing is to begin to check the scale. Have weigh sheets with pig numbers, sex, and previous weight ready for each pen or group of pigs.

When there are many pigs to weigh individually, have plenty of help. Three persons are sufficient in most situations. One person is needed for driving the pigs onto the scale and operating the entry gate, another for recording pig number, sex, and weight, and the third person for opening and closing the gate.

1. One individual opens the entry gate on the scale and, with the hurdle in hand, drives a single pig onto the scale.

2. If the scale has solid gates and sides, holding the exit gate partially open so that the pig coming onto the scale can see an opening helps in moving pigs onto the scale. A pig does not like to be driven into an enclosed area, because it evidently senses being trapped. A pig to be weighed usually will move onto the scale if he can see a route for leaving. Close and latch the exit gate as soon as the pig to be weighed is almost on the scale. Be alert, because some pigs may try to dart through the exit gate at high speed when they see an opportunity for escape.

3. The individual driving the pig onto the scale can read the ear notch or identification, or this can be verified by the person recording the weights.

4. Verify the sex and pig number and record the pig weight. If there is a previous weight to compare to, make a mental calculation and note whether the weight gain appears reasonable before the pig leaves the scale. Record the date and any sickness or abnormalities. Sick pigs should be given appropriate treatment.

5. Scale balance should be checked after each pen or group of no more than 15 to 20 pigs is weighed. Check also whenever there are indications something is not working, such as a scale weight indicator needle that moves abnormally or when a weight reading seems incorrect. Any buildup of solid or liquid excretia on the scale as weighing progresses may require frequent balancing of the scale to ensure accurate weights.

Postprocedural Management

After weighing, check the pigs that were weighed to make sure none are overheated or injured. Remove any sick or injured pigs to sick pens.

Page 273 provides a sample performance record for the sow and litter.

5.22 MOVING OR TRANSPORTING PIGS ON THE FARM

Pigs must periodically be moved from location to location on the farm because of different housing and environmental requirements at various stages of their life cycle. Examples of required moves are: moving gestating sows to the farrowing unit, moving newly weaned pigs from the farrowing building to a nursery building, moving lactating sows with their pigs to a sow–pig nursery, moving boars to the breeding pens, pastures, or lots, and the moving required to carry out such swine management practices as weighing and medical treatment.

The pig is less apt to follow a leader when being moved than are sheep and cattle. It is necessary, therefore, to maintain guidance and control over the pigs being moved. Providing an environment and equipment that facilitate the moving of swine will reduce stress to both pigs and those persons doing the job of moving.

Equipment Necessary (Options)
- Hand hurdles, panels, or gates
- Portable chute
- Permanently installed alleys, chutes, and gates that may be a part of the swine operation
- Small pens or boxes for pigs that attach to a three-point hitch on a tractor
- Two-wheeled trailers with which either a loading chute or hydraulic system is used to load and move hogs
- Farm trucks or pickups

Restraint Required
No restraint equipment is required, but control over the animal must be maintained either by hand or with equipment.

Step-by-Step Procedures
Loading, moving, and unloading animals should be carefully planned so that they will not become unduly excited, get too hot, or become injured. If strong, well-constructed equipment in good repair is used, moving hogs can be done rapidly and easily. There are four common moving situations with swine—driving pigs along a fence outside of a building, moving hogs with trailers or trucks, backing animals, and carrying young pigs.

Catching and Moving Pigs by Hand

Small pigs that weigh less than 30 lbs may have to be caught by hand and moved to another pen or the transporting vehicle.

1. Pigs must first be corralled or put into a small pen in which the handler can catch the pig without

Sample Record Form for Sow and Pig Performance

Breed (litter) _____   Sire (number or name) _____

   Litter number _____   Dam (number or name) _____

   Farrowed: Date _____   Farrowed: Date _____

Iron shots: Date _____   Sire: _____

Castrated: Date _____   Dam: _____

Weaned: Date _____   Age of dam _____

No. of boars: _____   No. of litters

No. of sows: _____   at this farrowing _____

Pig no.	Sex	Weight at— Birth	Weight at— 21 days	Days to 240 lbs	Back-fat	Loin-eye area Sonoray	Remarks:

Total no. of pigs _____

Avg. weight _____

Sows Record

Litter	Date	No. farrowed	Weaned
1			
2			
3			
4			
5			
6			

CAUTION: When moving or working with mature boars and gilts or sows with baby pigs, be sure to use a hurdle or make sure there is a partition between you and the animal to protect you should the animal try to attack or bite.

Do not try to separate fighting pigs with your arms or hands; you can be severely bitten. Use a hurdle or panel.

In hot weather, it is best to move hogs during the cooler hours of the early morning or late evening. Do not move older hogs, sows, and boars at all if the temperature is above 90°F with high humidity. Wait for cooler conditions.

If a moving process involves the mixing of pigs from different pens, move the group to a pen that is completely new for all pigs involved. Using the new pen as a new home to all pigs will reduce but may not eliminate fighting.

Never use an electric prod to move or load hogs. Use only a canvas slapper, a piece of plastic pipe, or a whip to prevent injury to the animal.

When moving pigs, it is best to make sure that the animal does not escape from you the first time. If he does escape, he will be even more persistent in his next attempts. The pig is smart and learns quickly.

Minimize disease transmission by cleaning and disinfecting all equipment after each use.

the pig getting away. It works best if at least two people work together in this task.

2. The first thing to do is to enter the pen of pigs. The pigs will usually run to the opposite end or corners of the pen, thus exposing their rear legs.

3. Grab one leg, lift the pig completely off the floor, and carry him to the new location or the vehicle in which he is to be transported. The last few pigs in the pen are the most difficult to catch because they have more room in which to escape; the job is easier with two persons.

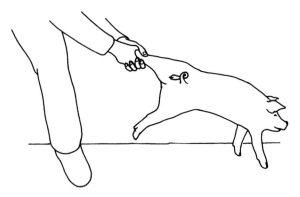

Driving or Moving Pigs in the Open

Older pigs, sows, and boars can be moved easily through the aisles within buildings. If pigs are

moved frequently between buildings, a permanent alley can be constructed to facilitate this process. Keep the alley narrow—18 to 20 inches—so animals cannot turn around. Sometimes when hogs are maintained in pastures or lots, it is necessary to move them without the use of a transporting vehicle. If a fence is present, a pig can usually be moved along it without difficulty.

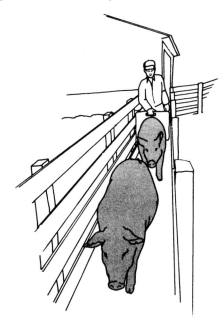

1. Use a handheld hurdle positioned either in front of you or to your side, keeping it ready to move in any direction to maintain animal control. The fact that the hurdle is solid will cause the animal to move away from it.

2. Follow the pig at about a 45° angle from the fence as it moves to its new location.

3. Do not rush the animal. Allow it to move at its own pace.

4. If the animal stops, gently nudge it to keep it moving.

5. Do not walk too far toward the head of the pig, as he will turn and go in the opposite direction. Be alert, as some pigs will want to return to the pen that is home to them. If the pig tries this, put the hurdle in front of its head and in contact with the ground. If possible, use some force to maintain control.

This method works best when only one animal is being moved, but with the help of several persons and additional hurdles and longer gates, several animals can be moved at the same time. It is best to use this method with older animals because they are slower and easier to move than young pigs. Young pigs are hard to control out in the open because they move faster and tend to dart and run away from you. It is best to move young pigs by transporting them in some vehicle or carrying them by hand to the new location. Females in estrus are often very difficult to move.

Moving Pigs by Vehicles (Hauling)

If animals are to be moved by hauling, there are several types of commercial equipment available. One is a box- or penlike structure which attaches directly to a three-point hydraulic hitch on a tractor. The pen can be raised or lowered into different positions so that animals can be loaded with the least amount of disturbance to the manager and the animal.

Another popular method of moving pigs is to use a two-wheeled trailer pulled by a tractor or some other vehicle. A chute may be used to load and move pigs in and out of trailers. Some trailers are equipped with hydraulic lifts; their beds can be set flat on the ground or raised to heights of 36 inches for loading and hauling. Farm trucks and pickups with stock racks are also used to move hogs on the farm.

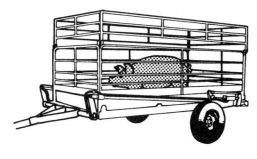

1. When moving hogs into a vehicle, use a hurdle or your knees to push and nudge at least one animal into the vehicle. In most cases, it will be advantageous to use bedding on the chute and in the vehicle.
2. Return to the pen and move another group toward the chute. Having at least one animal in the vehicle may prompt the remaining pigs to more easily enter the strange area.
3. Be patient but firm, and be careful not to injure the animals. Hogs, when being loaded, will tend to head back in the direction of the pen from which they came. Watch and anticipate the moves of the pigs already on the vehicle because very often one or all will attempt to leave.

Backing Pigs

Sometimes it is necessary to back pigs out of a stall or crate.

1. Use a solid object such as a hurdle, bucket, or shovel to do this.
2. Position and hold the solid object immediately in front of the head or eyes of the pig.
3. The pig will back up in its attempt to get away.
4. Keep moving the hurdle, keeping it directly in front of the pig until the animal has moved from the crate.

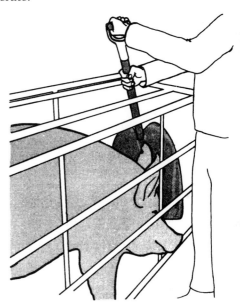

Postprocedural Management

After moving pigs, check them within 30 minutes to make sure that the animals are not overheated or fighting each other excessively. In hot weather, cool the animal if it is lying down and panting abnormally hard. When pigs are mixed, they will usually do some fighting until they tire. In rare cases it may be necessary to separate pigs by moving the aggressor to another pen.

Be cautious and observant when turning mature boars or sows from different pens together in the same pen because they may fight until overheated. Death could result.

5.23 TRANSPORTING PIGS OFF THE FARM

There are certain precautions and good management procedures that must be implemented when transporting animals off the farm to new locations. The size and weight of the animal, distance to be moved, season of the year, and temperature must be considered when planning to transport. With small (20 to 60 lbs) feeder pigs, for example, some pre-shipment treatment is practiced to avoid difficulties that may accompany the stress associated with pig movement.

While the pig is being loaded or is in transit, every precaution must be taken to avoid loss from bruising, crippling, heat stroke, and suffocation. The Livestock Conservation Institute reports that millions of dollars are lost annually as a result of bruises to swine slaughter animals. Most of the bruises occur in the ham area. Bruised meat on the carcass must be trimmed away; it cannot be used for human food.

Shrinkage or weight loss in transit is normal. A loss of 0.5 to 1% of the live weight is typical, but this increases with distance traveled and the number of hours the pigs are on the trucks. Short-distance shipments and careful loading and handling are helpful in minimizing shrinkage loss.

Proper techniques and equipment are required for the transportation of pigs off the farm.

Equipment Necessary

- Loading chute
- Transporting vehicle
- Bedding or sand
- Canvas backslapper or short pieces of plastic pipe

Restraint Required

No restraint equipment is required, but pens and chutes must be constructed so that hogs can be loaded and transported easily.

Loading Devices

Loading devices should be carefully designed and constructed to get the pig from the ground or building level into the transporting vehicle. If possible, load hogs off the level rather than up a chute. Loading chutes are required with most trucks. Loading chutes should be so constructed and in such a state of repair that there are no protruding nails, sharp corners, or other design features that might bruise or injure an animal. Many loading chutes are portable and can be adjusted so that there is a tight union between vehicle and loading chute. If there is a slot opening on the sides between the vehicle and loading chute, animals may try to escape through the opening, or they may get their feet caught in the slot on the floor, causing injury or breakage to the leg or foot. Telescoping panels on the sides of the chute can be used to block the gap between the truck and the chute, thereby preventing injuries caused when hogs try to escape through gaps.

A stairstep loading chute should have a $3\frac{1}{2}$ - to 4-inch rise and a 12-inch run for each step. The

slope for a permanently installed chute floor or ramp is 20°; for portable chutes it is 30°. If angles greater than this are used, the pigs have difficulty in moving up or down the chute. If stairs are not used, the chute should have cleats on the floor to aid in giving the animal footing.

The chute should be about 18 to 22 inches wide so that mature pigs cannot turn around once they start up the chute.

The Swine-Transporting Vehicle

If a truck is used for transport, the top of the exhaust stack should be above the top of the trailer, because diesel fumes entering the trailer can kill hogs.

Gooseneck and hydraulic trailers are increasing in popularity; they can be loaded and unloaded without a chute.

It is advisable to use partitions in the truck so that groups of 20 to 25 pigs are penned together. This will result in fewer losses from all causes. Livestock deaths do occur in transit, and most truckers have insurance to protect against such losses.

Summer. Bed the truck with a layer of moist sand. Load and move hogs during the cooler part of the day. During hot weather, allow 4 to 6 square feet of floor space in the truck for hogs weighing from 230 to 260 lbs. Larger hogs require more space, and those weighing less than 230 lbs require less space. Feeder pigs (20 to 60 lbs) require 1 to 1.5 square feet per head. When the temperature is 70°F or higher, allow more space per animal. Overcrowding pigs, especially during hot weather, may cause them to become overheated, suffocate, and die.

A rack with a well-designed top will protect hogs from the hot sun. Vent slots on the transporting vehicle should be open, or slats removed for ventilation.

Winter. In severely cold weather the sides of the truck should be 90% closed, with total closing at pig height. Slats should be replaced to protect the hogs from chilling effects of the wind when temperatures are below 50°F. Fresh straw or similar bedding material should be used in the truck to protect the pig from frostbite while in transit. Frostbite results in carcass losses. Pigs raised in environmentally controlled housing may have thin haircoats and be very susceptible to frostbite damage while in transit in cold weather. Keep hogs dry during cold weather to prevent chilling.

Step-by-Step Procedure

1. Sort hogs to be transported and put them in a separate pen, preferably near the load-out facility. The shipping of small pigs (20 to 60 lbs) or feeder pigs must be done with care. Some preshipping medication may be practiced to help the pig through the movement. Soluble antibiotics, electrolytes, and water-soluble vitamins may be administered in the water supply 36 to 48 hours before shipping.

2. In all seasons, it is advisable to partition the truck so that groups of 20 to 25 pigs of similar weight are penned together. The transporting vehicle should be clean (washed and disinfected) and bedded to reduce stress and the chance of disease transmission.

3. Have the loading chute (if required) and transporting vehicle in place. All equipment should be in good repair and tight so that animals cannot escape and will not attempt to do so.

4. Scatter fresh bedding on the loading chute. Sometimes the floor of the chute is constructed of the same flooring material on which the pigs are housed. For example, if pigs are maintained on slotted floors, the alleyway and sometimes the chute are also slotted. The pigs are familiar with the type of flooring and can be driven more easily onto the transporting vehicle.

5. With everything in place, you are now ready to load. Drive a group of pigs into the loading corral and then load a small number at a time. Do not rush the pigs. If you are lucky, one pig may take the lead and walk up the chute and onto the truck and the rest will follow. Keep crowding the pigs toward the chute with a hurdle to keep them moving. Hogs will often refuse to cross shadows or bright spots. Let the pigs investigate such spots. Do not rush them.

If pigs do not want to move up the chute, try using some persuasion with a canvas slapper or short plastic pipe (¾" to 1" in diameter) on those pigs near the entrance of the chute. Make sure the canvas slapper is not frozen; if it is, it will bruise the animal. Avoid the use of electric prods or hot shots, pitchforks, or other sharp objects. These instruments can cause injury or excite the animal, which can cause the animals to crowd and pile on top of one another, ultimately leading to carcass losses from bruising or deaths.

Try to force one pig onto the truck or vehicle by manually pushing and shoving until the pig is in the truck. Then quickly drive the rest of them toward the truck. Pigs will tend to move onto the truck when they see other pigs already on it.

Be patient, but firm. Hogs, when being loaded, will tend to head back in the direction of the pen they came from. Use a hurdle, if necessary, and move a few at a time.

Watch and anticipate the moves of the pigs already on the truck. Very often, one will attempt to leave the truck. If one succeeds, chances are they will all try to leave. Partitions in the truck help to control this. Do not overload, and allow sufficient space. Mature boars strange to each other should be penned separately in the truck.

6. If it is a long trip, check the load periodically en route to the new location. In hot weather, the vehicle should be kept moving once the animals are loaded. If it becomes necessary to stop en route, some effort should be made to sprinkle water on the pigs.

7. To unload the truck, do not rush the pigs. In most cases they will walk off the truck with little persuasion.

8. Clean and disinfect the truck before the next group of pigs is hauled.

Postprocedural Management

After transporting newly purchased feeder pigs, provide a dry, draft-free, well-bedded facility. Provide a special formulated starter ration with extra fiber and antibiotics, medicate the water the first week, observe the pigs often, and provide prompt and correct treatment for sick pigs.

For breeding stock, provide comfortable quarters, feed, and water. Be sure to isolate any breeding stock several hundred feet from the rest of the herd for at least 30 days. For slaughter hogs, provide comfortable quarters, feed, and water. Minimize stress.

5.24 BACKFAT AND LOIN MUSCLE DETERMINATIONS

The amount of fat that a pig has is an excellent indicator of the amount of muscle or meat that its carcass contains—the greater the amount of fat, the lower the amount of muscle. Muscularity or quantity of lean meat or fat in the carcass is heritable. By selecting lean and muscular breeding stock, improvement in this trait in slaughter swine can be made rapidly. Consumers of pork are interested in buying a cut of pork that has a high lean-to-fat ratio.

It is economically important, therefore, that swine producers emphasize this trait in producing pork.

Because it is impossible to easily measure and determine the amount of muscle in the live pig or the pork carcass, the backfat measure on both the live pig and carcass is used in swine selection programs to determine which are the leanest pigs.

There are three mechanical methods for determining backfat on a live hog. These are "A-mode" ultrasound imaging equipment, "B-mode" or real-time ultrasound imaging equipment, and the mechanical backfat probe.

Ultrasound Imaging: A-mode and B-mode

Ultrasound imaging machines use sound waves to create a visual image of a cross section of the animal or to determine individual site specific fat depths. Ultrasound imaging is not invasive and has numerous other uses in veterinary and human medicine. There are two types—A-mode and B-mode.

A-mode ultrasound machines are usually small, handheld units that are held in direct contact with the animal at the site or sites to be measured. Larger, more sophisticated machines may have a small handheld transducer connected to a portable console. A-mode instruments emit sound waves from a transducer and make use of the "pulse-echo" phenomenon to estimate tissue depth. Sound waves bounce off changes in tissue density (for instance, where fat changes to lean) and the elapsed time from sending the signal until the signal is returned to the transducer is measured. This time measurement is converted to some sort of distance scale or read directly from a screen.

A-MODE TECHNOLOGY

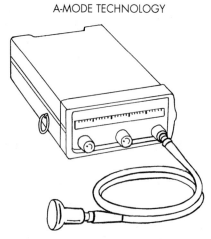

B-mode or real-time ultrasound machines equipped with a linear transducer are used to measure both backfat thickness and loin muscle area. In the B-mode technology, reflected sound waves are transcribed and presented on a television-like monitor as two-dimensional cross sections or images. From these, direct measurements of backfat depth and loin-eye area can be made or a computer program can be used to interpret the readings and provide measurement readings. These B-mode units, when equipped with various transducer types, enable the machines to be used for a variety of other diagnostic applications.

B-MODE TECHNOLOGY

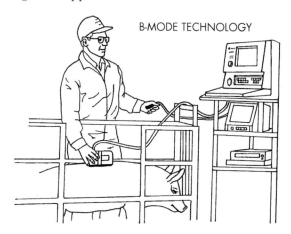

For pork production, B-mode ultrasound can be very accurate in determining carcass characteristics and predicting percent carcass lean, but does require a high degree of operator training and experience. B-mode ultrasound is also very expensive and is probably not affordable to the average producer.

A-mode machines are a practical ultrasound alternative for pork producers and can be used effectively to evaluate backfat and determine pregnancy status. Some systems interface directly with an electronic scale and computer to store and process the ultrasound data automatically.

Equipment Necessary

- Ultrasound unit
- Power source
- Vegetable oil
- Computer (optional)

Restraint Required

A small, narrow chute or scale is needed to secure the animal so that the backfat determination can be made. Animals to be pregnancy-tested can also be restrained in this manner or scanned unrestrained on a feeding floor or small pen.

Step-by-Step Procedure for A-Mode and B-Mode Ultrasound

1. Restrain the animal as described previously.

2. Apply a liberal amount of vegetable oil to the site to be scanned. The oil serves to transmit the sound waves from the transducer through the animal's hide into the underlying tissue.

3. Apply transducer and interpret data as instructed by user's manual. Always follow directions and make sure the equipment is working properly.

Mechanical Backfat Probe

A mechanical backfat probe is a simple, effective, and inexpensive technique for measuring backfat. While it is invasive, it is the most accurate method of measuring backfat for most producers. With a little experience, this technique can be a valuable tool in a swine selection program. When the cost of ultrasound technology is reduced, it will eventually replace the mechanical backfat probe. Until that time, the metal ruler will remain a valuable tool.

Equipment Necessary

- Disinfectant and cotton swab
- Scalpel
- Metal ruler (probe) calibrated in tenths of an inch
- Scale

Restraint Required

Hog snare

Step-by-Step Procedure

Backfat determinations on the live pig are taken at locations on the back of the animal. The first measurement is taken at the shoulder, or a point straight up from behind the front leg. The second is at a point up from the last rib. The last is at a point up from the stifle joint. Probes should be made from 1 to 2 inches on either side of the midline to avoid striking the vertebrae. At least two people will be required to do the backfat probe—one to hold the pig with the hog snare and the other to do the probing.

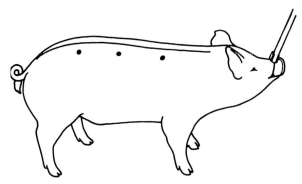

1. Weigh the pig to be probed.
2. Restrain the pig with a hog snare.
3. Disinfect the first site for probing.

4. With a disinfected scalpel, cut through the skin by making an incision about ¼" deep and ¼" long, with the incision perpendicular to and 1" to 2" off the midline. Wrap several layers of tape around the scalpel blade about ¼ inch from the tip to help prevent the blade from going too deep.

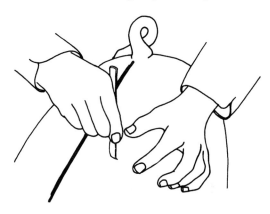

5. Measure the backfat by pushing the metal ruler down through the fat layer. Push the ruler until contact is made with the muscle in the back. Muscle will offer more resistance to the ruler than fat. With a little experience, one can tell when the muscle has been contacted. Be careful not to push the ruler into the muscle. The backfat probe (ruler) has a metal clip on it to facilitate the taking of the reading. Hold the ruler in place with one hand, and with the other hand push the clip snugly down to the pig's skin. Remove the probe (ruler) and read and record the number of inches of backfat. Disinfect the incision.

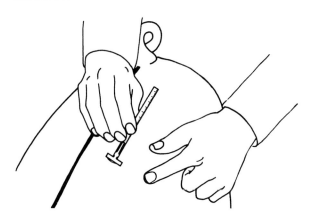

6. Repeat step 5 for the remaining two locations. The greatest fat depth is usually over the shoulder, and the least over the loin up from the stifle. The pig will feel the most discomfort in the rear location and will tend to jump forward during this probe.

7. Calculate the average of the three readings and normalize it to 230 lbs by dividing it by the pig's weight (in pounds) and multiplying the result by

230, or use a table of average values designed for this purpose.

8. To check the first reading, one can repeat the procedure on the other side of the pig.

Postprocedural Management

After probing, pigs may be returned to their pens.

Recovery Sequence

Little or no bleeding will occur at the incision sites. Incisions will scab over and heal in a short time.

5.25 ARTIFICIAL INSEMINATION IN SWINE

Artificial insemination (AI) is a relatively simple process that can produce tremendous benefits for the pork producer. AI provides cost-effective access to genetically superior sires. Artificial insemination, combined with on-farm collection, allows producers to use herd boars more extensively and allows the mating of heavyweight mature boars to gilts and smaller sows, decreasing the number of boars required in the herd and allowing boars to remain productive longer. AI facilitates the development of a closed herd where germ plasm enters via semen. However, the use of AI requires superior management skills on the part of the herdsman, including better record keeping, heat-detection skill and diligence, and ability to evaluate and select genetically superior animals.

Heat Detection

The first step in the AI process is determining estrus or standing heat. Proper timing of the inseminations is critical to achieving maximum conception rates and litter size. The normal estrous or reproductive cycle for typical females is 20 to 22 days in length, but varies from 18 to 25 days. The average length of estrus or standing heat is 24 to 48 hours for gilts and 48 to 72 hours for mature sows. Determining estrus is critical to the success of an AI program.

The chart below shows repeating 21-day cycles. Twenty-one days is the normal length of the estrous cycle of gilts and sows. The cycle of each animal is figured from the start of one standing heat [SH] to the start of the following standing heat [SH]. If you have a large herd, females will be starting and ending their cycles on every day of the month, every month of the year.

Sows in estrus exhibit certain behaviors and physical signs. For example, sows approaching estrus or in standing heat will often aggressively seek out a boar. Some sows, and more often gilts, will show evidence of swelling and reddening of the vulva. To determine whether a sow or gilt is in estrus, expose her to a boar through a fence and apply back pressure with your hands or by sitting on her back. Sows in estrus will assume a rigid stance and will repeatedly move their ears to an erect position. (The estrus response will be strongest if the sow is exposed to the boar only when estrus is being determined by the herdsman.) Sows and gilts should be checked for estrus status twice daily, in the morning and evening, and should be kept relaxed and comfortable. Do not check estrus just before feeding and do not excite or stress the females in any way. Accurate detection of initial standing heat is critical to the accurate timing of the inseminations.

Estrus detection is much more difficult if a mature boar is not available to stimulate the females. As many as 30 to 40% of all females will not exhibit an estrus response without exposure to a boar, and the estrus response of those sows that do show a standing heat will be much weaker if a boar is not present. For a complete discussion of "Heat Detection," see Section 5.3.

Insemination

Equipment Needed

- Paper towels
- Lubricant (nonspermicidal)
- Spirette
- Styrofoam cooler

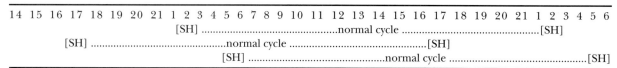

DAYS

14 15 16 17 18 19 20 21	1 2 3 4 5 6 7 8 9 10 11 12 13 14 15 16 17 18 19 20 21	1 2 3 4 5 6
[SH] ....normal cycle ....[SH]		
[SH] ....normal cycle ....[SH]		
[SH] ....normal cycle ....[SH]		

Standing heat [SH] is normally 48 hours long.

With once-a-day heat detection, breed each female on the first day they are detected and on each following day of standing heat [SH] that they will accept the boar. With twice-a-day heat detection, breed 12 to 24 hours after the females are detected to be in standing heat [SH].

Once the sow or gilt reaches standing heat, the timing of the inseminations can be determined. Most inseminations are made with fresh semen. Fresh semen can be effectively stored and maintained for 5 to 7 days, but is always best if used as soon after collection as possible. Fresh semen should be stored at 62 to 64° F and the bottles gently rotated four to five times daily during storage to resuspend the sperm in the nutrient extender. Avoid cold shock or change in semen temperature by keeping it at the proper temperature and transporting it to the sow in a styrofoam cooler or similar device. Avoid exposing the semen bottle to sunlight.

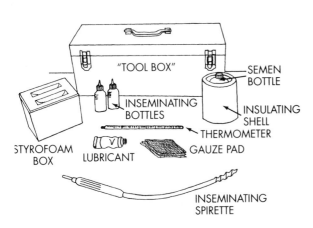

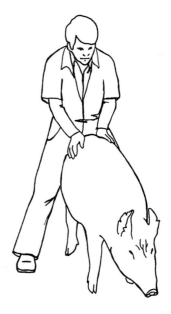

Frozen semen is an option, but requires more equipment and sophisticated techniques for storage and thawing. Frozen semen also produces lower conception rates and litter sizes compared to fresh semen.

Inseminate mature sows twice at 24 and 48 hours after the onset of standing heat. Inseminate gilts twice at 24 and 36 hours after the onset of standing heat. If three inseminations per sow are to be made, inseminate sows at 24 hours after the onset of standing estrus and again at 36 and 48 hours. Inseminate gilts every 12 hours, starting with the first insemination 12 hours after the onset of standing estrus.

Several different insemination tools are commercially available. Some are reusable; others are disposable and should be used only once. Many have a reverse-thread corkscrew design that simulates the anatomy of the boar penis. Others are designed to simply plug the cervix or vaginal tract, preventing semen backflow. Which spirette or catheter is best

for you is a matter of personal choice, convenience, and availability.

Semen is packaged in various containers for insemination, including bottles, bags, and tubes. Each should be stored and treated the same, but may require different coupling joints or insemination equipment. Most commercial semen containers include 80 to 120 ml of extended semen with 4 to 5 billion sperm per insemination.

When you are ready to inseminate a sow or gilt, start by stimulating the sow to exhibit a firm estrus response in a comfortable standing position, preferably opposite a mature boar. Maintain back pressure on the sow and wipe the vulva clean with a disposable paper towel. Apply a few drops of nonspermicidal lubricant to the tip of the spirette and insert the spirette into the vaginal opening with the tip in a slightly vertical position. Gently push the spirette forward while maintaining the slightly vertical orientation and begin to turn the spirette slowly in a counterclockwise direction. The slightly vertical presentation will prevent you from entering the opening of the urethra, which would produce severe pain in the sow. Slide the spirette along the top of the vagina until firm resistance is met. This is the cervix and it is usually 8 to 10 inches inside the vulva. Gently turn the spirette in a counterclockwise direction with your thumb and forefinger as it enters the cervix. The spirette should lock in the cervix as the sow clamps down. The spirette is locked properly when it will no longer turn and snaps back firmly when gently rotated. With the catheter in place, connect the semen container and

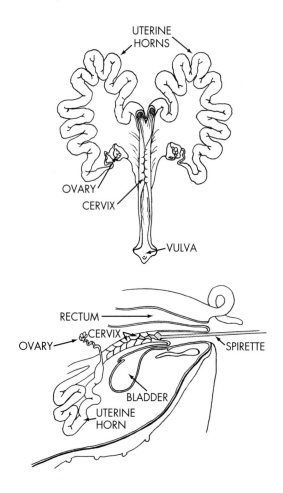

lift it to a position above the vulva to allow a downhill flow. Remember to maintain back pressure and apply stimulus to the sow by rubbing her underline and flank area. *The semen should flow into the sow with little or no pressure.* If the semen does not flow with gentle pressure or is running back and out of the sow, rotate the spirette further and make sure the tip is seated in the cervix. Be patient and gentle with the spirette. You may have to adjust it several times during the insemination. An insemination should take a minimum of 3 to 5 minutes with little or no backflow. When the insemination is complete, disconnect the semen container from the spirette and gently turn the spirette in a clockwise direction to remove it from the cervix. Once free of the cervix, pull it out directly. Maintain back pressure for a few more minutes and provide further stimulation to the sow by rubbing the flanks and underline.

5.26 ASSURING QUALITY PORK

The goal of the American Pork Producers is to produce high-quality, nutritious, and safe pork for the consumer. While *nutritious* and *safe* are relatively easy to understand and define as "good for you" and "won't make you ill," *quality* is much more difficult

to define, because it means quite different things to different people.

Quality can mean leanness, taste, appearance, color, water-holding capacity, intramuscular fat, nutritional value, safety, and/or any combination of these factors. It simply depends on whether you are defining quality from the perspective of the consumer, producer, packer, retailer, or USDA inspector.

While there is not a single definition for pork quality, the National Pork Producers Council (NPPC) has developed a series of pork quality targets. These targets represent minimums and, in some cases, ranges for quality parameters in fresh pork. The targets do not occur in isolation. Every pork production practice, from the selection of the breeding stock through the cutting and wrapping of the retail cut at the grocer's meat counter, impacts the end product quality.

The National Pork Producers Council has developed a *9-Point System for Assuring Pork Quality.* The 9 quality control points are all related to each other, and each is critical.

Quality Control Point 1—Genetic Input
Quality Control Point 2—Nutritional Inputs
Quality Control Point 3—On-Farm Hog Handling
Quality Control Point 4—Handling Hogs During Transport
Quality Control Point 5—Pre-Slaughter Handling
Quality Control Point 6—Stun, Stick & Early Post-Mortem Handling of Carcasses
Quality Control Point 7—Handling of Carcass During Evisceration
Quality Control Point 8—Chilling of Carcasses
Quality Control Point 9—Fabrication of Pork Cuts

This 9-point system, if implemented and adhered to, will help assure a high-quality meat pork product at the retail shelf level. However, in recent years, it has become apparent that in addition to high quality, the pork product must have an impeccable reputation for safety. If American pork producers are going to compete at the international level, all customers, domestic and international, must have total confidence in the safety of the pork product.

The Pork Quality Assurance program addresses the food safety issue in a comprehensive manner. There are 10 Good Production Practices (GPP) in Level III.

GPP 1 —Identify and track all drug-treated animals.
GPP 2 —Maintain medication and treatment records.
GPP 3 —Properly store, label, and account for all drug products and medicated feeds.

GPP 4—Obtain and use veterinary prescription drugs based only on a valid veterinarian client–patient relationship.

GPP 5—Educate all employees and family members on proper administration techniques and withdrawal times.

GPP 6—Use drug residue tests when appropriate.

GPP 7—Establish efficient quality production and an effective herd health management plan.

GPP 8—Provide proper swine care.

GPP 9—Follow appropriate, on-farm feed processing and commercial feed processor procedures.

GPP 10—Complete the quality assurance checklist annually, and recertify every 2 years.

AMERICAN QUARTER HORSE

Reprinted with permission from the ©American Quarter Horse Association.

ARABIAN HORSE

Courtesy of Arabian Horse Association, photographed by Jeff Janson.

LIPIZZAN

Courtesy of Department of Animal Science, Oklahoma State University

THOROUGHBREDS RACING

Courtesy of Department of Animal Science, Oklahoma State University

APPALOOSA

Courtesy of Appaloosa Horse Club Inc.

AMERICAN PAINT HORSE

Courtesy of American Paint Horse Association

MORGAN

Courtesy of Department of Animal Science, Oklahoma State University

STANDARD BRED TROTTER

Courtesy of United States Trotting Association

Horse Management Techniques

6.1 INTRODUCTION

Equus caballus, the modern horse, belongs to the family *Equidae,* which includes zebras, asses, and the horse, all of which are similar in physical appearance. What comes as a surprise to most people is that the entire equine family is a member of the order *Perissodactyla,* which makes the horse a close relative of the rhinoceros!

The horse as we know it today shows a wide variation in height, weight, and shape. Draft horses stand 18 to 19 hands tall, weigh 2,000 to 2,400 pounds, and are thickly made, heavy-boned animals. Their very essence exudes power. At the other extreme are the miniature horses, which stand only 7 to 8 hands tall and weigh no more than a medium-size dog. They appear very refined and almost fragile. The pleasure horse lies somewhere between these two extremes. Typically he is 14-3 to 16-0 hands tall,

weighs 900 to 1,300 pounds, and carries the image of an all-around athlete who is capable of speed or power as needed.

A Thoroughbred can cover a 1-mile distance at an average speed of nearly 40 mph, while an American Quarter Horse can dash short distances at speeds over 45 mph. Open jumpers clear heights exceeding 6 feet, and working hunters traverse the countryside for hours on end, taking fences, creeks, ditches, and brush in full stride. Horses of any breed can be trained to work as circus performers, as equine ballet virtuosos, or as movie stunt horses.

This same animal, *Equus caballus,* has also walked in dreary circles for long hours at grain and cane mills; he has plowed, tilled, and planted untold millions of acres of farmland; he has carried whole nations to battle; he was the enabling factor in the settlement of our country; he has run free and has fought human and animal predators to a standstill; and for a nickle or a dime, he carried your father, and perhaps even you, on his back at the town picnic.

It is a fortunate combination of psychological, physiological, and physical characteristics that enables one animal to adapt so well to the multitude of demands placed upon him and his kind—demands that require speed, power, endurance, courage, and aggression, as well as patience, intelligence, and memory. At other times in history, the horse has also been a source of meat, milk, companionship, and warmth for freezing warriors.

Handling horses safely and with enjoyment involves an awareness of their physical and psychological traits. They are large and quick, and come equipped with sharp teeth set in strong jaws. They have four legs, each with a hard, dense hoof. They are many times stronger than their human handlers. If you wish to have them submit to your wishes without undue force, you must understand their psychological needs. Such primary needs as hunger, fear, and gregariousness can become training stimuli in the hands of an intelligent human. The horse, while not the most intelligent of our domestic animals, does have an excellent memory—ranking second only to the elephant in the entire animal kingdom.

The horse reacts to danger today just as his ancestors did millions of years ago. His first impulse is to escape and run. If this fails, he will fight by kicking, biting, or striking out with his front feet. Danger to the horse need not be real, from your point of view, for him to react by bolting, kicking, biting, or striking. Surprise him by approaching without speaking, and he will react to the "danger" just as his ancestors reacted to the charge of a predator.

The horse does not interpret color as we do, and probably cannot see much beyond 400 yards. He is capable of seeing in a "monocular" and "binocular" mode. When he is seeing things to his left with his left eye and things to his right side with his right eye, and interpreting them simultaneously with his brain, he is using monocular vision. This is similar to our peripheral vision. When the horse is in the monocular mode, his ears will be doing different things. One might be forward, the other back, or they may be switching back and forth. With monocular vision, the horse is capable of detecting motion or movement. Watch the horse's ears: he is concentrating on the side that his ear is forward on. The horse can travel through an area and suddenly be spooked by an object from his left side that has been there the entire time. This is because he has been concentrating on those items on his right side.

To judge distance or size, the horse must use both eyes, look straight ahead, and adjust his vision, or focus, by raising or lowering his head. The horse is incapable of focusing by changing the shape of his eyeballs, as we do. When the horse is using binocular vision, his ears will be forward and he will raise and lower his head to get the proper "fix" on an object. For example, if the horse must negotiate a creek, cross a down log, or scramble up a ditch bank, he must move his head up and down so that he can focus in binocular vision mode and be able to judge how far away the creek is or how high the log is that is blocking the trail. We as riders must understand this binocular vision and "give the horse his head" as he attempts to negotiate the trail. Another good example of the horse moving his head to focus is when the horse, just learning to load, lowers his head to gauge the height of the step into the trailer.

The horse has blind spots in his vision. It is essential that we, as handlers, understand this. Remember that the horse is a poor visual concentrator. It is possible for us to walk into the blind spot cone of the horse and "surprise" him. If we do not speak to the horse, or otherwise make him aware of our presence as we enter the cone, we should not be surprised when he jumps away, jerks his head up, or kicks at us—all because we have frightened him.

The hearing of the horse is well developed, although not nearly as sharp as that of a game animal. The horse identifies his home, his grazing grounds, and his friends by their smell. Any new object or animal he encounters is investigated with his nose to determine if it is dangerous or acceptable. If the smell jogs the horse's memory into recalling a fearful event, the item or animal is either escaped from or attacked.

The sense of touch is well developed in the horse—literally from his nose to his tail. He is controllable because of touch receptors. His nose, mouth, ears, neck, withers, and rib area are especially sensitive.

A group of horses running as a band on pasture will most likely be led by an aged mare. She will be the boss horse, and all others, including the stallion, will follow her lead. Breeding season is an exception to this rule. During this time, the stallion becomes protective and dominant. Of stallions, mares, and geldings, the geldings are the least aggressive and least dominant.

Environmental factors such as thunder, lightning, wind, and precipitation have an effect on how horses behave. More important in this regard than any of these is a change in barometric pressure. Horses can sense the rise and fall in atmospheric pressure. They react differently on the day before a big weather change than they do during steady weather conditions.

The horse has no grazing pattern, as far as time is concerned. He will graze continuously if the weather is favorable. If it is extremely hot or cold, he may shelter himself from the elements during these times. If he is confined to a box stall and paddock, he is easily conditioned to a schedule of feeding. If you cause this conditioning to develop, expect him to show erratic behavior if you break the routine.

Horses can easily acquire several bad habits. Some horses are just plain lazy, while others will test you every time out. All horses will mimic the mannerisms of other horses. This can result in horses that are hard to catch, barn-sour horses, halter-pulling horses, buckers, strikers, kickers, biters, stall weavers, tail rubbers, and cribbers. All of these habits are extremely annoying vices. Each of them is also avoidable if you work to know, understand, and utilize the psychology of the horse.

In this chapter, you will be guided step-by-step through the most common management procedures involved in horse ownership. At every appropriate point, you will be cautioned to think ahead, look out for certain events, or be ready to react in a given manner. Cautions are inserted at proven danger points. They are not to be interpreted as the only danger points or as the only times you need to be safety-conscious around a horse.

Safety around horses is not a sometime thing. It must become a conscious way of life if you intend to be around horses and enjoy them to their fullest. Consider the following general cautions carefully before attempting any of the horse management techniques.

General Cautions

Horses are large, heavy, powerful, quick-reacting animals. They can injure you seriously without intending to.

The horse is an animal that, being incapable of human reasoning, will react to primary stimuli *when* they occur. The horse may be your best friend today, and bolt from you or over the top of you tomorrow, depending upon his reaction to a stimulus. He cannot determine whether a danger is real or imagined, nor is he going to remember, or care, that you may be between him and the only route of escape from some "spook."

Because of the way the horse sees, he has very poor frontal visual concentration. Approach him from an angle 45° off his shoulder. Never approach directly from the front or rear. (Refer to the illustration in Section 1.6 of Chapter 1.)

Speak to the horse as you approach him, and be certain that you have his attention. Horses doze standing up, and he may be sleeping when you speak.

A frightened horse needs time to settle down before you continue working with him. He may not be sufficiently aware of your presence if he is frightened, and you could get hurt.

Learn to read the horse's head, ears, and eyes. When he is about to "blow," the head will be high, rigid, and tense, with the ears either laid back or alternating from full-erect to full-back, and his eyes will be bugged, with the whites showing.

Never delude yourself into thinking you can stop a horse that is determined to leave with only a lead rope. You cannot! If you become entangled in the rope coils, you will be dragged; then, unless the horse knows what "whoa!" means, your shouting will only excite him more.

Keep your temper under control when working around horses. At times, they will need correcting, sometimes with harsh discipline, but never discipline with your heart; use your head instead.

6.2 APPROACHING, CATCHING, AND HALTERING

This job is either extremely easy or almost impossible, depending upon the horse's early training and acquired habits. If the horse was taught the meaning of "whoa!" and to accept the approach and handling of humans as a youngster, the job of catching him as an adult should not be difficult. If this was not taught to the young horse, or if it was and he was allowed to forget it or ignore it, then approaching, catching, and haltering the horse becomes very difficult and potentially dangerous.

Equipment Necessary

- Halter
- Lead shank

Restraint Required

No physical restraint is necessary to approach, catch, and place a halter upon a properly trained horse. You call upon the psychological conditioning acquired during the horse's earlier training when you command "whoa!" and expect the horse to stop, stand quietly, and allow you to approach and halter him. Most horsemen find that speaking to the animal while they approach calms the horse and helps to hold it in position.

Step-by-Step Procedure

You are most likely to be trying to catch a horse in one of three places: in his box stall, in a small corral, or in a large pasture. Each setting places different demands upon your horsemanship skills and is discussed separately. Regardless of which situation is presented to you, steps 1, 2, and 3 should be followed.

1. Attach the lead rope to the halter before you begin to go after the horse. This will prevent you from arriving at the horse with the lead rope in one hand, halter in the other, and a need for two more hands—one to hold the horse and one to connect the lead to the halter.

2. Enter the box stall, corral, or pasture only after alerting the horse to your presence.

3. Approach the horse while talking to him or while shaking his feed bucket. You should approach from a 45° angle off his shoulder, whether you approach from the front or from the rear.

CAUTION: Horses have a blind area immediately in front of them and to their rear. They literally cannot see anything in these areas, even something as large as a person. Couple this blindness to the horse's ability to sleep standing up and you have the potential for some dangerous moments for the nonthinking manager. (Refer to the illustration in Section 1.6 of Chapter 1.)

---

4. At this point you are approaching the horse, you have his attention, and you are about to attempt to place your hand upon his shoulder, withers, or neck. Almost all horses, except the very well-trained, will make some attempt to escape "capture" at this point. How you stop these escape maneuvers and control the horse depends upon whether he is in a box stall, small corral, or large pasture.

Box Stall. The horse will attempt to turn away from you and move in the direction opposite to your approach. If you persist in your initial move-

ments, you will wind up with a horse facing directly away from you, leaving you an excellent view of his rear end! To avoid this, step back as the horse turns away from your initial movement, and then move quickly (but not so quickly as to frighten the horse) into the direction in which the horse is turning. Horses can be outmaneuvered in this manner. Most horses will relent when you have arrived in their shoulder areas. If the horse persists in his turning away, do not lose patience and attempt a dangerous approach. Either do it safely, go for help, or use hay or grain to entice the horse to you. If when you attempt to enter the stall the horse presents his rear end to you with his ears laid back, and he shows no sign of changing this posture, do not enter the stall. The horse must be made to face you. The simplest ways are to call him or to entice him with food. A belligerent horse who still does not respond probably needs to be stung smartly on his hips with a whip until he turns to face you. This is no task for the novice or the hot-tempered. The novice will not be able to perform this task safely, and the hot-tempered person may forget that he is trying to teach the horse, not abuse him.

Small Corral. Many of the same problems exist here as in the box stall, except that the animal can move farther and faster. Your task, which is to stop the escape maneuvers, remains the same. Once again, the key is to gain the 45° approach angle and to keep it until you reach the horse. It may take just as many or more attempts outside as in the stall, but be patient and be safe. Should the horse constantly turn his rear end to you when you come near him, exactly the same procedures should be followed as when this occurred in the box stall. Keep in mind that this is a dangerous habit for the horse and it must be stopped. In the relatively open area of the corral, a wafer of hay or handful of grain in the feed bucket may be just the edge needed to "capture" the horse.

Large Pasture. Any area that is so large that you cannot cut across it and head the horse off by the time he reaches the next corner can be a real problem if the horse simply refuses to see it your way. Try "whoa," try talking, try anything you wish; you have a large problem if the horse is in a 5-acre field and refuses to let you walk up and touch him. It is to be hoped that "whoa," will work; if it does not, hope that the hay or grain does. If they do not, the horse must be moved to smaller quarters, caught, and retrained.

CAUTION: Whether in the box stall, corral, or large pasture, the horse may rapidly whirl his rear end toward you and lash out with his rear feet. You can minimize this risk by working hard to keep the 45° shoulder angle and by keeping alert.

---

5. With the lead rope in either hand and the halter in the other, place your hand upon the horse's withers, shoulder, or neck. Continue speaking to the horse and grasp some mane with the hand that is holding the lead rope while moving in closer to the horse.

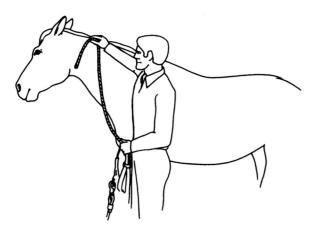

CAUTION: The horse may lunge and bite at you as you move your hand up along his neck. Admittedly, it would take an ill-tempered horse to do this, but to minimize the risk, keep your hand firmly on his shoulder or neck so that any movements he makes will be telegraphed to you.

6. Slip the lead rope over the horse's neck and fashion a handheld loop.

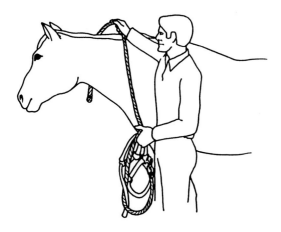

CAUTION: The horse may whirl and bolt, or try to test you by slowly turning away and attempting to leave. If he whirls and bolts, release him immediately to prevent losing your safe 45° angle. If he turns away slowly and tests your determination to control him, move with him as long as you are not losing your safe angle. If it becomes obvious that your lead rope or mane hold cannot stop him safely, release your hold, step away, and start over.

7. Place the halter upon the horse's head. If you are tall enough, you may choose to place your arm closest to the horse across his neck.

8. After the halter is in place, adjust it for a snug fit. Undrape the lead rope from the horse's neck. Remain in the safety area while leading.

CAUTION: If you leave the 45° safety zone by standing in front of the horse to adjust his halter or when crossing from one side to the other, the horse may strike out at you with his front feet.

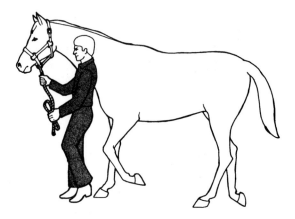

6.3 LEADING

Leading a horse from one point to another is a task that sounds simple. Most people would readily agree that they understand exactly what is expected. Unfortunately, not all horses have been trained to move in the direction and at the rate of speed chosen by the handler. The level of difficulty involved depends upon the amount and type of early training that the horse received as well as the horse's acquired habits.

Equipment Necessary

• Halter
• Lead shank

Restraint Required

Leading is not a technique that depends upon the use of physical restraint. The psychological conditioning of the horse's earlier training is the only useful restraint while leading. Prior training must have instilled in the horse a thorough understanding of what the "end of the rope" means, plus an immediate response to the command, "Whoa!" Only if these two aspects of ground schooling have been completed can the handler hope to control the horse physically. Even with the use of a top-quality nylon halter, bull-snap, and ¾" lead rope you will be unable to stop a frightened horse from breaking away. No one is able to stop 1,000 pounds of spooked or ill-tempered horse with only a lead rope and halter unless there is no place for the horse to run.

Step-by-Step Procedure

A horse should be capable of being led from either side; however, many horses are "one-sided," that is, capable of safely being handled from their left sides only. Until you are sure that your horse is "two-sided," lead him from his left side. Horsepeople refer to this left side as the *nearside*, with the right side being the *offside* of the horse.

1. Grasp the lead shank with your right hand, approximately 6" to 12" from where it attaches to the halter. (If you are leading from the horse's left side, reverse these directions.) Some horsemen prefer (and some horses respond better to) handling with a "loose" lead of 12" to 18" from the halter.

2. Fold the remaining length of lead rope into a figure eight in your left hand. Do not coil the extra rope into lariat loops.

CAUTION: The most common cause of mishap while leading is having the rope coiled around the manager's hand. If the horse should suddenly spook and bolt, the manager could be caught by the suddenly tightened coils and be dragged, or at least have his hand, arm, and shoulder damaged.

3. The handler should position himself approximately midway along the horse's neck and remain from 12" to 18" away from the horse's body.

4. The handler should not lead from in front of the horse, tugging on the lead, or allow himself to move too far back along the horse, allowing the horse to pull him along.

CAUTION: Being bitten is a common mishap while leading an ill-tempered horse. It is very easy for the horse to swing his head to either side and bite the handler, and it can happen too quickly for a day-dreaming person to avoid. Grasp the lead shank as shown and keep awake.

Another dangerous situation arises when the horse rears while you are leading him. This normally will occur while you are trying to force him to come forward. The horse will rear, perhaps flail his feet in the air, come down, and back up very rapidly. No one can stop the horse from rearing or backing up if he has a mind to. Don't try to! Immediately allow the folds of the lead rope to slip through your hand until the end piece is felt. This allows the horse to rear while keeping the handler out of reach. When the horse begins backing up, go toward him at about a 45° angle, keep a loose lead and command "whoa!" Most horses will not continue backing for more than a few yards unless the handler persists in pulling on the lead shank.

Most horsemen wear a light but durable pair of leather gloves while handling horses, especially green or ill-mannered horses. Gloves provide a surer grip on the lead rope and prevent your hands from being burned by rope slipping through them.

Keep in mind that a horse can move very quickly and that it is possible for him to whirl and kick before the handler realizes that the lead rope has been pulled from his hand.

Be alert for sudden striking (lashing out with the front feet), with or without rearing. If the handler stays in the safety zone, keeping a 45° angle off either shoulder, these dangers are reduced.

6.4 TYING

Tying the horse involves attaching one end of a lead rope to him and the other end to an immovable object so that he is forced to stay wherever you put him. This is a simple procedure, and to complete it with a well-trained horse, you must know only a few safety precautions, how to lead a horse, and how to tie the quick-release knot.

If the horse is young and unschooled or "spoiled" (negatively conditioned), the procedure becomes much more complicated. There are many more safety precautions to be observed, and a thorough understanding of horse psychology is required if you are to correct these problem horses.

Equipment Necessary

- Halter
- Tie rope
- Tie post, tie ring, cross ties, or tree
- Lariat
- Cotton rope: ¾", three-strand, loose weave, 25 to 30 feet
- Inner tube

Restraint Required

The entire procedure of tying a horse is one of restraint. For a well-trained horse, only the halter, tie rope, and a suitable place to which to tie the rope are necessary to keep him in one place. Unfortunately, many horses have either never been properly

taught to stand tied, or have picked up the vice of halter pulling at some point in their lives.

This is neither a book nor a chapter on horse training, but the problem of reschooling halter pullers is large enough to merit a discussion of a few techniques to assist the handler who must deal with such a horse. These same techniques can be used to teach a young, nonspoiled horse to stand tied.

Tying Well-Schooled Horses: Step-by-Step Procedure

A well-schooled horse will accept being tied with only his lead rope and halter secured to whatever tree, post, or tie ring you select. There are, however, several cautions that are pertinent to tying horses, no matter what their level of training or how gentle (or spoiled) they are.

CAUTION: Be sure of the quality and strength of all of the pieces in the tie system—halter, lead rope, snap, and post. Follow experienced horsemen's recommendations about the size and type of equipment. Tie your knot properly and be sure of the strength of what you are tying your horse to. Trees should be 6" in diameter, fence posts should have 6" tops, and any tie rings should be bolted firmly into or through solid, 6" material.

Be sure to tie the horse high and tight. The tie rope should be attached no lower than the withers of the horse. If possible, tie the horse about 24" from the post or ring. This gives your tie system the mechanical advantage. If the horse decides to pull, he will not be able to get a firm footing on the ground before he is snubbed off balance by the high, short tie.

Never tie to the rail or board portion of a fence, no matter how firmly it seems to be attached. The rails or boards, as compared to the posts, are the weak links in any fencing system.

Never tie a horse to a post that is part of any wire fencing. Horses may paw near a wire fence and entangle their feet, legs, or shoes.

Be sure that the post used for tying is tall enough and the tie short enough that the horse cannot rear and lunge forward onto the top of the post.

1. Lead the horse up to the tie post, tree, or tie ring. Stop when the horse's nose is about 2 feet from the post.

CAUTION: Do not pull the horse's head so close to the tie post that he cannot move it more than 6" in any direction. A horse must "have his head"; it is an essential part of his balance, and if normal movement is interfered with, the horse becomes fearful and reacts instinctively by trying to escape. If you insist on tying close, be alert. The horse may not wait until you have the knot secured before attempting to get away.

2. Wrap the lead rope around the post and secure it at about wither height. Use a knot that will not slip loose, will not become impossibly jammed if the horse pulls on it, and yet can be untied or released quickly in an emergency, even with the horse's weight pulling on it. The only knot fulfilling all of these criteria is the quick-release knot (see Chapter 1).

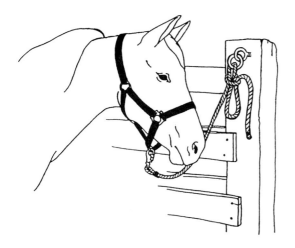

CAUTION: Do not tie lower than wither height or longer than 2 feet because the horse most certainly can and probably will get a leg *over* the tie rope under these conditions. Such a possibility, even with properly tied horses, is the reason that you should use "panic snaps" and quick-release knots.

3. After completing the knot, check it to be sure it is properly tied. Pull on the horse's end of the rope to jam the knot firmly against the tree, post, or ring.

CAUTION: It is easy to forget commonsense safety procedures while tying the knot. You must stay in the safety zone while tying and be alert to the horse's actions. Horses will often pull back, and if snubbed, immediately lunge forward, pinning you between them and the post, tree, or wall unless you are in the safety zone.

Watch your feet and legs. They are vulnerable to being struck or stomped upon while you are tying. You can minimize this risk by staying in the safety zone.

Many horsemen have their fingers or hands damaged while tying because either they were careless or learned sloppy technique in knot tying. There is no reason to place your hands or fingers between the rope and post or *into* any part of any knot while tying a horse. If you insist on doing so and the horse pulls back at that precise moment, you lose every time.

If you tie to a post, as compared to a ring or a rough-barked tree, you should always be sure that there are boards or rails above and below your tying point. If there are not, the loop may come off the top, allowing the horse to escape, or drop dangerously low, allowing the horse to become entangled in the rope. If you must tie to a smooth tree or post with no stops on it, use a clove hitch in combination with a quick-release knot.

Tying Problem Horses: Step-by-Step Procedure

Most halter pullers began as horses who "got away." To try to escape restraint is natural to all animals, especially to a fleet-footed, free-roaming animal such as the horse. Back when the now-spoiled horse initially got away, it probably did so because the handler failed to appreciate the horse's drive to be free. In all likelihood, the horse initially got away because he was restrained in a nonsecure manner or with weak, inferior equipment.

Strength-versus-Strength Technique

The keys to success here are a high-quality nylon web halter, a high-tensile-strength $\frac{5}{8}$" or $\frac{3}{4}$" tie rope, a well-made, heavy-duty bull snap, and a stout snubbing post, tie ring, or tree trunk.

1. Fit the halter snugly upon the horse, attach a tie rope to the halter, carefully approach the tie area, and use a quick-release knot to tie the horse properly to the post, ring, or tree.

CAUTION: Because this horse is a declared halter puller, you should exercise extra caution in all handling procedures. His bad habits and poor manners may have carried over to other areas. Particularly observe the cautions about being pinned, having your feet and legs stomped, and losing fingers while tying that are listed in procedures for handling well-schooled horses.

2. After being certain that the knot is properly tied, step back and let the horse fight the system. If, at that particular moment, he chooses not to be a halter puller, do something that will trigger his normal reaction. Normally, just waiting will trigger it; at other times, feet must be picked up or tails thinned (carefully!). Sometimes the horse must be tied a little shorter.

3. To be sure that the horse has thoroughly learned his lessons, tie him for 30 minutes to an hour each day for about a week. He finally will refuse to test the system. Tie him in a different place and in a different situation each day. You are trying to teach the horse that he must stand tied, whether in front of your snubbing post or to the side of your trailer.

Body- and Neck-Rope Techniques

These methods involve using the same wide, strong, nylon halter, the same lead rope, and the same stout post or tree that were used in the strength-versus-strength method, but in addition you will place a rope around the problem horse's rear girth or neck. If you know ahead of time that the horse is a real problem, this method is preferred over the strength-versus-strength technique. You will need a lariat, or 15 to 30 feet of ¾" cotton rope. In addition, you must handle the horse more to attach the body rope or neck rope. Sometimes this can be very difficult with problem horses.

1. Catch the horse, halter him, and lead him to a work area where someone should assist you by holding the horse's lead rope while you position the body rope around his rear barrel, about where the rear cinch would go. Position this rope snugly, leaving an inch or two of slack at the bottom where the knot is located. If you use the cotton rope, you do not want it to become tighter around the horse as he pulls back, so use a bowline knot to secure it. If you want the rope to draw tighter in proportion to how hard the horse pulls, use the lariat with its slip-through honda as your body rope. The lariat does release when the horse stops resisting.

2. After securing the body rope around the horse's rear girth, run the free end up between the horse's front legs, along his chest, up along the underside of his neck, and finally through the lead-rope ring on his halter. Pull out the excess, so that it now acts as a lead rope.

3. Tie the free end of the body rope to the stout post, tie ring, or tree. Tie it at about eye level and leave the horse 2 to 3 feet of slack. Now take a lead rope and attach it to the horse and post as you normally would, but adjust it so that it begins to restrain the horse just before the body rope really constricts his flanks. You time the ropes to pull in this sequence so that the horse associates the flank constriction with the pull on the lead rope.

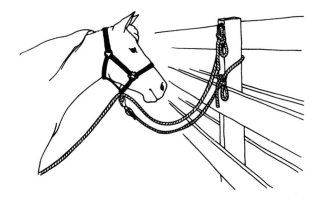

4. Now that both ropes are properly secured, step back and let the halter puller's bad habit cause him to pull. If necessary, you should cause him to pull by doing whatever normally triggers the reaction. The horse must fight the system, become uncomfortable with it, and teach himself the futility of pulling.

> **CAUTION:** The horse may lose his footing and go down. If this occurs, you may have to approach and release him by untying the ropes from the post. When he has scrambled to his feet, retie him as before. Do all of this very carefully.

5. The body-rope method can be changed to a neck-rope technique by replacing the rope around the flank with a rope around the horse's neck, just behind the jaws. In this case the cotton rope is preferred, and the bowline knot is essential. The neck rope must not constrict and choke the horse. You will need 12 to 15 feet of cotton rope.

> **CAUTION:** The precautions to be observed are exactly the same as those for the body-rope method and for tying in general.

6. Position the rope snugly around the horse's neck, just behind the poll and cheeks (jowls). Leave 1" to 2" of slack, and be sure of your bowline knot.

7. Run the free end through the lead-rope ring on the halter. Tie off this free end of the neck rope and a lead rope as you did in the body-rope technique. The rest of the procedure is the same as that described in steps 3 and 4.

Teaching the horse to stand tied with a neck rope pays double dividends, and all horses should be taught to accept it. It can be a cure for halter pullers, but just as important, it is especially useful on a trail ride. Normally, when you ride, the halter is removed. If you wish to tie the horse, you must replace the halter, which means removing the bridle. This is a hassle when you are out on the trail.

Some riders avoid this by tying with their bridle reins, but this is both insecure and very dangerous to the horse's mouth. It is much wiser to train your horses to accept the neck rope. It is convenient to carry a multipurpose rope along and simply slip it over the horse's neck when you want to stop for a few minutes.

Inner-Tube Technique

This is the best of the techniques available to teach a horse to stand tied. You must use the same quality and strength of pieces in this system that you did in the strength-versus-strength method, but by placing an inner tube into the restraint you remove the chance of poll injury and neck and back damage. Another plus is that the horse does not become psychologically dependent upon the presence of the neck or body rope to stand tied. This method is especially useful for teaching a young horse or foal to stand tied.

> **CAUTION:** The precautions to be observed are the same as those for the body-rope and neck-rope techniques and for tying in general.

1. Place a wide, nylon web halter snugly upon the horse's head. Attach the lead rope to a halter ring and select the strong post, tree, or tie ring to be used.

2. Double a single automobile or motorcycle inner tube (thus making it half as long as before) and attach it to the post, tree, or tie ring at a height of 6 to 7 feet from the ground. Be certain of this attachment by using ⅝" or ¾" rope to tie the inner tube.

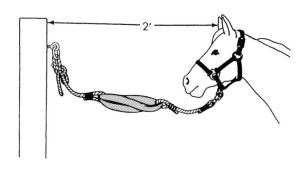

3. Attach the horse's rope to the inner tube so that he will have only about 2 feet of movement before he must "fight" the inner tube. Cause him to pull back by doing whatever caused him to do so before, or let his halter-pulling habit naturally come to the fore.

4. This method, like many of the others, must be repeated for several days in a row and, if possible, in different situations or locations. But the horse learns the lesson! He pulls, stretches the inner tube, and feels that he is winning, so he pulls more—but the inner tube never gives up! Soon he becomes tired, sore, and not so certain of his ability to beat the system. He will test it for a few more days, but eventually he will *learn*.

Cross-Tying

This is not normally considered to be a schooling or retraining technique, although it is often a very good one. The indirect pulls from the ties attached to each side of the horse's head prevent him from getting good footing if he attempts to break away.

Cross-tying is a management aid because you can place the horse in the cross ties and work completely around him without being tempted to duck under his lead rope or to squeeze between him and the tie post (both practices are dangerous).

1. Work in an area that is at least 10 feet wide.

> **CAUTION:** If you attempt to work in an area much narrower than this, you may be smashed into the wall, stepped upon, or forced to take unnecessary risks.

2. Attach two large, strong eyebolts or eye screws to a support post or structural beam. These should be placed at least as high as the head of the horse, and preferably 12" to 24" higher.

3. Attach chains, cables, or lead ropes to the eyebolts. They should be about 12" longer than the length required to reach the center of the cross-tie area at the muzzle height of the horse.

4. Attach bull snaps to the ends of the chains, cables, or leads, using ¼" S hooks for chain, clamps for cable, or backsplices for rope.

5. Hook a bull snap into the lower cheekpiece ring on each side of the halter.

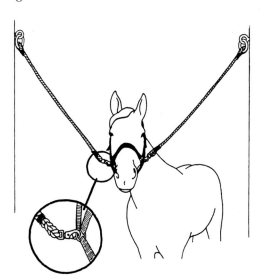

> **CAUTION:** This is a new situation for the horse. He may react to the newness or strangeness of it by attempting to bolt, pulling back, or even attempting to rear. Work calmly and with certainty, talk to the horse, and keep alert.

6.5 GROOMING

Grooming is the brushing, combing, and rubbing that is done to the body, mane, tail, legs, and feet of a horse. Its primary purpose is to remove the dirt, dead hair, and dust that have accumulated on the horse's skin and hair coat since the last grooming. A proper grooming session beautifies the horse and provides a stimulating massage to the skin, hair follicles, oil glands, and muscles. As an added bonus, making the horse stand and submit to your wishes and handling on a daily basis, even for only 20 to 30 minutes, is good horse-training technique. Your handling of the horse also ensures that you are aware of daily changes in the horse's condition and well-being.

Equipment Necessary

- Halter
- Lead rope
- Tie area
- Rubber or deep-massage currycomb
- Mud brush
- Medium brush
- Soft or finish brush
- Rag
- Mane and tail comb
- Hoof-pick
- Shedding blade
- Sweat scraper
- Grooming gloves

Restraint Required

While you are grooming the horse, it must be restrained in some manner. The restraint is normally a halter and lead rope or cross tie, but a body rope or neck rope should be used when dealing with a halter puller. In many cases, a horse can be safely and thoroughly groomed while standing free in the stall. You, as the horse's manager, should make this decision.

Step-by-Step Procedure

1. Assemble all of the equipment that will be necessary to complete the grooming procedure. It is recommended that each horse have his own set of grooming tools. This minimizes the possibility of spreading skinborne diseases from horse to horse. (Ringworm, a fungal skin condition, is a good example of such diseases.) A lightweight plastic or plywood grooming box or inexpensive bucket can be

labeled and used to hold these individual sets of grooming aids. If, under your management system, one common set of grooming tools is used for all of the horses, institute a strong program of external-parasite control to minimize the carrying of parasites from horse to horse. The only reasonable control program is to groom daily, observe closely, and immediately stop using the community brush when a problem arises.

2. Catch and restrain the horse in whatever manner is appropriate for you and your horse.

3. Begin the grooming session by briefly (about 30 seconds per side) rubbing the horse with your hands. This is not a massage session. Its purpose is to locate any swellings, regions of elevated temperature, localized soreness, or wounds that the horse may have picked up since the last grooming. Knit grooming gloves with rubber dots are especially useful for this. Your hands will stay cleaner, and loose hairs and dirt will be removed.

> **CAUTION:** If you start off by boldly and enthusiastically using the currycomb over an undetected sore spot, you may cause the horse to react with escape attempts, or worse, with such defensive attempts as kicking, biting, or striking.

4. If the horse is wet from sweating, rolling in the dew-laden grass, rainfall, or a recent bath, he will need to dry before you proceed to groom him. No harm will be done if you groom him while he is wet, but it is much more difficult to remove debris from, and put a shine upon, a wet coat.

To hasten drying, a vacuum sweeper can be reversed and directed upon the horse, or the hair can be brushed against its lay, standing it up and exposing it to the air.

> **CAUTION:** Condition the horse to the sound of the vacuum sweeper before blowing it upon his body. If you do not, you may frighten him into escape or defense attempts. One good way of conditioning the horse to the noise and blowing air is to first run it next to him, and then use it on him while he is eating.

5. It is essential that you use clean currycombs, body brushes, and mane/tail combs for the grooming process. It makes little sense to attempt to clean a horse with a dirty, gummy brush. During the grooming process, the currycomb and stiff brushes will "load up" with dirt and hair. This "crud" should be removed as often as necessary by sharply rapping the side of the brush handle against your boot or a fence post. Sometimes a close-bristled brush is best cleaned by brushing it clean against the edge of a board fence. The grooming equipment should be washed periodically in warm, soapy water, thoroughly rinsed, and placed, bristles down, in an airy place to dry.

6. Begin the actual grooming with the rubber curry or deep-massage currycomb. (In the spring of the year, when the horses are shedding, the first grooming tool to be used will more than likely be the shedding or hair-pulling blade—either a commercial variety or a homemade version.) The purpose of a currycomb is to lift to the surface dead hair, dry, flaky skin, and dirt. This lifting and depositing on the surface is best accomplished by using vigorous, quick, circular motions. The size of the circle is unimportant. Most horsemen will use small circles—about 8" to 10" in diameter. Establish a system to be sure of covering the horse's entire body, including the belly. The ears, face, and legs

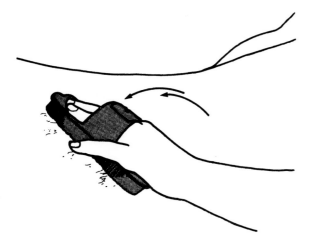

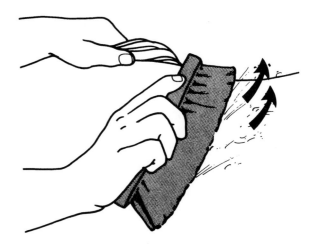

below the knee and fetlock are not covered by the hard-rubber currycomb. These are tender areas and should be brushed, not curried. It does not hurt to use the relatively soft-bristled deep-massage currycomb on the lower legs. In fact, a few extra moments spent massaging the coronet will pay extra dividends in stimulated hoof growth.

CAUTION: In the spring of the year, horses are especially thin-skinned and sensitive to being curried. Any currying, especially that to remove dead winter hair, should be tempered by common sense. Watch the horse's reactions for your cue to lighten up. Stepping away, laying the ears back, and aggravated tail twitching are the signs that your grooming efforts are too vigorous.

7. After you have curried the horse's entire body, use the stiffest brush in your kit, perhaps the mud brush, to flick away the dead hair, dirt, and other debris. Use plenty of pressure on the mud brush while you are scooting it along the horse's hair coat. End each stroke with a flipping or flicking motion to throw the hair and dirt from the body. If you neglect this flicking, you are simply moving dirt from one part of the body to the next. Each stroke should be no longer than 10" or 12". The use of longer strokes is not harmful, only less efficient. This mud brush can certainly be used on the entire leg and, with some care, on the face. The neck area, lying below the lay of the mane, is best curried and brushed by flipping the mane to the opposite side during the process. Try to be thorough as you use the stiff brush, but because you will be using at least one more brush, it is not essential that every last speck be removed.

8. After using the curry and stiff brushes, many horsemen will go immediately to a soft or finish brush to remove the last of the hairs and dust. It is a more thorough method to use a medium-bristle brush to follow the stiff brush, and then to move to the finish brush. With this method, the finish brush should not become soiled or filled with hair. Its purpose is to align the hairs and turn down the ends. The more aligned are the individual hairs, and the fewer the ends that are sticking up, the shinier the hair coat. Be certain to continue using the flicking motion and to thoroughly groom the face and head with these last brushes. Wipe the corners of the eyes and the insides of the ears and nostrils with a rag.

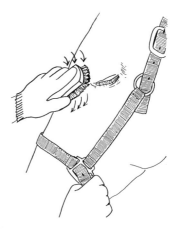

9. At this point, the forelock, mane, and tail should be taken care of. The term "taken care of" is used purposefully. Breeders of Arabians, Morgans, Saddlebreds, Tennessee Walkers, and other such breeds want long, flowing manes and tails. These manes and tails must be handpicked or gently brushed. Western horse owners usually are not so zealous in protecting the length of mane and tail, and certainly not the thickness of the mane and tail. These can be "taken care of"—deknotted, desnagged, deburred, and cleaned—by use of the mane and tail comb, deep-massage currycomb, coarse-toothed plastic comb, or stiff brush. You will lose some length and thickness with these tools, but, with some care, not enough to warrant handpicking. Spritz some silicone-based spray onto the tail and mane before attempting to untangle it. The silicone helps the brush, comb, and fingers to work the snarls free with a minimum of hair breakage. And, it will also add a welcome sheen to the mane and tail.

To begin working on the mane, stand at the horse's side and face the mane. Grasp the tool in one hand and the crest of the neck in the other. Start at either the poll or wither end. Take only a 2" or 3" bite at a time, starting from the bottom, and pull downward, at the same time pushing against the pull with the hand on the crest. With the next bite, move 2" to 3" farther up the mane toward the crest and pull as before, but continue down through that part that you just combed. To start at the crest instead of the bottom, or to take a larger bite, only results in hard work, sore hands, and an aggravated, sore-crested horse.

The tail is handled in exactly the same manner as the mane. Start at the bottom, use short pulls, and work upward.

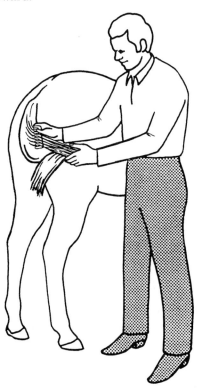

> **CAUTION:** Do not stand directly behind the horse to work on the tail. All horsemen appreciate the risk of getting kicked by a green or spooky horse, but even a sound, gentle horse can get fed up or surprised and kick out. Stand next to the horse's gaskin, facing the rear, reach for the tail, and pull it around toward you.

Sometimes dust and dandruff will fall out of the mane and tail and land on the polished horse. Use the finish brush to flick this away.

10. One step that really puts the finishing touch on the horse's hair coat is to hand-rub or rag-rub the entire body in the direction of the hair-lay only. This is for hair alignment with fewer ends sticking up and more sheen. Use some muscle in this procedure.

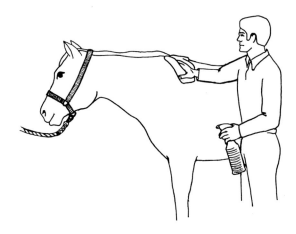

11. The final step in the daily grooming process is to pick up the hooves and remove any surface mud or dirt from the hoof wall. The proper procedure to use in picking up the foot and cleaning it is discussed in Section 6.6.

An old, stiff brush and some water or simply a wet rag will remove dirt and mud from the outside wall of the hoof.

6.6 PICKING UP AND CLEANING FEET

As part of the daily grooming procedure, and before and after each ride, the horse should have the bottom of his hooves thoroughly picked out or cleaned. To do so safely, the handler must be familiar with picking up and holding the legs of the horse.

Equipment Necessary

- Halter
- Lead rope
- Hoof-pick

Restraint Required

If the horse has been properly trained as a foal and weanling, he will yield to having his feet lifted and handled with no restraint other than someone at the end of the lead rope and halter. The horse should also be expected to stand tied to a post or ring during this procedure.

For those horses that will not peacefully submit to having their feet worked on, use a stronger restraint measure such as the nose twitch.

Step-by-Step Procedure

There are some horsemen who feel that the horse's feet should be picked up and cleaned in the same sequence each time. They feel that if you alter the sequence you will disturb or upset the horse. The most commonly suggested sequence is near forefoot, near hindfoot, off forefoot, and off hindfoot. Other horsemen feel that the horse should not dictate which leg gets lifted first, and they purposely start with a different one each time. This is good discipline for the horse, as he should submit to any of your reasonable demands. If the horse is new to you, common sense says to begin with the near forefoot. Most horsepeople, whether they intend to or not, will condition a horse to having his near forefoot lifted and cleaned first. It seems that this is traditionally the place to start.

1. Tie the horse securely to a post, place him in cross ties, or have an assistant control his head.

CAUTION: If someone holds the horse for you, it is your responsibility to see that they handle the horse properly. They should be positioned in the safety zones—45° off either shoulder. They should not be allowed to coil the lead about their hands or to stand and visit with friends while they should be paying attention to the task at hand.

While you are cleaning the front feet, the handler should stand on the side *opposite* you. In other words, if you are cleaning the near forefoot, the handler should be on the off shoulder. When you switch to the off forefoot, the handler switches to the nearside. This is done in case the horse should rear and attempt to escape. If the handler is opposite the footman when he is working on the front, it is a simple and safe matter for him to pull the horse toward himself and away from the footman. Were the handler on the same side, he would pull the rearing horse into the footman. You, as the footman, may have to insist on this. If the handler is doing his job correctly, he won't be able to see you work and will slowly but surely adjust his position so that he can see. When you are working on the rear feet, the handler must control the horse from the *same* side that you are on. When a horse rears and attempts to escape, the natural tendency, and the correct procedure, is for the handler to pull the horse into a circle toward himself. This maneuver swings the horse's rear legs away from both the handler and the footman.

It is also the handler's task to see to it that the horse does not bite you while you are working on the horse's front feet. While you are bent to your duties as the footman, your rear, back, and legs present nonmoving targets to the would-be biter.

---

2. Approach the horse from a 45° angle after alerting him to your presence. Continue talking to him as you draw near, and place your hand upon his withers, shoulder, or neck. The next step will be to pick up the horse's foot, so now is the time to place the hoof-pick in the rear pocket of your jeans—the one farthest away from the horse so you can get to it easily.

3. Begin with the near forefoot of the horse. Stand facing the rear of the horse. Your hip and shoulder should be next to the horse's leg and shoulder. Talk to the horse. Slide your left hand down the rear of the horse's foreleg, knee, and cannon until it is approximately 3" to 4" above the fetlock. Lightly pinch the tendon with your thumb, lean into the horse's shoulder with your shoulder, and say, "Foot!"—all at the same time. Only after the horse shifts his weight off the foot will you be able to lift it. Not all horses will respond immediately. Persist in this sequence until they do.

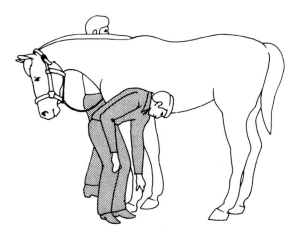

CAUTION: Some horses will lift their front feet very rapidly, actually jerking them up. This can catch you off guard and give your hand a good jolt, possibly even spraining your thumb. At other times the leg will come up and go back down so rapidly that you'll miss catching hold of it. When you lean into the horse, be certain that you do not move your left foot under him. If the horse jerks his foot up and down rapidly, your foot could be smashed.

There is an alternate method that may be preferred when you must lift the front leg of an unknown or ill-mannered horse. Approach him as before and stand next to him, facing the rear. This time, instead of sliding the hand nearest the horse down his leg, place it upon the horse's shoulder. Then, keeping it there to alert you to the horse's movements, slide your other hand down the rear of the horse's leg to the area about 3" or 4" above the fetlock. After the foot is picked up with this hand, it can be switched from hand to hand and examined as before.

4. After you are securely holding the horse's lower leg in the lifted position, place it between your legs just above your knees. To do so, you must momentarily switch the hands holding the horse's leg. Clasp the horse's leg securely with your legs. By positioning your feet in an exaggeratedly pigeon-toed manner, you will find it easier to maintain knee pressure. Sometimes it is not desirable to place the horse's leg between yours—when you are wearing good clothes, for instance. For these situations, it is correct simply to hold the leg with your left hand and clean the hoof with your right.

CAUTION: The horse may resist by rearing or backing. This may occur when you are lifting the leg, holding the leg, placing it between your legs, or when you are actually working on the foot. Some horses can be stopped in this attempt simply by growling something at them; others cannot. If your horse starts to

rear, and growling won't help, release the leg and step away. Use your leg and shoulder, which should be touching the horse's, as leverage to push away. Settle the horse and repeat steps 3 and 4 as necessary, but do not try to outpower him. If the horse persists in rearing or otherwise spoiling your efforts, consider a stern verbal reprimand accompanied by a sharp, openhanded slap to the rib cage, side, or belly. If this fails, resort to restraint procedures.

If your horse begins to lean on you by gradually shifting more and more of his 1,000-lb weight from his three legs to your 100- to 200-lb weight and two legs, you must nudge him, holler at him, or start over. There is no rationale for a 100-lb person to support a 1,000-lb horse. Do not allow this habit to develop.

5. Clean the hoof thoroughly, paying particular attention to the frog area and the areas where the bars and hoof wall meet. Considerable force should be applied to the hoof-pick. Dirt, rocks, and manure can become compacted almost to the point of being impossible to remove. The frog and sole of the hoof are quite durable, so bear down.

While you are cleaning the feet, check the amount of normal hoof growth and whether the shoes are still tightly nailed on. Also check for the presence of thrush, cracks, or sole bruises.

6. To pick up the rear legs and clean the rear feet, approach the horse in the same way that you did for cleaning the front feet. Stand at the horse's near shoulder, facing the horse's rear, and place your left hand on his withers or neck. After he has accepted your presence, move to approximately the horse's stifle area, keeping your left hand sliding along the horse's back. When you arrive at the stifle area, place both your hands upon the croup. Now move your right hand down the quarter to the gaskin, and finally over the hock to about 3" to 4" above the fetlock. Do not attempt to go from croup to fetlock in one motion. Rub each successive area once or twice before moving lower.

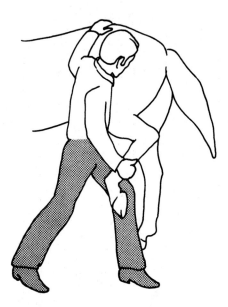

7. At this point, attempt to lift the leg. Your first direction of motion should be *forward with the leg,* not rearward or outward. Lift the leg and move it about. Determine at this point, before stepping under the leg, if the horse is going to resist.

with your left hand while you are anchoring it between your knees and to use your right hand to manipulate the hoof-pick.

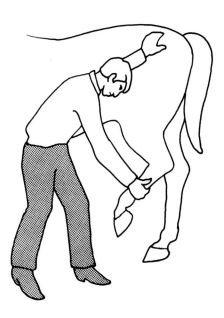

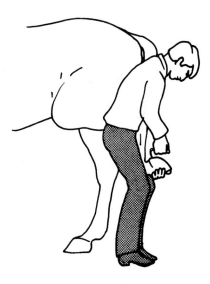

8. Step under the horse's leg as shown. Allow your left arm to slip rearward and down over the horse's hip. Do not be hesitant during this move. Start and complete it in one easy motion.

9. Now that you are under the horse's leg, it will probably be easier for you to help support the foot

10. For the offside legs, repeat steps 1 to 9 with the appropriate right/left switches. Keep in mind that some horses are not as well trained on their offside as they are on their near. This means that you must be more of a horseman and more patient.

6.7 THINNING AND SHORTENING MANES AND TAILS

There are two major reasons for thinning and shortening horses' manes and tails. First, it simply isn't *practical* to allow a horse's tail to become so long that it drags in the snow and mud, or to allow its mane to become so long that it fouls any harness you place upon it. Second (and this isn't quite as rational), we as horseowners think certain styles and lengths *look nice,* so we shape the manes and tails to suit us.

Fortunately, there is little effect upon the horse's well-being, regardless of the length and thickness of the mane and tail that we force upon him. With the exception of draft horses and certain ponies that have had their tails surgically bobbed for show purposes, there is always enough tail left for the horse to at least frighten flies and mosquitoes, even if he can't quite reach them.

Equipment Necessary

- Halter
- Lead rope
- Tie area
- Mane and tail comb
- Pocketknife
- Commercial mane-pulling tool
- Thinning comb

Restraint Required

The horse must be restrained in some manner. The halter and lead rope, tied securely, usually are all that is necessary.

Step-by-Step Procedure

Practically speaking, there is no method of shortening a mane or tail that does not thin it at the same time. As you pull out only the longest hairs to shorten the mane, you also reduce its bushiness or thickness by that amount of hair. Most of us are happy with this circumstance, because it's the rare horse indeed whose tail or mane is both too long and too thin.

Shortening Tails

1. Assemble all of the equipment that will be necessary to complete the tail work.

2. Catch and restrain the horse in whatever manner is appropriate for you and your horse.

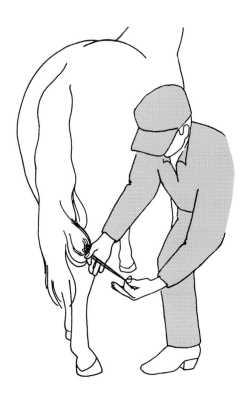

3. Comb out the tail so that all hairs are lying as straight and uniformly as possible. To shorten the tail, start at the bottom. Select five or six of the longest hairs and grasp them with the thumb and forefinger of one hand. With the other hand, push up and out of the way ("tease back"), for about 4", the shorter hairs in the immediate area of the five or six you are holding. You can do this teasing with your hand or with any kind of mane or tail comb.

4. Wrap these five or six hairs around your index finger (the one already holding the hair), squeeze them tightly, and pull straight down. Use a sharp jerk rather than a slow, steady pull. The hairs should pull out from the roots. If they seem to be breaking off instead, change your angle of pull slightly until they come from the roots.

> **CAUTION:** Horsehair is tough and securely rooted. If you grasp too many hairs at one time, or do not pull the five or six hairs sharply enough, they may cut your finger, or at least your finger will become quite reddened and sore. Some horsemen wear light, leather gloves for this task or wrap the five or six hairs around the teasing comb before pulling.

5. Comb out the teased-back hairs and once again grasp the five or six longest hairs in the same area. Repeat the preceeding tease, wrap, and pull sequence. To complete the tail, just keep taking the longest hairs from the bottom of the tail until you reach the desired length.

6. If the tail is really full and long, say 8" to 10" longer than desirable, it may be easiest to shorten it by "banging" it. Banging the tail refers to cutting it straight across the bottom, with a scissors, at or about the level of the fetlocks. To do this, simply grasp the tail, just above where you want to cut it off, use a sharp scissors, and cut the tail hairs as evenly as possible.

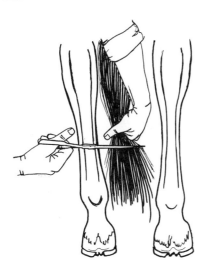

7. Release the tail hairs and once again thoroughly comb out the tail. Now, use the scissors to square off the tail bottom by cutting off the uneven hairs.

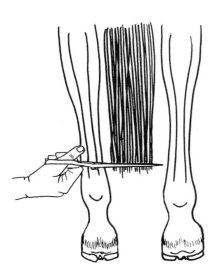

8. If you do not wish to have a square look to the tail bottom, repeat steps 1 to 5 from above, taking care to not shorten the tail more than desired.

Shortening Manes

1. Assemble all the equipment you will need to complete the mane work.

2. Restrain the horse in whatever manner is appropriate for you and your horse.

3. Comb out the mane so that all the hairs are lying as straight and uniformly as possible.

4. There are several methods that can be used to shorten the mane. They can all yield satisfactory, smooth-looking results when you have practiced how to use them properly. Mane-shortening methods include tease and pull, thinning shears, thinning comb, pocket knife, and electric clippers.

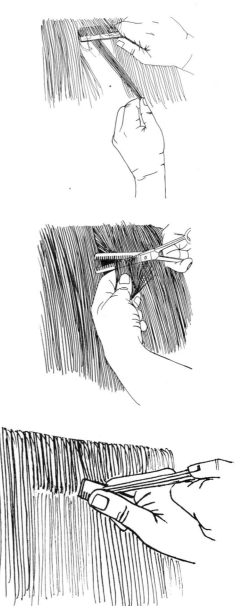

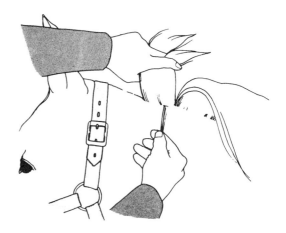

5. Whatever method you use, you will have to make a decision where to begin. There are as many people who insist on starting at one end of the mane or the other as there are those who insist on starting in the middle. Whatever spot you choose, carefully keep track of where you are heading so that you do not shorten one segment of the mane more than the next. If you do, there is little solution, if you want to even it up, beyond shortening it all to the shortest length.

Shortening the Forelock

1. The horse's forelock is handled in exactly the same manner as the mane and tail. A rough guideline for the forelock length is to shorten it to about the eye level of the horse.

CAUTION: Remain at the side of the horse. It is easier to stand directly in front, but it is also foolhardy and dangerous.

Thinning Manes

1. Comb out the mane so that all hairs are lying as straight and uniformly as possible. Start at the end that works best for you. Slip your strongest hand under the mane (between the mane and the neck) at the point at which you are going to start. Grasp five or six hairs from the bottom or middle layers of hair and pull sharply. For best results, these hairs should be grasped about 3" to 4" from their roots. Next, move about ½" along the underside, grasp five or six hairs, and pull as before. Continue this process along the entire length of the mane. The amount of thickness or thinness depends upon how often you grasp and pull hairs.

Be sure to select and pull hairs from the underside of the mane. Part of the reason for thinning manes is to make them lay better. If you pull from the bottom side, the top hairs can lie flatter because

there is less hair below causing them to bulge or stand up. Additionally, if you break some hairs, it is better to have them on the bottom than frizzing up on the top side.

Reconsider your inclination to use a thinning shears. Unless you are very good at it, the mane can look choppy and can actually lay worse than before thinning, because you now have short hairs, which are going to stand up straighter. There is no substitute for doing it correctly the first time.

2. There is an alternate way to thin manes and tails that appeals to many horsemen. It involves the use of a pocketknife to remove the excess hair. This method takes less patience than hand pulling, but unless you take great care in making uniform pulls and cuts, it can have the same shortcomings as the thinning shears.

In this method, instead of grasping the five or six hairs between the thumb and forefinger, they are pinched between your thumb and the opened blade of the pocketknife. This blade need not be extremely sharp; in fact, it's easier on your thumb if it isn't. After pinching this hair, pull the blade steadily toward you. (There is no need to jerk it.) Some hair will be cut, some pulled. Since you start from the underside of the mane, you should not wind up with "shorts" on top. If you are working carefully, the thickness

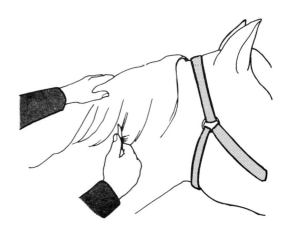

should be uniform throughout. In practice, there will be some "shorts" on top, and it is difficult to force yourself not to hurry. To be sure that you do not end up with a layered look, start some of your cuts at the roots, some an inch away, and so forth.

> **CAUTION:** Observe the same precautions as in hand-pulling the mane. There is the added problem this time of having to handle the knife. Keep it only moderately sharp, use only a 2" to 3" blade, and don't prick the horse's neck with it.

3. The horse's forelock can be handled in exactly the same manner as the mane. Your wishes and the way the horse looks are the only guidelines to forelock thickness.

> **CAUTION:** Do not perform these tasks while standing directly in front of the horse. Stay as close to the 45° safety zone as possible.

Thinning Tails

1. Comb out the tail so that all hairs are lying as straight and uniformly as possible. Find the point at which the tailbone ends. This will probably be about $\frac{1}{3}$ to $\frac{1}{2}$ of the way down the tail. Grasp the tail in your hand, leaving an inch or two of the tailbone extending below your hand. Manipulate this handful of tail so that you are holding only the bone and the long hairs coming out from it. You should release all of the shorter hairs flaring out from the sides of the bone or from above your hand. Tie a knot in this long hair, using the hair for the knot, or slip three or four rubber bands over it. This will prevent you from removing these longer hairs as you thin the tail.

2. Starting at the head of the tail, from either side, grasp five or six hairs between your thumb and forefinger. These hairs should be selected from the underside of the tail. Grasp them as close to their roots as is practical, wrap them around your finger or the comb, and pull sharply as in shortening the

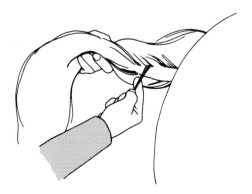

tail. Move down the tail an inch or two and repeat. Continue this down through the flaring hair that ends just above the tip of the tailbone. Do both sides. The ultimate thickness of the tail will depend upon how often you grasp and pull the five or six hairs.

> **CAUTION:** Use the same precautions as in thinning manes. There is the added danger that you may grow careless and move behind the horse to do this job. Stay near the stifle area, stand close to the horse, and stay alert.

3. Tails may be thinned by the pocketknife method. See the step-by-step procedure for thinning manes with the pocketknife.

> **CAUTION:** Be alert while working near the horse's rear legs and avoid sticking the pocketknife into the horse's tail. This is a touchy area for the horse and contains large blood vessels. Be careful!

4. There is something of a vicious cycle involved with thinning and shortening. It is difficult to separate the two in practice. You cannot shorten without thinning, and only the very practiced or very lucky can thin without shortening. The best advice is to complete these techniques in no fewer than two or three stages. In that way you can be sure of what the end result is going to be before you remove too much hair.

6.8 CLIPPING

Management Clip

Managers clipping is the technique of removing excess hair with an electric clipper. It does not include "show clipping," wherein all of the excess hair is removed to impress upon a judge just how capable you are with the clippers or how refined and breedy your horse is. Clipping for management purposes involves removing excess hairs from the coronet, legs, bridle path, ears, muzzle, and jaw. It should be considered the routine clip for health and normal, good appearance.

Two of the many reasons that a management clip is essential are: (1) long jaw or chin hair can be caught and pulled by the curb chain used as part of normal headgear arrangements, and (2) mud, snow, and ice can build up on and hang on to long fetlock hairs. Both of these occurrences are uncomfortable for the horse and the latter one can be unhealthy. Management clips should be repeated every 8 to 10 weeks.

Equipment Necessary

- Halter
- Lead rope
- Tie area
- Grooming equipment
- Electric clipper
- Extension cord (heavy-duty)
- Twitch
- Fuel oil or kerosene/blade wash

Restraint Required

Every horse owner would like to say that no restraint other than the normal lead rope and halter or cross tie is necessary when he clips his horse. Unfortunately, this simply isn't true. Many horses will resist being clipped. This behavior can vary from mild resistance, where you may have to use a nose twitch for a few minutes as you begin to clip, all the way to the extreme behavior of the horse that must be thrown and tied or chemically tranquilized before you can even get close to him with clippers.

The latter type, in its extreme, is rare, but not rare enough; nor is the horse who must be twitched or "eared" during the whole process of clipping. There is no excuse for any horseman to develop a horse who is clipper-shy. You may own such a horse someday; just be sure you purchased it, not ruined it yourself.

A young or green horse naturally will be apprehensive of the clippers. To the horse's mind, the buzzing and humming noises may represent flies, bees, rattling snake tails, or just another aggravation. Be patient. You do not have to win every battle with the horse. You can take several days to condition the young or green horse instead of "winning" by twitching him and doing the whole job on the first encounter. These first few sessions set the pattern for later behavior.

Step-by-Step Procedure

1. Assemble all of the equipment, including the fuel oil or kerosene, that will be necessary to complete the clipping task.

2. Arrange for an assistant to help you. It is always safer if there are two people present when doing work such as clipping. It is an absolute necessity when you are handling a young, green, or unruly horse.

CAUTION: Be sure the handler understands the basics of horse safety and where to stand in relation to where you are working (see Section 6.6).

3. Be certain that your electric clipper has a set of sharp blades of the proper size. An Oster Small Animal Clipper, Model A5, works well. The Number-10 blade is sized properly for the management clip. Before you start, lubricate the blades. Immerse them in the fuel oil or kerosene, while they are running, for a second or two. You may use a highly refined oil manufactured for this purpose. If you must use an extension cord, use a heavy-duty type, no smaller than 18 gauge. Household cords are too light, do not carry enough current, and cause the clipper motor to overheat.

CAUTION: Under no circumstances should you use gasoline to rinse and lubricate the clipper blades. Gasoline is of no value as a lubricant, and a single spark from the clipper motor will cause the gasoline to explode.

4. Groom the horse thoroughly. Nothing dulls clipper blades more rapidly than cutting through dirty, gritty hair. You are clipping only legs and heads, so the grooming need not extend beyond these areas.

5. You should begin the management clip on the horse's front legs. Most horses are less fearful of the buzzing about their legs than they are around their heads, muzzles, and ears.

CAUTION: Be certain to accustom the horse to the clippers, cord, and buzzing before you begin the trimming job. Even if the horse showed no concern the last time you clipped him, follow the sequence below.

1. Allow the horse to smell the quiet clippers and cord.

2. Touch and rub his neck, shoulder, withers, and forearm with the quiet clipper.

3. Start the clipper about 4 to 5 feet from his neck and shoulder area, *while he is looking at the clipper.* Touch the horse's neck with your free hand and speak to him. Bring the running clippers over to his shoulder and rub the clipper body over it.

4. Gradually move from there down the leg. Begin clipping at the coronet, keeping your free hand firmly on the horse's leg, and be alert.

5. If the horse should panic at any step, stop, back up, and then proceed slowly and patiently until he does accept the idea.

6. When it is obvious (say, after a half-dozen attempts) that the horse is not going to submit, twitch him and continue. As stated earlier in this section, young horses should be given several days' worth of chances before twitching, but an older horse should know better and probably has been spoiled by a previous owner.

Clipping the Legs

1. As you begin the specific clipping sequences, keep in mind that this is a management clip, with health, comfort, and neatness as the goals.

Examine the legs from the knees and hocks down. They may be dirty and gritty enough to need washing. If they are, do a thorough job of drying before beginning clipping.

Hair can be removed by clipping in either direction. More will be removed by clipping against the lay of the hair. When you are able to choose either direction, fewer mistakes will be made by clipping with the lay of the hair first, then against it for the finishing touches.

2. Begin the lower leg clip at the coronet. Place the clipper head parallel to the wall of the hoof and against the hoof. Start about 1" below the hairline and in the front center of the hoof. Move the clipper upward, keeping the same angle, and clip off the longer hairs sticking out over the hoof. Do not tilt your blade so that you follow the pastern line. This will put clipper marks on the pastern and make it necessary to clip the whole pastern to remove them.

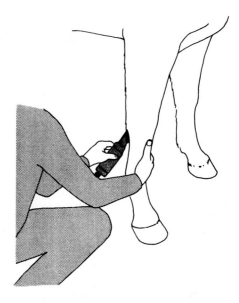

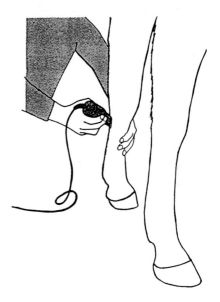

3. Move the clipper and repeat this procedure until you have completed the entire coronet. You will find it difficult to do the heel hairline, especially on the inside heel. One way to solve this is to reach across and do the inside heel of the leg opposite the one you are now clipping. Another is to lift the leg and clip the hairline of the heel while it is lifted.

4. The longer hairs on the rear of the pastern, fetlock, and along the back of the tendon should be removed. Do this by following the pastern line up from the heels, through the fetlock, and up along the tendon to about the lower one-third of the knee or hock.

It takes three or four clipper swaths to do the pastern and fetlock and only a single swath up the tendon. Any abrupt edges should be feathered by reversing the intended angle of the blades and combing the hair toward you.

Complete both front legs and then repeat the same sequence for the rear.

CAUTION: You are vulnerable, kneeling below the horse, and in a cramped position. It is easy to get careless. Remember to stay to the horse's side. Do not clip from directly in front of or directly to the rear of either set of legs. It is more convenient to do so, but very dangerous and foolhardy. Stay to the side and reach.

The horse may pick up his foot unexpectedly and hit you with it, either on the way up or on the way down. Perhaps he thought he had been cued to lift his leg, as when a farrier touches it, maybe flies were bothering him, or maybe he is just plain ornery and unruly. Whatever the reason or direction, it still hurts. Keep your free hand firmly on the leg being clipped so that you can react to his movements. You can cut the horse with the clipper blade, especially on the rear of the pasterns, if he should jerk his leg upward. Keep your hand on him and keep alert.

Anticipate the presence of the *ergot,* a horny, scale-like projection about the size of a pencil eraser, at the lower rear of the fetlock. Take care not to crudely jab into and dislodge this growth. It will not harm the horse, but it will aggravate him, cause some momentary distress, make him harder to work with next time, and identify you as a careless horseman.

Clipping the Head Region

1. As you begin these specific clipping sequences, keep in mind that improvements in the horse's health, comfort, and neatness are your goals.

Be certain that the mane is thoroughly combed out and lying approximately as you would prefer it. Use a rag to wipe out the ears and nostrils and to wipe off the face and jaws.

2. Begin this portion of the clip with the bridle path. After the legs, the next least sensitive region in the management clip is the neck. If there is a bridle path already present, and you like its dimensions, simply copy it. Be careful as you approach either end of it. A sudden movement of the horse can remove a lot of hair and create a whole new look to things—and it may be bad!

If there is no bridle path, you must start from scratch. Use these guidelines initially. Directly between the horse's ears, running from right to left, is a boney ridge called the *poll.* To the front of this "poll line" is forehead; to the rear is neck. Never go farther forward than the poll line with your clippers. When you begin, stop 1" or so behind this line.

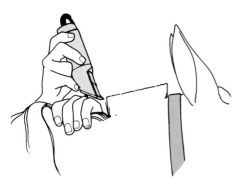

CLIPPING THE BRIDLE PATH

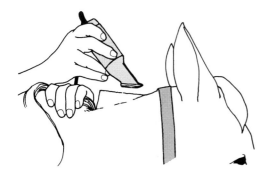

You can refine it later. An "ear's length" is a good estimate of how much mane to remove to the rear of the poll line. Stop a little short of a full ear length in your initial cut.

3. Step back and appraise your preliminary path. You'll note that it probably needs to be moved forward, up to the poll line, for it to look just right. Be careful as you cut hairs toward the poll. They cannot be replaced immediately. Your wishes determine the rearward extent of the bridle path.

As you make these cuts, press firmly so that you will obtain as short a clip as possible. If necessary, flip the clipper over and comb (feather) the body hairs on the neck into your clipped path.

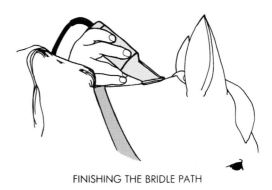

FINISHING THE BRIDLE PATH

CAUTION: To clip between the horse's ears, you must reach a long way up and crowd very close to the horse. You may be tempted to stretch the safety zone. Avoid doing this and pay particular attention to your feet, which are now very close to the horse's.

4. The ears should be clipped at this point. Keep in mind that this may be a very difficult task. Unless you have thoroughly and properly conditioned him, the horse will not allow this without twitching. Fortunately, the management clip demands very little in the way of ear treatment. Nature designed ear hair to help keep out wind, snow, rain, bugs, and flies. We should only trim it, not entirely remove it,

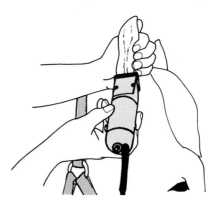

CLIPPING THE EAR

unless we are intensively caring for our horses by keeping them in a stall most of the time.

Gently grasp the ear in your free hand and fold it shut lengthwise so that the front and rear edges are lying next to one another. Run the clipper upward along these edges. This will remove the hair that stuck out and made the ear unsightly. Repeat this once or twice, perhaps clipping in a downward motion against the hair lay. Release the ear and run the clipper around the rim of the ear. Do not "bob" the tip. Shape it to your wishes. Dress the base of the ear as necessary to remove the unwanted longer hairs.

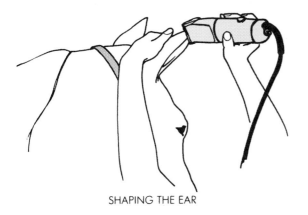

SHAPING THE EAR

CAUTION: Be as careful here as you would be in clipping any other area of the horse. If you have twitched the horse, be especially careful. Horses sometimes "blow up" when twitched, and not always immediately. A really mean horse will strike out when he is twitched. Stay in the safety zones and keep alert.

5. The last portions of the head region to be clipped are the muzzle hairs and underside of the jaw. The muzzle hairs will be the most aggravating for you to clip. The vibrating clipper tickles the horse's nose, and he continually twitches his muzzle, giving you a moving target to hit. To help accustom the animal to this vibration around his muzzle, start with the underside of his jaws. If the horse is well-mannered during clipping, the halter can be removed from the head and loosely connected around the neck. The lower jaw is V-shaped, with the base of the V being the lower teeth. The branches of the V extend back toward and to either side of the throatlatch. Depending upon the breed and size of horse, the widest distance between the branches is about 6". Clip the hairs close to the true bottom of the jaw, but avoid removing hair from the sides of these branches. As you handle the horse, this will become obvious.

If you are using a Number-10 Oster Clipper blade, do not indiscriminately clip against the lay of the

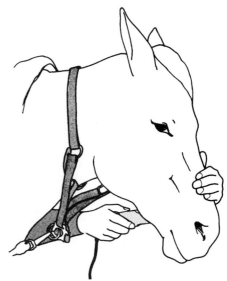

CLIPPING THE UNDERSIDE OF THE JAW

hair as you approach the outside of the branches of the jaw. You may remove too much hair and present too abrupt a stopping point. Either stop before the true bottom of the jaw branch or be prepared to flip the clipper over and comb (feather) the edge. Whatever your choice is regarding the edge of the jaw, be certain to remove the longer guardlike hairs from the underside. These start just behind the curb groove and extend to the throatlatch. Sometimes clipping with the hair will remove these; at other times you must clip against the lay. It is a slow and somewhat tedious procedure—be patient.

6. After you are satisfied with the jaw, move to the muzzle hairs. If you have a surgical blade, for a closer clip, switch to it at this point.

Start the muzzle clipping by placing the body of the clipper (if it is not hot) against the muzzle. Do this until the horse no longer twitches his nose vigorously. He probably will twitch to some extent at all times, but a couple of minutes spent getting him

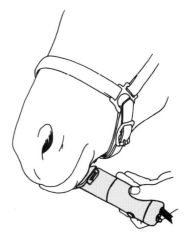

CLIPPING MUZZLES

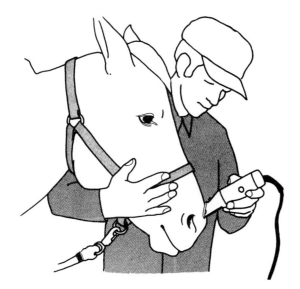

used to the tickle will yield large rewards in time saved. If you place your free hand, or even your whole free arm, over the horse's face (nose), it may help to calm his twitching. Begin clipping the long muzzle hairs from the underside of the lower lip, near the curb groove. Work your way up to the lips, and finally to the nostrils.

CAUTION: Do not attempt to hold the horse down with your free arm. You cannot do it, and to try is to invite injury.

Do not stand directly in front of the horse to clip the muzzle or jaw. Stay to the side and reach.

Take care not to poke the clipper into a twitching lip or nostril.

It is possible to clip a muzzle while the horse is twitched. Simply move the twitch once or twice and clip around it. If you bump the twitch chain with the vibrating clippers, the horse will usually be frightened anew and react by throwing his head upward. Be patient as you begin and hope that you can do it without a twitch.

7. There are hairs growing above and below the eye—not the eyelashes—that some people choose to remove. If you wish to remove them, cover the horse's eye with your cupped hand and pluck them out, snip them off with a blunt-tipped scissors, or use the clippers. Whatever your choice, be very careful not to injure the eye.

Clipping the Winter Coat

1. As you begin these specific clipping sequences, keep in mind that improvements in the horse's comfort and neatness and maintenance of his health are your goals.

As summer wanes and fall is approaching, the days grow shorter and the amount of sunlight diminishes. The decreasing day length triggers a response in the horse that causes him to prepare for cooler weather by growing a new winter hair coat. This new coat is an excellent insulator. It will keep him warm, but it will also make it a very difficult job for you to cool out your horse after exercise.

A resolution to this is to partially body clip the animal. The type of clip you give (amount and the location of body area you remove hair from) depends upon the environment you live in and the type of exercise you anticipate for your horses. If you remove a great deal of the winter hair coat from the horse's body surface, you will need to provide added warmth for the horse in the manner of a blanket or higher-quality shelter.

2. Prepare your clippers—the large shearing clippers. They should be newly sharpened, adjusted for quiet running (as possible), and oiled before you start. You will be removing a lot of hair, so have a stiff brush for cleaning out the grooves in the shears, a can of blade wash, an absorbent rag, and additional clipper lubricant.

3. If you are clipping in the fall, say late October to early November, the horse's hair coat will have thickened up a good deal. You will be removing a lot of hair, so the clippers may begin to run hot. Be certain to wash the blades often and keep the motor well oiled. Cooler clippers usually run more quietly. One clip in October–November should be enough for a partial body pattern. You will want to continue the 8- to 10-week management clips, however.

CAUTION: Just because you are trying to do your horse a favor by helping him control overheating during winter exercise, you cannot forget that your horse may vigorously object to the clipper noise, vibration, and smells. Approach the clipping carefully. Start the clippers while in his line of vision, well away from him—say 10 feet or more. Start and stop the clippers a few times. Then, approach him with the clippers while they are not running. Let him smell them. Touch his neck, chest, and withers with them. Move away again and restart the clippers. Approach the horse with them and touch those same areas with the clippers running. If he accepts them, fine; move to one of the clip patterns below and begin. If he does not accept them, keep working at it until he does. Even after you start, and he seems very calm about it all, be careful of your positions. Stay alert and stay in the safety zones.

Reread each of the cautions in this section. Horses can be unpredictable when they encounter clippers. Female horses are more sensitive, especially around their flanks, bellies, and udder area than males are. Take your time. Get help when it is necessary.

---

4. Groom the horse thoroughly. Get as much dirt, dust, and grime out of the hair coat as you can. The clippers will have to work much harder to clip a dirty horse, and they will run hotter and louder and need resharpening more quickly.

5. For even the simplest of the following clip patterns, use a stick of animal marking chalk to "lay out" the area of winter coat that you wish to remove. Balance and symmetry are easy to lose sight of when you are straining with the clippers, the horse is moving a bit, the wind is blowing the hair into your face, and you are sincerely wishing you had not started this whole process. Any farm supply store will have the chalk sticks.

6. *Chest-neck-throat clip.* Begin the clip on the chest floor, just where the legs tie into the shoulder/chest area. Clip the chest front, front of neck, and throatlatch. Also perform the management clip as explained previously in this section.

BELLY - CHEST - NECK - THROAT CLIP

8. *Harness/ranch clip.* If you work your horses hard in a sulky, training buggy, or sleigh, they are going to heat up and sweat, particularly under the harnessed areas. The shafts and traces of carts, buggies, and sleighs are just such areas. The harness clip will help reduce the heat up. It will most certainly shorten the cooling-out and dry-off period following the workout.

Ranch horses are worked in environments where they will be muddied; therefore the clip helps shorten the hair and the resulting cleanup in the body areas most likely to be dirtied. And, ranch horses will likely be kept outdoors on pasture during the colder months. This is why they are left with the hair coat intact on the back. Also perform the management clip as explained previously.

CHEST - NECK - THROAT CLIP

7. *Belly-chest-neck-throat-clip.* Instead of beginning on the chest floor, begin the clip near the inguinal area, just in front of the sheath of male horses and somewhere in front of the udder area of mares or fillies. You will be able to tell by the amount of hair cover. The amount of hair that you remove, to either side of the midline of the belly, depends upon your wishes and the amount of hair the horse has grown. The farther out you clip, away from the midline and up toward the sides of the horse, the more abrupt the edge will be where you do decide to stop. Finish this clip exactly as you did the check-neck-throat clip. Also perform the management clip as explained earlier.

HARNESS / RANCH CLIP

9. *Saddle or modified hunter clip.* The purpose of this clip is to provide for the horse that will be exercised vigorously and often across terrain that will most certainly muddy him. This is a clip for the horse that will receive much more than occasional use during the winter. This horse will be hunted through the fields several days per week. Note that the remaining hair coat closely resembles the outline of a hunt seat saddle. Also perform the management clip as explained previously.

SADDLE / HUNTER CLIP

Tips for Clipping Horses

- Reread each the cautions in this section. Horses can be unpredictable when they encounter clippers. Female horses are more sensitive, especially around their flanks, bellies, and udder area than males are. Take your time. Get help when it is necessary.
- Don't forget to lay out the areas to be clipped with animal chalk.
- Clip against the lay of the hair.
- Use overlapping strokes to reduce "clip lines."
- Do not clip close to the mane (from either side). Leave about 1/2 inch of hair to either side of where the mane hair starts. If you do not, the grow-back hair will look horrible.
- Pull loose skin taut as you clip over it. This will help with smoothness of cut.
- Hair whorls (cowlicks) will be present. Take your time and clip against the swirl from all directions necessary.

6.9 BATHING

Bathing the horse is a much disputed topic. There are those who never have and never will bathe their horses, and others who bathe their horses every week. Frankly, if both groups care for their horses equally well, you cannot tell the difference under normal circumstances. Whichever side of the argument you take, there is at least one very good reason for bathing a horse. That is to apply a medicated shampoo to treat a generalized skin condition. Whatever is involved in making the decision for your horses, do not make the mistake of transfer-

ring human enjoyment of the bath to the horse. A thorough daily grooming is just as invigorating and probably just as healthful as a bath.

Equipment Necessary

- Halter
- Lead rope
- Tie area
- Wash bucket
- Hose
- Shampoo
- Scrub brush, sponge, wash mitten
- Water scraper
- Grooming equipment

Restraint Required

Normally, only the halter, lead rope, and tie area or an assistant are needed. You could twitch, ear, or tie up a leg and wash the horse, but in practice, nobody does that on a continuing basis. If a horse is that ornery, stop showing him so you don't have to bathe him. For the most part, horses do not resist bathing. Some may pull or step away from rinsing, but this is a matter of a commonsense approach to getting the job done. Any animal, or any person for that matter, will step away from an abrupt blast of water.

Step-by-Step Procedure

1. Assemble all of the equipment necessary at the washing site. Ideally, this should be a hard-surfaced area with a drain or slight slope.

2. Arrange for water as close to the horse's body temperature, 101°F, as is practical. Realistically, no one uses a thermometer. They dip their hands in and, on a scale of cold to hot, blend until it is warm. There are a lot of horses washed with water straight from the well, cistern, or city water supply with no temperature modification, and they usually do just fine. The only problems are chilling the horse on a cold day and using bitterly cold well water on a hot day after a workout. Both of these can be avoided by using common sense. Don't bathe a horse with cold water on a cold day. On a hot day, use cold water only after a normal cooling-out period, and then be sure to start at the legs.

3. Comb out the mane and tail and curry out the worst of the dead hair and dirt from the horse's body. Use a stiff brush to remove excess mud and manure from the hooves.

4. Thoroughly wet the horse's entire body, mane, and tail. This is best done with a hose, perhaps with a rubber curry attachment, but a sponge and buckets will work. Some of the dead hair and dirt will be rinsed away by this wetting.

5. Dribble a lanolin-based livestock shampoo along the topline, or use your hand to apply it. Do not use an excessive amount; you can always add more as needed.

6. Start rubbing the horse's body with the scrub brush or wash mitten and work up a good lather. If it appears that you are not using enough shampoo, try more water instead. Some livestock shampoos need plenty of water to produce a good lather. Wash the horse's *entire* body, including the sheath or udder, recto-vaginal area, mane, and tail, but excluding the face, ears, and muzzle. The latter are best washed and rinsed separately.

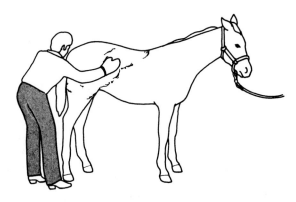

7. As you proceed, do not let the lather dry in one section as you do the next. Finish a section or two, then go back and spray one of the first sections so that it stays wet until you are entirely finished with the body.

8. Thoroughly rinse the horse from top to bottom. Keep rinsing until vigorous circular rubbing no longer brings lather bubbles to the surface. The underline, mane, tail, and legs are usually the hardest to rinse. If you do not remove all of the shampoo, the hair coat will be dull because of a gummy, dust-collecting residue.

9. Now is the time to apply a rinse coat-conditioner if you so choose. Because this is a management-type wash and not an often-repeated show wash, it really isn't necessary. Any natural oils removed from the hair by occasional shampooing will be replaced in a few days.

10. Use the sponge or a rag and clean water to wash the ears, face, and muzzle. If there is some particularly stained area, spot-clean it with some shampoo. Better yet, use one of the "wetless" shampoos for this task. Take care to avoid getting it in the horse's eyes.

11. Strip the excess water from the horse's body, mane, and tail. Use your hands for the mane and tail, and an aluminum sweat scraper, or better still, a section of broken fan belt, for the body. The piece of fan belt should be about 24 to 30 inches long. Because it is not rigid, it will mold itself to the horse's body contours.

12. Allow the horse to stand and dry; make sure he cannot lie down and roll. In chilly weather, or if time is critical, the horse can be force-dried with an animal vacuum/drier or with your own hair drier. Do not overuse heated air to dry the hair; this has a tendency to dry the oils from the hair and skin.

6.10 SHEATH AND GENITAL CLEANING

The sheath and penis of all male horses need periodic cleaning. There are sebaceous glands lining the sheath that produce an oil-like substance called sebum. When this mixes with dust, dirt, and sloughed skin cells, it forms a black crudlike substance called smegma. Smegma can remain soft and waxy or become dry and brittle flakes. Some horses can live with this and have no problems. In others the penis becomes irritated and so sore that urination is difficult.

Smegma can build up to such a degree that a "bean" is formed in the sheath, near the end of the penis. This might be small, on the order of a half-inch ball. In worst-case scenarios, it becomes the size of a golf ball. When that occurs, urination is extremely painful.

When the sheath is kept clean, the odor emanating from the horse, especially when he is heated up from exercise, is improved. And, since the cleaning removes the buildup of dirt and urine from the folds of the sheath, the horse's overall hygiene is improved. Sheath cleaning also reduces the likelihood of a urinary tract infection. The task is not especially difficult, but it is often neglected because the horse owner does not know how to clean the sheath or chooses to ignore the responsibility.

If the horse's sheath is swollen, he does not let his penis down to urinate, or he does let it down and you observe flakes of smegma clinging to it, it is high time to clean the sheath.

Equipment Necessary

- Pail, 3-gallon
- Latex or nitrile gloves
- Cotton balls or rolled cotton
- Mineral oil
- Animal-safe washing soap
- Water hose or basting syringe
- Halter and lead rope

Restraint Required

Normally, no special restraint is necessary. This is especially true if you have anticipated this day and have accustomed the young horse to being handled about his sheath and on the sensitive skin between his legs, while you were completing his normal daily grooming. Ideally, you might have actually done a "light" cleaning of the sheath on one or more occasions. If you have not done so, then you will need to be patient and careful. You most certainly will need an assistant to handle the horse while you do the washing and rinsing.

CAUTION: Watch your positioning during this procedure. It is easy to get caught up in the washing process and forget about the fact that you are within easy reach of the horse's hind leg, should he kick out. Keep your hands on the horse and pay attention to his body movements. When he begins to get "light" on his near rear leg speak to him, lean on him, or otherwise get his attention. This is foreign to him, but it is not hurting him. Be patient, but as long as you are safe don't quit because he is moving about a bit. Having just said that, if this is beyond your confidence zone, then get help from someone who has done it before.

Step-by-Step Procedure

1. Have someone handle the horse, if this is the first time for him, or for you, or if he naturally resists this cleaning every time it occurs. Otherwise, secure his head to a post or sound tie ring.

2. Put on the nitrile gloves, saturate some cotton in worm water and animal wash soap, and wash the belly, umbilical area, and outside of the sheath. The water should be tepid—not hot, not cold. Wash these areas well. They should be clean anyway, and it will get him ready for the actual sheath cleaning. Avoid splashing too much water onto his legs, as this can sometimes frighten the horse.

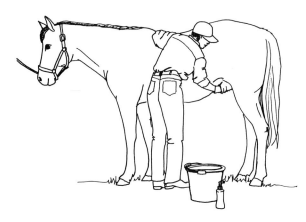

3. As the horse becomes accustomed to the external sheath washing, carefully slip your gloved and closed-fingered hand slightly into the inside of the sheath. This will be easy to do, if you glove and his sheath have soap on them. Don't try to go too deep right away. Just get the horse used to the intrusion.

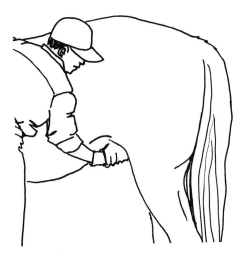

4. If the sheath does not feel like it is full of dried and rough smegma, the soap and water will likely be enough to remove the crud. If it is dry and rough, use the mineral oil. Pour some into your hand and work this into the inside of the sheath, or use the basting syringe to squirt it up into the sheath. Manipulation of the smegma will cause it to be softened by the oil. Then it can be removed with soapy washings. The oiling might have to be several times. Be patient. Realize that all of the crud does not need to be removed on this washing.

5. Washing the horse's penis is a matter of quickly and gently washing it with soap and water while he has it let down at the end of urination or while relaxing. Pay particular attention to any smegma-encrusted areas. It can be difficult to get more than a few seconds' time to complete this task. It is much easier to wash the stallion's penis prior to breeding. He is likely aroused because he recognizes the washing as part of the prebreeding ritual. Stallion or gelding, it is important to not be rough while washing the penis.

6. After the washing is completed, thoroughly rinse the inside of the sheath using wetted cotton, fresh tepid water, and the basting syringe to repeatedly squirt water into the sheath. It is important to keep rising until you are certain all soap residue has been removed.

Postprocedural Management

There are no essential postprocedural management practices necessary. However, if this has been a particularly difficult session because you have neglected to accustom the horse to the washing, then the time to start preparing for the next washing is now. Handle the sheath and scrotal area between the legs each day while grooming.

6.11 WRAPPING THE TAIL

This technique involves enclosing or wrapping the horse's tail in some sort of material. The materials available are many: elastic bandages, cotton "track" bandages, synthetic meshlike foam, terry cloth, and flannel strips. The normal widths for tail wraps are 3" to 6", and they should be at least 8 to 10 feet long.

There are several basic reasons for wrapping a tail, including breeding a mare, foaling a mare, veterinary surgery, to prevent damage to tail hairs from rubbing against the trailer butt-bar, to prevent soiling during shipping, and as a tail-training aid (spray and wrap the tail to make it lie flat and with no frizzies).

If you are wrapping for breeding or foaling, you will want to keep the tail dry as well as keep it out of the way. If so, consider using a plastic, full-arm inseminator's glove or bread bag. These can be taped to the tailhead and work quite well. The method certainly is much quicker than wrapping.

Equipment Necessary

- Halter
- Lead rope
- Tie area
- Tail-wrap material

Restraint Required

Usually only a halter, lead rope, and tie arrangement are needed. If the horse is a show-type animal, it is used to being handled. If the wrap is for surgery, and the animal is rank, wrap the tail after anesthetizing him. If the wrap is for breeding or foaling, and the mare won't accept the wrap, twitch her and then wrap the tail.

Step-by-Step Procedure

This sequence illustrates a wrap for management, not show purposes. If tail-training is your intent, other sources of information about what to spray with, how long to leave the tail wrapped, and so on, should be used.

1. Choose the wrap best suited to your needs and be certain that it is rolled properly. Many out-of-the-box wraps are reverse-rolled; that is, they have their Velcro tab or tie strings exposed for you to see. This is fine, because you can see what you are getting. Just be certain to wrap them into the inside, where they belong, before you finish the tail and can't figure out how to fasten it. A careful analysis and a trial run or two will solve the problem for you.

2. Hold the rolled-up wrap in one hand and the loose end in the other, with your hands about 10" to 12" apart. Place this single layer beneath the horse's tail and move it up as close as possible to the head of the tail.

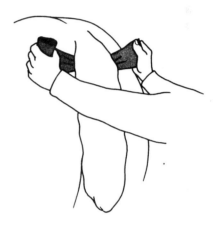

> **CAUTION:** Lift the tail when you do this, and also for the next few wraps, so that you do not rumple the hairs and have them kinked under the wrap. This will cause some discomfort and could cause the horse to rub the tail in an attempt to dislodge the source of aggravation.
>
> Do not stand directly behind the horse to wrap its tail. A well-trained show horse may not surprise you, but a mare in heat or one about to foal probably will. Stand to the side and reach.

3. Place the *end* of the wrap over the tail so that its end rests at about the middle of the tail. Holding this end snugly down with a single finger or two, pull the roll end of the wrap over it. As you cross the loose end, the material-on-material friction will take the place of your fingers. This first wrap or two is important, so do this carefully. It should not wind up being so loose that it immediately slips and falls away.

It makes no difference whether the roll is unwrapping from the top or bottom, provided that you have thought ahead and arranged for the Velcro tab or ties to come out right-side up. It is possible to

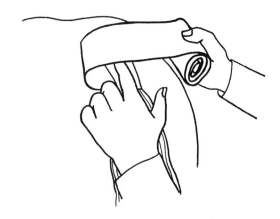

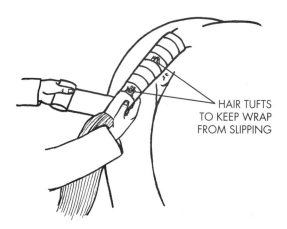

→ HAIR TUFTS
TO KEEP WRAP
FROM SLIPPING

CAUTION: Do not overreact to the possibility of a loose wrap by placing the tail in a stranglehold. The idea is only to have the wrap stay in place. Use only enough pressure to accomplish this. Practice a time or two, before the real need, so that your muscle, your wrap material, and your horse's tail are coordinated. There is a large blood vessel beneath the tail, especially prominent at its head, that you do not want to occlude.

have certain types of retainers come out wrong and be unable to fasten the wrap. Practice rolling these wraps until you get it right.

4. For the next wrap or two, do not move down the tail. In fact, if it is possible, actually wrap *toward* the tailhead. This will help to hold up the wrap.

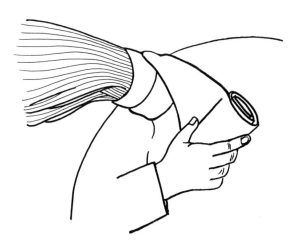

5. Begin to move on down the tail with your wraps. Keep the tension uniform and the degree of overlap constant. Only your own experience with your wrap material and your wishes about how far down to wrap can tell you how much overlap is best. It is probably a reasonable guideline never to leave less than an inch or more than half of the total width of the wrap showing.

6. Before you get too far down the tail, grab a pinch of hair (a dozen hairs or so) from just below your wrap and lay them upward onto the wrapped area of the tail. Pull your next wrap directly over these hairs. They serve as an effective "keeper" to prevent the whole wrap from slipping down and finally off the tail, much like a shed cocoon. It is probably a good idea to do this about every 6" to 8" of wrap. This will ensure the wrap's staying on, even on an up-and-down, nervous, foaling mare.

7. Decide how far down you are going to wrap. Your choices usually are to wrap to just below the end of the tailbone or to enclose the entire tail, including the switch. If you choose not to include the entire tail, the wrap is tied off with the velcro tab, strings, or clips, usually 2" to 3" below the end of the tailbone. The "switch," or broomlike end of the tail hair, is left free. This is the usual breeding and shipping wrap. If you enclose the entire tail, proceed as before up to the end of the tailbone, then fold the rest of the tail hair up onto the wrapped portion of the tail and enclose both with the remaining material by wrapping back up the tail. Tie this off as before.

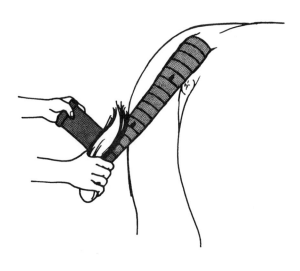

CAUTION: Do not leave the tail in this wrapped condition for long periods of time (more than 10 to 12 hours). At that point, remove it for 30 minutes to 1 hour, and then replace it as necessary. If you have applied the wrap extremely tightly, this "safe" time may drop to no more than an hour or two. Keep in mind the blood supply on the underside of the tail. Your efforts should be directed toward wrapping the tail only tightly enough to keep the wrap in place.

6.12 WRAPPING LEGS

This technique, sometimes referred to as *leg bandaging*, involves enclosing or wrapping the horse's legs in some sort of protective and supportive material.

There are several reasons for wrapping or bandaging legs, and the names of the various kinds of wraps indicate their purpose. They include sweat bandages, water bandages, exercise and running bandages, support bandages, and protection bandages. All of these are applied in essentially the same manner, and the first five are only specialized modifications of the last two.

Two types of material normally are used for wrapping legs, one for padding and the other for support. Padding material can include roll cotton, quilted pads manufactured expressly for padding, or disposable diapers, minus the plastic outside covering. Manufacture your own, but be certain that they are soft and lofty, have no harsh, cordlike edges, and measure approximately 15" × 20". The wrapping material is usually an elastic bandage, "track" bandage, synthetic meshlike foam, flannel strips, or gauze.

There is also a type of wrap designed to protect the horse's legs from flies or other buzzing insects. The brand name currently on the market is "Fly Wraps." They look like a velcro-attached shipping wrap. The attachments are velcro, but the material "protecting" the legs is a close-weave mesh. They are lightweight, durable, washable, and they work.

The discussion in this section covers wrapping or bandaging the leg for protection and support. A good example of when this type of wrap would be necessary is during transit. Protection, especially for the pastern and coronet areas, and support are especially important when the horse is being transported.

Equipment Necessary

- Halter
- Lead rope
- Tie area
- Padding
- Wrapping material

Restraint Required

Usually only the halter, lead rope, and tie arrangement are needed. Show animals, which routinely are hauled and therefore routinely wrapped, should present no problem. If an individual horse should continue stepping away or fidgeting beyond the limits of your patience or time, mild restraint becomes necessary. Consider lifting a leg, other than the one you are wrapping, off the ground so that the horse is nearly forced to stand still. If this fails, use the nose twitch.

Step-by-Step Procedure

1. Assemble the equipment before you begin. Be certain that the padding is adequate (three or four layers of roll cotton or its equivalent) and free of chaff and other debris. Position the roll of wrapping so that when you are finished with the leg, the tie device is positioned properly. Don't count on a new roll's being correctly rolled. Unwrap it, figure it out, and rewrap it (see Section 6.11).

2. Thoroughly groom the lower leg, starting from the knee and hock, and continue on down to include picking out the feet. Even though you are going to place a pad between the leg and the wrap, rumpled hair or a spot of caked mud will cause abrasion under the snugly wrapped leg.

3. This technique requires practice. If you persevere and practice the technique, you will discover that the only difficult, "takes-three-hands" part of the whole thing is the beginning. It is difficult to properly position the padding and keep it in place, properly position the first end of the wrap and keep it in place, and then wrap evenly up and down, preventing any slipping or wrinkling from occurring during this whole juggling process.

CAUTION: Take care that you do not become so absorbed in your work that you forget how to work safely around a horse. It is going to be much easier for you to "cheat" and crawl in front of, under, and behind your horse's legs. Do not allow this to happen. It is no less dangerous to you just because you are trying to be kind and helpful to the horse.

4. Place the padding around the lower leg so that it extends from just below the knee to about 1" *below* the coronet. It may touch the ground near the horse's heels. Arrange the padding so that its edges are on the sides of the legs.

CAUTION: Be sure that the edges of the padding stay on the sides of the horse's legs. If they should slip to the rear and place pressure on the tendon, a

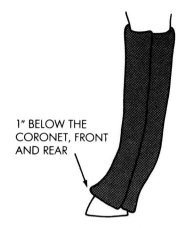

1" BELOW THE
CORONET, FRONT
AND REAR

"bandage-bow" or "tie-bow" can result. If they should move to the front and over the cannon, they will rub the bone and set up a painful localized inflammation. This same caution applies to the wrap fasteners; be certain that they stay on the outside of the leg.

The padding material must extend to below the coronet. If it does not, there is no protection over the coronet or heels. These are the areas that the horse most often injures during transit.

5. There can be a lot of discussion about where to begin wrapping. Some say always begin at the top, others say always begin at the bottom, and still others say start in the middle. While that discussion rambles on, start in the middle and be done with your support and protection wrap before the debate has subsided.

6. Open the padding slightly, being careful to keep it as tightly wrapped as you can, and insert 2" to 3" of the wrapping material under the flap. Close the flap and hold it in position over the end of the wrap. Take the roll of wrapping around the leg and back over itself on your side. The wrap insert should stay in place. Continue wrapping down the leg, leaving about half of each preceeding wrap layer uncovered by the next. The amount of tension to apply is a matter of experience. If the wrapping is

too loose, it gives no support; a wrapping that is too tight, with inadequate padding or wrinkles, also causes problems. The best advice is to pad thickly and smoothly and then snug the wrap up firmly and evenly. The only types of wrap that are likely to be too tight are the true elastic, athletic wraps, and perhaps some of the synthetic stretchable mesh types.

7. When the wrapping reaches the fetlock, go under and around the back of it, and then completely around the pastern once. Be sure to leave about 1" of padding unwrapped at the bottom. Then start back up from the opposite side of the pastern. Continue wrapping up the leg until you are within 1" of the top edge of the padding. From this point, if there is any wrap left, return back down the leg until it is used up. Tie the wrap or fasten it in the appropriate manner on the outside of the leg. Do not secure the fasteners more tightly than the rest of the leg is wrapped.

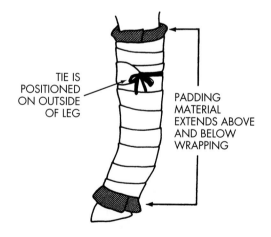

TIE IS
POSITIONED
ON OUTSIDE
OF LEG

PADDING
MATERIAL
EXTENDS ABOVE
AND BELOW
WRAPPING

CAUTION: If you tie on the front or rear of the cannon, you are inviting the bone and tendon problems mentioned in the CAUTION for step 4. If the fasteners are positioned on the inside of the leg, the horse will find a way to brush his legs together and unfasten them.

Leg wraps, even properly applied, are not meant to remain on the horse over long periods of time. Unless advised otherwise by a veterinarian when the wrapping is used for surgical bandaging or wound care, remove leg wraps for 30 to 60 minutes every 12 to 18 hours.

If you do not leave the padding extended above and below the wrap material, you run the risk of cutting off the circulation in the horse's lower leg. This is especially true if you use any kind of elastic wrap material.

6.13 TRIMMING HOOVES

Trimming the hoof refers to the removal of excess growth of the hoof wall and "dead" sole from the foot of the horse. Growth of the horse's hoof is a continuous process, and while normal activity of the horse, especially on dry, hard surfaces, keeps the hoof length near normal, you must occasionally intervene and remove what the horse has not worn away. At least, that is the sequence of events for the barefoot or unshod horse. Trimming unshod feet must normally be done every 8 to 12 weeks. This timing varies with climate, nutritional level, and the individual horse.

Trimming of the horse's feet also takes place in preparation for shoeing, to correct for a structural defect in a young horse (for example, splayfootedness), or to reshape a cracking or misshaped hoof. Each of these three types of trimming involves procedures and knowledge that are beyond the scope of this discussion. Horsemen are encouraged to learn these procedures and acquire this knowledge from their farriers. The intent of the present discussion is to prepare the horseman to properly trim the barefoot, or unshod, structurally normal horse.

Equipment Necessary

- Halter
- Lead rope
- Tie area
- Farrier's apron or chaps
- Hoof-pick
- Hoof knife
- Hoof nipper
- Hoof rasp

Restraint Required

Usually, no restraint other than the halter, lead rope, and tie area is required. It does seem, however, that if an otherwise normal horse is going to act up, it is usually at footwork time. The use of a knowledgeable assistant usually is better than simply tying the horse. (See Section 6.6 for handler positioning during footwork.) If restraint measures are necessary, the assistant should apply the nose twitch (see Chapter 1, Section 1.6).

Step-by-Step Procedure

1. Assemble all of your equipment before you begin. Be certain that the hoof nippers are well oiled at their pivot point and that the cutting edges meet and are not so badly nicked as to prevent cutting. The hoof knife should be as sharp as you can make it with a fine-toothed metal file. If the teeth of the hoof rasp are clogged with debris from any source, remove it with a wire brush. Foot trimming, even with sharp, efficient tools, is hard work. It makes no sense to use dull or damaged equipment.

2. Protect your clothing, especially the knees of your pants, by wearing a farrier's apron or chaps. You may also wish to wear a pair of leather work gloves. Your fingers and knuckles, especially the second joint of your thumb, are prone to being rasped or otherwise damaged.

CAUTION: Sew or tie a "weak link" into the belt and leg loops of your farrier's apron. With this weak link in position, the entire apron can easily be pulled free from your waist. You may never need this, but you cannot risk becoming entangled with the horse because he got his foot caught up in a leather garment that could not be ripped free when he started to jerk about. You are going to have open clinches and a partially removed shoe at some point in this technique; these can easily snag your apron. It is also possible for a horse to slip his leg into a particularly baggy leg loop of your apron. A workable weak link is a loop of rawhide of a very small diameter ($\frac{1}{16}$″).

If you wear the shotgun-type work chap, the weak-link idea doesn't work because of the zipper in the legs. If your chaps fit snugly, catchable or snaggable places are minimized, but the opened nail clinch and shoe end still could catch, tear, and poke through. If you have this type of chap, extra care becomes necessary. If you plan on doing your own shoeing someday, invest in the apron now.

3. Approach the restrained horse, grasp and lift a leg in the proper manner, and place the leg between your knees.

CAUTION: See Section 6.6 for the procedure to follow and precautions to be observed when lifting the horse's legs.

4. Clean the bottom of the foot thoroughly, using the hoof-pick or an old, worn-down, dulled hoof knife. Normally, most of the caked manure, mud,

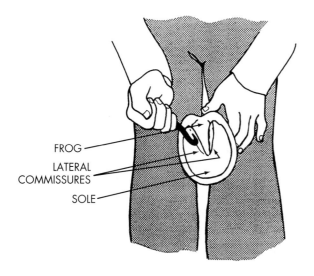

FROG —
LATERAL
COMMISSURES —
SOLE —

stones, and any other type of debris is easily removed. At any rate, it all must be removed. Pay particular attention to the lateral commissures and central cleft of the frog, to the sole at the junction of the bars with the hoof wall, and to the sole at the toe immediately behind the hoof wall. The commissures and cleft of the frog are often poorly cleaned because horsemen are fearful of bearing down on the hoof-pick. These must be thoroughly cleaned if thrush is to be prevented. When the sole of the hoof is clearly visible, and when the hoof-wall–hoof-sole interface is clearly defined, you are ready to proceed to the next step.

5. Grasp the sharpened hoof knife in either hand and examine it. You'll note that the handle and blade are curved. This is to accommodate the farrier in his attempts to trim the circular and concave (on the rear feet) soles of the horse's hoof. The knife may curve to either the right or the left; either

type of knife will work, regardless of whether you are right- or left-handed. Pick both up, handle them, and see which feels "right" for you. It boils down to which hand pulls best and which pushes best. Efficient use of the hoof knife involves two hands—one grasping the handle and applying the main force, the other guiding and pushing the blade with the thumb. Also note that as you begin to cut, the hooked tip of the knife blade must be kept pointed in the direction of the frog if efficient and safe cutting is to take place. Some people may choose to keep the knife in one hand for the entire hoof, others choose to switch at or near the toe. Switching hands does allow for more controlled blade placement, particularly at the sole–wall interface.

Begin paring away thin slices of the white, flaky, dead sole at one heel or the other. Pare away the bars until they are nearly flush with the sole. (This is to prevent their being torn away from the sole when the horse travels rough, rocky ground.) Continue around the sole of the hoof to the other heel. Be certain to create a sharp division between sole and hoof wall, because for the most part this overgrowth will serve as your guide to how much wall to remove. Enough sole has been removed when the flaky, white material changes to a nonflaky, bluish, and glossy textured "young" sole. Scoop out or cup the soles of the feet only a slight amount. In the natural state, the horse has nearly flat front feet and slightly cupped (concave) rear feet. Unfortunately, domestic horses do not have the strength and substance of sole that their wild ancestors had. Cupping or lowering the soles allows the horse's weight to be borne largely by the hoof wall rather than by the sole alone.

6. Using the hoof knife in one hand and guiding with the fingers of your other, trim up the ragged edges on the frog. Because of the normal wear and tear of the pasture, there may be no ragged edges; in that case, do not trim the frog any further. The

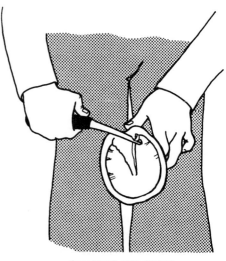

SHAPING THE FROG

purpose of trimming on the frog is to open up the commissures and cleft for proper cleaning—nothing more.

CAUTION: When working around the frog, you are getting close to a soft, easily cut area. The sole of the hoof can be hard as flint, and it is very likely that you will be pulling and pushing very hard to remove its excess. A slip of the knife at this point could remove a lot more frog than you really want to. This is why the tip of the knife, and not the butt, is positioned near the frog, why the two-handed push/pull technique is recommended, and why the knife should be very sharp. It also makes sense for you to stop when your knees begin to shake from fatigue. You'll rush things as you go beyond your endurance, thereby increasing the risk of injury. Remember that your hands and legs are also vulnerable to cuts; be alert and take care!

---

7. Place the open nippers so that the inside cutting edge is resting at the cleaned sole–wall interface. Tilt and position the handles of the nippers so that the cut you are about to make will be parallel to the bearing surface of the sole. If you position an imaginary line exactly in the middle of your open nipper handles perpendicular to this bearing surface, your cut will come out about right.

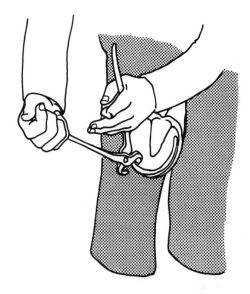

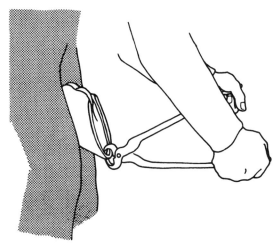

Firmly close the nippers until they click, signaling completion of the cut. For the next cut, move right or left and use the cut just made as your guide for continuing. Continue cutting to one heel or the other, then return to the toe and repeat the sequence toward the other heel. Do not tear away each "bite" with the nippers; rather, try to remove the excess hoof wall in one continuous "horseshoe." Besides looking nice and professional, this also shows that you have made clean cuts of even depth. This means much less work with the rasp to level the ground surface of the hoof.

CAUTION: First of all, don't push yourself beyond your physical ability. Two feet per session are plenty, until you become accustomed to the strain on your legs and back. When you push yourself too far, you will rush and become sloppy in your nipper cuts and actually create extra work for yourself. More important, you may remove too much wall, especially in the quarters of the hoof.

The toe of the hoof appears to grow more rapidly than the heel. Perhaps the real truth is that the heel and toe grow at the same rate, but the wall at the heel is thinner than at the toe, so the heel is simply worn away faster. At any rate, with normal foot wear, this means that more wall should be taken off of the toe than from the heel if you want the hoof to remain at the same angle it was before trimming. Most of us will want to do this. *If you begin at the toe, as outlined here, you must make a conscious effort to lessen your cut as you approach the heel.* If you begin at the heel, as some do, you must deepen your cuts as you move toward the toe. If you created a sharp sole–wall interface with your knife, this edge serves as a good guide for depth of cut.

---

8. The hoof should now be "sighted" and leveled as necessary with the coarse side of the rasp. *Sighting* refers to looking at the heels of the hoof, held between your legs, and gauging the depth or length of each heel. They should be equal. If one is longer than the other, the rasp must be applied more vigorously to that side.

The hoof should also be balanced when viewed from the front. If corrective trimming is performed to correct unusual hoof wear or an anatomical abnormality, the hoof may be left in an unbalanced condition. Corrective trimming should be attempted only by an experienced hoof trimmer or farrier.

10. Use the fine side of the rasp to remove or round the sharp edge of the leveled hoof wall. This is accomplished by placing the teeth of the rasp against the sharp edge at a 90° angle and "dubbing" (rasping) it back. If this is not done, the sharp edge will tend to bend, chip, or break away when the horse travels on rocky ground.

9. Rasp the entire ground surface of the hoof, being certain to achieve an absolutely level and balanced surface. The following guidelines will help you to achieve the levelness and balance necessary: (1) always rasp two points at once; (2) always rasp at least part of the heel and toe at the same time; (3) never rasp directly across the hoof, from quarter to quarter; and (4) be sure to rasp in both heel-to-toe and toe-to-heel directions.

6.14 SHOE REMOVAL

This technique is the removal of the iron rim or plate (horseshoe) from the bottom of the horse's hoof. Done properly, it involves more than simply pulling on the horseshoe. In fact, if "simply pulling" is the technique chosen, much harm could be done to the hoof itself.

It becomes advisable or necessary to remove a shoe for the following reasons: (1) one or more shoes have become loosened and the nail holes have deteriorated to the extent that they cannot be reclinched; (2) the horse has injured its foot, and the only way to treat it properly is to remove the shoe; and (3) it is your management decision to pull off all the plates and allow the horse to benefit from the barefooted exercise.

CAUTION: Even if there is an injury to only one foot or a loose shoe on only one foot, be sure to pull off the other member of the pair if it is going to be a week or more before replacement. The horse was not meant to have varying lengths, weights, traction, or flight patterns between paired feet.

Equipment Necessary

- Halter
- Lead rope
- Tie area
- Farrier's apron or work chaps
- Clinch cutter
- Hammer
- Shoe puller—old nippers, pull-offs, or pincers
- Rasp

Restraint Required

The halter, lead rope, and tie area usually are the only restraints required, but it does seem that if an otherwise normal horse is going to foul up, it is usually at footwork time. A knowledgeable assistant, on the head, is a better choice than simply tying the horse. (See Section 6.6 for handler positioning during footwork.) If restraint measures are necessary, the assistant should apply the nose twitch.

Step-by-Step Procedure

1. Assemble all of your equipment before you begin. Be certain that your seldom-used nippers open and close easily and that your clinch cutter has a modest edge. There is no reason for it to be sharp, but if its last use was as a steel chisel, it may need

CAUTION: Sew or tie a "weak link" into the belt and leg loops of your farrier's apron. With this weak link in position, the entire apron can easily be pulled free from your waist. You may never need this, but you cannot risk becoming entangled with the horse because he got his foot caught up in a leather garment that could not be ripped free when he started to jerk about. You are going to have open clinches and a partially removed shoe at some point in this technique; these can easily snag your apron. It is also possible for a horse to slip his leg into a particularly baggy leg loop of your apron. A workable weak link is a loop of rawhide of a very small diameter ($^1/_{16}$").

If you wear the shotgun-type work chap, the weak-link idea doesn't work because of the zipper in the legs. If your chaps fit snugly, catchable or snaggable places are minimized, but the opened nail clinch and shoe end still could catch, tear, and poke through. If you have this type of chap, extra care becomes necessary. If you plan on doing your own trimming and/or shoeing, invest in the apron now.

shaping up. Wear leather chaps or a leather farrier's apron to protect your jeans (and your hide).

2. Approach the restrained horse, grasp and lift his leg in the proper manner, and place it between your knees.

CAUTION: See Section 6.6 for the procedure to follow and precautions to observe when lifting the horse's leg.

Place the clinch cutter edge next to and, if possible, under the bent-over (clinched) portion of a nail. Hold the edge of the cutter as nearly parallel to the hoof wall as possible to prevent cutting into the wall. Rap the head of the clinch cutter sharply with a hammer until its edge and the force of the blows straighten out the bent-over nail. Repeat this for each of the nails. (There are usually three or four nails in both the inside and outside walls.)

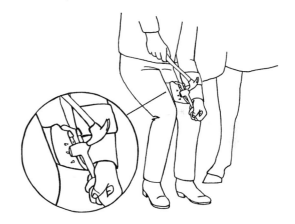

CAUTION: Be certain to count the number of clinches you open and to see that this number corresponds to the number of nailheads in the crease of the horseshoe. You do not want to leave a nail remaining in the hoof wall after you have pulled the shoe. Sometimes the head of a nail will pull through the shoe or break off when the shoe is pulled, leaving the long piece in the wall. If this should happen, get a pair of pliers and pull the nail out by the stem.

3. Insert the jaws of your pullers under either heel of the shoe. Close the puller handles so that the jaws are now nearly closed and are positioned between the hoof floor and horseshoe. Loosen the shoe by applying pressure downward and inward (toward the frog) on the puller handles. Use only one hand in doing this. The free hand should support the toe of the hoof to avoid damage to the coffin and pastern joints. This is especially important when using inward pressure because the torsion (twisting) may cause damage to a tendon.

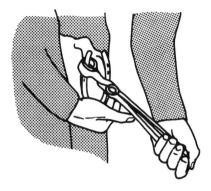

When the heel has been loosened, move the nippers up toward the toe an inch or so, and repeat the loosening. Now switch to the other heel and repeat for a couple of "bites." Keep moving from one branch of the shoe to the other and toward the toe until the shoe can be lifted free.

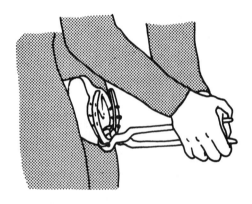

CAUTION: During the time you are loosening the shoe, try very hard not to have the horse jerk his foot away. You might get caught with the nails or shoe, and the horse could injure his foot when he places his weight upon the old shoe and protruding nails.

4. After you have removed the shoe, carefully count nails (the whole ones and the pieces) until you are satisfied that you have them all. Either salvage the pulled shoe or discard it, but be certain that it does not lie around causing clutter and being a hazard to safety.

5. At this point, you should pick-out the foot and properly trim it to withstand barefoot wear and tear (see Section 6.13).

6.15 EMERGENCY SHOE REPLACEMENT

This section describes the attachment (actually, the reattachment) of a horseshoe to the bearing surface of the horse's foot in an emergency situation. Normal shoeing, starting from scratch, involves science and artistry beyond the realm of everyday horse

management skills. Good horseshoers (farriers) spend many years and crawl under a lot of horses to perfect their skills. The horse owner or manager can hardly be expected to shoe skillfully, but he should be able to cope with the crisis of a thrown or pulled shoe.

If the horse pulls a shoe at home, the horse owner has the options of immediately calling the farrier for help, pulling the other front or rear plate, thereby getting rid of the imbalance, or replacing the missing shoe himself. If the horse pulls this shoe off out on the trail, 20 miles from the trailhead, on the second day of a week-long trip, the horseman has no option—he must replace the shoe himself.

Replacing a thrown shoe is within the capacity of any person who is willing to try and capable of following directions. Anyone who spends time on a horse's back, some distance removed from civilization's hustle and bustle, will carry a couple of extra shoes, one front and one rear, plus the tools with which to tack them on. Your farrier will sell you properly sized shoes and nails, and will even shape the extra plates if this is necessary. Explain to him why you want to learn, and he will show you how to nail the shoes on properly.

Saddlebags are an essential part of the gear for all trail riders. The entire shoeing paraphernalia will fit into these bags with room to spare. Clinchers and nipper handles can be shortened and a "shorty" rasp purchased for the trail. If this is done, the whole package will weigh no more than 4 pounds.

There is an alternative to replacing the thrown shoe that appeals to many horsepeople. It involves the purchase of an "easy-boot" or "overshoe" that is simply placed onto the bare foot of the horse much as you'd put on a boot. Straps and buckles hold it in position. The whole task takes no more than a couple of minutes and the "boot" can be carried in a saddlebag.

Equipment Necessary

- Halter
- Lead rope
- Trail chaps
- Horseshoes: one front, one rear (sized to fit your horse)
- Horseshoe nails
- Driving hammer
- Hoof nipper
- Hoof knife
- Rasp
- Nail clincher
- Easy-Boot® or other "temporary horseshoe" device

Restraint Required

Usually, no restraint other than the halter, lead rope, and steadying influence of the handler is required. Also recall that this is a saddle horse that should be trained to accept footwork. If restraint measures are necessary, the assistant should apply a nose twitch. (See Section 6.6 for handler positioning during footwork.)

Step-by-Step Procedure

1. Try to prevent the need to replace a shoe on the trail. It is not all that easy to shoe a horse, and your patience will probably wear thin, partially spoiling an otherwise nice day. Arrange for a routine farrier's visit shortly before the trail ride. You can check the shoes for obvious looseness by tapping lightly but sharply on them with a hammer. The sound should be a solid, dull thunk. If the shoe rattles loosely or gives off a slightly resonant ring, indicating that it is loose, it should be tightened before you leave.

2. As soon as you notice a shoe is missing, carefully examine the horse's foot. Check for any remaining nails or nail portions. Also check for missing portions of the hoof wall. It is not unusual to tear away small chunks of wall when a shoe is wrenched off without the clinches being opened. Clean the bottom of the hoof carefully and check for nail-puncture holes. The horse may have stepped on the shoe and protruding nails in the process of losing it. If you find evidence of puncturing, the holes must be cleaned and treated with iodine and the horse injected with tetanus antitoxin. If no punctures are present, nor any chunks missing, and if the horse is not lame from being footsore, you are lucky indeed.

3. Assemble your shoeing tools and secure the assistance of someone to help hold the horse. Be certain that this person understands where to stand during footwork (see Section 6.6). Take your hoof knife and begin to prepare the bare foot by picking, scraping, and cutting away any loose debris on the sole. The foot, since it was wearing a shoe until very recently, is likely to be nearly level. This is to your advantage and you should not change it. Take the rasp, and with a few strokes, heel to toe and then toe to heel, smooth out any raggedness that may have developed on the sole of the hoof. You must be certain exactly where the white line of the hoof is in order to nail on the shoe. This rasping will clarify that location for you.

4. Place the shoe in position on the sole of the hoof. Remember that your farrier previously sized and maybe even partially cold-shaped this shoe to your horse's foot. The shoe is in position if the wall of the hoof blends in all around with the hoof surface of the horseshoe. This fit may or may not be

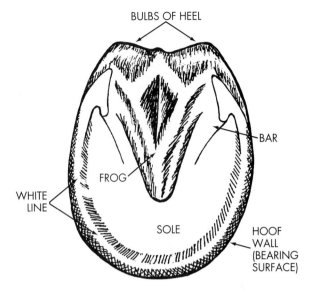

FRONT FOOT

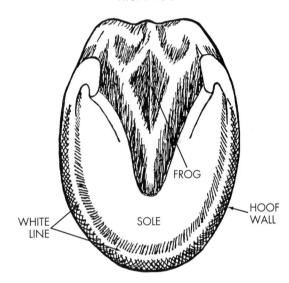

REAR FOOT

perfect. Make the best possible compromise. Another gauge of the shoe's being in position is that the white line of the sole is visible through the nail holes of the horseshoe. This is the more important of the two positioning gauges, and you should strive for minimum compromise here.

5. Hold the shoe in position with the edge of your left hand (for right-handed horsemen) and grasp the horseshoe nail between the thumb and forefinger of this same hand. Be certain that the checkered side of the nailhead faces the inside of the hoof. Position the nail in one of the heel holes with the point resting on or just slightly outside of the white line. Aim the point of the nail at a spot on the outside of the hoof wall about ¼" to ⅜" above the old nail holes. Tap the nail lightly several times to get it started. Now hold the toe and shoe in the

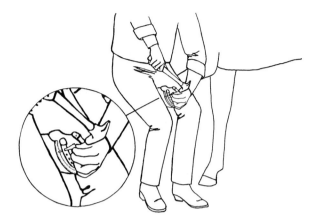

left hand (this helps to maintain the position of the shoe and helps you to control the horse's foot). Finish driving the nail until the head is firmly seated in the shoe and the point is protruding from the hoof wall. Use sharp blows to finish driving the nail so that the nail can "take" the direction of the bevel and come to the outside of the hoof wall.

6. Immediately bend the nail over by hooking the claw end of the driving hammer under the point and pulling it downward toward the shoe. Tap this bent nail toward the hoof wall so that it nearly lies against it. It is a sign of a practiced, professional farrier to drive each nail so that it will come out at exactly the same height on the hoof wall as the one before it. The goal for the emergency farrier is not quite the same. Your concern should be to have all of the nails come to the outside of the hoof wall—period! Most emergency farriers are too timid and will tap lightly for the entire length of the nail instead of rapping it sharply. Keep in mind that sharp raps are necessary for the nail bevel to be able to do its job: to orient the point of the nail outward. If the point does not surface, pull it back out, aim it a lit-

tle differently, and try again, this time using sharper raps. If this happens to you, watch for lameness and consider the administration of tetanus antitoxin—you may have quicked the horse. At the other extreme, if the nail exits too low on the hoof (say, less than ⅛" to ⅝" above the shoe), it should be removed and redriven. If you clinch in the nail over less hoof wall than this, you run the risk of the shoe's tearing away the wall when it is stressed by the moving horse.

CAUTION: There are several cautions here. The first is to appreciate the significance of the white line. This is the junction of the sensitive and the insensitive laminae of the hoof. You can nail *into* it, or you can nail to the *outside* of it, and the horse will feel no pain. If you nail to the inside of it, you "quick" the horse, causing sharp pain, quick reactions, and the likelihood of infection.

The next caution, actually related to the one above, involves the horseshoe nail. Recall that you placed it directly upon the white line, with the checkered side of the head facing inward. This checkered side of the head coincides with a beveled side of the point. The purpose of the beveled point is to help force the point outward, away from the bevel, as you drive the nail into the hoof. The point must exit the hoof wall if you are to anchor it and to be certain that you did not damage the internal portions of the hoof. Always double-check the orientation of the nail before driving it.

The last caution relates to the horseshoe nail that exits the hoof wall when you seat the nailhead into the horseshoe. Imagine what would happen to your leg if, just as you seated the nailhead into the shoe and had ¾" to 1" of sharp nail aiming directly at your thigh, the horse jerked free. Bend the nail over immediately!

---

7. Repeat the preceding steps for the other heel nail. In succession, following this nail, drive the rest of the nails, alternating side to side and working from heel to toe.

8. Take the short-handled nippers and snip off the bent-over nails, leaving about ⅛" of shank bent over next to the hoof wall.

BENDING THE NAIL POINTS OVER

CAUTION: Catch these clipped-off nail points. They can puncture the horse's hooves, the sole of your boot, or the tire on your pickup or horse trailer.

SNIPPING OFF BENT-OVER NAILHEADS

9. Take the head of the nippers (a farrier would probably use a small block of steel of dimensions 1" × 2" × 4") and hold it securely against the hoof wall and up against the ⅛" stub of the first nail you drove. While you hold your nipper head against the nail stub, strike a couple of sharp blows to the head of the nail. This will seat the nailhead in the shoe and begin to "clinch" or turn down the stub of the nail. Do this to each of the nails in the same order that you drove them.

CAUTION: There is some danger in hammering too hard and causing the wall under the stub to be strained and perhaps even torn away. A couple of sharp blows should suffice.

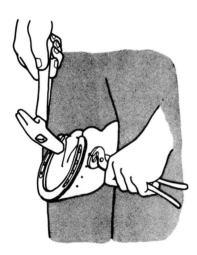

SEATING THE NAILHEADS

10. Take the horse's leg, pull it forward, and place the hoof upon your lower thigh so that you can work on the hoof in its normal position. Take the short-handled nippers and snip off each of the nail stubs until they protrude only about

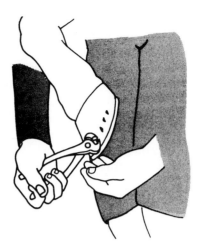

REMOVING NAIL STUBS

1/16" from the hoof wall. Again, catch all of the pieces of nail.

11. Place the curved jaw of the nail clinchers upon the nail stub and the flat jaw on the underside of the horseshoe. Gently close the clincher handles while at the same time gently pulling them downward toward your knee. Repeat this for each nail, alternating from side to side. Try to feel what is happening and clinch each nail by the same amount.

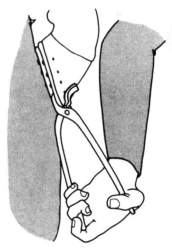

CLINCHING THE NAILS

CAUTION: Nails can be clinched too tightly, and a clinch-bound, footsore horse will be the result. Because this is an emergency shoe job, and your expertise at this may not be very great, try to clinch them snugly, not forcefully, and check them later.

12. At home, in the leisure of your stable alleyway, the farrier would now pretty-up the shoeing job (give it the professional look). It is not necessary for you to do this. The clinches need not be smoothed, nor the hoof–shoe interface delineated with the rasp edge. You have accomplished your task. The horse is once again rideable. Depending upon your confidence in your shoeing job, either call the farrier upon your return home or simply wait until his next routine visit.

13. Temporary horseshoes are excellent insurance for trail riders, and they are useful, even at home, for protecting damaged hooves or soles, for applying medical treatments, and for protecting trailer or stall doors from a determined kicker or striker. They are relatively inexpensive and simple to put on the horse, and they come in various sizes. Measure the width and distance from heel to toe of the hoof to be protected. Take these measurements with you when you purchase the temporary horseshoe.

6.16 AGING HORSES BY THEIR TEETH

The age of a horse can be determined by examining the overall profile and individual characteristics of his incisor teeth. In years past, when registration papers did not abound as they do today, the age of a horse had to be determined by examining the teeth. Our forefathers were therefore more proficient at this technique than we are. The ability to age horses accurately by briefly looking at their 12 incisor teeth increases in direct proportion to the number of horses' mouths you look at.

Years ago, when most horsemen were good at "reading the teeth," unscrupulous sellers had to physically and chemically change ("bishop") the teeth to deceive buyers. Today, because we are not so good at aging horses, the deceitful seller usually needs to alter only the truth, not the teeth.

A typical set of incisors can be used to categorize the actual calendar age of a horse into 12-month intervals from 1 year up to about age 10. After that, 2- to 3-year intervals must pass before discernable differences in the teeth can be seen. This accuracy is good enough for practical horse management. We may *like* to know if the mare is really $4\frac{1}{2}$ years old, as compared to 4 years, but we *must* be able to determine if she is 14 years old and not 4! Any horseman, equipped with the following information and guidelines, can become proficient at this skill if he or she will practice it.

Equipment Necessary

- Halter
- Lead rope

Restraint Required

Normally, none is required. A horse that will not allow you to open his mouth for inspection without rearing, backing violently, striking, or bolting is a poor candidate for purchase. You should expect the horse to try to turn away, however, to raise his head, or even to try to back away as you attempt to open his mouth, especially if you are rough-mannered and impatient. The presence of an assistant usually is of value.

Step-by-Step Procedure

1. You must become familiar with the dental nomenclature and timetable for eruption and wear of the horse's incisor and canine teeth (Table 6.1).

The horse has upper and lower sets of incisor teeth. The four teeth in the center (two upper, two lower) are termed the *central incisors;* the next four teeth, one on each side of the centrals, are the *intermediate incisors* (two upper, two lower); and the last four teeth are the *corner incisors* (two upper, two lower), situated laterally to (outside of) the intermediates. Four canines or *tushes* (tusks), two upper and two lower, appear at age 5 in the interdental space of all male and some female horses.

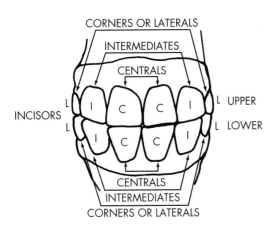

2. When you have become familiar with these guidelines, you are ready to approach the horse and examine his teeth. Approach the horse at a safe 45° angle from the nearside. Position your left hand upon the horse's upper lip and your right hand under his jaw, with your index finger positioned approximately in his curb groove. Take care not to partially occlude the horse's nostrils with your hand, as he will most certainly resist your efforts if you do.

CAUTION: Be certain that you do not stand in front of the horse to examine his teeth.

3. Use your thumbs—the left on the upper lip and the right on the lower—to part and lift or peel back the horse's lips. You do not need to use force for this procedure, nor do you have to expose a great deal of gum tissue. Lift the lips just far enough to see the incisors and to check for canine teeth. With the exception of the years 5 through 9 or 10, seeing the biting surface of the teeth is not critical to accurate aging. The recognition of permanent and temporary incisors and the progress of Galvayne's groove will suffice.

4. If you wish to examine the biting surface of the teeth, and you should do so to verify and refine your initial, closed-jaw estimate, slip the index and middle fingers of your right hand into the interdental space and place your right thumb on the horse's lower lip. Keep your left hand in its original position, and apply downward pressure upon the lower jaw with your right hand. The horse need only open his teeth from 1" to 1½" for you to see the biting surface, so don't use a tremendous amount of force. In fact, if you do, the horse will certainly resist. Be patient and firmly gentle.

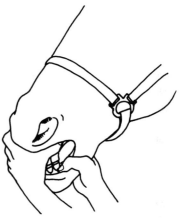

EXPOSING THE INCISOR TEETH

CAUTION: You have not been advised to grasp the tongue and pull it to the side to keep the horse's mouth open for examination. This is risky business. One wrong move and you can have painfully smashed fingers. Additionally, the horse's tongue is slick, and if you wish to hang on, you must nearly gouge it with your fingertips. You would also most likely wind up standing directly in front of the horse while you are holding on to his tongue. All this risk and pain are unnecessary, and you eliminate them if you will follow the directions just detailed. Be certain, even with this technique, that you keep your fingers down on the bars and away from the premolars and molars.

5. Use common sense before you stake your reputation on your guesstimate of the horse's age as indicated by his teeth. Check the length of the mane and tail. Are there gray hairs about the horse's muzzle? Does his overall appearance indicate youth or age? Are there signs that he is a cribber? Has he been pastured on sandy, sparse pasture? Some of these are general indications of age, and others are reasons that the tooth wear may not be an accurate estimate of age. They are things a good horseman should not overlook.

TABLE 6.1   Timetable for Eruption and Wear of Horses' Teeth

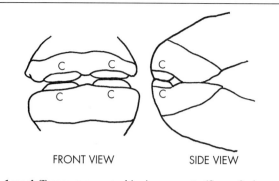

FRONT VIEW   SIDE VIEW

1 week Temporary central incisors erupt. (Some foals are born with these.)

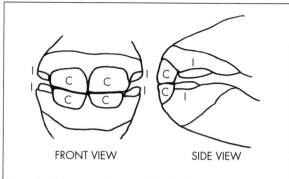

FRONT VIEW   SIDE VIEW

6 weeks Temporary intermediate incisors erupt.

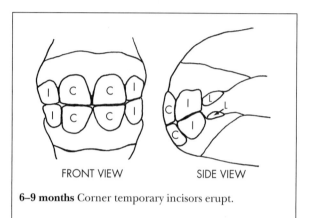

FRONT VIEW   SIDE VIEW

6–9 months Corner temporary incisors erupt.

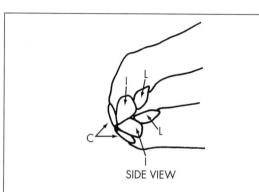

SIDE VIEW

1 year Central and intermediate temporary incisors are "in wear." (*In wear* means the upper and lower teeth are touching.)

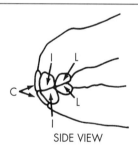

SIDE VIEW

2 years Central, intermediate, and corner temporary incisors are "in wear."

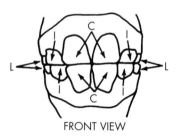

FRONT VIEW

2½–3 years Central permanent incisors erupt.

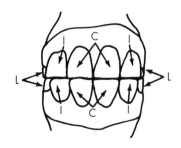

3½–4 years *Front view.* Intermediate permanent incisors erupt.

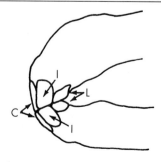

3½–4 years *Side view.* Intermediate permanent incisors erupt.

TABLE 6.1 (continued)

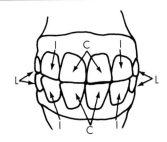

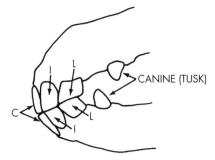

4½–5 years *Front view.* Corner permanent incisors erupt. Canines or tushes (tusks) have erupted (present in males, sometimes in mares).

4½–5 years *Side view.* Corner permanent incisors erupt. Canines or tushes (tusks) have erupted (present in males, sometimes in mares).

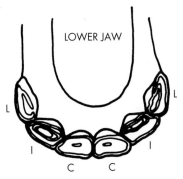

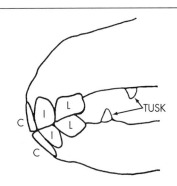

6 years *Top view.* Lower central incisor cups worn away.

6 years *Side view.* Corner permanent incisors are "in wear," tushes present in males and some mares, very slight beginning of "hook" on upper corner incisor.

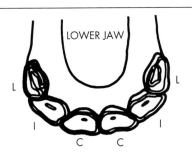

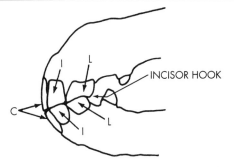

7 years *Top view.* Lower intermediate incisor cups worn away.

7 years *Side view.* Upper corner incisors develop "hook" on rear edge.

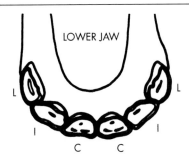

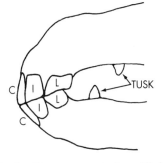

8 years *Top view.* Lower corner incisor cups are worn away. Lower central incisor shows "dental star."

8 years *Side view.* Upper corner incisor still shows some "hook."

TABLE 6.1 (continued)

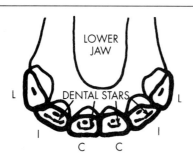

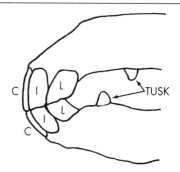

9 years *Top view.* Lower central incisors becoming rounded in shape. Lower central and intermediate incisors show "dental star."

9 years *Side view.* Upper corner incisor "hook" disappears.

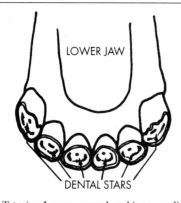

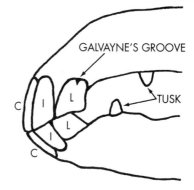

10 years *Top view.* Lower central and intermediate incisors becoming rounded in shape. Lower central, intermediate and corner incisors show "dental star."

10 years *Side view.* Galvayne's groove begins to appear at gum line of upper corner incisor. Angulation of "bite" (angle at which top and bottom incisors meet) is becoming less straight and more inclined or sloped to front of mouth. Teeth appear longer. Hence the term "long in the tooth" for an aged horse.

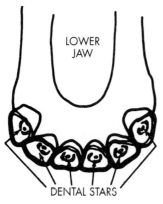

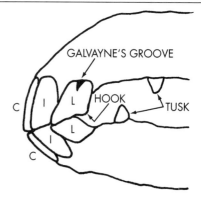

12 years *Top view.* Lower central, intermediate, and corner incisors are rounded in shape. Dental stars on all lower incisors are curved instead of straight line.

12 years *Side view.* Galvayne's groove is ¼ of the way down upper corner incisor. Hook has begun to return to the upper corner incisor. Angulation of "bite" becoming more pronounced.

TABLE 6.1 (continued)

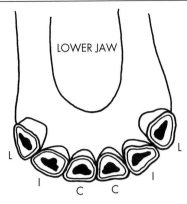

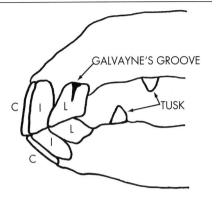

14 years *Top view.* Lower central, intermediate, and corner incisors becoming less rounded and more triangular in shape. Dental stars and pulp cavity blended on all lower incisors.

14 years *Side view.* Galvayne's groove extends halfway down the upper corner incisor. Angulation of "bite" becoming more pronounced.

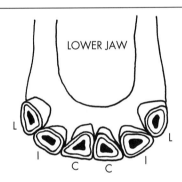

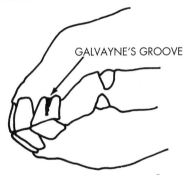

16 years *Top view.* Lower central incisors becoming triangular in shape.

16 years *Side view.* Galvayne's groove is $^3/_4$ way down upper corner incisor. Angulation of teeth is more pronounced.

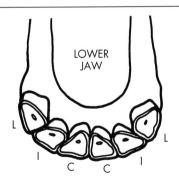

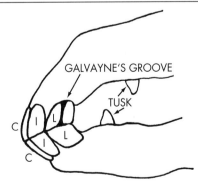

20 years *Top view.* All lower incisors are triangular in shape.

20 years *Side view.* Galvayne's groove extends full length of upper corner incisor. Angulation of incisors becoming more pronounced.

TABLE 6.1   **(continued)**

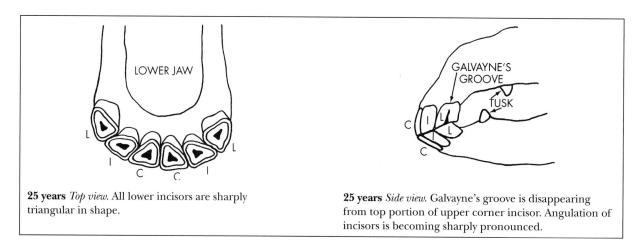

25 years *Top view.* All lower incisors are sharply triangular in shape.

25 years *Side view.* Galvayne's groove is disappearing from top portion of upper corner incisor. Angulation of incisors is becoming sharply pronounced.

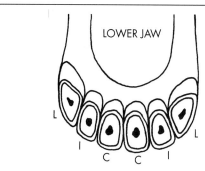

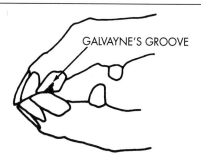

30 years *Top view.* Incisors are sharply elongated and show extreme wear.

30 years *Side view.* Galvayne's groove is present in no more than the lower $1/4$ of the upper corner incisor. Pronounced angulation of the teeth.

6.17 TOOTH CARE

The teeth of the horse are the first site of digestion. The mechanical digestion that begins in the mouth is very important because it is this physical chewing that breaks up the fibrous hulls of grains and smashes the coarse stems of hay. When chewing is efficient, the entire digestive process is off to a good start. If for any reason the grains or hays are not properly chewed, that is, if they pass through the mouth without having their hard outer coverings sufficiently broken up, the chemical digestion that should occur in the stomach and small intestine will be reduced. Food moves rapidly through the digestive system of the horse. Unless the hulls of grain kernels are broken up and the digestive enzymes allowed to enter the grain and do their job, a significant portion of the ingested grain will pass completely through the system and yield no nutritional value. It will be deposited on the ground in the feces.

Sharp edges on the premolar and molar teeth, retention of deciduous (baby) tooth caps, dental caries (tooth decay), and inflamed wolf teeth are the typical equine dental problems. A horse can have one, all, or none of these problems. Each of them will contribute to painful chewing. The horse will try to avoid this pain by reducing chewing time, by swallowing whole grain, by tilting his head, by altering his chewing motion, or by refusing feed. You can identify dental problems by watching for an increased number of grain kernels in the feces, large amounts of wasted grain dropped from the mouth to the ground, an unexplained loss of condition, or a sudden change to a sour disposition by a normally well-mannered animal.

Because this is a handbook of normal management techniques, the discussion in this section is limited to "floating" or rasping the sharp edges from the horse's upper and lower teeth. This is a technique that can be performed by most experienced horse managers, and one that mother nature does not normally take care of. Retained decidual tooth caps usually come off in their own time, and the extraction of wolf teeth and treatment of dental caries are best left to veterinarians.

Equipment Necessary

- Halter
- Lead rope
- Twitch
- Tooth floats: angle float and straight float
- Flashlight
- Bucket; fresh water

Restraint Required

Some horses, like some humans, have a high dentist-tolerance level. These horses may allow you to float their teeth with no more restraint than an assistant on the halter and your soothing, patient voice and manner. Others, and unfortunately these are in the majority, have less dentist tolerance and will resist your efforts to alleviate their tooth problems. This resistance usually is mild to moderate and can be overcome with a nose twitch in the hands of a capable assistant, patience on your part, and a gentle yet authoritative rasping technique. Working stocks seem to help some horses. Other horses will stop serious resistance when they are backed into or allowed to back themselves into a corner. On the severe side of the resistance scale are the horses that will fight you tooth and nail. If you have strong working stocks, strong rope, and strong, fearless helpers, you can usually float these horses; however, many equine dentists won't risk injury to this extent. They will use a mild chemical restraint to take the edge off.

CAUTION: This is one technique in which the inexperienced horseman can soon be in over his head. The horse is probably going to resist. The inexperienced person may actually encourage this resistance by his hesitancy and attitude of apprehension. It is hard work to float the teeth of horses, and even the best of equine dentists is going to run into horses who make it doubly hard by striking, rearing, and canting, lowering, or shaking their heads. If you attempt to be your own equine dentist, select the sanest of your horses needing the dental work, enlist an experienced horsemen to assist you, plan your procedure ahead of time, and be careful!

Step-by-Step Procedure

1. Be certain that the job needs doing. Not all horses need to have their teeth floated. It does seem, however, that if a horse needed it once in its lifetime, it will need it periodically thereafter. Observe the horse while he is eating and watch for head tilting, excessive grain falling from the mouth while chewing, and obvious signs of painful chewing (he is cutting both his cheek and his tongue).

2. The dental exam to determine the presence or absence of sharp tooth edges involves two people. One grasps the halter and steadies the horse from one safety zone. The other inserts the fingers of one hand into the interdental space (where the "bars" are) and grasps the tongue when the horse opens his mouth. It will take your entire hand—fingers, palm, and thumb—plus a good, strong effort to hold the tongue. Once you have a firm hold on it, move it farther outward and backward. This will place it between the horse's teeth on one side and he will not bite down.

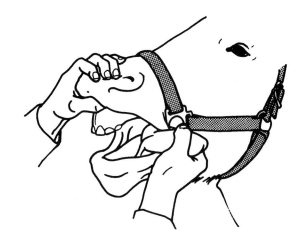

Use the flashlight to visually examine the premolars and molars. You are looking for sharp edges on the tongue side of the lower teeth and on the cheek side of the upper teeth. You can, if you wish, insert your fingers into the horse's mouth and feel for these edges. There is usually no reason to examine both sides of the jaw. The horse needs floating either on both sides or not at all.

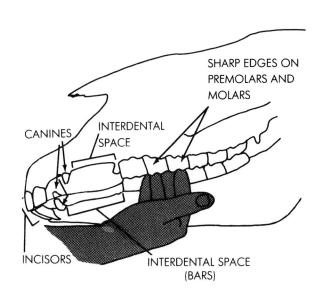

CAUTION: It does take a firm grip and some strength to hold on to the horse's tongue, but the tongue can be injured if you use your fingernails to assure your grip, and the web at the lower front of the tongue can be torn if you pull the tongue outward too forcefully.

It is very easy, while you are struggling with the horse's tongue and trying to shine the flashlight into his mouth, to forget where you are standing. You will be tempted to stand directly in front of the horse for the entire procedure. Appreciate the danger from striking and bolting when you stand in this manner. It should be avoided as much as possible.

Fingers can be painfully smashed by the teeth and jaw muscles of the horse. Be very careful when grasping the tongue and manually examining the teeth. If you are uncertain of how to perform these operations safely, request the assistance of an experienced horseperson.

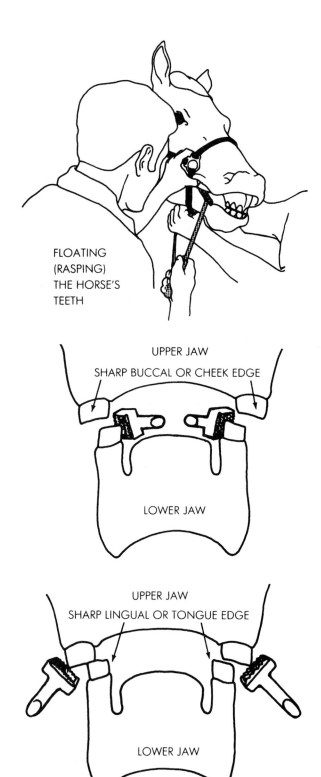

FLOATING (RASPING) THE HORSE'S TEETH

UPPER JAW

SHARP BUCCAL OR CHEEK EDGE

LOWER JAW

UPPER JAW

SHARP LINGUAL OR TONGUE EDGE

LOWER JAW

3. When it has been determined that the horse's teeth are in need of floating, establish the degree of restraint that you are going to use and your pattern of rasping. Twitching is advised for all horses, while the tongue hold depends upon whether the horse will keep his mouth open for the floating procedure without having his tongue held. If the tongue must be held, the sequence of rasping will be interrupted by your having to change the tongue from side to side. If you do not have to hold the tongue to rasp, the process will be quicker.

The rasp inserts on the float can be set to remove tooth edges when either pushed or pulled across the tooth. In addition, the float handle can be straight or offset. Horsemen who are performing these tasks for the first few times will find it easier to use an offset handle with the head set to push. A workable sequence, without tongue holding, is outside upper, then the diagonal inside lower, then switch hands and sides of the horse and repeat for the two undone edges. The hand not on the rasp is usually on the nose- or cheekpiece of the halter.

4. The actual floating of the teeth involves pushing or pulling the rasp across the sharp edges of the teeth until they are reduced to the degree that they are no longer sharp and painful to the horse. The rasp head should be held at such an angle that it contacts only the tooth edges. The length of stroke should be as long as practical, but care should be taken not to jab the float head into the cheek, gum, or tongue of the horse. The amount of pressure on the rasp will depend upon the prominence of edges to be removed. If they are small, heavy pressure can be used; if they are large, lighter pressure must be used or the float head will "stick" on the edges and not move freely. Moderate pressure, which will allow free rasp movement, is advised, even though more rasping movements will be required than when heavy pressure is applied. Additional benefits of lighter pressure are more control of the float and less resistance from the horse.

CAUTION: Control of the float is especially important at the extreme rear edge of the last molar and at the leading edge of the first premolar. Heavy-handedness in these areas will guarantee damage to the horse's mouth when your rasp slips from the teeth. Do not give in to the temptation to omit completely or do only a halfway job on these beginning and ending teeth. To skip them now only means that you will need to come back very soon to finish the job correctly.

The noise created by the grating of the rasp across the teeth is amplified by the hollow cavity of the horse's mouth and head. It is an unnerving sound to the inexperienced horseperson, particularly when you relate the whole procedure to your teeth! It is also unnerving to the horse who is inexperienced in this matter. When the noise begins, take extra care to anticipate reactions.

5. As you progress with the floating, the teeth of the rasp will become clogged with tooth particles. These should be periodically rinsed away in a bucket of clean water. On a younger horse (one under 5 years), you may, while rasping on the teeth, suddenly have what looks like a "grinding" tooth fall out of the horse's mouth. Do not panic. This is a "cap," the remainder of a deciduous or baby tooth, which is normally lost. It would not have been displaced by floating unless it were ready to come free.

6. Some blood in the saliva of the horse is to be expected as a result of your efforts to remove the offending edges on the premolar and molar teeth. The gums, cheek, and tongue will inadvertently be scraped, and when that happens they will bleed. As a sensitive human being, your obligation is to be as gentle and as careful as you can be while accomplishing the task.

7. When the rasping has been completed, reexamine, either visually or manually, the premolar and molar teeth. The sharp edges should no longer be prominent.

6.18 MEASURING HEIGHT

The height of a horse is the distance, in hands or inches, measured perpendicularly upward from level ground to the high point of his withers. A horse's height is normally recorded in hands, and a pony's in inches. One hand (h) is 4". Thus a 15.1h horse stands 61" tall, and a 53" pony is 13.1h tall. Note that 15.1h, 15.2h, 15.3h and 16.0h do not indicate decimal readings (15 and one-tenth, etc.);

rather, they signify the nearest whole hand, 15, and some fraction of a whole hand: .1 means ¼ and .3 means ¾.

Measuring the horse's height ought to be very simple, and physically it is. Unfortunately, the interpretation of the measured reading, which should be a simple, objective reading, is loaded with difficulties. It seems as if no one is satisfied with his horse's height. Pony owners always want to "shrink" their show animals to be able to show them to an advantage in the next-smaller class. Horse owners all want big athletic animals, so 14.3h horses become 15.2h when they are discussed. The "high point of the withers" keeps moving farther up the neck to accommodate these exaggerations. Thicker pads, thicker shoes, and longer feet can all cause a horse to "grow" very quickly.

Knowing the actual height of your horse is a must in properly exhibiting the animal. It also aids you in determining whether a young horse is growing properly and whether your breeding stock selection has been prudent. There are horsemen who will insist that they can eyeball a horse's height. It cannot be denied that, if you eyeball enough horses, you will become a good *guesstimator*, but only the measurer will know for sure. The same goes for "chinning" horses. Standing next to a narrow-made Thoroughbred of 16.3h and placing your chin on his withers is far different from doing this to a 16.3h stoutly built Percheron mare.

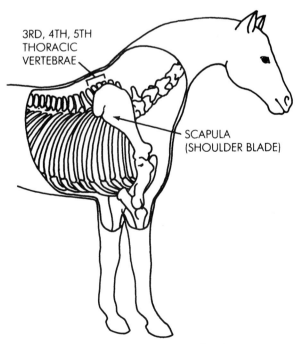

ANATOMICAL STRUCTURES INVOLVED IN DETERMINING THE HEIGHT OF A HORSE

Equipment Necessary

- Halter
- Lead rope
- Level, hard-surfaced measuring area
- Measuring stick or tape measure and carpenter level

Restraint Required

No physical restraint other than the lead rope and halter should be necessary, but measuring the horse does require that he stand still long enough for you to take an accurate reading. Once again, the value of "whoa!" becomes obvious. If you must measure the spooky animal who will not allow you to approach him with a measuring device, have your assistant use a mild form of restraint such as a nose twitch.

Step-by-Step Procedure

1. Establish a clear understanding of *where* the high point of the withers *should* be located before you attempt to measure the horse. As noted earlier, this is the area where the 3rd, 4th, and 5th thoracic vertebrae extend above the scapula of the horse. This point can be found by extending the shoulder lines along their natural slope, from both the right and left sides, upward and rearward until they meet.

Another method of locating this spot is to feed the horse and, when he lowers his neck while standing squarely and on level ground, marking the highest part of his withers area. This is exactly the spot to measure. However you arrive at it, determine and mark this spot *before* you try to measure the horse.

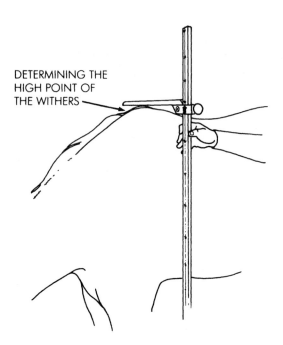

DETERMINING THE HIGH POINT OF THE WITHERS

2. Stand the horse up squarely and on level ground. Carefully approach him from the shoulder safety zone on his nearside. Take care not to frighten the horse by crudely waving the measuring stick around. If the horse should move away, be patient and simply try again.

3. Place the arm of the measuring stick over the horse's withers and lower it until it touches the predetermined spot on the withers. The arm of a measuring stick has a spirit level (floating bubble) to assist you in placing it in a level manner over the withers. Levelness is important if you wish to accurately measure the height of a horse.

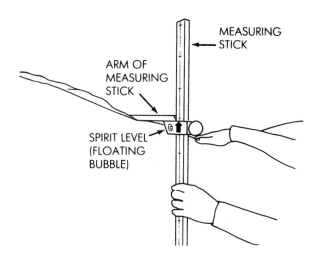

MEASURING STICK

ARM OF MEASURING STICK

SPIRIT LEVEL (FLOATING BUBBLE)

4. Immobilize the arm of the stick and read the scale on the upright part of the measure. This may be in hands or inches. If a third set of hands is available, a carpenter's level plus a steel tape measure makes an excellent "measuring stick." In this case, one person holds the horse, another holds the level, and the third person drops the tape from the level and reads it.

5. Record this measurement and the date on which it was made. Also record whether the horse was wearing shoes or pads when he was measured. For future reference and consistency, it would be helpful to measure the length of a front hoof (from the hair line, down the front center of the wall, to the ground).

6.19 IDENTIFICATION

A comprehensive system of permanent identification to distinguish one horse from the 8 to 9 million others in the United States is becoming increasingly important. Such a system may seem unnecessary to people who own only one horse, or even to the owners of groups of horses, each of which is distinguishable from the others on the basis of color or markings; however, the problem is not that simple.

Horses are shipped across state lines and even across the borders of countries. Many times, nothing more than a set of registration papers (mostly with no photo) or health papers accompanies the horse. A horse that is a brown Thoroughbred with no markings to you may be a bay Thoroughbred with no markings to some nonhorseman official. Theft of horses is a common problem, and a visible identifying mark of ownership helps to deter it. Fraudulent misrepresentation of one horse (usually inexpensive) for another (usually valuable) still happens in sale transactions.

Methods of Identification

Signalment

To identify a horse by signalment is to make note of and record all of his natural marks and colors. Items such as coat color, leg and facial colors, dorsal stripes, zebra stripes, chestnut shape, and hair whorls (trichoglyphs) are included. These are usually drawn upon generalized outlines of a horse and filed with a breed registry. The whole system depends upon good observers, good artists, equal color interpretation, and honesty. Unfortunately, this system does not work because it is too easy to have one ingredient missing from the above list.

Lip Tattoo

This method of identifying horses involves placing an indelible dye into the tissues below the skin on the inside surface of the horse's lip, usually the open lip. It has been used for many years and has served horsemen, particularly racehorse owners, well. It does have some shortcomings, such as a limit to the amount of information that can be stamped into the lip surface. Tattoos put into the lips of foals are even more limited with respect to space, and the tattoos fade out of young animals. Also, the possibility of disease transmission, especially equine infectious anemia (EIA) from repeated use of the pin-studded numbers (dies) is very real, and a laser beam can completely remove the entire lip tattoo.

Hot Branding

This method of marking a horse involves heating a branding iron to a nearly red-hot temperature (and color). The hot iron is then placed against the horse's skin until it has singed and branded the hide. Hot branding has the following definite limitations: it is painful to the horse, there is a limit to how much information can be branded onto the side of a horse in letters large enough to be readable after the healing process, and a hot brand can be unsightly to owners of pleasure horses. Hot brands are most useful in a within-herd or on-farm identification system.

Hoof Branding

Actually a form of hot branding, hoof branding involves burning an identifying number into the hoof wall of the horse. The burning or etching is accomplished with small branding irons. Other than the fact that it does not cause discomfort to the horse, hoof branding has the same drawbacks as hot branding. Additionally, a hoof brand must be redone every 5 to 7 months.

Trichoglyphs

The word *trichoglyph* comes from the Greek and means, literally, "hair carving." Today we use trichoglyph to mean a hair whorl or disturbance in the lay or stream of a horse's hair coat. These occur on the face, on the side of the neck near the crest, on the front of the neck, in the rear flank, and at several other locations. The frontal hair whorl, located approximately between the horse's eyes, will probably be the one to contribute most to a useful identification system. It is usable because no two animals have identical frontal whorls, and the whorls cannot be altered without leaving evidence that an alteration has occured.

To make a reading of a trichoglyph, a clay disc is pressed against the whorl or cowlick. The hair-whorl pattern is impressed upon the clay. This clay disc now becomes the mold for a rubber casting, which will be like the horse's "signature stamp." The unique whorl pattern will be stamped onto registration papers, health certificates, and other important papers. The drawback is obvious. Only a few people will have the clay, the rubber, the expertise to make the stamp, and the expertise to read the whorl and definitely declare the identity of the horse.

Implanted Microchip

The microchip, encased in a shatterproof glass capsule about the size of a grain of rice, is implanted with a syringe and hypodermic needle into the horse's nuchal ligament, which lies just under the horse's mane. A 12-gauge needle is used to implant the chip about halfway between the poll and withers. This is done by a veterinarian, using sterile technique. The microchip remains in the horse for its lifetime.

The system is entirely "passive," meaning that the chip does nothing until a scanner/reader is passed within a few inches of the chip. A signal is transmitted to the chip from the reader. This turns the chip on and it responds to the signal by sending out a signal of its own. The signal is a unique code that identifies the horse. Microchip technology is safe for the horse, inexpensive for the horse owner (about $10), permanent, cannot be altered, and it is reliable. However, it is not a theft deterrent because it is not easily seen and thieves know that few sale barns routinely scan horses for microchips. Also, the scanner or reader is expensive at this point in time.

Ocular Scans

There are two types of ocular scans—retinal scan and iris scan. The *retinal scan* uses a digital video camera and an infrared light to illuminate the retinal blood vessels in the horse's eye. The digital photo of these is interpreted by a computer and turned into a number code—not unlike the grocer's bar code—that is unique for the animal. The *iris scan* utilizes the same technology but creates the bar code, again unique for each horse, from the color, shape, and vasculature of the horse's iris. The iris scan is probably the direction of the future, because you can use the technology from a distance as compared to the retinal scan, which requires dilation of the horse's pupil.

Freeze Branding

Freeze branding involves the placing of a freezing-cold, copper branding iron to the skin of the horse. The cold destroys the pigment-producing cells of those hair follicles directly under the copper iron. The hair in the branding site will grow back white and will be very sharp in outline.

The numbering system used is not the conventional alphanumeric one. An Angle System has been developed by R. Keith Farrell, D.V.M., a USDA veterinarian, that produces a sharper brand than is possible with alphanumerics and increases the coding potential of the system (i.e., more horses can be identified without repeating numbers). The Angle System is easy to understand if you visualize it as two squares—one rotated 45° and superimposed upon the other. These two squares take care of the numbers 2 through 9. The numbers 1 and 0 and the alphabetic characters are just as simple to understand and use. The following two illustrations show the numbering code and two of the eight positions that the Angle System letter A can assume.

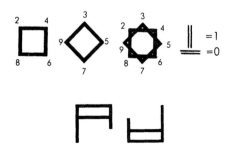

TWO OF EIGHT POSITIONS THAT THE ANGLE SYSTEM LETTER "A" CAN ASSUME

Equipment Necessary

The only method that is discussed here on a step-by-step basis is freeze branding. It is painless, does not require extreme horse restraint, requires only a modest investment in equipment, and offers the largest potential for development. It can be used with the alphanumeric or Angle System and for within-herd (on-farm) or nationwide identification purposes.

- Copper branding irons
- Dry ice or liquid nitrogen (1 lb per horse)
- Alcohol, 95% (1 quart per 5 horses)
- Electric clippers with surgical blade
- Styrofoam chest
- Cotton swabs
- Halter
- Lead rope

Restraint Required

This is a painless technique, and as such should require no special restraint. A knowledgeable assistant handling the horse should be sufficient, but you simply cannot predict exactly how a horse will react, so be prepared to use a mild restraint such as a nose twitch.

Step-by-Step Procedure

1. Restrain the horse, select the site to be branded, and clip the area as closely as possible. Surgical blades will clip closely enough. If you do not want the brand to show, flip the mane from its normal resting side to the opposite side and brand near the crest. When the mane is replaced, the brand will not show.

2. Determine the number to be placed upon the animal.

3. Select the appropriate iron or place the appropriate numerals in the brand holder, and place them into the styrofoam container.

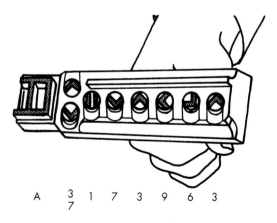

4. Place the alcohol and dry ice or liquid nitrogen into the styrofoam container. Only enough alcohol–ice mixture or nitrogen to cover the working part of the branding iron should be used at one time.

CAUTION: Painful freeze burns or tissue destruction can occur if you handle the dry ice, liquid nitrogen, or cold irons with bare hands. Carelessly pouring the nitrogen could result in having it land on or in your shoe. Be careful!

5. Scrub the clipped area with 95% alcohol until the hair stubble and skin are thoroughly wetted. The alcohol is used to draw out the body heat and to conduct the cold into the hair follicles.

6. Before the alcohol has evaporated, place the branding iron firmly against the hide of the horse. Hold it there for approximately 20 seconds. You needn't smash the horse with the iron; neither do you want it to move about once it has been placed in position. If the horse should move, the branding person should simply move with it, holding the iron in place.

Horses of lighter color—the grays and whites—need more time under the branding iron. White hairs on a gray or white horse are difficult to read. An additional 15 to 20 seconds will completely destroy the hair follicle and result in a balding of the numeral face. This open-face brand will be readable even on a light-colored horse.

7. Select the iron or numerals for the brand holder for the next horse and place these into the liquid nitrogen or dry ice. By the time the next horse is ready, the irons will be at the correct temperature. As the dry ice sublimes or liquid nitrogen boils away, add enough to fill the styrofoam container to the required level.

8. Welts will rise over the freeze-branded area; they will correspond exactly to the number branded onto the hide. They will rise in just a few minutes and will disappear in a few hours. They are painless and perfectly normal.

9. The white hairs will grow back in about 30 days. If the branding was done in the winter, the new, white hairs may not come in until the animal sheds his old hair coat in the spring. From that time on, the brand will remain. Each successive hair coat will grow in normally, except that the hairs will be white.

6.20 TAKING THE HORSE'S TEMPERATURE, PULSE, AND RESPIRATION

There are certain measurements or readings that can be taken from the horse that reflect his general health status. These vital functions include rectal temperature, pulse rate, respiratory frequency, skin pliability, mucosal color (gums, conjunctiva of the eye), composition of feces and urine, and gut sounds. There are normal ranges for each of these parameters, and any deviation from the normal range indicates that something has changed or is beginning to change in the normal functioning of the horse's system. Such a change is not always an indicator of infection. For example, the stress of hard work will elevate the temperature, pulse, and respiration (TPR), water deprivation will affect skin pliability, and lush grass or teeth in need of floating will alter fecal consistency.

The good horseman observes his horse's condition and behavior daily. When he notices any change in the physical condition, activity level, or behavior of his animal, he makes note of it and begins to check the TPR. Horsemen would like to look at each of the vital functions, but the measurement and interpretation of some of them are difficult. For this reason, most horsemen content themselves with learning to accurately measure and evaluate the horse's temperature, pulse rate, and respiratory rate. For the normal ranges of temperature, pulse, and respiration (TPR), see Chapter 10.

Equipment Necessary

- Halter
- Lead rope
- Tie area
- Veterinary thermometer with string and clip
- Watch with sweep-second hand
- Record book (sheet)

Restraint Required

Usually no restraint other than the lead rope and halter is required. The help of a handler is preferred for safety and because most horses will stand better for a handler than when tied to a post. The psychological restraint of a well-trained horse who knows what "whoa!" means and who has been conditioned to accept handling is most important. In fact, if you must fight the horse to collect the measurement, it probably will be invalid because the fight may have had a greater effect upon the TPR than any change in the horse's physiological systems. Be patient, and work calmly and gently about the horse.

Step-by-Step Procedure

Taking the Temperature

1. Take a veterinary-model thermometer and attach a 10" to 12" length of wrapping string to the loop in the end of it. To the other end of this string, attach some sort of alligator clip or clothespin. This allows the thermometer to be inserted with the string attached, and clipped to the tail hairs of the horse. If the thermometer is expelled, it will not fall

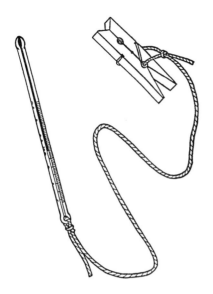

to the ground and break, nor can it be "pulled" into the rectum.

2. Shake down the thermometer so that it reads below 96°F.

3. Coat the thermometer with petroleum jelly or other nonirritating lubricant. This will allow for an easy insertion of the thermometer into the rectum. A dry thermometer will adhere to the anus and cause irritation to the horse.

4. Raise the horse's tail. This can be simple and safe, with no discomfort to the horse, if done correctly. Do not approach the horse from directly behind or stand directly behind the horse once you are next to him. Approach from the 45° safety zone off the shoulder, touch the horse, and then work rearward until you are standing next to his hip. From this position, reach to the tail and grasp it approximately 12" to 14" from the head of the tail. Pull or push it to one side, lifting only as much as is necessary to allow for the sideward movement.

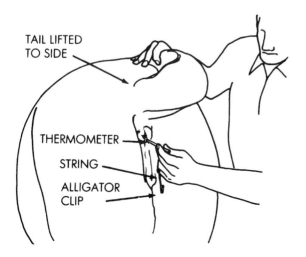

TAIL LIFTED
TO SIDE

THERMOMETER

STRING

ALLIGATOR
CLIP

CAUTION: Do not attempt to lift the horse's tail directly upward. It is uncomfortable for the horse, difficult for you to do, and will certainly provoke resistance.

5. Insert the thermometer into the rectum with a slight rotating motion. Attach the clip to a clump of tail hairs and test the attachment by tugging on the clip. The thermometer should be inserted so that about 1" to 1½" of the barrel is outside the rectum. If the thermometer is expelled, simply reinsert it. If the horse pulls it completely into his rectum, grasp the string and pull gently to retrieve it.

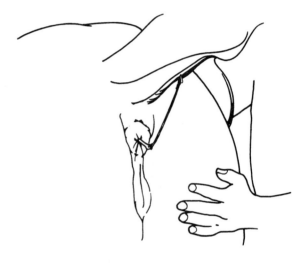

CAUTION: If the string becomes detached and the thermometer is lost into the rectum, you have two options. Put on an insemination glove and, by palpating rectally, retrieve the thermometer, or stall the horse and await the next bowel movement, which should expel the thermometer. Be certain that the thermometer has been removed from the horse, whichever method you use.

6. Allow the thermometer to remain in the horse for a minimum of 2 to 3 minutes. The tail can be released once the thermometer is properly positioned.

7. Remove the thermometer. Record the reading, date, and time of day on a record sheet. Wipe off the thermometer and shake it down for the next use.

Taking the Pulse

1. The pulse throb that is felt to establish the pulse rate results from the heart's contracting and sending out a wave or throb of increased pressure through the arteries. This pulse is detected by placing the fingers of your hand against an artery.

CAUTION: Be certain that your fingers only, and not your thumb, are in contact with the horse. Your thumb has a pulse of its own, which could be confused with that of the horse.

2. Select the artery at which you will measure the pulse rate. In the horse, this can be done at any of the following:

Facial artery: Back edge of the cheek, 4" to 5" below the eye.
Maxillary artery: Underside of the jaw, along the inner surface of the bones forming the "vee," one-third to one-half of the distance from the front.
Medial artery: Inside of the forearm, 3" to 4" above the knee.
Digital artery: Almost directly behind the knee.
Median coccygeal artery: Up under the tail, just about where the end of the thermometer protrudes while taking the temperature.

Select the site that is easiest for you and your horse.

3. Approach the horse in a safe manner from one of the safety zones. Place your fingers against one of these arteries, and when you can feel the pulse, begin timing with the second hand of your watch. Time for 30 seconds and multiply by 2, or time for 15 seconds and multiply by 4, and take the result to be the pulse rate in pulses per minute. You may find it difficult to take the pulse for 30 continuous seconds because the horse may move or you may lose your concentration. If this happens, simply try again. If you repeatedly lose the pulse, try for only 15 seconds and multiply by four.

4. Be certain to record the pulse rate next to the horse's temperature, the date, and the time of day.

Taking the Respiration Rate

1. The horse's respiration rate can be determined in three ways: (1) counting flank movements, (2) counting nostril movements or flares, and (3) holding a mirror 2" to 3" in front of one of the horse's nostrils and watching for and counting the vapor blurs on the mirror.

2. Be certain that you do not confuse muscular movements in the flank region for respiratory movements. One rise and the consequent fall of the abdomen–flank area should be counted as one respiration (this is also true for nostril movements and vapor blurs).

3. Choose your site, pick up the respiratory pattern, and begin to count. Time the respirations for

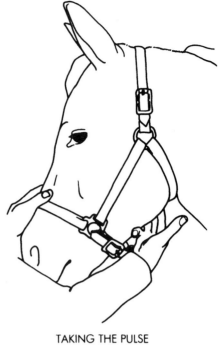

TAKING THE PULSE
(MAXILLARY ARTERY)

30 seconds and multiply by 2, or use 15 seconds and multiply by 4, for the rate in respirations per minute.

4. Record the respiration rate, pulse rate, temperature, date, and time of day on a permanent record sheet. Have this information available when you talk with your veterinarian.

6.21 PASSING THE STOMACH TUBE

This skill involves the insertion of a length of a small-diameter (½" to ¾" OD) flexible, plastic tubing into one of the nostrils of the horse, and the careful manipulation of it through the nasal passages, pharynx, and esophagus, and finally the actual passing of the tube into the stomach itself. The purposes of such a procedure include the following: the introduction of medicine directly into the stomach (as in parasite control or colic treatment), the feeding and watering of equine patients that either refuse or are unable to eat and drink of their own accord, and the treatment of esophageal choke in the horse. (See step 12 of the step-by-step procedure for this technique for an alternative oral method of introducing medication to the horse.)

Properly done, this technique is painless to the horse, and its effectiveness for its intended purposes is beyond question. It is not a technique to be taken lightly, however; it involves a great deal of skill and a thorough understanding of the anatomy of the equine respiratory and digestive systems. Competence in this skill requires practice. In other words, if you aren't going to do several horses, several times a year, you would be better advised not to mess around with this technique and possibly injure or destroy your horse.

Equipment Necessary

- Halter
- Lead rope
- Stomach tube: ½" to ¾" outside diameter rubber construction, 8' to 10' long
- Lubricant: petroleum jelly, K-Y Lubricant®
- Nose twitch
- Medication as indicated

Restraint Required

Some horses require no more restraint than a familiar handler on the end of the lead rope and halter. Others resist so violently that they must be tranquilized. The best procedure is to enlist the assistance of a knowledgeable, patient horseman who will steady the animal and back him into a corner or up against a wall so that he cannot continually keep moving away. The nose twitch should be applied off to one side and slightly more forward than usual to allow for passage of the tube. If all else fails, the administration of a tranquilizer will calm the horse enough that the tube can be safely passed.

Step-by-Step Procedure

1. Warm the tube to body temperature by placing it in warm water. This will lessen the resistance from the horse.

2. Lubricate the entire length of the tube that will be passed through the nasal passages.

CAUTION: If the tube is not lubricated, some bleeding will occur due to the friction of the dry tube upon the fragile nasal blood vessels.

---

3. Restrain the animal in whatever manner is appropriate. Note that this is an unusual and frightening experience for the horse. He must be restrained so that he does not injure you or himself. Be patient and do not overdo the restraint chosen. Consider using a strong, sure handler and an ear or nose twitch, and back the horse into a corner. As a final measure, chemical restraint can be employed.

4. Approach the horse and stand in the frontal safety zone, 45° off his shoulder. Place the thumb of your left hand in the upper interdental space ("bars") and the first two or three fingers inside the horse's nostril.

5. Insert the end of the tube into the nostril and hold it down with pressure from the fingers which were placed in the nostril in step 4. Push gently on the tube, keeping the natural curvature of the tube curving downward. After about 15" of the tube (mark the tube!) has been passed, its end is in the pharynx and ready to be "swallowed" into the esophagus. Now turn the tube so that any natural curvature is aiming upward toward the esophagus and away from the trachea (and lungs!).

6. Observe the horse and wait for the sight and feel of his swallowing. When he swallows, the tube will be pushed back toward the outside. When you see and feel this, immediately, but not crudely, push the tube back toward the horse and into the esophagus, which has been opened for swallowing.

7. When the tube has been inserted into the esophagus, *stop and check* before you go any farther. There is a risk of placing the tube into the trachea. Look for tube-end movement in the upper esophagus, usually 6" to 8" below the throatlatch. This will appear as a bulge in the area of the jugular groove on the left side of the horse's neck. Sometimes this is hard to see. If it is, push the horse's head to the right, which will stretch the left side of the neck and

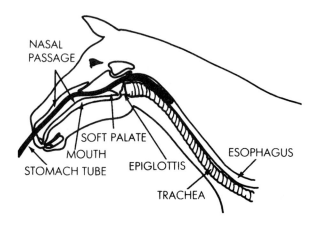

make movements of the tube more obvious. The esophagus occasionally lies on the horse's right side. When you think you see the tube, place your finger immediately below the bulge and push/pull the tube with your other hand. If the bulge moves against your finger, the chances are very good that the tube is in the esophagus.

8. Continue passing the tube until it has completely entered the stomach. This is somewhere about 5 to 6 feet along the tube (mark the tube!). As you push the tube into and through the horse's esophagus, a slight but constant resistance from the walls of the esophagus will be felt. If there is no resistance after the horse has swallowed the tube, it may indicate that the tube is in the trachea and headed for the lungs. Another indication that the tube may be in the trachea instead of the esophagus is that the tube stops moving approximately 18" to 24" before it should. That's about how much less tube it takes to reach the lungs instead of the stomach. *Mark that tube at the 5-foot mark!*

9. Have an assistant place his ear to the horse's stomach while you blow in the end of the tube. A rumbling, gurgling sound is further evidence that the tube is in the correct place. One final check is to grasp the trachea and shake it from side to side. If no rattling or vibration is felt, as there would be if the tube were in the wrong place, consider it safe to proceed.

10. Administer the medication slowly and according to directions. After the medicine has been delivered, but before withdrawing the tube, flush the inside of the tube with water. Blow the remaining contents of the tube into the stomach. Kink the tube or plug the end of it in some manner so that you can withdraw it without siphoning the medication back out of the horse's stomach.

11. Hold the tube down as you slowly withdraw it by placing your fingers in the horse's nostril. If there is any difficulty in withdrawing the tube, do not muscle it out. Work it back and forth and twist it gently until you have manipulated it free.

12. If all of this seems beyond your level of skill, or if you simply do not like the idea of tubing your own horse, consider using medication in paste form or on-feed powder or granules for worming. That still leaves the need for tubing for other medications and special treatments, but there are good alternative methods for routine worming. Paste wormers are administered from a tube and "gun" that closely resembles a caulking gun or from a large syringe-like plastic tube. The tip of the tube, which is about 2" to 3" long, is inserted into the interdental space and aimed upward and backward. The trigger handle is squeezed the appropriate number of times (see directions on the tube) or the syringe plunger depressed and the paste deposited upon the tongue. It is of such a consistency that the horse cannot "tongue it out." As he tries to get rid of it, he inadvertently swallows it. If the horse should resist insertion of the tube by backing up, move with him and continue squeezing the trigger until the correct dosage is delivered. After worming, hold the horse's mouth 6" to 12" higher than usual for a minute or so.

6.22 SADDLING

Saddling the horse is a skill that consists of placing a saddle blanket or pad, and then the saddle itself, in proper position upon the horse's back. Anyone who is brave enough to approach the horse and strong enough to lift a saddle can *throw it* across the horse's back; however, this is not the skillful saddling that is discussed in this section. There is a proper way to approach the horse, to position the blanket, to lower the saddle upon the horse's back, and to ensure that it is in the correct position. Finally, there is a proper way to cinch down the saddle.

Equipment Necessary

- Halter
- Lead rope
- Tie area
- Grooming equipment
- Saddle blanket or pad
- Saddle

Restraint Required

Usually, nothing other than the lead rope and halter is needed. There are times in the training of young or spoiled horses that hobbling or blindfolding is nearly an absolute necessity for placing a saddle upon their back, but handling the rank or untrained horse is not discussed here. There is no need for anyone who simply wishes to use or enjoy the horse to put up with a "hard-saddler." If your horse is of this type, have it trained (or retrained), or sell it and get another.

Step-by-Step Procedure

The procedure described here is applicable to the use of a Western-type saddle. There is no difference in the procedure for saddling with an English-type saddle, other than the shape of the pad, the shape of the saddle itself, and the method of cinching down the saddle. If you can properly saddle with one type now, only a few practice sessions will be necessary to make you skillful with the other.

CAUTION: Carefully examine the pad or blanket for burrs, gravel, caked areas, and wrinkles or hard spots. Burrs and gravel could cause bucking, resulting in danger to you. Caked, wrinkled, or hard spots will chafe the horse's back and cause him discomfort.

Examine the entire saddle for excessive wear, especially the rigging rings and plates, the girth billets, and the stirrup leathers. Check for leather deterioration by folding the leather. Replace any questionable parts.

1. Assemble all of the necessary equipment near the horse. The area should not be crowded with the saddle and grooming bucket so that you are stepping or stumbling over them constantly, but it is more efficient to have these things relatively close at hand. Stand the saddle and pad in the manner shown in the illustration. If you lay them down, there is a chance that they will pick up debris. This could be uncomfortable or dangerous to the horse and hasten the deterioration of your pad and blanket.

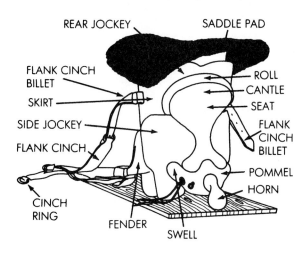

2. If time permits, groom your horse over his entire body. If time is short, at least be certain to thoroughly groom the area that will be covered by the pad, saddle, and cinches. This is to remove any dust, grit, or other debris that may be ground into the horse's hide by the prolonged pressure of the ride. Be certain to include the horse's hooves in this preride grooming.

3. Place the pad or blanket onto the horse's back, a foot or so farther forward than it should be during normal use. Then slide it rearward until it is properly positioned over the loin, back, and withers.

CAUTION: Be certain that you do not frighten the horse with the blanket. A good habit to get into is to always allow the horse to smell the blanket first, then slide it along his neck, and finally up and over his withers.

The pad is placed forward, then slid rearward, to assure that the hair below the pad is not rumpled and pressed in the wrong direction. If the hair is pulled and pressed against its normal lie, a sore, painful back could result. The final positioning of the pad is related to that of the saddle. For your purposes, figure that the gullet of the saddle will be placed directly over the withers. Place the pad so that it extends about 2" to 3" in front of the saddle. Before placing the saddle on the pad, be certain that there are no wrinkles and that the pad or blanket drapes equally to either side of the horse's topline.

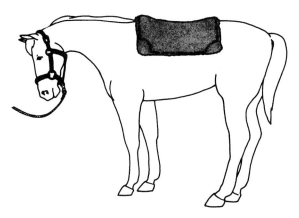

4. Grasp the saddle and rest it on your knee and upper leg long enough to hook the right stirrup over the horn and to place the front girth and rear cinch over the saddle's seat. If this step is ignored, the saddling process becomes much more difficult. Not only will the offside stirrup and both cinches be pinned between the saddle and pad, but the horse is likely to be spooked as they bang against his ribs. None of us is able to lift a fully rigged Western saddle to the height necessary and gently place it over the horse unless the stirrup and cinches are caught up in some manner.

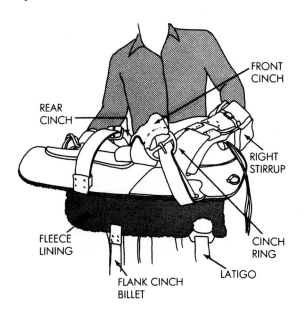

CAUTION: You must take care not to frighten or surprise the horse with the saddle. Refer to the cautionary note about blankets. While it would be difficult to rub the horse with the saddle, he should be allowed to smell it before you move to his side and plunk it down upon his back. Speak to him as you perform this task (more to be sure that he is awake and paying attention than to soothe his fears).

5. You are now ready to place the saddle upon the horse's back. Your right hand should be under the Cheyenne roll (or simply "roll"), and your left under the gullet or horn. Be certain you have the horse's attention. Lift the saddle high enough that the offside skirt clears the horse's topline. Gently lower the saddle into position.

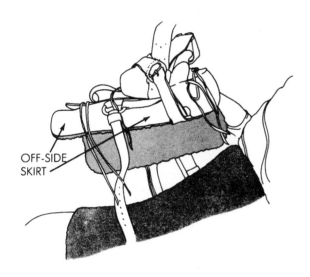

OFF-SIDE SKIRT

Adjust the saddle placement by lifting and repositioning it rather than trying to scoot it over the pad or blanket. Once again, proper placement is achieved when the gullet fits directly over the withers. Also keep in mind that the cinch rings should never be so far forward as to interfere with leg movement.

6. At this point, go to the horse's offside and lower the stirrup, cinches, and any leather ties to their free positions. Go back to the nearside, reach under the horse, and grab the front girth with your left hand. Cinch the front girth snugly, but do not attempt to tighten the cinch to the maximum at this point. If your saddle is not equipped with a belt-buckle or tackaberry-type girth attachment, see Chapter 1 for the correct method of tying the cinch knot. Cinch the rear girth in the same manner, but leave a 1" to 2" "belly" between the leather and the horse. Some saddles do not have a rear cinch.

CAUTION: There are "cinch-shy" or "cinchy" horses that are fine animals in all ways until you reach for the girth or try to tighten the cinch. At that point, they'll turn and bite you or reach up and kick at you with a rear leg. Watch out for this in any untried horse. Prevent this vice from developing by taking care not to pinch your horse during cinching and by not trying to "cut him in two" when you tighten the cinch. The cinch need only be tight enough to hold the saddle securely upon the withers.

The rear cinch should be adjusted until it nearly touches the belly, or it should be removed. Some horsemen like to let it dangle 4" to 6" below the belly. This is just enough room for the horse to stick his rear foot into it as he paws up at it. A serious accident could result.

7. Walk the horse around for about 30 seconds to allow him to relax and the saddle to adjust a bit, then recheck the cinch. It will probably need retightening. It is normal for a horse to "blow himself up" when he is being saddled. As you walk him around, he will relax and you can properly tighten the cinch.

CAUTION: Do not forget this step. Saddles that slip because of forgotten loose cinches are a major cause of horseback-riding accidents.

8. Unsaddling the horse involves reversing the sequence of steps in saddling. It is especially important that a rear cinch, if any, be undone first. Should you forget this step and the saddle slip after the front girth is released, the saddle could wind up being positioned under the horse's belly. Always release the rear cinch first.

6.23 BRIDLING

A bridle consists of a headstall, some type of bit, and a pair of reins. Bridling the horse involves simultaneously slipping the headstall upon the horse's head and placing the bit into its mouth. The reins can then be looped over the horse's neck and the whole bridle adjusted for proper fit. The key to skillful bridling is the coordination of hand and eye so that you are slipping the headstall on, opening the horse's mouth, and inserting the bit all at the same time in one smooth, sure, practiced motion.

Equipment Necessary

- Halter
- Lead rope
- Bridle

Restraint Required

There should be no restraint required to bridle a horse. All horses should accept having the poll strap or crownpiece of the headstall slipped over their ears and the bit inserted into their mouths. If a horse is worked with as a youngster and has its ears handled as part of the daily grooming process, there will be no ear problem. The horse that has been *crudely* "eared-down" (there is a correct way to do this!) in the past may become head-shy or ear-shy and refuse to allow you even to touch his ears, much less bend them over and slip them under the poll strap or crownpiece of the bridle. Another man-made problem is the horse who has had his teeth banged with the heavy bit one time too many. He simply refuses to accept the bit any longer.

Restraint really doesn't help either of these problems. The latter problem can be overcome with perseverance and the patience of Job; the former can be coped with by opening the buckle on the cheekpiece of the bridle and positioning the poll strap or crownpiece without going over the ears, just as you would apply a buckle-type halter.

Step-by-Step Procedure

1. Before you begin to bridle the horse, take time to consider a couple of things about the bridle. First of all, is it adjusted to fit the horse? Naturally, it will be if this is the only horse that you ride; however, if you have several horses and one favorite bridle, you will constantly be changing the adjustment. Learn to adjust by sight, or mark it in some way for each horse. A bridle that is too big goes on easily and can easily be adjusted. One that is too tight is a hassle for you and could be painful for the horse's mouth, eyes, and ears. Second, is the bit bitterly cold? If it is 10°F in your tack room, all the bits in there are also 10°F. Warm the bit in your hand or under your coat before shoving it into the horse's mouth. The bit lying in the bed of the pickup on a 100°F afternoon presents the reverse problem. It must be cooled before being placed in the horse's mouth.

2. Attach a lead rope to the halter of your horse, then remove the halter from his head and slip it over his neck and buckle it there loosely. Drape the end of the lead rope over his neck, keeping it off the ground and handy for you to grab should the horse begin to move off.

3. Grab the poll strap or crownpiece in your right hand and open it widely between your thumb and fingers. Drape the reins over your left arm at about the crotch of the elbow. Use your left hand to hold and position the bit with the thumb on one end of the mouthpiece and the fingers on the other.

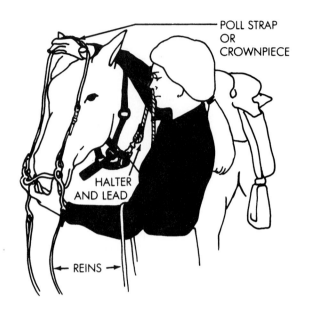

4. Approach the horse from the side, holding the bridle as just described. Stand at the horse's shoulder, facing him. Place your right hand and the poll strap or crownpiece directly in front of the horse's head at about eye level or a little above. With your left hand, gently place the bit against the point at which the horse's lips meet. Use the thumb and/or fingers of the left hand to gently press against the gums above the horse's teeth or against his bars to encourage him to open his mouth. As soon as the horse opens his mouth, gently but quickly pull upward with your right hand and hold the bit snugly in position.

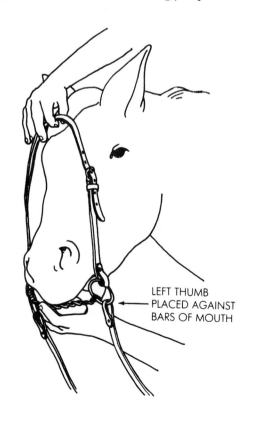

CAUTION: Even a gentle horse will hurt you if you allow him to smash your fingers between his teeth. Until the horse opens his mouth automatically when the bit is positioned against his lips, be careful that you keep your thumb and fingers on the gums or bars, not between the teeth.

Don't jerk the bit upward when the horse opens his mouth. If you rush it with a jerk, you will frighten the horse and possibly injure his mouth. You must move it quickly, however, and then hold it snugly in position or you'll have to repeat the whole process again after it has fallen out.

The presence of a curb chain or strap can be an added hassle when you are just beginning to acquire this skill. These are a light chain or half-inch leather or nylon strap attached to either the top rings of the bit or to separate rings just behind these. Their function is to enhance the action of the bit by leveraging the mouthpiece of the bit downward onto the tongue and bars of the horse. They must be placed in the curb groove, just about 4" up from the horse's lower lip on the bottom of his jaw. Sometimes these will hang up on the lower lip if you do not start them properly. It is easier to become confident in bridling a horse if you remove the curb strap when you are practicing.

5. Now that the bit is in the horse's mouth and held in position by your right hand, all that remains is to place the poll strap or crownpiece over the ears. Take the poll strap or crownpiece in your left hand and place your right hand on the horse's mane near the base of the ear. Without being rough-handed, rub the mane and base of the right ear until the horse allows you to bend the ear forward. When it is bent forward, slip the poll strap or crownpiece over it. In the case of a bridle with an ear loop, slip the ear forward into the loop. Repeat the poll strap or crownpiece positioning for the left ear. Sometimes it is easier to bend the second ear backward instead of forward.

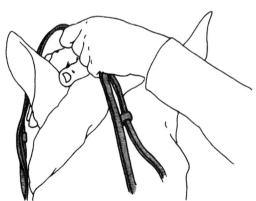

GENTLY BENDING EAR UNDER CROWNPIECE

CAUTION: In your efforts to place the ears under the poll strap or crownpiece while keeping the bit from falling out, you are very likely to poke the horse's eyes with the bridle's cheekpieces. If you are being gentle, the horse will react by simply closing his eyes and no harm will be done. If you are harsh and jerk the bridle, the horse will pull his head up and back to avoid the abrasion and a bad habit begins.

"Bending" the ears refers to *gently* folding them over—either forward or rearward. This action will not make your horse ear-shy. If the ear will not easily slip into the ear loop or under the poll strap or crownpiece, you probably have the bridle adjusted so that the bit will fit too tightly when you do get it on. Loosen the bridle by moving the buckle on the cheekpiece by a hole or two.

If the bridle is a split-ear or sliding-ear bridle, or in any other way has an ear loop, be sure to insert that ear first. More manipulation is necessary to insert the ear into the loop, and this is more easily accomplished if the other side of the bridle has not already been positioned behind the ear.

6. If the horse is ear-shy and you cannot bend the ears either forward or backward, open the bridle at the cheekpiece buckle on the left side and place it over the poll and behind the ears as you would a halter. You should not use an ear-loop bridle on an ear-shy horse.

CAUTION: This caution is appropriate for any horse, but especially so for a horse that already has exhibited a bitting quirk such as ear-shyness or head throwing. To avoid bitting, even if you found ways to stymie his head tossing, the confirmed escapist may strike, bolt forward, or whirl and bolt, with or without a kick. Anticipate these and keep alert.

7. Check to see that the horse's tongue is below the mouthpiece of the bit. It belongs in this position, and most horses will not persist in placing their tongues over the mouthpiece. Other horses do, unfortunately, and for them you must drop the bit down a few inches (not out of the mouth if you can prevent it) and try again.

8. To properly adjust the placement of the bit in the horse's mouth, open the belt buckle on the left cheekpiece of the bridle and move it up or down as appropriate. Most bits are properly adjusted when they create a "smile" in the corners of the horse's mouth. This smile consists of *one or two* folds of skin in the corners of the horse's mouth above the mouthpiece of the bit. The presence of more than two folds

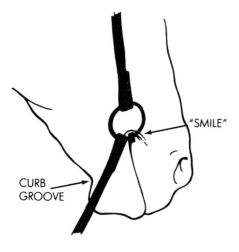

usually means that the bit is adjusted too tightly, while fewer folds indicate a bit that fits too loosely.

9. The curb strap on the bit must also be adjusted. This curb strap is the leverage point that causes the mouthpiece of the bit to exert downward pressure on the tongue and bars of the horse's mouth. If it is too loosely adjusted, it does not transfer the pressure; if too tightly adjusted, it has a tendency to numb the bars and tongue. In both extremes, the result is the same—no control. Most curb straps are properly adjusted if you are able to place your index and middle fingers side by side between the curb strap and the horse's jaw and then to rotate them so that they are positioned on top of one another. For most people, this results in about $1\frac{1}{2}$" of sag in the curb strap.

10. The last step is to secure the reins so that the horse does not step on them and damage his mouth. Tie them to the saddle horn or loop them around the horse's neck.

CAUTION: Never tie the horse with the bridle reins. The reins will break if the horse attempts to escape, but only after the horse's mouth has been damaged.

Always check to see that the reins are in good condition. In a panic situation, if you really pull back on the reins (to prevent a runaway, you'd haul back on one only), you do not want them to break, leaving you in an even worse situation.

6.24 CLEANING AND PRESERVATION OF LEATHER

Any high-quality equipment used for managing horses is expensive. This is especially true of leather equipment. Your saddle and leather headgear represent two of your largest cash expenditures. With an investment of $1,000 for a good-quality saddle and $5,000 to $7,500 for a leather headstall or halter, not counting silver or bits, it certainly makes sense to take proper care of them from the very beginning. This involves conditioning, periodic cleaning, and preservation.

Equipment Necessary

- Saddle rack or bridle hook
- Bucket of water
- Brushes (stiff)
- Sponges
- Toweling or terry-cloth material
- Saddle soap—paste and liquid
- Neat's-foot oil
- Leather conditioner

The procedures described are those for properly caring for a saddle. Bridles and halters should be disassembled and then conditioned, cleaned, and preserved in the same manner as the saddle.

Conditioning the New Saddle

1. Remove, or at least reduce, the squeaking that exists in a new saddle. Most of it is due to the pinching of the stirrup leathers between the skirt of the saddle and the bars of the saddle tree. This is normal and will eventually disappear as the new, stiff leather softens. To hasten the process, loosen the quick-change buckles on the stirrup leathers and drop the stirrups as far as you can without completely removing the leathers. Apply several light coatings of neat's-foot oil or saddle-leather conditioner to the stirrup leather, allowing each coating to dry before applying the next. Be certain to coat both the top and the underside of the leathers. To be certain of coating the entire stirrup leathers at any point where they will run over or under the saddle tree, run them up to the shortest position and repeat the coating process on any unconditioned segment of the leathers.

CAUTION: Neat's-foot oil has a tendency to permanently darken leather. Use a leather conditioner instead if you do not want this to happen.

---

2. Train the stirrup leathers and fenders to turn toward and remain in the proper riding position. The proper riding position is achieved when the stirrup leathers and fenders are twisted far enough to have the stirrups facing directly to the rear. As they come from the factory, they will face directly outward—a full 90° rotation from where they should be.

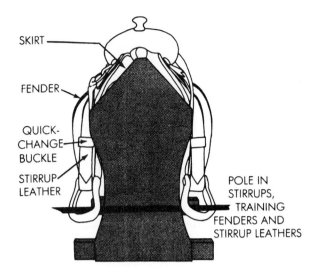

SKIRT

FENDER

QUICK-
CHANGE
BUCKLE

STIRRUP
LEATHER

POLE IN
STIRRUPS,
TRAINING
FENDERS AND
STIRRUP LEATHERS

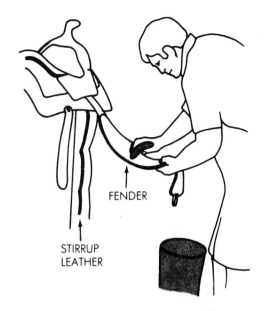

FENDER

STIRRUP
LEATHER

Use a sponge and plain water to wet the top and bottom sides of the lower half of the stirrup leathers. Keep wetting these areas until the leather begins to yield to your twisting.

3. Place the saddle upon the rack, turn the stirrups to the rear (perhaps even overturning them slightly so that they turn inward toward the horse), and place a broom handle through them. Allow the leather to dry in this position, then oil or condition the leather as described in step 1.

> **CAUTION:** Remember that neat's-foot oil or neat's-foot-oil compound darkens leather.
> Antique-finished saddles should not have neat's-foot oil applied to their topsides because the chemicals in the oil will attack and dissolve the finish.

Cleaning and Preserving the Used Saddle

1. Place the saddle on a rack and dismantle it as far as it is practical to do so. This usually involves removing the cinch, rear girth, and stirrups (but not the entire leathers and fenders).

2. Wipe off the entire saddle with a soft, dry piece of toweling. As you dry-wipe the saddle, pay particular attention to scratches, worn areas, and dry, cracking areas.

3. Wash the saddle thoroughly, using warm water, saddle soap, and a stiff-bristled brush. Don't be reluctant to wet the leather or really lean on the well-lathered brush. The idea is to clean the leather, and you cannot do that without soap, water, and elbow grease. Be certain to wash the underside of all leather, especially any parts coming into direct contact with the horse's body. Rinse each area thoroughly with a sponge to prevent the lather from drying on the surface and in the tooling cuts of the leather.

Take care not to wet the fleece lining of the saddle. If this were to happen and you allowed it to dry thoroughly, no harm would come of it. It does dry very slowly, however, compresses the fleece loft, and can be avoided with care.

Do not wash the seat of the saddle, even if it is a smooth seat as compared to rough-out. Most seats today are padded to some extent, and by washing them you run the risk of soaking the padding and discoloring the seat.

4. Allow the saddle to dry at room temperature. Do not permit it to dry next to a strong heat source or in direct sunlight. This drying may take from one day to one week depending upon how wet the saddle became during washing and upon how much humidity is in the air. It is not essential that the saddle become absolutely dry before oiling or conditioning, but if it is oiled before it becomes totally dry, the drying process is considerably prolonged.

5. When the saddle is dry, and the leather clean with the pores open, it is time to apply the oil or leather conditioner. Use a scrap of wool cloth or sponge, or better yet, a 3" × 3" square of wool fleece to apply the preservative. If the oil or conditioner is warmed slightly, it will penetrate the leather more quickly and deeply. Apply the preservative to the entire surface of the saddle, omitting the seat, and to the entire underside of the saddle, omitting the wool lining. Use a light, even coating. Allow this coating to dry before applying the second light coating. Set the saddle aside for at least 1 hour before continuing.

> **CAUTION:** Warm the oil or conditioner by setting it in the sunlight or on a heat register. A stove burner could ignite the oil and cause a fire.

6. It is now time to close the pores of the leather and to add some shine to the saddle. There are two ways of doing this that are recommended by leather workers. The first and easiest method is to apply a thin, but thorough, coating of a water-based polish to the entire upper surface of the saddle, omitting the seat. This will dry with a shine and will not clog the leather pores. The second way to finish preserving saddle leather is to rub in some more saddle soap. This time, use a drier lather and rub it until it disappears into the leather. When this second soaping is totally dry, use some toweling to polish the leather. Saddle soap does not polish to a high degree of brilliance, but rather takes on a soft sheen. There is no need to polish or saddle-soap the underside of the leather after it has been cleaned and oiled (or conditioned).

CAUTION: Do not use lacquer spray or wipe-on to impart the sheen to your saddle. Lacquer clogs the pores of the leather and makes it resistant to penetration by any type of preservative. Lacquer-treated (from the factory) saddles can be properly cared for and preserved for many years of use, but must be handled in a different manner. If you own one of these saddles, or are contemplating a purchase, contact your tack dealer for proper directions.

7. Rough-out seats should be cleaned with a brush designed for suede leather. This will remove dirt and make the leather fibers stand up. Smooth leather seats can be preserved with a *light* dressing of leather conditioner and water-based polish to seal the pores.

CAUTION: Too much preserver or polish on a seat can make "taking a good seat" almost an impossibility. It can also lead to stains on the seat and legs of your breeches.

8. Oil-tanned cinch straps and tie strings must also be cleaned and preserved. Some horsemen mistakenly think this latigo leather does not need care. It does feel oily most of the time, but that is the nature of latigo leather, not an indication that it does not need cleaning and preserving. Especially consider the cinch latigos. They are directly against the sweaty, grimy, salty horsehide and under constant stress. If they ever fail, the horseman is in for a "world of hurt." To clean them properly, use saddle soap and a brush. Scrub and rinse them several times and leave them hanging to dry. When they are dry, apply a couple of light, even coatings of leather conditioner.

CAUTION: Do not oil the latigos or overdo the conditioning. They could become sticky and prevent proper cinching.

9. If you are ever caught in a downpour or in an all-day soaking rain, there is no danger to your saddle if a few procedures are followed when you arrive home. Begin by wiping all saddle surfaces, stripping all latigos of excess moisture, and patting the sheepskin lining as dry as you can with toweling. Then hang your saddle up to dry by attaching a rope to the horn or through the gullet hole. Do not place a saddle with a wet liner upon a saddle rack. The fleece liner will mat down and is likely to mildew. Be sure to allow all latigos to hang straight and to turn the stirrup leathers to the rear, as when you initially trained them, so that they will retake the correct set. Finally, when the leather is dry, soap and oil or condition as before.

10. The cleaning and preserving procedures detailed here should probably be done by all horsemen at least twice yearly, even if their saddles are not heavily used. A hard-working saddle may need the whole process on a monthly basis. If the major purpose of cleaning is to remove grit and dust from the top surface and not to remove salt and sweat from the underside, the saddle-soaping and preserving steps for the surface leather may be all that is necessary.

6.25 LOADING

Loading the horse—getting the horse into the horse or stock trailer—can be an extremely easy and uneventful task, or it can be one of the most frustrating and patience-trying activities a horse-man can undertake. "Easy loaders" usually are brought up to the rear of the trailer, stopped while the handler places the lead rope over the horse's neck (to prevent his stepping on it), and then told to enter by either a verbal or touch command—all of which they alertly and immediately do. To finish this performance, they calmly stand while the handler positions the butt bar, closes the rear door, and fastens their heads. "Hard loaders" do none of these things. They resist with their entire 1,000 pounds and their total will at the first sight of the trailer.

Basically, horses become hard loaders because of negative psychological conditioning. Through some traumatic experience, they have learned to fear the horse trailer. Some really bad actors may simply resist any handling attempts. These present an entirely different problem and will not be dealt with in this text. Keep in mind that loading a problem horse presents real potential danger to the horse and to the handler.

It is easy to say to the owner of a bad-loading beast, "get professional help," or, "you should have done this or that," but that won't help the owner. To get the horse to the trainer you must load him, and what you should have done will help you only on the next horse, not this one. The following are tried-and-true ways of coping with hard-loading horses. One of them has always worked eventually.

Equipment Necessary

- Halter
- Lead rope
- Cotton rope
- Horse trailer
- Leather gloves

Restraint Required

This entire section involves restraint and coercion. The various methods are discussed in the text.

Equipment Checklist

Step-by-Step Procedure

1. Be absolutely certain that your trailer is safe and sound. Check the floorboards for soundness by pounding on them with a hammer; check to see that the hitch and all door latches and hinges work properly and lubricate them; check to see that your trailer is firmly attached to the hitch on the towing vehicle. *If possible*, select a tall (6'6" to 7'), wide (with moveable center partition), open-appearing trailer with a brightly colored interior. A two-horse trailer with dark-wood interior and darker, contrasting padding may look good to us, but it looks like a big, dark cave to the horse!

2. Check your restraint equipment. Select a nylon web halter of the highest quality and strength. It should be fitted snugly to the horse's head. The halter seams, buckles, and snaps should not show signs of weakening from wear. Lead ropes and snaps should be checked. Any frazzled or rotted leads and any balky or rusted snaps should be discarded. The ¾" cotton rope should be fresh, loosely woven, three-stranded, and 25 to 30 feet long.

3. The tie rings in the trailer's mangers should be firmly welded to the front of the trailer and smooth-edged to prevent frazzling and weakening of the rope.

4. Assemble your help. Plan enough ahead that you can enlist the assistance of at least two more handlers. They should be people who respect the danger involved and are comfortable working around horses.

"Easy" Hard Loaders

The "easy" hard loader is a horse that will load with minimal coercion but always shows resistance and always needs some coercion (the inducement of feed in a bucket in the manger has no effect on this horse). The resistance may last for 1 to 10 minutes; then something "rings" in this horse's head and he enters the trailer more or less of his own accord. If you allow time for this display, it is only mildly patience-trying; if you are pushed for time, it can be frustrating.

1. Lead the horse up to the trailer's rear without jerks, threats, or any other actions that may induce resistance. Have the rear trailer doors, the front escape door, and any front windows open. The more open the trailer looks, the easier the loading will be. Decide ahead of time which side of the trailer you'll haul the horse on. If you are hauling singly, place the horse on the high-side of the trailer (the side nearest the center of the road), as this will usually allow for best roadability of the trailer. If you are hauling double, put the hard loader on the side with the

escape door. Arranging matters this way permits the good loader to be put into the trailer first as an example or for reassurance.

> **CAUTION:** Never attempt to load a horse by entering the horse area if there is no escape door. If the horse should jump in all at once, which many green or spooky horses will do, you could be trapped and smashed or stepped upon. You cannot escape into the manger quickly enough.

2. Approach the trailer with a show of outward confidence and try not to stop. Attempt to walk right in with the horse in hand. Be certain to instruct your assistants to position the butt bar quickly if this should occur. If the horse should stop or begin pulling back, do not attempt to use brute force to pull him in with the lead rope. You *cannot* do this, and the pulling and jerking only reinforce the resistance. Talk to the horse; coax it closer and closer to that first step. Often this first step solves the problem and the horse continues on in. Give-and-take pulling on the halter also helps. Your goal at this point is to bring the horse right up to the trailer doors, so that his legs are about 12" from the rear of the trailer.

3. Instruct one of your assistants to lift one of the horse's front legs and place it up onto the trailer floor. The horse may pull it back out immediately. If he does, the assistant should persist in replacing the foot in a patient, positive manner. When the foot is in the up-and-in position, repeatedly tug and release on the lead rope. The other assistant should place his hand on the horse's rump and give some mild pushing-type encouragement.

> **CAUTION:** Be certain that this horse is not a kicker, because the assistant at the rear is vulnerable. The horse may also resist by rearing, so the foot-lifting assistant should be doubly cautious. The handler of the horse in any loading situation is liable to get rope burns if he attempts to hold onto the lead rope of a horse that is determined to leave. Care must be taken to keep your fingers out of lead loops and from between trailer partitions and the lead.

4. If the above technique did not work, have your assistants clasp hands behind the horse and place their locked arms against the rear of the horse's gaskin area. They should then combine their strengths in a lifting–pushing motion as the handler tugs and releases on the head. On a mildly resistant horse, the method usually works.

> **CAUTION:** If the horse is a kicker, the assistants are in danger. The kicking horse can, will, and does kick at people who are handling him in this manner.

5. If the horse is a kicker, or if the assistants choose not to clasp hands behind the horse, a cotton rope can be placed against the animal where the assistants' arms would have been. The assistants combine their strengths in a lifting, pushing motion as before, which produces the same effect as does the use of their arms. Doubling the rope is advisable because it gives the horse a greater feel of coercion and keeps excess rope from becoming entangled in the handlers' feet.

> **CAUTION:** The horse may swing to one side or the other as he is being pushed or pulled into the trailer. If this occurs, you probably will have to stop and realign him for safety's sake. The corners and edges of trailer doors are a definite injury site for the hard loader.
>
> At some point in this butt-rope–hand-clasp technique, the horse may lose his footing, rear, or blow up and begin frantic scrambling. Release your pressure from both ends when this occurs and start over. There is nothing to be gained by your being dragged out of the trailer and your assistants being trampled.

6. Perhaps your horse has been taught to load with a whip or broom. If you have a whip-trained or broom-trained horse, you'll have to accept that fact and load him accordingly. *Lightly* tapping the fetlocks and rear of the cannons with the whip is all that is necessary if your horse has been whip-trained. A broom-trained horse will respond to the shaking of the broom, touching of the broom to the ground, or touching of the broom fibers to his fetlocks, cannons, hocks, or quarters.

CAUTION: Do not resort to flogging horses with the whip or banging on them with the broom. The idea is to find the right stimulus to cause them to load of their own accord, not to beat them into submission.

Some horses resent the brushing of the whip or broom against their bodies and respond with some swift, short kicks. Take care that you do not turn a non-kicker into a kicker, and take care that you are not caught by one of those retaliatory kicks.

"Hard" Hard Loaders

The "hard" hard-loading horse is a real problem in that he must be physically forced to go into the trailer. There is no sudden relenting on his part to indicate that he has seen the light. He is the type of horse that will offer maximum resistance until the last latch is bolted. The type and even the intensity of resistance may be no different from those offered by an "easy" hard loader, but this horse just never stops resisting. The fact that he weighs 1,000 pounds or more, and you and your assistants only half as much, presents some very real problems.

1. Try the techniques described in steps 2 through 6 of the procedure for handling "easy" hard loaders, but note that with this type of horse, you cannot afford to give back even one inch of gain!

2. The handler on the head should run the lead rope (or neck or body rope if you prefer—see Section 6.4) through a tie ring in the front of the trailer in such a manner that he can hold it for whatever time is necessary without allowing it to slip. As ground is gained from picking up feet, butt ropes, clasped hands, and whips or brooms, it should be taken up by the handler and not given back unless the horse loses his footing and goes down.

3. As an added source of leverage for the assistants, the butt rope can be tied to a ring on one side of the rear of the trailer, run behind the horse, through the other side ring, and back to one of the assistants, who is standing in a secure position. Once again, remember that nothing gained is ever given back, so run this rope around the ring in a manner that makes it easy to control.

CAUTION: The horse can easily lose his footing and go down. If this happens, the tension on the lead rope and the butt rope must be released immediately and until the horse has regained his footing.

There is some danger that the horse may scrape his front cannons on the rear of the trailer. This is another reason that a horse's legs should be wrapped when he is to be loaded and hauled (see Section 6.12).

Do not tie any rope to the center post at the rear of the trailer, or for that matter, use the center post as a leverage point. It is nothing more than a doorjamb on most trailers and can be easily torn away by a determined horse. Be certain that the trailer is firmly attached to the hitch of the towing vehicle. An unhitched trailer will be pulled around the lot by a hard loader.

4. Leverage is the key to getting this type of horse loaded. You simply must outmuscle him to get him loaded. Ratchet-type "come-alongs," calf pullers, and fence stretchers have all been used when necessary. If you must live with a confirmed hard loader, mount the come-along ratchet permanently in front of the trailer and modify openings to accept the come-along cable. Be certain to place a quick-release snap somewhere in the line of pull if you use such a device. Some of them are difficult if not impossible to release under a strain. In using a come-along, treat it as the lead rope or body rope and slowly but relentlessly take up on it. Have your assistants use a butt rope to position and guide the horse's rear end.

"Untrained" Hard Loaders

This hard-to-load horse is a problem, but he is not a problem horse. He simply does not know what is expected of him and has no desire to enter a box or cave on wheels! Patience is the key to success with this horse. He must be taught to load easily and with enthusiasm. If you abuse him, or if he is hurt while loading, you will create a permanent hard loader.

1. For this horse, try feed as an inducement, try coaxing, or try butt ropes or hand clasping, but do it with a minimum of grief to the horse. He must learn to enjoy his sojourn in the trailer.

2. The best of all methods for this green, untrained horse is to allow him to teach himself. This places the burden of planning ahead squarely on the owner's shoulders. The horse will teach himself, but it will take two or three days. The horse should be confined in a *small* corral, a corner of the barn, or his stall. All food, water, and edible bedding should be removed from this area. The trailer should be placed into the corral, adjoining the corner of the barn, or against the door of the stall. The food and water should be placed in the manger of the trailer and the doors *removed* from the trailer (or firmly secured) so that the horse has safe, free access to the interior of the trailer. Hunger, thirst, and time, usually less than a day, will soon have the horse entering and leaving the trailer without hesitation.

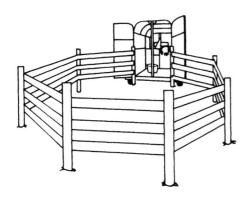

CAUTION: The trailer must be firmly attached to a towing vehicle. This is to ensure that it is completely stable and will not tip when the horse enters.

3. There is one additional method of allowing the green horse to teach himself. It is of particular value to the owner or manager who cannot use or does not have a facility to back the trailer up to as in step 2. It is still possible to allow the horse to teach himself to load if he understands standing tied.

To use this method, firmly attach the trailer to a towing vehicle. Position the trailer in a shady area and remove the rear doors if possible. If this is impossible, at least tie them firmly into the open position. Place feed in the manger and water on the floor of the trailer at the very front of the unit. Use a rubber tub for this purpose. Catch the horse, lead him to the trailer, fit him with a strong, nylon web halter, and cross-tie him to the tie rings in the trailer mangers (two snaps and two leads to the horse—one tied to the left manger ring and one tied to the right).

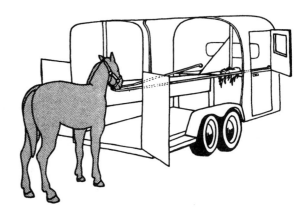

The horse should be exercised two or three times a day and returned to his stall at night. Do not allow feed or water at any time except in the trailer.

CAUTION: Do not leave the horse unobserved. Most horses will relent within a few hours, and by the end of 2 or 3 days of this schooling they are freely entering and leaving the trailer.

6.26 TRAILERING (HAULING) HORSES

Trailering (hauling) horses is a fact of life. The overwhelming percentage of horse owners live in towns and cities, clustered into groups with fellow Homo sapiens, far removed from the public lands, horse shows, gymkhanas, rodeos, and trail rides that are the destination sites of the hundreds of thousands of trailers crisscrossing the countryside each weekend. Adding to this are the specialized veterinary practitioners and farriers who insist that you bring the horses to them instead of them making farm calls. Horse owners are taking their horses with them for vacations, sometimes clear across the country. Mare owners think nothing of hauling their females hundreds of miles for the opportunity to breed to a well-known champion stallion.

It all seems pretty normal and commonplace . . . except for those horse owners who are plagued with hard loaders or who have a mechanical breakdown en route. It is easy to forget that for the horse, a plains animal used to open spaces and the ability to run from enemies, entering a 7' × 7' × 9' cave on wheels is entirely unnatural. We have just completed the discussion of loading techniques (Section 6.25). Following are suggestions that will minimize the stress of hauling for the horse (and the horse owner). Follow these suggestions and an easy loader will remain an easy loader because he knows there is nothing to fear from the cave, even if it is speeding down the highway at turnpike speeds. Overlook too many of the suggestions that follow, and you will sour the horse on being hauled and create all sorts of problems, starting with hard-loading horses and continuing through nursing them through the recovery from the stress-induced illness following a traumatic hauling session.

Equipment Necessary

- Horse trailer or stock trailer
- Pulling vehicle, properly sized
- Feed and water
- First aid kit
- Halters and lead ropes

Restraint Required

In one sense, trailering horses requires maximum restraint; in another, minimal restraint. The horse is in the trailer, totally restrained from going anywhere the trailer is not going. Within the trailer, the horse is usually tied to the tie ring at trailer's front, near the feed area. There are horse owners who insist on hauling "loose" (the horses are not tied to the front of the trailer). If your horse is not accustomed to hauling loose, tie his head to the front of the trailer.

If you are tying the horse during the trip, tie him loose enough to allow some freedom of head movement, but not enough for him to turn his head back past the divider or to get it down below the manger in front of him. Many people prefer a leather halter to haul with because they think it will break if the horse loses his footing and goes down in the trailer. Leather is certainly less strong than nylon, but don't count on either one breaking. Always use panic snaps to tie the horse. Then, if the horse does go down, a quick jerk on the panic snap will release his head. The snap must be kept in a rust-free con-

dition so that it will function properly. A quick-release knot is the next best option, if the panic snap is unavailable.

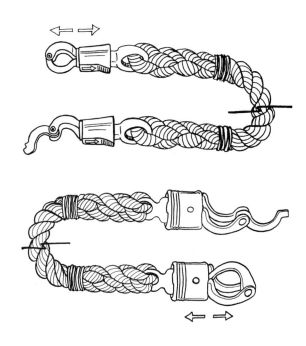

Step-by-Step Procedure

1. Be certain that your trailer is totally safety-checked: lights, brakes, flooring, latches, tires, hitch, safety chains, spare tire, etc. This check needs to take place about one week before the trip. Last-minute safety checks result in, "Aw heck, those tires aren't that slick . . . let's go one more time with them."

2. Wrap your horse's legs—all four of them. Wrap them properly for both support and protection (see Section 6.12).

3. Provide for adequate ventilation in the trailer. "Adequate" means enough air movement to keep the trailer from becoming stifling hot and stuffy and, at the other extreme, minimizing the draftiness that can result from a trailer with poor-fitting doors and windows. If hauling takes place during the heat of the day, temperatures inside the trailer will be 10° to 15° warmer than the outside temperature.

4. If it is going to be a trip during cold weather, bring a horse blanket along. Standing in the trailer for hours on end can result in a chilled horse.

5. Travel with a good first aid kit. Include bandage materials, antiseptic spray or ointment, thermometer, current antibiotic (on ice), syringe, and needle. Consult with your veterinarian about what to include for your horses.

6. Drive sensibly. Stopping distances increase with a loaded horse trailer behind. Trucks do not turn the same with the drag or push of the trailer. Jackrabbit starts and screeching stops are tough on the horses being trailered, as well as being tough on your equipment. Accelerate slowly and stop slowly, allowing your horses to balance themselves.

7. It may be an old wives' tale, but many horse owners feel more comfortable hauling a single horse or the heavier part of the load on the "high side" of the trailer. The high side is the side nearest the center or white line of the highway. Most roads have a crown in the center, making the center the high side and the edge nearest the ditch the low side. There is good logic to this. If a wheel of the trailer is allowed to drop off the edge of the pavement, it will certainly cause less of a lurch coming back on if the 1,000-lb horse is not on the side trying to jump back onto the pavement.

8. Carry your health papers and your insurance papers. Be certain your trailer and vehicle tags are current. Obey the speed limits and wear your safety belt.

9. In addition to being certain that your horse's health *papers* are in order, be certain that your horse's actual health is in order. The stress of traveling, coupled with the stress of coming nose-to-nose with horses carrying different strains of "bugs" for your horse to catch, make it important that you do all that you are able to immunize your horse before traveling. Consult with your veterinarian six weeks in advance of your planned trip.

10. Stop every 3 to 4 hours for gas or coffee and allow the horse to stand quietly in the trailer for 20 to 30 minutes. As weather and temperature dictate, you might want to open a window or upper rear door. This rest will do both you and the horse a lot of good. Certainly, check the horse at each stop, immediately after you shut the engine down and just before you take to the freeway again. Provide fresh drinking water at each stop. A wisp or two of hay he is accustomed to will also make for a nicer trip for the horse. Some rest stops have horse exercise rings. If your horse is an easy loader, take him from the trailer and use it. If he is a hard loader, leave him in the trailer. You would be surprised just how quickly a crowd will gather to watch the loading circus!

6.27 BREEDING HORSES

The standard procedure employed by most horse-breeding farms to get their mares "in foal" is to hand-mate the animals. This process involves a mare handler, one or more handlers for the stallion, and a designated breeding area. Practically, it means taking the stallion and mare in hand (at the end of a halter and lead rope) and controlling both during the breeding process. Hand mating is the alternative to pasture mating and artificial insemination.

Equipment Necessary

- Halters
- Lead ropes
- Teasing area
- Nose twitch and/or breeding hobbles
- Washing, rinsing, and drying materials
- Tail-wrapping material
- Breeding area
- Disposable rubber gloves

Restraint Required

If the mare is truly in heat, she is physiologically and psychologically ready to accept breeding from the stallion. If the stallion has been properly conditioned with regard to the behavior patterns you will or will not accept during the breeding process, he should do nothing to frighten the mare or to injure her- or himself. If the stallion and mare are psychologically prepared for the breeding, the only restraints required are control devices. Halters, lead ropes, and such mild restraints as nose twitching and breeding hobbles are all that are necessary. If the breeding is being forced upon the mare, or if the stallion is rank and ill-mannered, stronger restraint measures may be necessary. Forced breedings and ill-temper in your stallions should be avoided.

Step-by-Step Procedure

1. Determine the most opportune time to breed the mare. This timing coincides with the greatest development of the ovarian follicle. The follicle contains the egg to be fertilized and produces a hormone, estrogen, which causes the mare to psychologically accept the stallion. Your task is to determine when she is most ready to accept (stand for) the stallion; that is, when she is most strongly in heat.

The signs of being in heat include:

a. An obvious desire for company. She will seek the close companionship of other mares, geldings, or even her handlers. This is not always seen and should not be used as a definite sign when observed. It may occur 2 or 3 days before standing heat.

b. Blinking of the vulva. Normally the mare does this several times after urinating, but spontaneous and continued blinking is one of the more certain signs of approaching estrus.

c. Squatting, frequent urination, lifting of the tail, *and* blinking of the vulva are sure signs of standing heat.

You will be unable to observe these signs of heat to determine when the mare is ready for breeding unless you have developed a "teasing" program. Unlike some other farm species, mares typically do not "ride" or accept being ridden by other mares, or become loose and drip from the vulva. The only sure way of determining the time to breed, short of daily rectal palpation of the ovaries, is to tease.

2. Teasing is literally what its name implies. A stallion that will not be allowed to breed is brought close to the mare being teased. Handlers for the stallion and mare observe her actions for the signs that indicate she will accept the stallion (that she is "in heat," or simply, "in").

There are many ways to tease mares; most tried-and-true methods involve some variation of the techniques listed below.

Teasing Bar or Chute. The mare is kept on a lead shank and positioned across a 4-foot-high solid barrier from the stallion, who also is on a lead shank.

Aisle Teasing. The mares are left free, either in their stalls or in a long working chute, and the stallion is brought to their location on a lead shank.

Pen Teasing. The stallion is allowed to run without restraint in some type of enclosure, while the mares are running freely in an adjoining enclosure. The mares are free to visit with the stallion across the barrier separating the two pens.

Open Teasing. The mare and stallion come into contact without a separating barrier. The stallion is always on the lead shank, while the mares may or may not be restrained. This is the least desirable method because of high risk of injury to the handler, and it is not recommended.

Employing any type of teasing method involves a good knowledge of horse handling and psychology. One must observe very carefully, tease frequently so that small changes in response can be noted, and record the status of each mare in a teasing record book.

Teasing the "Wet" or Nursing Mare. Very often, the mare to be teased will have a nursing foal at her side. Since her maternal instincts will still be very strong either at the foal heat or for the first several months following birth, it may be difficult to get a good read from teasing the mare. She will be "worrying" about the foal either left behind in the stall or the corral or running about the teasing area. There are a couple of good ways to resolve this problem. You can have enough people power and assign someone to hold the foal in front of and in the direct vision of the mare. Or, you can erect a small foal pen in front of the teasing area with the net result being the same—the mare sees the foal and does not fret. A recently introduced commercial product that holds promise is the *"Easy-Breeder."* It protects the stallion, the mare, and the foal, and it allows the mare to see and be in close proximity to the foal.

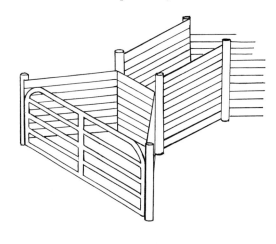

CAUTION: During the teasing process, the stallion will become excited and want to mount and breed. You are only teasing the mares, so you will have to deny this urge. The excitement and subsequent frustration, mare after mare, day after day, can lead to an ill-mannered, rambunctious stallion. There is a real risk of being bitten, struck at, pawed, or kicked by such an animal. Be assertive in the handling of your teasing stallion and be alert for any deviant behavior.

When you determine that the mare is in heat, both she and the stallion should be prepared in the manner set forth below.

3. The stallion's penis and entire genital area should be cleaned. For this task, the stallion is best handled by two people—one to hold the lead shank, the other to do the actual washing. Each should thoroughly understand his role ahead of time.

To wash the stallion, put on the rubber gloves, saturate a handful of cotton in warm wash water (around 100°F), and carefully place it on his umbilical, scrotal, and sheath area. Wet these areas several times, using new cotton each time, then insert some into the sheath to wet the inside and the penis. Avoid splashing the water or dribbling it down his leg; this can frighten or aggravate the stallion, thus making him difficult to manage.

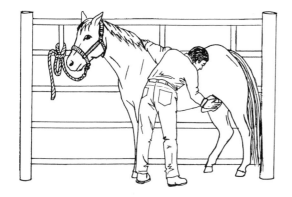

Put soap on wads of cotton and wash twice the umbilical and scrotal area and the outside of the sheath. For the inside of the sheath and penis, soap more cotton and wash repeatedly until the area is completely clean of dirt, scale, other detritis, and odor. If the buildup of "crud" is too great or too crusty to be removed easily with soap and water, coat the "crud" with baby oil and let is soak in for 24 hours.

After the washing is completed, use cotton and clean, warm water to rinse the area until all soap and residue are gone. Thoroughly dry the genital area; water has a spermicidal effect. Although the washing process would be easier if the stallion had an erection, it is not essential to doing the job properly.

The washing of the stallion's genital area should be done in the same place in the building each time, preferably in the breeding area at a distance of approximately 5 to 20 feet from the place where mating will occur.

CAUTION: Both handler and washer must be extremely cautious and alert during the entire washing process. Anticipation of the breeding act makes the stallion both powerful and unpredictable. Although the animal may already have learned to accept this handling without dancing and "talking," the handlers should stand in the safety zone about 45° off the horse's shoulder, and the washer should approach at about 45° from the front of the horse.

---

4. If enough competent help is available, it is best to wash the mare at the same time as the stallion. If both animals cannot be prepared at one time, wash the mare first, in her stall, then the stallion, in the breeding arena; finally, bring the mare to the stallion for breeding.

Wash, rinse, and dry the mare's vulva and hips, and the area between her legs. Remember that incomplete washing may permit introduction of bacteria, which can lead to resorption of the fetus, and that incomplete drying will allow water to come in contact with the sperm and kill them. The water

temperature, type of soap, and washing and drying materials should be of the same types as those used for the stallion.

5. Some stallion owners who stand their studs to outside mares will allow the mare owner to tease the mares and bring them to the stallion when they are in heat. This reduces the overall costs for the mare owner and the labor requirement for the stallion owner. If that is how your farm operates, it is possible that the first time you observe the mare's vulva is while you are washing her prior to breeding. You may notice that she has a Caslick's. If she does, it needs to be opened prior to breeding.

A Caslick's procedure is a simple surgical procedure performed to prevent pneumovagina or wind sucking. Wind sucking is the aspiration of air into the vagina of the mare. This interferes with the normal reproductive processes by allowing bacteria to enter. These bacteria will very likely cause inflammation of the vagina (vaginitis). The inflammation will then extend to the cervix, and finally to the uterus (endometritis). All of these conditions greatly reduce the possibility of a successful breeding.

In the Caslick's procedure, the upper edges of the vaginal lips are sutured together. A small $1\frac{1}{2}$" to 2" opening is left at the bottom of the vagina through which the mare can urinate. The Caslick's needs to be opened for breeding, redone following breeding, and finally opened for foaling. If the mare skipped a year following her last foal, it is possible she arrived at the breeding shed with the Caslick's in place. Call the veterinarian to have it opened.

If you study the following drawings, you can get a good idea of how the procedure is performed and what the finished product looks like. Local anesthesia, perhaps a mild sedative, and a set of stocks are all that is needed.

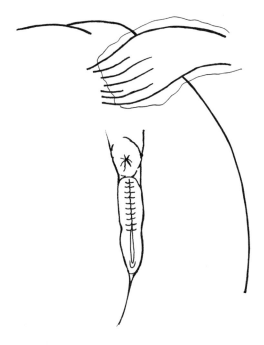

6. Wrap the mare's tail at least to the end of the tail-bone, and tie the remaining switch back onto the wrapped area. Tail wrapping reduces the possibility of hair cutting the head of the penis during intromission and prevents the penis from passing through the tail hair, where it could pick up filth (see Section 6.11).

7. After the animals have been prepared for breeding, the mare should be restrained in the manner selected. It is best if a nose twitch is sufficient restraint. The mare's handler stands in the safety zone (about 45° off her near shoulder) with one hand on the lead shank and the other on the twitch.

> **CAUTION:** The mare's handler should remain alert during the entire breeding process for two undesirable reactions: (1) the mare bolting ahead or striking out with a foreleg, and (2) the stallion flailing a hoof during mounting and dismounting.

8. The stallion's handler then *walks* him to the mare, approaching at a 45° angle from the rear on her nearside (not directly from the rear). Never allow the stallion to charge or become unruly. If he should charge, stop him immediately and reprimand sharply with the riding whip that you carry for that purpose. Since he is undoubtedly excited, make sure that you have his attention during the reprimand and that he understands exactly what you want. Be especially careful when handling an aggressive young stallion that he does not suddenly whirl and kick when approaching the mare.

The stallion may "visit" or "talk" to the mare as he approaches. This is fine and should not be discouraged. A visiting interval may range from 15

to 20 minutes for a totally "green" stallion to only 15 or 20 seconds for an experienced one. He is most likely to visit from the area of the mare's withers back to her hips, until he is ready to mount. (You will learn to sense when that time is as both you and the stallion gain experience.) At that point, he should be allowed to move to the mare's rear.

9. Don't let a young stallion mount until he has a full erection. It will only cause turmoil and aggravation for both him and the mare if he is not truly ready for intromission. If he seems ready to mount before achieving a full erection, change his mind by tugging on his lead shank *before* he shifts weight to his rear legs.

The stallion should be allowed to mount from any angle of not more than 45° from the nearside of the mare, as long as he is nearer her hips than her withers. Don't try to line him up, but rather let him do his own maneuvering into the correct position.

When the stallion begins to rear up on his hind legs to mount, the handler must allow the lead line (purposefully made longer than usual) to be slack in his hands, or he will inadvertently stop the mount or distract the stallion by tugging on his head. Unfortunately, this commonly happens in the excitement of a mount. Both handlers must also be alert at this point because the stallion rises and flails his forelegs somewhat and the mare may strike or bolt.

The inexperienced stallion may have to have his penis guided for proper intromission. It is not uncommon for him to direct it into the rectum or to an area below the vagina. Simply grasp the penis, guide it toward the vagina, and step out of the way. Do this only after he has mounted, has finished flailing, and you are confident that the mare won't kick.

> **CAUTION:** Do not permit nipping or biting. For most young stallions, slight but persistent tugs on the lead shank will prevent this habit from getting started. In a few cases, however, where the habit seemingly is set, it may be necessary to protect the mare's withers and crest area with leather or blanket material, or simply to muzzle the stallion.

10. The actual time of intromission usually is about 45 seconds and no more than a couple of minutes. A good sign that the stallion has ejaculated is the spasmodic and spontaneous "flagging" of his tail. Allow the stallion to dismount on his own.

As soon as the dismount is completed, remove the mare's twitch, pull her quickly to her own near-side, and lead her away at a brisk trot. Moving her in the same direction as the stallion prevents her from kicking out and prevents him from whirling and kicking her.

11. The mare should be walked for 15 to 20 minutes after breeding and not allowed to urinate or stop and strain. She should then be washed, have her tail wrap removed, and finally be returned to her stall or to an outside paddock.

12. The stallion should be thoroughly washed and rinsed at his prebreeding "washing spot." It's psychologically good to reward the stallion after breeding. Give him about a pound of oats during the postbreeding washing and another pound or so when he's back in his stall. Once started, continue the practice because he will come to anticipate it.

13. About 10 days after breeding (or on day 16 or 17 of her estrous cycle), teasing the mare should commence just as before and continue through her upcoming potential heat period. If she is pregnant, she will not come into heat. Teasing should continue daily for 6 to 7 days, just as in the previous heat period. If she shows no inclination to accept the stallion, she may be considered in foal, but she can be checked for pregnancy manually at 35 to 42 days if early diagnosis is desired.

> **CAUTION:** Great care should be taken with rectal palpation at this time; development of the early embryo can easily be disrupted by rough handling. If no palpation is desired, another teasing session through her next heat period would add to the confidence in the diagnosis of pregnancy.

14. From approximately day 45 to day 150, there are hormonal pregnancy tests and ultrasound evaluations that can be conducted. The best recommendation is to complete one of these at 75 to 90 days. Occasionally, a mare will resorb an embryo between day 45 and day 90. Testing by day 90 allows the mare owner to try breeding again during the current season if he so wishes. Many times, late-return mares are not noticed because teasing has logically ceased, they are put into the back pasture, and you are relaxing because they palpated safe-in-foal at 40 days.

15. After breeding, the mare should be allowed to exercise freely and be able to work in any manner to which she is accustomed. This level of activity can continue well into pregnancy. Only jumping, galloping, and activities fraught with the danger of slipping should be avoided in late pregnancy.

6.28 FOALING

The normal gestation period in the mare is 330 to 340 days. It has been documented, however, that live foals have come as early as 315 days or as late as 370 days after breeding. During this gestation period, the unborn foal has developed and grown from the fertilized egg, through the embryonic stages, and finally arrived at pregnancy's end point as a full-term fetus. During the early stages of this growth and development, observable changes in the mare were minimal, with the exception of an ever-enlarging abdomen. During the last month of pregnancy, however, there are certain signs that the mare is drawing closer to delivery. These signs are obvious to the careful manager. They should be watched for and their occurrence and dates noted in a record book.

The signs of approaching parturition are listed in Table 6.2. Keep in mind that these are average figures. They may or may not be accurate for your mare.

As the time for foaling draws near, say 12 hours or less, the following behavior of the mare will be noted: she will seek seclusion from other horses if she is running in a band; in the stall, she will walk the perimeter and at times act confused about what is going on; she will paw the floor of the stall as if trying to decide on a nest location; she may break out

TABLE 6.2 Signs of Approaching Parturition in the Horse

Sign	Time before Parturition
Enlargement of mammary gland (udder)	4 weeks
Relaxation of muscles surrounding the tailhead	7 to 10 days
Foal "drops" backward and downward in abdomen	7 to 10 days
Nipples of the udder fill with colostrum	4 to 6 days
Vulva loosens and opens slightly	2 to 4 days
Nipples wax	1 to 2 days

into a mild or profuse sweat; she will continuously glance back at her abdomen, first on one side and then on the other; and there may be a rather continuous dripping of colostrum from the nipples during the final hours.

Equipment Necessary

Normal Birth

- Clean, well-bedded stall
- Halter and lead rope
- Flashlight
- Bucket
- Household scale
- Tail wrap

Abnormal Birth—Assistance Required

- Obstetrical chains with handles
- Clean, dry towels
- Mild antiseptic

Restraint Required

If all progresses well, and the foal is born unassisted, possibly even unattended, no form of restraint is required. A halter and lead rope for the mare will become necessary if any examination is required to determine if delivery is progressing normally. Psychological restraint, that is, the verbal control that you can exert over the mare because of previous conditioning, is certainly most valuable when you want to enter the stall before, during, or after the birth. This could be important to you, because some mares become quite protective of their newborn foals and will challenge anyone entering the stall.

Step-by-Step Procedure

1. Decide ahead of time where the mare will foal. The two obvious choices are out on pasture or in a foaling stall. Pasture foaling is a good choice, and many foals are born on good, clean pasture. It makes sense to have this pasture close to your area of activity so that the mare can be observed as you go about your daily routine.

If the choice is to foal in a stall, be certain that it is as large as possible and free of *all* sharp edges and projections. Before foaling, it should be disinfected and bedded with straw. Excessively deep bedding is not recommended as it will hinder the foal's attempts to rise. The foaling stall should always be free of drafts and out of the mainstream of activity if at all possible.

2. Decide whether you are going to try to be present for the delivery or whether you will be content with finding the new arrival out in the pasture or in the stall on the morning after. Most foals are born between 11 P.M. and 7 A.M. It takes a tremendous commitment of time and effort to be present for the birth of a foal. It also means that you must make use of a foaling stall, even if the pasture may be more hygienic, safer, and therefore a better choice on your farm. If it is your choice to attend the mare during delivery, you must familiarize yourself with the normal sequence of events, as well as the abnormal, so that you can help the mare, either by not interfering or by actually assisting when necessary.

Steps 3 through 12 of this procedure are not management functions in the strictest sense. They are stages in the normal delivery process. The attendant must know and understand the sequence. He also must know whether he can remedy the abnormal situation or whether he must call for immediate and professional assistance. Abnormal births and steps to assist these are presented in steps 14 through 19.

3. In the following sequence, assume that the mare has been placed in her stall, her tail wrapped, and her vulva and buttock area washed with a mild antiseptic soap.

CAUTION: During the washing and tail wrapping procedure, the foaling attendant should double-check to be certain the mare's vulva has not been stitched partially closed during a Caslick's operation to prevent wind sucking. A veterinarian should have been summoned to open these sutures a week ago, but we are occasionally forgetful of or surprised by this condition. If the mare needs "opening," wait until the water has burst and feeling in the vulva is at a minimum, then cut the vulva along the central seam to its original size. This is accomplished by inserting a rounded-end scissors or round-tipped knife into the opening of the vulva, at the bottom, and cutting carefully upward, holding the blades upward and pressing outward as you cut. If you allow the stitched mare to deliver uncut, she will tear the sutured vulva and require extensive veterinary repair.

As the first stage of delivery, which involves the dilation of the mare's cervix so that the foal can pass through, begins, the mare will become increasingly restless. She may lie down, only to immediately rise again. She is likely to repeat this sequence time and time again. She may roll as the pain increases and repeatedly reach for her abdomen with her muzzle. She will carry her tail elevated, stretch out as in an attempt to urinate, and evacuate her bowels repeatedly. She may be in a profuse sweat. She will, all in all, act like a mare with colic. This is normal and should not cause undue concern to the attendant. If you fear that the mare may actually have colic, offer her a wisp of good hay or a handful of her favorite grain. If she eats it, her rolling is due to foaling pains, not colic.

4. The next event that will occur, and it sometimes causes consternation to the inexperienced attendant, is the rupturing of the water bag. The water that will literally gush out is the fluid environment in which the foal has been floating for the past 11 months. As the mare's uterus continues to contract and to exert outward pressure upon the foal and fluids, the water bag is broken. It must be if the foal is to be delivered. The fluids will lubricate the birth canal and prevent the trauma of a dry birth. It occurs from 10 to 30 minutes before birth.

The mare may be standing or lying down when the water breaks. The gush probably will frighten her so that if she was lying down, she will immediately rise. If the water breaks and the mare rises and remains standing for a few seconds, take the opportunity to remove and re-bed the wetted straw. This will take only a few minutes and will assure the foal of a dry bed into which to be born. Do not disturb the mare to re-bed. If she is lying down, is particularly nervous, or if you do not have help to control the mare, skip the rebedding until later.

5. After the water breaks, the mare will lie down and begin to strain to give birth to the foal. As soon as she lies down, observe the situation and be certain that there is room between her buttocks and the wall of the stall for the foal to be delivered. If there is not, enter the stall, get her up, and let her lie down again, hopefully in a better position. Continue this until she is positioned properly.

6. Watch for the forefeet of the foal to emerge from the vulva. This should occur within 10 minutes after the water breaks and the mare begins to strain. The feet may be covered with the amnion or free of any membranes. Either way, it is of no concern at this point. What is of concern is that *two* feet do appear shortly after the water bag breaks. They should be positioned side by side, or better yet, one just in front of the other. The soles of the foal's emerging

feet should face *downward*. A slight sideward rotation is no real problem; however, if they are upside down, the birth is going to be complicated.

7. By the time the legs have emerged to the knees, the foal's muzzle should be seen lying between the cannons. With another contraction or two at this point, nearly all of the head will be visible. The first real obstacle to be encountered is the foal's shoulders. They are an abrupt "stop" as compared to the slender and angular legs, head, and neck. The mare probably will have to strain in earnest to pass them through the birth canal. Her voluntary straining will coincide with the involuntary contractions of her uterus.

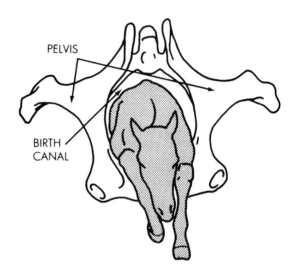

8. After the shoulders have passed the birth canal, the foal will literally slide out, up to the hips. If the mare is large and the foal small, even the hips will pass freely and the foal will be completely through the birth canal. If the foal is large and the mare small or fatigued, the foal may become hung at the hips or "hip-locked." In this situation, the mare should be helped, quickly.

The shoulder-stopped and hip-locked foal is not really considered an abnormal birth because all of the legs, head, and neck are properly oriented in the birth canal. Immediate assistance is valuable because it prevents fatigue of the mare and unnecessary wear and tear on the foal. Grasp the forelegs of the foal, using the towels to assure your grip, and pull outward and downward. Swinging (more correctly, exaggeratedly wiggling) the foal from side to side will help to maneuver the shoulders or hips through the birth canal (also called the *pelvic arch*). You should time your pulls and your resting to coincide with the mare's pushing and resting. It does little good for you to pull on the foal unless she is pushing at the same time.

9. When the foal is born (i.e., completely through the birth canal), the attendant should immediately make sure that no fetal membranes are covering the mouth and nostrils of the foal. If they are, and are not removed, the foal will die of suffocation. The attendant should grasp the membrane with one or both hands and pull it free of the entire face. This is a simple task and should be done quickly and quietly without disturbing the mare and foal. After the membranes are removed, step back out of the stall and observe. If you made note of the time at which the water broke, you'll find that the whole delivery process was completed in 20 to 30 minutes.

10. After a period of lying on the straw and resting, the foal will begin struggling to gain his feet. Allow him to do this on his own. His neuromuscular coordination must develop, and this is the first step in the process. As the foal struggles for his feet, he will free himself of the placental membranes and break his attachment to the umbilical cord. This is normal and should not be a matter of concern.

CAUTION: Care should be taken to not disturb the resting mare and foal immediately after birth, with the exception of clearing the foal's face. There is fetal blood in the placenta that is being pulled back into the foal while they are lying there resting. If you disturb the foal and cause him to scramble about and prematurely break the umbilical attachment, some of this blood will be unavailable to the foal. It should not be considered abnormal for the mare and foal to rest for up to 30 minutes after foaling.

11. The last stage in the birth process is for the mare to shed the placenta, or afterbirth. The foal may already have struggled free of it; however, it is still attached to the inside of the uterus of the mare. When the mare rises from her prone foaling position, the placenta may fall free. If it does not, the mare must expel it, with additional uterine contractions, in the same manner as she expelled the foal while giving birth. Normally, the afterbirth is shed within 30 minutes to an hour after birth. Up to two hours, and possibly three, is not considered abnormal. Longer retention times than this are abnormal and require veterinary attention.

12. The placenta should now be picked up, placed into a bucket, and weighed. The placenta itself will normally weigh from 12 to 15 pounds. Record this weight and notify the veterinarian if the placenta is lighter or heavier than this range. The placenta should also be laid out on a floor and examined to see that it is complete and to ensure that there are no pieces remaining attached to the uterus of the mare.

The liverlike, bean-shaped pieces of material that may be present on the floor or caught up in the placenta are called *hippomanes*. These can be 4" to 5" long, 2" wide, and ¾" thick. They look like an organ of some sort that has been misplaced. They are perfectly normal and are a collection of sloughed-off cells and salts from the fetus.

13. There is a list of things that must be done for the foal during the first minutes and hours after his birth. These are most important to his very life and to his future well-being, and are discussed in Section 6.30.

CAUTION: After the expulsion of the placenta, the uterus will continue to contract so that it involutes (returns to near normal size) more quickly. These contractions may be so strong that they cause the mare to exhibit signs of colic. This is postpartum colic and may occur from 15 minutes to 15 hours after foaling. If the mare begins to fret and toss about, the foaling attendant should enter the stall and control the mare so that she does not unintentionally injure her foal. Postpartum colic will last from 10 minutes to an hour. It may be necessary to contact the veterinarian if the pain becomes severe or lasts more than an hour.

Prolapsed uterus is the final concern in managing the recently foaled mare. This is a serious situation, and any time the mare attendant sees any part of the uterus protruding from the vulva he should contact the veterinarian immediately. The prolapsed uterus should be wrapped in a clean, wet sheet to prevent it from lying in the straw and debris while waiting for the veterinarian.

If the weather permits, the mare and foal should be turned out into a private paddock, corral, or pasture for exercise on the day after delivery. Thirty minutes to an hour is appropriate.

14. Beginning with this step, suggestions will be presented about how to recognize the start of an *abnormal delivery* and then how to cope with it. The foaling attendant should not sit by and merely observe the foaling, nor should he attempt to remedy situations beyond his experience *if he has a choice.* His job is to be alert to the total delivery process and to immediately observe any difficulty that the mare may be experiencing. If the mare is in trouble, he should make an intelligent, careful examination and then, depending upon the circumstances, he should remedy the situation according to the directions here, or call the veterinarian for assistance.

CAUTIONS: There are several general cautions that must be stressed before giving any specifics. They are important and should be remembered and observed.

Be patient. Be certain help is needed before you rush in and start assisting. If the foaling is progressing normally and you intervene, you may upset the natural progression of events and *create* problems.

Do make an examination when you feel that trouble is appearing. Know what you are looking for and interpret carefully what you see and feel. Act only after considering the problem and the steps to remedy it.

Never attempt an examination of a standing mare without an assistant controlling her head. If the mare is lying down, it is possible for the attendant to examine her by himself, although it is still recommended that an observer be present in case of mishap.

If, after your examination, you decide you cannot or will not attempt to handle the problem by yourself, call the veterinarian immediately. If you examine the mare immediately, decide quickly, call for help quickly, and the veterinarian responds quickly, most problem situations can be salvaged.

If you have summoned help, do not allow the mare to go down and strain or roll and toss about the stall. Keep her up and moving easily until the help arrives.

Keep your wits about you, especially if no one is coming to assist. Most malpresentations of the foal can be coped with if you just try to visualize what should be as compared to what you are presented with.

With the exception of an upside-down foal, do not attempt to change the overall orientation of the malpresented foal. In other words, do not attempt to turn a breech (hind legs first) foal around for a normal frontal delivery.

15. If *one foot only, one foot and the muzzle, both forefeet and no muzzle,* or the *muzzle only* are presented, with the mare attempting to deliver more of the foal, prepare immediately for an examination. Quickly, but calmly, wash the mare's vulva and anal area and your hands and arms with an antiseptic soap. Insert your hand and arm gently into the vulva, following whatever is showing (foot, muzzle, etc.) inward. Feel about until you have found the problem. Look for: a leg or legs bent backward behind the pelvic arch; the head bend downward, backward over the top of the foal's own back, or off to one side; a leg folded over the top of the neck; or a foot poked through the uterus into the rectum or abdominal cavity. Take the time necessary to accurately diagnose the problem.

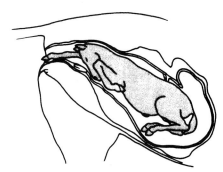

MALPRESENTATION: LEG AND HEAD DOWN

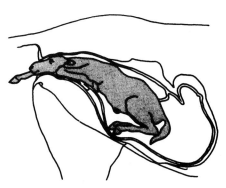

MALPRESENTATION: ONE LEG BACK

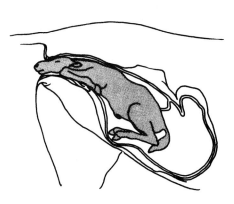

MALPRESENTATION: BOTH LEGS BACK

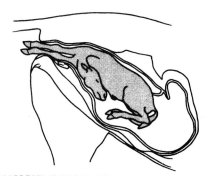

MALPRESENTATION: HEAD AND NECK BACK

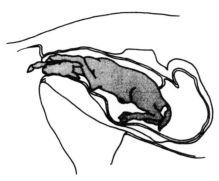

MALPRESENTATION: LEG OVER NECK

Remedying these malpresentations involves the obvious. The leg, legs, or head must be put into the correct position. Your efforts at bending the foal's knees, elbows, or neck into the proper orientation will be greatly complicated by the straining of the mare. She is a powerful animal and your muscle is no match for hers. Do your repositioning between her contractions. If you must push the foal back into the mare (for example, to gain more room to move a leg or neck), do the pushing between her contractions.

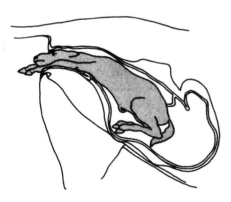

CORRECT PRESENTATION POSITION.
MALPRESENTATIONS REPRESENTED ABOVE
MUST BE ADJUSTED TO THIS.

Before pushing the leg or legs of the foal back into the mare and running the risk of losing them, slip the obstetrical chains over them. It is a great psychological boost to know that you can always retrieve them. The chain, slipped around a bent foreleg, may also be used to provide the leverage necessary to straighten it and pull it into the birth canal.

CAUTION: Birthing the problem foal is pure hard work. Muscles you've never used before will become fatigued and you will become panicky and try to brute-force it into position without thinking. Prevent this at all costs. Brute force you will need, but

you must continue your thinking as well. A thorough understanding of the anatomy (not long names, just which bones are where and in which direction they can move) surely helps. The use of a clean Turkish towel or a pair of clean, rough-textured gloves helps to hold the slippery legs and neck as you strain and pull for position. It is perfectly acceptable (in fact, most often necessary) to use both arms and hands inside the mare at one time.

16. If you are presented with *two feet with the foot pads facing upward* instead of downward, you must immediately determine whether they are front feet or rear feet. This is no time to try to decide by the shape of the feet. You must scrub as before, reach in, and examine the situation. If you reach in and feel hocks, hips, and a tail, you are faced with an "uncomplicated" breech delivery.

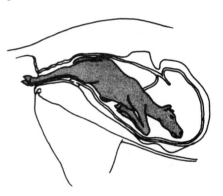

MALPRESENTATION: BACKWARDS

Allow the mare to deliver by herself as long as things are moving rapidly—that is, with no obvious hangups. If the mare appears to be in trouble, grasp the foal's rear legs, using a towel for help in gripping, and pull outward and downward. Wiggling the foal from right to left will help to clear the bulk of the hips through the birth canal. Continue your efforts to help the mare, pulling as she pushes. The unassisted foal can die of anoxia (lack of oxygen) in this position if the umbilical cord becomes smashed against the floor of the birth canal, so once you decide to assist, move quickly.

If the examination reveals that the upturned feet are the front feet, the situation is a lot more serious. The foal is being presented upside down. Feel for the head and neck and see how they are oriented. Also test the resistance of the uterus to pushing the foal back in just enough to turn the foal over. If the foal can be turned, you can deliver it by assisting as you would a fatigued mare delivering a normally presented foal. Pull when she pushes, outward and downward.

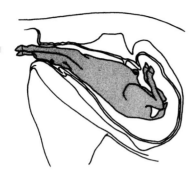

MALPRESENTATION: UPSIDE DOWN

If the foal cannot be turned, orient the head to face outward and assist the delivery of the foal upside down. Be aware that the foal's spine won't bend very far in the wrong direction. Your pulling will be mostly outward and somewhat upward. Wiggling the foal from side to side will help.

17. If you are confronted with three feet, your task is to determine whether there are twins (quite rare in horses) or whether one of the hind feet is up in the birth canal. If it is a hind foot, it must be pushed back into the uterus. The head and neck should be aligned and the foal delivered as in a fatigued mare.

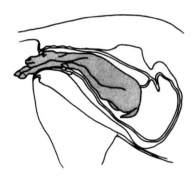

MALPRESENTATION: THREE OR FOUR LEGS FORWARD

Twins must be delivered one at a time. It is very difficult to sort the legs, necks, and heads when you can't see them and when you have only an extremely fatigued arm to sort with. Work at it until you have two legs that belong together. Put the obstetrical chains over these so you won't lose them, then push the foal back in until you have the head and neck oriented correctly and started into the birth canal. Now exert a pull on the chains until the feet, legs, and muzzle are coming as they should. Allow the mare to assist you with her contractions. When the first foal is delivered, examine the position of the second. If it is normal, and the mare is not fatigued, allow her to at least start the delivery. Assist her as you would a fatigued mare delivering a normal foal.

18. A *full-breech presentation*, with the hind legs under the body, is very difficult to correct. It is especially difficult if you allow the mare to strain for any length of time and thereby really jam the foal's buttocks well up against the pelvic arch. If the mare ap-

pears to be struggling and nothing is showing at the vulva, examine immediately for the possibility of a complicated, full-breech delivery.

MALPRESENTATION: BREECH

The hind feet must be brought up into the birth canal before the foal can be born. It is very tight in the uterus, and the foal's legs and hocks bend only in one direction, so anticipate a great deal of hard work to correct this situation. Basically, the hocks must be bent to a position under the foal's buttocks, as if he were sitting on them, and from there the legs can be bent backward and pulled into the birth canal. Keep in mind that the mare is going to be straining during all of this and that you will have enough room to do this maneuvering only when she is relaxed and you can push the foal back into the abdominal cavity.

19. If you should ever be presented with a *foot or both feet protruding from the anus* (not from the vulva as they should be), immediately intervene by inserting your arm into the rectum and pushing the leg or legs back through the hole. Then quickly wash and insert your hand into the birth canal and pull the leg or legs back into it. (Note that the birth canal also has a hole in it because the foot had to go through it in order to poke into the rectum.) From this point on, handle the delivery as one of the previously described situations.

CAUTION: A veterinarian must be called to sew up the tears in the mare's vagina and rectum.

20. There are other possible types of malpresentation, but it is impossible to cover them all here. If you can prepare yourself to mentally and physically handle those covered here, you will, with the same determination and knowledge, be able to handle other situations.

CAUTION: At this point, the casual, would-be equine midwife should be thoroughly frightened and intimidated. The stakes here—the life of a mare and foal—are too great to attempt something casually. If the veterinarian can be summoned, the best advice is always to summon him. If none is available when you need the help, you are out of the realm of a casual midwife and into a very real do-or-die situation. If you have prepared for such an emergency, and if you will keep your wits about you, disaster is not a forgone conclusion. You must determine *if* a problem exists, decide *how* to remedy it, and then simply (or not so simply) do it!

6.29 RESUSCITATING THE ASPHYXIATED FOAL

Today, cardiopulmonary resuscitation (CPR) is more descriptively called cardiopulmonary cerebral resuscitation (CPCR). CPCR is the restoration of a heartbeat and the consequent maintenance of brain (cerebral) function. There is a need to perform CPCR when the foal has been deprived of oxygen (asphyxiated). The usual time when asphyxia presents is following a difficult birth. Usual signs of asphyxia include the foal not breathing, gasping for breath, breathing at a rate less than 10 respirations per minute, heartbeat is absent or less than 40 bpm, or there is no response to stimulation.

Equipment Necessary

There is no specialized equipment necessary for foal resuscitation, although a foal resuscitation mask and pump would make the process easier if you have been instructed in their proper use.

- Towel (rough-textured)
- Nitrite gloves
- Basting syringe (6" to 10" length)

Restraint Required

There is no need for restraint of the foal. The mare will need to be restrained by an attendant or even better she should be removed from the stall where the foal is being resuscitated. She will fret a good bit, but better that she does not endanger you and the foal as she charges about tying to free herself from the attendant.

Step-by-Step Procedure

1. Clear the airway of the foal. This means to remove all membranes and mucus from the nose and nasal passages. Use your gloved hand, the towel, and the basting syringe. It is necessary to suction only the nasal passages. Horses do not breathe through their mouths. Any mucus remaining in the mouth will not be a problem. There will be plenty of time to remove it later.

2. Rub the foal vigorously with the rough towel. This may stimulate breathing. Do not shake the foal, hang it by its feet, or swing it about.

3. If the umbilical cord remnant is bleeding more than a few drops, tie it off or clamp it with hemostats.

4. If breathing has not started in 30 to 60 seconds, have an attendant call the veterinarian while you continue trying to save the foal.

5. Place the foal on its side, on a firm surface. Straighten the neck so that there is not an obvious bend in the airway. Clamp one of your hands onto the downside nostril of the foal, thereby closing off that airway. Place you mouth over the upside nostril and blow breaths into it at the rate of 1 breath every 2 to 3 seconds. You are trying to achieve 25 to 30 breaths per minute. The foal's chest should expand with every breath you administer. Continue breathing for the foal for about 1 minute, then stop to assess whether there is spontaneous breathing. If there is not, breathe for the foal for another minute and assess again. Continue this until the veterinarian arrives. If the heart rate establishes at 70 to 80 bpm and breathing is spontaneous, you can stop the CPCR.

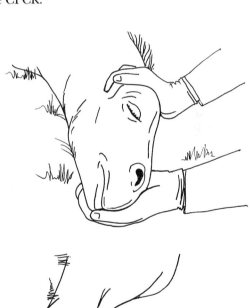

6. If the heart rate is nonexistent or is below 40 bpm, chest compressions should be established. The lack of heart rate indicates that the blood is not circulating. Chest compressions will mimic the beating of the heart and provide a semblance of normal circulation.

7. Breathe for the foal for about 30 seconds and check if the heart has started. If it has, then wait briefly to see if the heart rate reaches 40 bpm, if it does not, begin the compressions.

8. Kneel beside the foal with your chest and shoulders over the foal. Place the heel of one of your hands onto the foal's shoulder area, just behind the foal's elbow. Place your other hand on top of the first hand. Use some of your weight to compress the chest of the foal. You are trying to compress the heart so that it will mimic the natural beating. Use some thoughtful care. If your thrusts are too easy, you will not compress the heart. If your thrusts are too hard, you will crack the ribs and bruise the heart. You are trying to compress at the rate of two compressions every second. Yes, that is rapid!

9. Continue compressions for 3 minutes then check for vital signs of heartbeat, heart rate, and pupil dilation (wide dilation is not good).

10. Prepare yourself for the worst. CPCR is only successful 10 to 15% of the time in humans. Horse CPCR is even less successful . . . approaching 10%.

Postprocedural Management

The foal that has survived the trauma of asphyxiation and CPCR is in need of professional veterinary care. There was a reason that the foal's systems were shut down. That causative agent or problem must be diagnosed and dealt with.

6.30 CARE OF THE NEWBORN FOAL

After the delivery, both the mare and foal will probably show a need for and desire to rest. They will remain lying on the straw-covered stall floor for 30 minutes or more. The foal has the typical bedraggled and bewildered look about him that only the equine mother can call handsome. The mare looks weary, quite maternal, and certainly well pleased with herself. This is as things should be, and you should not disturb them until after one or the other rises and breaks the umbilical cord.

The severing of the umbilical cord, which normally occurs the first time the mare stands up following delivery, signals the beginning of the next series of management techniques that the horse owner must perform if he intends to assure the well-being of the foal.

Equipment Necessary

- Lead rope and halter for the mare
- Foal halter
- Flashlight
- Tetanus antitoxin
- Combiotic (penicillin–streptomycin)
- Syringes and needles
- Tincture of iodine, 7%
- Baby-food jar
- Twitch
- Prepackaged enema fluid

Restraint Required

There are two restraints required, one for the mare and one for the foal. The mare must be restrained if you are to be able to work freely upon the foal. Her maternal instincts will be very strong, and she may challenge or actually attack you as you enter the stall. You can best cope with this situation by entering the stall with the halter and lead rope in hand and talking in a soothing, moderate voice without a show of anxiety. Be sure to be careful, but walk right in, go directly to the mare, halter her, and proceed. She may lay her ears back, bare her teeth, and bluff a charge by swinging her head and neck at you. Call this bluff by proceeding as indicated, but be careful not to come between her and her foal or to allow yourself to be maneuvered into a corner.

Once the mare is haltered, have an assistant hold her while you attend to the foal. Allow the mare to remain in the stall, but position her against one wall. This leaves most of the stall open for work and prevents the foal from running to the other side of the mare for safety. Allow the mare to "have her head." You want to restrict her large movements, not her observation of her foal.

As you begin to catch the foal, keep in mind that you are teaching him his first lesson. He is either about to learn that trying to escape is futile, and besides, they don't hurt you anyway, or that all you need to do to escape is to scramble quickly and fight strongly. You must calmly move about until the foal

is cornered and not grab or lunge at him until you are certain of latching onto and holding him. The foal should be held as shown in the illustration. At this point, work can begin on the neonatal management of the foal.

Step-by-Step Procedure

1. While the foal is being held in the manner just illustrated, have the third person involved begin neonatal care by dipping the navel stump. Pour an inch or more of the 7% tincture of iodine into a clean baby-food jar. Place this opened jar over the navel stump, hold it firmly against the foal, and shake it vigorously for about 30 seconds. Remove the jar and allow the remaining cord and stump to drip dry. Any iodine stains on the abdomen or legs will disappear in a few days. You may notice an umbilical hernia while dipping the navel. Treatment for this is discussed in step 11.

Tincture of iodine is a strong 7% (harsh) compound. It does cause an irritation to the navel stump, and there has been effort to find a less harsh, but equally effective, replacement. Tincture of iodine also stains skin and clothing quite readily. Some owners are using "tamed iodine" compounds in an effort to cut back on the irritation and the staining. Chlorhexidine is another option, although the recommendation is to apply it for 30 seconds for five consecutive days to achieve maximum benefit. Reports for chlorhexidine are promising, but many owners will not have the opportunity to treat for five consecutive days. Tincture of iodine has worked well for many years, and it is indeed difficult to fix something that is not truly broken.

2. Examine the foal's inguinal area to determine its sex. The presence of a very small udder and two half-inch teats in the female, as compared to a scrotum and penis in the male, must be accompanied by a vulva and anal opening in the female, as compared to an anal opening only in the male. Your examination may seem unnecessary, but there are developmental problems in fetuses that seal over the anal opening or vulva. Your quick examination will reveal any problem in this area. The testes may or may not be descended into the scrotum of the colt at the time of this early examination. It is of no real concern at this time, but it does provide a reference point for interpreting any problem with cryptorchidism that may crop up. A large mass of tissue in the scrotum indicates a scrotal hernia, about which a veterinarian should be consulted.

3. Examine both eyes to be certain that the condition known as *entropion* (inverted eyelids) does not exist. In this condition, the eyelashes, usually the bottom ones, are constantly rubbing the eyeball. If you look closely, you can see this with your naked eye. If the eyelashes look normal, they are normal. If the eyelashes are not visible, they are inverted against the eye or rolled completely inward and downward. If they are inverted, you must manipulate the eyelids with your fingers and evert them. Continue this manipulation for a minute or so per affected eye, and repeat three or four times daily. If the condition does not correct itself in 3 or 4 days, summon the veterinarian, who will perform a minor surgical technique. If you ignore the problem, loss of sight is an almost certain consequence.

4. Check the lips, gum pad, and roof of the mouth. In the vast majority of cases, these areas are developed and closed as they should be. In a very few others, the lip, gum pad, and roof of the mouth may be incomplete, or "cleft." If you should find such a condition, contact your veterinarian for advice.

During this examination, check also for overbite or underbite and note for your own reference whether teeth are present or not. Some foals have two front incisors at birth, while others pick them up in the first 10 days. Undershot or overshot jaws are not a real concern at this point, but your examination now provides a reference point for later exams.

5. Inject the foal with a full dose of tetanus antitoxin. Many horsemen will also inject both the foal and mare with appropriate dosages of a penicillin–streptomycin combination antibiotic and repeat the combiotic injections on the following 2 days for both the mare and foal. The tetanus antitoxin is cheap insurance against tetanus, and there is an open avenue for the entry of tetanus spores after foaling. Combiotic injections should help both animals to resist bacterial infections during the stressful first few days.

6. Run your hands over every inch of the foal's body, checking for lumps, cuts, swelling, and bone or tendon abnormalities. Make a written note of anything unusual that you find. Pick up each of his feet, but only briefly, and note the gristlelike pad on the bottom of the hoof. This is called the "golden hoof"; its purpose is to prevent uterine damage from the foal's hooves. This handling is also good for the foal's "people-izing."

During this step, you should closely examine any gross abnormalities that are present. Some of these are extreme enough to warrant humane destruction of the foal. You and your veterinarian should make this decision. Others, such as crooked legs, knock-knees, contracted tendons, and totally relaxed pasterns will either spontaneously improve with time or respond to appropriate nursing, veterinary, and orthopedic care. Record all of your observations during this initial exam. You will need a reference point from which to evaluate progress.

7. Remove the halter from the mare, release the foal, and step back out of the stall. Continue observing the pair until you see the foal nurse. This early milk, or colostrum, is essential for the foal's disease resistance. The earlier he receives it, the better off he is. If the mare should not allow the foal to nurse, she must be restrained, usually just by holding the halter, or at most by applying a twitch to her nose. Most mares who won't allow nursing are simply sore-teated from the engorgement of the udder. Once the foal has nursed, the pressure and soreness is gone and the foal is allowed to nurse at will.

This first nursing will usually occur within the first hour. If 2 hours have elapsed without nursing, assist the foal by holding him in position. Your management treatments usually will stimulate the foal's system sufficiently to cause him instinctively to seek something to nurse.

Over the next several hours, you must also note whether the foal nurses both sides of the udder. Occasionally, a foal uses only one side while the other grows painfully engorged. When he does bump it or attempt to nurse it, he may be rejected. Handle this rejection in the same manner that you handled the first nursing.

8. Your next concern is whether the foal passes the *meconium* and has his first bowel movement. The meconium is a black, tarlike substance that plugs the rear of the digestive tract of the foal during gestation. The colostrum is usually laxative enough to cause this to be loosened and passed. If it is not passed by the end of the first day, the foal must receive an enema.

Catch and restrain both animals as before. Use a commercially prepared, prepackaged, 4-ounce bland enema. Carefully insert the 2" tip of the container into the foal's rectum and administer the enema gently. Results are sometimes immediate, so watch where you are standing! If a second enema is indicated by the foal's acting colicky or straining, by all means administer another. If the problem still exists, contact your veterinarian.

The normal physiological parameters for the foal and the normal times of occurrence are as follows.

Physiological Parameter	Normal Measurement or Time of Occurrence
Heat rate—lying down	40–80 bpm
Heart rate—standing	100–150 bpm
Heart rate—1 day old, stable	70–90 bpm
Respiration rate	60 breaths/min
Umbilical cord ruptures	15–30 min after foaling, when mare stands
Foal stands	1–2 hrs
Foal nurses	1–2 hrs
Temperature—rectal	98.6°F–101°F
Meconium passed	Usually 4–6 hrs. If not passed by 24 hours, give enema.

9. The first week in the life of a foal is critical to him and he must be watched carefully for any changes in his overall attitude. Sudden lethargy or dullness, sudden hyperexcitability, or a lost desire to nurse may indicate a potentially serious problem that should be checked immediately. Always contact your veterinarian at the first sign of unexplained trouble with a foal. Certain birth injuries, septicemias, "joint-ill," *atresia coli, atresia ani,* isohemolytic disease, and immune deficiency can all affect the foal during this first week to 10 days.

10. Diarrhea is a common problem in the foal that probably will crop up during the second week of life. Its onset will more often than not coincide with the mare's foal heat. Foal heat, or the mare's first heat after foaling, usually occurs somewhere between 5 and 11 days postpartum. Because of the timing, this attack of diarrhea is termed the *foal-heat diarrhea.* Although researchers are uncertain about exactly what does cause this diarrhea, they are convinced that it is not related to the hormone estrogen that is present during heat periods of the mare. Ingestion of "lochia," a disrupted feeding schedule, milk overload from constant nursing, and the strongyloid family of internal parasites have been suggested as causes.

Whatever the cause, management of foal diarrhea is important. Persistent diarrhea will quickly result in dehydration of the foal if it remains unchecked and untreated. A first step for the foal manager, following his initial observation of scoured feces on the tail, underside, and buttocks of the foal, should be to wash and dry the scoured areas. Vaseline should then be generously rubbed over the previously scoured area. If this is not done, the caustic nature of the scoured feces will cause the foal's hair to be sloughed off in this area. In most instances, the diarrhea will pass in 3 or 4 days. Observe the foal closely for signs of illness (see step 9). If the diarrhea does not pass, or if it worsens, administer 2 to 4 ounces of a kaolin–pectin mixture orally. A dose syringe can be used to administer the product. Choose a flavored product and the foal will swallow it without hesitation. The tube on the dose syringe should be inserted into the interdental space at the side of the mouth and directed toward the throat. Depress the plunger in small increments so that you do not overwhelm the foal. It is best if he swallows it himself, but if he is reluctant, tip his head upward slightly, use small depressions of the plunger, and allow the fluid to drain toward the foal's throat.

CAUTION: Do not shove the syringe tube way back into the throat or tilt the head at an extreme angle. You may cause the foal to aspirate the fluid into his trachea and lungs. Immediate strangulation or pneumonia is the result. The best advice is to use a flavored product and allow the foal to swallow it himself. If the diarrhea does not respond, contact your veterinarian.

11. Umbilical hernia is another problem that your foal may have at birth or develop shortly thereafter. The hernia occurs very close to the stump of the navel and appears as a lump below the skin. It may be as small as a walnut or as large as an orange. It should not be a cause for concern beyond noting that the condition exists. It should be palpated (and found to be soft) and manipulated (to see if it can be worked back into the abdominal cavity) periodically. If it enlarges or becomes hardened, veterinary attention is needed. Umbilical hernias normally repair themselves by the time the foal is 6 months old. If the condition still exists at that time, the veterinarian will correct it surgically.

12. The immunization program and parasite-control measures should begin early for the foal. Tetanus toxoid, Eastern-Western encephalomy-elitis vaccine, rhinopneumonitis vaccine, and other vaccines or bacterins commonly administered in your area should be started by the time the foal is 3 months of age.

Ascarids (roundworms) and strongyles (bloodworms) are a particular problem for the young horse. By the time the foal is 2 months of age, it should be wormed for ascarids. Repeat this worming at 3 months, coupled with a product effective against strongyles. For the rest of the first year, worm the foal every 60 days with a product or products aimed at ascarid and strongyle control. Once, in December or January, add a boticide to the program.

13. An orphan foal can be reared successfully if close attention is paid to a few guidelines. The likelihood of raising an orphan foal is greatly increased if the foal received colostrum. If he does not, he will

not be resistant to the multitude of infections that he must face in his first few days of life. Ideally, the foal should receive colostrum within the first 6 hours of birth, but it continues to be of value to him for up to 24, and perhaps 36, hours. After that time, the foal's system can no longer absorb the antibodies in the colostrum. This means that if the mare should die after delivery, she should immediately be milked out as thoroughly as possible. Wrapping the udder in warm, wet towels will help some. Partition the colostrum into three or four servings and refrigerate them. These should be given to her foal during the first 6 to 12 hours.

If the mare cannot be milked out, colostrum from another mare can be used. A neighbor, large breeding farm, or your veterinarian may be a good source. The mare selected for donating colostrum must have foaled within the past 2 or 3 days. If she foaled before this, her fresh milk will no longer contain the necessary level of antibodies. Many breeders are milking excess colostrum from their mares with foals at side and freezing it for just such an emergency. Year-old colostrum, if it has been properly frozen, is still valuable.

14. There is a slight possibility that an orphan foal can be grafted or transferred to a foster mother. This foster mother may have lost her foal at birth. Whatever the situation, much patience and ingenuity is required to "fool" or convince the mare to take a strange foal. There is also the possibility of buying a fresh nanny goat and teaching the foal to nurse from her. This may sound far-fetched, but after the conditioning period, the nanny learns to climb up on a box or boxes so she is high enough for the foal to reach.

15. If you decide to bottle-feed the foal, you will need a large soda bottle, a couple of lamb's nipples, and a recipe for a foal formula. Straight cow's milk or straight condensed milk is too rich and must be modified. The following formulas have proven successful:

Formula 1

1 pint condensed milk
1 pint warm water
2 tablespoons syrup
4 tablespoons limewater*

*Limewater is 1 part lime (calcium hydroxide) mixed with 700 parts water. A drugstore will mix this for you.

Formula 2

1 pint 2% cow's milk
4 ounces limewater*
1 teaspoon sugar

Formula 3

20 ounces whole cow's milk
12 ounces limewater
2 ounces sugar

Formula 4

FOAL-LAC (Borden) (as directed)

Consult Table 6.3 for feeding amounts and schedules.

TABLE 6.3   Feeding Information for Bottle-Fed Foals

Age of Foal	Feeding Frequency	Amount of Formula per Feeding
First 7 days	Hourly	6–8 oz
2nd week	@ 3 hours	16–24 oz
3rd and 4th weeks	@ 4 hours	24–32 oz
5th week	@ 6 hours	32–48 oz
Weaning*	@ 6 hours	No more than
	Until weaning	48 oz per day

*See step 17 for a discussion of weaning.

If it is more convenient for you, a foal can be taught to drink from a pail instead of a soda bottle and nipple. Whatever your choice, the containers must be kept clean by washing and drying them after each feeding. No more than a 24-hour supply of formula should be mixed at any one time.

16. All foals, especially an orphan foal that will be "early-weaned," should be introduced to creep feed at an early age. Most horsemen take care of this task during the first 2 to 3 weeks of age. If the foal has a mother, he will see her eating and learn the process by mimicking her. He will probably eat with her out of her feed box or tub. If she will not allow this, build him a separate feeder from which she cannot eat.

Select a high-quality creep ration and start the foal out by placing just a handful of the ration into the feeder. Gradually increase the amount until it reaches 1 pound per day. Hold it there until the foal is 4 weeks of age. At that time, start increasing it again by quarter-pound increments until it reaches 2 pounds per day. Hold the amount at 2 pounds until the foal is 6 weeks of age. Continue this progression of 1-pound increases every 2 weeks, until you reach a point where the foal does not quite clean up the amount fed in a 24-hour period. Feeding him at that level will maximize his genetic growth potential. The idea is to keep as much feed in front of him as he will eat. High-quality legume or grass–legume hay should also be fed at the rate of 1 pound of hay per 100 pounds of body weight. If at the very beginning the foal is reluctant to eat, sprinkle some extra sugar or syrup on a handful of the hay and forcefully (very gently!) place it into his mouth. Do not hand-feed any longer than necessary to start him eating.

17. There are differences of opinion about when to wean foals. Some say 6 months, some 4 months, others 2 months. The foal should stay with the mare as long as she continues to supply a significant amount of nutrition to him; that is, as long as her milk production remains at a high level. (In the case of the orphan foal, weaning at 8 to 10 weeks is advised. Be certain that the foal is eating hay and grain at the suggested levels.) In the large majority of mares, milk production has declined significantly by the fourth month of lactation. Whatever your choice, start the weaning process 5 to 7 days before the actual separation. Cut back on the mare's grain ration by 10 to 15 percent daily until it reaches one-third of what it was before you started the cutback. This will help to reduce her milk supply. On the day before weaning, remove one-half of the mare's water and keep it reduced until the second day after weaning. When you separate the mare and foal, do not give in to their incessant fretting and reunite them.

Do not milk the mare out. She will swell and become sore in the udder, but the nonremoval of milk will help to shut down its production. Milking her out only prolongs the inevitable. In 3 or 4 days, her udder will begin to reduce in size.

When you separate the mare and foal, an attempt should be made to move them out of sight and sound of each other. Put the foal into a pen with an older gelding or dry mare so that he is not physically *and* psychologically alone. You must protect him from himself during these first few days. Fences should be high and tight and all imaginable hazards removed from his environment. If it is necessary to pasture the mare and foal together again you may do so, but not before 4 to 6 weeks have elapsed since weaning.

Equine Genetic Diseases

Glycogen branching enzyme deficiency (GBED)	Homozygous recessive trait	Carrier unaffected. Mare *and* stallion must be carriers of the recessive gene.
Hereditary equine regional dermal asthenia (HERDA)	Homozygous recessive trait	Carrier unaffected. Mare *and* stallion must be carriers of the recessive gene.
Hyperkalemic periodic paralysis (HYPP)	Homozygous dominant trait	Carrier unaffected. Mare *and* stallion must be carriers of the dominant gene.
Junctional epidermolysis bullosa (JEB)	Homozygous recessive trait	Carrier unaffected. Mare *and* stallion must be carriers of the recessive gene.

Equine Diseases of Genetic Origin

Most foals are born healthy and vigorous and in a few short days and weeks develop into active, energetic young horses. Some, however, seem to never get started, languishing from the very outset, no matter how carefully you attend to neonatal care. There is a growing awareness that these foals may be presenting some early stages of diseases or defects of genetic origin. Above are some currently common equine diseases of genetic origin, and a brief explanation of the inheritance mode for these afflictions.

Genetic Principles

Homozygous means the forms of the gene are the same.

Heterozygous means the forms of the gene are different.

Genetic Basis of Homozygous Recessive Trait Disease

G = dominant form of the causative gene

g = recessive form of the causative gene

GG = homozygous dominant—individual is genetically normal and appears normal

G g = heterozygous for the gene—carrier of recessive gene (g)—appears normal

g g = homozygous reessive—carries both recessive genes—has disease

Genetic Basis of Homozygous Dominant Trait Disease

G = dominant form of the causative gene

g = recessive form of the causative gene

GG = homozygous dominant—individual carries both dominant genes—has disease

G g = heterozygous for the gene—carrier of dominant gene (G)—appears normal

gg = homozygous recessive—individual is genetically normal and appears normal

In the table above, there are three diseases listed that present themselves when a foal is born in the *homozygous recessive* condition. That is the foal must be "gg" for the gene(s) that cause the disease. In the other disease, just the opposite is true. For the *homozygous dominant* disease, the foal must be "GG."

Keep in mind that for the diseases in the table, neither the "gg" nor "GG" foal survives to breeding age. Study the schematics below to see how the disease is passed from normal-appearing parents to offspring.

Homozygous Recessive Trait

Stallion × Mare (both normal)	Stallion × Mare (one normal, one carrier) (carrier appears normal)	Stallion × Mare (both carriers) (both appear normal)
GG × GG	GG × Gg	Gg × Gg

[Above are the *genotypes*, the genetic makeup, of the stallion and mare.]

[Below are the gametes, the sperm and egg, each with one-half of the genetic makeup of the stallion and mare.]

G or G x G or G   G or G x G or g   G or g x G or g
GG GG GG GG   GG Gg GG Gg   GG Gg Gg gg

Homozygous Dominant Trait

Stallion × Mare (both normal)	Stallion × Mare (one normal, one carrier) (carrier appears normal)	Stallion × Mare (both carriers) (both appear normal)
gg × gg	gg × Gg	Gg × Gg

[Above are the *genotypes*, the genetic makeup, of the stallion and mare.]

[Below are the gametes, the sperm and egg, each with one-half of the genetic makeup of the stallion and mare.]

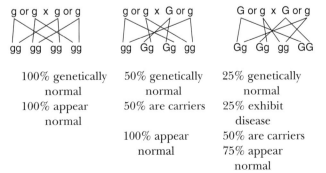

g or g x g or g   g or g x G or g   G or g x G or g
gg gg gg gg   gg Gg Gg gg   Gg Gg gg GG

100% genetically normal	50% genetically normal	25% genetically normal
100% appear normal	50% are carriers	25% exhibit disease
	100% appear normal	50% are carriers
		75% appear normal

6.31 CASTRATION

Castration involves the surgical removal of the testes, epididymides, and a portion of the spermatic cord. In the horse, it is always accomplished under at least local anesthesia accompanied by a general sedative, or under general anesthesia.

Most colts are castrated between the ages of 12 and 24 months. Castration helps to prevent the crestiness and heavy-frontedness found in intact males and it may improve an unmanageable horse's disposition. It almost certainly will increase a horse's monetary value because knowledgeable horsemen are always anxious to buy a good gelding.

Castration is always indicated if the stallion in question has a heritable conformation defect. An excellent example of this is the cryptorchid horse, sometimes called a "ridgling," "original," or "cryp." The cryptorchid horse carries one (unilateral cryptorchid) or both (bilateral cryptorchid) testes abdominally instead of descended into the scrotum where they should be. This *is* a heritable trait and will be passed on from father to son should the stallion be capable of settling mares.

If the stallion is bilaterally cryptorchid, his chance of settling mares is slim because of the reduced production of normal sperm in the abdominally carried testes. In the unilateral cryptorchid, the descended testis can and does produce normal sperm and such a stallion can settle mares. As a mare owner, *you* should verify that any potential mate for your mare has both testes descended into the scrotum.

There is an additional concern in managing the cryptorchid horse. While the normal sperm production of abdominal testes is reduced, the production of the male sex hormone testosterone is *increased*. This means that the cryptorchid stallion is likely to be more aggressive and show more desire to breed than a normal stallion. This is because the elevated temperature of the abdomen, as compared to the scrotum, increases the rate of activity of the cells that produce testosterone.

CAUTION: Castration of a horse is a far different consideration than castration of a calf, lamb, or pig. Calves weigh from 100 to 500 pounds at castration and have a descended scrotum. Lambs weigh from 10 to 30 pounds when they are castrated, if they are castrated at all, and they, too, have a descended scrotum. The young pig comes closer to the horse in that it carries its scrotum tight to the body, but it differs from the horse in that it is castrated in the first days or weeks of its life.

The horse is almost always of nearly mature size when you reluctantly admit that he is not the next world champion stallion and decide to geld (castrate) him. This is no place for the amateur surgeon. There is a very real risk to you (kicking and striking) and to the horse (bleeding and overdosing of tranquilizer) when you try to handle an animal of this size. Experienced veterinary surgeons often need all of their knowledge, tools, and skills to save a patient that is bleeding postoperatively.

For these reasons, it is recommended that you contact your veterinarian and have him perform this task, once you have made the decision to geld your stallion. It should be pointed out that the earlier you perform this operation the easier it is on you, your veterinarian, and your horse. There is no benefit to leaving your colt intact beyond the February or March following his birth year.

6.32 PARASITE CONTROL PROGRAM

The parasite load in horses can be controlled and held to a low, minimum-impact level through a carefully constructed and closely followed management plan. If internal parasites are not controlled, they will cause irreparable internal organ damage to the horse, cause the horse to look shabby and unkempt, and cause you to spend a great deal of money as you try to treat the organ damage or purchase supplements to improve the "bloom" of the horse. For you and the horse, it is a far better approach to have a parasite control program and to follow it.

There are considerably more than 100 internal parasites that affect the horse. Fortunately, if the horse manager focuses control efforts on the *large and small strongyles (bloodworms), ascarids (roundworms), bots, pinworms, tapeworms, and strongyloides (threadworms)*, most of the others will also be controlled.

Control of internal parasites involves more than the administration of one of the marketplace dewormers. While there are excellent products on the market today that will do exactly what they say they will in the advertisements, your control plan must follow some basic principles for maximum product efficacy to occur.

Step-by-Step Procedure

1. Recognize that there are at least seven different classes of deworming medications for horses:

SIMPLE HETEROCYCLIC COMPOUNDS:
 Piperazine
BENZIMIDAZOLES:
 Thiabendazole
 Mebendazole
 Fenbendazole
 Cambendazole
 Oxfendazole
 Oxibendazole

IMIDAZOTHIAZOLES:
  Levamisole
TETRAHYDROPYRIMIDINES:
  Pyrantel pamoate
  Pyrantel tartrate
PHENYL-GUANIDINES:
  Febantel
AVERMECTIN AND MILBEMYCINS:
  Ivermectin
  Moxidectin
ISOQUINOLINE-PYROZINES:
  Praziquantel

2. Select three or four deworming medications from the marketplace *that represent different classes of medications* from the list. Rotate these dewormers throughout the year.

CAUTION: Dewormers must be rotated so that the parasites in your horse do not develop a resistance to the medication. Be certain that you are not rotating dewormers selected from within one of the classes from the previous list. If that were the case, resistance would result and the products would not control the parasite load.

3. Realize that certain classes of deworming medication may be especially effective against certain internal parasites, but nearly useless against others. Your parasite control plan must take this into consideration. For example, organophosphates are especially good against bots, but ineffective against roundworms and strongyles. If your horse develops tapeworms, only the pyrantels and praziquantel will control them. Strongyles, roundworms, strongyloides, and pinworms are controlled by the other classes. The avermectins have the broadest spectrum of efficacy.

4. Be certain that the correct full dosage reaches the stomach of the horse. This implies that you have read the directions, know the weight of your horse, know the pregnancy status of the filly or mare, are using the proper administration technique, and that the horse has not rejected the product after you stopped watching.

Top-dress dewormers, those that are placed into the feed bucket with the daily grain, are notorious for being rejected by the horse. It may be the smell or it could be the taste of the product that cause the rejection. Whatever the cause, if the horse does not consume the grain with the dewormer laced into it, the horse has not been dewormed. Likewise, if a paste dewormer is used—and they are the most popular today—be certain that the correct dosage is deposited into the rear of the mouth and that the horse swallows it. After pasting the horse, observe it for a few minutes to be certain that he does not spit the paste out after holding it in his mouth for a few minutes.

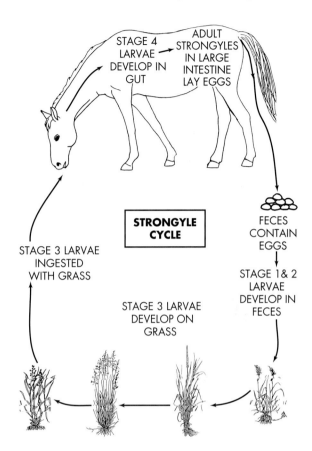

STAGE 4 LARVAE DEVELOP IN GUT

ADULT STRONGYLES IN LARGE INTESTINE LAY EGGS

STRONGYLE CYCLE

STAGE 3 LARVAE INGESTED WITH GRASS

FECES CONTAIN EGGS

STAGE 1 & 2 LARVAE DEVELOP IN FECES

STAGE 3 LARVAE DEVELOP ON GRASS

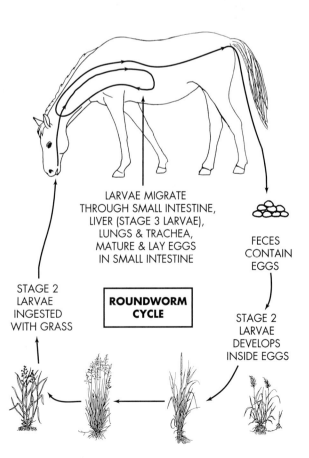

LARVAE MIGRATE THROUGH SMALL INTESTINE, LIVER (STAGE 3 LARVAE), LUNGS & TRACHEA, MATURE & LAY EGGS IN SMALL INTESTINE

FECES CONTAIN EGGS

STAGE 2 LARVAE INGESTED WITH GRASS

ROUNDWORM CYCLE

STAGE 2 LARVAE DEVELOPS INSIDE EGGS

5. Manage the environment in which the horses are maintained.

  a. Pick up and dispose of manure daily.

  b. Compost the manure, if it is to be spread back onto the pasture areas.

  c. Prevent fecal contamination of the feed supply and water sources.

  d. Feed hay from a manger and grain from a bucket, instead of feeding from the ground.

  e. Rotate pastures. This has a double benefit: it provides greater amounts of grass for the horses and it allows time for the sun and elements to destroy parasite eggs and larva.

  f. Do not overstock your pastures.

  g. Mow and drag (chain- or tooth-harrow) pastures to break up manure and expose eggs and larva to the elements.

  h. Isolate new horses for a sufficient length of time for you to perform a fecal examination and deworm as indicated.

  i. Deworm all horses that have contact with one another, at the same time.

  j. Remove bot eggs from the legs and body of the horses. If they are allowed to remain, the horse will lick them off and ingest them. This starts the whole bot life cycle. A safety razor or bot block works well to remove the eggs.

6. Schedule the administration of the deworming medications for maximum benefit. Examples for a quarterly deworming program and for every 2 months follow. Your exact parasite control program should be an integral part of your comprehensive herd health program. Consult with your veterinarian for a custom program.

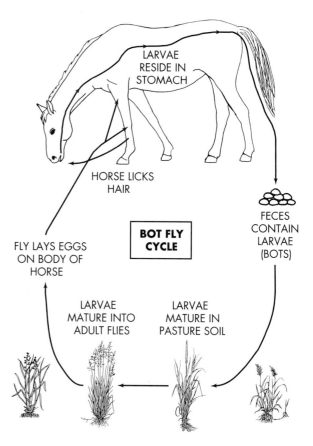

Every 2 Months Deworming Program

Month	Dewormer Class Product
January	Benzimidazole
February	
March	Avermectin
April	
May	Pyrantel Pamoate
June	
July	Benzimidazole
August	
September	Milbemycin + Praziquantel
October	
November	Pyrantel pamoate
December	

Quarterly Deworming Program

Months/Quarter	Dewormer Class Product
January—1ST qtr	Benzimidazole
February	
March	
April—2ND qtr	Avermectin
May	
June	
July—3RD qtr	Benzimidazole
August	
September	
October—4TH qtr	Milbemycin + Praziquantel

CAUTION: Pay particular attention to the deworming program for pregnant mares, foals, and young horses. The foal will be seen to nibble on feces as soon as he is bold enough to wander about the stall. Licking on his mother and stall facilities will also provide an early parasite infestation. Parasites can rob a most significant amount of nutrients from the foal's system, if they are not controlled. And they can cause permanent damage to the liver, lungs, arteries, and stomach. The horse owner will certainly notice the loss of growth and thriftiness and the diminished resistance to disease. The horse continues to grow well into his third year of life. To maximize the genetic potential and your monetary investments, be very attentive to managing the parasite control program for your breeding herd and young horses. Many horse owners will deworm their young horses on a 30-day schedule during the entire first year of life.

7. A list of the *signs and symptoms of a poorly controlled parasite infestation* in your horses includes the following items. However, much more time should be spent planning and administering your deworming program than learning the signs and symptoms of not deworming. This is clearly a case where an ounce of prevention is worth a pound of cure.

Roughened hair coat	Delayed spring shedding
Dull hair coat	Tail rubbing
Unthrifty, dull appearance	Anemic gums, sclera of eyes
Weight loss	Reduced appetite
Dehydration	Diarrhea
Pot belly appearance	Frequent colicky symptoms
Nasal discharge	Cough
Pneumonia	

6.33 FLY CONTROL

Equipment Necessary

- Fly repellant (spray, roll-on, gel)
- Grooming equipment
- Some type of physical barrier to prevent flies from landing on horse

Restraint Required

You are going to groom the horse, then apply a spray to legs, body, and head. Normally, this would not require a great deal of restraint. But, if the horse is new to you, or you already know he spooks at the sound of the spray, you will need to restrain him in whatever manner is necessary to get the repellant applied. A lot of patience can often overcome the fear of the spray.

CAUTION: The horse has no sense of the fact that you are trying to improve his comfort level. If he spooks from the spray, and you are careless about safety zones, you could get stepped on kicked or run over. Be alert and stay in the safety zones.

Step-by-Step Procedure

1. Restrain the horse, preferably in cross-ties.

2. Thoroughly groom the horse. Fly repellant will sit on top of dirt, grime, and dust, if it is not removed prior to application. The repellant must reach the hairs of the horse, perhaps even the skin, if it is to be effective. Pay particular attention to removing the heavy dust.

3. Pour some repellant onto a cloth and apply it in this manner, before attempting to spray it onto the horse. Sometimes it is the smell of the repellent coupled with the sound of the spray that spooks the horse. Start on the shoulder and stay where he can see you. Gradually move from there to the rest of the body and down the legs, but save the head for last.

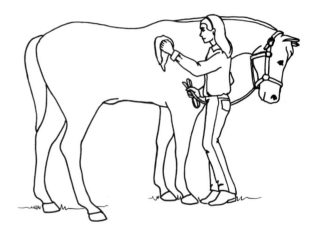

4. Apply the repellant to the head and neck with extra care. Remember to apply it to the ears . . . and well down inside them. Resist getting it too close to the eyes of the horse, unless it states on the product container that it does not irritate the eyes.

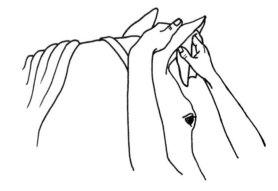

5. At some point, you will want to use a spray repellant. Sometimes a spray is quicker and more convenient to apply at the last moment before an event. Patience is the key to accustom the horse to the "hissing" sound of the spray. Start on a part of the body where he can watch and see you during the whole process. Spraying does not hurt the horse, so once he has accepted the "hissing" noise as something that is associated with you and not some feared critter, he will allow it. Be especially careful around ticklish areas such as the face, flank, and belly.

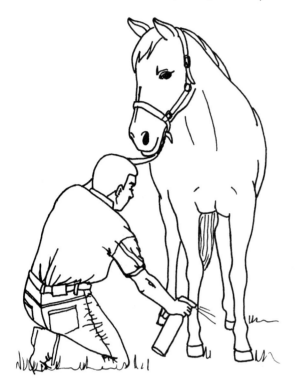

6. Sometimes a physical barrier to prevent flies landing on the horse is the best solution. It is especially good for horses that are maintained on pasture, with a minimum of daily grooming and spraying. If gnats are a nuisance in your area, bonnets that cover the ears are an excellent choice. Horses get used to these bonnets more quickly than you would think.

CAUTION: Be careful when you put a bonnet on your horse for the first time. It will be strange for him. Stay in a safety zone and keep alert.

6.34 BODY CONDITION SCORING

If you feed your horse too much pasture, hay, or grain, he or she will gain weight and become what we humans call "fat." The reversal is also true; too little feed and the horse becomes thin. The concept of fatness or thinness varies from person to person. You may think a horse is fat, while your neighbor considers the same horse to be conditioned "just right," and perhaps a bit on the thin side. Part of the problem lies in the lack of a precise definition of "fat," "thin," or "just right." Body condition scoring is a step toward defining the condition of horses. And, it will make it much easier to converse with one another and to make recommendations for managing various classes of horses. For example, a pregnant mare that is about to deliver and begin lactation should be at a BCS (body condition score) of 5 or 6.

Equipment Necessary
- Body condition scorecard
- Written explanation of body condition scores from 1 to 9

Restraint Required
There is no restraint required for most of the process. During the learning curve, you may want to actually handle the horse and feel the condition as well as see it. In that case, a halter and lead rope are required.

Step-by-Step Procedure

1. Secure a small card showing the skeletal anatomy of the horse. Study this so that you are aware of which bones are supporting the muscle, fat, and skin that you can see.

2. Commit the following body condition scores to memory or create a pocket-sized version to carry with you as you assess your horses. After you do a few assessmental it will not be necessary to have the cards.

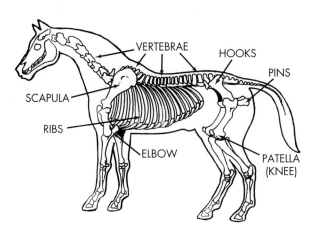

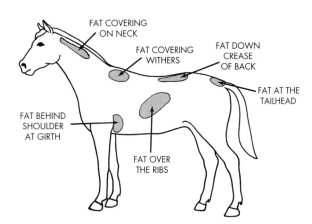

Score	Terminology	Description
1	Poor	Horse is emaciated. Spinous processes of the backbone, ribs, tailhead, and hipbones are prominent. Bones underlying the shoulders, withers, and neck are visible. There is not fat cover on any part of the body.
2	Very thin	Horse is emaciated. Spinous processes might have a small amount of fat cover. Ribs, tailhead, and hipbones are still prominent. Bones underlying the shoulders, withers, and neck are not as visible as BCS 1.
3	Thin	Fat cover is beginning to develop on the spinous processes. There is a slight covering of fat over the ribs, although they are still easily seen. Withers, shoulders, and neck bones are still prominent, though less so than BCS 2.
4	Moderately thin	There is still a slight spinous process ridge along the back. Ribs are faintly visible. Some fat is beginning to develop around the tailhead. Hip joints are no longer visible. Bones underlying withers, neck, and shoulder are no longer visible.
5	Moderate	Back is flat. There is no ridge of spinous processes and there is no back crease due to fat. Ribs cannot be seen, but they are easily palpable. Tailhead is soft and spongy. Withers are rounded. Shoulders and neck are blended smoothly into the body.
6	Moderately fleshy	Slight crease might develop down the back. Can still feel ribs, but there is substantial fat cover over them. Tailhead is soft. Withers have fat along sides. Neck has fat along sides. There is fat behind the shoulders.
7	Fleshy	Slight crease down back. Ribs can be felt, but substantial fat covers them and fat can be felt between the ribs. There is more fat around withers, shoulder, neck, and tailhead.
8	Fat	Obese horse. Definite crease down the back. Ribs felt with difficulty. Tailhead fat globs are very soft. Withers and neck are unnoticeably thickened due to fat deposits. Fat is deposited along inner thighs.
9	Extremely fat	Obese horse. Obvious crease down back. Patchy fat over ribs. Pones of fat are bulging at the tailhead. Withers, neck, and shoulders are showing pones of fat. Fat on inner thighs might touch from side to side. Flank begins to fill with fat.

Postprocedural Management

Body condition scoring of horses, on any species, is a non-invasive technique. So, careful postprocedural management practices are not essential.

However, abrupt dietary changes, either up or down, should always be avoided when feeding horses. Reductions can be made over a 3- to 4-day period. Feed increases should be made more slowly, involving 5 to 7 days.

SHEEP BREEDS

SUFFOLK

Courtesy of American Sheep Industry Association

DORSET

Courtesy of American Sheep Industry Association

RAMBOUILLET

Courtesy of American Sheep Industry Association

MONTADALE

Courtesy of American Sheep Industry Association

COLUMBIA

Courtesy of American Sheep Industry Association

HAMPSHIRE

Courtesy of American Sheep Industry Association

POLYPAY

Courtesy of American Sheep Industry Association

DORPER

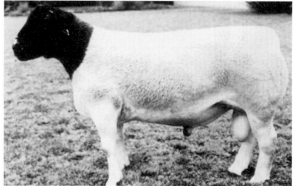

Courtesy of American Sheep Industry Association

chapter SEVEN

Sheep Management Techniques

7.1 INTRODUCTION

To be successful at managing a sheep flock you must like sheep. If you do not, do not attempt to work with them, but select another species, because sheep, perhaps more than most other livestock, respond to the type of management they receive.

Characteristics of Sheep

You can develop a liking for sheep by learning to understand their habits and characteristics. For example, sheep are gregarious and tend to flock together or follow the leader when they move from one location to another. If this trait is understood they can be handled easily. Sheep have limited means of self-protection and are usually rather timid and unaggressive. As a result they tend to run from danger and are easy prey for predators. Through the centuries of their domestication they have come to depend upon man for protection, which must be afforded them in times of danger.

Sheep are very particular in their eating habits. Unlike most other animals, they will refuse to eat or drink from dirty or contaminated feed troughs or water containers. They need fresh feed and water daily, unsoiled by feces or contaminated materials. They have a highly developed sense of smell and will refuse to eat feeds with strong odors, such as fish meal or other meat products. They have a high requirement for salt, which should be supplied daily along with plenty of fresh water. They prefer slowly moving streams rather than stagnant pools, but sheep will not enter water unless forced to do so. They will avoid wet, marshy ground and mud, both of which help to spread foot rot.

Sheep have three sets of glands not found in other ruminants. One set, the suborbital face glands, is located below the eyes. Another set is located in the groin on either side of the udder, and a third set, the interdigital pouches, is located between the toes on each foot. All secrete fatty or oily materials. If the glands between the toes become plugged with mud and their secretions retained, sheep may develop a type of lameness that resembles foot rot.

Sheep are superior to cattle in utilizing grasses and other forages and can obtain 80 to 90% of their yearly nutrients from forages alone. For this reason they should be limited in the amount of cereal grains and other concentrates consumed. Mature sheep have a low glucose tolerance; their blood contains only 40 to 50 mg of glucose per 100 ml. Excessive consumption of concentrates results in enterotoxemia, founder, diarrhea, and bloat, and can be fatal in many cases.

Sheep prefer to graze in the early morning and late evening. They consume the shorter grasses, so pastures with more than 6 to 8 inches of growth are unpalatable to them. They will destroy 90% of the weeds with which they come into contact. They prefer to graze on higher ground, which may be beneficial in spreading manure on less fertile areas. Sheep manure has a higher fertilizing value than that of any other farm animal except poultry.

Most sheep are subject to infestation by internal parasites, which have a secondary life cycle on pastures. Without an adequate program of parasite control, sheep may become heavily parasitized on pastures. This can result in reduced weight gains in lambs, anemia, diarrhea, and eventual death.

Because of their ability to utilize forages, sheep can spend a large part of the year on pastures. This can help to reduce feed costs, especially for the mature sheep. Housing will be necessary during the lambing period, but need not be expensive or elaborate. Protection from severe storms, snow, and wind should be provided for young lambs or newly shorn sheep during the winter months.

The critical temperature, or ambient temperature below which the sheep must speed up its metabolic rate to maintain a constant body temperature of 102.3°F, is 28°F for a ewe in full fleece on a maintenance ration. For the same ewe, newly shorn, this increases to 50°F.

Because of the high cost of land and competition with grain farming, many sheep producers in the midwestern and eastern parts of the United States are finishing lambs in confinement on slotted floors. This has the advantages of eliminating predators, reducing the parasite load, and eliminating the need for bedding and fencing, as well as increasing the number of sheep on a given area of land. A high level of management is required for this type of operation, but it usually results in a shorter finishing period, higher rates of gain, greater feed efficiency, and a reduction in labor if the operation can be automated.

Be a Good Observer

Once a person has learned to like sheep and has decided to raise them, he or she must become a keen observer. The development of this ability will result in fewer problems and a greater productivity in the flock. A noted Scottish sheep owner once said, "Every sheep has a message to give, but only the best shepherds can read it." A person who is familiar with the appearance of a healthy sheep can detect one that is ill or having problems. Sheep that are not alert, allow their ears to droop, have a glassy look to their eyes, segregate themselves from the rest of the flock, refuse to eat, or do not conform to the normal habits of sheep usually are showing evidence of illness or disease. If one can detect this unusual behavior early, the old adage that "a sick sheep is as good as a dead one" can be avoided, and treatment can be started before it is too late. Most deaths result from the fact that the disease is not detected early enough for the treatment to be effective.

Because of the gregarious nature of sheep, a sick one, out of habit, will stand at the feed trough with healthy sheep and appear to be eating, while in effect she is not. Young lambs that wiggle their tails and appear to be nursing may not be getting milk and may die of starvation. Daily observation is necessary, therefore, during the period of late gestation and early lactation to detect unusual behavior and evidence of illness. Less frequent observation is required during the rest of the year, but one will find that it is time well spent.

Timing of Management Operations

The timing of management operations is perhaps more important with sheep than with other species of livestock. Jobs that are delayed or not done can result in poor productivity throughout the year.

Because sheep are seasonally polyestrous in their breeding habits (that is, most breeds normally mate in the fall months only), a delay in the breeding season results in a delay in the lambing season the following spring. This can result in increased housing and feeding costs, poor utilization of pastures, and a lower price for the lambs that are marketed during September and October, when the price of slaughter lambs usually drops.

Timing is very important in developing an effective program for control of internal parasites if the program is to be effective when damage by parasites is greatest.

Such simple management jobs as docking, castrating, weaning, vaccination against enterotoxemia, and a treatment of nutritional deficiencies must be completed on a definite time schedule for best results. Each sheep producer should develop a management calendar for each month of the year to be sure that each job is completed on time and not overlooked.

Profit Potential of Sheep

The profit potential of the sheep enterprise as compared to alternative sources of income should not be overlooked. In many small flocks, profit has not been a major consideration; but in larger flocks, where sheep are the main source of income, it must be considered. In any given year, reasonable prices are received for market lambs sold, but the price paid for wool has fallen sharply. In addition to the drop in prices for wool, the wool subsidy payment from the USDA has been eliminated. In the mutton breeds, typical of farm flocks, the price to hire a shearer for the sheep is not covered by the value of the wool shorn. In many occasions, it literally costs you money to shear and sell the wool! There is still some profit in the wool clip from white face range sheep flocks. White face ewes shear a much heavier fleece. It is of higher quality and therefore brings a higher price than fleeces from mutton breeds. Predator losses, increased feed costs, housing, land, transportation, interest on captial investments, depreciation, labor, and marketing also contribute to the difficulty of making a profit with sheep.

An effort should be made to reduce feed costs, which comprise more than 50% of all costs, by maximizing utilization of forages, crop residues, winter pastures, limited feeding during maintenance, detection of nonpregnant ewes, and other methods. Housing costs can be reduced by changing the lambing season and by dividing the flocks into smaller units to utilize the same facilities more than once. Land costs can be reduced by intensifying operations through confinement, feeding programs, and fertilization and rotation of pastures, and by double-cropping land with high-yielding pasture crops. In cases where it is difficult to hire qualified labor, it may be more economical to substitute capital for labor through automation, especially in the feeding and manure-removal operations. Transportation and marketing costs may be reduced by cooperative efforts on the part of sheep producers through lamb and wool marketing pools and tele-auctions.

The profit potential of smaller flocks of sheep should not be overlooked. Profitable small-flock operations can often be instituted by farmers with an excess of forages and pasture, part-time farmers with limited labor and facilities, young farmers with limited capital, and youths—including 4-H and FFA members.

In any situation, the greatest single factor affecting the profit potential of sheep will be the number of lambs marketed per ewe per year. The present level of approximately 1 lamb marketed per ewe in the western states and Texas and from 1 to 1.2 lambs marketed in the Midwest and East is not high enough to realize a good profit from the industry. With increasing costs, this figure should be at least 1.5 lambs marketed per ewe, and with excellent management it should be 1.8 or higher. Through the infusion of the genetics available from some of the more naturally prolific breeds, such as the Finnsheep, or from some of the composite breeds, such as the Polypay or Texel, and with accelerated lambing programs in which ewes lamb every 8 months, producing 3 lamb crops in 2 years, it is possible to market 2 lambs per ewe per year. This will more than double the profit now realized in most flocks.

Future Opportunities in Sheep Raising

The increasing need to conserve energy and feed grains puts sheep in an enviable position compared to other meat-producing ruminants. They are more efficient than cattle in converting forages into food and can obtain from 80 to 90% of their yearly nutrient requirements from this renewable natural resource without seriously competing with humans for cereal grains or for energy.

Sheep are amazingly efficient in converting grass into protein, which can be stored in meat and in wool. The world protein shortage is expected to worsen in the future. In the United States, more protein is produced in wool each year than in sheep and goat meat. Protein in wool can be hydrolyzed and used as food for humans and animals if necessary. No proteins can be stored more easily for long periods of time than they can in wool.

Sheep have other advantages that make them attractive to farmers with an excess of forages and pastures. Some of these advantages are as follows:

1. The investment cost per animal unit is low, making sheep suitable for those with limited capital.

2. They have greater flexibility in breeding and management than do cattle, and so they better meet the needs of part-time farmers.

3. Returns come quickly because of a short gestation period (148 days) and a short finishing period (150 days), a total of less than 1 year from mating to market.

4. Sheep produce two major products, meat and wool, and from one to three lambs yearly.

5. Sheep adapt more readily to seasonal feed supply and drought and have lower water requirements than cattle.

These advantages will cause an increasing number of farmers to consider sheep as a part-time or major source of income on many farms and ranches in the future.

7.2 BREEDING

Most breeds of sheep are seasonally polyestrous; that is, they will undergo a series of estrous cycles only during the fall. The onset of the breeding season is triggered by the decreasing light intensity of the shorter days and cooler nights.

Some breeds such as the Dorset and Rambouillet will breed at other times of the year, including the spring months of March through May, when most breeds are undergoing seasonal anestrus and have no estrous cycles. These out-of-season breeders can be used to produce either one lamb crop in September and October on a 12-month lambing interval, or three lamb crops every 2 years on an 8-month lambing interval. In the latter program, which is known as *accelerated lambing*, ewes produce lambs in January, September, and the following May.

Management of ewes and rams is nearly the same for any of these systems. Adequate preparation before the breeding season is important.

Equipment Necessary

- Shearing equipment
- Dose syringes
- Anthelmintics
- Paint brands
- Ewe-marking harness with different-colored crayons
- Vitamins A, D, and E
- Syringe, needles, and disinfectant
- Wool-branding paint

Restraint Required

Restraint will be required to "crotch-out" the ewes, to worm, and to trim feet. Rams will require restraint for shearing, and if a marking harness is used they will have to be held by the head when it is placed on them or when the color of the marking crayon is changed.

Step-by-Step Procedure

1. Tag ("tagging") all ewes by removing the wool and dung tags by shearing around the dock and vulva. If this were the tagging process to prepare for the lambing season, we would also shear inside and along the front of the rear legs and completely shear out the udder so that the lambs could have free and open access to the nipples. Now, in preparation for the breeding season, is an excellent opportunity to tag out the eyes, that is, free them from any tags or overgrowth of wool. In certain areas of the world, this tagging process is termed *crutching* and sometimes *crotching*.

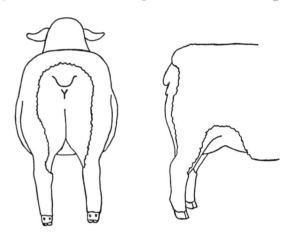

2. Trim the feet of all ewes and inspect them for foot rot. Treat them for foot rot if necessary.

3. Deworm all ewes with a recommended anthelmintic. (See Section 7.12.)

4. Use wool-branding paint to stencil all ewes with numbers corresponding with their permanent flock numbers so that they can easily be identified at a distance when bred.

5. Two weeks before you wish to start the breeding season, "flush" all ewes by feeding them approximately ¼ to ½ pound of shelled corn or other high-energy concentrates in addition to pasture. This will put them on a rising plane of nutrition and cause them to shed more ova, thus increasing the number of multiple births.

CAUTION: Do not graze ewes on any legume pasture; such plants contain estrogens that prolong or prevent conception. These plant estrogens interfere with fertilization and cause embryonic mortality. Use grass or nonlegume pastures during the breeding season.

---

6. Shear the ram completely two weeks before the start of the breeding season. This will cool him, make him more active, and help to increase his fertility.

7. Trim the feet of the ram and worm him.

8. If possible, obtain a semen evaluation by a qualified technician on all rams to identify infertile or low-fertility rams.

9. Place a marking harness on the ram a day or two before the breeding season starts to allow him to get used to it. This harness contains a removable crayon that will mark the ewes on the rump as they are bred. The color of the crayon should be changed every 16 to 17 days to identify ewes that conceive during the first estrous cycle, those that are rebred during the second cycle, and those that do not breed. Start with light-colored (yellow, orange) crayons in the harness and progress to darker as the cycling continues.

10. To start the breeding season, place the ram with the ewes. In hot weather (temperatures over 80°F), he should be removed during the day and allowed to be with the ewes only at night to prevent him from becoming overheated. In cooler weather he can remain with the ewes throughout the day,

but shelter should be provided to protect him from high daytime temperatures.

11. In purebred flocks, the sires of the lambs must be known. One ram should be placed with 25 to 50 ewes for two estrous cycles, or 32 to 35 days. He can be followed by another cleanup ram for an additional 16 to 17 days to complete the breeding season and ensure that late-cycling ewes are bred.

12. In commercial flocks where identification of the sire may not be necessary, more than one ram can be used simultaneously. Three rams are better than two, because two rams tend to fight each other, which may delay breeding. If two rams are used, they can be rotated every 12 hours, allowing only one ram with the flock at any one time.

13. Remove the rams from the ewe flock after two estrous cycles (32 to 35 days), or three estrous cycles if a cleanup ram is used to breed the late-cycling ewes or slow breeders.

Postprocedural Management

At the close of the breeding season, the ewes can be placed on a maintenance diet until 4 to 6 weeks before parturition. This diet can consist of pasture, corn fields after the corn is picked, or a ration of 2 to 3 pounds of hay or other roughage per day.

During the last 4 to 6 weeks of gestation the ewes should receive from ½ to ¾ pound of shelled corn or other energy-equivalent concentrate in addition to a roughage ration of 3 to 4 pounds of hay or equivalent. This will supply additional energy for the rapid growth of the fetuses during this period and help to prevent ketosis or pregnancy disease (lambing paralysis). The ewes should also receive plenty of exercise during this period.

If the ewes are receiving a low-quality roughage, haylage, or silage that has not been sun-cured, or if they are on a winter fescue pasture, they should receive vitamins A, D, and E during the last 4 to 6 weeks of gestation. These vitamins can be supplied as supplements to the ration or they can be injected intramuscularly in the shoulder, forearm, or the inside of the rear legs in a wool-free area. Use the dosage recommended by the manufacturer. If possible, clean and disinfect the area to be vaccinated and do the same to the needles after each use to prevent infection.

After the breeding season has been completed, the rams should continue to receive from ½ to ¾ pound of concentrate in addition to pasture or 2 to 3 pounds of hay equivalent until they have regained the weight and flesh lost during the breeding season. They can then be placed on a maintenance diet until cold weather arrives or until they are again used for breeding.

7.3 DIAGNOSIS OF PREGNANCY

Pregnancy can be detected in ewes between 40 and 120 days after the breeding season. The diagnosis can be made with an accuracy of from 90 to 95% when the ewes are examined by an experienced shepherd. Pregnancy diagnosis provides information that can be used to cull nonpregnant "boarder" ewes, to select the more fertile ewe lambs that cycle early for replacements, and to rebreed nonpregnant ewes at an earlier date in accelerated lambing programs. This results in the development of a more productive flock and in considerable savings in feed costs.

The methods that are currently used to detect pregnancy include: (1) the use of a vasectomized ram with a ewe-marking harness; (2) the use of a fertile ram fitted with a breeding (prevention) apron and a ewe-marking harness; (3) ultrasonic scanning; (4) the Doppler instrument to detect movement, such as blood circulation to the uterus, fetal movement, uterine fluid movement, and perhaps even the fetal heart beating; and (5) udder development, determined by palpation.

Ultrasonic scanning and the Doppler instrument method are more costly, due to the electronics involved, but they can be cost-effective and their use justified when the flock is large, when the equipment ownership is shared with several other flock owners, or when the equipment is owned by a traveling technician who "preg checks" your flock for a per-head fee.

Equipment Necessary

- Vasectomized ram with marking harness
- Fertile ram with apron and marking harness
- Dose syringe
- Doppler instrument with rectal probe
- Sonascope with transducer
- Ewe-holding cradle
- Tilting squeeze chute

Restraint Required

No restraint is needed with the vasectomized ram or fertile ram fitted with an apron. With the Doppler instrument method, the ewe must be placed on her back in a comfortable horizontal position with one assistant holding her head and another her hind legs, or she can be placed in a ewe-holding cradle with attachments to secure the hind legs, or the ewe can be placed in a tilting squeeze chute so that she can be turned on her back. The sonascope can be used with an assistant holding the ewe by her head. Palpation of the udder requires only slight and momentary restraint.

Step-by-Step Procedure

Vasectomized Ram

1. Fit the ram with a ewe-marking harness containing a crayon of a different color than that used during the breeding season. This can be done from 2 to 7 days after the breeding season is completed and is continued for 16 to 17 days.

2. Record the number of any ewe that is marked during this period.

CAUTION: Be sure that the marks are of sufficient depth of color to denote breeding. False breeding marks are sometimes made during an attempted breeding, but these ewes may be pregnant. A nonpregnant ewe in heat will stand for the ram, allowing a deep imprint to be made when the ram mounts.

---

3. Ewes that have been thoroughly marked are usually not pregnant and can be separated from the pregnant ewes and retained for future breeding or marketed.

Fertile Ram with Apron

1. Fit an apron made of burlap or polyethylene material under the ram's belly, covering the penis, and tie it with string or straps across his back.

2. Place a ewe-marking harness equipped with a marking crayon on the ram and proceed as with a vasectomized ram. Some rams will not attempt to breed ewes in estrus under these conditions and should be replaced with another fertile ram or a vasectomized ram.

3. Record the numbers of marked ewes. (See the cautionary note above.)

Doppler Instrument

1. Restrain the ewe on her back, on the ground, in a tilting squeeze chute, or in a ewe-holding cradle.

2. Inject 6 to 8 ounces of a soapy lubricant into the rectum with a dose syringe.

3. Insert the Doppler probe into the rectum. Orient it toward the ewe's backbone so that it rests under the uterus.

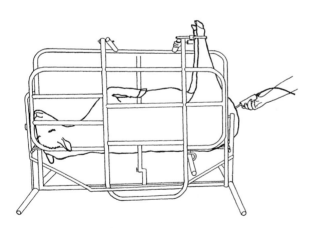

4. As the probe reaches the middle uterine artery, the heartbeat of the ewe can be detected by the instrument. (The pulse rate should be 90 to 110 beats per minute.) As the probe moves into the rectum, the blood flow in the placenta can be heard as a sound similar to that of wind blowing through trees. This sound will be followed by that of the fetal heartbeat (130 to 160 beats per minute). The tone of the fetal heartbeat will not have the depth or resonance of the maternal heartbeat. If no fetal heartbeat is detected, the ewe can be presumed nonpregnant.

5. After the diagnosis is made, slowly remove the Doppler probe and then clean and disinfect it.

Ultrasonic Scanning

Ultrasonic waves of about 2 million hertz (cycles per second) can pass through living tissue and be used to determine pregnancy in ewes. These sound waves are painless and have no harmful effects on the ewe or fetus. The working part of the sonascope is the transducer, which transmits and receives sound waves. A crystal in the transducer converts electrical energy into sound energy, which is sent into the animal.

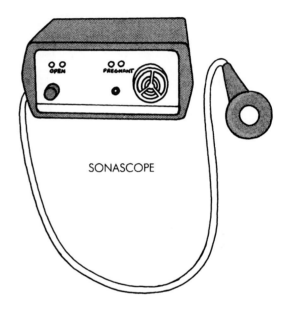

SONASCOPE

These sound waves do not travel through air; air must be sealed out with a nondetergent motor oil or mineral oil.

When the sound waves encounter different kinds of tissues such as skin, fat, muscle, or connective tissue, they are reflected back to the transducer, converted to electrical energy, and analyzed. If the instrument receives signals indicating pregnancy, a specific light goes on and a "beep" is sounded. This is what happens when the sonascope is used. Next-generation ultrasound instruments, such as the ultrascan Pregnancy Visualization Unit, convert the signals to an image showing the embryos in the uterus.

As soon as the ewe becomes pregnant, the uterus begins to fill with fluids. The weight of the pregnant uterus causes it to sink downward to the bottom of the abdominal wall. As gestation progresses, fluids continue to expand the uterus forward and to the ewe's right side. The uterine fluid can be detected by ultrasonic techniques at about 70 days of gestation. After 120 days the fetus fills the uterus, making ultrasonic detection less accurate.

1. Be sure that the battery has been charged and that the instrument is working properly. Follow the directions of the manufacturer of the sonascope.

2. Restrain the ewe to be tested by having an assistant hold her or secure her in a chute or stand. Allow her to stand in a normal position.

3. Apply the nondetergent motor oil or clear mineral oil to the transducer.

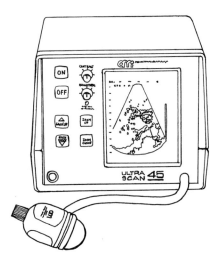

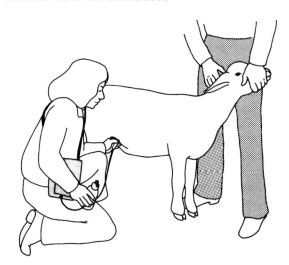

4. Hold the transducer flat against the non-wooled area below the right flank, 2 to 3 inches in front of the right nipple of the udder. Aim the transducer so that the ultrasonic sound waves are beamed forward at a 30° angle from the horizontal and at a 45° angle toward the opposite side of the ewe behind the last rib, about halfway from the top of the back. This positioning will avoid the urinary bladder, which is behind the uterus and can give a false reading.

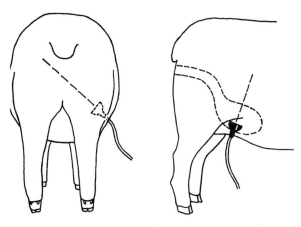

5. Pregnancy is indicated by a light, an audible tone (beep or buzz), or by an image on a cathode-ray screen (TV-like screen). The type of signal will vary depending upon instrument manufacturer. Whatever the signal, it will occur very shortly after the transducer makes contact with the flank skin of the ewe.

6. Nonpregnancy will be signaled by a blinking light, a different-colored light, a different audible tone, or by a different image on the screen. Read and follow the directions carefully so that you understand what the machine is indicating to you.

> **CAUTION:** Good contact of the transducer with the skin is essential. If the instrument light flickers, apply more oil for better contact. If the skin contains foreign substances or is dirty, clean it with alcohol or water and wipe it dry.

Bagging, or Udder Palpation

Bagging, or udder palpation, which means checking the udder of the ewe shortly before lambing, is certainly the oldest and still the most common method used to determine if a ewe is carrying a lamb or lambs. Udder palpation relies on the fact that the udder undergoes development and enlargement in the pregnant ewe as gestation progresses. Udder development is particularly noticeable during the last weeks of pregnancy. Accuracy is quite high when using this method.

Despite its accuracy, udder palpation as a pregnancy determiner has drawbacks. Some ewes will not develop an udder until gestation is nearly complete, making it likely that the shepherd will designate the ewe as open and place her into the open-ewe pasture instead of giving her the close-up ewe management and attention she needs. Also, even if the shepherd makes the correct call and the ewe is open, this method does not enable sorting of the ewes in time to manage the open ewes to the best economic and biological advantage.

Experienced operators should be able to diagnose pregnancy with 90 to 95% accuracy using any of the previously described methods.

Postprocedural Management

Ewes diagnosed as nonpregnant should be separated from the pregnant ewes and fed at a lower nutritional level to save feed. Nonpregnant ewe lambs could be marketed while they are still of market size and weight. In accelerated lambing programs using an 8-month lambing interval, those ewes diagnosed as nonpregnant between 60 and 120 days can be placed in an alternate flock and rebred in 4 months. This prevents the ewes from being barren for a full 8 months.

7.4 LAMBING

Pregnant ewes have a gestation period of from 144 to 152 days. Most medium-wool breeds will average 145 to 147 days, but those with Finn breeding will average 142 to 145 days, while those with fine-wool breeding will average from 147 to 150 days.

Adequate preparation in advance of parturition will increase the survival rate among newborn lambs. The lambing area should be completely cleaned. After cleaning, ground limestone or superphosphate (0-46-0) can be spread over the floor to help disinfect the lambing area and prevent accumulation of moisture. The area should be well bedded with 3 to 4 inches of bedding. Lambing pens should be set up.

Most lambing pens or jugs consist of two hinged hurdles, 4 feet long, 5 feet wide, and 3 to 4 feet high, but large ewes with twin lambs require 5' × 5' hur-

dles. If a warm barn is available, the ewes can be completely shorn at 1 to 4 weeks before lambing. Shorn ewes require less space, the barn remains drier because wool absorbs and holds moisture, the lambs can nurse better, and the ewes will not lie on their lambs and smother them. Shorn ewes will not expose lambs to unfavorable weather conditions and chill them. If ewes cannot be completely shorn, they should be faced by shearing the wool from around the eyes and on top of the head, tagged by removing the manure locks around the rear parts, and "crotched out" by removing the wool from around the vulva and the udder.

As a ewe approaches parturition, she will become less active than formerly and her udder will become full and distended. She may have a hollow appearance in front of the hips and above the dock. Within a few hours of lambing she will become restless, segregate herself from the remainder of the flock, and lie down and get up frequently. This is accompanied by frequent urination and a pawing of the bedding. As she starts to labor, her head will be pointed upward as she strains, and the placental membranes (water bag) will protrude from the vulva. A normal parturition will occur in 30 to 45 minutes. If the ewe is carrying more than one lamb, they may be born from 10 to 20 minutes apart.

Difficult births are most apt to occur in overfat ewes, ewes that have not been exercised daily, and in ewe lambs or yearlings. Older ewes that have lambed once or twice usually will have very little difficulty.

Equipment Necessary

- Mineral oil
- Prepared lubricating disinfectant
- Stout cord
- Lambing instrument
- Disposable gloves

Restraint Required

Most ewes will require little help if the presentation of the lamb is normal. They should not be placed in the lambing pens or jugs before lambing because the lamb may be injured in delivery or stepped on by the ewe, and the placental fluids keep the pens wet and stimulate the growth of disease-producing

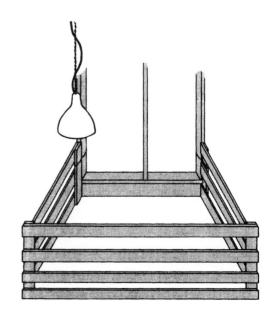

agents. After they have lambed, they should be placed in a lambing pen with their lambs for 1 to 2 days. This is especially important if they have more than one lamb, because the firstborn lamb may wander away and be rejected by the ewe. If assistance at parturition is anticipated, it may be desirable to restrain the ewe by placing her in a lambing pen before lambing.

Step-by-Step Procedure

Normal Presentation

1. Observe the ewe for type of lamb presentation as parturition progresses. Normal presentation is with the lamb's head lying on the two forelegs. The feet usually appear first. If the feet and head pass through the pelvic arch without assistance, the lamb will usually be born without help.

2. Remain near to render assistance if necessary.

3. If the presentation is normal, the ewe may not need help; however, if the lamb is large, assistance should be given by pulling the forelegs down toward the ewe's hocks as the ewe strains in labor. Make sure that the head follows the forelegs.

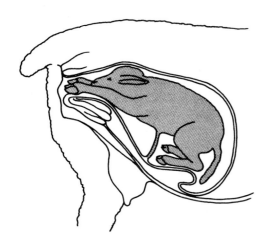

4. After the lamb is born, immediately clean the mucus from the nostrils and mouth, making sure that the lamb starts to breathe. A piece of straw inserted into the nostrils will usually cause the lamb to sneeze and start the breathing process.

5. If the lamb does not breathe, stimulate the process by pressing gently on the ribs or by moving the forelegs vigorously. Sometimes it may be necessary to hold the lamb by the hind legs to drain the fluids and mucus from the mouth and throat.

6. After the lamb is breathing normally, place it where the ewe can lick it dry. She identifies her lamb by smell, and it is important that this be done as soon as possible.

7. Observe the ewe for at least 20 to 30 minutes for a second lamb or multiple births and assist if necessary. If in doubt, examine the ewe for addi-

tional lambs by inserting your hand into the uterus while the cervix is dilated.

8. Remove the placental membranes (afterbirth) from the lambing pen after they are expelled by the ewe. Do not attempt to remove them from the ewe by force.

9. Disposable plastic gloves can be used when delivering lambs or examining the ewe for additional lambs.

> **CAUTION:** If it is necessary to assist the ewe in parturition, be sure to wash and disinfect your hands before entering the uterus. A mild disinfectant mixed with mineral oil will help in entering the uterus, and plastic gloves will help reduce infection. If it is necessary to pull the lamb, pull downward only as the ewe strains to avoid tearing the cervix or vagina. If the umbilical cord does not break when the lamb is born, scrape or tear it apart, but do not cut it, as it will sometimes hemorrhage. Do not attempt to remove the placental membranes (afterbirth) by force, as this can cause hemorrhaging in the uterus. The afterbirth should break loose within 30 to 60 minutes after parturition. If it has not broken loose after 24 hours, obtain help from your veterinarian.

Abnormal Presentations

If the presentation of the lamb is not normal, it usually will be necessary to assist in the delivery.

1. If the *head appears first with one or both forelegs back*, it will be necessary to gently push the head back behind the pelvic arch and find the forelegs. This can be done without suffocating the lamb if the umbilical cord is not broken, because the lamb receives its oxygen from its mother. A snare in the form of a stout cord or lambing instrument may be helpful in holding the head while searching for the forelegs. The snare should be placed over the head and behind the ears. After the forelegs are placed in

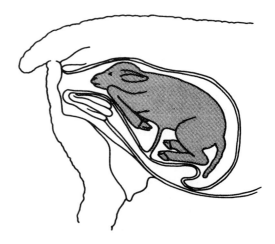

the vagina and the head follows, the lamb can be removed as in a normal presentation.

2. If *the forelegs appear and the head is back,* gently force the legs back until the head can be found and guided into the vagina. If the ewe has been in labor for some time in either this or the previous situation, the lamb's head may be swollen because of poor circulation. This is where a snare with a stout cord or lambing instrument may help in positioning the head or pulling it through the pelvic arch.

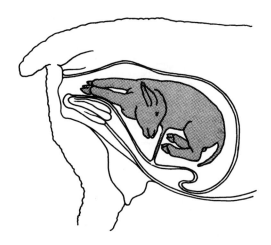

3. In *a breech presentation, the tail will appear first* and one or both rear legs will be caught at the hocks behind the pelvic arch. Again it will be necessary to gently force the lamb back into the uterus and find the rear legs. Do not attempt to pull the legs by the hocks, as they will straighten out and puncture the uterus. Rather, place the hand under the rear feet and lift them over the pelvic arch. In a breech presentation, it will be necessary to remove the lamb rather quickly to prevent suffocation if the umbilical cord is broken. This must be done gently; you are pulling against the lamb's rib cage and the ribs can be fractured.

4. If *more than one lamb is being born at one time,* find the legs and head of one lamb, gently force the second lamb back into the uterus, and remove the first one.

5. If *the cervix has not dilated,* parturition will be very difficult and the danger of tearing the cervix will be great. If upon examination it appears that the lamb cannot be expelled, a veterinarian should be contacted. He or she can inject the ewe with hormones to help dilate the cervix.

CAUTION: The same cautions that apply to normal presentations also apply to abnormal ones; in addition, good judgment and patience will pay dividends, especially in breech presentations, where injury to the lamb is greatest.

Postprocedural Management

In all parturitions where the lambs are alive and breathing, they should be placed where the ewe can dry them by licking. If the delivery has been difficult, both the ewe and the lambs should be allowed to rest until they want to stand. The lambs will nurse better if allowed to get somewhat hungry, but they should not become chilled. Additional care of the ewe and lambs is discussed in Section 7.5.

7.5 NEONATAL CARE

The first few hours after parturition can be critical for both the ewe and the lambs, especially during very cold weather. Under these conditions, a warm area or heat lamps should be used to prevent chilling of the lambs. Additional care is necessary to ensure that the lambs nurse properly and are treated to prevent navel infection.

Equipment Necessary

- Hinged lambing pens (4' × 4', 4' × 5', or 5' × 5', 3'–4' high)
- Iodine solution: 7%, contained in a squeeze bottle
- Ear tags
- Paint brands numbered 0–9
- Scales
- Lambing record sheets
- Ewe stanchion
- Heat lamp
- Water buckets
- Stomach tube for lambs

Restraint Required

The ewe and her lambs should be restrained in a lambing pen made of 4' × 4' hurdles for small- or medium-size ewes, and 4' × 5' or 5' × 5' hurdles for larger ewes. The pen should be 3 to 4 feet high. Stanchions can be used for ewes that refuse to accept their lambs.

Step-by-Step Procedure

1. Milk a small amount of milk from both teats of the udder to start the milk flow and to remove the plugs from the teats.

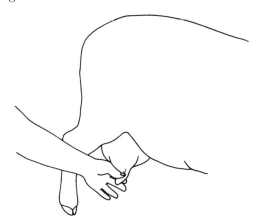

2. Be sure that the lambs receive the colostrum or first milk of the ewe within the first hour after birth. If the ewe does not have milk or is slow in letting her milk down, the colostrum from another ewe or from a dairy cow can be used. This colostrum should be collected in advance and frozen for such emergencies.

3. Weak lambs that cannot nurse should be fed colostrum via a stomach tube; the colostrum should be warm to the touch. This can be done by stretching the lamb's neck backward so that the tube can be inserted directly into the esophagus.

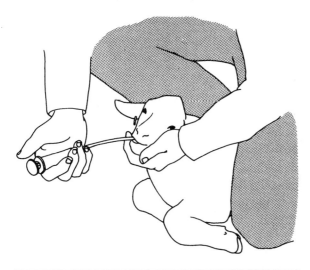

CAUTION: The stomach tube may enter the trachea. If this occurs, the lamb will show signs of strangulation. Remove the stomach tube, let the lamb recover, and reinsert the tube by stretching the neck backward.

4. Apply 7% iodine solution to the navel. This will prevent infection, which will cause a swelling of the joints known as "navel ill."

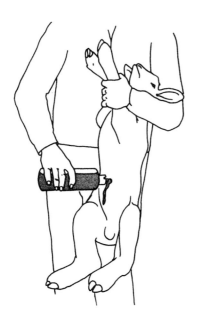

5. Obtain the birth weight of each lamb. This can be done with a small utility scale.

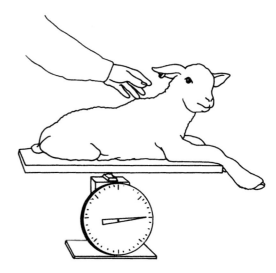

6. Identify each lamb with a metal ear tag as discussed in Section 7.8.

7. Record the tag number of each lamb, its sex, dam, and birth weight, and any remarks about its birth in the lambing records. This should be done for all lambs, living or dead.

8. After the ewe expels them, remove the placental membranes (afterbirth) from the pen.

9. Do not start feeding grain to the ewe too soon after lambing. This may stimulate too much milk production and cause digestive problems. Let her have 3 to 4 pounds of hay or other roughage for the first 24 hours.

10. The ewe will need some water after parturition. It should have the chill removed and be offered in a partly filled bucket, which should be removed from the lambing pen after she drinks.

11. After 24 hours have passed, the ewe can be offered some grain, the amount of which should be increased gradually until she is at full feed after the second day. Ewes with one lamb can receive from 1 to 1½ pounds of grain. Those with twins should receive up to 2 pounds.

12. The ewe and her lambs can remain in the lambing pen for 1 to 2 days, depending upon the need for the pens and the condition of the ewe and lambs.

13. Clean the lambing pens after each use, disinfect them, and rebed them with 3 to 4 inches of bedding.

14. Before the ewe and her lambs are removed from the lambing pen, they should all be stenciled (paint branded) with the same number for easy identification when they are placed in a mixing pen with other ewes and lambs.

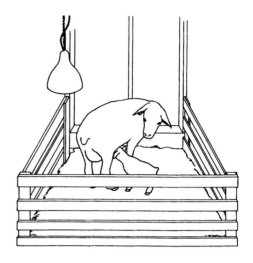

15. Ewes that refuse to accept their lambs, or those that lose their lambs, can be forced to accept their own lambs or orphans by restraining them in a stanchion until they do so. The stanchion should be so arranged that the lambs can nurse and the ewe can be fed and watered and get up or lie down.

16. A ewe that loses one or more lambs at birth may be persuaded to accept an orphan by immersing her lamb and the orphan in a strong, warm, salt solution or by rubbing her placental fluids onto the orphan lamb.

CAUTION: Do not place a full bucket of water in a lambing pen where small lambs can fall into it and drown. If a bucket is left in the lambing pen, be sure that it contains 6 inches of water or less.

Nervous ewes should be restrained with a halter or stanchion to prevent injury to the lambs.

Observe the ewe and her lambs often to be sure that the lambs are nursing properly. Lambs that continually bleat or appear weak are usually not getting enough to eat. Help them to nurse or raise them artificially.

If heat lamps are used, be sure that they are placed 3 to 4 feet above the bedding to prevent starting a fire and so that they do not scorch the ewe's back. Electrical outlets should be placed horizontally above the lambing pens so that the plugs are disconnected if the heat lamp falls. Do not hang the lamps by the electrical cords, but use a chain or wire to support them.

Postprocedural Management

Some ewes and lambs may require additional management after parturition. These include ewes with retained placentas, infections caused by injury during parturition, diarrhea, failure to milk, mastitis, or those that have had difficult parturition (dystocia).

Lambs that are weak, fail to nurse properly, develop diarrhea because of infection or overeating, and lambs with pneumonia or other infections should be retained in the lambing pens or area and treated accordingly. Consult a veterinarian if necessary.

7.6 DOCKING

The removal of the tail in lambs is known as *docking*. That part of the tail remaining on the body is referred to as the *dock*. This process is necessary in most breeds for the following reasons: (1) to improve sanitary conditions, since the long wool on the tail will become saturated with feces and urine and become a target for fly strikes or screwworm infestation; (2) to increase productivity in ewes, in which the tail may interfere with breeding and lambing; (3) to improve the appearance of sheep for exhibition in the show ring; and (4) to increase the value of market lambs. Most buyers pay less for long-tailed lambs because the tail is inedible and may weigh several pounds.

For best results, lambs should be docked before they are 2 weeks of age. Big, strong lambs can be docked within 24 hours of birth, before they leave the lambing pen. Smaller, weaker lambs should be docked later when their survival is assured. Lambs can be docked when they are more than 2 weeks old, but the trauma and bleeding will be greater.

Docking is accomplished by severing the tail, preferably between the vertebrae. If done properly, very little bleeding will occur; the tail contains no large arteries or veins, and the spinal cord does not extend into the tail. The tail should be removed at the end of the caudal folds on the underside of the tail, 1 inch from the body. Lambs docked shorter than this are subject to rectal prolapse. Short docking is one of the major causes of this problem because nerves in the rectal area can be severed or damaged.

Docking can be done in a number of different ways depending upon the preference of the shepherd. If help is available, one person can hold the lamb and one can do the docking, but some procedures require only one individual to carry them out. The most commonly used methods, and the advantages and disadvantages of each, are discussed in this section.

Equipment Necessary

- Sharp pocketknife
- Emasculator
- Emasculatome (Burdizzo®)
- Elastrator
- Hot docking irons
- All-in-one pliers
- Pruning shears
- Docking cradle
- Iodine in squeeze bottle—7% solution
- Stout string
- Tetanus antitoxin
- Fly repellant
- Syringe

Restraint Required

If two people are involved, one should hold the lamb in a vertical position with the back of the lamb against the holder's body. The front and rear legs on each side should be held in each hand. This allows the person doing the docking to feel for the vertebrae in the tail and do a straighter job of removal.

If only one person is doing the docking, the lamb can stand and be held by the neck between the legs or placed on its rump in a docking cradle.

Step-by-Step Procedure

Sharp Pocketknife with Assistant Holding Lamb

1. Set the lamb up so that the tail is positioned on a firm base against which to cut.

2. Locate the junction of the caudal folds with the underside of the tail (about 1 inch from the body).

3. Force the loose skin surrounding the tail toward the body so that it will cover the end of the dock after the tail is removed.

4. Place the knife blade on the tail and cut between the vertebrae with a forward and downward movement.

5. Draw the knife blade toward you, severing the skin on the bottom (upper side of the dock when the lamb stands).

6. Use a squeeze bottle to apply a 7% iodine solution to the dock.

7. If excessive bleeding occurs, pinch the dock between your fingers until the bleeding subsides or coagulation occurs.

Sharp Pocketknife without Additional Help

1. Hold the lamb, which is in a standing position, by placing its neck and/or rib cage between your legs.

2. Use your left hand to hold the tail in a horizontal position. Hold the knife in your right hand and place it 1 inch from the lamb's body on the underside of the tail.

3. Remove the tail by cutting upward.

4. Apply 7% iodine solution to the dock.

5. Bleeding can be reduced by pinching the dock between your fingers.

6. If a docking cradle is used, proceed as though another person were holding the lamb.

The knife method requires the least expensive equipment and can be done quickly. A clean cut will usually heal faster than will one in which the tail is crushed or burned.

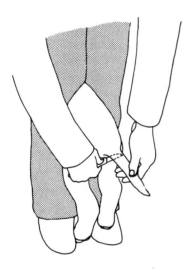

> **CAUTION:** Bleeding may occur, so lambs should be observed for a period of time after docking. If bleeding persists, a stout string can be placed on the dock as a tourniquet. It should be removed within an hour to prevent swelling of the dock.

Emasculator

This instrument is normally used to castrate bulls or stallions. It has both a cutting edge and a crushing surface. Be sure that the crushing surface is toward the lamb, so that the blood vessels on the remaining dock are crushed.

1. Place the emasculator around the tail, with the crushing surface toward the lamb.

2. Remove the tail by slowly compressing the handles of the emasculator.

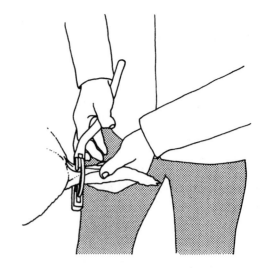

3. Hold the emasculator on the dock for 5 to 10 seconds to allow the blood to coagulate.

4. Apply a 7% iodine solution to the dock.

This method will reduce bleeding, but may take longer to heal because some of the tissue and possibly the vertebrae are crushed in the docking process. The cost of this instrument is greater than that of a pocketknife.

Emasculatome (Burdizzo®)

This instrument also is used for castration and has two blunt edges that pinch the tail, thus reducing bleeding.

1. Place the emasculatome on the tail 1 inch from the body and close the handles.

2. With a sharp knife, cut the tail off *inside* the jaws of the emasculatome, leaving the jaws clamped on the dock for 5 to 10 seconds.

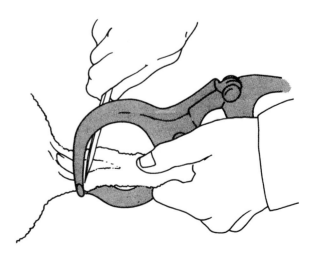

3. Remove the emasculatome and apply a 7% iodine solution to the dock.

There will be practically no bleeding with this method, but it will not work well with large lambs over 4 weeks old because it is difficult to close the emasculatome on large tails. The cost of the emasculatome can be justified for large flocks.

Small emasculatomes can be sprung after continued use in docking lambs and should not be used for castration.

Elastrator (Rubber Rings)

This instrument can be used for both docking and castration.

1. Place the rubber ring on the prongs of the elastrator.
2. Spread the rubber ring so that it is large enough to slip over the tail.
3. Slide the rubber ring over the tail to a point 1 inch from the lamb's body. Keep the prong toward the lamb's body so that the elastrator can be removed.
4. Release the tension on the elastrator and leave the rubber ring on the tail.

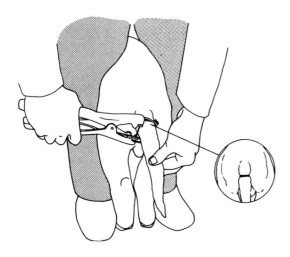

This method is bloodless; the rubber ring shuts off the circulation in the tail and causes it to slough off in 1 to 2 weeks. It may cause some pain for an hour or so until the tail becomes numb. Lambs docked by this method are vulnerable to tetanus, which enters the body through the dead tissue in the tail. For this reason, some shepherds prefer to cut the tail off after 24 hours, leaving the rubber ring on the dock. This method requires double handling of the lambs.

Hot Docking Irons

This method is used to dock lambs when they cannot be checked daily as closely as they can be in smaller flocks. It is bloodless if done properly, and the heat cauterizes and disinfects the wound in one operation.

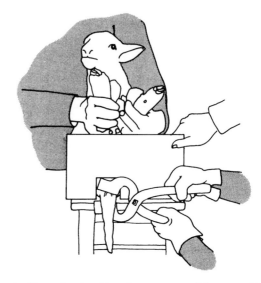

1. Heat the docking irons to a red-hot condition.
2. Have an assistant hold the lamb in the manner used for the pocketknife method.
3. Place a 1-inch-thick notched board over the tail next to the body to protect the lamb from the hot irons.
4. Clamp the hot irons onto the tail, which is burned off at 1 inch in length.

This method requires more equipment than other methods, but is safe and requires very little checking for bleeding.

All-in-One Pliers and Pruning Shears

The all-in-one pliers are designed for both docking and castration. The blades are straight, but the points can be used for pulling. Pruning shears have a curved blade that fits well around the tail.

1. The lambs can be held by an assistant (see the pocketknife method) or held by the operator between his legs.
2. Place the shears on the tail at 1 inch from the body and cut the tail off cleanly.

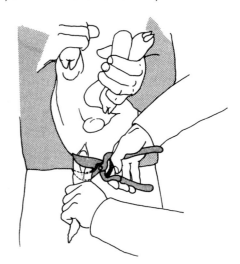

3. Apply a 7% iodine solution to the dock.
4. Observe lambs for excessive bleeding.

These shears are easy to use, and the cost is only slightly greater than that of the pocketknife.

CAUTION: For those methods in which excessive bleeding may occur, pinch the dock for a few minutes to allow for coagulation. If bleeding persists, place a string or baling twine around the dock above the cut as a tourniquet, but remove it in 1 hour.

Lambs showing evidence of anemia, as evidenced by pale lips and eyelids, or weak lambs, should not be docked until their conditions improve.

To prevent rectal prolapse, do not cut the dock shorter than 1 inch.

If tetanus has occurred previously in the flock, vaccinate all lambs with tetanus antitoxin as they are docked. This is a safety measure if the elastrator and rubber rings are used.

Postprocedural Management

Use clean bedding in pens where lambs are placed after docking to reduce the chance for infection. Allow the lambs to rest after docking until the trauma of the operation has subsided. In cold weather, a heat lamp placed over the creep will encourage the lambs to rest and also prevent chilling. Docking should be done on a bright, sunny day, if possible, as the lambs recover more quickly under such conditions than they do on cold, wet days. The docks should scab over and heal in about a week. Overcrowding may cause the lambs to bump each other and prolong the healing process; it should be avoided. If there is danger of a fly strike or maggots, apply a recommended fly repellant as the lambs are docked.

If tetanus infection is likely, 200 to 300 units of tetanus antitoxin should be injected either intramuscularly or subcutaneously according to the manufacturer's directions. Intramuscular injections can be given in the muscle of the shoulder or forearm or on the inside of the rear leg in a wool-free area. The needle should be inserted into the muscle and the syringe plunger pulled slightly backward to determine if the needle is in a blood vessel. If blood appears in the syringe, pull the needle back slightly and complete the injection. If a subcutaneous injection is used, set the lamb on its rump and lift one foreleg. Inject the antitoxin under the loose skin in the wool-free area behind the foreleg. In either case, the area to be injected should be cleaned with isopropyl alcohol before the injection is made. Clean the needle after each use with isopropyl alcohol to reduce infection in other lambs.

7.7 CASTRATION

Castration involves the removal of the testicles from the male lamb. The castrated male is called a *wether.* Castration is usually done when the male lamb is between 2 and 4 weeks of age. It can be done earlier or at the same time the lamb is docked, but most shepherds prefer to castrate at a later date to avoid too much trauma or stress at one time.

In commercial flocks, castration is not recommended if the ram lambs can be marketed before they reach puberty at 5 months of age. Research has shown that ram lambs grow faster, have greater feed efficiency, and produce a carcass with less fat and more lean meat than do wethers. Many markets will accept ram lambs with no reduction in price up to about 5 months of age. Under these conditions, castration cannot be justified economically. After this age, intact rams sell at a lower price.

In purebred flocks, only a few of the best ram lambs are retained for breeding purposes. The remainder are sold as market lambs or wethers. The purebred breeder has very little information upon which to select the best ram lambs if he castrates between 2 and 4 weeks of age. If the flock is enrolled in a performance-testing program, this information is not available until the lambs are 90 days of age, when castration of ram lambs can cause a severe decrease in their rate of growth. The purebred breeder would be justified in not castrating those ram lambs not selected for breeding, but in marketing them as intact males.

Castration is justified when ram lambs cannot be marketed by 5 months of age, as in range production, or when wethers are required for carcass or market lamb shows. It may be justified when it is impossible to separate ram and ewe lambs, which would result in ewe lambs being bred at too young an age.

Equipment Necessary

- Castrating knife
- Emasculatome (Burdizzo®)
- Elastrator
- Rubber rings
- Iodine, 7% solution
- Tetanus antitoxin
- Syringe
- Fly repellant
- Isopropyl alcohol

Step-by-Step Procedure

Castrating Knife

1. An assistant should hold the ram lamb in a vertical position with the lamb's back against the holder. The fore- and rear leg on each side should each be held with one hand.

2. The operator should examine the lamb for evidence of scrotal hernia, which is manifested by an enlargement of the scrotal area. He should determine also if both testicles are in the scrotum. If a hernia is present or if both testes are not descended, the lambs should not be castrated by this method. Consult a veterinarian or leave the lamb uncastrated.

3. Grasp the scrotum by the bottom and stretch it away from the body. While you are stretching the scrotum away from the body, you will be able to feel the two small testes inside the scrotum, sliding upward away from your fingers and toward the body.

4. Using a castrating knife, cut off the bottom one-third of the scrotal sac. Do this by carefully placing the knife blade against the scrotum, above your fingers, and with firm pressure, make the cut. Done properly, the testes will be above the knife blade and your fingers below it.

5. With one hand, force the scrotum toward the body, exposing the testicles. Press the scrotum against the body wall to prevent rupture of the tissues.

6. Grasp one testicle firmly between the thumb and forefinger of the other hand and pull it out of the scrotum slowly, breaking the spermatic cord in the process.

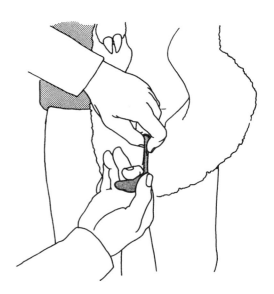

CAUTION: If any portion of the spermatic cord is left exposed, pull it out; it can be an avenue for infection to enter the body cavity.

7. Grasp the other testicle and remove it in a manner similar to the first.

8. Apply a 7% iodine solution to the severed edges of the scrotum.

9. Set the lamb down gently into a clean, well-bedded pen. Do not drop him into the pen; doing so can cause a hernia.

Emasculatome (Burdizzo®)

This method can be used on larger or older lambs, for which castration with the knife is dangerous and causes more trauma. It can also be used if a scrotal hernia is suspected. In such a case the spermatic cord is clamped below the hernia.

1. Have the assistant hold the lamb as in the castrating-knife method.

2. Check the lamb for scrotal hernia or for a testicle retained in the body cavity.

3. Work one of the testes down into the scrotum, well away from the body, and place the emasculatome on the scrotum so that it will clamp the spermatic cord inside the scrotum. This is best done if the spermatic cord can be forced to the outer edge of the scrotum.

Do not clamp across the entire scrotum. It is necessary that the septum remain intact to carry the blood supply that keeps the scrotal tissue alive. If the scrotal tissue dies, it will slough off, exposing the testicles.

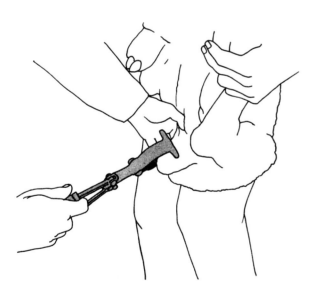

4. Clamp the scrotum and hold the clamp in position for 10 seconds.

5. Repeat this procedure for the other testicle.

6. If the procedure is done properly, the emasculatome will sever or damage the spermatic cord so that it will not transport the sperm. The testicles will atrophy or shrink in size from reduced blood supply, but there is no visible damage or bloodletting.

7. Examine the lambs periodically after this method has been used to determine if the testicles are shrinking in size. If they are not, the procedure may need repeating.

8. The scrotum may be swollen for 1 to 2 days after castration.

Elastrator

This method employs strong rubber rings that are placed on the scrotum to reduce the blood supply. This causes the testicles and scrotum to atrophy and slough off in 2 to 3 weeks.

1. Place the rubber ring on the prongs of the elastrator and stretch the ring so that it will slip over the scrotum and both testicles.

2. Be sure that the prongs of the elastrator are pointed toward the lamb's body.

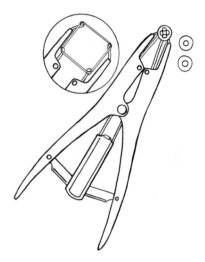

3. Hold the scrotum and both testicles in one hand and slip them into the rubber ring.

4. Position the rubber ring between the testes and the body and release the tension on the ring. Remove the elastrator, leaving the ring on the scrotum.

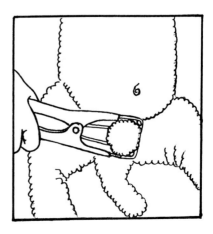

5. If one testicle slides above the ring, cut the ring with a sharp knife and repeat the process so that both testicles are below the ring.

6. Set the lamb down gently when finished.

7. This method will cause some pain for 30 minutes until the tissues become numb from the pressure. The lamb will lie down and may kick.

CAUTION: Do not attempt to use a castrating knife or elastrator to castrate a ram lamb that has a scrotal hernia. Use the emasculatome, consult a veterinarian for help, or leave the lamb uncastrated. Be sure to identify this as a ram lamb in your records.

Most castrated lambs will be stiff in the rear legs for 1 to 2 days after castration. If this condition persists, or if there is a history of tetanus on the farm, immunize against tetanus all lambs that have been castrated with the castrating knife or the elastrator. This may not be necessary if the emasculatome is used, unless there is a break in the skin.

If tetanus infection is likely, 200 to 300 units of tetanus antitoxin should be injected either intramuscularly or subcutaneously according to the manufacturer's directions. Intramuscular injections can be given in the muscle of the shoulder or forearm or on the inside of the rear leg in a wool-free area. The needle should be inserted into the muscle and the plunger of the syringe pulled slightly backward to determine if the needle is in a blood vessel. If blood appears in the syringe, pull the needle back slightly and complete the injection. If a subcutaneous injection is used, set the lamb on its rump and lift one foreleg. Inject the antitoxin under the loose skin in the wool-free area behind the foreleg.

In either case, the area to be injected should be cleaned with isopropyl alcohol before the injection is made. Clean the needle after each use with isopropyl alcohol to reduce the incidence of infection in other lambs.

Postprocedural Management

Use clean bedding in the pens where lambs are placed after castration, especially those castrated with the knife. This will reduce the chance of infection. Allow the lambs to rest for 30 minutes after castration until the trauma of the operation has subsided. In cold weather, a heat lamp placed over the creep will encourage the lambs to rest and prevent chilling. Choose a bright spring or summer day for castration as the lambs will react more favorably than they will on a cold, wet day. Observe lambs castrated by the knife for evidence of infection and administer antibiotics if infection is present.

Check lambs that have been castrated by the emasculatome for atrophy of the testes. If this does not occur in one or both testicles, repeat the procedure.

If there is danger of fly strike or maggots, a recommended fly repellant should be applied when the castrating knife is used.

7.8 IDENTIFICATION

Proper identification of sheep and lambs is necessary if any records are to be kept on the flock. It is necessary to determine offspring and parents, to make selections of breeding stock and sale animals, and to register purebred animals with breed associations.

A number of systems can be used for identification. Some, such as stencils, grease sticks, or spray paint, are temporary and are meant to be used until more permanent methods are instituted; others, such as ear tags, are more permanent, unless they are lost or torn out of ears. Tattooing or ear notches are permanent.

Equipment Necessary

- Paint brands; 2½-inch for lambs, 4-inch for sheep
- Wool-branding paint
- Metal ear tags
- Ear-tagging pliers
- Ear punch
- Plastic ear tags
- Applicator for plastic tags
- Ear notcher
- Tattoo outfit
- Antiseptic
- Grease crayons
- Spray paint cans

Restraint Required

To apply stencils, ear tags, ear notches, or tattoos in the ears requires only that the animal be held by the head. If tattoos are placed in the flank, it will be necessary to set the animal on its rump.

Stenciling Lambs and Ewes

This temporary method of identification is used on young lambs shortly after birth. As the lambs and their dam are released from the lambing pens it is easy for them to become separated, with the result that some lambs suffer from malnutrition. The lambs and their dam can be given the same number, using the small stencils for the lambs and the larger stencils for the ewes. It is then easy to reunite the lambs with their dams or quickly to identify any lambs that may need special treatment. Wool-branding paint should be used because it will wash out of the wool when it is scoured or washed in processing and not damage the wool as will ordinary paint.

Step-by-Step Procedures

1. Pour a small amount of wool-branding paint into a container to saturate a piece of cloth or burlap. This will prevent the drippage that will occur if the stencils are dipped in a full can of branding paint.

2. Place the stencil in the container until the figure is covered with branding paint.

3. Place the stencil on the back of the lamb or ewe so that it can be read from front to rear while standing on the left side, or from left to right while standing at the sheep's rear. It may be necessary to rotate the stencil to conform to the contour of the animal's back.

4. Allow the paint to dry before moving or handling the animal.

5. Clean the equipment after use.

Ear Tags

Ear tags are usually made of metal or plastic with the numbers either stamped or painted on them. They can be purchased prenumbered from livestock supply firms. Some breeders add a farm or family name to the tag. Metal tags are either self-piercing or require that a hole be punched in the ear. Some sheep breed associations furnish metal registration tags to be used on purebred sheep. Many of these require that a hole be punched in the ear and the tags closed by bending or clamping.

The use of plastic ear tags usually requires that a hole be punched in the ear, either with an ear punch or an applicator that inserts the tag at the same time. In some cases it is necessary for the owner to paint the number on each tag.

Applying Self-Piercing Tags

1. Insert the self-piercing ear tag into the pliers.

2. Find the area on the inside of the ear that has the widest space between the ribs of cartilage. There

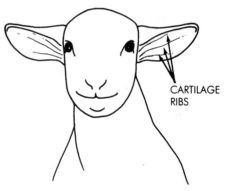

CARTILAGE RIBS

will be one rib below the area and two above it. Allow space for the ears of young lambs to grow when using metal ear tags. Leave at least $\frac{1}{2}$ inch between the edge of the ear and the ear tag.

3. Place the ear tag in the ear with the number facing forward. Clamp it tightly, so that the tag is sealed to prevent its loss from the ear.

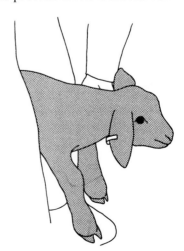

Applying Nonpiercing Metal Tags

Nonpiercing metal tags will require that a hole be punched in the ear. This can be done with an ear punch designed for this purpose.

1. Locate a spot in the widest part of the ear between the rib nearest the bottom and the two ribs at the top.

2. Punch a clean hole through the ear, removing any loose cartilage that remains.

3. Place the ear tag in the ear with the number facing forward.

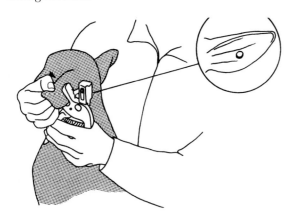

4. Clamp the ear tag together with a pair of pliers or bend the small tab at the end of the tag to secure it in the ear.

Applying Plastic Tags

1. Select a tag style. Both single (number tag and "button" keeper) and double (two number tags with one serving as the keeper) plastic tags can be used successfully. The double is more versatile because with one piece being used on each side of the ear, the tag can more easily be seen, regardless of which way the animal is facing.

2. Select the tag size to be used. Keep in mind the reason for using a visible tag is so that you can read it from a distance. If you will never want to read the tag until you have restrained the animal, then a small tag is acceptable. If you will want to read it from a few feet away, then select the largest tag appropriate to the size of the animal. At first, a lamb's ear may droop with a large tag, but as it grows, the ear will become strong enough to support it.

3. Be certain to select contrasting ink and tag colors. If you are uncertain, a safe bet is always a yellow tag with black numerals and letters. Additional color combinations can be used to determine breed, sex, year of birth, or individual sire lines.

4. Select a numbering system for the flock. The first number might be the last digit of the year of birth, with the following numbers being the sequential birth order.

5. Select either prenumbered or blank tags and lettering ink. Prenumbered are more convenient, but less adaptable to your system and flock. If you do use the blank tags, use the type of ink recommended by the tag manufacturer. The secret to long-lasting readable ear tags is using an ink that will not fade. Good ink and tag combinations bond to each other. Avoid the Magic Marker type of coloring tool that merely marks on the surface of the tag. Number the tags at least 1 day before you insert them into the ear. If you use a single tag and button keeper, number both sides of the tag.

6. Insert the ear tag into the correct applicator pliers. Two-piece tags require that the male portion of the tag be placed over the pin on the applicator and the female portion be inserted into a clip. Before you attempt to place this into a lamb's ear, check that each tag you have inserted is properly aligned by starting to close the pliers but stopping just as the two pieces begin to meet. If they meet properly, they are properly aligned and ready to be placed into the lamb's ear.

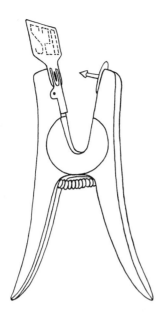

7. Select the ear to be tagged. This will depend upon your working facilities, your preferences, and whether the breed registry has some guidelines for you to follow.

8. Select the tagging site on the ear. For lambs, it is best to locate a spot on the widest part of the ear and above the lowest rib on the ear, but below the top two ribs.

9. Hold the ear of the lamb with one hand while using the other hand to insert the tag.

> **CAUTION:** Be certain to restrain the head securely while piercing the ear, and continue to do so until you have removed the pliers from the ear. The animal will often throw its head from side to side, and while the lamb is not strong enough to escape, the shaking could tear the ear if you are not securely restraining the lamb.

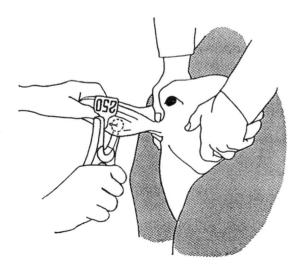

10. Treat the pierced area of the ear with an antiseptic spray, iodine, or other animal wound dressing to help prevent infection and fly irritation.

11. Release the animal and allow it to return immediately to its flock mates.

Ear Notching

Many purebred breeders do not like to use ear notches because it detracts from the appearance of sheep that are exhibited in the showring. Commercial breeders will find it the most economical method of identification because it requires less equipment than other methods. For best results, it should be used on sheep that have no wool on the ears, because the wool makes the notches difficult to see.

A simple system of notching that can be read and interpreted easily should be used. The systems used in swine identification can be used with sheep. A notch in the top of the left ear of the sheep represents the number 1; in the bottom of the left ear, 10; and in the end of the left ear, 100. A notch in the top of the right ear represents the number 3; in the bottom, 30; and in the end of the right ear, 300. Thus, sheep number 135 would have one notch in the bottom of the right ear, one notch in the top of the right ear, two notches in the top of the left ear, and one notch in the end of the left ear. This system can be used to number 999 animals. If more numbers are needed, a hole can be punched in the center of the left ear for 1,000 and in the center of the right ear for 3,000.

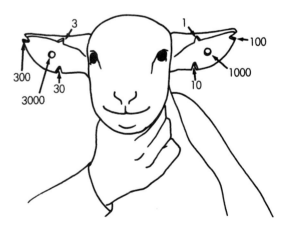

1. Hold the sheep by the head and use a pair of ear notchers that remove a V-shaped amount of tissue from the edge of the ear to notch the correct number in the ear.

This method may cause some bleeding. The notch should be treated with iodine or some other antiseptic, especially during fly season.

Tattooing

This is a permanent method of identifying sheep if it is used properly. It requires a tattooing outfit, which consists of a pair of pliers and a set of numbers or letters made in the form of dies with sharp, pointed, needlelike projections that pierce the skin. A tattoo ink or paste is forced into the puncture and remains visible after the puncture wound heals. The tattoo ink or paste can be placed on the dies and pressed into the skin by the pliers or it can be rubbed into the puncture wound after the pliers are used. The pigment in the ink heals into the skin, leaving a permanent number or letter.

Tattoos can be seen best on sheep that have white faces and ears. Those with black pigment in the ears can be tattooed inside the rear flank, which usually contains less pigment than the ears.

Do not tattoo young lambs. The space in the ear may not be large enough if more than two numbers are used. As the ear grows, the tattoo grows also, and it may spread out so that it cannot be read. Use a temporary identification until the lamb is at least 6 months of age, and then tattoo it.

1. Place the number dies in the tattooing pliers in the proper order. This will be in reverse order as you look at them. To prevent error, press the dies into a piece of cardboard to determine if the number is in proper order. Some pliers have the numbers on a small drum that can be rotated to the correct number. Most pliers have space for four or five numbers.

2. Locate the widest spot in the ear between the rib nearest the bottom and the two ribs at the top. Clean the area with alcohol.

3. Press the dies into the skin of the ear.

4. If the tattoo ink was placed on the dies, remove the pliers and rub the ink into the punctures.

5. If no ink is placed on the dies, apply it to the ear and rub it into the punctures.

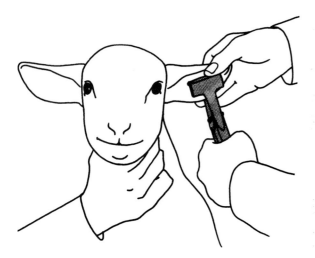

6. The same procedure can be used on the inside of the flank. It will be necessary to set the sheep on its rump in the manner described in Section 7.14. The tattoo should be applied to the inside of the flank where there is no wool growth.

7. Clean the equipment after use.

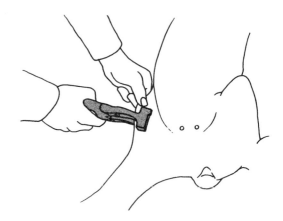

Postprocedural Management

Sometimes the ears of sheep or lambs identified with self-piercing ear tags or plastic tags become infected and swollen. If such is the case, clean the area around the tag and treat it with iodine or some other antiseptic until cured.

7.9 FOSTERING LAMBS

Lambs may need fostering under the following conditions: (1) when the mother dies, (2) when the mother has no milk, (3) when the mother refuses to allow the lamb to nurse, (4) when the mother gives birth to more than two lambs, or (5) when the mother loses one or more lambs at birth or shortly thereafter.

Under most conditions, grafting or fostering of lambs will be more successful if it can be done at birth or within a few hours after birth. The ewe recognizes her lambs by smell. Once she becomes accustomed to her lamb's odor, she will not accept other lambs. It is necessary, therefore, to give the foster lamb the same odor as her own lamb or to mask the odor of the foster lamb in some manner.

Equipment Necessary

- Bucket, 10- to 12-quart
- Salt solution
- Sharp knife
- Fostering stanchion
- Lambing pen
- Iron stakes
- Wire
- Ewe halter

Restraint Required

If the ewe refuses to accept the foster lamb, it may be necessary to place her in a fostering stanchion until the graft is accomplished. The stanchion should be constructed so that the ewe can eat, drink, and lie down, but cannot injure the lamb or refuse to let it nurse. It should be equipped with sides 8 to 10 inches above the floor to keep the ewe from moving away from the lamb. Most ewes will accept the lamb in from 3 to 6 days.

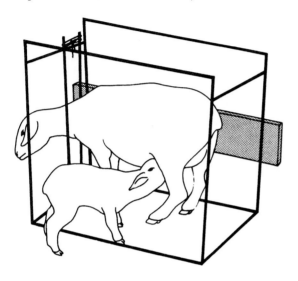

Step-by-Step Procedure

1. If the lambs are dead at birth or the ewe loses one lamb and is capable of raising two lambs, a foster lamb or lambs should be placed with her immediately.

2. Rub the foster lamb with the fluids contained in the afterbirth of the dead lambs or with the dead lamb itself, especially around the head and tail where the ewe smells for identification.

3. If the ewe refuses to accept the foster lamb, immerse the lamb in a warm solution of salt in water to

mask its odor and do the same with her own lamb so that the lambs smell alike. The salt solution can be prepared by dissolving a cup of iodized salt in 10 quarts of warm water. This procedure is called a "wet" graft.

4. If the wet graft fails, skin the dead lamb and place the skin over the back of the foster lamb. Cut holes in the skin of the dead lamb and place the forelegs of the foster lamb through them to hold the skin in place. Be sure to leave the tail of the dead lamb on the skin.

5. If all these methods fail, place the ewe in a stanchion with her own lamb and the foster lamb, or with two foster lambs, until she accepts them.

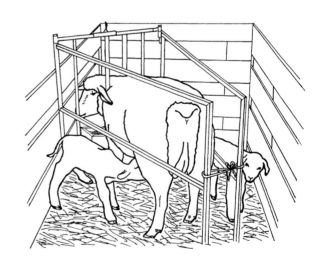

6. If a ewe stanchion is not available, two iron stakes can be driven into a dirt floor. The stakes should be 6 inches apart and 3 feet high. The ewe's neck can be placed between the stakes and a wire can be placed above her head to prevent the stakes from spreading apart. A hinged lambing pen should be placed around the stakes to prevent the lambs from wandering away from the ewe.

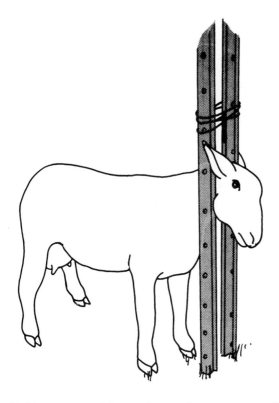

7. If a ewe stanchion or iron stakes are not available, place a halter on the ewe and tie her to the lambing pen. She should be tied in a manner that will prevent her from jumping out of it. This method is not as satisfactory as other methods because the ewe cannot be restrained enough to permit the lamb or lambs to nurse.

8. Some ewes will accept a foster lamb if a dog is tied near the lambing pen; they apparently accept it to protect it from the dog.

Postprocedural Management

Observe the ewe with foster lambs to be sure that she has accepted them and that the lambs are nursing. If none of the fostering methods are successful, the lambs should be raised artificially on cold liquid milk replacer.

7.10 WEANING AND MASTITIS CONTROL

Lambs can be weaned satisfactorily at as early as 5 weeks of age if necessary, but most are weaned when they are from 8 to 12 weeks old. Milk production in the ewe reaches a peak after 4 weeks of lactation and declines steadily until the tenth to twelfth week of lactation.

If the ewe is milking well, there appears to be no advantage to weaning lambs before they are 8 weeks old, because their weight at this age is highly correlated with the milk production of the dam. Earlier weaning may be justified if the ewe has mastitis in one or both sides of the udder, or if her milk supply is low. Later weaning subjects the lambs to greater infestations of internal parasites if they are grazing on the same pastures as their mothers.

Both ewes and lambs must be adequately prepared before weaning is done to reduce the milk supply of the ewe and to protect the lamb from enterotoxemia (overeating) when the ewe's milk is no longer available. Abrupt weaning will cause less confusion than will gradual weaning, in which the lambs are returned to the ewes to nurse them out.

Equipment Necessary

- Pens—separated for ewes and lambs
- Vaccine—*Clostridium perfringens,* Types C and D bacterin with tetanus toxoid
- Syringe and needles
- Balling gun and sulfamethazine boluses

Restraint Required

For weaning, ewes should be separated from the lambs and placed in separate pens or pastures where the ewes and lambs cannot see or hear each other. Mastitis treatment involves restraint by hand or with the aid of a squeeze chute.

Step-by-Step Procedures

Weaning

1. Two weeks before weaning, vaccinate all lambs to be weaned against enterotoxemia (overeating disease) with *Clostridium perfringens,* bacterin, Types C and D, as recommended by the manufacturer. Set the lamb on its rump and lift one of the forelegs. Use isopropyl alcohol to clean the area to be injected and inject the toxoid or bacterin under the loose skin with a syringe graduated in milliliters or cc's. Clean the needle with isopropyl alcohol after each injection to reduce the incidence of infection in other lambs.

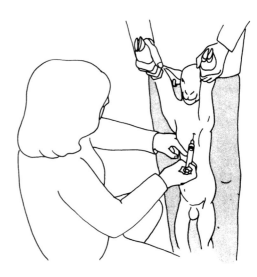

2. Two days before weaning, discontinue the feeding of concentrates to ewes from which lambs will be weaned.

3. One day before weaning, remove all feed and water from the ewes to reduce milk flow.

4. Wean the lambs abruptly by separating the ewes from the lambs and moving them to an area where they cannot see or hear the lambs. Leave the lambs in familiar surroundings with the same creep feed they were receiving prior to weaning.

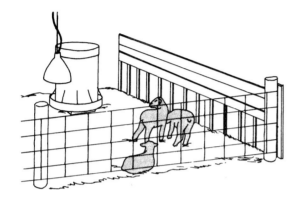

5. The ewes can receive water and a maintenance ration of roughages on the second day after weaning, but no concentrate should be fed.

6. Lambs may be revaccinated against enterotoxemia at 14 days after weaning if that is recommended by the manufacturer of either the toxoid or bacterin. This will give additional protection to the lambs.

Mastitis Control

Mastitis may develop in ewes during lactation or following the early weaning of lambs. It is caused by staphylococcus, streptococcus, coliform, or pasteurella organisms.

The most severe form of mastitis is the gangrenous type, sometimes called "blue bag." It develops rapidly, and the ewes develop a high fever. The udder becomes hard, reddened, and swollen, and eventually develops gangrene, which gives it a characteristic blue color. The ewes are in considerable pain and refuse to let the lambs nurse. Death may occur in 25% of the affected ewes in 4 to 5 days. If the ewe recovers, the udder remains nonfunctional and the affected portion may slough off.

Nongangrenous mastitis may go unnoticed. The udder becomes hard, swollen, and inflamed. Milk production is reduced and the udder or half of the udder becomes nonfunctional. Ewe survival rate is greater than with the gangrenous type, but the lambs will either die of starvation or toxemia or attempt to nurse other ewes. In so doing they spread the infection from their dam to other ewes.

It is important, therefore, to separate any ewe that has mastitis, and her lambs, from the flock to prevent the spread of the disease organisms. The lambs can be raised artificially on a milk replacer and the ewes should be marked for culling.

1. Separate ewes with possible mastitis and their lambs from the rest of the flock.

2. Check the ewe's temperature rectally; 102.3°F is normal.

3. Wide-spectrum antibiotics, especially streptomycin, can be injected intramuscularly into the ewe according to the manufacturer's directions.

4. Sulfamethazine can be given orally in a 15-g bolus to affected ewes as soon as mastitis is noticed, with a second 15-g bolus given 24 hours later.

5. If the surface of the udder sloughs off, a veterinarian should be called to treat the affected ewe.

Postprocedural Management

Heavy-milking ewes should be checked for mastitis following the weaning of lambs. Distended udders can be partially milked by hand to relieve some pressure, but the udder should not be emptied, because the pressure in it reduces milk secretion. Those ewes with mastitis should be treated with antibiotics.

7.11 ARTIFICIAL REARING

Lambs can be raised successfully on a liquid milk replacer if they cannot be raised by their natural mother or grafted onto a foster mother. The best results are obtained if the lambs can be placed on artificial rearing before they are 24 hours old and after they have received colostrum. Colostrum is rich in nutrients needed for survival, has a laxative effect on the lamb, and contains immunoglobulins (antibodies) that provide protection against disease and activate the sucking instinct. If colostrum is not available from the natural mother, it can be provided from another ewe or from a dairy cow. The dairy

colostrum should be collected in advance and frozen until needed. It should be thawed at room temperature. Do not heat it, because heating destroys the immunoglobulins.

Lamb milk replacers are available commercially. Most are prepared from a milk source and contain casein, the milk protein, and lactose, the milk sugar, which are produced exclusively by the mammary gland. Milk replacers are prepared in powder form and when mixed with water have the approximate composition of ewe's milk. Ewe's milk contains about 8.2% fat, 5.8% protein, 0.92% minerals, and 80% water. It has more total solids and therefore more energy than the milk from either the cow or the goat.

The composition of lamb milk replacers should be within the following percentage range on a dry-matter basis: crude fat, 30 to 32%; crude milk protein, 22 to 24%; crude fiber, 0 to 1%; lactose, 30 to 35%; and ash, 5 to 10%. When the milk powder is mixed with water it forms a liquid milk replacer approximating the composition of ewe's milk. Milk replacers prepared for calves contain only 20% fat, a concentration that is not adequate for lambs.

The powdered milk replacer should be thoroughly mixed with hot water according to the manufacturer's recommendations until the powder is evenly distributed throughout the mixture. It then should be cooled to 32 to 34°F in a refrigerator. Small quantities can be mixed with hand mixers or blenders, but larger quantities will require larger mixing equipment. A used spin-dry washer is satisfactory and contains a pump that can be used to transfer the mixture to a container for refrigeration. In any case, the equipment must be thoroughly cleaned and disinfected after each use. Disinfectants used for cleaning milking equipment in the dairy industry can be used. This applies to the nursers as well.

For best results, the liquid milk replacer should be fed cold (less than 40°F) to the artificially reared lambs. The low temperature reduces bacterial activity, prevents overeating by reducing the intake of the milk replacer, permits self-feeding, and reduces such digestive disturbances as colic. The replacer can be kept cold in the nursers by freezing a plastic container of water and placing it in the nursing container. Bacterial activity can be reduced in warm weather by adding 1 ml (cc) of formalin (37% formaldehyde) to each gallon of liquid milk replacer during the mixing process. Once the lambs have been trained to nurse the cold milk replacer they can be self-fed until weaned. They should consume small amounts of the replacer frequently, which simulates the natural nursing process on the ewe. One nipple is required for each three to five lambs. The nipples should be between 12 and 15 inches from the floor.

Research indicates that lambs can be weaned from the liquid milk replacer at 21 to 28 days of age, provided they have become accustomed to a dry creep ration. Abrupt weaning appears better than gradual weaning or dilution of the milk replacer with water. There will be some growth-check at weaning until the lambs consume enough dry feed to meet energy needs. Part of this growth-check is due to dehydration, so a source of fresh water must be supplied with the dry creep feed. The concentrate should contain from 20 to 24% crude protein and have a high-energy content. Pelleting will increase consumption and reduce dustiness.

Equipment Necessary

- Lamb milk replacer (powder)
- Mixing equipment
- Nursers equipped with lamb bar or nursette lamb nipples
- Creep feeders
- Water containers
- Training and mixing pen
- Disinfectant
- Formalin
- *Clostridium perfringens,* Types C and D bacterin with tetanus toxoid
- Bucket and nipple
- Bottle and nipple

Restraint Required

After the lamb has been removed from the ewe or fed colostrum, it should be placed in a training pen and taught to nurse. The lamb should be held to the nipple at first, but care should be taken not to hold its head or it will become dependent upon this help. Two or three sessions usually are adequate, especially if the lamb is hungry. An older lamb that has been trained will help in leading new lambs to the nurser. Once the lamb is trained to nurse, it can be self-fed until weaned.

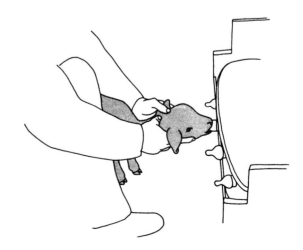

Step-by-Step Procedure

Large Number of Lambs

1. Mix the liquid milk replacer according to the manufacturer's directions.

2. Refrigerate the milk replacer and feed it cold from the nurser.

3. Train the new lamb to nurse from the nurser and check periodically to see that it is not hungry.

4. Remove the lamb from the training pen after it is nursing properly and place it in a mixing pen with older lambs until weaned. The population of the training pen should be limited to 10 lambs, but the mixing pen may contain 20 to 30 lambs.

5. Provide a high-energy creep ration and an adequate supply of water.

6. Wean the lamb abruptly from the liquid replacer at 21 to 28 days of age. The lambs should be left in the same surroundings until they have adapted to dry feed. Lambs weaned gradually by reducing the milk replacer or by diluting it with water do not adapt as quickly to dry feed as do those weaned abruptly.

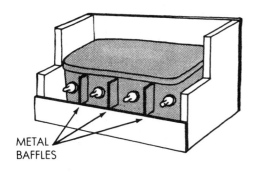

METAL BAFFLES

CAUTION: Some lambs develop the vice of trying to nurse other lambs. This usually involves tails, navel cords, or ears. Such lambs may not be getting enough milk and should be forced to nurse the replacer more frequently. If this vice cannot be corrected, remove them from other lambs and handle them separately.

Some lambs may develop the habit of chewing the ends off the nipples, thus wasting the replacer. This can be prevented by placing a 1" × 3" plastic or metal baffle between the nipples so that the lambs cannot chew them with their molar teeth.

At 2 weeks of age, the artificially reared lambs should be vaccinated against enterotoxemia (overeating disease) with *Clostridium perfringens* Types C and D bacterin. Set the lamb on its rump, raise one foreleg, and inject the vaccine under the skin (subcutaneously) in a wool-free area behind the leg according to the manufacturer's directions. Use isopropyl alcohol to clean the area to be vaccinated. Clean the needles with isopropyl alcohol after use to reduce the incidence of infection in other lambs. The injections should be repeated at approximately 5 to 6 weeks of age.

Clean and disinfect the milking and nursing equipment once a week to reduce bacterial growth. Keep the milk between 33 and 40°F, especially in warm weather, by use of a plastic bottle containing ice.

If bedding is used in the training and mixing pens, it should be removed and replaced each week. Lambs that develop diarrhea and intestinal infections from eating the bedding can be treated with antibiotics or sulfonamides. The development of these problems can be prevented if no bedding is used or if the lambs are placed on wire floors above the bedding.

Small Number of Lambs

If only two or three lambs are to be raised artificially, it is preferable to attempt to graft them onto a foster mother. If this is impossible, a small nurser can be made by boring a hole near the top of a plastic bucket and fitting a lamb bar or nursette nipple and tube to it. The lambs should be trained and handled in the manner previously discussed. If the owner insists on hand-feeding a small number of lambs on a bottle, the following schedule can be used as a guide:

Age of Lamb	Frequency of Feeding	Amount per Feeding
First 24 hours	Every 2 hours	Not over 2 ounces
2nd to 4th day	Every 3 hours	3–4 ounces
5th to 7th day	Every 4 hours	4–5 ounces
2nd week	Every 4 hours	5 ounces
3rd to 8th week	Every 6 hours	8–10 ounces

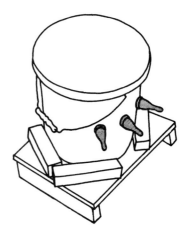

The feeding of cold milk after the first week will reduce colic and digestive disturbances in the lambs.

Postprocedural Management

Artificially reared lambs should not be placed with naturally reared lambs until they reach the same level of maturity. Lambs weaned from the artificial nurser at 21 to 28 days cannot compete with naturally reared lambs weaned at 56 to 60 days. Instead, artificially reared lambs should be kept in a mixing pen and continued on a creep ration of 20 to 24% crude protein until they are 56 to 60 days old. The protein level can be reduced to 15 to 17% at that age.

Artificially reared lambs will be a few pounds lighter than naturally reared lambs at 56 to 60 days of age, but this difference usually disappears by the time both groups are 90 days old.

7.12 WORMING TO CONTROL INTERNAL PARASITES

Internal parasites cause considerable losses in sheep flocks through death, reduced weight gains, lower wool production, and poor reproductive efficiency. Parasitism is most damaging to young lambs and to poorly managed animals that may suffer from malnutrition.

Prevention is as important in controlling internal parasites as treatment. Most internal parasites have a life cycle in which the adult stage develops in the true stomach (abomasum), the small or large intestine, or the lungs. Eggs from adults are eliminated in the feces. Under proper conditions of moisture and temperature, the eggs develop into larvae that migrate onto the blades of grass and are consumed by grazing sheep and lambs to repeat the cycle. Methods used to break the life cycles of internal parasites include (1) pasture rotation, (2) early weaning of lambs before they can pick up the larvae from the pasture, (3) separation of lambs

from ewes during grazing, (4) weaning of lambs and finishing in drylot or on slotted floors, and (5) the use of clean pastures for lambs that are finished on grass.

Treatment of sheep infested with internal parasites consists of drenching (worming) with approved anthelmintics (drenches) to eliminate or reduce the body load of adult parasites or to reduce egg production. This should be done on a regular schedule with the entire flock on a year-round basis.

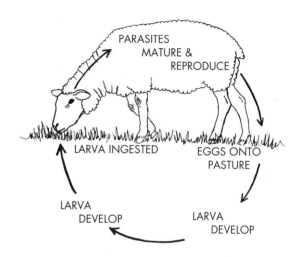

1. Worm all mature sheep at least four times a year: (1) in late fall at the close of the pasture season; (2) in the spring before turning the sheep onto pasture; (3) in early summer; and (4) before the breeding season in late summer or early fall.

In addition to this general guideline, add the following: (1) Treat ewes within 7 to 10 days after lambing, then repeat in 7 to 10 days, and for flocks with a history of heavy infestation, follow this with a third worming in another 7 to 10 days. Turn the ewe-lamb combinations onto clean pasture. Remember, your goal is to sell as many pounds of top-quality lamb as you can, as quickly as you can. You can do this only if the lamb growth is not hindered by internal parasites. Spending a couple extra dollars on wormings is a good investment. (2) Treat the lambs about 14 to 21 days after they begin effective grazing; that is the time when they will be ingesting parasite eggs and larvae from contaminated grass. (3) Consider altering the general guidelines above (see item 1) and treat the entire flock if the weather has been particularly wet and warm—conditions that favor rapid buildup of parasites in the ground and forage. (4) Treat all rams before turning them in with the ewes. Breeding season is demanding and a heavy parasite load will contribute to a rapid loss of condition, vigor, and libido in the ram.

2. Worm all lambs at weaning time. Those remaining on pasture should be wormed every 3 to 4 weeks until marketed. Those placed in drylot or on slotted floors may be wormed as needed.

3. If infestation is heavy in mature sheep, they should be wormed more often than four times a year as needed.

4. Pregnant ewes should not be wormed within 2 weeks of lambing.

The more common internal parasites affecting sheep are the large stomach worm (*Haemonchus contortus*), nodular worm (*Oesophagostomum columbianum*), tapeworm (*Moneizia expansa*), lungworm (*Dictyocaulus*), and under certain conditions liver flukes (*Trematodes*) and coccidia (protozoa). Others may be present also, depending upon the section of the country and environmental conditions.

The external symptoms of internal parasitism include loss of weight, unthriftiness, diarrhea, coughing, anemia, edema under the jaw and along the underline, and loss of appetite. These symptoms must be recognized at an early stage and treatment begun at once. Fecal examinations can be made by a veterinarian to determine which of the parasites are present.

There are several brand-name anthelmintics on the market today, but they are each based upon the two currently sheep-approved products, levamisole and thiabendazole. Of the two, levamisole is the broader spectrum. The phenothiazines, lead arsenates, copper sulfates, heloxons, nicotine sulfates, carbon tetrachlorides, and sulfonamides of times past are no longer approved.

CAUTION: Lambs and mature sheep should be wormed prior to marketing only in compliance with federal regulations regarding the use of anthelmintics on animals that will find their way into the human food chain. Follow label directions carefully!

Equipment Necessary

- Anthelmintics
- Dose syringe, 2- or 4-ounce
- Automatic drencher
- Chutes or holding pens (for restraint)
- Wool-marking chalk
- Forceps or balling gun
- Coccidiostat

Restraint Required

It is necessary to restrain each animal as it is wormed. A sheep-handling chute will make handling easier for large flocks. Individual sheep can be caught, straddled by the person administering the drench, and backed into a corner of a pen. Mark each sheep when worming is completed to prevent redrenching.

Step-by-Step Procedure

1. Prepare the worming material (anthelmintic) as directed by the manufacturer.

2. Catch and restrain the sheep.

3. Place your hand under the jaw of the sheep and control its head.

4. Place the tube of the drenching syringe into the mouth of the animal at the side between the inscisors and the molars and extend it over the base of the tongue.

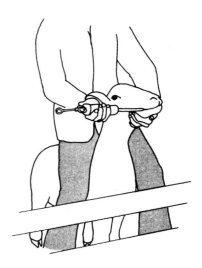

5. While holding the head level so the animal can swallow, slowly force the drench into the throat.

CAUTION: Care should be taken not to force the drench into the lungs.

---

6. Mark the drenched sheep with wool-marking chalk.

7. If an automatic drencher is used it will have a longer curved tube, which can be placed farther down the throat.

8. The head should be raised to permit the longer tube to enter the esophagus.

9. The nose can be grasped to prevent breathing and the drench administered faster than with a syringe.

10. Most automatic drenchers are attached to a bag containing the drench, which can be worn on the operator's back. They are usually calibrated for automatic delivery of the correct dose.

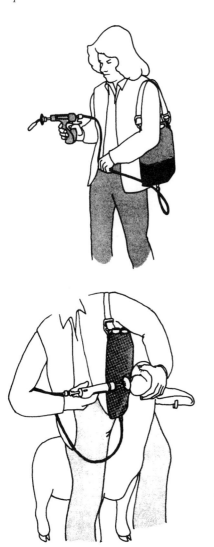

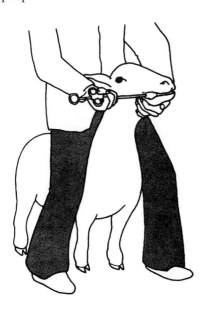

11. Some anthelmintics are prepared in the form of a bolus. This should be placed on the back of the tongue with a pair of forceps or balling gun made for this purpose.

CAUTIONS: Follow the manufacturer's directions for mixing and administering each anthelmintic; these directions vary with the size or weight of animal and with different products. Do not confuse ounces with milliliters (cc's).

Follow the manufacturer's (label) directions on administering anthelmintics to market animals.

Do not worm pregnant ewes within 2 weeks of lambing. The anthelmintic may be dangerous to the fetuses and rough handling can cause abortion.

Postprocedural Management

Observe all animals after worming for reactions to the drench. Labored breathing may indicate that some of the drench has entered the lungs and a veterinarian should be consulted.

Some animals will show symptoms of greater parasite infestation than others. These should be dewormed more often.

Coccidiosis Control

Coccidiosis is a disease, not a condition caused by internal parasites. However, it is appropriate to consider it here because it is a very important disease that can determine profit or loss in your operation, and the control of coccidiosis is best timed when it coincides with internal parasite control.

Prevention is the key to successfully managing coccidiosis. Begin treatment of all sheep classes—lambs, ewes, and rams—*before* periods of stress. These stress events include forecasted periods of prolonged severe weather, lambing, weaning, switching lambs from grass to feedlots, long breeding seasons, poor nutrition for ewes and rams, and any time there are heavy parasite infestations.

For *preventive* type of treatment, sulfonamides and lasolocid are approved. Check with your veterinarian for a coccidiostat that works well in your environment.

Most times, it is good management to deworm and administer the coccidiostat simultaneously. Parasite loading and the added stress of one of the above events is a certain recipe for a health "wreck."

7.13 CONTROL OF EXTERNAL PARASITES

The two most common external parasites of sheep are sheep "ticks" or sheep keds (*melophagus ovinus*) and sheep-biting lice (*bovicola bovis*). Sheep "ticks" are not ticks at all but wingless flies. They have only six legs, while most ticks have eight legs. They are bloodsuckers and have a dorsal sac that becomes engorged with the blood of the host animal. They can be detected by parting the wool and finding them attached to the skin. They produce living young in a pupal case that is attached to the wool fibers until the young ked emerges. The pupal case of the sheep ked cannot live except on the wool fibers of sheep.

Sheep lice, both chewing and sucking, are harder to detect because they are much smaller and are well camouflaged by their wool-like color. They feed on the skin, causing irritation, or cause blood loss from the sucking.

In both cases, the symptoms are rubbing or scratching by the host animal, resulting in a loss of wool. (Wool in the fence is an indication of the presence of external parasites.)

A more serious but less common external parasitic disease of sheep is scabies (not to be confused with *scrapie*), which is caused by the sarcoptic mange mite. This mite causes large scabs to form on the skin, with extensive loss of wool and reduced growth or production. Fortunately, a national eradication program was undertaken a few years ago and this disease has been eliminated from the United States. An outbreak of scabies must be reported to a veterinary office.

Sheep do become infested with ticks, real ticks, not the sheep "tick" or ked, especially in the summer. Western range ewes are especially prone to this problem. These ticks are of the six-legged larval, eight-legged nymphal, or eight-legged adult variety. Irritation-caused wool rubbing, weakness and unthriftiness from blood loss (sucking), and occasionally paralysis are symptoms of a tick infestation. A possible severe complication can be the transmission of anaplasmosis and tularemia from the tick. The same tick species can infect humans and spread tularemia and spotted fever to us.

Treatment for sheep keds, sheep lice, and ticks involves either dipping or spraying with an insecticide approved by the FDA. These include pyrethroids, organophosphates, and ectrin (for keds). They can be purchased from livestock supply houses, and care should be taken to mix them properly.

CAUTION: The concentration for dipping is *one-half* the concentration recommended for spraying.

Dipping is considered to be more effective than spraying because the material more completely penetrates areas behind the flanks and other hard-to-reach places. The dip can also be reused if the severity of the infestation requires a second treatment in 14 days. For best results, sheep should be shorn before either treatment to allow the insecticide to reach the skin and to save the liquid that would remain in the wool.

Equipment Necessary

- Approved insecticide
- Dipping vat or spraying equipment
- Hurdles or fencing

Restraint Required

Sheep do not like to enter water or a spray and must be forced to do so. Holding pens must be positioned both ahead of and behind the treatment area. Sheep will enter the dipping tank or spraying area more readily if the approach chute is curved such that they cannot see ahead.

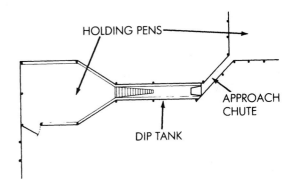

The dipping tank can be constructed of metal or concrete. It should be 12 feet long at the top and 6 feet long at the bottom, with a cleated incline for the other 6 feet. The tank should be 2 feet wide at the top, sloping to 1 foot at the bottom, and it should be 6 feet high. The sheep should be completely immersed in the liquid (including their heads and ears).

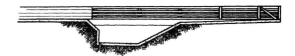

Spraying can be done in a holding pen with a high-pressure sprayer equipped with a nozzle, or the sheep can be walked through a spraying device equipped with nozzles on each side as well as the top or bottom. In either case, the use of a pen is recommended to allow the treated sheep to drain briefly to save the liquid and prevent contamination of soil or pastures by the insecticide.

Step-by-Step Procedure

1. Dipping or spraying is best done 1 week after shearing in the spring to allow any cuts from shearing to heal.

2. Lambs in short fleece can be treated without shearing.

3. Prepare the liquid insecticide according to the manufacturer's directions.

4. Choose a bright, sunny, spring day so the treated animals will dry quickly and the insecticide will not be washed off or diluted by rain.

5. Fill the dipping vat to 1 foot from the top.

6. Force the sheep through the dipping vat or spray, but allow them to pause long enough to permit the insecticide to be effective. Allow only two sheep to be in the vat at one time.

7. Keep the sheep in a holding pen for at least 5 minutes after treatment to be sure that they drain properly.

8. In severe cases of infestation, repeat the treatment again in 7 to 14 days to kill the newly developed or hatched keds or lice.

9. If equipment is not available for dipping or spraying, arrangements can be made with custom dippers who have portable equipment that can be brought to the farm.

CAUTION: Put the required amount of water into the vat or sprayer and mix the insecticide according to directions. A solution that is too strong will burn the skin, and a solution that is too weak will not be effective. Remember to use one-half the spray concentration for dipping and to follow the manufacturer's directions.

Young lambs under 1 month of age should not be dipped because of the danger of drowning. They can be sprayed if they are heavily infested. Older lambs can be dipped if they are separated from mature sheep and handled separately.

Unweaned lambs will become infested with external parasites from their mothers. If the ewes are shorn and not dipped or sprayed shortly thereafter, most of the parasites will migrate to the lambs. Both groups should therefore be treated at the same time after shearing.

Do not store shorn wool that is contaminated with external parasites near the shorn sheep where the parasites can crawl back onto the sheep or lambs.

Do not return treated sheep to the barn from which they came, because some of the parasites or eggs will be in the bedding. The barn should be completely cleaned before it is reoccupied.

Examine sheep for external parasites before purchase. Newly purchased sheep should have no contact with animals on your farm until they have been examined and proven free of infestation or treated if infested.

Sheep exhibited at fairs or shows can become infested with keds or lice by contact with infested sheep. Isolate these sheep from the flock upon return to the farm and treat them if necessary.

Sheep or lambs that are destined for market should be treated with the insecticides indicated only with strict compliance with manufacturer's directions regarding withdrawal times.

Postprocedural Management

One or two treatments 7 to 14 days apart each year at shearing time are usually sufficient to eliminate external parasites. If the sheep are exposed to other infested sheep at shows or sales, they should be treated before they are returned to the flock.

7.14 FOOT TRIMMING

Foot trimming in sheep has two functions: (1) to properly shape the foot in young, growing animals and (2) to control foot rot in older animals. Foot trimming should be done at least twice per year—at shearing time in the spring and again before breeding time in the fall. If the foot has become misshapen or the animal develops lameness, it should be done more often as needed.

Equipment Necessary
- Sheep foot trimmers
- Sharp knife
- Pruning shears
- Tilting squeeze chute

Restraint Required

It will be necessary to restrain the sheep in one of two ways. The most common method is to set the sheep on its rump and hold it between your knees. This can be accomplished by standing on the left side of the sheep and holding the sheep by the head with your left hand. Move your right

hand to the sheep's right hip. Hold the jaw tightly and bend the sheep's head sharply over its right shoulder. Press down on the sheep's right hip, forcing it to the floor, and raise the front feet and head so that the sheep rests on its rump and leans against you.

The second method of restraint is to use a tilting squeeze chute in which the animal is rotated on its side while being kept under the pressure of the chute.

Step-by-Step Procedure

1. Restrain the animal in the manner selected.

2. Grasp the foot by the fetlock and use your thumb to separate the toes. Use the foot trimmer to cut away the bottom of the exterior hoof wall, which has overgrown the sole, until it is level with the fleshy center portion of the toe. This should be done on the outer and inner portions of the toes on all four feet.

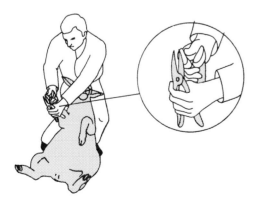

3. In the absence of a foot trimmer, a sharp knife can be used in the same manner, but care should be taken not to cut your finger or thumb.

4. If there is evidence of foot rot the foot should be treated at this time.

5. After all four feet have been trimmed, the sheep should be allowed to stand on a level surface. Examine the job, and if the toes are uneven they should be retrimmed to level them up.

> **CAUTION:** It is possible to trim the foot too closely, causing it to bleed. It should be treated with a 7% iodine solution if this happens.
>
> If ordinary pruning shears are used to trim the feet, be sure that the cutting blade is always on the inside cutting outward toward the outer wall of the foot to keep from cutting too deeply.

7.15 TREATMENT OF FOOT ROT

Contagious foot rot is one of the most troublesome diseases plaguing the sheep industry. The disease is as old as the art of shepherding sheep. The industry has never been without sheep infected with foot rot and likely never will be. The condition is currently a major production problem in many areas of the United States.

Foot rot is an economically devastating disease of sheep, though it seldom causes death. Losses are incurred through the reduction of growth and gains in lambs and the impaired grazing ability of older animals whenever the outbreaks occur. Shepherds know that it is far easier to prevent the flock from getting the disease than it is to control or eradicate it once the animals have become infected.

Foot rot is caused by an interaction between two types of anaerobic bacteria (bacteria that grow in the absence of oxygen). *Bacteroides nodosus* can live only in the animal's foot, and *Fusobacterium necrophorum* lives in soil and sheep manure and is always present where sheep are raised. The interaction of these bacteria destroys the connective tissues between the hoof wall and the sole of the foot, causing them to separate and allowing a pronounced lameness to develop. *B. nodosus* can be eradicated from the flock, because it will live for no more than 2 weeks outside of the sheep's foot. If the shepherd can accomplish this eradication, or if the flock is currently clean, meaning that no sheep are currently carrying *B. nodosus* in their hooves, foot rot will not occur, because one of the two necessary bacteria is missing.

For animals in the flock to start showing symptoms of foot rot, it is necessary that the skin between the toes become wet, bruised, scratched, and infected by *B. nodosus*. *B. nodosus* then becomes established in the deeper layers of the skin and produces an enzyme that liquefies tissue protein around it. The infection separates the heel, sole, and wall of the hoof from their attachments to the foot, causing inflammation, lameness, and the characteristic odor associated with foot rot.

B. nodosus may live for months in the flesh of the sheep's foot. Such sheep become carriers and spread the *B. nodosus* bacteria throughout the farm or ranch through soil and bedding. Recall that these bacteria, thus spread about, are viable and infective for about a 2-week period.

Foot rot in sheep is not the same disease that affects cattle; it is caused by a different organism, causes different types of lesions, and responds to different types of treatments. Proper treatment of foot rot in sheep requires repeated examinations and observations, proper foot trimming, proper use of disinfectants, isolation of infected animals, and removal of carriers from the flock.

Equipment Necessary

- Sheep foot trimmers
- Sharp knife
- Tilting squeeze chute
- Footbath
- Copper sulfate solution
- Zinc sulfate solution

Step-by-Step Procedure

1. The first steps in the control of foot rot are to catch and restrain the sheep in the manner described in Section 7.14 and to examine them for evidence of the disease. This is usually done by setting the sheep on its rump or placing it in a tilting squeeze chute.

2. Foot rot can be detected by soft spots in the hoof wall, separation of the outer wall of the hoof from the sole, and inflammation in these areas. It may also appear as an ulcer between the toes. The foot may be enlarged and feel hot at the junction of the skin and the hoof. When excess amounts of moist, soft, or dead tissue are trimmed away, the foot has a strong, foul odor characteristic of foot

rot. Exposure of pockets of infection to the air has a detrimental effect on the growth of the anaerobic organism that causes the disease.

3. Trim the feet of the sheep as described in Section 7.14. To maximize the effect of whatever medication you select to treat foot rot, the foot must be thoroughly trimmed to expose the infected tissue. Diseased, necrotic (dead), and undermined hoof areas must be pared away to allow medication and air to reach the infective bacteria. Topical medications should be sprayed or painted on immediately after trimming.

> **CAUTION:** Foot trimmers should be cleaned thoroughly and allowed to soak in a disinfectant bath after trimming an infected foot.

4. After the feet have been trimmed, they should be treated either individually or in a footbath. Individual treatment is usually employed when only a few sheep in a flock are affected. It consists of submersing the foot in a container of disinfectant, pouring the solution over the foot, or painting the solution onto the hoof. This should be followed by a second treatment 7 to 10 days later. Ten percent zinc sulfate in water (8 pounds of zinc sulfate dissolved in 10 gallons of water), 10% copper sulfate in vinegar, or two parts copper sulfate in one part pine tar have been shown to be the most effective topical treatments. Antibiotics in alcohol solution have also been reported to be from 50 to 90% effective where feet were thoroughly trimmed, medication was applied once, and sheep were held in a dry area. The above medications (except the pine tar–copper sulfate mixture, which must be painted on) can be applied with a hand sprayer; 5 to 10 ml will be needed per infected foot.

5. Footbaths are used when a large number of the animals or whole flocks are treated. The footbath should be 12 to 16 feet long, 8 inches wide at the bottom, 12 inches wide at the top, and 6 inches deep. The use of walk-through footbaths containing a 10% zinc sulfate or 10% copper sulfate solution every fifth to seventh day will greatly reduce the spread of foot rot to normal animals. Enforced daily self-foot-bathing with 10% zinc sulfate also reduces the spread and will help most infected animals recover. The footbath will not "cure" carriers or those with severe deterioration of the hoof unless their feet are carefully trimmed. Ten percent zinc sulfate in lime, mixed dry and kept dry, can be placed in a box between feed and water to reduce spread of foot rot. This is particularly useful during freezing weather when a solution would freeze. This method is not effective for treatment of animals already infected. The footbath can be placed in a gate through which sheep are forced to walk daily, between their food and water, or it can be placed in a cutting and handling chute to slow their movement through it.

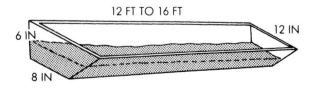

6. The sheep most severely infected with foot rot should be separated from the rest of the flock, placed in a pen or drylot, and treated individually until completely cured. They can then be returned to the healthy flock.

7. Treatment of infected sheep with systemic antibiotics is becoming more popular. Today, oxy-

tetracycline (200 mg/ml) (LA200) is being used. As with all antibiotics, follow the label directions for proper dosage.

> **CAUTION:** Copper sulfate solution is quickly inactivated by organic material. It should never be used in metal containers or metal footbaths because the copper will plate onto the metal and reduce the strength of the solution. Only enamel, wood, glass, or epoxy-painted metal should be used as containers for copper sulfate.

Postprocedural Management

Once foot rot has been controlled or eradicated in a flock, certain management procedures should be followed to prevent another outbreak or reintroduction.

1. Trim the feet of all sheep and examine them for foot rot at least twice a year, once at shearing in the spring and again at breeding time in the fall. If the summer is unusually moist, a midsummer examination may be necessary.

2. Isolate any lame sheep from the rest of the flock and treat them individually.

3. Do not force sheep to use muddy lots, lanes, or pastures for long periods of time.

4. Keep areas around feed bunks and waterers as clean and dry as possible.

5. Do not use cinders, fine gravel, or finely crushed stone as footings for sheep. These materials tend to injure the feet and provide avenues for bacterial invasion of the hoof.

6. Never buy sheep from a flock that is known to be infected with foot rot. Don't even purchase apparently clean sheep from a flock with this reputation. It is also a high-risk venture to buy sheep at sale yards where clean and infected sheep are grouped together, even temporarily.

7. Insist on proper cleaning and disinfection of trucks prior to transporting sheep—either to or from your operation. Clean trucks coming to your farm or ranch will help prevent the introduction of foot rot organisms onto your place. Clean trucks leaving with a load of your sheep will protect your buyers and your reputation.

8. Assume all new additions to your flock are infected with foot rot and arrange for a quarantine area that will be available for at least 14 days, with 21 days being a better recommendation, if it is possible. Always trim all feet immediately upon arrival, treat the feet of all sheep following trimming, and reexamine periodically during the quarantine period.

9. In a range setting or grazing association, avoid common-use trails and corrals where infected sheep have traveled or have been penned during the preceding 2 weeks.

10. Use of vaccines, such as Foot Vac, can decrease the spread of foot rot in flocks where the causative organisms are the same genetic strains as those contained in the vaccine. Usually, two doses are given subcutaneously behind the ear, 4 to 6 weeks apart. If possible, time the last dose for about 14 days before a seasonal outbreak is anticipated. Vaccines are most beneficial when used in conjunction with other foot rot control measures.

11. The use of systemic antibiotics or antibiotics injected into the foot is expensive and of limited success in controlling the spread of foot rot.

12. This disease can be best controlled by using various combinations of all of the treatments previously discussed.

13. Eradication will require the use of all of the methods outlined here. In addition, it will require commitment, perseverance, and a willingness to cull animals that do not respond. An initial increased expense in labor and working facilities will be required but will pay off many times over in the long run. To eradicate foot rot, it will be essential to have quarantine facilities for keeping new animals isolated and a "sick pen" to keep infected animals isolated.

7.16 TREATMENT OF BLOAT

Bloat usually occurs when sheep are grazing on young, lush legumes such as alfalfa or the clovers, although it can occur in drylot as well. Some animals appear to be predisposed to bloat on any feed, and susceptibility to bloat may be heritable.

Bloat is caused by an excessive accumulation of frothy gases in the rumen or paunch and the reticulum. Free gases are normally eliminated under proper nerve response by eructation or belching through the cardiac opening of the rumen into the esophagus. When the gases are mixed with rumen contents, the foam that forms does not bring about the proper nerve response and bloat occurs. This condition is known as *frothy bloat.*

Bloat is characterized by an abnormal distension on the upper left side of the abdomen; in extreme cases this distension may develop on the right side also. The bloated animal will not eat, stands with the nose elevated and the legs spread apart, and often kicks at its stomach. There is rapid breathing and evidence of stress. There is great pressure on the diaphragm, which eventually causes suffocation from the inability to breathe.

Equipment Necessary

- Knife
- Canula and trocar
- Dose syringe
- Hose; 2' to 3' long, ½" to ¾" in diameter

Restraint Required

Bloated animals should be caught for treatment. Because they have difficulty in breathing, they should not be chased or forced to move long distances. Move them quietly and quickly into a pen or barn and allow them to stand during treatment.

Step-by-Step Procedure

Treatment of Bloat

1. In the early stages of bloat, the animal can be drenched with half a pint of warm water to which 1 tablespoon of formalin or bicarbonate of soda has been added, or it can be drenched with a quart of warm milk to reduce the formation of additional gas.

2. An old method of treating bloat is to fasten a round stick in the sheep's mouth. Chewing on the stick may cause belching.

3. A stomach tube or piece of hose ½" to ¾" in diameter and about 2' to 3' long can be passed into the rumen to remove some of the gases. The front feet should be elevated so that the gases in the rumen are nearer the cardiac opening. After the gases have been removed, the hose or tube can be used to administer 6 to 8 ounces of mineral oil directly into the rumen to prevent further bloating.

CAUTION: Do not attempt to drench the animal with mineral oil without using the tube, as there is a danger of getting some of the oil into the lungs.

---

4. As a last resort if the above methods fail, the rumen can be punctured with a small knife or a trocar and canula to remove the gases. The stab wound should be made through the body wall directly into the rumen, midway between the posterior rib and

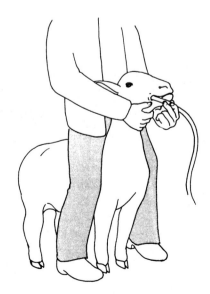

the hip bone. Sheep treated in this manner should also be treated immediately with a large injection of an antibiotic, followed by daily injections until the wound heals and the danger of infection has passed. The advice of a veterinarian is recommended.

Prevention of Bloat

1. Bloat can be prevented by using a legume–grass mixture in pastures or waiting until legumes attain some maturity, because the more fibrous plants stimulate belching.

2. Fill sheep with dry hay before pasturing them on legumes, and continue to feed the dry hay as long as the danger of bloat exists.

3. Put the sheep on legume pastures for the first time when the forage is dry to reduce the weight of the ingested forage.

4. Provide an adequate supply of water so that the sheep will drink often and not consume large amounts of water at one time.

5. Provide adequate salt or a salt–mineral mixture.

6. Experience has shown that it is better to leave all sheep on the pasture once bloat is encountered than to remove them for part of each day to reduce their consumption of legumes.

7. Sheep or lambs that bloat in drylot and chronic bloaters should be removed from the others and confined by themselves where their feed intake can be controlled or reduced.

7.17 AGE DETERMINATION

The ages of sheep from 1 to 4 years old can be determined by observing the changes that take place in the incisor teeth on their lower jaws. At best this can only be an approximation of the exact age. Examination of the teeth may be necessary when the exact date of birth is unknown and the age must be determined for the purchase of commercial ewes, for checking the ages of animals in shows and sales, and for culling older animals or those with malformations of the teeth and jaw.

The change in teeth after 4 years of age will depend upon the longevity of the breed and the management and nutrition they have received. Fine-wool breeds such as the Rambouillet may show a full mouth until they are 10 to 12 years of age, while medium-wool breeds may begin to lose their teeth at 6 to 7 years. Sheep that graze on sandy soils or woody plants will wear their teeth down more quickly than those grazing under more favorable conditions.

Equipment Necessary

None.

Restraint Required

Hold the animal by the head and part the lips to expose the incisor teeth for examination.

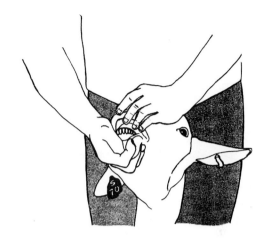

Step-by-Step Procedure

1. A lamb has eight temporary incisors that begin to erupt at approximately 2 months of age and remain until about 12 months of age.

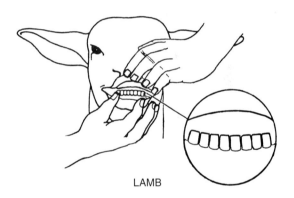

LAMB

2. At approximately 12 months of age, the two center temporary incisors are replaced by two larger, wider, permanent teeth. They may emerge at about 10 months or may not be full grown until 14 months, depending upon the growth rate and maturity of the animal.

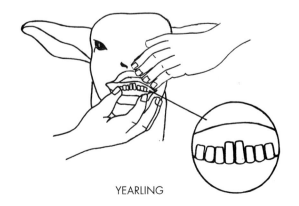

YEARLING

3. At 2 years of age the second pair of temporary incisors, one on either side of the center pair, are replaced by permanent incisors. The mouth now has four permanent teeth in its center.

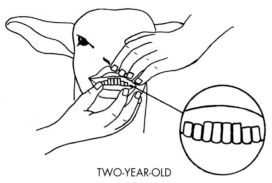

TWO-YEAR-OLD

4. At 3 years of age, the third pair of temporary incisors, one on each side of the four permanent teeth, are replaced by permanent teeth. There are now six permanent teeth in the center of the mouth and two temporary incisors on the corners.

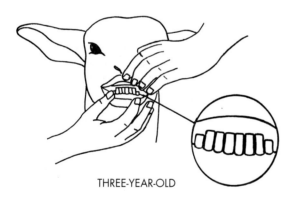

THREE-YEAR-OLD

5. At 4 years, all of the incisor teeth have been replaced by permanent incisors and the mouth shows eight permanent incisors. These animals are said to have a full mouth.

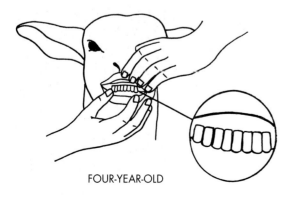

FOUR-YEAR-OLD

6. As the animal ages beyond 4 years, the teeth appear to be longer due to a receding of the gums, and there is greater space between the teeth. Such an animal is referred to as a *spreader* if no teeth have been lost, but its exact age cannot be determined.

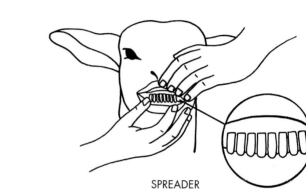

SPREADER

7. At 7 to 8 years of age, some teeth may be lost. Sheep with missing teeth are referred to as *broken mouth* sheep.

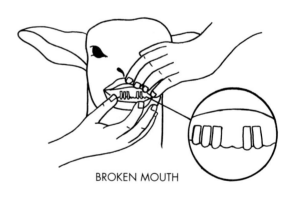

BROKEN MOUTH

8. In sandy areas the teeth may be worn down by grazing, and sheep in which this has occurred are referred to as *short mouth* animals. This can occur at any time after the sheep is 4 years old.

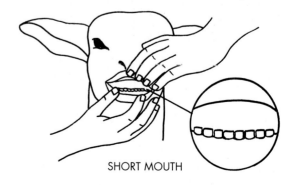

SHORT MOUTH

9. Sheep that have lost all of their incisor teeth are referred to as *gummers.* They have difficulty in grazing forage and usually show the effects of poor nutrition.

10. Lambs that have malformations of the teeth and jaw should be culled because this is an inherited defect. These malformations include *undershot jaw,* in which the incisor teeth on the lower jaw extend beyond the dental pad on the upper jaw, and *overshot jaw,* in which the lower jaw is shorter than the upper jaw.

11. Older sheep in which the condition of the teeth appears to interfere with their ability to utilize feed should also be marked for culling.

7.18 PREPARING FOR SALE OR EXHIBITION

The purpose of preparing sheep for sale and exhibition is to improve their appearance by accentuating the strong points in their conformations. A skillfully done job will improve the sheep's appearance, but a poorly executed one can make the animal less attractive. Following correct procedures can help to improve the sheep's appearance and avoid unsightly errors.

First, familiarize yourself with the general characteristics of your breed. Study those with good conformations, and also study pictures of well-prepared sheep so that you know what they should look like. One of the best ways to learn trimming is to watch an experienced shepherd prepare an animal. Then examine each animal with which you intend to work and determine its strong and weak points so that you can accentuate the positives.

Yearling breeding sheep of the medium-wool breeds—Dorsets, Suffolks, and Hampshires, for example—should be completely shorn 6 to 8 weeks before the sale or show so that they are exhibited in short fleece. The back and belly can be reshorn just before washing, which should be done 2 to 3 weeks before showing.

Yearling breeding sheep of the fine-wool breeds such as Corriedale, Columbia, and Rambouillet should be shorn 3 to 4 months before showing and should not be reshorn. Spring lambs of these breeds should not be shorn.

Market lambs of all breeds should be completely shorn from 2 to 4 weeks before showing. Very little trimming is needed. The back, head, and belly can be reshorn just prior to a sale or show, if needed.

Equipment Necessary

- Shearing equipment
- Hand sheep shears with $6\frac{1}{2}$- to 7-inch straight blades
- Brush
- Circular currycomb
- Wool card
- Water bucket
- Soap or mild detergent
- Woolen cloth
- Sheep-trimming stand
- Sheep blanket
- Sheep-dip

Restraint Required

The sheep must be placed on a sheep-trimming stand with its head held in a yoke, which can be adjusted to the size of the animal.

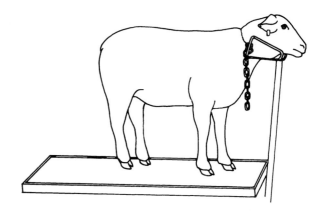

Step-by-Step Procedures

Cleaning the Fleece

1. If the fleece is dirty, it should be cleaned either before you begin trimming or as you trim. The fleece of Corriedale, Columbia, Rambouillet, and other fine-wool breeds should not be washed, but it can be cleaned during the trim by using the circular currycomb and wool card.

2. The medium-wool breeds can be washed by making them stand or sit in a tank of water to which mild detergent has been added. Use a combination of your hands, a currycomb, and a brush to thoroughly clean the fleece. Allow 2 to 3 weeks after washing for the natural yolk to return to the fleece.

3. If you use soap, it should be thoroughly rinsed out of the fleece.

4. If a tank is not available, you can use a hose with a spray nozzle. The pressure of the water will clean the outer ends of the wool fibers.

5. The legs, rear parts, brisket (chest), and chin will usually require the most washing.

6. Allow the fleece to dry before beginning to trim it. The remaining dirt can be removed during the trimming if you dampen the wool, rake it with the currycomb, card it, and then trim it.

7. Once the sheep is clean, be sure to provide clean bedding and to cover the fleece with a clean blanket to keep from having to repeat the job.

Trimming the Fleece

One of the secrets of good trimming is to keep the blade of the hand shears that is next to the fleece stationary while the other blade pulls the fibers over it. This technique is developed with practice, and a much smoother job is obtained than when both blades are moved. Start at one end of the area to be trimmed and move slowly and deliberately to the other end without lifting the shears from the fleece. Nicks in the fleece occur when the shears are lifted from the surface.

1. Place the sheep on the trimming table and brush the fleece to remove any dirt.

2. Dampen the fleece with water by using the brush. Do not soak the fleece. A mild solution of a sheep-dip will help to clean and straighten the fibers.

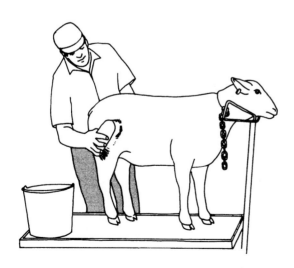

3. Comb the entire fleece with the circular comb to remove the dirt on the outer ends of the fibers.

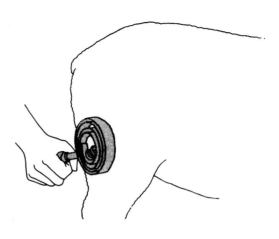

4. Card the fleece with the wool card to straighten the fibers and break up the fleece.

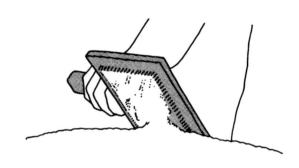

5. Trim off the rough, carded wool until a smooth surface is obtained. The fleece will trim more easily if it is damp.

6. Trim from the top to the edges of the sheep's back, making sure that the top is level.

7. Trim the sides until the proper shape is obtained.

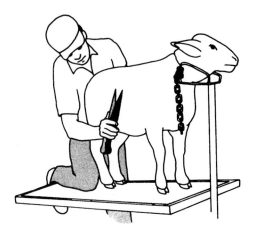

8. Square the dock and closely trim the twist and the rear legs.

9. Trim the brisket and shoulders.

10. Trim the head and neck according to accepted breed practices.

11. Pack the fleece with the back of a dampened wool card after trimming is completed.

12. Place a clean blanket on the sheep until the next trimming. Two or more trimmings may be necessary.

Showing the Sheep

1. In the show ring, hold the sheep by the wool and loose skin under the chin. You may place a hand on the dock (tail area) when moving the sheep.

2. Set the sheep up with all four legs squarely under it and do not allow it to stretch so that the back becomes weak and sags. Gentle pressure on the neck will help to strengthen the line of the back when the judge handles the sheep.

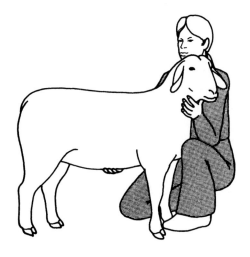

3. When showing, keep the sheep between you and the judge. Stay far enough away from other sheep that the judge can see your animal. Do not move completely around the sheep or step over its back. Hold the head slightly high or in a normal position. Always be alert to follow any directions the judge may give. Attempt to develop an easy manner of showing that gives an appearance of confidence and does not excite the animal.

Postprocedural Management

It will be necessary to go over the sheep two or three times to do a smooth job of trimming. Do not remove a lot of wool at any one time; a number of short trimmings is preferable. The last trimming should be done either on the day of the show or the day before, after the fleece has had time to set from the first trimmings.

7.19 CATCHING, HANDLING, MOVING, AND WEIGHING

An understanding of some behavioral traits of sheep will make the catching, handling, weighing, and moving of the animals much easier. Most breeds of sheep are naturally gregarious; that is, they stay close to one another, partly for protection. This trait is called the *flocking instinct*. A sheep that remains apart from the rest of the flock is usually ailing in some respect. They tend to follow a lead sheep in moving. They do not like to move toward direct sunlight, but will move toward light when in a darkened area. They will not approach an unfamiliar area without being forced to do so, but will usually enter an open gate or door. They do not like to walk in water or muddy areas, or on sharp gravel or cinders. They will usually crawl under a fence rather than jump over it.

Sheep have little defense against an attack by strange or wild dogs or predators and will usually run from them. However, a well-trained sheepdog can be invaluable in moving and herding sheep. Single sheep will usually face the handler or attacker and can be backed into a corner for catching. Sheep must be caught, handled, and moved so that the owner can perform many of the management skills necessary for profitable production. This can be done easily if the step-by-step procedures are followed.

Equipment Necessary

- Holding pens
- Scales
- Cutting chute
- Hurdles

Restraint Required

This is a technique of movement, and as such requires no restraint in the usual sense. The boundary restraint of fencing is usually employed in farming areas. In range areas, only the psychological restraint that the shepherd has over his flock, or a good pair of sheepdogs, is used to control the animals.

Catching Sheep

1. An individual sheep can be caught by forcing the group into the corner of a barn or pen. Move easily and quietly. Once you get the group "started" toward the corner, they will all flow into it. This certainly works easier if you have an assistant or a good dog.

2. As you move into the group of sheep to catch the identified animal, be alert. The individual sheep can be caught in a variety of ways. If it is facing you, grasp the head and neck in both hands and move quickly to its side. From that position, it is easy to hold the head and neck in an upright position and thereby control the animal. If the sheep is passing in front of you, from one side to the other, it can be caught under the chin, using either hand. Once that hand is in place, quickly move into position where you can use both hands to control the head and neck. Some people grab a handful of skin and fleece as the animal passes across the front of them. Avoid this if possible. Fleece will pull free from the skin, if enough force is used. If you are nearer to the

rear of the sheep than to its head, as it passes across in front of you, it can be caught by the loose skin in the flank just ahead of the rear leg. Immediately use a hand in both flanks if necessary, and get to the head and neck for control as quickly as possible.

3. A shepherd's crook is a very handy tool and can be used to great advantage in catching sheep. All is takes is a little practice and a good understanding of sheep behavior. The bottom line is that the crook extends the reach of your arm by about 5 feet. A crook allows you to reach across the outer layers of a crowd of sheep and catch the desired animal. Crooks can be made of wood or metal. Mostly, crooks are designed to catch the sheep about the neck. There are, however, leg crooks. Most people find the neck crook more useful.

CAUTION: Be absolutely certain that the crook, either the neck or leg variety, is not used for a club. Sheep will be hurt, the crook bent or broken, you will look and feel the fool . . . and the sheep will have escaped by the time you regain control of yourself.

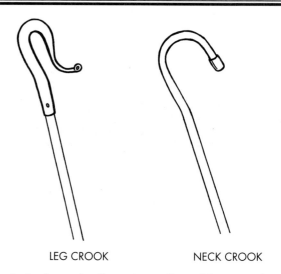

LEG CROOK   NECK CROOK

4. A through discussion of catching, moving, and handling sheep can be found in Chapter 1, Section 1.7.

Moving Sheep

1. Keep the flock together when moving them.
2. Move the flock slowly and quietly without exciting them.
3. If one or two sheep break away from the rest of the flock, try to hold the flock together; the strays usually will return.
4. If the flock is to be turned into a barn or pasture, send an assistant or a dog ahead of them to accomplish this.
5. If sheep are to enter a cutting chute, provide a wide entrance with gates that will close behind them. They will usually enter the chute faster if they can see an opening at the opposite end. This can be accomplished if a wire gate is used instead of a solid one.
6. Gates at the entrance and exit of a cutting chute should pull upward, but the cutting gates should open inward to direct the sheep into separate pens. The cutting procedure is best accomplished if the sheep can move continuously through the chute; when they stop they tend to back up.
7. Backing up can be prevented by placing ¾-inch rods or pipes through the chute, 9 inches from the bottom and spaced 4 feet apart. The sheep will jump forward over the rods, but cannot lift their rear legs high enough to back up.
8. If a cutting chute is not available, sheep can be crowded into a pen or a temporary pen made of hurdles and sorted or handled individually. They can be caught more easily if they are forced together in a group so they cannot break and run. Hurdles can be used to force them together as the number in the pen decreases.

Weighing Sheep

1. Lambs of 70 lbs or less can be weighed by holding them while standing on bathroom or platform scales and subtracting the weight of the holder.

2. For heavier lambs or sheep, a crate with an entrance and exit gate can be attached to the scales. The weight of the crate should be subtracted from the total weight.
3. For large flocks, the scales can be placed in or at the exit of the cutting chute. A crate with an entrance and exit gate must be used to detain them long enough to be weighed.
4. Portable scales equipped with a crate, gates, and either a dial or beam for quick weighing are available for large flocks.

CAUTION: All scales should be balanced after 25 sheep or lambs have been weighed to adjust for accumulations of manure or bedding.

---

5. In small flocks, sheep can be caught and handled individually if they enter a barn or enclosed area. They can be trained to enter this area by placing the salt and water supply there or by feeding hay or grain there once or twice a day. Portable hurdles can be used to crowd the sheep into a small area or to separate them in the handling process.

7.20 SHEARING

In the United States, sheep are normally shorn once a year in late winter or early spring when the cold weather moderates. Some producers prefer to shear pregnant ewes before lambing if they have buildings adequate to protect the shorn ewes, while others in warm climates may shear the ewes a second time before breeding in late summer.

Shearing before lambing has certain advantages, among which are those listed below.

1. Shorn ewes require less floor space and less feeding space than do wooled ewes.
2. The bedding stays drier with shorn ewes because wool absorbs and holds moisture from the outside.
3. Shorn ewes will stay inside the barn during a snow- or rainstorm and not chill newborn or young lambs that might accompany them outside the barn.
4. Shorn ewes usually will not smother newborn lambs by lying on them, as sometimes happens with unshorn ewes.
5. Young lambs can find the nipple more easily and will not nurse on wool tags.
6. Shorn ewes have a higher metabolic rate than wooled ewes, making them less susceptible to ketosis. The critical temperature for shorn ewes is 50°F, and for wooled ewes of similar size it is 28°F. This is

the ambient temperature below which the ewe must increase its metabolic rate to maintain a constant body temperature of 102.3°F.

If ewes cannot be completely shorn before lambing (and before breeding), they should be "crotched out" or tagged by shearing the wool from the udder and dock area and from the head and eyes of those breeds with wool on the face. This will make lambing and nursing easier.

Shearing before breeding in late summer tends to increase heat loss from the body and cools the ewe and ram. This will induce estrus in the ewe and stimulate spermatogenesis in the ram.

Certain fundamental procedures must be followed if one is to do a good job of shearing sheep.

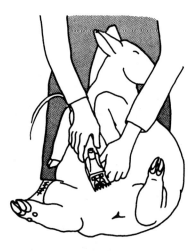

1. The sheep must be held properly in a comfortable position to prevent its struggling during shearing. Most shearers use the Australian method, which is described in this section.

2. The skin should be stretched so that it is smooth in the area being shorn.

3. Wool fibers should be cut only once next to the skin to avoid "second cuts" or short fibers of reduced value.

4. Belly wool, leg wool, and tags have a lower value and should be kept separate from the higher-quality wool from the back, neck, and sides.

5. The fleece should be removed in one piece so that it will remain together when tied.

6. The fleece should be tied only with paper wool twine; binder twine (sisal) and nylon twine will contaminate the wool.

Adequate preparation is very important to a good job of shearing. The cutters and combs should be sharp and should be cleaned and resharpened after each job of shearing.

The shearing floor should be clean and free of straw or chaff. It should be swept clean of second cuts and manure tags after each shearing. Do not drag straw and manure onto the shearing floor with the sheep to be shorn. Sheep must be dry before shearing. If sheep are shorn when they are wet, the wool will heat and become discolored from the moisture.

It will be necessary to clean cutters and combs after two or three sheep are shorn. Lanolin from the wool will cling to the teeth of the instruments, which should be cleaned with a stiff brush and warm water.

CAUTION: Do not immerse the shearing head in water. Disconnect the electric power when cleaning the cutters.

Equipment Necessary

- Sheep shears
- Shearing mat; canvas, about 6' × 6'
- Lubricating oil
- Extra cutters and combs
- Stiff brush
- Paper wool twine
- Sharpening equipment
- Glove for left hand

Restraint Required

In the Australian method, the sheep is set upon its rump as described in Section 7.15. Additional restraint methods are described below.

Step-by-Step Procedure

The procedure described below is the Australian method of shearing. The instructions given here assume that the shearer is right-handed; left-handed shearers should transpose the directions.

1. Start the motor of the shearing machine.

2. Set the sheep upon its rump and support it firmly between your knees.

3. With the sheep facing forward, rest its right foreleg on your left side to tighten the skin of the belly.

4. Make the first shearing stroke downward, starting in the right foreflank and following the body contour to the right rear flank.

5. Make the next three or four strokes (however many are necessary to open the entire upper breast) about 1 foot in length and also straight downward, but position them directly upon and to either side of the breastbone.

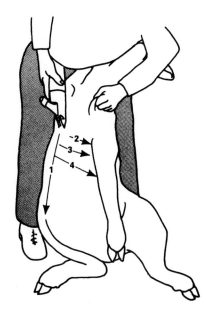

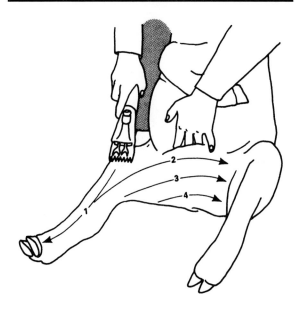

tinue upward and around the udder or scrotum. The strokes should end at the sheep's left leg where it joins the body.

> **CAUTION:** Be extremely careful when shearing around the udder that you do not cut or damage the nipples.

6. Change the position of your left arm so that you are holding the sheep's left front leg with your left wrist. This will leave your left hand free to tighten the skin on the sheep's belly. Finish shearing the sheep's belly by using strokes that start at the opened area created by the first strokes, finishing them well over into the left flank. In contrast to the previous strokes, these should be parallel to the ground.

8. The shearer should now move his right leg back a few inches and allow the sheep's right front leg to drop between his legs. The sheep's right leg will rest somewhere near the back of the shearer's right knee.

9. Straighten the sheep's left hind leg by applying pressure on its left stifle joint with your left hand.

10. Shear the top and inside of the left hind leg, starting the strokes at the stifle and continuing toward the toe.

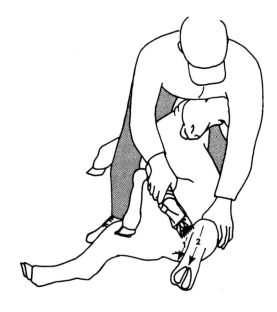

7. Reposition yourself by backing off slightly. Bend over the sheep and use your left hand to tighten the skin in the crotch area. Now shear out the top inside of the sheep's right rear leg, starting the stroke at the stifle and continuing to the toe. The next three or four strokes (whatever is necessary to shear out the entire inside of the right rear leg and the crotch area) begin at the toe and con-

11. The next stroke starts at the left toe of the sheep and runs just to the outside of the top of the sheep's left leg. The stroke is completed at the sheep's left flank. An additional stroke or two may be necessary to finish shearing the sheep's left flank.

> **CAUTION:** Be careful not to cut the loose skin of the flank.

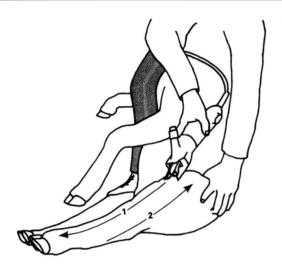

12. Reposition yourself slightly by moving your left foot an additional 6 to 8 inches closer to the sheep's hip. The sheep's right leg is still behind your right leg and its head is still in front of your knees.

13. Lean over and apply pressure on the sheep's left flank with your left hand. This will straighten the sheep's left leg so that you can finish shearing the outside of the left leg.

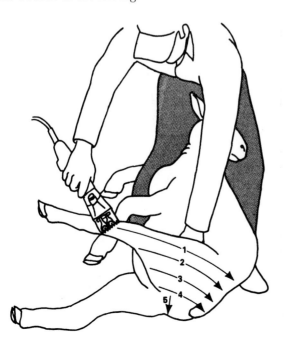

14. Shear the outside of the left leg by starting the strokes from the toe and moving them along the curvature of the leg to the midline of the sheep's back. Continue shearing down the sheep's left hip until it is clean of wool. A stroke or two should be made below the rectum or vulva; these strokes will clean part of the upper right leg.

> **CAUTION:** Slide the shearing head at an angle to avoid cutting the tendon above the hock (the hamstring).

15. Lean way over, perhaps even move your left foot another 6 to 8 inches toward the sheep's rump, and change the angle of your wrist at the same time so that the head of the shears is positioned at an upward angle. Shear all of the wool from the tail by stroking forward along the backbone to the point at which your previous strokes have ended.

16. Straighten your back, slide your left foot back toward the sheep's head by about 6 to 8 inches, but

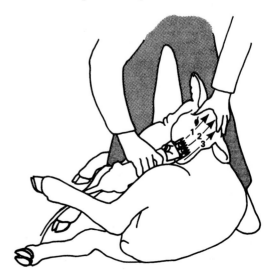

do not change the position of the sheep. Grasp the sheep's head in your left hand and use strokes from the nose toward the ears to clean the face and top-knot of all wool. Stop the strokes at a line connecting the two ears.

17. Reposition the sheep by straightening it up, but keep it facing to your right. Place your left foot close to the sheep's hip or rump and the heel of your right foot into the crotch between the sheep's rear legs. When you are properly positioned, the sheep's right legs will be behind your right leg and its left legs will be in front of it.

18. Grasp the sheep by its muzzle with your left hand and stretch its neck backward and over your left knee. Help secure the sheep's position by squeezing its brisket between your knees.

19. Start a stroke at the point of the brisket and shear up the neck and under the sheep's chin. This stroke is stopped only an inch or so from your hand, so be careful not to allow the sheep to pull your left hand into the handpiece. At the completion of the stroke, flip the handpiece to the right. This will cause the fleece to break and enable you to see where to make the second stroke. Make a second stroke along the neck and chin just below the first (to the sheep's left side).

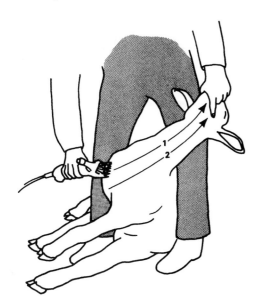

20. Gently twist the sheep's head so that its nose is pointed directly upward, with the right side of its face resting against the front of your left leg. Shear the sheep's left, upper shoulder and the left side of its neck by taking strokes that begin on the unshorn part of the shoulder. Travel up and along the neck, along the sheep's jaw, and stop about 1 inch from your left hand.

CAUTION: Be careful not to cut your left hand.

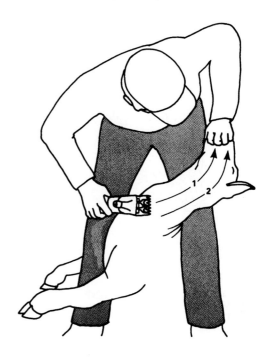

21. Finish shearing the left side of the neck with strokes that end at the base of the left ear.

22. Hold the sheep's head against your left thigh by grasping the sheep's left ear. Shear around the left ear and across the top of the head. Two or three strokes across the head are all that are necessary.

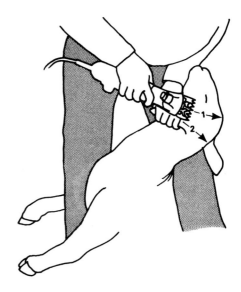

23. Keep the sheep's body between your knees as before and hold its head against your left leg with your left elbow. Stretch the skin on the sheep's left shoulder with your left hand. Clean the sheep's left shoulder with strokes starting at the left knee. Continue these strokes upward until the shoulder is free of wool and tags.

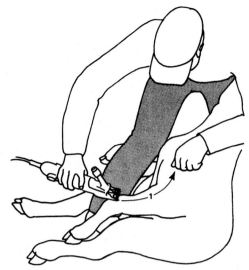

24. Use downward strokes to shear the sheep's left front leg below the knee.

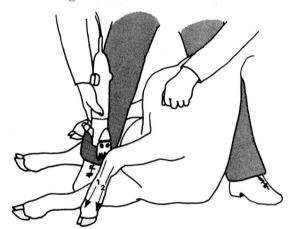

25. Use your left hand to hold the sheep's left leg extended next to your body, shear the outside of the sheep's left leg, and clean out the left foreflank. The strokes start below the knee and end where the leg joins the shoulder.

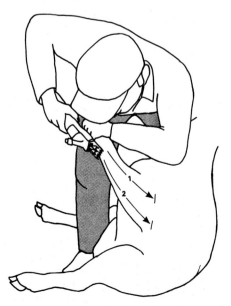

26. Reposition the sheep by keeping your right foot in the sheep's crotch and allowing the sheep to slide down your left leg and onto its back. Keeping your left foot under the sheep's right shoulder prevents it from getting up. The sheep's right leg is pinned against its body by your lower left leg. Your left hand should be pulling the sheep's left foreleg toward its head so that the skin on its left side is stretched.

27. Shear the full length of the sheep's side with three or four strokes, traveling from the tail to the shoulder, until you reach the backbone.

CAUTION: Keep combs close to the skin on each stroke and follow the contour of the body. Avoid reshearing as much as possible, because this produces "second-cuts."

28. Reposition yourself by placing your right knee on the sheep's left hip. Keep your left toe under the sheep's right shoulder to keep its forefeet off the floor.

29. By applying downward pressure on the sheep's head and at the same time pulling it toward your rear with your left hand, stretch the sheep's neck around your left leg. Slide your left foot backward 3 or 4 inches to allow the sheep to "roll up" toward you.

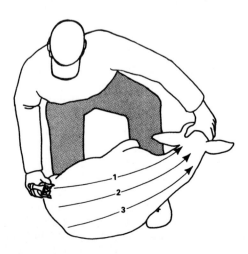

30. Shear two strokes past the backbone. This is called the *long blow* and is shorn at this point to save time on the last side. On the last stroke of the long blow, hold the sheep's right ear with your left hand and shear under it and out along the jaw to finish cleaning the face.

31. Reposition yourself and the sheep by moving your right foot forward and rolling the sheep slightly more onto its back. Clamp the sheep firmly between your lower legs, which should be positioned across the sheep's rib cage.

32. Clear the ear and face of tags; then shear out the right side of the neck with downward strokes to the point of the shoulder. Use your left hand on the neck to stretch the skin. After the last stroke clearing the right side of the neck, allow the sheep's right foreleg to come forward from behind your left leg.

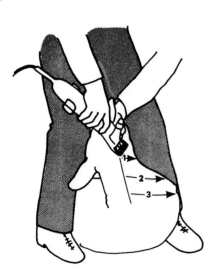

33. Reposition the sheep by pulling it up onto its rump and allowing its head to fall backward between your legs. This will stretch the skin on the right side of the sheep.

34. Push downward on the point of the sheep's right shoulder with the palm of your left hand to

straighten the sheep's foreleg. Shear the right shoulder and foreleg with strokes starting at the leg–shoulder junction and moving toward the toe. On the final stroke, tighten the skin of the leg by pulling it upward with the fingers of your left hand. Place the handpiece under the leg and shear it with a stroke moving toward the toe.

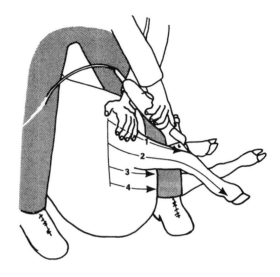

35. Hold the sheep firmly between your legs. Its head should also be between your legs and resting on its own left side. This stretches the skin on the right side of the sheep. Shear around the sheep's right side from top to bottom with two or three strokes, allowing the wool to fall downward as it is shorn.

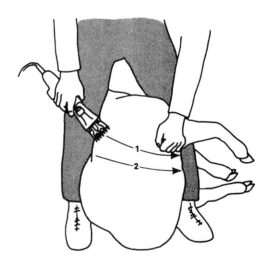

36. After the last stroke is taken, reposition yourself and the sheep by moving your left foot to the other side of the sheep's hind legs. Raise its head and place it upon your left knee. The left side of the sheep's body should now be resting against the front of your legs. Use your left arm to clamp the sheep's head against your knee.

37. Continue to shear the right side of the sheep with slightly downward strokes that start at the edge of the long blow and continue to the right toe. Apply pressure to the stifle joint with your left hand to straighten the leg, and slide the handpiece around the hock to keep from cutting the tendon.

38. Pull the sheep's head toward its right side and backward toward its backbone. Clamp it into this position with your left forearm. Move your left foot back so that the sheep lies on its left side, but keep your toe under its shoulder. This allows you to more easily complete the last strokes, which are the final cleaning of the rump and dock.

39. Allow the sheep to stand, examine it quickly, and return it to the pen with the other sheared sheep.

40. Separate the belly wool from the leg wool and pack them separately.

41. With the sheared side out, roll the sides and tail area into the center of the fleece and fold the back and head area over it.

42. Tie the fleece loosely with paper wool twine, place it in a wool bag (large, burlap, 8 feet long, 30 inches diameter), and store it in a dry, mouse-free area until it is marketed.

It is best to shear white face sheep before shearing the black face, because if you mix the sequence, the white face wool will very likely be contaminated with black fibers from the black face wool. Shear the black face breeds last.

Bundled and tied fleeces should be bagged by face color of the sheep. For the best marketing advantage, place white face fleeces and black face fleeces in separate bags.

CAUTION: Follow the Australian method of holding the sheep so that it is resting and does not struggle. Under these conditions, the sheep can be shorn more quickly with fewer cuts, and the fleece will be removed in one piece.

Do not grasp the hind feet, as the sheep will react by pulling away. Use pressure on the stifle joint to straighten the legs.

Never lift the unshorn fleece with the left hand and attempt to shear it off. This lifts the skin as well, which will be cut in shearing. Instead, use your left hand to stretch the skin away from the handpiece.

Do not force the shears through the wool. If you need to force the shears, either the cutters and combs need sharpening, or you are shearing against a wrinkle or other obstruction and will cut the sheep.

Be extremely careful when shearing around the udder, scrotum, sheath, and wrinkles not to cut or damage them. The same care should be used around the loose skin in the flanks and the tendons of the legs. Some nicks and cuts are unavoidable, but care should be taken to hold them to a minimum.

Use a shearing glove on the nonshearing hand to protect yourself against injury from the handpiece.

Keep the cord of the cutters behind you so it cannot be cut, producing an electric shock.

Do not rush through the shearing procedure in an attempt to increase speed. Speed comes with practice. The appearance of the shorn sheep (which should have a minimum of cuts) and the condition of the fleece are as important as speed. An experienced shearer can shear a sheep in 5 minutes or less, which works out to about 100 sheep per day. An amateur should be content with half this number until he has mastered the correct procedure.

Do not shear sheep when the wool is wet. Wet wool will heat when stored and become discolored.

Use only paper wool twine to tie the fleeces. Bailer twine (sisal) and nylon will become entangled in the wool fibers and will not assume the dye when processed into cloth.

Postprocedural Management

Newly shorn sheep should have protection from the cold and rain until they have had time to regrow some wool cover. Lambs that are not marketed by July 1 should be shorn, because they will gain better in hot weather. These lambs will sunburn if they are exposed to the direct rays of the sun. If a large number of sheep are to be shorn, the shearer should purchase a power grinder and learn to sharpen the cutters and combs. All shearing equipment should be cleaned and sharpened after each use and properly lubricated.

7.21 ANNUAL MANAGEMENT CALENDAR

The "order" or sequence of activities will essentially be the same for all parts of the country. However, the actual seasons, or months, of the year will vary by climate and geography. There will also be adjustments based upon differences between purebred and range flock operations.

Physiological Season of the Sheep	*Management Activities/Concerns*
Pre-Breeding (Late Summer/Early Fall)	Body condition score ewes Nutritional management—base upon desired body condition score Cull problem ewes External and internal parasite control Breeding soundness—rams Vaccinate ewes—"abortion" diseases Flush ewes—increase ovulation rate "Tag" ewes Trim feet—treat foot rot—ewes and rams Shear rams—to keep cooler
Breeding (Fall)	Purebred flocks—1 ram per 25–50 ewes—for 2 estrous cycles (32–35d) Range flocks—2 rams per 100 ewes—for 2 estrous cycles (32–35d) If using marking harness, change crayon color at 16–17 days Replace rams if lame or lacking libido
Gestation (Late Fall/Early Winter)	Monitor body condition score of ewes—adjust nutrition level Verify pregnancy status at 40–100 days—ultrasound and sort Vaccinate ewes Internal and external parasite control Shear ewes Prepare lambing ground and "jugs"
Lambing (Winter)	Allow ewes to lamb naturally whenever possible Place ewe and lambs in "jugs" for mothering Identify lambs See to colostrum consumption Dip navel in iodine Be prepared to foster orphan lambs
Postlambing (Late Winter/Early Spring)	Dock all lambs Castrate ram lambs Permanently identify lambs Internal and external parasite control
Weaning (Spring/Summer)	Vaccinate lambs against enterotoxemia Remove grain from ewes 2 days before weaning lambs Reduce amounts of all feed and water last day before weaning Remove ewes from lambs and place out of eye and ear, if possible Closely observe ewes for signs of mastitis. Treat aggressively. Internal and external parasite control

ALPINE

Courtesy of American Dairy Goat Association

LA MANCHA

Courtesy of American Dairy Goat Association

NUBIAN

Courtesy of American Dairy Goat Association

SAANEN

Courtesy of American Dairy Goat Association

OBERHASLI

Courtesy of American Dairy Goat Association

TOGGENBURG

Courtesy of American Dairy Goat Association

ANGORA

Courtesy of Texas Sheep & Goat Raisers' Association

BOER

Courtesy of Texas Sheep & Goat Raisers' Association

Goat Management Techniques

8.1 INTRODUCTION

Worldwide, the goat population exceeds 350 million animals and there are over 60 recognized or "official" breeds. These are multipurpose animals that produce milk, meat, fiber, and skins.

In the United States, where there are between two and three million goats, there are currently nine breeds that are enjoying increasing popularity. They include the Alpine, La Mancha, Toggenburg, Nubian, Oberhasli, and Saanen (the dairy breeds), the Boer (a meat breed), the Angora or Mohair goat, and the Pygmy. The Pygmy has become popular as a pet in the last few years.

There are several reasons for the increasing popularity of goats in the United States. One of the most important reasons is the ever growing determination of Americans to be self-sufficient. A small herd of goats may be the only unit of livestock that can allow the small, city-working, country-living landholder to really help himself toward self-sufficiency.

The goat is uniquely suited to this type of agriculture in that it is relatively inexpensive to purchase and reproduces efficiently—often rewarding your efforts with two or three offspring per year. Goats do not demand specialized housing or feedstuffs, yet they yield a wide and generous variety of products. Because of their size and temperament, goats can be raised on small acreages and handled by even the younger members of the family. For many families who are attempting to build a more basic and meaningful existence for themselves and future generations, the goat may be the animal whose time has come again.

Another valuable contribution of goats is the control of brush. In most of the United States, this is not a huge concern, but in developing nations, one of the major reasons for keeping goats is to control the woody vegetation that grows in the countryside. Goats work well in this regard.

One should not get the impression that goats are owned only by inhabitants of underdeveloped countries, homesteading families, return-to-the-landers, or hobbyists. There are literally hundreds of financially successful dairies across the United States that keep from 50 to 150 does each. Ten times that number of goat owners generate some level of income from the sale of dairy products, meat, or skins. These small producers (hobbyists) maintain 10 to 30 does each. Really big successful goat dairies milk from 500 to 900 does each day. Admittedly, there are very few states with more than one or two of these operations, but they do exist and are models of modern-day mechanization.

The six most popular breeds of goats in the United States—Toggenburg, Saanen, French Alpine, Nubian, Oberhasli, and La Mancha—are kept primarily for their milk production.

The *Toggenburg*, a breed native to northeastern Switzerland, is the oldest breed of dairy goat in the United States. It is also the smallest breed; a typical doe stands 26" at the withers and weighs 120 to 125 pounds. Bucks average 33" at the withers and weigh 165 to 175 pounds in breeding condition. The Toggenburg is solid-colored, with the body color varying between fawn (very light brown) and chocolate (deep, dark brown). The breed has white ears with a dark spot in the middle, white stripes on its face extending from above its eyes to its muzzle, front and rear legs that are white on the inner side and totally white from the hocks and knees downward, a white triangular rump patch on each side of the tail, and either white wattle roots or white spots where the wattles would have been if they are absent. The head is concave (dished) in profile with small, upright ears. As a breed, they do not milk as heavily or have as high a butterfat test as the Alpines or Saanens. They are noted for their large udders, which are carried high and strongly attached.

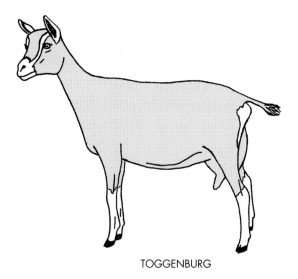

TOGGENBURG

The *Saanen* is a quiet, not easily excited breed of Swiss origin. It enjoys a well-deserved reputation for high total milk production. More precisely, the reputation is for persistency and longevity of lactation, not for extremely high peaks in daily production. The breed is large, with an average doe standing 30"

at the withers and weighing 150 pounds. The Saanen buck stands 36" and weighs from 175 to 200 pounds. The breed is white or cream in color, but may show black skin spots (skin only, never black hairs) on the nose, ears, and udder. While the head has the concavity and the forward-pointing, erect ear of the Toggenburg, it is a larger, coarser head, which is in keeping with the overall larger Saanen body. The udders are typically large and strongly attached.

SAANEN

The *French Alpine* is a large, variously colored breed of goat that was developed originally in the French Alps and refined into the breed as we know it today in lowland France. During the breed's development in France, the goats were selected for body size and milk production, with no attention paid to color pattern. No colors or patterns of color are truly predictable in the French Alpine goats; the body colors range through white, gray, fawn, cinnamon, red, brown, black, and piebald, with such varied patterns as white neck (Cou Blanc), black neck (Cou Noir), tan neck (Cou Clair), wild Chamois (Chamoisee), and black with Toggenburg markings

FRENCH ALPINE

or black with white underbody (Sundgau). While the size also varies greatly, the average doe usually exceeds 30 at the withers and weighs 150 pounds or more. Mature bucks may stand 40" at the withers and weigh 200 pounds. French Alpine does can be excellent milkers and typically have large, correctly shaped udders with properly placed teats.

Goats of the *Nubian* breed give the impression of being the aristocrats of goatdom because of their graceful, upright, and proud carriage and way of moving. The breed derives its name from its ancestral home, Nubia, in northeastern Africa. As a whole, they are a tall, long-legged group that is considered to be hardy. It is the most numerous of the five dairy breeds in the United States. Mature does should average 30" at the withers and 150 to 160 pounds, while bucks stand 35" and weigh 210 pounds. The Nubian is distinguished by long, wide, pendulous ears and a pronounced roman nose. The ears hang close to the face and should be free of folds. Any color or colors, either solid or piebald, are acceptable. Black, brown, red, tan, gray, cream, and white are typically seen; they may be accompanied by lighter or darker markings, especially about the ears, face, muzzle, and underbody. The Nubian doe produces less total milk than the Swiss breeds, but its milk usually has a higher (an additional 1% or more) butterfat test than the milk of other breeds. The duration of lactation also tends to be shorter than in the Swiss breeds. The Nubian has the reputation of fleshing out quite readily, and as such would be a good choice for chevon production.

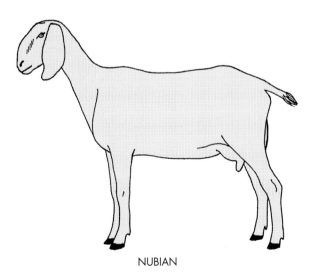

NUBIAN

The American *La Mancha* breed is a newcomer to the scene. It has been recognized as a distinct breed by the American Dairy Goat Association only since 1958. The breed was developed in Oregon from short-eared goats found in Spain, especially in the La Mancha region. First imported in the late 19th century, the short-eared Spanish goat was crossed with outstanding animals of the four major dairy breeds. The resultant hybrid always carried the short ear. Any body color and markings are acceptable. The size and weight are similar to those of the Swiss breeds. The face is intermediate in profile between the concave Swiss face and the convex Nubian face. The extremely short ear is the distinguishing characteristic; in fact, the breed is often colloquially called the "earless goat." Gopher, Cookie, and La Mancha are three categories of ear shape based upon the folding and bending configuration of the ear stump. Few La Manchas have outstanding, "world-type" milk or butterfat records, but as a whole the breed milks not too differently from the average does of the other dairy breeds.

The *Oberhasli*, like the Saanen and Toggenburg, has as its origin the mountains and valleys of Switzerland. This Swiss dairy goat is of medium size, hearty, vigorous, and alert, with erect and inquisitive ears. While not "flighty," it is not a lethargic breed.

OBERHASLI

Officially, the color is "chamois," with solid black being allowed in the does. Chamois is best described as being the color of a bay horse, with body colors ranging from a light sandy hue to a deep, rich, reddish brown. The straight face has two black stripes carrying downward from above and inside the eyes to the black muzzle. The forehead is often totally black. There are also two black stripes, one from the base of each ear, running to a point behind the poll of the head, where they meet, connect, and carry down the back line as a dorsal stripe that terminates in a black tail. The ears themselves are black on their interior surface and bay on the exterior. The stomach and udder are black, as are the legs below the knees and hocks. Bucks typically show more black, especially on the shoulders, lower chest, and back.

Roman noses, a desirable characteristic of the Nubian breed, are discriminated against in the Oberhasli. Oberhasli goats can be either polled or horned.

Of the breeds of goat kept for other than milk production, the Angora is most numerous. Between 400 and 600 Angora goats were imported into the United States from Turkey and South Africa in the middle part of the nineteenth century. Today there are more than 1 million Angora goats in the United States, most of them in central West Texas.

The *Angora* is a small-bodied goat that is managed for the production of mohair—the name given to its long, silky, high-luster hair coat. Does normally weigh 120 pounds, while bucks may attain weights of 150 pounds. The hair coat grows from ¾ to 1 inch per month, and the average total clip per year (two shearings) is 7 to 9 pounds.

The milk of the Angora is not normally used for human consumption because of its extremely low volume—barely enough to raise its own kids. Because of this low milk production and selection for the genetically uncorrelated trait of hair growth, the kids grow slowly.

The Angora is becoming increasingly popular today as a homesteader's goat because of its mohair production. The hair can be homespun just as wool can, but the fiber is stronger and withstands more abrasion. Mohair will take an infinite variety of dye shades and retain its natural luster and sheen.

The Boer goat was developed in the eastern Cape region of South Africa as a product of the worldwide interest in goat meat. The breed meets the following criteria, which are essential for a meat-type animal: good conformation, fast growth rate, good carcass characteristics, and a high fertility level. In addition, the ability to adapt to severe and varied climates was essential.

Boer goats are large-framed animals. Mature males will weigh between 260 and 380 pounds, while the females will scale between 210 and 265 pounds. Under performance-tested conditions, males averaged 80 pounds at 3 months of age, 160 pounds at 8 months, 222 pounds at 12 months, and 313 pounds at 25 months. Females averaged 63 pounds at 3 months, 139 pounds at 12 months, and 220 pounds at 24 months. The Boer goat is capable of attaining an average daily gain of over 0.88 pound daily in feedlot situations. In addition, the Boer's gain potential on pasture or rangeland offers possibilities for improving the growth rates of crossbred or milk-type goat herds. The Boer goat's dressing percentage exceeds 50%.

BOER

The ovulation rate for Boer goats ranges from one to four eggs per doe, and a kidding rate of 200% is common. The Boer goat can be considered a prolific breed. It reaches puberty early. Males will average about 6 months of age and the females will be about 10 to 12 months old when they first mate. The Boer goat has an extended breeding season, and it is possible to achieve 3 kiddings every 2 years. Boer goats are good milkers, which enables them to successfully raise their multiple offspring.

ANGORA

Goats are intriguing animals. The more you associate with them, the more you read and learn about them, and the more you talk to other "goat people" about them, the more you are tempted to jump right into the thick of it and start keeping goats! This may be the correct sequence of events for you, but before you begin keeping goats in earnest, you should consider several aspects of human and goat psychology.

Keeping goats requires a tremendous commitment of your time and a willingness to be tied to the land. Goats are kept for their milk-producing ability, and they must be milked twice daily, rain or shine, Saturday and Sunday, Christmas and Easter. The moments spent quietly milking are very rewarding and refreshing to your spirit, but you must be willing to pay the price.

The goat can be an extremely frustrating and trying animal. It can literally cause you to tear your hair while it blithely hops or climbs your new and expensive fence, tramples your newly seeded lawn, bounces onto the hood of your vintage sports car, and proceeds to eat the buds from your favorite rosebush. But this is the same animal that looks to you for help in her hour of need at kidding time and, more important, looks at you afterward as if thanking you for your assistance. For that particular feeling that comes from salvaging perhaps two animals' lives from a complicated kidding, the goat owner should be willing to sacrifice several rosebushes.

You must also realize that the goat has a set of behavioral characteristics that are quite unique. They basically will not be driven ahead of man or beast if they choose to stand and face the "enemy." If they do bolt from danger, they do not do so as a group; they scatter in any and all directions. They are intensely curious and will make all possible efforts to investigate anything that piques their interest. They are excellent jumpers, able to clear a 48" fence easily if not heavy in kid or pendulous of udder. They are browsers by preference, not grazers. They are excellent climbers and need only a narrow ledge to perch upon for the next jump. (Couple their jumping and climbing ability with their curiosity and desire to browse and you have many potential problems in controlling goats.)

Goats are very impressionable, and so are easily conditioned to a routine. But here again the uniqueness of the goat rises to the fore. Most animals, once thoroughly conditioned, will keep this imprint in their psyche, and years later a given cue will elicit some of the once-familiar responses. Not so the goat! Either you persist in a routine or they dismiss it from their behavior pattern. All of this leads some behaviorists to believe that the goat is perhaps more intelligent than our other farm animals.

For all of this purported intelligence and hardiness of makeup, the goat absolutely requires the assistance of his keepers if he is to cope with the demands and restrictions that we place upon him in today's production systems. The techniques discussed in this chapter were designed to assist you in controlling your goats and to help you carry out your production program.

8.2 CATCHING, HANDLING, AND CONTROLLING GOATS

Goats usually are not difficult to catch and handle. They are not especially large or quick-moving, and they are not mean, vicious, or dangerous (with the possible exception of a spoiled buck). Goats are creatures of habit and are extremely curious and affectionate (needing to both give and receive affection). Any problems you may have in catching, handling, and controlling your goats are likely to stem from a lack of understanding of the psychology behind their habits, curiosity, and need for affection.

As the techniques of catching, handling, and controlling goats are discussed, only enough psychology will be introduced to validate the logic behind a given step. You are encouraged to study the overall behavior of your goats and to read and learn about other goatherds' techniques and experiences. You are likely to find that as your knowledge grows, your appreciation for and enjoyment of your goats will grow also.

Equipment Necessary
- Feed bucket
- Fencing materials
- Catch pen
- Snow fencing
- Feed bunk
- Stock dog (well controlled)
- Collar (with ring)
- Neck chain
- Bullstaff

- Tether stake
- Tether chain

Restraint Required

The restraint required to catch a goat is mostly psychological in nature rather than physical. In other words, you must outsmart the goat by capitalizing on one or more of its habits until you can get close enough to grab it.

Handling the goat requires only moderate strength and agility. As your understanding of goat behavior increases, you will be able to anticipate their escape maneuvers and will find it ever easier to control them. A collar, neck chain, or halter is all that usually is necessary to handle a goat.

CAUTION: The adult male goat is an aggressive, strong, and potentially dangerous animal. He should be handled only by an experienced adult.

Some sort of range limitation is necessary when controlling the goat. In its most lenient forms, it can be the perimeter fencing around your property, the interior fencing surrounding the goat pasture or lot, or a small exercise run adjoining the shed. In its most stringent form, the range limitation can be the length of a tether chain. Control of the animal in each of these range categories is a function of the quality and height of the fence and the length and strength of the tether.

Catching

The procedures used and the degree of difficulty involved in catching goats are directly related to the management system under which the goats are maintained. They can be either intensively managed or free-roaming, without routine.

Intensively managed goats include the milking does, young animals whose feed is being supplemented with a daily grain ration (for example, replacement doe kids, bred doe kids, and doelings that are approaching kidding), breeding bucks or replacement bucks of all ages, and any animals being fed for butchering.

The underlying management factor that is common to all of these categories is the routine gathering together of the goats at a given point and for a given reason at least once a day. Whether this gathering is for milking or feeding, you are creating a habit—a conditioned response—in these animals whenever you bring them together. The cue that elicits the response—coming together at a given time and place—can be the "clock" of a full udder

for milking does or the sight of the feed buckets for the others. The wise manager will also condition his goats to the sound of his voice or the tone of his whistling.

The physical catching or grabbing of goats so conditioned involves only moderate effort from the manager. After the animals are assembled in a holding corral or a small pen, the gate can be closed and the animals worked as necessary.

Step-by-Step Procedures

1. Establish a management routine. Milking or feeding time can be the major cue, but be certain to use as many additional or secondary cues as you can for any routine. Examples of such cues are feeding or milking at the same time each morning and evening, always calling for the goats or whistling them up in the same manner each time, calling or whistling even if they are standing in the feed pen or milk corral waiting for you, and even going as far as to wear the same pair of coveralls and hat each time you handle them.

2. Construct a pen, just large enough to comfortably hold the number of animals you have, around the feed bunks or hay feeder or adjoining the milking room. Be certain to allow for a free-working 4- to 6-foot gate.

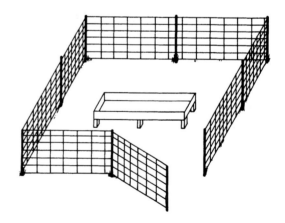

3. Assemble the equipment necessary for the management technique you wish to perform before you assemble the animals.

4. Wait until as near a routine time as you can to assemble the animals. (If you use the cues to which the goats are accustomed, they will respond by assembling in the customary place several hours ahead of the normal time.)

5. Perform the normal activity first; that is, milk them or feed them before vaccinating or trimming feet.

6. Upset the animals as little as possible by performing the management tasks as quickly, proficiently, quietly, and gently as you can.

Free-roaming, without-routine goats are quite a different problem for the would-be goat manager. These animals are not conditioned to assemble at any time for any reason. Goats in this category include pet goats that are now neglected (feed, minerals, and water are provided, but the animals are no longer milked or used to pull a cart), goats that are kept solely as four-footed brush clearers, or a goat flock that is kept only to produce chevon but otherwise allowed to run without restriction.

Catching and handling these goats will tax your strength and endurance and strain your patience and basic good humor to the last thread. The best advice is not to allow this situation of free-roaming without a routine to develop. If free-roaming fits your operation, then use it, but keep the animals coming back to the work area on a once-a-day basis. (A handful of their favorite feed or minerals or water can be the inducement.)

If you are stuck with the task of rounding up the wild ones, some combination of the following steps or suggestions may be of value.

1. Anticipate the need to handle the goats as far in advance as you can and attempt to condition them so that they will come to the feed bucket. At first, go to them, slowly. Talk to them, and if they scatter, stop moving for a bit. Get as close as you can, squat down, keep talking, and pour the feed on the ground in an appropriate number of small piles. In a couple of days, if they like the feed, they may come to you and your bucket. In a few more, they may follow you to the bunk. Allow several days of bunk feeding before attempting to work them. This program works best in times of short supply of browse and pasture.

CAUTION: A mature herd buck may challenge your right to approach and move among his flock. Such bucks can move quickly and are capable of hurting you. Be careful.

2. You can attempt to drive the goats ahead of you, up to and then along a fence line, and finally into a waiting catch pen. This sounds fine, and it may work for you, but some problems to be considered are: (1) goats don't drive as easily as sheep do—they stand and face the unknown rather than running from it; (2) even if you can goad them into running, they won't bunch and run from you, they will scatter and run; (3) they fight dogs and will not move ahead of them any better than they will move ahead of you (perhaps worse).

If you are determined to drive them, amass as large a crew of drivers as you can. Form a human

fence and approach them slowly and steadily. If they stop and face you, stop and wait until they turn and start moving before you begin again. If you push them too hard from too close, they are likely to bolt back through your ranks. Carrying a burlap sack in each of your hands sometimes helps the forward movement to continue.

Move them to and then along a fence line. Keep them moving slowly. Channel the goats from the fence into a catch pen. A roll or two of snow fencing can be the channel fencing, and it can also become the catch pen if it is quickly circled about the goats or pulled across the corner of a pasture.

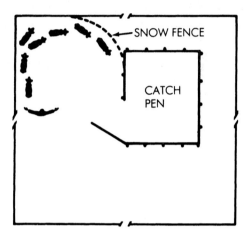

3. A properly trained and controlled dog can save you some steps and mental strain with a herd of goats that is undecided about whether to come to your call and feed bucket. Using the dog to advantage does require that the goats associate you or the barn with safety. If they were ever or are now coming to feed, milking, or water, this method may work.

You and the dog should approach the herd until they are aware of your presence. Stop walking; send the dog well around the herd and down it. Let your

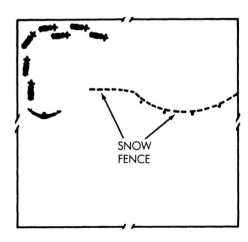

call or whistle to the goats be the cue to the dog to push the goats toward you. Your calling and moving to the barn or pen may offer the goats an alternative to facing the barking, scampering enemy. You will be "pulling" the herd while the dog "pushes" it. As long as the dog stays 50 to 100 feet from the herd, he may not cause the goats to scatter.

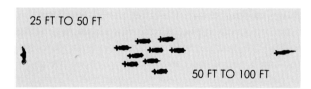

25 FT TO 50 FT

50 FT TO 100 FT

Handling

Handling goats fortunately is not a large problem for an adult-sized person, even if they are not tame from being handled daily. With the exception of a mature buck, the goat is neither strong enough nor aggressive enough to escape your grasp.

1. If a collar or identification chain is present, use it to handle the goat. Most goats will stop struggling to escape if you grasp the collar or chain and slip it up toward the base of the skull.

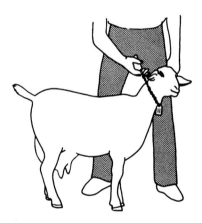

2. If the goat must be tied or continues to struggle in escape attempts, place an adjustable rope halter over its head. Any goat, with the possible

exception of one who lowers his head and charges, can be controlled with a halter.

3. The overaggressive buck can be controlled with a halter, collar and ring, and shortened bullstaff with a spring snap on the end of it. Depending on the individual animal, an assistant may be necessary to help place the halter and staff. If the buck is really mean, the bullstaff must be connected first, then the halter put in position while the assistant controls the staff. If the animal can be approached and haltered safely, but then quickly runs out of patience and charges, one person can do both the haltering and positioning of the staff.

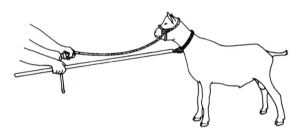

4. Many truly tame goats can be handled with nothing more than one hand under the chin and the other hand behind the poll on the upper neck.

CAUTION: Whatever the method of handling, the buck goat should be respected. He is dangerous and unpredictable, particularly during the breeding season. He is most likely to charge and butt you or to rear up and flail out at you with his front legs and feet.

The horns of any goat of any age should be considered dangerous. They are hard and sharp. An animal that has them will know how to use them to advantage. Zoos that choose to exhibit horned goats commonly place short lengths of hose or rubber balls over the horn tips for the safety of the handlers and other animals.

The most common escape maneuver will *not* be to try to break away to the side or rear as you are controlling the head and upper neck. The goat is more likely to wait for the moment when your grip or attention has faltered just a bit, then half jump, half lunge ahead—either directly over or just to the side of you. If you are not anticipating this, you will be off balance at the lunge and down on the ground just before the goat escapes.

Controlling the Goat

To control the goat is to limit his range of movement. This means that either fences or a tether system become necessary. Each goat handler is faced with a unique set of circumstances that precludes the giving of a set of recommendations that will work for all situations. Consider the following as steps in the control technique, as suggestions for you to consider, or as cautions describing pitfalls that you should avoid.

1. Construct all fences so that they are a minimum of 48" tall. If it is possible, 54" is a much better height. Any fence lower than 48" to 54" will be jumped when the need arises.

2. Barbed-wire fencing should not be a first choice for controlling goats, no matter to what height it is constructed. Even when it is dead tight, it will catch and tear the goats' hides as they run into it, lean against it, or stick their heads through it. When the strands are loose, entanglement becomes almost certain as the goat attempts to climb through the strands or walk them down.

3. Rail, pipe, or board fences become very expensive if they are to be 48" to 54" tall and have rails, pipes, or boards close enough together to be kid-proof (6" to 8").

4. Electric fence can be used as cross-fencing to limit pasture access or to ensure proper pasture or animal management practices. Recommendations are to place live wires at 12", 27", and 42" above the ground. "Flag" these well with tape or pieces of rag so that the goats see them, are attracted to them, and get "stung" by them. The goats thus become aware of the wires and avoid them.

5. Woven-wire fencing, sometimes called *hog fencing* or *stock fencing*, can be used, but it has the drawback of requiring corner braces, which the young goats will use as a bridge to climb over the fence. Wire-net fencing is also easy for the goats to stand up against and too inviting for them to use as a back rub. As this is a type of fencing popular for other livestock, it must be made to work for goats also. Using the heaviest-gauge wire you can afford, place the wire on the goat side of the posts. Running a strand or two of electric or barbed wire along the goat side of the fence, at 12" and 24" from the ground, will help to make it goat-proof.

6. Chain-link or Cyclone fencing, 4' to 4½' high, with posts at 6' centers, has the fewest drawbacks. It is also the most expensive, and therefore normally is used only by the wealthiest of goat owners. It should always be used for the buck pen. For this purpose, it should be from 5' to 5½' tall and well stretched and supported.

7. Whether you select chain-link, woven-wire, or board fencing for the buck, it *must* be tall (5' to 5½'), well constructed, and escape-proof. The gate should be capable of swinging inward or outward as the need arises. A mound of earth that is 5' to 6' tall and 5' × 5' at its top should be constructed in the center of the buck pen. This provides much-needed exercise and a throne for him to perch upon as he surveys his domain. Much boredom and such pen vices as fence walking or walking down of fences can be avoided by providing such a mound for the buck. Be certain to construct the mound at least 8' from the fence line so that the buck cannot jump out of the pen from it.

8. Tethering the goat involves the use of a collar, ring, several lengths of chain, and a tether stake. The stake, which has a freely revolving ring at its top, is driven into the ground. A length of chain is attached to it and to the collar ring on the goat. As the day progresses and the grass is grazed away, another length of chain can be added to the existing one. Tethering can be inhumane if the goat is left without shade, water, or protection from dogs. Goats are not usually tethered outside at night.

An alternative form of tethering involves the use of two stakes, from 10' to 100' apart, with a stout wire or cable stretched between them. A short chain (2' to 3' long) is clipped to it and then to the collar ring on the goat. This arrangement is called the *running tether* and is usually better for the goat in that it provides more exercise and free-choice sun or shade.

When given the opportunity to choose, always allow the goat to roam freely. Tethering is a very poor alternative to freedom of movement. However, if the fences are down or otherwise incapable of retaining the goat, tethering is an excellent alternative to a 10' × 10' drylot or stall.

9. The *Spanish halter* is a device commonly seen in the European countryside. It starts with a web or leather strap encircling the rib-cage. Attached to this "belt" is another strap which comes up from between the front legs. Finally, this strap from between the front legs attaches to a headstall or halter. The value of the Spanish halter lies in the fact that it allows goats the freedom of grazing grasses, forbs, and brush growing in orchards, without the mobility that would allow them to reach up and damage the tree branches.

8.3 GROOMING AND CLIPPING

Grooming a goat involves essentially the same procedures that are involved in grooming any other animal. It may be easier with goats because they are, by nature, perhaps the cleanest species of farm animal. If a milking doe is provided with a clean, grassy pasture and a shed with clean bedding, she will remain stain-free at all times. If you as the goatherd will provide a minute or two of grooming each day at one of the milkings, not only will your goat be stain-free, clean, and slick-coated, but she will very quickly become a pet as well.

Equipment Necessary

- Brush
- Mud rag (toweling or sack)
- Rub rag
- Currycomb
- Hoof-pick
- Electric clipper

Restraint Required

Grooming is a routine management practice only for goats such as milking does or show goats that are handled on a daily basis. For these animals, restraint involves only a neck chain or collar, a halter, or a grooming stanchion (similar to the milking stanchion). For goats that are maintained as brush clearers only, and are not milked, the restraint for their once-in-a-while grooming will almost certainly involve the grooming stanchion. Buck restraint depends upon your working relationship with the buck and upon the time of the year. If he is a worked-with, well-mannered beast, restrain him as you would a milking doe. If he is mean, or if it is breeding season, restrain him with the respect he is due.

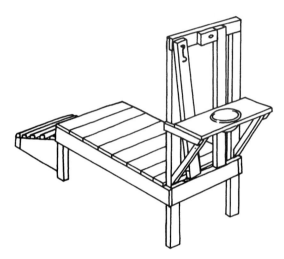

Step-by-Step Procedure

Body Grooming

1. Assemble the necessary equipment.

2. Catch and restrain the goats in whatever manner is necessary.

3. Begin grooming with a stiff-bristled and not too densely bristled brush. Begin brushing with the lay of the hair on the side of the goat's neck. Work upward and rearward from there. Short, flicking motions,

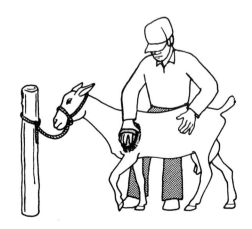

instead of long sweeping ones, are most efficient in removing dead skin, dirt, dead hair, and chaff.

4. Brush the entire body, including the face, udder, and legs. If the lower legs and feet are muddy, do not use the brush. Wipe these areas with an old towel or burlap sack. Brush them when they are dry.

> **CAUTION:** Goats enjoy grooming. Do not destroy this enthusiasm by brushing harshly around the ears, eyes, muzzle, or teats.

5. During the spring of the year, when the goat is shedding its winter coat, preface the brushing with a thorough currycombing. Use either a rubber, deep-massage currycomb or a steel-toothed version. In addition to removing the long outer hair, these tools will also remove the dense, woolly undercoat that was present during the winter.

> **CAUTION:** Do not use the currycombs on the face, udder, or lower legs.
> Currycomb gently. Use only enough force to lift the scurf to the surface. Skin is especially sensitive during this spring shedding.
> Currycombing or even deep, vigorous brushing should be curtailed during the fall and winter. The woolly undercoat is important for the goat's winter warmth.

6. Wipe the ears, eyes, nostrils, and muzzle with a clean rag.

7. Complete the body grooming with a quick, brisk rubdown with the rub rag. This is always rubbed with the lay of the hair and is not used on wet, muddy areas. This step puts a high gloss on the coat and assures that you are aware of any swellings, minor cuts, or localized temperature changes on your goat.

8. The final step in the routine grooming procedure is to clean the feet. This is a very simple and quickly performed task, but an important one. It is important to your goat's overall comfort and it assures that you become aware of excess hoof growth or disease problems.

A hoof-pick (as a horseman would carry) or pocketknife can be used. Lift the leg and brush the easily dislodged dirt and debris from the bottom of the foot and from between the toes. Then use the pick to remove the embedded debris from these areas. Pay particular attention to the area between the toes and to any portion of the foot that may be growing over the sole.

The Management Clips

Milking Clip. Goat milk is a high-quality, healthful product. It is becoming increasingly popular as more and more people rid themselves of biases about how goat milk will *probably* smell or taste. The simple truth is that goat milk, properly produced and handled, is entirely free of any touch of goaty odor or taste. There are several management guidelines that can help to produce a product of this quality. One of the most important is to keep the milking does clipped.

In a warm environment, the entire body can be clipped if you so desire. In areas of the country where cold weather is a problem, clipping the entire body is not reasonable, as it could be detrimental to the goat's health. Regardless of the temperature, the udder clip or milking clip is essential if you wish to produce clean, low-bacteria-count milk that does not smell or taste goaty.

1. Place the milking doe on the grooming stand and restrain its head. Some goat owners will use the milking stand for this procedure. This practice is not recommended because the milking stand should be kept as clean and free of hair, chaff, and dirt as possible.

2. Accustom the goat to the noise and feel of the vibrating clipper. This can be done by holding the running clipper in one hand and petting the goat with the other. Gradually ease the clipper body against the outside of the upper rear leg and allow the goat to become accustomed to the vibration.

3. Clip around the tailhead, extending the clipped area up along the rump for about 3 inches. From the tailhead, clip a path diagonally forward and downward into the flank area, to a point about 5 inches in front of the udder. Clip everything to the rear of this path, including the sides and rear of the udder itself and the entire rear legs. It is far more efficient to run the clipper against the lay of the doe's hair.

MILKING CLIP

CAUTION: It is possible to inflict a severe cut on the doe's udder with the electric clipper. This danger can be minimized by stretching the skin of the udder with your free hand and by having an assistant hold his hip or shoulder against the doe.

4. The hair on the belly of the doe should also be clipped in a swath about as wide as the udder and extending 6 to 8 inches forward. Special care should be taken as the clipper head is moved into the region of the udder's tie-in with the belly.

5. Repeat this process as often as necessary to keep the hair in the clipped area short and the udder easy to wash before milking.

Buck Clip. The purpose of the buck clip is to further aid in controlling the characteristic odor of the buck goat. A large portion of the buck odor should have been eliminated shortly after birth when the kid was disbudded and deodorized. It is practically impossible to render a buck completely odorless because of a natural behavioral trait he possesses. When aroused, the buck goat will urinate upon his front legs, brisket, and beard. As you might imagine, this habit can create quite an odor and wears heavily upon your appreciation of buck keeping. Fortunately, the following technique will largely eliminate this problem.

1. Catch and restrain the buck in whatever manner is appropriate for his temperament. He may stand quietly while an assistant holds him in a show-goat pose or if he is haltered, or he may best be restrained on the grooming stand.

CAUTION: It is much more difficult to handle the buck if you wait until the start of the breeding season. Clip him and perform any other routine management tasks in late summer before his change to breeding-season temperament.

2. Accustom the buck to the noise and feel of the vibrating clipper in the same manner as you did the milking doe. Instead of using the rear leg as a conditioning site for the vibration, use the buck's shoulder.

3. Closely clip the hair from the buck's brisket and front legs. In addition, completely remove the buck's beard with the clipper. He may still urinate upon himself in these areas, but with the short hair, the retained odor will be noticeably less. An additional improvement in the aroma of the buck will be noted if you bathe him after this clipping.

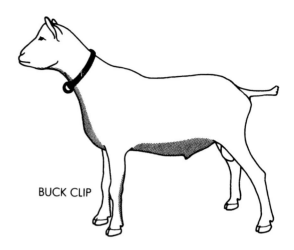

BUCK CLIP

4. Repeat the clipping procedure as often as necessary. The excess hair in these areas is of no value to the domesticated male goat.

8.4 BREEDING

There are three systems available for breeding goats: pasture or pen breeding, artificial insemination, and hand mating.

Pasture (Pen) Breeding

In the pasture-breeding system, the buck is allowed to run freely with the does. It makes no difference whether the "pasture" is a drylot, a 1-acre exercise yard, or a 20-acre field.

This type of breeding system involves very little labor and does result in a large percentage of does conceiving. It also has the following disadvantages: you have no control over when the does are bred, and consequently none over when they kid; your doe kids, only 4 to 5 months old, are quite likely to be bred; the buck may chase one particularly attractive doe for 2 or 3 days while allowing several others to come into and go out of heat without being bred; the buck will mark the does as belonging to him by rubbing his scent glands against them daily.

Artificial Insemination

In this system, frozen semen is thawed and deposited into the reproductive tract of the doe by a trained technician using a plastic inseminating pipette.

This system is growing in popularity and will continue to do so as long as suppliers continue to offer a selection of goat semen.

There is quite a bit of hassle involved with keeping a buck, and it is expensive if you try to keep a genetically superior herd improver around for only a few does. For a fee of $10 to $20, you can purchase semen, breed your doe, and then 5 months later have offspring from some of the best bucks in the country if you use artificial insemination. The disadvantages include adjusting your schedule on breeding day to that of the inseminator if you do not do your own inseminating, and the need for you and your family to become very adept at spotting does in heat.

Hand Mating

In this system, the buck is kept in a lot of his own and is never allowed to run with the does. As the does come into heat, they are taken to his pen, either singly or two or three at a time, and the buck breeds them there. The breeding is usually observed, and as the doe is bred, she is removed from the pen.

This system has the advantages of giving you precise breeding dates and of allowing you to skip breedings until a later heat period if you wish to stagger the freshening dates of your does. The disadvantages include the need for a separate buck shed and lot, the considerable amount of time involved, and, unless you have many does, the high buck cost per doe.

Buck kids and doe kids are capable of breeding, or of being bred *and conceiving,* as early as 12 weeks of age. Certainly, this is not to be encouraged, and from 3 months of age, the sexes must be separated. Young does should be bred in October, November, and December, when they are 6 or 7 months old, to freshen in March, April, or May as doelings. Young bucks that are 6 to 7 months of age can be penmated to a dozen or so does without being harmed.

Eighty percent of all does come into heat between September 1 and December 31, which means that 80 percent of the kids are born between February 1 and May 31. Because they breed in the fall, goats are termed *short-day breeders.*

Recently, several progressive goat breeders have been using artificial lighting (20 hours per day during January, February, and March, followed by 14 hours per day after that) to fool the does into believing that the days are growing shorter, just as in the normal fall breeding season. Even though it was spring outside the barn, and the days were actually lengthening, the does under the lights inside the barn responded by coming into heat, breeding, and conceiving in April. This assures freshening does in September and October, when normally bred does are nearing the end of their lactations. For producers with appropriate facilities, artificial lighting to induce breeding is a technique that can become a useful management tool.

As do other farm animals, the goat has a repeating estrous cycle. This means that every 20 (18 to 22) days, the female goat is in heat (estrus) and is capable of being bred. During estrus, her ovaries have responded to hormones by producing an egg or eggs ready for ovulation and fertilization, and the hormone levels in her body have been adjusted so that she will accept the buck's advances. Psychological and physiological readiness must be coordinated if the breeding is to be successful.

Only hand mating is discussed further in this chapter. Pasture mating is out of your control, and artificial insemination is in the realm of the trained technician. The only difference in your management techniques for hand mating and artificial insemination is that in the former method the does are penned with the buck, and in the latter they are penned for the breeding technician.

Equipment Necessary
- Collars—one each for the buck and doe
- Lead rope
- Pocket record book
- Bullstaff
- Nose ring

Restraint Required
The buck pen, where the breeding will take place, should be a stoutly constructed, escape-proof enclosure. The fences should be 5' to 6' high so that the buck cannot jump out. Individual animal restraint involves a collar and lead rope for the doe and, if necessary, a collar, nose ring, and bullstaff for the buck.

Step-by-Step Procedure
1. Learn the signs of heat (estrus). They include restlessness, almost constant bleating, some loss of appetite, a continuous tail twitching, a redness and swelling around the vulva, and a thin mucous discharge from the vulva (sometimes). Occasionally, a goat in heat rubs up against her herdmates. Goats do not "ride" as often as cattle do, but some does will accept being ridden or will ride other goats when they are in heat.

There are two relatively surefire tests to determine if the doe is in standing heat. *Standing heat* is that stage of estrus in which the doe willingly stands and accepts the buck. This can last for as little as 12 hours for some does and as long as 2 days for others. One test for standing heat involves running your hand firmly along the doe's back and observing her tail. If it twitches even faster than before, as if in response to your rubbing, the doe is psychologically ready to breed.

The other test involves taking a rag, about diaper size, and rubbing it all over a buck goat's scent glands, beard, brisket, and front legs until it smells like he does. Place this rag in a jar and seal it. Take it to the doe and open it by her nose. If she tries to get into the jar next to the rag, she is ready to breed.

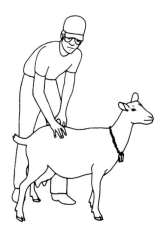

If you check heat twice daily and are doing a good job of it, you can take advantage of a basic physiological phenomenon to enhance the likelihood of breeding success. If you spot the doe in heat in the morning, wait until evening to breed her. If you spot her in the evening, wait until the following morning. This times the ripening and release of her egg more closely to the depositing of the sperm in her reproductive tract.

2. Set up a system to observe the does for signs of heat. This cannot be done on a hit-and-miss basis. For your milking does, there is no problem; you can observe them in the holding lot and in the milking stanchion twice daily. For the doe kids or yearlings not yet bred, a concerted effort must be made to observe them twice a day. Allow 15 to 20 minutes for each observation period. It will take half that long for the youngsters to get over the newness of your presence and return to their normal activity.

Another method of detecting heat is to place the buck on one side of a good, buck-proof fence, while the does to be bred are held on the other side. You must still observe the interaction, but you can do it from a distance and not run the risk of disturbing normal reactions of the does or buck.

In case there is not a buck present on your farm (you are planning to detect heat and then haul the does to the neighbor's buck for breeding), use a variation of the buck-rag-in-a-jar technique. Go to the neighbor and prepare the rag by rubbing it over the buck's head, belly, chest, and genital area . . . just like you did for the rag to place in the jar. Hang this rag in a tree, on a post, or on a fence adjoining the does. Does in heat will come to the rag just like they did to the buck on the other side of the fence.

As the does come into heat, they will look for the company of a buck and be attracted to the rag. This system works fairly well if you keep the rag "fresh."

3. Record in your pocket record book the identification of the does that are in heat. Note what stage of lactation they are in and calculate ahead to see which does you want to dry off in 60 to 90 days. Drying off is a standard procedure, to allow a 2- or 3-month rest period between lactations. If a certain doe is a fine milker and carries well out past 300 days in milk, you may decide to skip breeding her during this heat period and wait another 20 days.

A doe is normally bred in her seventh or eighth month of lactation. After breeding, she is milked for another 60 to 90 days, putting her in about the tenth month of lactation (300 days), and then dried off and allowed to rest for 60 days. With this system, the doe will kid and freshen every 12 months.

The easiest and safest way to dry off a doe is to start by skipping the evening milking for a week. After this, milk only every other morning for a week, and then stop milking altogether. Turn the doe into the gestation group and feed her as correctly as you can. She is pregnant and represents next year's milk supply. There are some successful goat breeders who dry off their does abruptly and rely on the engorged, swollen udder to shut itself down. It will do this, but there is some risk of mastitis and considerable discomfort for the does.

During the dry period, the doe should have access to a lot where she can exercise freely.

Consider breeding some does in each of the prime months (September through December) and others out of season by using artificial lighting so that your milk supply will be evened out and continue well into the following winter and, possibly, throughout the entire year. This ability to schedule is perhaps the greatest advantage that pen breeding and artificial insemination have over pasture mating.

CAUTION: If you do not schedule your breedings over a period of several heats, you will be breeding in the fall, freshening in the spring, oversupplied with milk in the summer, and buying milk in the winter.

4. After you have determined that the doe is in standing heat and that you do want her bred this period, place a collar with a 5- to 6-foot length of rope on her. Take her to the buck pen and place her in with the male.

Remain with the doe during breeding. Hold the lead rope during the breeding, but not so tightly as to pull on the doe, and in no way physically interfere with the normal breeding. The buck will sense that the doe is in heat and attempt to breed her in a matter of minutes. The mount of a goat is very quick; the buck spends no more than 10 seconds atop the doe.

After the doe is bred, she may arch her back as if to urinate and discharge a mucus-like substance from the vulva. Don't worry about this if it occurs; the mucus does not represent all of the semen the buck deposited.

After breeding, immediately remove the doe from the buck pen.

CAUTION: The male goat becomes quite possessive of his does. You may be challenged when you try to remove the doe. This is why you have the collar and rope on the doe. As soon as the breeding is over, tug the rope and leave with the doe.

There is the occasional buck that will not breed the doe while you remain in the pen, no matter how strong her state of estrus. With a buck of this type, you must leave and come back for the doe later or observe from a distance.

A small-framed or young buck may need a bale of straw to stand upon so that he will be tall enough to breed the doe. Placing the rear legs of the doe into a depression will serve the same purpose.

If you wish to cover (breed) the doe a second time (all of the doelings should be), it is better to return her to the buck 12 hours later than to let her run freely in the pen with the buck. A vigorous buck with highly fertile semen can easily breed five or six does per week, but no more than two in any one day. It is not unusual for a well-known, herd-improving buck to have 100 does brought to him each season.

If the doe should ever escape your grasp and the buck does challenge you, be certain that you pay him his due respect. He is large enough, strong enough, and quick enough to injure you. Use the bullstaff, which is a stout pole 5' to 6' long, to fend him off as you go after the doe. There is no reason to want to hit the buck, but you should not hesitate to do so if his bluffs turn into actual charges. Always try to entice him into his shed with a favorite feed before resorting to a show of force.

CAUTION: It is difficult to imagine a buck good enough to keep around if he is dangerously mean. If you feel that you must keep such a buck, at least ring him as you would a dairy bull and learn to use the bullstaff properly. A 2-inch self-piercing ring works fine for the nose ring, and a spade handle with a spring snap makes a good bullstaff.

5. Record the breeding date and the fact that you did observe the breeding. Continue to watch the doe for any signs of heat during the next weeks. Pay very close attention at 18 to 24 days from the breeding date, because that would have been her normal heat period had she not been bred.

8.5 KIDDING

Parturition, the birth of the kids, occurs after a gestation period of 145 to 155 days. The average length of pregnancy is 149 or 150 days.

To any livestock producer, the birth of the young marks an important time of the year. For the goat breeder it's the time of the year that milking does freshen after a period of being dry, and your source of milk, either for home use or sale, is renewed. Next year's milking-doe replacements arrive and your dreams of expansion are rejuvenated. The buck kids and excess females are marked and managed for slaughter or sale. All in all, it should be a refreshing time of the year.

Adequate preparation is essential if the kids are to have a high survival rate. The kidding does themselves can be lost if you have not managed them properly during the gestation or if you are not prepared to assist them, should they need it, during delivery.

Equipment Necessary

- Kidding stall
- Box or carton with bedding
- Towels
- Antiseptic soap, warm water, washbasin
- Gloves (cloth, rough texture)
- Obstetrical chains
- Heat lamps

Restraint Required

There is no restraint required during the kidding. In fact, care should be taken to ensure that the doe will not kid while she is tethered or stanchioned. The doe should be moved to a kidding stall or pen, no smaller than 6' × 6', for the actual delivery. She should be allowed to move about as she sees fit during labor and delivery. It becomes necessary to restrain the doe only if she is having difficulty in delivering the kids and you must examine her to determine the problem. Having an assistant hold her head, or tying her to a post, is usually sufficient.

Step-by-Step Procedure

Preparturition Preparation

1. Prepare the kidding area. This should consist of a draft-free, well-bedded stall at least 6' × 6' in size. If possible, the doe should be moved into this pen about 2 weeks before her due date. She should be allowed access to an outside exercise lot during a portion of the day. If you live in a warm climate and have the extra pasture space, the doe could be allowed to kid in a clean, fresh pasture lot.

2. Plan on being present at the time of delivery. It will mean a loss of sleep, but it could also mean saving the life of the kid or doe. Chilling is one of the causes of death in newborn kids. As a precaution, in case you miss the delivery, install a heat lamp above the kidding stall. This does not take the place of your being present at birth and drying and warming the kids, but it may keep them alive on a cold night until you arrive.

3. Have your plans made about how you are going to manage the feeding of the kid—including the very first feeding. Your choices are to allow the doe to raise the kid, in which case your goat milk supply will be greatly reduced, to allow her to feed the kid for the first day or two and then take over with the bottle or pan, or to take the youngster immediately after birth, milk the doe, and bottle- or pan-feed the kid from that time on.

4. If you allow the does to raise their own kids, you will need a kidding pen for each doe unless they are several days apart in due dates. In this system, the doe and kids are normally kept in the kidding pens for 2 or 3 days. If the doe is going to be allowed to feed the kids for only a couple of days, fewer stalls are necessary, but a nursery is needed for the kid raising. The last alternative, removing the kids at birth, calls for the least number of kidding stalls but the most carefully thought-out nursery.

5. Assemble the equipment necessary for kidding. This includes towels or sacks for drying off the newborn, a carton or box that contains bedding in which to place the kids, and soap, warm water, washbasin, and towels to clean the doe and your hands and arms after delivery. If you must assist at delivery, a pair of rough-textured, cloth gloves or a set of obstetrical chains is a big help.

6. Observe the doe closely. As the delivery date draws closer, once-daily observation is not adequate. Learn the signs of approaching parturition.

Signs of Approaching Parturition

Keep in mind that an accurate set of flock records, including breeding dates, is the most accurate indication that kidding is rapidly approaching. (Some does have not "read the book" and show none or only a few of the signs.)

1. Two or three weeks before kidding, and possibly even earlier, you will be able to see and feel the kid. Look for it on the doe's right side.

2. About 2 weeks before kidding, the doe will show a softening and relaxing of the muscles and ligaments on both sides of her tail. The udder will begin gradually to enlarge and fill from this time on.

3. About 3 or 4 days before kidding, the doe's udder will appear quite large and actually get in her way as she gets up and lies down. A first-kid doe may totally mislead you on this as her udder may not fully develop until 2 weeks or more after birth of the kids.

4. The last 1 or 2 days will be marked by the doe's being nervous, restless, and lying down a great deal. She will discharge a thin mucus from the vulva that will gradually thicken as parturition approaches.

5. The last 12 hours will be the most trying for the doe, and for you. She will fret and stew continually. Bleating will be almost continuous, especially when you enter the barn to check on her. She may carry her tail straight out or slightly elevated. The first several vertebrae of the spine, just in front of the tailhead, will seem to be standing higher and taller than usual. As these last 12 hours wind down, the doe will repeatedly lie down and get up, look back at her flank, and perhaps even kick up at it. The mucous discharge from her vulva will thicken and become gelatinous.

Three Stages of Parturition

While most animal managers consider birth of the young to be one big process, it is in reality broken into three stages. The first stage is the positioning of the fetus and dilation of the cervix. The last 12 hours covered above take care of this stage. At the transition from stage 1 to stage 2, the water will break. Stage 2 is the actual delivery. It is the presentation of the newborn kid. Stage 3 of parturition involves the expulsion of the afterbirth and the beginning of the involution, or return to normal size, of the doe's uterus.

Normal Delivery

1. After the water has burst and thereby lubricated the birth canal, the actual delivery of the kid should be complete in 30 minutes or so. Keep in mind that it is the straining of the doe and the pressure from the unborn kid's feet and head that cause the water bag to break.

2. The normal position of the kid in the uterus just before parturition is right side up, with the front feet first and the head lying on or between the fetlocks and knees. The normal presentation (what you see) as the kid is being delivered should be the two front feet with the pads turned downward, with the muzzle of the kid lying at about his knees. If any variation of this presentation occurs, the delivery is going to be abnormal and will most likely require your assistance.

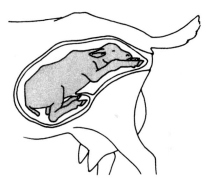

3. As the doe attempts to deliver the kid, there are several points at which she must strain in earnest. The head presents the first enlargement that must be passed through the birth canal. Then the shoulders are encountered and the cervix must be restretched. This can take several minutes, especially if the kid is large. The rib cage and abdomen come easily and provide a brief respite for the doe just before the last hurdle. The hips are the last enlargement that must pass through the birth canal, and often these are the worst moments for the doe. When the hips do pass, however, it is all over and the kid literally slides or falls out all at once.

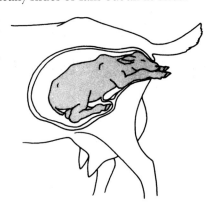

4. You should be present during this sequence and prepared to assist if necessary. If the position of the kid in the uterus and the presentation are normal, as described in step 2, and if the kid is of normal size, no assistance will be necessary as the doe begins to present the kid. As a guide to how long to allow the doe to work at trying to deliver the head, shoulders, and hips herself, you should think in terms of continuous progress, not in minutes. If the doe does not deliver more of the kid every 5 or 6 strains, once it is visible from the outside, she should be helped.

5. Consider the following as merely assisting the normal delivery. Nothing is visibly wrong—the forefeet and head are visible and properly oriented in the birth canal. Perhaps the doe just isn't strong enough to expel the kid, or the kid could be larger than normal. The cardinal rule in assisting the delivery is to pull only when the doe is straining to expel the kid. Grasp the feet and pasterns of the kid, using some toweling or the gloves so that you can grip the slick legs. Pull downward and outward as she pushes. Don't jerk, just pull firmly while she is pushing. Stop when she does. Sometimes it helps to pull the kid outward, downward, and to one side or the other. This procedure causes the shoulders and hips to come through the birth canal one side at a time. If you start assisting the doe, keep helping her until the kid is entirely delivered. Remember to pull only when she pushes.

6. After the kid is born, immediately clean the mucus from the nostrils and mouth, making sure that the kid starts to breathe. If the kid does not breathe, stimulate the process by pressing gently on the ribs or by moving the forelegs vigorously. Sometimes it may be necessary to hold the kid by the hind legs to drain the fluids and mucus from the mouth and throat. Swinging the kid while holding it in this position may also aid in establishing normal breathing.

CAUTION: After the kid is delivered and you are sure he is breathing normally, step back and do not disturb the mother and offspring. The umbilical cord is probably still attached, and the fetal blood that is in the placenta will be pulled back into the kid over the next few minutes. Do not deprive him of this blood by causing a premature severing of the cord. Sometimes the cord breaks immediately at birth. If this occurs, there is nothing you can do about it and it should not be a cause of concern.

If the umbilical cord does not break when the kid is born or when he or the doe begins to move about,

scrape or tear it apart 2" to 4" from the kid's body. Do not cut it with a sharp instrument. Blood does not clot well in a sharply made cut, and excess blood loss may occur.

7. When the umbilical cord has been separated from the kid, pick him up and thoroughly and briskly dry him. Use the toweling or burlap. After the kid has been dried, place him in the holding box.

8. Observe the doe for at least 20 to 30 minutes for a second kid or multiple births, and assist if necessary. If in doubt, wash your hand and arm with a disinfectant soap and examine the doe for additional kids by inserting your hand into the uterus and probing gently.

9. Remove the placental membranes (afterbirth) from the kidding pen after they are expelled by the doe.

CAUTION: Do not attempt to remove the placenta (afterbirth) from the doe by force. Allow her to shed it unassisted. This will normally occur in an hour. If it is still attached after 12 to 18 hours, call the veterinarian.

Assisting the Abnormal or Difficult Delivery

The directions given earlier state that a doe will deliver her kid or kids about 30 minutes after the water bag breaks. You were also advised not to intervene and assist the doe as long as every five or six strainings expelled more of the kid—once she started to deliver it. This is accurate information and works fine if you were there to observe the water bursting. Most often you will not be; rather, you will catch the doe delivering or trying to deliver kids on one of your periodic barn checks. The following step-by-step procedures and cautions are designed to help you cope with that situation.

CAUTION: Be patient. Be certain that help is needed before you rush in and start assisting. If kidding is progressing normally, and you intervene, you may upset the natural progression of events and create problems. Do not make an examination until you have observed the doe straining unproductively for 30 minutes to 1 hour. If the doe is growing visibly weaker as you observe, disregard this time frame and examine immediately.

Do not attempt an examination of a doe without an assistant present to restrain her or, if you are alone, tying her head. She will be frightened and confused by the complicated kidding, and if you have to chase her around the pen for the examination, you'll make things worse.

Before entering the doe for the examination, wash her anal and vulval areas with an antiseptic soap. Also wash your hands and forearms and lubricate one of them with a sterile lubricant if one is available. If it is not, use a small amount of the soap as a lubricant.

As you make the examination, keep your wits about you. Most malpresentations of the kid can be coped with if you are able to visualize what the problem is inside the uterus and compare that situation to what it should be. This caution implies that you have done your homework and know both what the normal presentation should be and what the abnormal presentation is most likely to be.

Always keep in mind that the goat is capable of multiple births. Sometimes the legs of the two or three unborn kids become entangled. At the onset of labor, one leg from each kid may enter the birth canal.

1. *The doe has been straining unproductively for 30 minutes and no part of the kid is visible.* In this situation, wait no longer. Wash yourself and the doe as indicated, place your hand in the vagina, and make the examination. Expect to find a full-breech delivery, a set of twins that have their legs entwined, or just a very poorly positioned fetus. In these cases, no part of the kid has entered the birth canal.

Full Breech. For the full breech, your task is to push inward on the buttocks and reposition the youngster to the easier-to-deliver uncomplicated breech presentation.

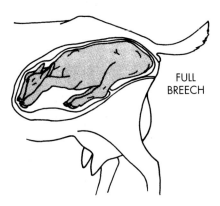

FULL BREECH

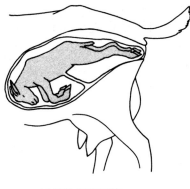

UNCOMPLICATED BREECH

CAUTION: In repositioning the kid, place your hand over the edge of the hoof as you pull it rearward and upward. If you do not, it could poke through the uterus and into the abdominal cavity.

Do not attempt to push the kid back into the doe while she is straining to expel it. Push when she relaxes; try to hold your position when she strains.

Once you start assisting the uncomplicated breech delivery, move rapidly. As the youngster enters the birth canal and the umbilical cord is pinched against the floor of the pelvis, he could suffocate if you delay.

Multiple Births. Your task in examining the doe who is trying to deliver two or more kids is to sort out pairs of legs. This takes some concentration. Don't worry if they are front or rear legs. Start a pair of them into the birth canal and attach an obstetrical chain to them, then determine whether they are front or rear legs (bottom of feet up = rear legs; bottom of feet down = front legs). If they are rear legs, assist delivery by pulling outward and downward when the doe strains. If they are front legs, you must be certain to position the head into the birth canal before assisting.

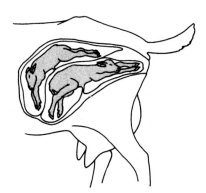

After the first kid of a multiple birth is born, the second will many times be delivered unassisted. Give the doe a few minutes (10 to 15) to start things on her own before you intervene.

Improperly Positioned. See the following steps for directions to remedy the most common malpresentations.

2. *Two feet, pads oriented downward, are visible and the doe is straining unproductively to deliver more of the kid.* This abnormal presentation is not unusual. It is usually the result of the head and neck of the kid being turned to one side or downward, between his legs, instead of into the birth canal. Your task is to push the kid back into the uterus and to reposition the head onto the front legs and subsequently into the birth canal.

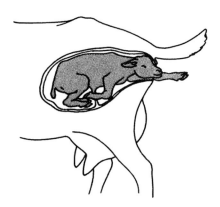

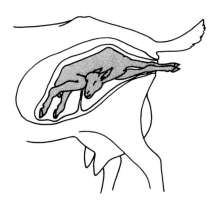

Place the obstetrical chains, one loop per leg, over the legs of the kid before you push the kid back in. You may, in your struggle to turn the neck and head, lose the legs if you do not attach the chains.

It will take some strength to turn the head and neck your way, so don't hesitate to pull. Grasping the lower jaw or an ear sometimes helps to assure your grip. A second chain can be placed over the head and used as an aid for repositioning.

CAUTION: It takes some real concentration and considerable strength to get this job done. When your arm fatigues and your mind starts to panic, do not lose control. Stop, reassess the problem, and start again.

---

3. *Head only, or one foot and the head, are visible, with the doe straining unproductively to deliver more of the kid.* This is not an unusual situation. It involves one or both of the legs not being properly positioned in the birth canal. Your first task is to palpate and be certain that you are not dealing with a multiple birth (where the head and leg may belong to different kids). To correct this malpresentation, slip the obstetrical chains over whatever is visible and then push the kid back in to gain enough room to bring the leg or legs up into the birth canal.

CAUTION: Try to keep your hand over the edge of the hoof as you pull it upward. It could pierce the uterus and protrude into the abdominal cavity.

---

4. *One or two feet, pads oriented upward, are visible; the head may or may not be in the birth canal.* Immediately investigate. Sort out the legs and head to determine if multiple births are further complicating this situation. Since the pads are facing upward, you can be certain that the delivery is either going to be an uncomplicated breech type or the kid is upside down. Examine by palpating and determine if the feet are on the hind legs (uncomplicated breech) or on the front legs (upside-down delivery). If the delivery is breech, follow the directions for assisting a full-breech delivery. If the kid is upside down, you must decide whether to turn him or not. If the legs and head are in the birth canal, go ahead and pull the kid without turning him as you would any other, but pull straight outward, not outward and downward.

If any part of the kid is missing from the birth canal, the best procedure is to push the kid back into the uterus, turn him right side up, position his legs and head, and assist the delivery by pulling as the doe pushes.

After the delivery has been assisted and the kid is breathing normally, see steps 7, 8, and 9 in the subsection on "Normal Delivery."

Postprocedural Management

Both the kids and the doe must be attended after either an assisted or unassisted delivery. For the procedures for caring for the kids and doe, see Section 8.6.

8.6 NEONATAL CARE

In practice, the goat breeder must not separate the delivery of the kids from the management procedures that should be performed on them within a few minutes after birth. After the delivery, the kids are checked for normal breathing and then immediately placed in an open-top, well-bedded, cardboard or wooden box. While you attend the doe and assist at any subsequent delivery, the kids are warm and safe in the box.

After half an hour or so, the doe will be finished with kidding and should be left to rest as you switch your attention to the kids. Care should be taken to perform the following procedures carefully and completely. They are critical to the survival and subsequent well-being of the kids.

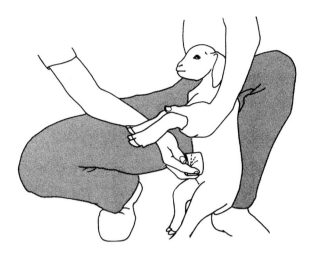

Equipment Necessary

- Toweling or burlap
- Tincture of iodine (7%) in baby-food jar
- Milking pail
- Bottle and lamb nipple
- Antiseptic soap and warm water
- Small utility scale
- Identification tools
- Disinfectant
- Fly repellant (wound-safe)

CAUTION: Do not delay in performing this technique. Joint ill or navel ill can result if you do. For best protection, it should be completed in the first half hour following birth. If you find a several-hours-old "surprise" in the kidding pen some morning, you should still dip the navel, in spite of the fact that the surest protection is achieved in the first minutes after birth.

Restraint Required

For the most part, none of the neonatal management techniques for either the kids or the doe requires strong restraint. It will certainly be easier if there is another set of hands present, but even this is not absolutely required. The doe will probably have to be tied while she is milked in the kidding pen.

3. Identify each kid with whatever identification system is necessary and appropriate for your operation.

Step-by-Step Procedure

1. After the doe has finished delivering the kids, or while you are observing the delivery of the second or third kid, remove the firstborn from the carton and thoroughly rub it dry with the toweling or burlap. This will finish drying it, and will also warm it and stimulate its neuromuscular systems.

2. Treat the navel by placing the baby-food jar containing fresh tincture of iodine over the umbilical stump (and any remnant of the cord). Pick the kid up and hold him securely in one of your arms. Use your other hand to place the iodine jar firmly over the navel. With a combination of shaking the jar and rotating the kid, saturate the navel for 20 to 30 seconds. Remove the jar and reseal it for use on the next kid. Allow the navel to drip dry.

4. If it is important to you, obtain the birth weight of the kid. Place a bucket on the platform of the utility scale and turn the dial to zero. Place the kid in the bucket and read the dial when it has stopped moving. Record this weight in your record book.

5. By this time, you have handled the kid for several minutes and have observed its sex and general vigor and well-being. Carefully examine now for specifics. Are its eyelashes turned out normally or are the lower lashes turned in and lying against the eyeball (entropion)? If entropion is present, contact an experienced goat or sheep breeder or veterinarian. The treatment involves a simple manual eversion of the eyelash for most kids or, in nonresponsive cases, a veterinary-installed suture in the lower eyelid.

CAUTION: If you neglect this, in four or five days a white cloudiness (scar tissue) will form on the eyeball under the lashes. Loss of sight is the certain result.

---

6. Are the visible sex organs normal? Are the anus, vulva, and mammary gland present and formed normally in the female kid? Are the penis, scrotum (with two testes), and anus normal in the buck kid? In the doe kid, are supernumerary (extra, more than two) or malformed teats present? If there are abnormalities, contact an experienced dairyman or veterinarian for advice.

7. Determine whether the kid will be polled (naturally hornless) or whether you will have to disbud it in 3 or 4 days. See Section 8.8 for the procedure to determine if horn buds are present in the newborn kid. If the kid is a female and you are selecting for the polled condition, the likelihood of *hermaphroditism* (intersexing, or the presence of both male and female organs in the same animal) is increased. To examine for this condition, open the lips of the vulva and observe its lower end. If an enlarged clitoris (round, pea-sized swelling) is present, the doe kid may be hermaphroditic and she should be watched during her development to see if your early diagnosis was correct. If it was, remove her from consideration as a breeding doe. Grow her out and butcher her as you would a wether.

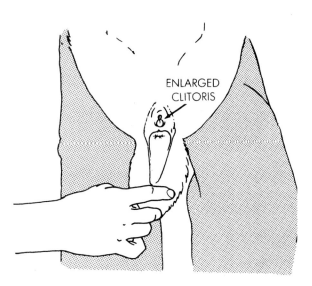

ENLARGED CLITORIS

8. You must now make the final decision about how you will rear the young. Will the doe be allowed to nurse the kids or will they be separated immediately and hand-reared? This is a weighty decision and should not be made lightly. The labor is much reduced if normal rearing is practiced, but the kids will need extra handling if they are not to grow up wild and wary of people. Some does are terrible mothers and some kids just never catch on to the system. If you allow natural rearing, your supply of milk is used for the kids.

Hand rearing is laborious. At first, you should feed the kids three to four times per day—every day. This does taper off, but even early-weaned kids depend on you for their total care for 5 to 6 weeks. Advantages of hand rearing include friendlier kids, and subsequently friendlier milking does, and more goat milk for your consumption (you will use a milk replacer after the first week). It is also a great way to teach care and responsibility to your children.

If your choice is to rear naturally, wash the doe's vulva and udder with antiseptic soap and warm water. Strip a stream or two of colostrum (first milk) from each teat. Turn the kids and doe into a holding pen much like a sheep "jug." See that the kids each nurse during the first 30 minutes after being transferred. Hold the doe and kids in this pen for 2 to 3 days before turning them out with other family groups.

9. For our purposes, we will assume that you are going to hand-rear the kids and that you are going to separate the doe and kids immediately, not ever allowing the kids to nurse. We will also assume that the kids are going to be fed from a nipple rather than from a pan.

CAUTION: Care must be taken to keep the bottles, nipples, self-feeders, and pans meticulously clean. Bacteria grow quite well in milk residue and are the primary cause of diarrhea (scours) in hand-reared kids.

10. Assemble your milk pail, milk bottle and nipple, soap, wash water, and towels. Restrain the doe in the kidding pen if necessary. Allow her to nuzzle the kids (still kept in the carton) if she shows interest in doing so. Wash the vulva, legs, and udder if she has shed the placenta. If she has not, wash only the udder at this time and the rest after she has shed it.

After drying the udder, milk the doe into the pail, transfer the colostrum to the feeding bottle, and feed from 2 to 4 ounces to each of the kids. This first feeding may be slow, but stick with it until the colostrum is consumed. It is important that the kids receive full feedings of it during their first hours. There is no need to milk out the doe completely. Remove only enough colostrum to feed the kids and relax the tension on the udder. After this initial

milking, switch to twice-daily milking, and at the second milking on the third day after kidding, milk out the doe completely. Excess colostrum should be refrigerated and fed during the first 2 days. If you warm the colostrum, do it in a double boiler or very slowly in a single pan of water. If it is warmed too rapidly, it will "set up" and take on a custardlike consistency. If it is overheated, the immunoglobulins in the colostrum will be destroyed.

11. Remove the kids from the doe immediately after this initial milking and place them in the nursery with other kids that are within a few days of their own age. If a nursery is unavailable, a pen in a draft-free portion of the barn is satisfactory. The doe will fret less if the kids are out of sight and sound. If it is particularly cold, heat lamps can be placed in a safe position (see manufacturer's cautions) over these pens.

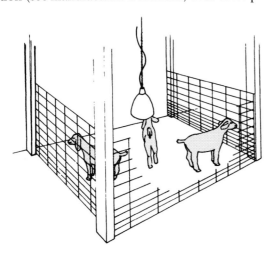

After these early management procedures have been performed on the kids, return your attention to the doe. She should be given a pail of lukewarm water (containing a cup of molasses, if you like) to drink. This will help to replace fluids lost during kidding. Place a rack or manger of good hay in the pen and keep it before her at all times. Wait until the second or third day after kidding to begin feeding grain. Take a full 2 weeks to have her on a high-grain diet. Always feed grain in proportion to the doe's milk production.

12. Allow the doe to rest in the kidding pen for a week before turning her in with the rest of the milking string. After the third or fourth day, the milk returns to its normal quality and can be saved for human consumption.

13. For the first 2 weeks, feed the kid three times each day—as close to every 8 hours as is practical—10 to 12 ounces at a feeding. The milk should be warmed to 100°F before bottle feeding. If you wish to group-feed the kids, construct a lamb-bar-style feeder that can handle 10 or 12 kids at one time.

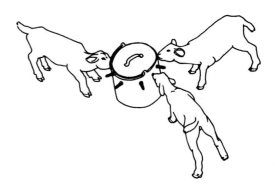

CAUTION: At the first sign of diarrhea (scours) in the kids, cut back on the amount of milk being consumed by about 50%. Hold it there for as many feedings as it takes for the diarrhea to stop. If the diarrhea persists for more than a couple of days, contact the veterinarian.

Actually, the better advice is to hold the amount of fluid intake constant but to dilute the richness or concentration of the milk by 50%. Do this by mixing milk and water on a one-to-one basis, then feeding the normal amount of this liquid.

During the first 2 weeks, the kids were fed goat milk. Because one of the main advantages of hand rearing is to maximize *your* utilization of goat milk, it is now time to switch the kids over to a milk replacer. Select one of the brand-name ewe or cow milk replacers and follow the manufacturer's directions for mixing. Take about 5 days to gradually replace the goat milk with the substitute. If you accustom the kids to it, cold milk replacer is satisfactory and palatable.

Allow the kids to consume 36 to 48 ounces of milk replacer per day by the end of the third week and perhaps even 64 ounces by the end of the fourth. Hold at this level until 3 days before weaning. Wean the kids by limiting them to 24 ounces on each of the last 3 days, then none from that point on. Weaning can occur, if you manage it properly, after the fifth or sixth week.

Kids can be reared by using the lamb-bar system, cold milk replacer, and free-choice feeding from the second day after birth. This system offers promise to the labor-poor producer.

Keep a small flake of high-quality hay before the kids at all times, starting at about 1 week of age. Feed this in some sort of manger or rack so the kids do not soil it. A good lamb or calf starter should be introduced at this time. Feed only small amounts of it to keep it fresh, until the kid consumes from $\frac{1}{4}$ to $\frac{1}{2}$ pound of it daily. Increase the quantity to suit your wishes. Your goal should be to grow the youngster, not fatten it.

Offer water to the kids twice daily—this is preferred to free-choice water availability. Some kids will consume water and be filled with it instead of milk, hay, and grain. A block of salt should also be in the nursery at all times.

14. Other management procedures that should be carried out during the first 3 or 4 days of the kid's life are disbudding and deodorizing.

15. Keep the kids in the nursery until they are 4 to 6 weeks old. At that point, they can be turned out with the herd during the daylight hours if the weather is warm.

16. Castrate the buck kids by the time they are weaned—certainly, no later than 8 weeks of age—unless you intend to market the bucklings without their being castrated. If you choose to do the latter, the buck kids and doe kids should be separated from the time they are 3 months of age.

8.7 MILKING

Milking does freshen each year, usually in the spring, and then for approximately the next 300 days will produce an average of 3 to 4 quarts of milk per day. If you come from a dairy background and think in terms of pounds of milk instead of quarts or gallons, this translates to somewhere between 1,800 to 2,000 pounds. This is a respectable figure when you realize that the goat that produces this weighs 150 pounds. This would extrapolate to a 1,500-pound dairy cow producing from 18,000 to 24,000 pounds of milk.

The top milking does from each breed—Saanen, Toggenburg, Alpine, La Mancha, and Nubian—will produce over 4,000 pounds of milk each year. The all-time top milk production by a goat exceeded 6,800 pounds in 365 days! (Most milk production

figures will be for 300 days, not 365, since the does are usually allowed a 60-day dry period to rest before freshening again.)

Dairy goats should be milked twice daily at approximately 12-hour intervals for maximum milk production. Regularity of schedule, however, is far more important than the 12-12 schedule. If a 10-14 schedule (for example, 6:00 A.M. and 4:00 P.M.) fits your lifestyle better, use it; just be regular about it.

For maximum milk production, dairy goats must come from good genetic stock, receive excellent management from the point of view of health practices, and be maintained on a top-level nutritional program.

Equipment Necessary

- Milking stanchion
- Milking pail (stainless steel, partially hooded)
- Milk strainer
- Filter discs
- Chlorinated udder wash
- Wash bucket
- Paper towels
- Strip cup
- Dairy wash

Restraint Required

The milking doe should be placed in a milking stanchion whenever possible. An elevated stanchion keeps the doe in position until you are finished, keeps the milk pail and your clothes off the ground, and spares your back. It is possible to milk the doe on the ground if she is tied or being held by an assistant.

Step-by-Step Procedure

1. Place the doe in the milking stanchion. Brush her quickly but thoroughly, paying particular attention to the inside of the legs and the belly and udder area. If the does have been given a "milking clip," this step is much more effective in reducing contamination of the milk. Place a pound or so of grain in the feed pan and allow her to start eating this while you prepare for milking.

2. Mix a pail of warm udder wash solution according to the manufacturer's directions. Wash your hands with some of this solution and then wash the doe's udder. Do a thorough job of this and use water that is comfortable but very warm to your touch (120 to 130°F). Not only are you cleansing the udder, you are also stimulating the doe to "let down" her milk. After washing, dry the udder and your hands with the paper towels.

Rinse (sanitize) the milk pail, milk strainer, and any other utensils used for milking, transfer, or storage with a chlorine solution. Do not dry these utensils after they are sanitized or wash them off with tap water. Allow them to stand and drain for a few minutes.

3. In step 5, you will be given directions about how to milk the doe. This will involve trapping milk in the teat, squeezing your fingers in sequence, and causing the milk stream to be ejected from the teat.

The first three or four streams of milk from each teat must be directed into the strip cup. This does two things. First of all, it causes you to discard the first bit of milk from each teat. This is important because it is always heavily loaded with bacteria, which would raise the overall bacterial count significantly. Second, it allows you to check for mastitis. The screen in the strip cup is black, and stringy, flaky, lumpy, mastitic milk can be spotted easily. Always use the strip cup before each milking. If the milk should appear mastitic, contact your veterinarian immediately. Do not use milk from a doe suspected of having even a mild case of mastitis.

CAUTION: Mastitis can be transmitted from goat to goat. Always milk "suspect" goats last. Contaminated bedding can spread the disease (so don't just pitch the milk from the strip cup onto the floor). After milking a mastitic goat, always disinfect your hands.

The milk from mastitic does should always be discarded—never kept for human consumption.

4. Sit on the platform attached to the milking stanchion. This is usually constructed so that you are milking from the doe's right side. Position the milking pail under the doe so that the stream of milk from each teat can be directed into it. Maneuver yourself so that you can comfortably grasp the doe's right teat in your left hand and her left in your right. Notice that you can control, or at least signal your wishes to the doe by using your right shoulder and forearm on her side and belly and your left hand against her right rear leg.

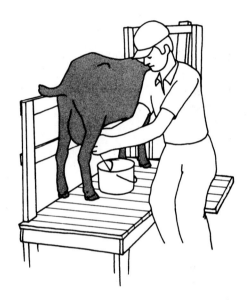

5. Grasp the teat by encircling it with the thumb and forefinger of your hand. Note that this grip is positioned at the junction of the udder and teat.

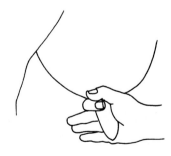

CAUTION: If you grasp above the teat—actually, on the lower part of the udder—you will destroy some of the supportive and secretory tissue and cause the premature deterioration of the udder.

Squeeze your thumb and forefinger together firmly. You are trying to trap the milk that is present in the teat with this pinching action so that as you squeeze with the rest of your fingers, it will be forced outward, not upward and back into the udder.

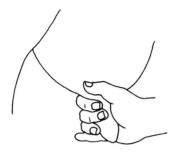

6. Maintain a firm grip with your thumb and forefinger and firmly squeeze the teat with your middle finger. This forces the milk even farther down the teat. Next, use your middle finger to squeeze, and finally use your little finger. If all has gone well, milk should have been ejected from the teat and into the pail. Now release your grip completely and allow milk to refill the teat.

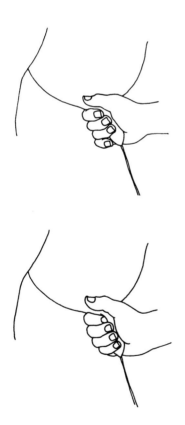

7. Repeat the same sequence with your other hand. Eventually, work up to a smooth rhythm in which one hand is squeezing while the other is relaxed. Speed is not as important as a smooth rhythm, but you will soon be surprised at just how rapidly you can perform this technique.

8. When you think you have milked the doe out (depleted her udder of milk for this session), do two more things to assure that you actually have. First,

bump the udder firmly but gently three or four times with the back of your hands. This imitates the nursing young and stimulates the letdown of any remaining milk. Second, strip the teat until milk is no longer ejected. Stripping is accomplished by grasping the teat up near the udder junction between your thumb and forefinger and running them down the teat without allowing the milk to escape back up into the udder.

Stripping is of value because only by removing every last bit of milk can you stimulate the goat to produce to her maximum potential. In addition, the last bits of milk are the richest in butterfat.

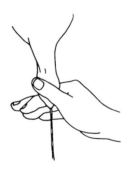

9. As soon as you have finished milking out the doe, pour the milk from your pail through the strainer and into the storage container. Place this container into an ice water bath and bring the temperature down as low and as rapidly as possible. In a commercial dairy milking system, milk goes from body temperature to 40°F in 60 seconds! Your goal should be to get it to 40°F in 1 hour. A refrigerator is a poor choice for this in that it cannot draw heat out as rapidly as is necessary to maintain maximum milk quality.

Remove the doe from the milking stanchion and turn her back into the rest of the herd. Repeat the previous steps for the remainder of the milking string.

10. Rinse all milking equipment in lukewarm water as soon as possible after use. After rinsing it, wash all of the equipment in warm, soapy water using a brand-name dairy wash and brush. A final rinse with boiling-hot water completes the cleanup. The utensils should be set aside to air dry.

CAUTION: If your first rinse is with cold water, it will not liquefy the butterfat. If it is with hot water, it will coagulate the protein and make it adhere to the surface of the pail.

11. Goats can be milked by machine just as efficiently as cattle can. Some of the largest goat dairies milk up to 500 does twice daily with either modified

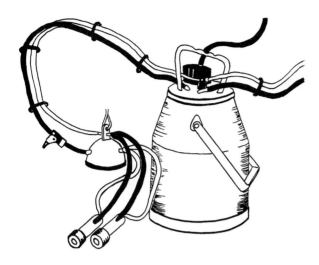

ELIZABETHAN COLLAR

cattle milkers or with specialized goat claws. Machine milking of dairy goats is usually done with a pipeline or milking parlor setup. Goat-milking machines and parts are offered for sale by most dairy supply houses.

12. Milk picks up off-flavors quite easily. Observe the following precautions to minimize this.

> **CAUTION:** Avoid feeding "strong" feeds such as onions or silage within 4 hours of milking.
>
> The milking room should be separated from other parts of the barn by a set of doors. This is to minimize the manure and bedding odor picked up by the milk.
>
> Do not smoke while milking. Smoking in a barn is a fire hazard and tobacco smoke taints the milk.
>
> Do not allow the buck to run with the milking does. He will mark them with his scent and thereby taint the milk.

13. Self-sucking is one management problem that will drive you to distraction until you sell the offending goats or utilize a restraint method to stop the habit. Self-sucking goats have perfected the technique of bending their heads around to their sides, back toward the udder, and sucking on their

own teats. However the habit may have started, you are now faced with coping with it or allowing the goat to continue drinking her own milk. Use either the Elizabethan collar or the anti-suck harness to control self-sucking goats.

8.8 DISBUDDING

Disbudding is the practice of removing the horn buds from the very young kid. Not every kid is destined to have horns as an adult, because some goat breeders are selecting lines or families that are naturally polled (hornless). As a practical management procedure, however, every kid—male and female— should be examined for horn buds.

Some people feel that a goat with horns is a beautiful animal; others feel that the sharp horns and sharp hooves are the goat's only defenses against the ravages of wild beasts. The first of these arguments cannot be denied. The horned goat is a beautiful and impressive animal—especially the mature buck. Consider, however, the following disadvantages of attempting to cope with the horned goat in your day-to-day management practices. First of all, there is the real danger to you and your family. If you maintain a herd buck, he will challenge you from time to time, especially during breeding and kidding seasons. Sharp horns can inflict terrible bruises, punctures, and lacerations. How about the danger to the nonhorned or less aggressive flock members? Horn removal puts all animals on an equal-armament basis. Do you ever plan to show your goats? Show regulations prohibit the exhibition of horned animals. Couple the above with horns entangled in fences and the aggravating modifications that must be made to all goat equipment plans to make them work with horned animals, and horns soon lose their beauty.

A wild or feral goat does need to use its horns and hooves for defense. To round up a wild goat flock and dehorn it would be a sadistic, cruel act, because it would leave the flock at the mercy of its predators. The farm flock of milking and meat goats is not in

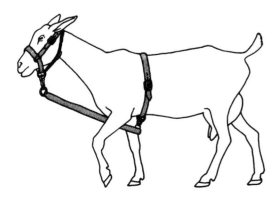

ANTI-SUCK HARNESS

the same situation, however. It is the responsibility of the goatherd to protect his animals from marauding predators.

Equipment Necessary

- Disbudding iron
- Scissors
- Caustic paste
- Petroleum jelly
- Band-Aids (round, patchlike)
- Antiseptic
- Fly repellant

Restraint Required

The key to successful disbudding is to perform the technique on young kids, preferably when they are 3 or 4 days old. If the kids are of this age, restraint is

not a difficult task. The kid may be held by an assistant, held firmly between your own thighs, or placed in a holding box.

Step-by-Step Procedure

Chemical Method

1. Determine whether your kid is going to grow horns and therefore needs disbudding. The time to

do this is at birth. On the still-wet newborn, the hair on the head is smoothly matted down, with the exception of two tufts or swirls of hair where the horns will erupt on the kids that need disbudding. When you spot these tufts at birth or in the next few days thereafter, mentally mark that youngster for disbudding.

2. Make a further examination at 3 or 4 days to confirm that the kid needs disbudding and that it is not going to be naturally polled. Do this by using scissors to carefully clip away the hair tufts where the horns appear destined to erupt. Clip an area the size of a half dollar as closely as you can, and then examine the tuft area. If the kid needs disbudding, a small, $\frac{1}{4}$-inch bare spot will appear over the horn bud. If the hair is even, the kid is going to be naturally hornless. One final check is to manipulate the bare spot with your fingers. If it moves freely, the kid is genetically hornless; if the skin is tight and immovable, consider this as further evidence that disbudding is required. Some owners disbud bucks between 4 to 8 days and does between 5 to 10 days.

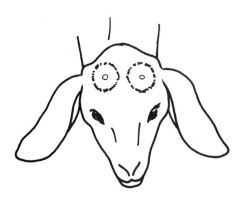

3. If you have only one or two does kidding per year, the chemical method of horn-bud removal is likely to be your choice. It is efficient, economical, and easy to administer. After you have restrained the kid and clipped the horn-bud area, place the round, Band-Aid patch directly over the horn bud. Then smear petroleum jelly in a circle around the patch. Use a heavy coating about $\frac{1}{2}$" to $\frac{3}{4}$" wide all

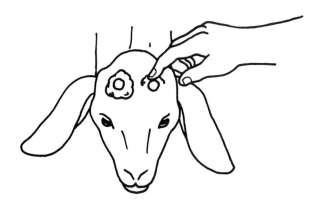

the way around each patch and rub it well into the hair. This is to prevent the chemical-disbudding compound from running or oozing away from the horn-bud site and possibly damaging the eyes or ear tissue. The patch will prevent the accidental protection of the horn-bud site.

4. Remove the round patch and apply the chemical dehorning paste, liquid, or stick directly to the horn-bud area. Follow the manufacturer's directions.

> **CAUTION:** Be certain to protect your hands during this procedure. These chemical compounds are caustic and can cause a painful burn if allowed to remain on your skin. Soap and water, and perhaps a vinegar rinse, will help to alleviate any discomfort.
>
> Disfigurement of ear edges and eye damage can occur if the chemical is allowed to run from the application site. Be certain to use a wide band of petroleum jelly to help prevent this.

5. Hold the kid in your arms or in a holding box for about 30 minutes to an hour after treating him with the chemical. The burning sensation will have stopped by this time and he will no longer attempt to rub it from his head.

> **CAUTION:** Do not bypass this step. These chemicals burn the horn bud and the kid will try to rub it off. He could smear the chemical into his eyes, or onto other goats. You will not achieve the desired results if he is turned loose and allowed to rub, and disfigurement of flockmates is probable.

Disbudding Iron

1. An excellent alternative method for disbudding kids involves the use of a disbudding iron. The iron is heated to a red-hot temperature and color, either electrically or with an external heat source such as hot coals or a small propane torch, and then placed over the horn bud. To perform this technique, the kid must be restrained and the hair tuft clipped as in the chemical disbudding method.

2. Heat the disbudding iron until it glows with a cherry-red color. Place the end of the iron directly over the horn bud. Press it firmly against the head of the kid and hold it in position for 10 to 15 seconds. This will sear the hair and skin and destroy the epidermis of the horn bud. Reheat the iron and apply it to the other horn bud.

3. The head of the kid will now show two rings branded into the hair, each with a horn bud in the center. Reheat the iron and gently sear this horn bud with the iron's edge.

> **CAUTION:** Use an appropriate amount of pressure on the disbudding iron. Only enough pressure to hold it in position without slipping is required. More important, the head of the kid must be restrained as nearly without movement as possible.
>
> Do not use an iron that is inadequately heated. To do so is cruel to the kid because the treatment will be ineffective and necessitate dehorning at a later date.

4. Deodorizing of *all* buck kids should be performed at the same time that the horned buck and doe kids are disbudded. It takes only an additional minute or less and prevents another handling and subjecting of the kids to additional distress.

Postprocedural Management

With the exceptions already noted in the step-by-step procedure, there are no special postprocedural management measures necessary. However, kids disbudded by either the chemical or hot-iron method will sometimes develop a scab in the treated area. This should be considered absolutely normal and cause you no concern unless it is fly season. If flies are buzzing, use an antiseptic fly-repellant dressing until the scab is sloughed off and the area healed over.

A drop or two of blood may ooze from the disbudded site when the scab is sloughed. This is of no concern. Treat with the antiseptic fly repellant as before.

The horn root or growth plate widens very quickly after birth. It is possible to miss an edge of this plate with the disbudding iron. If this occurs, scurs will develop from the remaining horn root. These will be ugly, misshapen horn masses. As you notice the start of scur development, re-disbud the kid immediately with a hot iron. Scurs of this nature can be eliminated with the disbudding iron.

8.9 DEODORIZING

The buck goat has a well-known and not easily overlooked characteristic of smelling bad. In addition, he has the equally well-known habit of labeling his territory and everything in it with his buck-goat aroma. The labeling is done by rubbing his head—specifically, the horn area—against the objects to be claimed or marked. Unless you are keeping the goats strictly for something to look at or as four-legged defoliating agents, and will handle them only the minimum amount to keep them healthy, the odor becomes a real problem.

Your milking does will soon smell like a buck goat if the buck is allowed to run with the flock. Fresh milk picks up odors very easily, and it too will assume a goaty aroma and flavor from the marked does. (Not all goaty-flavored milk can be blamed on the buck, however; some does also develop scent or musk glands.)

The buck's shed, fence posts, feed buckets, and anything else he touches will soon become saturated with this odor. The gloves, coveralls, and boots you wear when feeding him will also be tainted—so much so that if you do not wish your house to acquire the essence of goat, you'll leave them in the garage.

The musk or scent glands are located immediately behind and along the inside edge of each horn base. In a polled kid, the glands are in the same area but the absence of horn buds negates their use as reference points. In an older, polled goat, the scurs or horny boss of the animal are used as the reference point instead of the horn or horn bud. In both young and old animals, the glands appear as an area of shiny, thick, dark skin. When the glands are actively producing musk (seasonal peaks), they appear extra thick and corrugated or folded.

It is a simple task to deodorize a kid, and the procedure can be performed while disbudding. No additional equipment or handling is required.

Equipment Necessary

- Disbudding iron
- Scissors

Restraint Required

Goats are most easily deodorized as kids, at the same time that they are disbudded. Because this is done at 3 or 4 days of age, restraint is not a difficult task. As in disbudding, the kid may be held by an assistant,

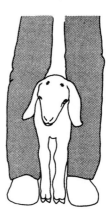

held firmly between your own thighs, or placed in a holding box.

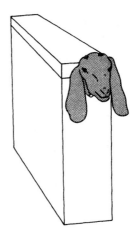

Step-by-Step Procedure

Because deodorizing the kid is an extension of disbudding, the techniques should be performed one after the other, with disbudding being first.

1. Restrain the kid in the manner selected and clip away the hair tufts over the horn-bud area and the hair immediately inside of and behind the hair tufts.

2. Disbud the kids with the hot-iron technique (see Section 8.8).

3. Reheat the iron until it glows with the necessary cherry-red color. Apply the hot iron about $\frac{1}{2}$ inch behind and toward the center from the disbudded horn buds for approximately 10 seconds. This will make two sets of overlapping circles.

> **CAUTION:** Try not to overlap the rings from the right and left sides and thereby destroy (burn through) the band of skin that lies between the horn buds. If this should occur, disbudding and deodorizing are no less effective, but healing will be delayed.
>
> Use an appropriate amount of pressure on the disbudding iron. Only enough pressure to hold it in position without slipping is required. More important, the head of the kid must be restrained as nearly without movement as possible.
>
> Do not use an iron that is inadequately heated. To do so is cruel to the kid because the treatment will be ineffective and necessitate dehorning at a later date.

4. Naturally polled kids should also be deodorized. While there are no horn buds or hair tufts to use as reference points, the gland is present in the same position that it is in horned kids. The skin of the scent gland is shiny and darker than the surrounding skin.

5. It is possible to de-scent an older goat, even a mature buck or doe. If the animal is hornless, the task is no more difficult than is de-scenting a young kid—with the exception of the restraint. A mature buck or doe cannot be held between your legs; nor can it be placed in a holding box. If the animal is a horned buck or doe, the de-scenting is more difficult to perform. It can be performed as an additional step along with dehorning, or the animal can be surgically dehorned first and then, after recovery, descented with a disbudding iron.

6. If you wish to de-scent an older polled goat, the animal should be thrown (gently!) onto its side.

This procedure is the same as "throwing" a mature ram or ewe. The step-by-step restraint procedures are described in Chapter 1.

7. Two capable assistants should hold the goat to the ground while you tie its hind legs together. Place a folded blanket or a couple of feed sacks under the goat's head to prevent injury and to present a better angle for de-scenting. One handler should manage the abdomen and rear quarter while the other controls the neck, shoulders, and head.

8. Clip or shave the hair over the scent glands. Recall that they are shiny and darker than the surrounding skin.

9. Heat a disbudding iron until it glows a cherry-red color. Place the iron over one of the scent glands and firmly hold in position for 10 to 15 seconds. Reheat the iron and repeat the procedure for the other gland.

> **CAUTION:** Observe the precautions given in step 3 of this procedure.

10. Some goats carry small patches of musk glands on locations other than the head. A common place for these to occur is on the inside of and just above the hock joint. If a goat still smells bad after routine deodorizing, wash it and search with your nose for the remaining sources of scent. Because of the danger of damage to connective tissue of the hock, consult with your veterinarian before removing these musk glands with a disbudding iron.

Postprocedural Management

Unless you hold the iron on the kid or mature goat for longer than the recommended 10 or 15 seconds, there should be no adverse effects from the deodorizing. If the iron is held on for too long a time, the skin may burn and scab. If this occurs, the scab should be managed to prevent infection or fly aggravation as is done in disbudding.

Deodorizing or de-scenting will not affect the male's fertility, his libido, or the doe's attraction to him.

8.10 CASTRATION

Castration of the young buck kid involves the removal of the testes, epididymides, and a portion of each sperm duct. The castrated male goat is called a *wether*.

For most goat owners, there is no reason not to castrate each and every buck kid. The genetic quality that is normally bred into a small, low-income flock of a half-dozen does is not great enough to merit keeping the buck kids for potential herd sires. You can purchase a high-quality buck kid or buckling from a larger breeder and probably do it more cheaply than you can raise your own.

Buck kids should not be allowed to remain intact past 12 weeks unless you have a lot or pen in which to house them separately from the females. Buck kids are capable of breeding the females from the time they are 4 to 5 months of age.

Intact male goats make very poor and dangerous pets. They become spoiled and unmanageable and are quite capable of hurting even an adult by butting (with or without horns) or by rearing up and striking out with their front feet. All buck kids that are not of herd-buck quality should be castrated.

Castration should take place at the youngest age possible. The amount of shock to the kid's system and the degree of growth-performance setback increase as he grows older. Most goat owners choose to castrate somewhere between 2 and 4 weeks of age. The best advice is to castrate as soon after birth as you can grasp both testes in the scrotum. (This may be as early as 2 or 3 days after the kid's birth.) There is no harm or even great apparent pain if you wait until months later, but neither is there a reason to wait.

The wethers can run with the milking flock if you so wish, but if *chevon* is your goal, they will grow faster and be of higher quality if you keep them separated and feed them some extra grain.

Equipment Necessary

- Emasculatome
- Elastrator
- Knife
- Emasculator
- Disinfectant
- Antiseptic
- Antiseptic soap
- Fly repellant, wound-safe
- Syringe and needle
- Tetanus antitoxin
- Antibiotic (injectable)

Restraint Required

If the goat is castrated as a buck kid, the normal restraint is to have an assistant set him up on his lower back or rump with his hind legs pointing forward. The kid's back is vertical or nearly vertical and held against the body of the holder. The rear legs are lifted upward and they and the front legs are held together in the hands of the holder. The kid will be more conveniently located for the castrator if the holder places the rump on a table or fence rail.

Animals more than 3 or 4 months old cannot be held in this manner because of their strength. They can be restrained in a stanchion or cast to the ground and held by two assistants.

Step-by-Step Procedure

Knife and Emasculator

Castrating with the knife and emasculator is preferred by many livestock producers. It is quick, inflicts pain for only a very brief time, and you can be absolutely certain that both testes have been removed. Done properly and at an early age, it is nearly a bloodless procedure.

CAUTION: A kid that has a scrotal hernia should not be castrated by any cutting method. The emasculatome procedure is recommended in this situation but should be performed only after taking great care to assure that the cord-crushing jaws are positioned below the hernia.

1. Be certain that you have all the necessary equipment and the required number of assistants assembled. The castrating equipment should be sanitized with a disinfectant.

2. Catch and restrain the animal in the manner appropriate to his size and to the number of assistants available.

3. Wash the scrotal area, plus a couple of inches immediately surrounding it, with an antiseptic soap. Wash and disinfect your hands in the same manner.

4. Grasp the lower end of the scrotum in your fingertips and cut off the lower one-third of it. Use a cut that runs parallel to the ground. The testes are maneuvered to a position above the cut line since there is no need to cut them. At the completion of this step, the testes are visible as two discrete organs protruding from the lower end of the remaining scrotum.

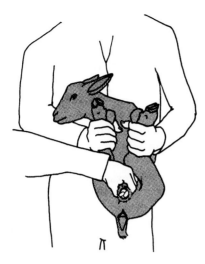

Some producers make a separate incision in each side of the scrotum to remove the testes. This procedure will work fine if the two side cuts are extended to a point low enough on the scrotum to allow for fluid drainage and to eliminate any blind pocketing effect (and probably infection) in the lower scrotum. Removing the lower one-third of the scrotum has been chosen as our castration procedure because it promotes the necessary drainage and eliminates the pocket for infection.

5. Using your fingers, grasp each testis, one at a time, and pull downward. The testes are slick and difficult to hold onto, so grasp them firmly. Pull slowly but continuously until the testis comes free from its attachment near the top of the scrotum. There will be a segment of cord attached to each testis that is removed at the same time.

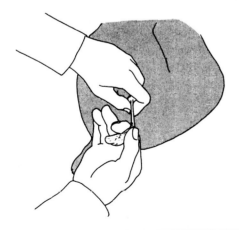

CAUTION: If the animal is more than 4 or 5 weeks old, use your knife to sever the cord instead of pulling it until it breaks. Do not cut the cord with the knife blade; instead, scrape it until it abrades through. If you pull the cord or cut it cleanly, excessive blood loss could occur.

If the buck is mature, the cord will bleed excessively even if abraded instead of cut. Use the emas-

culator on a cord of this size. Place the crushing jaws toward the animal's body and the cutting edge away from it so that the cord is crushed, and bleeding stopped, at a point above the cut.

Do not excite the animal by chasing him about, being unnecessarily loud, or subjecting him to a rougher-than-necessary handling. All of this causes the heart to beat faster and blood pressure to rise, both of which will cause extra bleeding.

If a segment of the spermatic cord is visibly protruding below the end of the scrotal stump, it must be removed. Left exposed, it acts as a wick to pull bacteria up into the body. Pull it free or abrade it with the knife.

6. Douse the castration site with antiseptic and administer an injection of tetanus antitoxin.

Postprocedural Management

Place the castrated kids into pens that have been cleaned and freshly bedded. The cleanliness will help to reduce the possibility of infection. Prevent drafts from blowing upon the kids, and in extremely cold or damp weather, place a heat lamp over their pen. Check the surgical site daily, and if evidence of infection is present, administer antibiotics for 3 days according to the drug manufacturer's directions. If no infection is present, the kids can be turned out in 3 to 5 days.

If flies are present during the recovery period, spray the surgical site with a wound-safe repellant daily.

The kid will probably appear stiff in the rear legs for 2 or 3 days after any castration procedure. If the stiffness persists beyond this time, consult your veterinarian.

Emasculatome (Burdizzo®)

Use of the emasculatome in castration is preferred by many livestock producers. Because it involves no cutting and, when done properly, does not even break the skin, it is entirely bloodless. Consequently, there is no chance for postprocedural infection.

Critics of the method point out that you are never sure that the animal has been properly castrated until you see both testes in hand and out of the scrotum. It does take special care to be certain that both spermatic cords are crushed, but it can be done, and tens of thousands of lambs, kids, and calves are castrated annually with the Burdizzo.

1. Be certain that you have all of the necessary equipment and the required number of assistants assembled. The emasculatome is not a cutting instrument, so strictly speaking it does not have to be sanitized. Sanitization will not damage the instrument, however, and there is always a chance that the skin may be broken.

2. Catch and restrain the animal in the manner appropriate to his size and to the number of assistants available.

3. Douse the upper portion of the scrotum, near where it attaches to the body, with the antiseptic liquid.

4. Grasp the scrotum in one hand and manipulate it until you have one of the testes well down into the scrotum and the cord leading to it held between your fingers. Place the jaws of the emasculatome onto the upper scrotum, just below the rudimentary

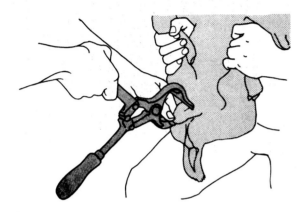

teats. Position the jaws of the emasculatome so that about two-thirds of the scrotum is below the crush site when the jaws are closed. Leave the instrument in the closed position for 15 to 20 seconds.

Open the jaws, lower the instrument about ½ inch, and crush the other side of the scrotum.

CAUTION: The spermatic cord is very elusive when you attempt to crush it between the jaws of the emasculatome. Be certain that you feel it within the confines of the jaws before and after the jaws are closed.

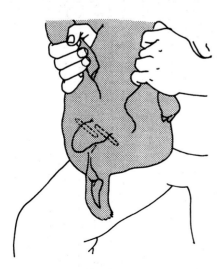

Postprocedural Management

Even though this is supposed to be a bloodless method of castration, occasionally the skin of the scrotum will be broken by one of the corners of the emasculatome jaws. For this reason, examine each kid carefully after the procedure. If any skin was broken, liberally douse it with antiseptic and give the kid an injection of tetanus antitoxin.

Swelling of the scrotum will occur in the next 24 hours and last for 2 to 3 days. This is normal, and the animal should be examined for infection only if the swelling persists. The scrotum (cod) will remain visible for the animal's lifetime, but an examination will disclose that the testes have degenerated and disappeared.

Elastrator

In this method of castration, a heavy rubber band or ring is placed over the scrotum near its upper end. The rubber ring stops the circulation of the blood to the testes and scrotum. In 10 to 14 days, the scrotum and testes will slough off. This procedure is quick to perform and causes pain for only about an hour until the scrotal area has become numbed. It is most effective when performed on young animals whose scrotal muscles and other tissues have not become well developed.

1. Be certain that you have all of the necessary equipment and the required number of assistants assembled. Castration with the elastrator is bloodless. It is not necessary to disinfect the elastrator or rubber rings.

2. Catch and restrain the animal in the manner appropriate to his size and to the number of assistants available.

3. Douse the upper portion of the scrotum, near where it attaches to the body, with antiseptic.

4. Place the rubber ring on the prongs of the elastrator. Orient the prongs so that they face toward the kid's body. Expand the ring by squeezing the handles of the elastrator. Place this expanded

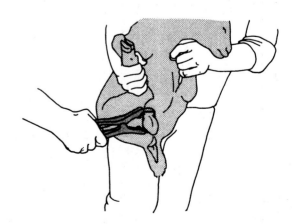

ring over the scrotum and testes and position it as close as possible to the kid's body without interfering with the rudimentary teats.

5. Grasp and manipulate the scrotum until you are certain that both testes are descended and are controlled by your hand.

6. Press the trigger lever that displaces the ring from the prongs, thereby positioning it on the testes.

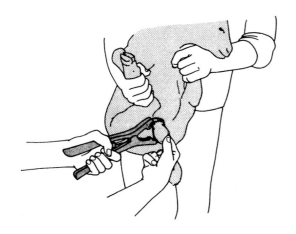

CAUTION: Double-check to be certain that both testes are below the ring. If they are not, carefully cut the ring and start over.

7. Administer an injection of tetanus antitoxin. Even though this is a bloodless, nonsurgical procedure, the tetanus organism can gain entry through the irritated tissue that will develop around the rubber ring.

Postprocedural Management

In 5 to 7 days, an area of tissue irritation will be obvious around the rubber ring. This is a likely area for an infection to develop. Catch all of the elastrator-castrated kids and wash this area with disinfectant soap, rinse, and re-disinfect. Spray the area with a wound-safe fly repellant if the season warrants it. In another 7 to 10 days, the scrotums and testes will begin to slough off. Reexamine the kids at this time and check for infection. Cleanse and disinfect the wounds and administer antibiotics as necessary.

8.11 IDENTIFICATION

The proper identification of goats is essential for all aspects of efficient goat production. It allows for the orderly proof of and transfer of ownership. It is re-quired if the goat owner wishes to register purebred animals with appropriate breed registries. It enables the producer to keep comprehensive records of milk production, multiple births, and health problems. From these records and any others that you may choose to use, all of which require a permanent identification system, prudent selection of breeding stock can be made.

There are several systems of identification that can be used. The system you select will depend upon the size of your herd, the environmental conditions under which they are maintained, the primary purpose for identifying individual animals, rules and regulations of breed-governing bodies, and equipment that you have available.

Keep in mind that there are two basic types of identification—the breed association number, which suits their records, and the within-herd or home numbering system, which is designed for your needs. Registered animals that you wish to sell as registered or to exhibit at breed shows must bear an association number wherever they tell you to place it. Your home herd of milking does, whether grade or registered, should carry an additional set of numbers. These could be simply a way of telling one animal from another, or they could tell you the year of birth, birth sequence, level of milk production, or other items. They should tell you this at a glance, which means that you should use large numerals and perhaps depart somewhat from convention. Do not dismiss the use of freeze branding, for instance, just because in the past someone could not figure out how to make freeze branding work on a Saanen.

Equipment Necessary

- Tattoo pliers and digits
- Tattoo ink
- Neck chain or collar; tags
- Ear notcher
- Disinfectant
- Widemouthed jar or can
- Squeeze bottle
- Cotton swabs
- Freeze-branding irons and equipment

Restraint Required

The restraint required will vary with the age and strength of the animal as well as with the type of identification system chosen. Regardless of the system chosen, a young animal (up to several months of age) can be held in the arms of the assistant. Older animals should be placed in a grooming stanchion or thrown and held to the ground by two assistants.

Tattooing

This is a permanent method of identifying goats if used properly. It requires a tattooing outfit consisting of a pair of pliers and numbers or letters made of sharp-pointed, needlelike projections that pierce the skin. A tattoo ink forced into the punctures remains visible after the puncture wounds heal. The tattoo ink can be placed directly on the digits and pressed into the skin by the pliers, or it can be rubbed into the puncture wounds after the pliers are used. The pigment in the ink heals into the skin, leaving a permanent number or letter.

Tattooing is usually done on the ears, and tattoos can be seen best on goats that have white ears. Those with black pigment in the ears can be tattooed, but the numerals and letters are more difficult to read. On black-pigmented goats (for example, the Alpine), use green tattoo paste instead of black.

Step-by-Step Procedures

1. Assemble the necessary equipment. Select the first number to be tattooed into the ear and place these digits into the tattoo pliers. Lock them in securely.

Before piercing the ear with the tattoo needles, test to be sure that the number is correct. A piece of cardboard makes a good test site. Remember that the numbers are formed by a series of needles only and that they are formed backward. It's easy to make a mistake.

2. Place this first setup and the remaining digits into the jar or can and add enough disinfectant to cover the needlelike projections.

3. Catch and restrain the animal in the appropriate manner. Keep in mind that the goat's head must be held motionless at the moment of tattooing.

4. Examine the ear of the animal. You will find ribs of cartilage with spaces between them. You must insert the tattoo numerals into the widest of these spaces without damaging the cartilage.

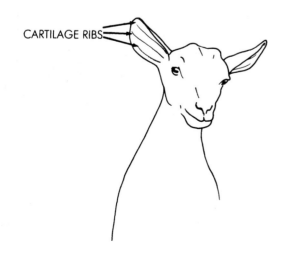

CARTILAGE RIBS

CAUTION: If the cartilage is pierced by one or more of the needles, there is some danger of damage to the cartilage, which could prevent the ear from being carried in the characteristic position. A bigger concern is the fact that the tattoo ink is not as visible in the cartilage, which will make positive reading more difficult.

5. Clean the front and back of the ear with disinfectant in the area to be tattooed. There is little likelihood of a serious infection developing, but it takes so little time for this extra precaution that it is foolish to skip it.

6. Place the tattoo pliers over the edge of the ear and quickly, but smoothly and forcefully, press the needle-point numerals into (possibly through) the ear by squeezing the handles of the pliers.

7. Release and carefully remove the pliers from the ear. Place them back into the disinfectant.

CAUTION: If the needles have not come completely free of the ear before you attempt to remove the tattooer, you could scratch or tear the ear. Proceed slowly and carefully!

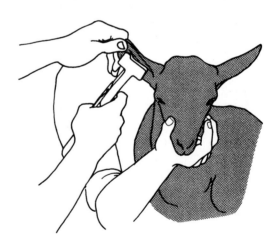

8. Take some tattoo ink or paste and *thoroughly* rub it into the needle punctures. Be certain to apply ink to both the inner and outer surfaces of the ear if the tattoo needles pierced completely through.

Some producers choose to place the ink or paste onto the needle points before tattooing, the theory being that the ink penetrates along with the needles. They then rub the tattoo site to enhance penetration. In practice, the surer way to a dark tattoo is the first method or a combination of the two.

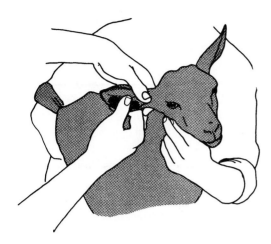

9. Wait until the kid is 5 or 6 months of age, if you can, before tattooing it. There is more space for the numerals and the bulk of the ear growth is completed by this time. If the kid is tattooed at a young age, growth of the ear will spread the tattoo out. Coupling that to a thickening of ear tissues and an increase in skin pigment results in an "unreadable without a flashlight behind the ear" tattoo.

These young kids should have a temporary identification such as an ear tag or neck chain.

10. The earless La Mancha presents the unique problem of where to place the tattoo. Breeders have selected the tail web. It can be tattooed on both the right and left sides. Many breeders will use one side for the farm number and the other for the Association number. The techniques involved in the actual tattooing are exactly the same. The secret to highly readable tail-web tattoos is to pull the skin on the side of the tail tightly out to the side. This provides a wide, even surface to tattoo.

Neck Chains

There is not much discussion necessary with this identification system. It consists of a chain or cord around the neck with an identification tag attached to it. The tags can be made of metal, nylon, or plastic.

There is one decision that must be made. How strong should your chain be? At one extreme is the high-tensile chain that will never break. At the other is the plastic or nylon chain that breaks free each time the goat snags it, no matter how slightly. The best advice is to compromise. It should not come free at every tug, nor do you want to lose a goat because it becomes snagged on a branch or post and cannot break free. The chain should fit snugly with little or no free play when it is resting against the neck–chest junction. If young, growing kids are identified this way, they must be checked at weekly intervals to see if the chains have become too snug and need loosening.

Ear Notching

Ear notching is the approved method for marking Angora goats. There is no reason, other than an unwillingness to break tradition, that this system of identification would not work equally well on meat and dairy goats.

The system and procedure are identical to those used for sheep. Refer to Chapter 7 for details.

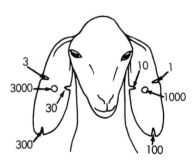

Branding

Branding is not yet a widely accepted method of identifying goats. However, there is no reason for this. The same techniques that are used on cattle and horses can be used, and just as effectively, on goats.

Freeze branding should be especially appealing to goat producers because it is painless, easy to apply, and provides a permanent, highly visible, within-herd recognition system. Refer to Chapter 2 for a detailed discussion of the procedure.

8.12 DEHORNING

Dehorning goats involves the removal of the horn tissue after the horn has erupted through the skin and protrudes in a recognizable form. This may involve a horn of from 1 to 3 inches in an immature animal or a rack of 10- to 12-inch horns in a mature buck.

It must be repeated here that if the decision to allow horns to develop or not is under your control, that decision should always be to disbud the kid at 3 to 5 days of age and to deodorize him at the same time. There is no reason to allow a milk- or meat-breed buck or doe to sport a set of horns.

If the youngster is disbudded properly, the pain and reduction in performance are minimal. When you must dehorn a mature breeding buck or milking doe, the pain is more substantial, a prolonged recovery period is involved, and milk production is reduced for varying lengths of time. The best advice is to disbud 3 to 5 days after birth. If that has not been done, dehorn as soon as possible thereafter. If you are stuck with a mature horned doe or buck, thoughtfully evaluate its value to your operation before undertaking the dehorning. If the doe is a poor specimen who produces little milk and breeds irregularly, sell her and buy another. If the buck is out of poor milk-producing stock, does nothing to improve your breeding program, or is mean, sell him. Consider using someone else's buck, using artificial insemination, or buying a good hornless buck.

Equipment Necessary

- Dehorning tools: hacksaw, wire saw, elastrator
- Forceps or hemostat
- Scissors or electric clipper
- Rope
- File or fine-toothed rasp
- Clotting powder
- Antibiotic (Combiotic)
- Tetanus antitoxin
- Tetanus toxoid
- Needle and syringe
- Bandaging material
- Antiseptic ointment
- Fly repellant

Restraint Required

To dehorn an older goat, the animal should be thrown (gently!) onto its side. This procedure is the same as throwing a mature ram or ewe (see Chapter 1).

Two capable assistants should hold the goat to the ground while you tie the hind legs together. Place a folded blanket or a couple of feed sacks under the goat's head to prevent injury and to present a better angle for dehorning. One handler should manage the abdomen and rear quarter while the other controls the neck, shoulder, and head.

If you are dehorning a young goat, perhaps only a couple of months old, this degree of restraint may not be required. Two assistants are still necessary, but tying is not. One assistant controls the youngster's body, the other immobilizes the head.

CAUTION: Keep in mind that this is a difficult task that does cause the animal to suffer some pain. A veterinarian can use a local or general anesthetic and perform the task painlessly and efficiently.

General Procedure

If you decide to dehorn the animal yourself, observe the following guidelines: (1) do not dehorn during or less than 30 days before the fly season, and (2) do not dehorn a pregnant doe—the stress of the procedure may cause her to resorb, or "slip" (abort) her kid. Expect the milk production of the doe to be reduced for a week to 10 days after dehorning.

Make a commonsense decision, based upon your experience, about which dehorning procedure will be used. This decision should involve the age of your goat, the size of the horn at the base, the tools you have on hand, and the help you can muster for the task.

Saw Dehorning

1. Assemble your tools and assistants. If any member of the crew is involved in dehorning for the first time, discuss with that person exactly what to expect and when to expect it.

2. Clip the hair from the base of the horn as closely as possible, creating a clipped ring approximately 1 inch wide all around the horn base. If your goat resists this, restrain him before completing the clipping.

3. Restrain the animal.

4. Saw off the horns, one at a time, along with about ¼" of the skin growing up from the base of the horn.

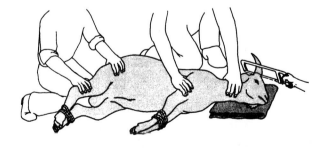

CAUTION: If you do not saw off this portion of the skin (and the skull underneath it), there is a strong probability that the horn will begin to grow back.

The inexperienced dehorner can remove too much skull along with the horn base. While sawing, check the orientation of both ends of the saw.

5. If your reason for dehorning is to shorten the horns so that risks of injury are reduced or day-to-day management practices are made easier, take the easy way out—for you and your goats—and simply dub the horns off. The entire process is the same, except that a 1- or 2-inch horn stub is left attached to the skull.

6. While still restrained, the animal should be given injections of tetanus antitoxin and antibiotics.

> **CAUTION:** It is of no use and may actually be detrimental to give only one injection of antibiotics with no follow-up therapy. It is best to repeat the injection twice, making a total of three injections, over the 36 to 48 hours after dehorning.

7. Carefully examine the horn stubs for continued bleeding. If you sawed the horns off very close to the skull, there will be arterial bleeding which you will see as a stream of spurting blood. Use the forceps or hemostat to grasp the vessel from which the blood is spurting and pull it outward. It will break, leaving the end down in the horn stub below your field of vision. The spurting will now be reduced to a welling-up of a pool of blood within the horn stub. This will cause a clot to form and stop the bleeding.

8. Sprinkle a liberal amount of clotting powder upon the horn stubs.

9. Bandage the horn cavities or horn stubs. Use a 4" × 4" nonadhering pad dressed with an antiseptic ointment over each horn site. Wrap with 3- or 4-inch-wide gauze directly over this. The wrapping should go over the horn sites, around the head between ear and eye, under the jaw, and then back over the top again. To help hold them in position, some wrappings should be placed behind the ear.

> **CAUTION:** Avoid the tendency to overtighten the bandage. It is possible to wrap so tightly that swelling could occur and make breathing difficult.
>
> Do not omit this step. It is important that dust, chaff, and dirt stay out of the open horn cavity. The bandage can be removed after 5 or 6 days.

Elastrator Dehorning

This method uses a very strong ring or band of rubber to dehorn the goat. It can be used on any size horn and on any age animal. Due to the nature of the goat horn, it works best on mature or nearly mature horns. It has a tendency to slip from a young, smooth, tapered horn.

1. Assemble your tools and assistants. If any member of the crew has not handled goats before, discuss with that person exactly what to expect during the technique.

2. Restrain the animal. (If your doe or buck will accept the handling described in steps 3 through 6 without throwing and tying, there is no need to do so. There is no pain involved in these steps, so strong restraint measures may be unwarranted.)

> **CAUTION:** If you attempt this procedure with minimal restraint, beware of the horns. They are the goat's defensive weapons, and a quick lunge, twist, or thrust could injure you or one of your assistants.

3. File a groove or notch about ⅛ inch deep and ⅛ inch wide completely around the base of each horn. This should be as close to the skull as possible without filing into the flesh of the scalp.

4. Use the elastrator to stretch a rubber ring, and place it over the goat's horn. Release the trigger of the elastrator when the rubber ring is over the groove you filed into the horn. If the ring is not placed exactly where you want it (into the groove), move it with your fingers until it is. These rings are tight, and moving them may take some concerted effort. Repeat the procedure on the other horn.

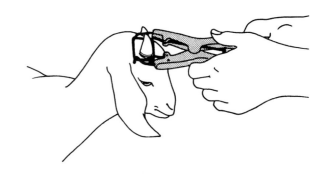

5. If the groove is not very deep and it appears that it will not hold the rubber ring securely, place several wraps of adhesive tape onto the horn directly above the band. They will prevent the rubber ring from riding upward on the horn. This is also necessary when working with the smooth, tapered horn of a young animal.

6. Administer an injection of tetanus toxoid at this time. It will take 4 to 6 weeks for the maximum immunizing effect of this injection to develop. This interval coincides with the time that the horn finally sloughs off and an open wound exists.

7. Observe the horns periodically during the elastrator dehorning process. It will take 4 to 8 weeks for the horns to slough off, and the rings may need replacing (after being broken or rubbed off) one or more times.

8. The stub will bleed if the horn is prematurely sloughed or rubbed off. When this occurs, apply a clotting agent.

CAUTION: The goats will rub their heads and horns against any available object in an attempt to rid themselves of the rubber ring aggravation. They should be checked periodically to make sure that they do not become entangled in fences.

The total dehorning process (4 to 8 weeks) should be completed before fly season. If the fly season does sneak up on you, apply a fly repellant to the area of the rubber ring.

Scur Removal

Horned Goats. When scurs develop on a previously dehorned goat, it means that a complete job was not performed in the initial dehorning process and that part of the growth plate or horn root was left intact. The resulting regrowth is an ugly, misshapen, dangerous horn mass that should be removed.

The step-by-step procedure for removing the scurs from an adult goat is the same as that for dehorning goats. It should be noted, however, that the time to handle the problem is when the horn scurs *start* to develop on a young kid. If a hot iron is taken to the scur when it is first noticed, the problem will develop no further. The step-by-step procedure for hot-iron scur treatment is discussed in Section 8.8.

Naturally Polled Goats. The scurs that may develop in the horn area of a polled goat are, of course, not the result of an inadequate dehorning. They are nothing more than an outgrowth of the goat's skin,

very similar to the chestnuts or "night eyes" on a horse's legs. They can be easily distinguished from horn-regrowth scurs, because skin scurs are not attached to the skull. Skin scurs can be moved about (as you can move the skin over your own knuckles), while horn-regrowth scurs are as rigidly attached as the original horn.

Skin scurs will often wear away and their size be kept to a minimum as the goat browses or rubs itself. If, for appearance's sake, you wish to remove these, use a sharp knife and carefully pare them away. They can be shaved down until ⅛ inch or less remains without causing them to bleed. If you should cut too deeply, the scur area will bleed, but no worse than would normal skin if you nicked it with a cutting instrument. A scur area of this sort will continue to produce the horny, scaly growth during the goat's entire life.

Postprocedural Management

Goats of any age who have had their horns sawed off should be penned for several hours after dehorning. This is to make sure that your measures to stop the bleeding have been successful. If they have not, you must repeat them, and if again unsuccessful, summon the veterinarian.

Goats chill and succumb to physiological shock very easily. The holding pen should be well bedded and free of drafts.

If the fly season should arrive ahead of schedule, you must treat any remaining scabs or open wounds with an antiseptic and fly repellant.

Remove any bandaging after 5 or 6 days. Take care when removing it that any bandaging stuck to the wound does not remove the scab that may have started to develop. If the dressing has stuck to the wound, apply a coating of light mineral oil and massage gently until it is free. Treat the wound site with antiseptic ointment and fly repellant if necessary, and leave it unbandaged.

8.13 FOOT TRIMMING

Growth of the goat's hooves is a normal and continuous process. In the wild state, the goat usually searches out the most brushy, harshest, and rockiest landscape for his home range. This environment keeps the hoof growth worn away and there is no need for human intervention. In a domesticated flock, managed in drylot or on a minimum of range, the new hoof growth cannot be worn away, and the goatherd periodically must trim the feet of his flock. This procedure is usually repeated every 6 to 8 weeks.

Trimming the feet of the goat involves removing the excess growth that occurs at the toes, heels, and sidewalls of the hooves. This improves the appearance of the goat. Nothing detracts more quickly from the eye appeal of any hooved animal than a set of neglected, overgrown hooves. More important considerations for hoof trimming involve improved mobility and comfort for the animal, the leg-straightening effect it can have when properly done on 2- to 3-month-old animals, and a reduction in the incidence of foot rot.

Equipment Necessary

- Foot-trimming shears
- Pocketknife
- Hoof rasp
- Grooming stand, halter, or collar
- Rags
- Iodine or other foot-rot remedy
- Leather gloves

Restraint Required

When having their feet trimmed, goats are usually restrained by controlling their heads. This can be done by placing them on a grooming stand that has a built-in stanchion, by haltering or collaring them and then tying them to a post or rail, or by having an assistant present to hold the haltered or collared animal. Alternative methods of restraint include the use of a tilting squeeze chute and throwing and tying the goat. The latter method is usually reserved for nervous or ornery individuals that will not stand for footwork.

Step-by-Step Procedure

1. Assemble the necessary equipment.

2. Restrain the goat in the selected manner.

3. The goat will resist foot trimming less if you begin the task with a front leg. Stand at the goat's shoulder and face toward the rear. Bend over and grasp the lower leg near the fetlock or pastern. Bend this leg backward and upward so that it is flexed at the knee.

CAUTION: It will be uncomfortable for the goat if you bend the knee beyond a 90° angle. If you do not overbend the knee, however, it is hard on your back.

Even if you are careful not to overbend the knee, the goat may attempt to pull the foot away or to move entirely away from you. You should retain your grip on her leg, and on your patience, and wait a moment or two until she settles down. Do not attempt to hold the leg perfectly still.

4. Use the point of the knife or foot-trimming shears to clean out the hoof. Pay particular attention to the area between the two toes and to the areas *under* any overgrown hoof wall. If the feet are muddy, use the rag to wipe them dry before beginning to clean them.

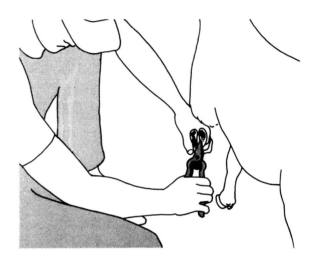

5. Trim any excess rim (folded-over hoof wall) from each of the toes with the shears or pocketknife. At this step, cut the sidewalls down to a point even with the bearing surface, or sole (frog), of the foot.

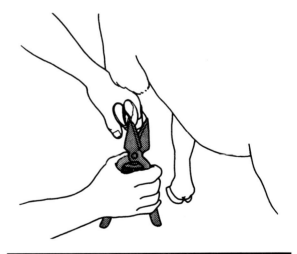

CAUTION: The goat may attempt to jerk her leg away or to jump free when you least expect it. Control the knife or shears at all times so that you or the goat are not injured. When you use the knife, minimize the number of cuts made by pulling the knife toward you. A pair of medium-weight leather gloves will be further insurance against accidental cuts or scrapes.

There will be less chance of injury to yourself or to the goat if you use the shears instead of the knife. This is especially true if the hooves are dry, hard, and brittle. If you use ordinary pruning shears, be certain to keep the cutting blade on the inside of the hoof, and cut outward toward the sidewall. If you do not, you will inadvertently cut too deeply into the sole.

6. Level the bearing surface of the foot. When you are finished, the entire sole of the foot should be flat on the ground, and the hairline of the pastern above the hoof wall should be parallel to the ground.

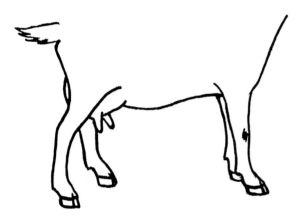

As you begin to level the foot, you will note that there is more growth at the heel and toe than in the center or frog area of each hoof half (toe). Start your leveling at these high points. Do not remove any growth from the center area of the hoof half until the heels and toes are cut down level with it. Make your cuts in thin slivers. Take two or three thin slices from the heel, then two or three from the toe. Keep repeating these alternating cuts until the

slices show a slight pinkish hue instead of the grayish white color of the insensitive hoof. The toes will usually need to have more removed than the heels. At all times, keep in mind the shape that you want the hoof to have when you are finished. If the feet are dry and hard, a hoof rasp or carpenter's Surfoam® is a better choice for removing excess growth from the heel and toe.

CAUTION: When you arrive at the pink tissue, you are immediately above a source of blood and sensitive tissue. Stop! If you don't like the shape of the hoof at this point, wait 30 days and try again to shape it to your ideal.

If you happen to cut too deeply and draw blood, don't panic. You are not the first person to accidentally do this. Help to stop the bleeding with pressure from your fingers. Disinfect the wound with iodine and bandage the hoof to prevent dirt and manure from entering. Give the goat an injection of tetanus antitoxin. Keep the animal in a small paddock for a day or two until you are certain an infection has not developed. A worsening limp and fever in the affected foot are symptoms of infection.

7. Move to the other front leg and repeat the same procedures. When you are finished, the two front feet should be as nearly identical as possible.

8. Move to the goat's rear (you should still be facing rearward), grasp the pastern of a rear leg, and position it over your knee or calf. Your exact

position will depend upon whether you have the goat restrained on a grooming stand or tied to a post. The procedure for trimming the rear feet is the same as that for trimming the front feet.

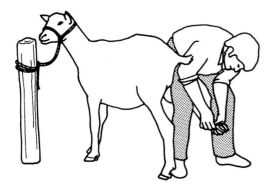

9. Therapeutic, corrective foot trimming can be performed on some conformationally imperfect animals. This is especially true of young animals whose bones are still growing. In a dairy goat, the most likely conformational defect is a "cow-hocked" condition. In this defect, the points of the hocks turn inward and are closer together than are the fetlocks. If you think about this for a moment, you will realize that the animal must also be splay-footed for the condition to occur.

To correct the cow-hocked condition, you must correct the splayfooted condition. Trim the outside toe of each rear hoof shorter than the inside toe. This will allow the hoof and leg to turn inward and cause the hock to roll outward.

CAUTION: If the feet of the kids are badly affected, this correction should take place over several trimmings. The same caution applies when changing the hoof angle by selectively lowering the heels or toes of each hoof.

8.14 TREATMENT OF FOOT ROT

Contagious foot rot is a troublesome disease plaguing the goat industry. The disease is as old as the art of herding goats. The industry has never been without goats infected with foot rot and likely never will be. The condition is currently a production problem in many areas of the United States.

Foot rot is an economically important disease of goats, though it seldom causes death. Losses are incurred through the reduction of growth and gains in kids, reduced milk production of the does, and the impaired grazing ability of older animals whenever the outbreaks occur. Goat owners know that it is far easier to prevent the herd from getting the disease than it is to control or eradicate it once the animals have become infected.

Foot rot is caused by an interaction between two or more types of anaerobic bacteria (bacteria that grow in the absence of oxygen). The interaction of these bacteria destroys the connective tissues between the hoof wall and the sole of the foot, causing them to separate and allowing a pronounced lameness to develop.

For animals in the flock to start showing symptoms of foot rot, it is necessary that the skin between the toes become wet, bruised, scratched, and infected by the foot-rot bacteria. When the bacteria become established in the deeper layers of the skin, they produce an enzyme that liquefies tissue protein around the infection site. The infection separates the heel, sole, and wall of the hoof from their attachments to the foot, causing inflammation, lameness, and the characteristic odor associated with foot rot.

Foot rot in goats is not the same disease that affects cattle; it is caused by a different organism, causes different types of lesions, and responds to different types of treatments. Proper treatment of foot rot in goats requires repeated examinations and observations, proper foot trimming, proper use of disinfectants, isolation of infected animals, and removal of carriers from the herd.

Equipment Necessary

- Foot-trimming shears
- Pocketknife
- Hoof rasp
- Grooming stand, halter, or collar
- Rags
- Iodine or other foot-rot remedy
- Leather gloves
- Zinc sulfate solution

Step-by-Step Procedure

1. Catch and restrain the goat in the manner described in Section 8.13.

2. Foot rot is characterized by soft spots in the hoof wall, separation of the outer wall of the hoof from the sole, and inflammation in these areas. It may also appear as an ulcer between the toes. The foot may be enlarged and feel hot at the junction of the skin and the hoof. When excess amounts of moist, soft, or dead tissue are trimmed away, the foot will have a strong, foul odor characteristic of foot rot. Exposure of pockets of infection to the air has a detrimental effect on the growth of the organisms that cause the condition.

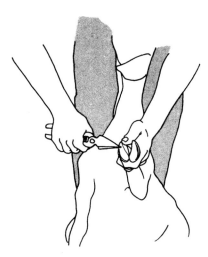

3. Trim the feet of the goat as in routine foot trimming. See Section 8.13.

To maximize the effect of whatever medication you select to treat foot rot, the foot must be thoroughly trimmed to expose the infected tissue. Diseased, necrotic (dead), and undermined hoof areas must be pared away to allow medication and air to reach the infective bacteria. Topical medications should be sprayed or painted on immediately after trimming.

CAUTION: Foot trimmers should be cleaned thoroughly and allowed to soak in a disinfectant bath after trimming an infected foot.

---

4. After the feet have been trimmed, they should be treated either individually or in a footbath. Individual treatment is usually done when only a few goats in a herd are affected. It consists of submersing the foot in a container of disinfectant, pouring the solution over the foot, or painting the solution onto the hoof. Ten percent zinc sulfate in water, 10% copper sulfate in vinegar, or two parts copper sulfate in one part pine tar have been shown to be the most effective topical treatments. Antibiotics in alcohol solution have also been reported to be from 50 to 90% effective where feet were thoroughly trimmed, medication was applied once, and goats were held in a dry area. The above medications (except the pine tar–copper sulfate mixture, which must be painted on) can be applied with a hand sprayer. Each infected foot will require 5 to 10 ml.

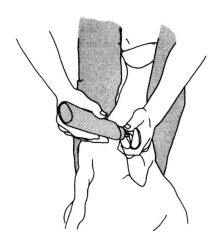

5. Footbaths are used when a large number of animals or whole herds are treated. The footbath should be 12 to 16 feet long, 8 inches wide at the bottom, 12 inches wide at the top, and 6 inches deep.

Walk-through footbaths containing a 10% zinc sulfate or 10% copper sulfate solution, used every fifth to seventh day, will greatly reduce the spread of foot rot to normal animals. Enforced daily self-footbathing with 10% zinc sulfate also reduces the spread and will help most infected animals recover. The footbath will not cure carriers or those with severe deterioration of the hoof unless their feet are carefully trimmed. Ten percent zinc sulfate in lime, mixed dry and kept dry, can be placed in a box between feed and water to reduce spread of foot rot. This is particularly useful during freezing weather, when a solution would freeze. This method is not effective for treatment of animals already infected. The footbath can be placed in a gate where goats are forced to walk through it daily, between their food and water, or it can be placed in a cutting-and-handling chute to slow their movement hrough it.

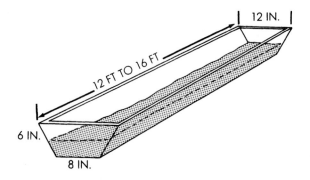

CAUTION: Copper sulfate solution is quickly inactivated by organic material. It should never be used in metal containers or footbaths because the copper will plate onto the metal and reduce the strength of the solution. Only enamel, wood, glass, or epoxy-painted metal should be used as a container for copper sulfate.

1. The goats infected with foot rot should be separated from the rest of the herd, placed in a pen or drylot, and treated individually until completely cured. They can then be returned to the healthy herd.

2. Trim the feet of all goats and reexamine them for foot rot at least every 30 days.

3. Do not force goats to use muddy lots, lanes, or pastures for long periods of time.

4. Keep areas around feed bunks and waterers as clean and dry as possible.

5. Do not use cinders, fine gravel, or finely crushed stone as footings for goats. These materials tend to injure the feet and provide avenues for bacterial invasion of the hoof.

6. Never buy goats from a herd that is known to be infected with foot rot. Don't even purchase apparently clean goats from a herd with this reputation. It is also a high-risk venture to buy goats at sale yards where clean and infected goats are grouped together, even temporarily.

7. Insist on proper cleaning and disinfection of trucks and trailers prior to transporting goats either to or from your operation. Clean trucks and trailers coming to your farm or ranch will help prevent the introduction of foot-rot organisms to your place. Clean transport leaving with a load of your goats will protect your buyers and your reputation.

8. Assume all new additions to your herd are infected with foot rot and arrange for a quarantine area that will be available for at least 14 days, with 21 days being a better recommendation if it is possible. Always trim all feet immediately upon arrival, treat the feet of all goats following trimming, and reexamine periodically during the quarantine period.

9. Avoid common-use trails and corrals where infected goats have traveled or have been penned during the preceding 2 weeks.

10. The use of systemic antibiotics or antibiotics injected into the foot is expensive and of limited success in controlling the spread of foot rot.

11. This disease can be best controlled by using a variety of combinations of all of the treatments discussed here.

Eradication will require the use of all of the methods outlined. In addition, it will require commitment, perseverance, and a willingness to cull animals that do not respond. An initial increased expense in labor and working facilities will be required, but this will pay off many times over in the long run. To eradicate foot rot, it will be essential to have quarantine facilities for keeping new animals isolated and a "sick pen" to keep infected animals isolated, and to cull animals with recurring foot-rot infection.

8.15 WATTLE REMOVAL

Wattles are globules or 2-inch pendants of skin that are frequently found hanging from the sides of the necks of goats. They are nonfunctional and are thought to be remnants of gill slits that all mammals shared somewhere back down the evolutionary tree. They are found on both sexes and on all breeds.

They are a frequent source of injury and site of infection. Kids will suck on them, causing an irritation and possible infection. They should be removed whenever they are large enough to be easily managed. Because of the ease with which they are removed, this procedure is usually coupled with another, such as worming or foot trimming.

Equipment Necessary

- Surgical scissors
- Fly repellant
- Antiseptic

Restraint Required

The head and upper neck region must be restrained (motion-free) for 5 to 10 seconds while the wattle is snipped from the skin. An additional 15 seconds of restraint is required for disinfecting, but this need not be motion-free.

If the animal is young, an assistant can hold it in his arms. Older animals can be held in a stanchion, pinned into a corner, or thrown and held to the ground.

Step-by-Step Procedure

1. Restrain the animal in the appropriate manner.

2. Thoroughly disinfect the skin area surrounding the site at which the wattle is attached.

3. Disinfect the blades of the scissors and, while holding the wattle outward from the animal, snip it free at its "neck." This is the thinnest point of the wattle and is usually found at its point of attachment to the skin.

4. Disinfect the removal site with an antiseptic. Treat with a wound-safe fly repellant only if flies are a serious problem.

Postprocedural Management

The procedure is so simple and nontraumatic to the animal that observation, "just to be sure," is the only follow-up procedure necessary. If the wattle is clipped free at its thinnest point, usually not even a drop of blood will ooze out.

8.16 AGE DETERMINATION

The ages of goats from about 3 months to 4 years old can be determined by observing the changes that take place in the incisor teeth on their lower jaws. Certainly, this method is only an approximation, but it is necessary when the exact date of birth is unknown and the age must be determined for purchasing breeding does, for checking the ages of show or sale animals, and for culling older animals or those with malformations of their teeth or jaws.

The change in teeth after 4 years of age is a function of wear rather than eruption of new teeth. Breeds of goats will vary in the wear patterns of their teeth; and the management practices, nutritional plane, and environmental conditions the animals are maintained in will also cause differences in wear patterns after 4 years of age. Still, estimating the age of older sheep, by evaluating the wear patterns of the incisors, is a skill worth learning, and at times it is absolutely essential.

Equipment Necessary

None.

Restraint Required

Hold the animal by the head and part the lips to expose the incisor teeth for examination. Use the minimal amount of restraint necessary. Usually, all that is necessary is to halter the animal or simply hold it by the head and chin.

Step-by-Step Procedure

1. A kid has eight temporary incisors. At birth, the central temporary incisor and often the second temporary incisor on each side have erupted. The third temporary incisor erupts at about 3 months. The fourth and last temporary incisors normally erupt within a month after the third.

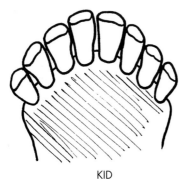

KID

2. At approximately 15 months of age, the two central temporary incisors are replaced by two wider, larger, permanent teeth. Actual timing of eruption may vary by 2 months on either side of the 15-month average.

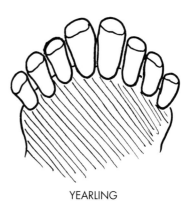

YEARLING

3. At 24 months of age, the second set of temporary incisors is replaced by the permanent incisors. The mouth now has four permanent teeth in its center.

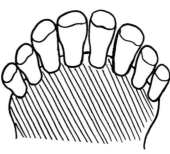

TWO-YEAR-OLD

4. At 3 years of age, the third pair of temporary incisors, one on each side of the four permanent teeth in the center of the mouth, are replaced by permanent teeth. There are now six permanent teeth in the center of the mouth and two temporary incisors on the corners.

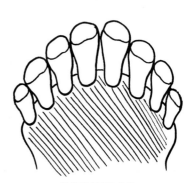

THREE-YEAR-OLD

5. At 4 years of age, all the temporary incisor teeth have been replaced by the permanent incisors. There are now eight permanent incisors in the mouth. The goat is said to have a "full mouth" at this time.

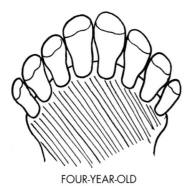

FOUR-YEAR-OLD

6. As the animal ages beyond 4 years, the teeth appear to be longer because of receding gums. The bearing surfaces of the incisors also begin to show wear after 4 years of age.

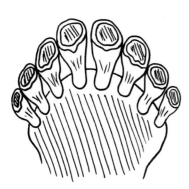

7. In addition to the incisors, the goat has 12 molars in the upper jaw and 8 molars in the lower jaw. There are no incisors in the upper mouth of the goat. Instead, there is a toughened area, a horny plate, in the gum area where the upper incisors should have been.

8.17 MAINTAINING HERD HEALTH

Goats by their very nature are healthy and hardy animals, and they are relatively easy to maintain that way in spite of the rather harsh demands placed upon them by today's production systems. While these modern production systems do provide us with out-of-season breeding, 300-day lactations of 6,000+ pounds of milk, milking herds of 600 to 700 does on 20 acres, and wethers capable of growing at 0.8 to 0.9 pounds per day, they are not without their penalties. The tampering with hormone cycles, the increased nutritional loads, the stress of total confinement—all of these can be predispositions to "dis-ease" problems.

No single production stress, or any combination thereof, can cause a breakdown in the goat's normal defense mechanisms as long as we bolster them with herd health management techniques. In this section, we discuss the broad technique of maintaining a healthy herd, which includes the mini-techniques of observing animals for signs of "dis-ease," taking temperature, pulse, and respiration, routine injections, drenching, dosing, blousing, worming, controlling bloat, and treating mastitis.

Equipment Necessary

- Pocket record book
- Thermometer, with string and clip
- Alcohol
- Injectables—as appropriate to the production unit
- Syringes and needles
- Drench bottle
- Dosing syringe
- Watch with sweep-second hand
- Balling gun
- Paste
- Deworming gun
- Deworming compounds

Restraint Required

Daily, routine observation of each animal in your herd is a technique in its own right. To properly observe animals in this manner requires no extra restraint. In fact, if you impose an artificial restraint upon them, your observations cease to be routine. When it is time to employ a physical technique such as drenching, taking temperatures, or deworming, the animals must be restrained in stanchions or crowded into a corner of a pen. See Chapter 1 and the individual techniques that follow for these restraint procedures.

Step-by-Step Procedures

Observing the Animal

1. On each and every day, as you go through the routine of daily chores, the following indicators of your goats' state of health should be observed and mental notes made of their condition: hair coat, skin condition, amount of fat over the ribs and vertebrae, grain and roughage consumption, cud chewing, water consumption, consistency of the fecal pellets, level of milk production, disposition or temperament, presence of bloat, presence of swollen extremities, peculiar posture, lameness, cough, reluctance to move, paleness of gums or conjunctiva of the eye, and any change in the normal behavior or activity patterns.

The list is long, and at first it does take an effort to consciously observe each indicator and relate it to the way it looked yesterday and the way it looked a week or a month ago. Keep in mind, however, that the lists of diseases and disease symptoms are even longer and that a whole lot more effort is required to cure a sick goat than to prevent a healthy one from becoming diseased.

2. There is no practical way for you to establish standards for each indicator, and fortunately there is no need to do so. Your task is to observe change. At first, you will be able to notice only the obvious, quick-changing indicators such as a drop in milk production, loss of appetite, or a change in consistency of feces. This should not concern you. As your experience broadens and your powers of observation grow, the little, subtly changing indicators such as hair coat, skin condition, and fat cover can be picked up.

3. When you do notice a change in a health indicator, the affected animal or animals must be given assistance immediately. Some indicator changes warrant immediate veterinary attention if the goat's life, or at least its productive life, is to be saved. Other problems can be solved by adjusting certain aspects of your management program.

CAUTION: There is no way to say, for your unique management system, which indicators are most important and which changes need professional attention. Until your experience in *managing* goat health problems is increased, content yourself with the perhaps more important task of *preventing* goat health problems.

Taking Temperature, Pulse, and Respiration

The ability to accurately measure the body temperature, the rate of heartbeat, and the breathing rate and depth is a technique that must be learned if you are going to prevent a serious outbreak of diseases in your production unit. A change in any one of these is an indicator of an impending problem. The measurement of them is the logical and immediate second step to the observation that something is wrong with the overall well-being of the individual or herd.

1. The goat's temperature is taken by placing a well-lubricated thermometer into its rectum. Use petroleum jelly or mineral oil as the lubricant. It is important for the goat's comfort that the thermometer be rotated while it is being inserted. Insert the thermometer until only 1 inch of it remains

visible outside the rectum. Attaching a string to the ring at the end of the thermometer and then using a clip at the other end of the string for attachment to the tail hairs will prevent the thermometer from being expelled and broken on the ground or sucked into the rectum of the goat.

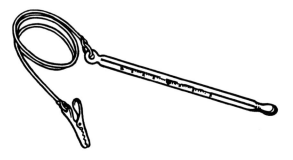

Remove the thermometer with a twisting-pulling motion after 3 to 4 minutes. Wipe it clean, read and record the temperature, shake it down to below 96°F, and clean it with alcohol for the next use. The normal body temperature of a mature goat is 102.5 to 103.0°F.

2. The pulse of the goat can be felt by placing the index and middle fingers of one hand upon a major artery wherever it crosses a bone or large muscle. In the goat, this is most easily done by using the artery just below and slightly to the inside of the edge of the jaw, about two-thirds of the way rearward from the muzzle. There are palpable arteries also under the goat's tail and behind the knees.

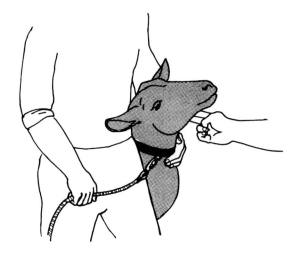

Pulse rate is recorded in beats per minute. Using your watch's sweep-second hand, count the beats for 30 seconds and multiply by 2. If it is too difficult to maintain a constant finger contact for 30 seconds, hold for only 15 seconds and multiply by 4. The normal pulse rate of a mature goat is 70 to 80 beats per minute.

> **CAUTION:** Avoid using your thumb to measure the pulse or to help control the animal. It has a pulse of its own, and if it touches the animal you could confuse the pulse rates.
>
> Do not measure the pulse of the goat immediately after chasing him or casting him to the ground. The excitement could cause a greater elevation than any suspected disease.

3. The rate of respiration (breathing) can best be measured by watching the movement of the goat's flank. If you are to make an accurate count, the goat cannot be moving around and masking the flank movements. For that reason, many recommendations will direct you to make these counts while the goat is lying down. This is not necessary if the goat is in the milking stanchion or resting quietly while standing.

Observe the flank from either side of the goat. Consider each rise (inspiration) and fall (exhalation) of the flank as one complete respiratory movement. The respiration rate is recorded in breaths or respirations per minute. Count for 30 seconds and multiply by 2, or for 15 seconds and multiply by 4. The normal respiration rate is 21 to 27 breaths per minute for a mature goat.

Routine Injections

Refer to Chapter 10 for a step-by-step explanation for giving injections. It is the purpose of this section to point out to you that there are injectable "tools"

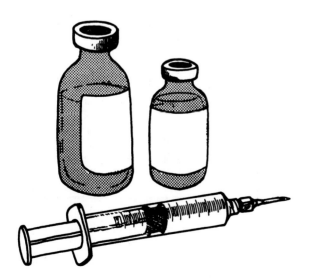

that you should use in maintaining the health and production levels of your herd. The proper use of these tools is no less a technique that needs learning than is the use of the hot iron at disbudding and deodorizing time.

You must come to realize that the use of these injectables is no guarantee of success. Indeed, they can be misused and cause a large measure of harm. Only your experience and consultation with a know-ledgeable professional can tell you when their use is warranted. The following injectable products can be used as aids in managing the health of your goats.

Clostridium Perfringens, Types C and D, Toxoid. Used as an aid in preventing enterotoxemia (overeating disease).

Vitamin Supplement. Includes vitamins A, D, and E.

Selenium-Vitamin E Supplement. Used as an aid in preventing neonatal losses of kids.

Vaccines, Bacterins, and Toxoids. These are specific to given diseases. Some of them are standard and available in any part of the country; others are developed as an aid in combating local combinations of microorganisms.

Antibiotics. Use only as warranted by the severity of the situation. Indiscriminate use can result in a buildup of drug resistance.

Today, most injections can be given subcutaneously.

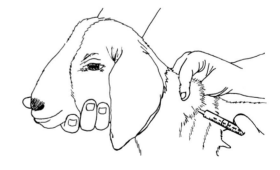

Drenching and Dosing

Drenching and dosing are two methods of administering moderate amounts of liquids or emulsions to the goat. For dosages of up to 8 or even 10 ounces, it is quicker, easier for you, and less uncomfortable for the goat than using the stomach tube. Mineral oil, kaolin–pectin combinations, anthelmintics, and antacid preparations can be given in this manner.

1. Back the goat into a corner and crowd its left side against a wall. Encircle its upper neck with your left arm. It helps if there is an assistant available to help hold the goat into position along the wall.

2. Insert the thumb of your left hand into the goat's mouth, and with a combination of prying and pushing, open its mouth wide enough to allow you to insert the neck of the drench (soda) bottle.

CAUTION: Goats have sharp teeth and strong jaws. If you position your thumb or fingers between the premolar and molar grinding teeth, they will be pinched and cut.

3. Insert the neck of the drenching bottle into the corner of the goat's mouth. It is only necessary to place 1½" to 2" of the bottle into the mouth. At the same time that you are inserting the bottle, lift or tilt the goat's head so that it is positioned as if the goat were looking upward at a 45° angle.

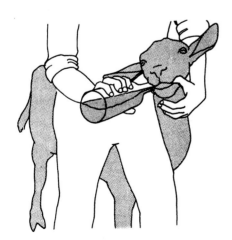

4. Pour the liquid slowly onto the rear of the tongue and allow the goat to swallow it.

CAUTION: Do not pour too rapidly. You may cause the goat to gasp for air and aspirate some liquid into its lungs.

Some producers feel that if the mouth of the bottle is placed against the roof of the mouth, the goat will close its windpipe in a reflex reaction. This would prevent fluid from entering the lungs.

5. Using the dose syringe requires exactly the same techniques of restraining, head positioning, and slow administration of drug as does drenching. The only difference is that the 2"- to 4"-tip of the dose syringe, rather than the neck of the bottle, is placed into the corner of the goat's mouth.

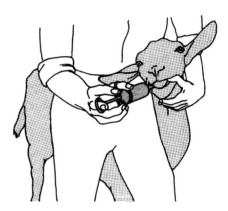

Bolusing

Sometimes referred to as "balling," this technique involves using a bolus gun or balling gun to place a tablet or bolus on the rear of the goat's tongue. Several popular anthelmintics are dispensed in bolus form. The size and shape of the bolus or tablet will vary. For easiest administration, the bolus should fit snugly (without falling out) into the head of the balling gun. Balling guns can be purchased with extra heads so that a bolus of almost any size can be used.

1. Back the goat into a corner and crowd its left side against a wall. Encircle its upper neck with your left arm. It helps if there is an assistant available to help to hold the goat in position along the wall.

2. Insert the thumb of your left hand into the goat's mouth, and with a combination of prying and pushing, open its mouth wide enough to allow you to insert the head of the balling gun.

CAUTION: Goats have sharp teeth and strong jaws. If you position your thumb or fingers between the premolar and molar grinding teeth, they will be pinched and cut.

3. Thoroughly coat the bolus with petroleum jelly or lubricate it with a thick mineral oil or molasses before you place it into the head of the gun. This will help the animal to swallow the otherwise dry, large bolus.

4. Insert the balling gun into the corner of the goat's mouth. It is necessary to insert enough of the gun into the mouth that the bolus can be deposited back near the base of the tongue. At the same time that you are inserting the gun, lift or tilt the head of the goat so that it is positioned as though she were looking upward at a 45° angle.

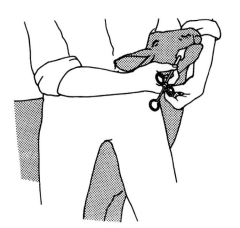

5. When the head of the gun is in position in the goat's mouth, fully depress the plunger. This will dislodge the bolus. Remove the gun, hold the goat's mouth closed, keeping the jaw elevated to 45°, and stroke the neck from jaw to chest floor. This will encourage the goat to swallow the bolus.

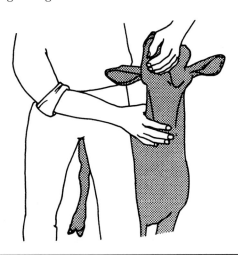

CAUTION: While the bolus should be placed as far back on the tongue as possible, care should be taken to avoid shoving the head of the gun back into the pharynx of the goat. This is unnecessary, painful, and may result in your placing the bolus into the windpipe of the goat, which could cause suffocation.

6. Observe the goat for several seconds to be certain that she has swallowed the bolus and does not spit it out as soon as she is released. If this should happen, catch her again and readminister the same bolus (or pieces of bolus, if she has managed to smash it with her teeth by this time).

Worming

Goats should be wormed on a routine schedule. There are several good anthelmintics available, and the wise producer makes use of all of them on a rotating basis. Continual use of only one drug results in the parasites becoming resistant to it.

An anthelmintic can be administered as a drench from a bottle or dose syringe, as a bolus or tablets from a balling gun, as a powder or granules sprinkled on the feed, or as a paste delivered into the mouth by a caulking-gun-like applicator. Drenching, dosing, and bolusing are discussed earlier in this section. The following technique is used to administer paste wormers.

1. Paste wormers have the consistency of peanut butter and are packaged in a tube $1\frac{1}{2}$" to 2" in diameter by 10" to 12" long. The animal end of the tube has a $\frac{1}{2}$" × 3" tip for insertion into the mouth. The operator end of the tube is screwed into an applicator gun equipped with a plunger and squeeze handle. As the handle is squeezed, the plunger enters the operator end of the tube and forces out a precalibrated amount of wormer from the tip. Releasing the handle recocks it for the next dose, or part thereof, depending upon animal size and age and upon the manufacturer's directions.

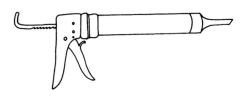

2. The animal is restrained, held, and positioned in exactly the same manner as for bolusing, dosing, and drenching.

3. Insert the tip of the applicator into the corner of the goat's mouth. Point the tip to the rear and upward toward the roof of the mouth. It is not necessary to force the tip up into the roof or back into the pharynx—just aim it in the general direction.

4. While holding the goat's head slightly elevated, administer the appropriate number of clicks of worming drug. It is not necessary to tilt the head as sharply in this technique as in dosing or drenching. The paste adheres to the roof of the mouth, and as the goat tries to tongue it out, she partially dissolves it and inadvertently swallows it.

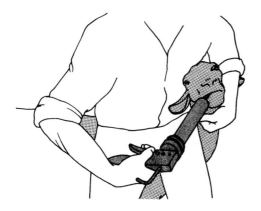

5. After the applicator is removed, keep the goat's head up for 10 to 15 seconds to assure that the wormer is not lost from the mouth.

Bloat Control and Mastitis Treatment

These techniques are covered in detailed step-by-step sequences in separate sections of Chapters 3 and 4.

Postprocedural Management

Drenching, dosing, bolusing, or "pasting" in themselves require no special handling after the techniques are performed, but the drug manufacturer's recommendations must be followed postprocedurally. These will vary with the drug. Be certain to read the directions.

8.18 HARNESSING YOUR GOAT

Clearly, harnessing your goat, teaching it to pull a cart or a travois, is not a necessary goat management skill. It is, however, a wonderful hobby, an excellent pastime, for you and your family. You will find the goat to be an apt pupil. In a very short time, you will be aboard the cart with Nanny, Billy, or "Ole Wether" stepping out smartly in front.

Equipment Necessary

- Halter
- Lead rope
- Surcingle
- Long lines
- Travois
- Goat harness
- Cart or wagon
- Driving whip

Restraint Required

There is no real restraint, per se, involved with harnessing your goat. However, it is necessary for the goat to know how to lead, to be willing to respond to your voice, and to be used to standing tied.

Step-by-Step Procedure

1. Teach the goat to stand tied. Place a strong nylon web halter onto the goat and use a strong 5/8" lead rope with a stout bull snap attached to it. Tie the goat, at about head height, to a post or small tree that will not break. Allow 2 or 3 feet of rope between the halter and the post. Stand back, relax with a cup of coffee, and allow the goat to find out for himself that it is futile to fight the situation. Don't be alarmed if the goat pitches a large fit. The whole idea is for him to teach himself that all the fighting will only serve to wear himself out. Goats are smart, and it will take only a few sessions for the goat to learn what is expected of him.

2. Teach the goat to lead. Using the same halter and lead rope, start out by simply trying to take the goat for a walk. This will work best with the younger, smaller goat. If the goat is confused or reluctant to follow you, gentle tugs, some snacks, and verbal encouragement will often cause the goat to move out and follow your lead.

3. If you are fortunate, this will be all that is necessary. However, the goat might not respond. It may drag its feet, pull back on the lead, and act as if it will never learn to follow you. This is where the smaller goat is a good idea. Pull a little, stop a little. Try short tugs, try steady pulls. Use snacks or an encouraging voice to get the goat to follow. With any luck at all, the goat will learn this lesson in a few sessions. You will need to decide if you want the goat to follow you on the lead or whether you wish for it to walk at your shoulder. Whatever your choice, teach it to the goat right from the start. Constant, unrelenting repetition will soon have the goat abiding by your wishes.

4. The worst case scenario, during these training sessions, is that the goat will resist by laying down! That is the strongest display of bull-headedness that the goat can offer. If the goat is small, pull upward and forward on the lead rope. It is essential that the goat receive no positive reinforcement from lying down. Get the goat up and moving using strength, voice, snacks—anything your imagination can muster. Stay at the process, in a gentle and humane manner, and you will be rewarded. If the goat is large and heavy, too heavy for you to move, and absolutely will not respond to voice, tugs, or snacks, sell it quickly! Life is too short to fight with a goat that will lie down and refuse to respond whenever it feels like it. Some people have had good luck in resolving this problem by using a shock collar designed for use in training dogs.

5. The next logical step, after the goat will lead, is to teach the goat to pull or drag the travois. This is especially easy to teach if the goat trusts you. The lesson here is to do a thorough and patient job of teaching the goat to lead. Do this part correctly and the goat will trust you and make your next steps easier. For the goat to be able to pull the travois, you will need a surcingle about the heart girth of the goat (padded where it crosses the withers) and a chest strap around the front of the neck. The actual travois needs to be nothing more than a couple of long sticks, a couple of cross-braces, and some means of attaching the travois to the surcingle.

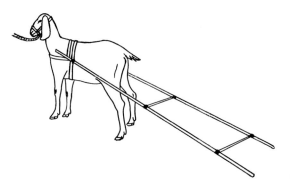

6. Place a halter and lead rope onto the goat. Try to have a helper—someone the goat is accustomed to—assist you during the first days of the travois work. While one person handles and soothes the goat, the other should place the travois over the back of the goat and attach it to the padded area of the surcingle. The travois is very light, and the goat will hardly notice the pull when she is led ahead. Begin the teaching process by leading the goat and talking to it for reassurance. Soon, the goat will move ahead and disregard the travois dragging behind.

7. Negotiating corners, adding weight to the travois, and having the goat respond to "hup" and "whoa" will take but a few additional lessons, if you have a plan, are patient, and are entirely consistent in your commands and expectations. It will be necessary for the goat to learn "hup" and "whoa" so that you can eventually move to the rear of the goat and still control its movement. Those two commands can easily be taught while leading the goat.

8. The next step in the process is to teach the goat to respond to pressure from the reins so you can drive it from behind with a set of long lines. "Hup" and "whoa" are used to make the goat go and stop. The long lines and rein pressure make it go left and right. One tried-and-true method to accomplish compliance to rein pressure is to have a trusted handler lead while an assistant handles the long lines from behind the goat.

9. Lead the goat out in a straight line. Coordinate the activities of the handlers so that the person from behind applies rein pressure in one direction

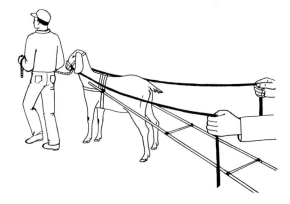

or the other, say to the right, just before the person on the lead line leads the goat in that direction. Normally, the goat will resist the unaccustomed long line pressure, but immediately respond to the lead line. As the lead line pressure is being applied, the long line pressure should be applied continuously and gently, perhaps with a bit of give-and-take pressure. Be certain to repeat this sequence in both directions an equal number of times during each session. Gradually, the lead line person will not be needed. Be certain to practice "hup" and "whoa" from the long line person in the same way as the directional control.

10. Teaching the goat to pull a cart or wagon is accomplished in exactly the same way. There is a bit more harness involved and a more complex device for the goat to pull, but if the goat can handle the travois, it will be an easy step up to the full harness and wagon or cart.

8.19 PACKING YOUR GOAT

Packing with your goat, like harnessing your goat, is clearly not a necessary management skill. However, like hitching and driving, it is an excellent hobby and pastime for you and your family. Packing with your goat has an added bonus. It causes you to become physically active. Hiking with your goat into mountainous or backcountry terrain is an excellent aerobic exercise. You will return from the pack trip more physically, and mentally, fit than you were before you left.

Most of the things the goat needs to learn to be a pleasure for you to pack are the same things that you must teach if you want to hitch and drive your goat. Certainly, it must learn to stand tied, it must lead without resistance, it must accept the restraint of the surcingle, and it must accept weight on its back—much like the weight of the travois.

Equipment Necessary

- Halter
- Lead rope
- Travois
- Saddle pad
- Pack saddle
- Panniers

Restraint Required

There is no real restraint, per se, involved with packing your goat. However, it is necessary for the goat to know how to lead, to be willing to respond to your voice, and to be used to standing tied.

Step-by-Step Procedure

1. Teach the goat to stand tied. Place a strong nylon web halter onto the goat and use a strong 5/8"

lead rope with a stout bull snap attached to it. Tie the goat, at about head height, to a post or small tree that will not break. Allow 2 or 3 feet of rope between the halter and the post. Stand back, relax with a cup of coffee, and allow the goat to find out for himself that it is futile to fight the situation. Don't be alarmed if the goat pitches a large fit. The whole idea is for him to teach himself that all the fighting will only serve to wear himself out. Goats are smart and it will take only a few sessions for the goat to learn what is expected of him.

2. Teach the goat to lead. Using the same halter and lead rope, start out by simply trying to take the goat for a walk. This will work best with the younger, smaller goat. If the goat is confused or reluctant to follow you, gentle tugs, some snacks, and verbal encouragement will often cause the goat to move out and follow your lead.

3. If you are fortunate, this will be all that is necessary. However, the goat might not respond. It may drag its feet, pull back on the lead, and act as if it will never learn to follow you without being drug. Getting past this point can be frustrating. Pull a little, stop a little. Try short tugs, try steady pulls. Use snacks or an encouraging voice to encourage the goat to follow. With any luck at all, the goat will learn this lesson in a few sessions. You will need to decide if you want the goat to follow you on the lead or whether you wish for it to walk at your shoulder. Whatever your choice, teach the goat to work in that position right from the start. Constant, unrelenting repetition will soon have the goat abiding by your wishes.

4. The worst case scenario, during these training sessions, is that the goat will resist by laying down! That is the strongest display of bull-headedness that the goat can offer. If the goat is small, pull upward and forward on the lead rope. It is essential that the goat receives no positive reinforcement from lying down. Get the goat up and moving using strength, voice, snacks, anything your imagination can muster. Stay at the process, in a gentle humane manner, and you will be rewarded. If the goat is large and heavy, too heavy for you to move, and absolutely will not respond to voice, tugs, or snacks, sell it quickly! Life is too short to fight with a goat that will lie down and refuse to respond whenever it feels like it. Some people have had good luck in resolving this problem by using a shock collar designed for use in training dogs.

5. The next logical step, after the goat will lead, is to teach it to pull or drag the travois. Even if you will never harness the goat, and your only intention is to pack with it, teaching the travois is a good idea. Dragging a travois is a good way to transport longer, bulkier loads into a camp site, especially if the terrain is relatively open.

The travois is especially easy to teach, if the goat trusts you. The lesson here is to do a thorough and patient job of teaching the goat to lead. Do this part correctly and the goat will trust you and make your next steps easier. For the goat to be able to pull the travois, you will need a surcingle about the heart girth of the goat (padded where it crosses the withers) and a chest strap around the front of the neck. The actual travois needs to be nothing more than a couple of long sticks, a couple of cross-braces, and some means of attaching the travois to the surcingle.

6. Place a halter and lead rope onto the goat. Try to have a helper whom the goat is accustomed to assist you during the first days of the travois work. While one person handles and soothes the goat, the other should place the travois over the back of the goat and attach it to the padded area of the surcingle. The travois is very light and the goat will hardly notice the pull when she is lead ahead. Begin the teaching process by leading the goat and talking to it for reassurance. Soon, the goat will move ahead and disregard the travois dragging behind.

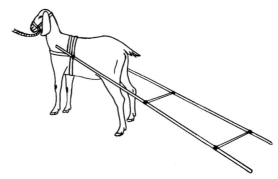

7. Negotiating corners, adding weight to the travois, and having the goat respond to "hup" and "whoa" will take but a few additional lessons, if you have a plan, are patient, and are entirely consistent in your commands and expectations. It will be necessary for the goat to learn to respond to the voice commands of "hup" and "whoa" so that you can eventually allow the goat to travel down the pack trail without a lead rope and still control its movement.

8. The next step in the process is to teach the goat to accept the pack saddle. At this point you will be happy that the goat is accustomed to accepting the surcingle that the travois was attached to. If you chose not to teach the travois, it will not take a long time for the goat to accept the pack saddle cinch. However, do be patient. It does restrict a goat's movement and may take a few times before much else can be taught besides simply wearing the saddle.

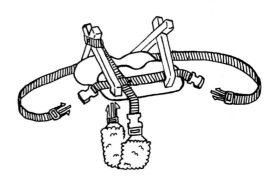

9. Have a friend help you with the first few saddlings. If no one is available to help, tie the goat before you attempt to saddle it. As you prepare to saddle the goat, use the following sequence, unless you have a better one. There is a reason for each step and for the sequence of the steps.

a. Brush the back and heart girth of the goat, then use your hands to rub down the area of the back that will be covered by the pad and the area that will be covered by the cinch. Check for embedded dirt, burrs, or any other sort of chaff.

b. Brush the underside of the pad. This is not as much to clean the pad as it is to be certain there is nothing there large enough to be an irritant for the goat's back.

c. Place the pad onto the goat's back by placing it too far forward, up over the withers, and sliding it rearward until it is in the correct position. This smooths the hairs in the direction of their natural lie. Never slide the pad against the lay of the hair.

d. Place the actual pack saddle onto the pad.

e. Place the butt strap or breeching under the tail.

f. Buckle the cinch.

g. Place the breast collar around the front of the chest.

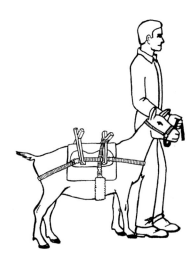

10. Introduce the goat to the pack load. Most normal pack load items can be carried in panniers. Panniers are canvas, wood, metal, or fiberglass containers, usually having some sort of closure or lid. It is the easiest way to carry the odd-sized assortment of items a pack trip needs. Whatever the material you select for your panniers, they will usually come in pairs. It is essential that the panniers be nearly exactly matched in weight and closely approximate in bulkiness. A normal weight for a mature dairy breed goat to carry is 50 to 60 pounds total. If a 50-pound load is selected, there should be 25 pounds in each pannier. Weight the packed panniers, so that you know for certain what weight the goat is carrying. Start with empty panniers on the goat until it is accustomed to moving down the trail with the added bulk of the panniers. After the goat is carrying the panniers comfortably, weight can be added gradually until the full load is being carried.

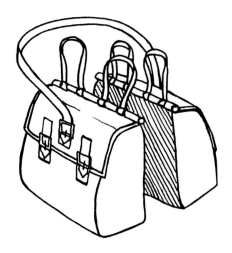

BARRED ROCK

BLACK ORPINGTON

BUFF COCHIN

RHODE ISLAND RED

WHITE WYANDOTTE

BLACK TURKEY

BROAD-BREASTED BRONZE TURKEY

WHITE HOLLAND TURKEY

All photographs on this page are courtesy of WATT Publishing Company. Prints may be purchased at www.wattcountrystore.com.

chapter NINE

Poultry Management Techniques

9.1 INTRODUCTION

Poultry can be defined as large birds that were domesticated centuries ago and are now raised for food in the form of eggs and meat. The best examples are chickens, turkeys, ducks, and geese. Other birds that are classified as poultry, although they are raised mostly as ornamental birds, include the guinea fowl, pea fowl, swans, and pigeons.

The chicken and the goose were domesticated over 3,000 years ago. In terms of sheer numbers raised for food, the chicken is the main kind of poultry. It is said to have originated in Southwest Asia from one or more of the jungle fowl species that still inhabit that region. The one species of poultry that is thought to be native to the North American continent is the turkey.

Today there are more than 225 breeds and varieties of chickens, turkeys, and ducks. Most of these are raised for show or exhibition. Only a few are of importance in the commercial production of meat and eggs.

A description of the recognized breeds and varieties of poultry is given in the *American Standard of Perfection,* published by the American Poultry Association. It is often referred to as the exhibition poultryman's bible because it contains all the details of size, shape, and color of feathers, comb, face,

legs, body, and tail that are the standards to which the breeds and varieties should conform.

The raising and showing of fancy or exhibition poultry is a popular avocation for many people. No one knows for certain the number of fancy poultry raised each year, but it is substantial and represents an important segment of the total poultry population.

The other part of the poultry population, and the most important economically, is that used to produce meat and eggs for human food. The birds used for the commercial production of meat and eggs originated from those breeds and varieties that are described in the *American Standard of Perfection.* Whereas the show breeds and varieties are purebreds, a commercial bird is usually a crossbred produced by crossing purebreds, varieties, or strains.

All white-egg-laying hens are crosses between strains of White Leghorns that have been selected for many years for efficiency in the production of eggs. Similarly, most commercial brown-egg hens have been produced by crossing genetically improved Rhode Island Red cockerels with Barred Rock hens. There is a special reason for using a Red rooster on a hen with barred feathers. This is because this cross produces a black pullet and a barred-feather-patterned cockerel. At hatching, the pullets and cockerels can be separated by the color

497

of the chick's down, resulting in a saving of 2¢ or 3¢ per chick over having a skilled operator use a method known as *vent sexing*.

Today, virtually all commercial breeders incorporate a genetic trait into the gene pool of their chicks that causes the wing feathers of the female to grow more rapidly than those of the male during the incubation period. Thus, sexing (determining males from females) of the chicks is possible based upon the length of the feathers, or "down," of the chicks. However, feather sexing must be completed during the first 24 to 48 hours after hatching, as the feather growth difference diminishes very rapidly with time. Feather sexing is much faster than vent sexing and can be readily and quickly learned by hatchery workers.

Broiler chicks are crossbreds produced from genetically pure White Rock hens that have been mated to males synthesized from several breeds, but that are basically White Cornish. The commercial turkey industry uses crosses, although not all turkeys raised for meat are crossbred.

The commercial poultry industry has chosen crosses over pure strains or breeds because of the crossbreds' increased egg production, improved feed efficiency, and accelerated growth rate. The use of crossbreds to produce eggs or meat does not eliminate the need for purebreds. The crossbred comes from the crossing of two "pure" breeds, varieties, lines, or strains. Superior performance in the crossbred depends upon superior performance in the purebred stock.

Almost all the commercial poultry used for egg or meat production are white-feathered. Typical examples of breeds and varieties that are raised commercially are White Leghorn, White Rock, White Cornish, Large White turkeys, Small White turkeys, and White Pekin ducks. The reason that white-feathered birds are raised is simply that they produce a cleaner, more appealing-looking carcass than do birds with colored feathers. This did not happen by chance, but was a result of a combination of (1) the consumer demanding the more pleasing-appearing carcass in the marketplace, (2) the processor paying more for the white-feathered bird, and (3) the producer responding by raising white-feathered birds.

Commercial poultry is a far cry from the fowl that used to roam the barnyard. While the barnyard hen laid her eggs mainly in the spring and early summer, and then only enough eggs to hatch a dozen chicks, her modern counterpart now lays all year round and hardly ever thinks of sitting on eggs to hatch them. Most commercial flocks average at least 240 eggs per hen per year, with some approaching the magical number of 1 egg per day per hen for 12 months. Similarly, the barnyard cockerel, which

used to furnish a family Sunday dinner at 3 or 4 months of age, now takes only 56 days to grow to 4 or 5 pounds, using 2 pounds or less of feed per pound of live weight. Commercial poultry is skillfully selected and bred for efficient production and fed precisely balanced diets. The birds are housed in a near-optimum environment and kept as free of disease as possible through programs that include blood testing of breeders to eliminate carriers of disease, vaccination to produce immunity to disease, and the use in the feed of chemicals that control serious diseases such as coccidiosis and blackhead.

Poultry differs markedly from other domestic farm animals. Some of the differences are easily recognized. They have feathers instead of wool or hair, walk on two instead of four legs, fly if necessary, and lay eggs, to name a few. Other differences are not as apparent, but nonetheless have a profound effect on the management practices required. For example, poultry feed is very finely ground, partly because poultry have no teeth. In lieu of teeth, nature provided them with a gizzard in the digestive tract. The gizzard has a tough lining and thick muscles that, along with bits of rocks and stones that collect therein, grind large particles of food into smaller ones. For many years it was common practice to provide sand, gravel, or crushed rock to a flock to assist the gizzard in grinding large particles of corn or wheat to a more digestible size. Recently, research results have shown that the feeding of crushed stone is unnecessary as long as the feed is in a finely ground form before it is fed to the poultry.

Poultry rations are low in fiber, with most rations containing only about 3% of such roughage. The digestive tract is short. Food can pass through it in 6 hours or less. There is no provision for microbial digestion as there is in the horse, cow, sheep, or goat. Most poultry rations are formulated with this in mind, and as a result are high in energy and low in fiber.

The young of poultry develop outside of the body of the hen. A hen does not have to wait for the young to be born before she can lay another egg. She ovulates, puts the correct quantity and quality of food to support the development of an embryo around the ovum, wraps it all up in a shell, and lays an egg. She can and often does ovulate and lay an egg every day. She used to store these eggs in a nest until a convenient nest-full had been laid. She then stopped laying, became "broody," and went to work hatching her chicks. The development of the embryo outside the body of the hen has given rise to the incubator and the hatchery industry. It has been the job of the geneticist to change the genes of the hen so that she no longer has the "broody" instinct that causes her to stop her laying cycle and to sit on a nest full of eggs to hatch her young. In the proper light and nutritional environment, a hen may lay continuously for 12 months or even longer. The production of many eggs per hen per year has given rise to the table-egg industry.

Anatomically, a bird has some features that link it to its evolutionary past. Scales on its legs link it to the reptiles. Scales are modified skin found in the lower orders of animals but absent in the higher orders. A *cloaca,* a chamber just before the anus, which receives the products of the digestive, urinary, and reproductive systems, is present in chickens and reptiles, but not in the horse, cow, goat, sheep, or pig.

Many of the bones of poultry are fused to enable the bird to walk on two legs with its backbone parallel to the ground and to fly with its weight carried on its wings (modified front legs). Many of the vertebrae in the thoracic, lumbar, and sacral regions are fused to one another. The vertebrae in the sacral and lumbar regions are also fused to the bones of the pelvic girdle, which in turn are fused to each other.

The bird has air sacs that are connected to the lungs and that are found in the thoracic and abdominal cavities. Some of these air sacs have connections with certain bones such as the humerus. The function of air sacs has been debated for many years. The most realistic function that has been proposed has to do with increasing the efficiency of respiration.

Poultry behave differently from other farm animals. They naturally flock together. They are timid and shy and will run when approached by a person. They are easily spooked, and chickens and turkeys will pile up in the corner of a pen when frightened, causing the birds on the bottom of the pile to smother in a very few minutes. Turkeys are curious. A pail left in a pen of poults (baby turkeys) will end up full of poults that have jumped into it out of curiosity.

Poultry flocks have a very strict social order commonly called the *pecking order.* The top hen in the pecking order will peck and "boss" all other hens in the flock. The next hen, although bossed and pecked by the top hen, can boss and peck all those beneath her in the pecking order, and so on down the order to the bottom hen, which bosses no one but is pecked by all others. Roosters in a flock establish a similar pecking order entirely independent of the hens.

The pecking order is established by fights between pairs of hens. Each hen challenges or fights every other hen. Any hen may challenge any other at any time. When this happens, the two hens may fight it out to see who is boss. A well-defined social order seems to contribute to the well-being and stability of the flock. Although people may object to the severity with which poultry discipline each

other, or feel sorry for the low bird in the pecking order, our efforts to thwart or skirt this social order by all sorts of management gimmicks probably contribute little to the total good of the flock.

The total volume of knowledge about poultry is great. It is probably the best researched of the domesticated farm animals. Much of this information is important to the successful management of a poultry enterprise and is available in books, bulletins, and scientific literature.

The poultry industry has achieved a high level of efficiency not only because of the research that has been and is being done, but also because of the progressive attitude of the commercial poultrymen who are willing and eager to add to the basic knowledge they already have and to apply it in the management of their poultry enterprises.

9.2 PARTS OF THE FOWL

Most of the names on the illustrations at the beginning of this chapter are important to people who breed and show fancy breeds and varieties of chickens. To them, the color of the wing bow, wing bay, or wing bar may be extremely important. The color, shape, or number of points of the comb or the smoothness or color of the shank may be the difference between a grand champion and an also-ran. A commercial producer of broilers or eggs, on the other hand, may have little interest in the shape of the comb or the parts of the wing. Nonetheless, to be able to talk intelligently to someone about chickens, even the commercial producer finds it useful and helpful to know the names of the various parts. You will also find it extremely useful to know the nomenclature of the fowl so that when you read about one of the parts, or hear it mentioned, a clear-cut mental image of what and where that part is will appear immediately.

Equipment Necessary

- Live hen and rooster
- Nomenclature chart

Restraint Required

Catch a hen and a rooster. Put them in a cage or crate so that you can examine them individually. Refer to Section 9.3 for techniques to catch poultry.

Step-by-Step Procedure

1. Refer to the charts in Appendixes K—M and identify the various parts listed. Learn the parts on both the hen and rooster. Notice how they feel: cold–warm, soft–rough, sharp–dull. Notice the relationship of one part to another—the comb and the head, the comb and the wattles, the ear lobe and the ear, the spur and the shank.

2. Spread the wing of the bird to expose the large, main wing feathers. Identify the large secondary and primary wing feathers and the smaller axial feather that separates them. The axial feather should be opposite the elbow of the wing. It can be identified by holding the front edge of the spread wing in the palm of your hand, with your fingers pointing to the rear. By pushing up from the underside with your middle finger, you should be able to cause the axial feather to surface. It is about half as long as the primaries and secondaries. Relative to the axial feather, the secondaries are in the direction of the body and the primaries are in the direction of the tip of the wing. Count the primaries. The normal number is 10 mature feathers in the adult. Anything less than 10 indicates that the bird is molting (a shedding of the old and a growing of new feathers). You can tell the difference between the old and new feathers. Old ones will often be longer, duller, worn, dirty, and frayed. New ones may be in any stage of growth—from just poking through the skin to full grown—and are cleaner, smoother, brighter looking, and more rounded at the tip. (Note that molting is not an abnormal condition. It is normal for young poultry and at certain seasons of the year in adults. When a pullet starts to lay, she is completing the growth of her third set of feathers.)

3. Learn the normal condition, color, shape, position, and location of the parts. Learning the normal characteristics of poultry will take time, and most poultry that you examine will be normal. Defects and abnormalities show up infrequently. If you are unsure of what is normal, ask someone who is an authority on poultry. Keep in mind that there may be some variations in shape, color, and condition among live chickens of the same age, sex, and breed. This is acceptable. The color or condition of some parts may change with age or when egg production starts or stops. After you learn to identify the normal, it will be easier to tell when something is wrong or when a part is abnormal.

4. Identify the differences between the sexes. Notice that most of the external parts of the male and female chicken are the same except for the larger size of the spur, comb, wattles, feather configuration, and body size in the male. Observe the different feathers of the area in front of the tail and in the hackle. The area in front of the tail is called the *cushion* in the female and the *saddle* in the male. Note that the male has sickles, which are the predominant long and pointed feathers in his tail, whereas the female lacks these.

Inspect the feathers of the saddle and the hackle of the male, and note that these are narrow and pointed at the tip, in distinction to the round, blunt-tipped feathers in these regions of the female. (The feathers of the hackle and, to a lesser extent, the saddle of the male of some breeds are superior material for use in making fishing flies.) The most reliable indicator of the sex of a male chicken is the presence of these pointed feathers in the saddle, back, and hackle regions. They are often glossy in appearance. In young chickens of some breeds, at 6 to 8 weeks of age the only way that males can be separated from females is by looking for these sharp, pointed feathers. As chickens get older and approach sexual maturity, around 5 to 6 months of age, other body differences can help to identify pullets and cockerels. All males will show marked spur growth after 6 months of age, while the female spur remains undeveloped. The body size of the male is about 25% larger than that of the female. Males grow much larger combs, wattles, and ear lobes than do females. They crow, dominate the females, and in general are much more aggressive and belligerent.

In the absence of external sexual parts, you must learn to use the shape of the feathers, spur growth, and the size of comb, wattles, and body to differentiate between male and female chickens. Some females will grow spurs, while others may crow or even mate with other females. Some males will sit on a nest and pretend that they have laid an egg. Don't let them fool you. If they have round, blunt-tipped feathers in the hackle, back, saddle, and tail, they are females. If these feathers are narrow and pointed at the tip, they are males.

5. Examine the diagrams of the turkey tom and hen (Appendix M). Notice the similarities between chickens and turkeys with respect to their feathers, wings, beaks, shanks, scales, spurs, and carriage of the body, neck, and head. Notice the dissimilarities, or things that distinguish a turkey from a chicken: the shape of the body, carriage of the tail, snood, throat wattle, caruncles, beard, and absence of comb.

6. Identify the difference between the turkey tom and hen. Notice that there are no external sex organs, so we must use such secondary sex characteristics as the beard, the snood, color of head furnishings, voice, and strutting to identify the sex.

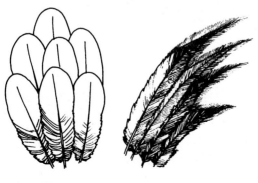

FEMALE   MALE

The most obvious sex identifier in the tom is the presence of the beard, which is a tuft of wiry, hairlike feathers projecting from the front of the breast. All adult toms have a beard; no hens have beards. The beard appears at about 3 months of age, so that is the earliest that sex can be identified by external signs. In most commercial flocks of turkeys, the males and females are identified as soon as they are removed from the incubator by a specially trained person known as a "sexor" who everts the cloaca and looks for the male and female organs which are located inside it.

Another distinguishing feature of the sexually mature adult tom is the large size of the snood. The tom's snood may be 3 to 4 inches long, hanging down over the beak. He possesses the ability to lengthen or shorten it or to change its color. The female's snood sits just behind her beak on top of her head, projecting about $\frac{1}{2}$ inch upward and forward. It is small and visibly unchanging in color and size.

Toms are larger than hens, weighing about 25% more. Toms weigh about 25 to 30 pounds at 6 months of age, while hens weigh 16 to 25 pounds at this age. In general, the larger birds in a flock are the toms and the smaller ones are the hens.

Toms may also be distinguished by their characteristic strut with tail spread in a characteristic open-fan position; the striking and vivid coloring of the caruncles, snood, and face, which changes to different hues of reds, blues, and white as the tom struts and performs for the females; and the characteristic "gobble" voiced by the tom. The hen, on the other hand, seldom struts, has a very plain, unchanging red head, and emits a simple chirping sound.

9.3 CATCHING, HOLDING, AND CARRYING POULTRY

Most poultry are small enough that they can be caught and held by hand. Chickens, turkeys, and ducks do not like to be caught and handled; they will struggle, scratch, and flap their wings in an effort to free themselves from your hold. By making use of certain equipment and applying certain knowledge, the catching and holding job can be made easy. Experience and practice will perfect your skill. Outsmarting the animal will remove much of the frustration from the job. The younger the bird, the easier it can be caught and held. The more often you handle your poultry, the tamer they become, until one day you will be able to walk into their pen and pick them up without resistance.

Equipment Necessary

- Catching hook
- Folding wire panel

Restraint Required

Under most conditions, the poultry to be caught will be confined to a pen, a wire cage, or, in the case of poults, chicks, or ducklings, to small compartments in a box. You won't need a halter or a rope and you won't need to wrestle the bird to the floor with your body. Bare hands will do it.

Step-by-Step Procedure

Newly Hatched Poults and Chicks

1. Chicks and poults are caught and handled more easily at hatching than at any other time. Reach down and grasp the baby with one hand, holding it as you would an egg.

2. Position the chick or poult in your hand by turning it with the fingers of the same hand or by changing it to the other hand. Holding it in one hand allows you to maneuver the other hand while performing any one of a number of management skills to be described later.

3. Try different placements of your fingers around its fuzzy body to find one which will give you complete control of its movement.

4. Position the chick or poult against the palm of your hand, its head toward your fingertips, with your little finger between its legs and your middle and ring fingers wrapped loosely around its body, leaving your index finger and thumb free to pinch a fold of skin on its head or spread its wing for banding. This technique will be useful in skills described later in this chapter.

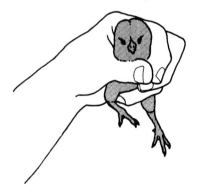

CAUTION: Be gentle! Don't squeeze, choke, or smother the chick. Remember that, like other babies, it is very tender and fragile.

Using a Folding Wire Panel

As chickens and turkeys become older, catching and holding them becomes more difficult. Their natural impulse is to escape from whatever is chasing them.

They will run, fly, hide—anything to escape the pursuer. In wide-open spaces, it is almost impossible to catch a chicken or turkey, because there is no place to corner it. The smaller the area available for the bird to run in, the easier it is to corner it. Small, temporary pens within larger ones may be used to assist in the catching process.

1. Wire catching fences can be bought from a commercial poultry-equipment company or made by hinging a series of three or more stiff-welded wire panels together so that they can be folded or expanded. The main requirement is that the hinged wire panels when expanded remain erect and standing. Each panel is approximately 2 feet wide by 3 or 4 feet high. The younger the bird to be caught, the lower the panels can be. A 2-foot-high panel is satisfactory for 2- or 3-week-old chicks or poults.

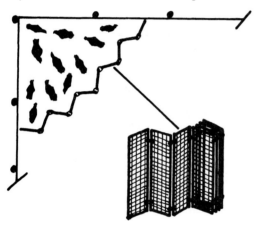

2. Make a smaller pen in one corner of the large pen or field by using folding wire panels and the two sides of the larger pen. The smaller pen should be just large enough that standing room only will be available to the birds. Make sure that the wire panel is held or anchored securely so that it will not collapse from the weight of the birds pushing against its sides.

3. Open the wire panel along the longer side of the pen.

4. Herd the chickens or turkeys into the open end of the smaller pen.

5. When most of the birds have been driven into the pen, enter it, closing and locking it or holding the panel behind you.

6. Work as fast as you can to empty the small pen. Grab a bird's legs from the back; pull the legs from under its body, forcing it to drop to the floor; and lift the bird, feet first, head hanging down, up and over the wire panel. Someone on the other side of the panel can take the bird from you and place it into a crate.

7. A second drive may be necessary to catch the stragglers, or they can be caught with a hook.

> **CAUTION:** When driven into a small enclosure, chickens and turkeys will pile up against the fence or wall. The ones on the bottom of the pile may smother and die if left for more than a minute. Be on the lookout for pileups. Spread them out into a single layer. Empty the pen as quickly as possible.

Using a Catching Hook (Chickens)

If only one or two birds from a large flock are to be caught, it may be easier to run them down individually than to catch the whole flock with a catching fence. Hooking a running chicken requires good hand–eye coordination and body control. It is an art in which you can become proficient with practice.

1. A catching hook can be made from stiff wire. It should be about 4 feet long, with a handle on one end and the other end bent back on itself to form an S-shaped hook. The open end of the hook should be just wide enough to allow the shank of the bird to slip through, and the bottom of the S hook flared just enough to allow the foot to slip in and be held. Hooks can be purchased from some local feed stores.

2. Locate the bird to be caught.

3. Use one hand to hold the catching hook by the handle. Extend the hook out toward the floor in front of you. The open end should be 3 or 4 inches above the floor, facing the outside of the bird's shank.

4. The trick is to pull the open end of the hook over the shank of one of the chicken's legs, drawing back on the hook when the shank has been caught. Pull the leg back and up from under the bird. During the hooking process, the shank should slip into the bottom of the hook, and the bottom of the hook then slip down the shank and snare the foot.

5. Pull the bird toward you. Reach down and grab both legs in one hand. Disengage the foot from the hook.

CAUTION: Most chickens will not stand still and allow themselves to be hooked. It is best if you can catch the bird on the first attempt before it becomes aware of what you are doing. Many times the first pass with the hook will be a miss; the bird runs, and you run after it. So it becomes a question of moving in close enough on the run to slip the open end of the hook over the shank. Be aware of sharp protruding objects in the pen that could injure you or the fowl as you chase it.

There are situations in which you might want to catch chickens that are confined to a small wire cage: (1) in a commercial cage-layer or grower operation housing thousands of birds under one roof with several birds to each cage, where the birds are to be processed or caught for moving to other quarters, and (2) at a poultry exhibition in a judging situation where small numbers of individually caged birds, perhaps 4 to 10, are to be evaluated on the basis of breed type or egg production.

In the first situation, it is important that you handle as many birds at a time as fast as you can. In the second situation, you have only a few birds to handle and each is handled and inspected individually. You may be working under the scrutiny of a judge or inspector and want to exhibit the finer points of catching and handling.

In Small Wire Cages—Commercial Setup

In the commercial setup, cages are small and located so that you can reach inside and easily catch all the birds.

1. Open the cage door cautiously. By standing close, block the open door with your body or arms so that none of the birds can escape.

2. Work with both hands inside the cage. Catch the legs with both hands, pull the legs out from under the bird, and drop its body to the floor of the cage. Hold the legs between the fingers of one hand and continue to catch birds with the other hand until you can hold no more.

3. Remove each handful of birds through the cage door feet first. Do not remove more birds at one time than can pass through the opening without injury.

4. At this point, depending on the job to be done, either place them into a chicken crate, continue to hold them while someone else does something to them, or carry them out of the building.

5. Several chickens can be carried in each hand at one time. Carry them with their feet up, shanks between your fingers, hock joints extended, and heads toward the floor.

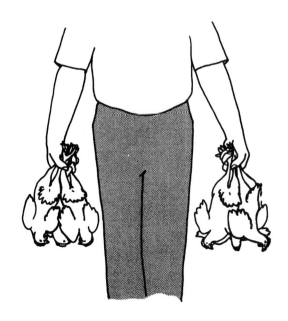

> **CAUTION:** Be very careful when pulling a handful of birds through the open cage door not to catch their wings on the door frame or tear their skin on sharp, protruding wire. Catching wings on objects may result in broken bones. Carry chickens by both legs. One-leg carrying results in crippled birds. Gentleness and care in handling are as important here as in any other operation to prevent needless injury to the bird.

In Small Wire Judging Cages (Chickens)

There are certain situations in which live birds must be handled individually and removed head first from the cage. Two examples are: (1) judging in show competition and (2) evaluating birds for egg, meat, or breeder traits. In both situations you are removing caged birds under the watchful eye of a judge or instructor who is evaluating how well you can handle a bird. In most cases, the cages will be at table height so you can stand normally to catch the birds.

1. Approach the judging cage slowly and open the door. Reach in, and with your right hand maneuver the chicken until it stands with its head facing to your right.

2. Extend your right hand over the back of the chicken and grasp the wing that is away from you,

firmly holding it so that it remains folded against the side of the bird.

3. Pull the chicken by the wing you have grasped so that its head is pointed toward the cage door. With your left hand, keep the chicken's right wing against its body.

4. Slide the fingers of your left hand, palm upward, under the bird's breast. At the same time, keep your left thumb on the right wing to hold it against the body. You may have to pull some of the wing feathers down with the downward movement of your hand to enable your thumb to keep the wing against the body.

5. With your right hand grasping the left wing and your left hand under the bird's body, lift the chicken from the cage head first. Then slide your right hand back under the bird to grasp its legs. Keep the palm of the right hand up.

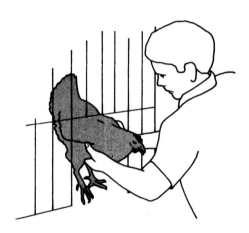

6. Hold the legs extended straight out to the rear at the hock joint, with your index finger between the legs and your thumb over the back of the hock joint. With the legs extended, you are in control of the bird and it cannot kick or scratch to escape.

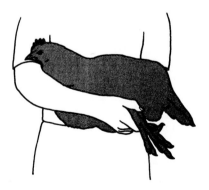

7. Always keep the chicken's head toward you. At this point, its legs are held by your right hand and its body is resting on your right forearm.

8. To look closely at the bird's head, place your left hand under the front of the breast. Continue to hold the legs with your right hand. Raise the bird to your eye level.

9. This technique of holding the bird can be used to examine wings, pubic area or vent, legs, feet, breast, back, and plumage.

10. Always put the bird back into the cage head first. It may help if you stand with your left side toward the coop and toss the bird gently through the door, allowing it to land on its feet.

The method described here is the proper way to hold and handle a chicken. Do it the right way, smoothly, and easily, and others will recognize you as a professional. So will the chicken. It will feel secure and allow you to move it around and hold it in many positions. Handle the chicken the wrong way, and it will quickly sense that it has a good chance to get loose and will try to do so. Remember to keep the chicken's legs extended at the hock joint.

Large Turkeys

The method for catching turkeys is much the same as that for catching chickens. Turkeys usually are raised in floor pens in houses (confinement rear-

ing) or in fenced lots and fields. As with chickens, it becomes a question of herding the turkeys into small enclosures within the pen or field where they can be cornered with a minimum of chasing and effort. Catching pens for turkeys must be fastened and anchored more securely to existing structures than is the case for chickens because the turkeys are heavier. In addition, it is of great importance, especially when catching turkeys for marketing, to avoid having them crawl or climb over each other. Only a few turkeys should be herded into a catching pen on each drive to avoid the piling tendency that occurs with large numbers of the birds. For example, no more than 10 toms or hens should be caught in a 4' × 6' catching pen. Commercial growers have many different schemes for herding their birds into a narrow space so that they can be caught easily and quickly with a minimum of scratching and flapping.

1. Corner the turkey to be caught.

2. Grab both legs from the rear. Use both hands.

3. Pull the legs out from under the bird, dropping its body to the floor. Avoid situations where the breast (keel) might strike a sharp object such as the edge of a roost or feeder.

4. Lift the turkey by both legs, with its feet up and its head down.

5. If the bird tries to right itself by bringing its body up, transfer both legs to one hand and use your free hand to push its body down. Keep the hock joint extended.

6. Do not let the turkey flap its wings. If necessary, hold both legs in one hand and use your other hand and arm to press the turkey's body against your leg, forcing its wings against its body. If the flapping cannot be controlled by this maneuver, place the turkey back on the floor with its breast

down, head up, legs grasped in your hand, and hocks extended. When it is resting in this position it should stop flapping immediately.

> **CAUTION:** The flapping wings of a large turkey can hurt people. Handle the turkey firmly but gently. Do not let the turkey beat its wings against the sides of the pen, the floor, or other objects. Severe bruising or broken wing bones may result. Do not let turkeys climb over each other. Their sharp toenails scratch the tender skin of other birds. Scratches, scabs, bruises, and broken bones produce a low-grade carcass and a much lower price per pound.

9.4 IDENTIFICATION

Chickens and turkeys can be marked in several ways to tell them apart. One of the simplest markers is colored ink. Felt-tipped pens of different colors can be used to mark the feathers on the back or side of the bird. The mark is easily visible, but temporary, because it wears off in a short time.

> **CAUTION:** Marking a single bird in a pen sets it apart as being different. Others, seeing that the bird is different, begin to peck it and drive it away. In an extreme case they may keep it from eating and drinking or peck it to death. The solution to this is to mark all of the birds. None will be different. Avoid red marks because the mark may be mistaken for blood by the other birds and give rise to cannibalism.

More permanent types of identification are the toe punch, wing band, wing badge, and leg band. Bands and badges are made of aluminum or plastic and can be stamped or inked with any combination of letters and numbers. They have the additional advantage of not marring the body or detracting from the looks of the bird. Identification can be applied at almost any age. Certain procedures such as toe punching and wing banding ordinarily are done to baby chicks as they are removed from the hatching trays or are still in chick boxes. Wing badges and leg bands are used when the bird is almost full-grown (5 or 6 months of age).

Equipment Necessary

- Toe punch—for chicks or poults
- Wing bands—aluminum or plastic; lettered and numbered; sizes for poults and chicks
- Wing badges—plastic, lettered and numbered
- Leg bands—aluminum or plastic; lettered and numbered; sizes from bantam hens to mammoth turkey toms
- Leg bands or rings—spiral, celluloid, or plastic; various colors; sizes from bantams to turkeys
- Pliers—for sealing aluminum wing bands
- Pliers—for sealing aluminum leg bands

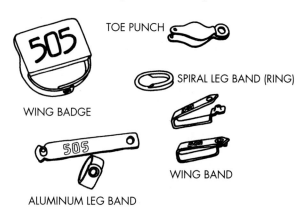

WING BADGE   TOE PUNCH

SPIRAL LEG BAND (RING)

ALUMINUM LEG BAND   WING BAND

Restraint Required

Chicks, poults, chickens, and turkeys must be held individually. Banding or toe-punching chicks and poults is a one-person job. With adult birds, banding and badging can be done by one person, but two people can do it easier and faster. One catches and holds while the other applies the band or badge.

Step-by-Step Procedures

Toe-Punching Baby Chicks or Poults

This is the simplest of the permanent identification methods.

1. Hold the chick or poult with one hand, using your thumb and index finger to steady the leg and shank and expose the web between the toes for punching.

2. With your free hand, apply the toe punch to the web between the toes. Center it in the web. Make a clean-cut hole.

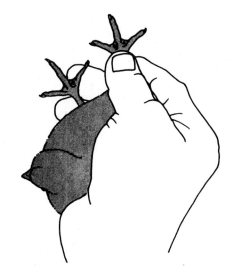

3. Remove the punched-out skin from the hole. If the punched-out skin is not completely removed, the hole may heal over and close.

4. Using both webs on each foot, in all possible combinations, 15 identification patterns are possible. These could be used to identify sire or dam families, different hatches, different strains, different years, or other variables.

5. As the bird grows in size, the hole also grows and can be easily seen. It is possible for the hole to tear out when it is snagged on a protruding object, but even if the hole is torn, the evidence that the web has been punched should remain.

Chicks and poults are toe-punched most easily at 1 day of age. There will be little or no bleeding. If they are kept in chick boxes for an hour or two after punching, any evidence of blood will have disappeared, so there should be no special attraction for others to pick at the toe or foot. At older ages, the danger of a drop or two of blood running from the punched-out hole is greater, and the opportunity to hide the blood from the chicken's eye is less. Thus the danger of triggering an outbreak of toe pecking and cannibalism is greater.

Wing-Banding Chicks and Poults at 1 Day of Age

The wing band is one of the most widely used types of identification. It is applied at 1 day of age and encircles the front edge of the wing web. With identification and separation of eggs in the incubator, chicks and poults can be marked to identify their sires and dams. As each chick or poult is removed from the hatching tray, a numbered band is inserted in the wing web. This identification number is then recorded under those of its sire and dam. This procedure is called *pedigree banding*. The wing band identifies the chick or poult for life.

Wing bands are lightweight, made of aluminum or plastic, and can be stamped with any combination of letters and numbers. One end of the band is pointed to make it easier to push up through the underside of the web. This end also contains a round hole to receive the rivet, which is located on the other end. After the band is inserted, the two ends are brought together around the front edge of the wing and sealed with special banding pliers. When placed properly, the band will encircle the tendon that goes from the bird's shoulder to its wrist. This tendon is relatively tough, and it will hold the band in place without restricting muscle, skin, bone, or feathers.

CAUTION: Some chicks inadvertently catch their bands on an object in the pen or hook their toe in the band and are unable to free themselves. If not found and freed quickly, the chick will eventually free itself by tearing the band from its wing. The band will be lost and identification of that chick is impossible if more than one band is lost. No more than 1 or 2 percent of bands that are properly placed should get lost.

---

1. Pick up the chick or poult with your left hand, with its head up and pointed toward your fingertips. Position its body in the palm of your hand, with its head up between your middle and ring fingers. Use your ring and little fingers to hold the body, with your little finger between its legs. Place your middle and index fingers over the chick's back and over the top of its wing so that the web is under your fingertips. Use your thumb and index fingers to grasp and spread the wing to expose the web.

This positioning is used for banding the chick's right wing. If the left wing is to be banded, the head should be up and held between your thumb and index fingers. Your index and middle fingers should be over the top of the wing, and your thumb and middle finger used to grasp and spread the wing. Notice that the fingertips of the middle and index fingers are over the web so that when the pointed end of the band is inserted from the underside, it will come out between them. This position gives

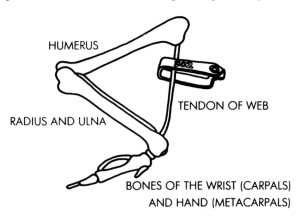

HUMERUS

RADIUS AND ULNA

TENDON OF WEB

BONES OF THE WRIST (CARPALS) AND HAND (METACARPALS)

support to the web so that the band does not tear out during the insertion process.

2. Grasp the band with the rivet and bent end between the thumb and index finger of your free hand, with the pointed end free and facing up and away from the thumb, ready for insertion in the web.

3. With the pointed end, come up through the web from the underside, aiming the point between your index and middle fingers, which lie on top of the web.

> **CAUTION:** The band should be inserted about halfway between the front edge of the web and the elbow joint, nearer the bird's shoulder than its wrist. This will keep the band from slipping over the wrist of the young chick, where it might interfere with normal growth. The skin of the web will resist puncture by the point, but increased upward pressure with your fingers will overcome this resistance. With practice, you will learn just how much pressure must be applied. Avoid setting the band too close to the bone (humerus). A heavy muscle will develop later in this area.

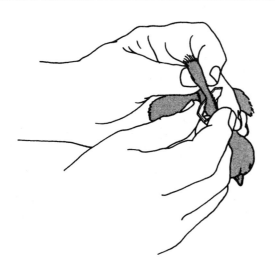

4. Bring together the open ends of the bands so that the rivet goes into the hole in the pointed end. The thumb and index finger of the hand holding the chick can be used to press and hold the open ends together until the rivet is set.

> **CAUTION:** Avoid pulling on the band. The tendon that is the main holding force keeping the band on the wing is still soft and weak in newly hatched chicks.

5. Use the hand that inserted the band to pick up the banding pliers and flatten the head of the rivet so that it cannot slip out of the hole.

> **CAUTION:** With the open-type rivet, the projection on the inside of one of the jaws of the banding pliers goes into the open end of the rivet. Apply just enough pressure with the banding pliers that a definite mushrooming of the rivet occurs. This will seal the band. Too much pressure can crush the rivet through the hole, in which case the band is not sealed.

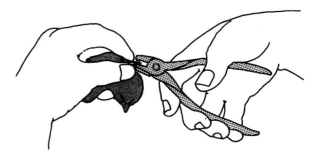

6. To help in finding the bands on the birds later on, band all chicks or poults on the same wing.

> **CAUTION:** Occasionally a band will tear out of the wing during the banding process. When this occurs, there is not much that can be done except to band the other wing.
>
> Bands may cause trouble by sliding over the folded wing at the wrist. If this is not corrected, the inflexible band will cut into the muscle and bone as the wing grows, resulting in an ulcerated sore and deformity of the wing. Make a practice of checking all wing bands at 10 days of age when bands locked over the end of the wing can be slipped over the wrist and back into a normal position with a minimum of difficulty and damage. When the band cannot be slipped over the wrist because the wing has grown too large, about the only thing that can be done is to cut off the band with a wire cutter or scissors and replace it.

Wing-Badging Adult Chickens

Plastic wing badges are available in different colors, lettered and numbered. They are used mostly to mark hens that are being trapnested so that the identity of the hen can be marked on the egg that she laid. In contrast to wing bands, wing badges do not puncture the wing, but fit completely around it at the shoulder. They are much larger than wing bands, stand upright on the shoulder, and can be read at a distance of 6 to 8 feet without difficulty.

Because the numbers on wing badges are much more easily seen and read than those on wing bands, a badge is used in addition to the wing band in large flocks of trapnested hens in which identification of individual hens is made almost daily.

Badges increase the speed and accuracy with which identification is made, which justifies the cost of the extra badges and badging. For pedigreed chickens that are not trapnested and that are identified only a few times in their lives, the extra cost of a wing badge is not justified, and the wing band alone is adequate identification.

Although holding and badging can be done by the same person, it is much quicker and easier if one person catches and holds the hen while the other slips the clasp of the badge around the wing and locks it.

1. One person holds the hen by its legs in one hand, with the palm of the other hand under its keel. Hold the bird out in front with its body upright and facing a badger at a height convenient for him to work on the wing. (If an empty chicken crate or similar object is available, set it on end. The holder can then lay the bird's body on the crate, controlling its movement by holding its legs stretched to the rear.)

2. The badger slips the clasp of the badge over the wing at the shoulder so that it can be read from the side (numbers face away from the side of the bird).

3. Lock the ends of the clasp together.

4. Adjust the feathers of the wing so that they fit under the badge and clasp in such a manner that the numbers are not hidden.

CAUTION: In extremely cold weather, plastic becomes very brittle and the clasps will break more frequently under the stress of being stretched over the wing. Badge during warm weather if possible, or plan to keep the badges warm when applying them in cold weather.

Leg-Banding Adult Chickens and Turkeys

At least two types of leg bands are available. Aluminum ones, carrying different types of information such as the year, type of program, letters, and numbers, have been widely used in the past in poultry-improvement programs. They are sealed on the leg and generally stay until cut off with a wire cutter or scissors. The other type is the celluloid spiral ring, available in various colors and widths, which is easily slipped over the shank. Both types are available in sizes that vary from small enough to fit a 1- to 1½-pound bantam hen to large enough to accommodate a 30- to 40-pound tom turkey. It is important to use bands that fit snugly enough that they will not slip over the foot, yet loosely enough that they will not cut into the shank as the bird grows.

As with wing badging, leg banding can be either a one-person or a two-person job. If the only job to be done is leg banding, a two-person approach is easier, quicker, and recommended. If leg banding is to be one step in a chain of jobs being done on the same birds at the same time (such as drawing blood for blood testing, reading wing bands, recording data, or vaccinating, where the birds are already caught and an extra pair of hands may not be available to hold), it may be necessary that the bander also hold the bird. In the following example, it is assumed that you will hold the bird while banding it.

1. Sit with the bird in your lap facing you, its legs stretched to its rear and its hock joints positioned above one knee.

2. Cross one of its legs over the other, bringing it down between your knees. Hold the shank and foot of this leg with your knees.

3. The other leg remains stretched out across your knee and is held in place by the crossed-over leg.

4. Slip the spiral band over the shank as you would slip a key onto a spiral key ring.

5. With an aluminum band, wrap it around the shank and put the rivet of one end in the hole of the other. Hold the two ends together with the fingers of one hand. Squeeze the rivet with banding pliers just enough to cause the rivet to mushroom and form a seal.

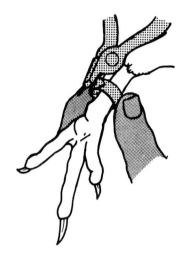

Celluloid spiral bands go on easily, but they are also very easily removed. Over a period of 3 to 4 months, most celluloid bands will get caught by a toenail or some protruding object in a pen and be ripped off. They are at best good only for a period of a week or two before they begin to get lost. Aluminum bands rarely get lost if they are of the proper size and are sealed properly. To remove them, cut them with a wire cutter or tin snips.

9.5 INSPECTING FOR DEFECTS, ABNORMALITIES, EXTERNAL PARASITES, AND MOLT

Inspecting chickens or turkeys individually for evidence of disease, infestations of parasites, or abnormalities of one kind or another is becoming a lost art. There are so many signs of trouble that can be picked up from a 30-second examination that a good poultryman will periodically pick up a bird from his flock and give it a quick once-over.

There are numerous instances in which chickens or turkeys should be inspected individually. Any bird, male or female, that is to be used for breeding should be examined carefully. Only those that are free of defects and abnormalities and those that conform closely to some standard that you have set should be used in matings.

Choosing the best birds from your flock to enter in a poultry show requires careful inspection. The *American Standard of Perfection,* a book published by the American Poultry Association, describes completely what each of the recognized breeds and varieties of fowl should look like from comb to toe. A show judge will be constantly evaluating each bird by the standards set forth in this book. Defects and deformities of the plumage, comb, beak, legs, toes, and other body parts also are described in the book. Some of these will disqualify a bird from competition; others will detract from its overall excellence, but still allow it to compete. Many of these defects and abnormalities can be found only by individual handling and inspection.

Judging for past and present egg production requires an inspection of the vent, pubic bones, abdomen, keel bone, shanks, beak, and eye ring. Pullets with deformities or abnormalities that are known to reduce egg production should be removed from egg-type breeder flocks. Commercial production of broilers, roasters, turkeys, or egg layers requires periodic handling and inspection of a few randomly selected individuals from a flock for evidence of disease, body condition, weight, and other indicators of abnormal conditions. It should become automatic to walk through the pens once or twice a day, pick up a bird now and then, and look it over.

Restraint Required

It is assumed that a flock of chickens or turkeys will be confined to a pen or cage. They can be caught by using one of the techniques described in Section 9.3. Some birds may be tame enough that they will not run away when you enter the pen, but will allow themselves to be picked up and inspected at any time.

Step-by-Step Procedure

See Section 9.3 for the proper way to hold and handle chickens.

1. Use your right hand to grasp the hocks of the bird in the correct manner, with its breastbone lying along your right forearm and its head toward you. Examine the beak, eye, comb, wattles, face, and mouth. Look for such abnormalities as a crossed beak, deformed eye opening, pupils that are other than round, an iris that is other than orange, or sores on the face, comb, or wattles (other than those caused by fighting or accident).

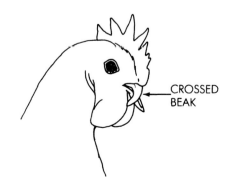

CROSSED BEAK

2. Starting with the neck, spread the feathers and look at the skin on various areas of the body. Look for new, immature feathers (pinfeathers), which indicate that a molt is in progress. A molt is the shedding of the old and the growing of a new feather. In a young, growing chicken, molting is a natural and continuous process. At 6 months of age, a pullet before it starts to lay should have completed its final molt. At approximately 5 to 6 months of age, a male should stop molting and be covered with adult feathers. During the first season of lay, which in a commercial egg-laying flock lasts from 12 to 14 months, the hen (and rooster) should not molt. Molting is undesirable in a laying flock because it indicates a lowered rate or cessation of egg production. It is a sign of an adverse happening that upset the hen enough to trigger the molt. Because a complete molt from head to tail can take from 6 weeks to 6 months, a good poultryman would want to seek the cause and correct it as quickly as possible. Some common causes of molt are disease, going without feed or water for a day or two, moldy feed, failure of lights to turn on at the normal time in a lighted laying house, and moving birds to strange quarters. It pays to know what is going on in your flock.

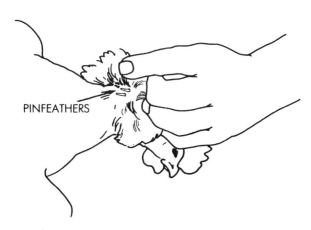

PINFEATHERS

Not all molting is unnatural or abnormal. Nature provides for a yearly change of feathers in adults in all avian species by way of a molt. An annual molt is usually triggered by a shortening of the day in late summer and fall. The presence of an occasional molting bird in a flock of several thousand is not something that should be cause for alarm. The point being made is that in a flock of first-year layers that has been genetically selected for egg production, a general molt is an unwanted condition because it indicates lower egg production.

3. Run your hand along the bird's back. It should be flat and straight. Any curvature or humpy condition is not normal. Continue moving your hand over the hips to the end of the tail. Occasionally, the rump (tail) will be missing.

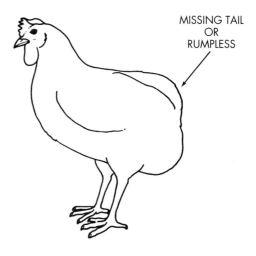

MISSING TAIL
OR
RUMPLESS

CAUTION: Birds with defects of the back and tail should never be used for breeding because the condition may be hereditary, and the abnormality may interfere with mating and cause a bird to be infertile. It is not likely that these defects would reduce the number of eggs laid, but they might reduce reproductive efficiency.

4. Examine the breastbone (keel) by running your hand along it from front to rear. It should be straight and free of curvature. The skin over the keel should be smooth and free of blemishes. A common abnormality that occurs just under the skin along the breastbone of broilers, roasters, and turkeys is a

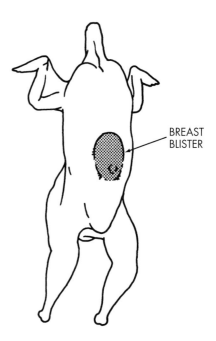

BREAST
BLISTER

breast blister (a raised, fluid-filled bubble) measuring an inch or two along the front end of the keel. In mild cases the blister will amount to no more than a slight puffiness. In severe cases, the blister may become infected and ulcerated and scab over.

CAUTION: Do not use birds with crooked breastbones or breast blisters for breeding. With the elimination of roosts in commercial production, a crooked breastbone is a rarity. Breast blisters are common, however, and seem to be associated with heavy, fast-growing birds, certain strains or stocks that are very susceptible, and such management conditions as wet litter. In broilers or turkeys being raised for market, either abnormality will downgrade the carcass.

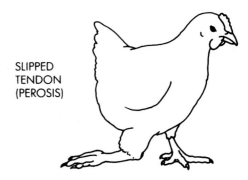

SLIPPED TENDON (PEROSIS)

5. Examine the vent and abdomen. Hold the bird by the legs so that its feet are up and its head down. Use your free hand to spread the feathers around the vent and look at the skin. Look at several areas above and below the vent. The skin and feathers should be clean. Look carefully for scabs on the skin or blackened feathers, which would indicate an infestation of lice or mites. The northern fowl mite, a tiny, black parasite, is a frequent invader of both hens and roosters and can be seen crawling on the skin and feathers of the vent and abdomen. As an infestation becomes more severe, the feathers become black with mites. Mites bother chickens and cause decreases in egg production, body weight, and fertility. Mites can be found crawling over your hands and arms after you handle infested chickens, but they are harmless to humans and can very easily be brushed or washed off. Treatment of infested flocks is recommended. Consult your county agent or university poultry or entomology extension agent for the latest in control methods.

Lice, which are much larger than mites and have brown bodies, can be found crawling over almost any part of the chicken, but are most evident in the vent area. They bother the bird, cause reductions in egg production, body weight, and fertility, and should be eradicated. Control methods change periodically, so it is best to check for the latest recommendations.

6. Examine the legs. A good pair of legs, placed squarely under the body, is essential for carrying the bird to the feed trough, for carrying its body weight, for proper mating, and for escaping from other birds that may want to do it harm. A bird that is crippled won't be able to grow or lay eggs efficiently. Such birds will not produce efficiently and take up valuable space. Hold the legs together in front of you where you can get a good look. The legs should be straight, not bowed. The hock joint should be trim, not swollen. There is a condition called *perosis*, or *slipped tendon*, in which the shank and foot are thrown out to the side. Perosis is a complicated abnormality that may be caused by a deficiency of certain minerals or vitamins. It is also considered to have a hereditary basis. Birds that have deformed legs should not be kept for any purpose. They should be culled immediately.

7. Examine the shanks. They should be thin, smooth, and covered with tightly fitting scales. Roughness of the shank or a lifting of the scale away from the shank is not normal and you should look for a cause. There is a mite that burrows under the scale and causes a condition known as *scaly leg*, which can be controlled by proper treatment. Check with the county or university inspection people for the latest recommendations.

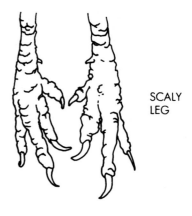

SCALY LEG

8. Examine the feet. The toes should be straight—three pointing to the front and one to the rear. Crooked toes of varying degrees of severity are common in broiler and turkey stocks. In certain cases the toes may be so severely crooked that they interfere with walking or standing. Breeders with crooked toes should be culled immediately, and keeping a bird so afflicted is questionable at any age. It should be culled.

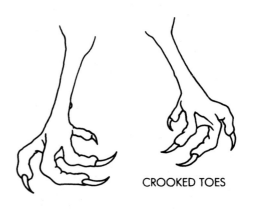

CROOKED TOES

Last, but not least, examine the ball or pad of the foot for infection or swelling. A severely infected foot can cripple a bird and render it useless. Because there is no good treatment for this condition, the bird should be culled. Birds that carry any of these abnormalities, no matter how slight, should never be used as breeders.

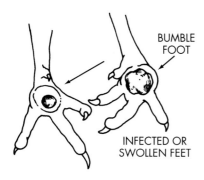

BUMBLE
FOOT

INFECTED OR
SWOLLEN FEET

9.6 PRODUCING AND INCUBATING FERTILE EGGS

Reproduction is one of the most essential parts of poultry production. Without the production of fertile eggs and their incubation to produce chicks, poults, and ducklings, the industry associated with these species would not exist. Research findings are constantly adding to our knowledge and understanding of the optimal conditions of mating and incubation to help us maximize reproductive efficiency. The art and skills associated with using this knowledge to successfully manage a breeding flock or an incubation system are also essential to the efficient production of quality chicks or poults.

Mating

The recommended age for the mating of cockerels and pullets is 6 months. Cockerels are sexually active as early as 4 or 5 months, but for maximum fertility, the placing of males with the females is usually delayed until they are 6 months of age. Pullets start to lay eggs when they are around 5 or 6 months of age, but the first eggs are small and their hatchability is low, so delay saving eggs for hatching until they are of medium size (23 ounces per dozen).

Only fertile eggs will hatch. To produce fertile eggs, place males in the pen with the females and leave them together as long as fertile eggs are required. A flock of 15 or fewer pullets requires one rooster. Any flock larger than this requires two or more roosters. Within a day or two, all females will have mated with a rooster and fertile eggs will be produced. A mated flock should produce at least 95% fertile eggs.

The recommended ratio of cockerels to pullets in a breeding flock depends on the breed and is as follows:

- Egg type
  Leghorns—1 male to 15 females
- Meat type
  White Plymouth Rocks—1 male to 12 females
  Langshan, Cochins, and Large Breeds—1 male to 10 females

Young males and females are more fertile and produce eggs with better hatchability than do old ones. The highest fertility (95%) and hatchability (90 to 95%) occur in flocks during their first laying

---

CAUTION: The presence of too many males in a flock may reduce fertility because of excessive fighting and interference of one male with another during a mating. Hold excess males in a separate pen and use them for replacements as needed. It is better to replace all males at once than to add a male now and then to replace one that died or is incapacitated. If you must put a new male in a breeding flock that contains several males, do it after dark to minimize fighting. Regardless of when you do it, much fighting between old and new males will take place as a new pecking order is developed.

---

season (usually from 9 to 12 months after the onset of egg production). Both hens and cocks are less effective during their second year. Fertility from 2-year-old cocks is often very disappointing, sometimes dropping to 50% or less. Hatchability from a 2-year-old flock is also less than from a 1-year-old flock. Mating 2-year-old or older hens with 6-month-old cockerels will give fertility and hatchability that are much superior to those obtained from older cock birds. Even in flocks in their first season of production, fertility and hatchability gradually decrease until they reach a level at the end of the season that is uneconomical. Some flocks, notably broiler strains of White Rocks, reach such a low level of hatchability after 9 months of production that they should be marketed and replaced for economic reasons.

Because sperm are stored in the oviduct in sperm nests, a single mating can produce fertile eggs for as long as 30 days. This means that even after all males have been removed, the hens will continue to produce some fertile eggs for several weeks. When changing males under conditions in which it is essential to know the sire of the chick (assuming that the hens are trapnested and the hen number is recorded on the egg), the old male should be removed and replaced with the new male at least 2 weeks before eggs are saved for incubation. This is sufficient time for the new sperm to replace most of the old sperm in the sperm nests. The new sperm will fertilize the eggs because they will be much more numerous and vigorous than the old.

Artificial Insemination

There are times when natural mating to produce fertile eggs cannot be used or relied upon. In the commercial production of turkey hatching eggs, artificial insemination (AI) is used to supplement natural mating. Even though toms and hens are kept together in the same flocks and allowed to mate naturally, all hens are inseminated artificially once every 2 or 3 weeks to ensure that fertility is kept as high as possible.

Natural matings sometimes produce a disappointing 40 or 50% fertility, which can be improved to 90% by supplementing with artificial insemination under some circumstances. The reason that natural mating often produces low levels of fertility is that turkeys have been selected and bred to have exceedingly wide and broad breasts. The male has become front-end heavy, and in attempting to mate with the female he often loses his balance and falls forward without completing a successful copulation.

There are several instances in which artificial insemination of chickens might be necessary. If fertile eggs are desired from hens caged individually in small wire cages, AI is the most economical and successful way to get the job done. Semen collected from males kept specifically for this purpose is inseminated into caged hens. Weekly insemination produces fertility in the 90% range.

Artificial insemination can also be used to improve fertility from matings that involve males in their second and third mating seasons. Two-year-old and older roosters often do not perform satisfactorily when mated naturally, in some instances producing zero fertility. If chicks are desired from these old sires, AI is often the answer.

Obtaining semen from a turkey tom or a chicken rooster and inseminating it into the oviduct of a hen is a skill that can be learned best from someone who is skilled in the art of AI. To become proficient in AI takes much practice under the supervision of a good teacher who can correct errors in placement of and pressure applied by the hands on the body of the bird, timing of stimulation, and holding of the bird. Describing the procedure is not the way to teach this technique. It must be shown. It is recommended that if you wish to learn the art of AI, you have it demonstrated sev-

eral times, then practice it under the watchful eye of an experienced inseminator.

Saving Eggs for Incubation

Produce clean eggs. To do this, the hens must lay in a nest that contains a good supply of such nesting materials as straw, hay, sawdust, and shavings. Eggs that are laid on the floor, cracked, misshapen, too large, or too small do not hatch well and should not be incubated.

Many hatch poorly because of poor internal environment. Dirty eggs may be contaminated with bacteria, which not only can be lethal to the embryo, but also may inhibit the proper exchange of gases and evaporation of water through the shell. Floor eggs in most cases have been prematurely laid and are deficient in thickness of shell. In addition, the germ needs the full time spent in the oviduct to develop properly.

Peewee eggs often have no yolk, and hence no embryo. Jumbo eggs often carry two yolks, and although each yolk has a germ cell that can start to develop, the nutrients in the egg usually are not sufficient or not in proper balance to support development of two chicks. (Twin chicks have been reported in the literature, but their occurrence is rare.) Why cracked eggs do not hatch is a mystery. If you patch the crack with cellophane or adhesive tape sometimes the chick will hatch. This seems to indicate that the crack in the shell does something to the internal environment of the egg that is lethal to the embryo.

Eggs should be gathered from the nests frequently. Three times a day is a minimum; more often is better. Once-a-day gathering results in more broken and dirty eggs, and even worse, it fosters the bad habit of egg eating.

Provide one nest for every three to five hens. Too many hens per nest results in more broken and dirty eggs, and hence fewer usable eggs.

Eggs that are to be incubated should be stored in a cool room with high humidity (60°F, 60 to 70% relative humidity). Most commercial hatcheries have specially built rooms that are cooled and humidified for ideal storage of hatching eggs. In the absence of a special room, a corner of a basement away from the water heater or furnace is satisfactory.

CAUTION: Do not store hatching eggs in a normal refrigerator. The temperature is too cold and reduces hatchability drastically.

Eggs should be incubated within a week after they are laid. If eggs are stored for more than 1 week, a 5 to 10% reduction in hatchability can be expected.

It helps to change the position of the eggs daily during the second week of storage so that the germ cell is put in contact with a different area of the egg contents each day. This is done by tipping cases from an angle of 45° in one direction to 45° in the opposite direction each day.

Sources of Fertile Eggs for Hatching

The best source of hatching eggs may be your own flock. If you do not have your own flock, contact a hatchery in your area or a farm that maintains a breeding flock. Your county extension agent may be able to suggest a source.

Eggs bought in a store (sold for eating purposes) will not hatch. Most are produced by flocks without roosters and most are coated with a thin layer of light mineral oil to preserve the interior quality by preventing the escape of air and moisture through the shell. (The passage of air and moisture through the shell is necessary for development of the embryo.)

Hatchability

Commercial hatcheries regularly achieve at least 85 chicks from every 100 eggs set. A setting hen may do even better. She may hatch 11 or 12 chicks from 12 eggs. With home-constructed incubators and small ready-made ones, a 50 to 70% hatch can be expected, which is an indication that somewhere in the incubation period the small incubator fails to provide the optimum conditions of incubation.

Incubation Period

The times required to hatch eggs from different kinds of poultry under optimum conditions of incubation are given below.

Type of Egg	Incubation Period (Days)
Chicken (including bantam)	21
Turkey	28
Duck (all except Muscovy)	28
Muscovy duck	35–37
Goose	28–34
Guinea	28
Pheasant	23–28
Pea fowl	28–30
Bob White quail	23–24

Incubators

An incubator is an environmental chamber designed to maintain constant temperature, humidity, and air quality. It provides a source of heat, moisture,

and a means of changing air. It can be very simple—for example, a cardboard box with a lightbulb for heat, a pan of water for moisture, and air-holes for ventilation—or it can be very complex, with heating coils, thermostats, humidistats, air inlets, air exhausts, fans to circulate the air, automatic egg-turning devices, alarms to signal trouble, and temperature- and humidity-recording devices.

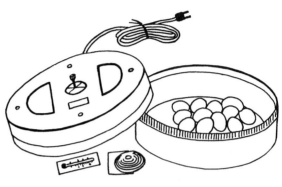

UTILITY INCUBATOR
(SMALL-SCALE)

Incubators are classified as either *still-air* (gravity circulation of air) or *forced-draft* (fans circulate air evenly and constantly throughout the machine). The main difference in the recommendations for operating the two types is associated with temperature.

In operating a ready-made incubator, follow the instructions provided by the manufacturer. Incubators differ, and each company has carefully researched the conditions under which its machine produces the best results.

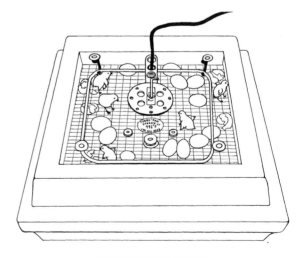

TEACHING INCUBATOR

If you make your own incubator, be sure that it can provide the optimum conditions discussed in the following section. These should be followed as closely as possible. Experience with your particular incubator will indicate adjustments that you must make to assume a maximum hatch.

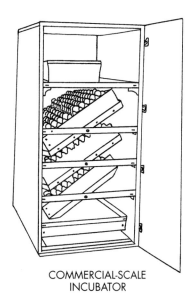

COMMERCIAL-SCALE
INCUBATOR

Location of the Incubator

The incubator should be located in a room in which the temperature remains fairly constant and that can be ventilated without creating drafts. A basement often meets these requirements better than most rooms in a house. The exact temperature of the room is not critical, but 70 to 75°F is desirable. Sunlight shining on an incubator can cause an undesirable rise in incubator temperature. Even short exposure to a temperature of 105°F is lethal to avian embryos.

Equipment Necessary

• Incubator

Small incubators can be purchased from Lyon Electric Co., P.O. Box 81303, San Diego, CA 92112; Fairsrew Hatchery-FVT, Inc., 18795 S 580 W, Remington, Indiana, 47977, (800)440-1538 and McMurray Hatchery, Inc., *www.mcmurrayhatchery.com*. Incubators can also be constructed from a number of different materials. Plans can be obtained from Feather Site at *www.feathersite.com/Poultry/BRKIncubation*. State University Cooperative extension offices are also excellent sources for plans.

Step-by-Step Procedure

1. Pretest the incubator. Be sure that it is in good working order before setting the eggs. Close the incubator and operate it empty until the temperature holds at the recommended level (99 to 100°F for forced-draft, 103°F for still-air).

2. When you are satisfied that the incubator is operating properly, set the eggs in the trays and place them in the incubator.

3. The optimum temperature for incubation in a forced-air incubator is 99 to 100°F. In a still-air

incubator, it is 101°F during the first week, 102°F during the second week, and 103°F during the third week.

Place the bulb of the thermometer in a still-air machine at the level of the tops of the eggs, but not touching them, and preferably in the center of the tray. Because all parts of the chamber of a forced-draft incubator should be at the same temperature, place the thermometer where it can be read conveniently.

> **CAUTION:** Check the different areas of the incubator trays for hot or cold spots. If these cannot be corrected, move the eggs to different places on the tray each day to equalize the conditions for all the eggs during the incubation period.

4. During incubation, the eggs should lose water slowly through the pores of the shell. A rapid loss of moisture is harmful. To prevent this, add moisture by keeping a pan of water under the egg tray at all times. The water pan should be approximately the size of the egg tray. Humidity is maintained by evaporation of water from the pan. Use warm water to fill the moisture pan. Do not let the eggs contact the water.

Maintain a 50 to 55% relative humidity during the first 18 days. Raise it to 65 to 70% during the last 3 days. A wet-bulb thermometer is used to measure the relative humidity. Psychrometric charts are available that translate the thermometer reading to relative humidity.

If a wet-bulb thermometer is not available, the proper humidity can be estimated by candling several eggs and comparing the size of the air cell at several stages of incubation with the standard shown here. If the air cells of the eggs are larger than the standard, increase the humidity; if the air cells are smaller, decrease the humidity. Although this is a trial-and-error method, it yields fairly good humidity control.

Humidity can be controlled also by the air openings in the incubator. If the humidity is too high, open the vents wider to move more air and increase

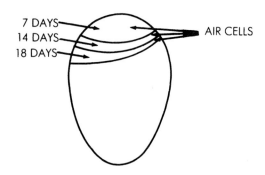

the evaporation of water from the eggs. If the humidity is too low, close the vents to cut down on airflow. Never close the vents completely.

CAUTION: Every time the incubator is opened, the temperature and humidity in the chamber drop quickly. They return to their proper levels slowly. If the incubator is opened and closed too often, the incubation period could be lengthened because of the low temperature. Avoid unnecessary opening of the incubator.

The humidity in the incubator is influenced by the humidity of the incubator room. Incubator humidity can be raised by increasing the humidity in the room by the use of vaporizors, by placing pans of hot water in the room, by washing the floor with warm water (if the floor is made of concrete and can be drained), and by other methods.

5. The developing embryo in the incubating egg needs oxygen and gives off carbon dioxide (CO_2) and water. Without ventilation, the oxygen level of the incubator air falls, while the concentrations of CO_2 and water increase. Ventilation in the incubator is a must. Provision must be made to let the stale air out and the fresh air in. Vents must be partially open at all times.

In a still-air incubator without a thermostat, temperature can be controlled by opening the vents wider to lower it and closing them to raise it.

As chicks begin to hatch, more air will be needed and the vents should be opened wider. As a matter of fact, when chicks begin to hatch the incubator temperature may rise sharply because of their body heat. Increased ventilation will be necessary to keep the temperature from going too high.

Humidity should be kept as high as possible during the hatching process to keep chicks from drying and sticking in the shell before they hatch. Some of the needed moisture is provided by the drying of the newly hatched chicks.

Ventilation during the hatching process is a compromise between ventilating to hold the temperature down and maintaining humidity as high as possible. If the temperature rises above the recommended level, increased ventilation may be the only way to bring it down. If this becomes necessary, the humidity must be increased by other methods, such as raising the humidity in the room, or increasing the surface area of the water in the incubator by using sponges or hanging pieces of cloth so that they will touch the water in the pan and act as a wick. Some incubators or incubator rooms are air-conditioned to control temperature.

6. Eggs must be turned at least three times per day during the first 18 days of incubation. In commercial machines, the eggs are turned automatically. In most still-air machines, the turning is done by hand.

Set the eggs on their sides in the trays. Mark an *X* with a pencil on one side and an *O* on the opposite side. If the Xs are up when the eggs are to be turned, roll the eggs with the palms of your hands so that the Xs are under and the Os are up. This way it is possible to tell whether all the eggs have been turned.

A simple turning schedule is first thing in the morning, again at noon, and the last thing at night. Eggs may be turned more often than three times per day, but the number of turnings should be an odd number so that the eggs do not spend every night on the same side.

In small incubators, move the eggs to different parts of the tray to offset variations in temperature that occur in different sections of the tray.

7. Do not turn the eggs during the last 3 days of incubation (days 18 through 21). Take steps to increase the relative humidity to 65 to 70%. Watch the temperature closely. Increase the ventilation if the temperature begins to rise. Open the incubator only to add water or make necessary adjustments. The chicks will start to pip the shell at about the 18th or 19th day. Those that are going to hatch should have done so by the end of the 21st day.

As soon as all the chicks have dried and fluffed up completely, they can be removed from the incubator. They should be placed temporarily in a covered chick box, 25 to a compartment, and held in a 70°F room where they will be allowed to rest and "harden" for no more than 24 hours before they are placed in a brooder house with feed, water, and heat. The hardening process can be omitted and the chicks placed directly into a brooding setup.

CAUTION: If the temperature runs low during incubation, it is possible for the hatch to be delayed a day or two. The longer eggs are held before incubation, the longer it takes them to hatch.

---

8. Incubating eggs can be checked for fertility and the presence of a viable embryo by a process called *candling*. Candling is done in a darkened room by passing a strong beam of light through the egg and observing the outline and color of the internal parts. The degree of embryo development will determine what can be seen in a fertile incubating egg. With white-shelled eggs (White Leghorn eggs), definite embryo development is seen around the fourth or fifth day of incubation. With brown-shelled eggs, the pigment in the shell makes it difficult to see much before the seventh or eighth day.

If the egg is fertile and the embryo is alive, the embryo can be seen as a small dark spot floating in the egg with a network of blood vessels extending from it in all directions. The blood vessels are bright red and give a faint reddish hue to the egg contents. If the egg is not fertile, it appears clear and only the faint, round outline of the yolk can be seen floating inside the egg. Dead embryos can be spotted because they are smaller than live ones and the red blood vessels are missing. In an egg that contains a dead embryo, the reddish hue is absent and instead the contents reflect a straw-colored hue.

Toward the end of the incubation period, the embryo has grown in size so that it fills the egg. The egg contents appear black because the embryo is blocking the passage of the light. Careful observation will reveal movement of the embryo within the egg—a kicking of its legs or a jerking motion of its body.

One of the dominant features of an egg in the last week of incubation is a large air cell. This can be plainly seen in the large end of the egg. The large air cell is necessary for the chick to hatch because when the chick is ready to hatch it breaks into the air cell with its beak, taking its first breath of air from the air cell. The air cell also provides room for movement of the chick's head and beak as it pips its way out of the egg. The size of the air cell at various stages of incubation is indicative of the evaporation of water from the egg and can be used to check that the proper level of humidity is being maintained in the incubator (see step 4 for details).

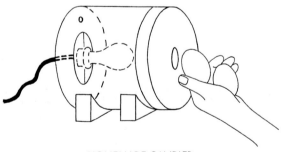

HOMEMADE CANDLER

A candler can be purchased from a poultry supply company, or one can be made from a shoe box, wooden box, or coffee can. Cut a hole about 1 inch in diameter in one end and mount a 40-watt bulb inside the container. Hold the egg in front of the opening with the large end up. Rotate the egg gently to cause the contents to float close to the surface of the shell where they are more visible.

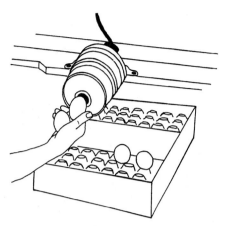

"BLACK-LAMP" CANDLER

CAUTION: Handle the eggs with care. A sudden jolt or jarring of the egg can be detrimental to the embryo. If you crack the shell, the embryo will die. Do not keep the eggs outside the incubator for a long period of time. They cool rapidly. Ten to fifteen minutes away from the heat is long enough. Although prolonged cooling of a developing embryo will not kill it, its development is slowed. Several of these long cooling periods may delay your hatch by several hours to a day.

9.7 BROODING CHICKS

There are two types of brooding: natural and artificial. In nature, the mother hen takes care of the brooding. She will provide temperature control, lead the chicks to the correct types of feed, and provide physical protection from injuries. Who has not seen the picture or drawing of a "broody" hen with her wings spread and feathers fluffed in a squatting position over her chicks, who are peeking out from under at all angles? It is a brooding system that has been going on for eons, because it works very well. The small poultry raiser who would like to brood some chicks is well advised to contact a poultry hobbyist and borrow a broody hen to take over when the incubation period has ended. The other option is artifical brooding. The remainder of this section will cover that topic.

The newly hatched chicks that have just come out of the incubators are at a critical stage in their lives. They require precise levels of temperature control, the correct amount of feed, and protection from injury. Fortunately, while the parameters for baby chick temperature are narrow and the new chicks do need the correct type of feed, it does not take a fancy facility to meet all of these needs.

Equipment Necessary

- Brooder
- Source of heat
- Bedding—wood shavings or sawdust
- Disinfectant

Step-by-Step Procedure

1. Chicks can be brooded in commercial brooding batteries, in a cardboard or plywood box, or on the floor under an electric or gas-fired brooder. Whichever type you choose, be certain to have the area prepared ahead of time so that the transition from incubator to brooder is as stress-free as possible.

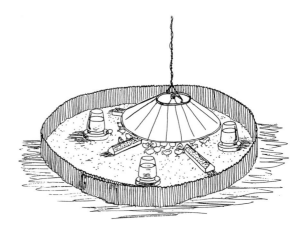

2. Clean and disinfect the brooding area and/or equipment a day or two ahead of the arrival or transfer of the chicks.

3. After the disinfectant is dry, bed the entire area with pinewood shavings or sawdust.

CAUTION: Avoid the use of hardwood shavings or sawdust. Hardwood bedding allows the growth of fungi that can cause the poultry disease *aspergillosis,* or "brooder pneumonia."

CAUTION: If a rodent problem exists, use a mesh subfloor below the shavings or sawdust. This will prevent rats from burrowing up into the brooder from below. A few rats can decimate the chick flock in a hurry, so be careful.

4. Turn on (or light) the brooders and adjust the temperature. The temperature under the brooder should be 90 to 95°F, about 2" to 2 ½" above the bedding.

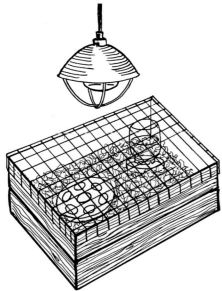

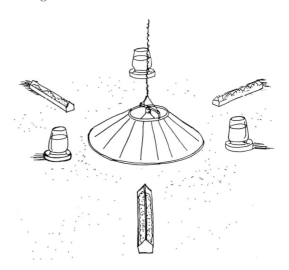

5. Place feeders and waterers in the brooding area before the chicks are introduced. Newly acquired chicks are easily frightened and will scurry

TOO HOT

TOO COLD

DRAFTY

CORRECT TEMPERATURE

about and bunch up on one another if they are disturbed by introducing the feeders and waterers after the chicks are in the brooder. One hundred baby chicks will require two or three 1-gallon waterers and 3 or 4 linear feet of baby chick feeders.

6. Stabilize the temperature and establish and follow a schedule for temperature control. Use the position of the chicks in the brooder area to determine whether the temperature is too high or too low. The schedule for brooder temperature reflects a 5°F decrease per week. After 5 or 6 weeks the chicks are capable of regulating their own body temperature.

Age of Chicks	Brooder Temperature
1st week	90 to 95 degrees F
2nd week	85 to 90 degrees F
3rd week	80 to 85 degrees F
4th week	75 to 80 degrees F
5th week	70 to 75 degrees F
6th week and beyond	70 degrees F

7. *Feeding chicks.* The best ration for chicks, from day 1, is a commercial chick starter (usually about 23% protein). Select one from a trusted feed supplier and follow the directions on the bag. Some poultry raisers place a few shallow lids, filled with starter, into the brooder area, so that the chicks will be enticed to start eating more quickly. Remove these after 4 or 5 days. Once the chicks are eating from the feeders, take care not to run out. Abrupt changes in ration can cause digestive problems. Each chick you have will eat about 10 pounds of starter ration in the first 10 weeks of its life.

CHICK FEEDERS

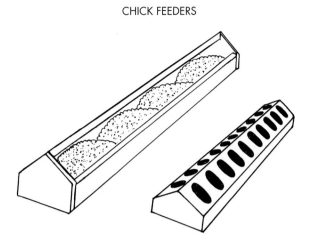

8. *Space requirements.* There are several formulas for calculating how many square inches of brooder space are required for each chick. If you have the space, allow 1-square foot per chick, from day 1 to 8 weeks of age. For 100 chicks, this will require a 10' × 10' area.

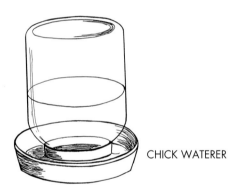

CHICK WATERER

9. *Vaccination program.* Most chicks that you will purchase from a commercial hatchery will have been vaccinated against Marek's disease at hatch. Unless you have a disease problem that is endemic to the area, there is not a real strong reason to consider vaccinating a broiler flock against other diseases. A good starter ration contains a coccidiostat and reduces the probability of coccidiosis in the chicks. If your broilers become sick, get a diagnosis from a good veterinary diagnostic center and treat according to their recommendations. If you are raising layer chicks, consider vaccinating for Newcastle disease, infectious bronchitis, and fowl pox in addition to Marek's disease.

10. *Sanitation program.* Sanitation involves more than using a disinfectant to clean equipment. It also includes feeding only fresh and nonmoldy feed, cleaning and rinsing the waterers daily, cleaning the feeders as necessary, and isolating the chicks from any other birds on the premises.

11. *Lighting program.* Chicks do not grow well in the dark. Especially during the first 2 days in the brooder, the chicks should have continuous light. This will help them become accustomed to eating and drinking. Most brooders will have an "attractor" light to help accomplish this. After 2 days, if there is natural light entering the brooder house, the attractor light can be turned off. If you are producing layers, there is a scheduled lighting regimen you can follow to help stimulate egg production in the pullets.

9.8 DUBBING AND DEWATTLING CHICKENS

Dubbing refers to the removal of the comb, and *dewattling* to the removal of the two wattles in chickens. Dubbing can be done most easily shortly after the chicks have been removed from the incubator. The wattles, however, are so small and hard to find at this time that it is impractical to remove them until they have grown more. Wattles are seldom removed from pullets, but in certain cases males are dewattled. Dubbing is sometimes practiced in flocks of White Leghorn pullets that are to be kept in cages for the production of table eggs. Some commercial hatcheries recommend that their pullets be dubbed and provide the service for a cent or two per chick at the hatchery. Not everyone believes that dubbing is a desirable practice, and some hatcheries do not recommend it. Those that do say that it increases egg production, and some research reports indicate that this is true. The comb of the White Leghorn hen that is kept in a cage grows abnormally large and is susceptible to cuts and scratches from parts of the cage and from the toenails and beaks of her cage sisters. As a result, the hen becomes very sensitive around the head and comb and hesitates to eat and drink if it means brushing her comb across parts of the cage. This may discourage her from eating and drinking at a rate necessary for maximum egg production. For this reason, some poultrymen believe that it is better to remove the comb and eliminate this possible source of stress.

Cockerels that are to be kept for breeding in a commercial breeding flock may benefit from dubbing and dewattling. Again, such cockerels usually are a large-combed breed (such as the White Leghorn) in which combs and wattles may interfere with eating or drinking or freeze in extremely cold weather. Dubbed and dewattled cockerels are more vigorous and better able to defend themselves against domineering males of the flock.

Dub and dewattle cockerels when they are 4 to 6 weeks old. At this time their combs and wattles are easily seen and grasped, cutting is easy, bleeding will be minimal, and a clean, close cut can be made. At 1 day of age, dubbing is a bloodless, simple operation. As the bird grows older, dubbing and dewattling become more difficult, with more bleeding, until at sexual maturity the operation is almost a major one—the comb is thick, the wattles are large, and the bleeding profuse. It is recommended that cockerels be dubbed and dewattled at as young an age as is consistent with getting a good, clean removal of the comb and wattles.

Equipment Necessary

- Small scissors or toenail clippers (for day-old chickens)
- Larger scissors (for older birds)

Restraint Required

A day-old chick is held in one hand. For 4- to 6-week-old birds, one person holds the bird while the other operates the scissors.

Step-by-Step Procedure

Day-old Chicks

1. Hold the chick in one hand with its body against your palm. The chick's head should be positioned upward and toward the fingertips. The middle, ring, and little fingers should be curved around the body, with the index finger and thumb steadying the head.

2. Use the fingers of your free hand or the open blades of the scissors to separate the down in the middle of the head. Look just behind the beak and nostrils for the small, yellow comb, which is about ½ inch in length.

3. Starting at the front of the comb and using the tips of the scissor blades and a succession of short snips, shear the comb off as close to the head as possible.

CAUTION: If the comb is not cut off close to the head, much of the value of the operation will be lost because any comb tissue remaining will enlarge at sexual maturity and a large comb will result.

Cockerels—4 to 6 Weeks of Age

It is easier to dub and dewattle if two people are working together. One person restrains the bird while the other performs the operation.

1. Have someone hold the male by the shanks and wings, with its head up and pointed toward you.

2. Grasp the top of the comb with your thumb and index finger. With the scissors in your other hand, cut the comb off as close to the head as possible.

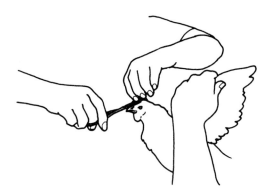

3. Grasp each wattle and cut it as close to the neck as possible.

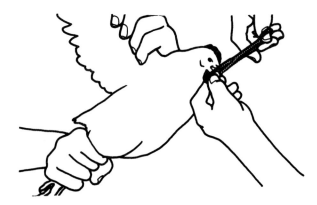

Postprocedural Management

No special treatment is required for chicks. The operation will not affect them. After an hour or so in chick boxes, they can be provided feed and water in the brooder house.

For older birds, no special treatment is required. Bleeding, if any, will stop quickly because of the rapid clotting inherent in chickens. Healing will be rapid. Keep dubbed and dewattled birds separate from nondubbed birds until healing is complete if only part of the flock is dubbed or dewattled.

CAUTION: The body temperature of the chicken is 107°F, and because of this high temperature it is not susceptible to infections by pus-forming bacteria that might be carried by human hands or the scissors used in the operation. To illustrate this, Louis Pasteur tried to infect chickens with the anthrax bacillus but could not do so because the bacteria could not tolerate a temperature of 107°F. Antiseptics or disinfectants are of little value if applied to the comb before cutting or to the cut surface afterward, and their use is not necessary. It is much more important to disinfect the scissors after they have been used on a group of birds, or after use on each bird, to prevent the transmission of avian diseases.

9.9 DESNOODING OF TURKEY POULTS

Desnooding refers to the removal of the snood. This structure is a tubular, fleshy appendage located on top of the head, immediately behind the beak. It bears the same relationship to the parts of the head on the turkey as does the comb on the chicken. It is plainly visible at hatching, resembling a small bump or tuft of down.

SNOOD

As the turkey grows older, the snood grows, becoming a prominent feature of the male's head— 5 to 6 inches long and ¼ to ½ inch in diameter, hanging down over the beak. The snood of the tom elongates and contracts with the activity and mood of the turkey. In the hen, the snood remains small, projecting upward about ½ inch above the head.

No specific function has been found for this structure. Because it is much larger in the male and reaches its maximum size and is most colorful in the mating season, it is likely that its function is related to the courtship of the female by the male, perhaps enhancing his masculinity and attractiveness. It is sometimes referred to as a secondary sex characteristic.

Because it is very large and prominent in the male, the snood is easily grabbed by other males during their numerous fights. Cuts, scratches, and bruises are common on the snood, providing a convenient entrance into the bird's body for disease organisms. This is considered to be an important means of spreading such diseases as erysipelas.

Because the snood appears to have no important function, and because of its possible role in the spread of disease, it has been a standard recommendation for many years to remove it in those turkeys that are grown for market. This operation is easiest to do on day-old poults by pinching it off with your fingers or cutting it off with a small scissors. The greatest advantage accrues to the toms; it is doubtful that desnooding holds much advantage for the hen. If your poults have been identified by sex, desnood the toms and leave the hens alone.

Equipment Necessary

- Small, sharp scissors

Restraint Required

The day-old poult will be confined to a poult box. Confine older turkeys to a small space within a larger pen.

Day-Old Poults

1. Pick up the poult with one hand. Hold it with its head up, beak pointed toward the tips of your fingers, and body enclosed by the palm of your hand and your middle, ring, and little fingers. Use your thumb and index finger to steady the head.

2. With your other hand, pinch off the snood by squeezing and twisting strongly with your thumbnail and index finger as you would remove a strawberry stem.

Poults of up to 3 Weeks of Age

1. Hold the poult in one hand in the manner described above.

2. Use small, sharp scissors to clip off the snood close to the head.

CAUTION: The younger the poult, the easier is the job and the less the effect on the poult. When a poult is 1 day old, desnooding is simple and bloodless. As the turkey grows older, it becomes harder to hold, and bleeding and shock are more likely. Plan ahead. Desnood when poults are 1 day old.

Postprocedural Management

After desnooding, let the poults remain in their boxes for an hour or two to allow the small area of exposed tissue to dry.

With older poults, pen the desnooded birds separately from those that have not been operated on.

9.10 TOE CLIPPING

Clipping the toes of day-old poults is a common practice among turkey growers. It prevents scratches and tears of the skin at later ages. Turkeys confined to a house or shelter, in contrast to those raised in large open fields, are more nervous, excitable, and crowded, and hence more apt to scratch and tear each other's skin. Thus it is almost a necessity to toe-clip confined birds. Turkeys reared on range also can damage each other with their toenails when they are being caught and loaded for market. Toe clipping is therefore good insurance to lessen the number of undergrade carcasses resulting from scratches and tears.

The inside and center toes of each foot are clipped. Clipping is done best and most easily, causing a minimum of bleeding and inconvenience to the poult, when the poult is 1 day old.

Baby chicks can also be toe-clipped. Some researchers recommend it for the prevention of "hysteria" in young pullets being grown in cages. Pullets may develop a sensitivity to the tearing and scratching of the claws of their cage mates, which occurs when they climb and crawl over each other while trying to escape when frightened or disturbed. Eliminating the causes of scratches and tears calms the birds. The clipping of the inside and middle toes of day-old birds has been shown to reduce or eliminate this hysteria.

Equipment Necessary

• 4- or 5-inch surgical scissors, or toenail clippers

Restraint Required

Baby chicks and poults will be in boxes and can be handled easily without restraint equipment.

Step-by-Step Procedure

1. Hold the day-old chick or poult in one hand, with its feet up and its head pointed toward your body. Grasp the legs between your thumb and index finger, with the bird's breast down and resting on the palm of the hand. Restraint is supplied by gently wrapping your middle, ring, and little fingers around the bird's body.

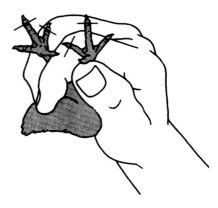

2. Use the scissors to cut off the tip of the toe just inside of the toe pad area. Include all of the toenail and a bit of the toe pad in the cut.

3. Remove both the inside and middle toes of each foot.

Postprocedural Management

The toe will bleed at the point of the cut. Keep the chicks or poults in their boxes for an hour or two after cutting to allow bleeding to stop and the wound to dry. There is very little danger of infection.

CAUTION: Clipping toes with an electric debeaker (described in Section 9.11) that cauterizes as it clips will prevent bleeding, but is much slower than the scissors method. Most turkey producers prefer to toe-clip with the scissors at 1 day of age because of the savings in time. Cauterizing is desirable and often necessary on older turkeys and chickens to eliminate bleeding.

9.11 DEBEAKING

In recent years, animal welfare activists have voiced opposition to the use of the term *debeak*. It is their contention that this word implies removal of the entire beak, which of course would be totally inhumane treatment. They have suggested an alternative term, one they feel is more descriptive of the actual process: *beak trim* or *beak trimming*. While their concern is well intentioned, the word *debeaking* is so ingrained in the management literature and language of the entire industry that it is unlikely to change in the foreseeable future.

Debeaking is the removal of a part of the end of the beak. Chickens, turkeys, ducks, pheasants, and quail are debeaked to prevent cannibalism. Debeaking is insurance against cannibalism because there are few flocks of poultry kept in confinement that will not feather-pull or cannibalize to some degree.

The loss from cannibalism of 1 or 2 birds in a flock of 10 or 20 is just as serious to the owner as the loss of 1,000 or so from a flock of 10,000. In either case, there is no way to predict when cannibalism will start or if it will stop by itself. Once it gets started in a flock, it never seems to stop.

In addition to losses from cannibalism, the habit of feather pulling (feather eating) leaves the carcass scabbed and covered with stubs of feathers, which results in a severe downgrading of the carcass. The actual and potential economic loss in flocks of turkeys, broilers, and ducks that are raised for meat is great.

There is no research that shows cannibalism is the result of a disease or a deficiency of vitamins or minerals. Disease is more likely to make the flock so listless and lethargic that the birds do not want to pick. In fact, one of the suggested cures for cannibalism is to feed a high level of salt in the ration for several days. This stops cannibalism temporarily, but only because a high salt level is toxic to poultry and makes the birds sick—too sick to be interested in this vice.

A vitamin or mineral deficiency may aggravate cannibalism, but research indicates that boosting the level of any specific vitamin or mineral in the ration does not prevent cannibalism.

Most commercial poultrymen accept the fact that any poultry kept in confinement—even for a short time and even if provided with adequate space, water, and amounts of a balanced ration—sooner or later get bored and look for mischief. The most likely expressions of this mischievousness are feather picking and the resulting cannibalism.

High temperatures, high levels of light, sunlight shining into a pen, and high stocking densities seem to induce or trigger cannibalism. Certain strains and breeds are more prone to cannibalism than others. Practices that help to control cannibalism are reducing the room temperature by increased ventilation, reducing the level of light (if you can read a newspaper in the pen, it's too light), blocking sunlight with burlap bags or paper over the windows, spreading the birds into more space, and keeping them busy and entertained by feeding whole oats or other grains in the litter several times a day. Hang green feed such as kale, alfalfa, or grass high enough off the floor that the birds have to jump to get a bite. As you can see, these recommendations are impractical in certain situations. In others, they would be impossible to implement because of the type of house or system in use or the large number of birds involved.

The most practical cannibalism control—the one that is effective under all conditions and for the life of the bird—is debeaking. Debeaking is a management practice that removes a potential source of trouble from a flock. Don't wait until cannibalism occurs to include debeaking in your management plans. Do it when it is best for the birds and best for you.

The factors discussed below must all be considered in debeaking.

Age. A chicken, turkey, or duck of any age can be debeaked, but as with other management practices, the undesirable effects upon production traits are fewer if it is done early in the life of the bird. Many poultrymen do not like to debeak when the birds are 1 day old. They believe that early feeding and drinking by a chick or poult is critical and nothing should be done that might interfere with the first bite or drink that the bird takes. Most commercially grown poultry are debeaked between 10 days and 3 weeks of age, before picking starts but after the bird has a good start in life. Debeaking of 1-day-old chicks has been shown to be feasible, however, and it is a practice used by some egg and broiler producers to reduce labor input.

Type of Poultry. Broilers, roasters, turkeys, egg-type pullets, layers, pheasants, ducks, and any other poultry kept in confinement may benefit from debeaking. Roosters are debeaked to reduce the damage that they may do to each other by fighting. Show poultry (fancy breeds and varieties kept only for exhibition) are not debeaked, because this detracts from their beauty.

Regrowth of the Beak and Redebeaking. The objective of debeaking is to prevent cannibalism for the life of the flock. To do this, enough of the beak must be removed to prevent regrowth to its normal length during the lifetime of the bird, but not enough to reduce the growth rate in a bird raised for meat or the production potential in an egg producer. It follows then that a broiler that will be marketed in 7 or 8 weeks will need less of its beak removed than will a turkey that goes to market at 6 months of age. The task of controlling beak length in a layer kept for 1½ years is much more difficult. In broilers, about one-third of the upper beak is cut off, whereas in turkeys, half of the upper beak is removed.

California workers have devised a method called "precision" or "block" debeaking for use on egg-type chicks. This procedure employs a Lyons Electric Debeaker and, when done according to directions, will permanently debeak the bird. It has the advantage of killing the growing points of the beak so that re-growth does not occur.

If redebeaking is needed in egg-type pullets or in pullets that are to be used as breeders, it can be done when they are handled to move them from the growing house to the layer house.

Equipment Necessary

- Pocketknife or paring knife
- Dog toenail clipper (can be purchased at any store where dog and cat supplies are sold)
- Lyons Electric Debeaker (available from NASCO-Fort Atkinson, 901 Janesville Ave., P.O. Box 901 Fort Atkinson, WI, 53538-0901, [920]563-8296).

The Lyons Electric Debeaker is a machine that cuts off and cauterizes a portion of the beak. It consists of a cutting blade that is heated to a cherry-red color, a support bar on which the beak is held, and a foot pedal that operates the blade. Depressing the foot pedal brings the red-hot blade down through the beak. The debeaker can be fastened to a table and operated in a sitting position, or a tripod can be purchased to hold the debeaker at standing height for the operator.

Debeaking of baby chicks is made easier by a guide plate which takes the place of the beak support. The guide plate is equipped with holes of three different sizes, the choice of which depends upon how big the chick is and how much of the beak you wish to remove. The chick's beak is inserted into this hole and the hot cutting blade moves down immediately behind the guide plate, cutting off the end of the beak.

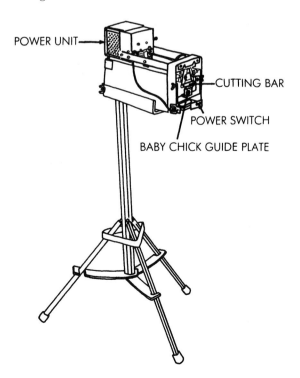

Another attachment that makes chick debeaking semiautomatic is a power unit that attaches to the debeaker. The power unit is activated by pressing a bar on the front of the debeaker, which moves the blade down behind the guide plate and through the beak. The power unit automatically holds the cutting blade in a down position for 2 seconds to cauterize the cut face of the beak. The blade then returns to its starting position.

Restraint Required

The birds are confined to a chick or poult box, small pen, or cage so that they can be caught easily and held in your hand.

Step-by-Step Procedure

Commercial Egg Layers or Breeders

Debeaking Chicks 1 to 10 Days of Age with the Lyons Electric Debeaker, Baby Chick Guide Plate, and Power Unit.

1. Hold the chick with its head up in one hand, using your palm and middle, ring, and little fingers to restrain the chick. Place your thumb on the back of its head and your forefinger under its throat. You will use your thumb to push and hold the beak in the hole of the guide plate. With your forefinger, pull back gently on the throat to retract the tongue to avoid burning it and to withdraw the lower beak slightly.

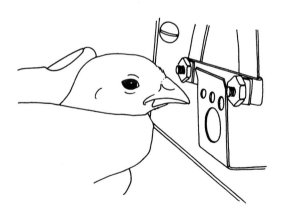

2. Insert the chick's closed beak into one of the holes in the guide plate (use the hole that will allow about one-half of the beak to be removed). Apply sufficient pressure with your thumb to hold the chick's beak in the hole.

3. The cutting blade should be heated to a glowing cherry red (approximately 1,500°F). Push the time bar on the front of the debeaker to activate the power unit. This will cause the cutting blade to come down behind the guide plate and cut off the portion of the beak that is protruding from the back side of the hole. The blade stays in the down position for 2 seconds. Continue to hold the beak against the hot blade for these 2 seconds to cauterize the cut surface.

CAUTION: Do not let the chick pull its beak away from the cutting blade. Keep the pressure on the back of its head with your thumb. The 2-second cauterization time regulates the amount of beak tissue killed in back of the cut, which is extremely important in preventing regrowth of the beak. If done properly, the debeaking is permanent.

Hold the chick's beak level in the guide-plate hole while debeaking. If the beak is tipped up, too much of the lower and not enough of the upper beak will be cut off. If the beak is pointed down, too much of the upper and too little of the lower beak will be removed. The cut should be square and perpendicular to the inside edges of the upper and lower beaks.

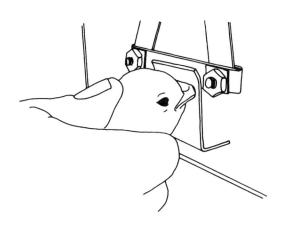

4. When the blade returns to its up position, drop the chick back into its box or onto the floor of its pen or cage.

5. Use your free hand to pick up and hold another chick, which can be passed to the debeaking hand when needed.

Debeaking Pullets 2 Weeks of Age and Older, Adult Hens, and Roosters with a Lyons Electric Debeaker, Minus the Chick Guide Plate and Power Unit.

1. Hold the bird by the legs in one hand, with its head up.

2. Use your other hand to grasp the bird's head between your thumb and forefinger, with the back of the bird's head in the "V" formed by your thumb and forefinger.

3. Force the tip of your forefinger between the upper and lower portions of the beak. Keep the beak open by sliding your forefinger between the upper and lower beaks.

4. Press the upper beak over the beak support of the debeaker.

5. The red-hot blade is brought down through the beak by pressure on the foot pedal, cutting off and cauterizing half of the upper beak. Leave the lower beak intact.

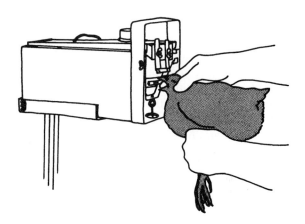

6. Hold the cut surface against the hot blade until cauterization occurs and bleeding is prevented. One or two seconds should be sufficient.

7. Debeaked birds can be returned to their pens or cages immediately. They will start eating and drinking quickly.

CAUTION: Because the upper beak has been removed, its natural wearing action on the lower beak will be absent. In time the lower beak may grow abnormally long and interfere with eating, especially in pullets kept for over 6 months after debeaking. This can be corrected by trimming the lower beak back to its original length with the electric debeaker. To avoid rehandling of birds that are to be held for more than 6 months, the lower beak can be cut back when the upper beak is removed. The lower beak should be left ⅛" to ¼" longer than the upper beak.

Debeaked birds cannot pick up feed from the bottom of the feeder. Keep at least ¾ inch of feed in the feeders at all times.

Debeaking of layers may cause a temporary slowing in the rate of lay, but the flock will return to a normal rate in a few days.

Emergency Debeaking with a Knife

In some flocks, chickens or turkeys may have to be debeaked immediately to prevent further deaths from cannibalism. An electric debeaker or dog toenail clippers may not be available. A sharp knife, either a pocket or kitchen variety, will be adequate.

The operation is simple and consists of peeling off the hard outer layer of the tip of the beak.

1. Make a short cut across each side of the upper beak, about ¼ inch back from the tip and just deep enough to cut through the hard outer layer.
2. Loosen the tip edge of the cut with the knife-point so that it can be grasped between the sharp edge of the knife blade and your thumb, or with a pliers.
3. Peel off the hard outer covering by pulling it toward the tip and the opposite side. There is little or no bleeding with this type of debeaking, and eating is not interfered with. The exposed soft tissue on the tip is sensitive enough to deter feather pulling and cannibalism. The hard covering of the beak will grow back in 2 or 3 weeks and cannibalism may start again. Redebeaking with an electric debeaker or dog toenail clippers may be necessary to effect a more permanent control.

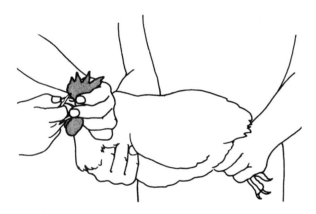

Debeaking Turkeys

Poults 2 or 3 Weeks of Age. Turkey poults, like chicks, are cannibals under conditions of confinement and often show cannibalistic tendencies at a very early age. Poults should be debeaked at an early age before cannibalism has a chance to start, but only after they have learned to eat, drink, and find the source of heat. Poults are more difficult to "start" than chicks, hence the reluctance to place any kind of stress on them until they have learned to eat and drink. Most turkey producers debeak poults at 2 to 3 weeks of age. If the young flock is not eating and growing well or is showing an abnormally high mortality rate, debeaking may be postponed until the situation improves.

Either an electric debeaker or a dog toenail clipper is used. Most turkey growers prefer the dog toenail clipper. Only the upper beak is cut off.

1. Hold the poult in one hand. The back of the poult's head is placed in the angle formed by your thumb and forefinger, which are placed along each side of the poult's head. Force the bird to open its mouth by pushing the tip of your forefinger between the upper and lower parts of the beak, near the angle of the jaw.
2. Place the upper beak in the cutting part of the clipper, so that a square cut can be made. Cut off about one-half of the upper beak (do not cut the lower beak).

CAUTION: Some bleeding will occur, but it should stop in 5 to 10 minutes. If all the birds are debeaked, they will be too busy taking care of themselves to bother others.

Most beginners are hesitant about debeaking and fail to remove enough of the beak. When this happens, the beak grows back to its normal length before the bird is ready for market. This may force another debeaking when the birds are too big to be handled easily, or so close to the market date that handling may do more harm than good.

Postprocedural Management

The cut surface of the beak needs no special treatment. It will heal quickly. Return debeaked birds to their pens as soon as the job is done. Allow them unlimited access to feed and water. Debeaking decreases the birds' ability to pick up individual bits and pieces of feed or to drink from shallow levels of water. Feed and water depths should therefore be kept above the ½- to ¾-inch level.

9.12 CAPONIZING

Although caponizing is a relatively simple operation, it requires a thorough knowledge of male avian anatomy, careful work, attention to detail, and many hours of hands-on, supervised practice. It should be attempted on your own only after thorough instruction from a skilled practitioner and numerous supervised operations.

Caponizing is the castration of male chickens. Turkeys, ducks, chickens, and any other type of poultry can be castrated, but only chickens gain enough in meat quality to make the practice worthwhile. The word *capon* refers to the castrated chicken.

Capons are kept until they are about 6 months of age, when they weigh 8 to 10 pounds. The statement is often made that they grow faster and are larger than normal cock birds, but this is not so. They do, however, put on more fat as they grow, and their meat is more tender. At 6 months of age they weigh more than their uncastrated brothers, mainly because of the excess fat they have accumulated.

The benefits of caponizing have been described as producing a more tender, tastier, sweeter, and juicier meat than is found in other types of chickens. Because of their large size and the fact that it costs more to produce them than it does to produce other meat birds, capons are considered a specialty item, produced for a particular market. Historically they have been most popular at Easter, with lesser demand at Thanksgiving, Christmas, and New Year's.

The consumer demand for capons has diminished, probably because turkey and chicken roasters are being produced that challenge the capon on size, tenderness, tastiness, and juiciness and cost the consumer less per pound of meat.

Capon production has costs that the raising of other types of meat birds does not, such as the cost of doing the caponizing. The capon producer's time and skill is worth something and must be charged to the capon. About 1% of the birds will not survive the operation. Then, too, because of the additional length of time that capons must be kept, they have poorer feed efficiency than roasters or turkeys. They have also been shown to be more susceptible to certain diseases (such as leucosis) because of decreased resistance caused by the removal of the testes and the resultant lowering of the levels of male sex hormone. Problems such as leg weakness and breast blisters plague them because of their weight. The net result is that the risk in raising capons is high. Because of this, capons should bring a higher market price than other meat birds. Such has not been the case in recent years because the capon market is very limited. Nevertheless, the demand to know the art of caponizing remains high.

Here are some factors that must be considered in capon production.

Breed or Variety to Use. Choose a breed or variety that is genetically capable of the efficient production of a high-quality meat. The White Rock is widely used. Any cockerel can be caponized, but a breed such as the White Leghorn, which is small in size and inefficient for its production of a low-quality, stringy meat, would be a poor choice. Certain strains of fast-growing White Rocks, although superior for broiler production, also do not work out well in capon production. They show a susceptibility to leg weakness and breast blisters when they are grown beyond the broiler age. Some hatcherymen offer a strain of White Rock that excels in capon production in that they are slightly slower-growing than most broiler strains, but much more resistant to leg weakness and breast blisters. It pays to shop around and locate a strain that is recommended for capon production.

Bird Health and Vigor. Caponize only healthy and vigorous cockerels. Weak, sickly birds will not survive the operation. A cockerel with an active case of respiratory or intestinal disease such as bronchitis or coccidiosis should never be caponized. Runts and undersized cockerels make poor capons.

Age. The best age for caponizing is 2 to 4 weeks, at a weight of $1\frac{1}{2}$ to 2 pounds. Caponizing tools are made to remove the testicles (which are no larger than a kernel of wheat) cleanly and in one piece. Cockerels that are $1\frac{1}{2}$ to 2 pounds in size carry testes that are the proper size for the tools to be used. When the size of the testis is larger than can be completely enclosed in the forceps or spoon, it is difficult to remove all of the organ in one piece and the danger of leaving a small piece of the testicle in the bird is great. Even the smallest piece of testicular tissue left in the bird will regenerate into a functional testis and result in a normal male. These birds are called *slips* and bring a much lower price per pound than capons.

CAUTION: Slips should be removed from a flock of capons when they are recognized because they try to mate with capons, scratching and tearing the tender skin of the capon in the process. This will downgrade the carcass and result in a lower price per pound.

Lighting. It is extremely important that the inside of the body cavity be illuminated enough during the operation that the operator can see the testicles. Set up your operating table on a clear day on the north side of a building, out of the direct sunlight. Direct sunlight makes it extremely difficult to see into the body cavity. In a building, an abundance of illumination from indirect lighting is desirable. The operator will have to experiment with light placement to obtain the desired illumination in the opened body cavity.

Good lighting, proper instruments, and adequate restraint of the bird are very important in producing a successful operation.

Equipment Necessary

- Table or barrel top on which to perform the operation
- Sharp knife or scalpel
- Spreaders (used to open the hole cut into the abdomen)
- Depressor and probe with hook
- Forceps for testes removal

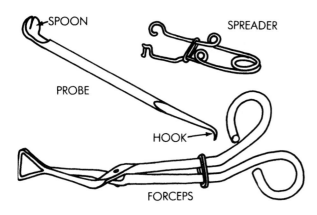

SPOON

SPREADER

PROBE

HOOK

FORCEPS

Restraint Required

The cockerel is placed on its side on the operating table, held by its legs and wings. An assistant operator holds the bird or straps are used to fasten the legs and wings to the table.

Step-by-Step Procedure

1. Cockerels should have no feed or water during the 24-hour period before the operation. This will empty the intestines so that they will fall away from the testes and the upper part of the body cavity. By placing the head somewhat above the feet (tipping the restraining board), the intestines can be made to fall even lower in the body cavity. If this is done, the operator's view of the testes will not be obstructed when the bird is opened, and the chances of puncturing the intestines with a knife or probe are greatly reduced.

Cockerels will eat their own droppings during this period of starvation, thus filling their intestines. Place the cockerels on a wire floor sufficiently high off the solid floor that their droppings will fall through and be inaccessible to them.

2. Fasten the bird to the operating table. The bird must be lying on its left side, with its wings together and above its body. The legs should also be held together and stretched to the rear to expose the sixth and seventh ribs. (Note: The muscles of the thigh normally cover the sixth and seventh ribs

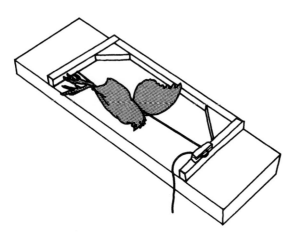

[the last two ribs of the rib cage]. The legs must be stretched to the rear to move this muscle away from the area of the incision.)

3. Pluck the feathers from the skin over the sixth and seventh ribs.

4. Use alcohol to clean the bare skin in the area of the sixth and seventh ribs.

5. Locate the last rib with your left hand, making sure that the thigh muscle is not covering it. (Note: The last rib does not have as much spring or arch as the sixth. As a result, it takes a noticeable drop toward the inside of the bird. This offset can be felt with your fingers to identify the area to be cut.)

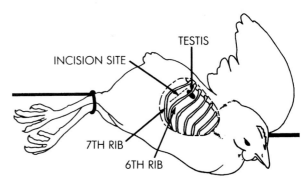

TESTIS

INCISION SITE

7TH RIB

6TH RIB

6. Hold the scalpel in your right hand and make a cut about an inch long between the last two ribs, cutting through the skin, muscle, and other tissue. Start the cut slightly below the upper limit of the abdominal cavity. The cut must be deep enough to expose the thin membranes of the air sacs and those covering the intestines.

CAUTION: The ventral part (bottom) of the seventh (last) rib is attached to the bottom portion of the sixth. If you feel your knife cutting through a rib, it is a signal that the cut is being made too low. Cease cutting toward the keel and enlarge the cut toward the backbone. If the cut is made too low, trouble may be encountered in finding the testes. If the cut is started too high, there is a danger of nicking the kidney with your knife and causing considerable bleeding.

7. Insert the spreader in the cut. Spread the opening to reveal the thin membranes covering the internal organs.

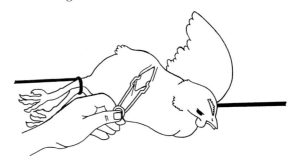

CAUTION: Be sure that the cut has been made deep enough that when the ribs are spread, the clear, colorless membranes are visible. Enlarge the opening to about 1 inch in diameter.

8. Use the hook on the probe to pick a hole in the membrane. Widen the tear in the membrane to expose the intestines and the testicles. Look toward the backbone—the testicles are found in front of the kidneys, attached closely to the tissue of the upper part of the cavity on either side of the backbone. They are usually yellow in color, but may be white, gray, or partially to completely black. They resemble grains of wheat, although occasionally they will be more elongate and narrow.

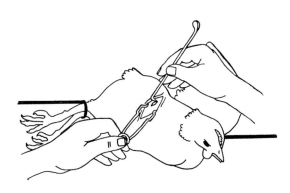

9. Remove the near testicle first, clamping the jaws of the forceps under it so that the whole testicle is enclosed in the jaws of the forceps. With a twisting, pulling motion, remove it through the opening. Another instrument that can be used to remove the testicle is a spoon. The bowl of the spoon should be at a right angle to the handle and split down the center. The spoon is slipped under the testicle, and the testicle is removed with a twisting, pulling motion.

CAUTION: When you remove the testicles, you grasp them, one at a time, with the forceps and twist. In doing so, you break them free from the cord they are attached to. As you do this, be certain that you grasp the entire testicle and that a part of it does not get left behind. In addition, take care that you do not accidentally grasp the artery that lies immediately behind the testicle. This artery can be ruptured as the forceps are twisted. If it is ruptured, the bird will bleed to death.

Every bit of the testicular material must be removed. That is why it is so important that the testicles not be any larger than can be clamped in the jaws of the forceps. Be careful when removing the testicle that it doesn't drop back into the body

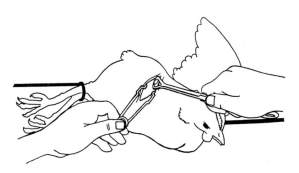

cavity. If the testicle remains anywhere in the body cavity, it will continue to grow. The cockerel will then develop masculine characteristics and be classified as a "slip."

10. The left testicle lies on the other side of the backbone in essentially the same position as the right testicle. To reach it from the same incision, push the intestine aside with your hook and spoon. When you have located it, use the forceps to remove this testicle, just as you did the other. If you cannot locate the left testicle through the initial incision, turn the bird over and make a second incision just as you did on the first side.

11. The incision does not need to be closed with stitches. When the rib spreader is removed, the skin of the incision will come back together on its own. However, no harm will be done if you wish to pour hydrogen peroxide over the wound. Some poultry raisers also coat the wound area with Neosporin antibiotic ointment. The bird can be released immediately.

Postoperative Procedures

Immediately after the operation, the caponized birds are returned to full feed and water. They can be penned together and they will not bother each other's wounds. Capons should not be mixed with other birds but should be kept as a separate flock until marketing time.

No special treatment is recommended for the wound. It will heal quickly and rarely, if ever, becomes infected. The wound will heal in about 10 days. Because an air sac, which is connected directly to the lung, was punctured in making the cut between the ribs into the body cavity, air will push up beneath the skin from the air sac and form a "wind puff." This is not serious, and will eventually disappear when the tissue between the ribs mends and closes the cut. Wind puffs can be relieved temporarily by puncturing the skin with a knife and forcing the air out past the blade. This relief will be temporary and the air will return under the skin until the cut between the ribs heals.

Although caponizing is a simple operation, it requires careful work and a knowledge of the chicken's abdominal anatomy. It is desirable for beginners to perform their first caponization under the guidance and direction of an experienced operator. Skill and dexterity can be developed by practice. If dead birds are available, practice on them.

9.13 PROCESSING POULTRY FOR FOOD

Production and processing go hand in hand when it comes to supplying an edible poultry product to the consumer. Premium-quality meat birds can be produced on the farm, but unless they are processed carefully they could end up inedible because of contamination with intestinal contents, too tough to eat because of improper handling or chilling, or off-flavor because of contact with chemicals, gall, or feces. Similarly, a superior job of processing is not going to improve a bird that was inferior while it was alive. Processing cannot fatten a thin bird, remove tears, scratches, bruises, or cuts, purify a carcass contaminated with insecticides or disease, or tenderize an old bird. Ready-to-cook poultry should be produced and processed with quality as the number one priority.

Poultry of any age can be prepared for human consumption, but keep in mind that the meat from poultry that is older than 6 months is more restricted in its use than that of younger birds because of a loss of tenderness with age. For example, Leghorn hens that have spent a year in a laying house or roosters that have been used in breeding flocks for 9 months to a year are generally processed into diced meat for chicken soup because their meat is too tough to be eaten fried or baked. White Rock hens that have spent 9 months producing hatching eggs will be processed and sold as stewing hens because it takes a boiling pot or pressure cooker to make them tender enough to eat. Broilers, roasters, and turkeys are grown specifically to produce a product that is tender enough to be barbequed, fried, or roasted.

Processing poultry for human food involves a number of specific steps from the live bird to the ready-to-cook product. The processing of broilers is discussed here. Modifications to these steps can be made as needed for processing other types of poultry.

Facilities and equipment for processing poultry vary widely. The simplest is a backyard setup in which the bird is killed by cutting off its head with an axe, scalded in a pail of boiling water, hung by the feet from a tree, defeathered by hand, eviscerated, and cut up on the kitchen table with a butcher knife. The more complex setups are aimed at volume and efficiency. They include virtual assembly lines of poultry moving through the many stages of processing from killing to eviscerating under the critical eye of the USDA inspector. The scalding, picking, and evisceration are automated, and all weighing, counting, and recording devices are electronic.

To simplify the discussion, only a small processing facility, with mechanized equipment and a line operation such as you might find in a university teaching plant, is considered here.

Equipment Necessary

- Sharp killing knives
- Metal killing cones
- Scalding tank—with thermostat to maintain a constant water temperature of 138 to 140°F and timer to aid in keeping the birds submerged for 30 to 75 seconds.
- Batch-type mechanical picker with timer—capable of picking the feathers from 10 to 12 broilers at one time. The picker is equipped with a set of rotating rubber fingers and literally scrubs the feathers from the birds as they bounce into the fingers.
- Evisceration trough—to receive the viscera and waste
- Gas-jet burner—to singe long hairs from skin
- Water sprayer—to wash and rinse carcasses

Restraint Required

The bird is placed in a killing cone to restrain it completely.

Step-by-Step Procedure

1. Grasp the live bird by its legs with its head down and place it into the killing cone, head first.

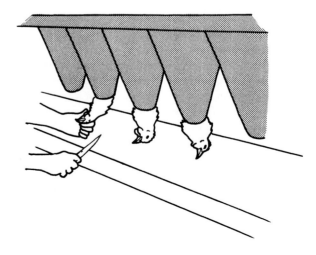

Do not allow the bird to tuck its legs under its body, but rather extend them to the rear so that they are pointing out the top (large end) of the cone. The bird should be wedged into the cone so that it can neither flop around nor escape.

2. Reach down and hold the bird's comb lengthwise between your thumb and index finger with the top of its head resting in the palm of your hand, throat facing up and toward you.

3. Push a killing knife into the soft tissue of the throat at the base of the skull. Insert the knife all the way through the throat, from side to side, as close to the skull as possible, with the cutting edge facing

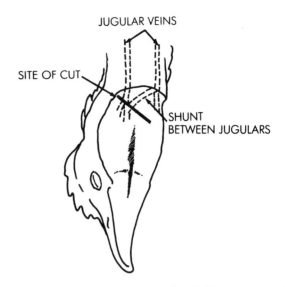

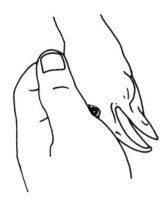

away from the skull. Cut through the tissue to the outside. If the cut is made properly, the bird will bleed freely because the jugular vein and its shunt have been cut.

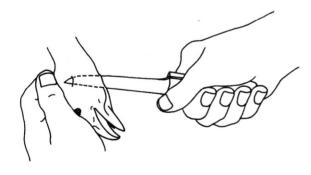

CAUTION: If the bird is not bleeding freely, recut deeper toward the base of the skull until you sever the jugular. If the jugular and the shunt that connects the right and left veins are not cut, bleeding will be incomplete and many of the veins of the skin will remain filled with blood, resulting in a darkened carcass with low keeping qualities.

---

4. Leave the bird in the cone until bleeding stops and struggling ceases.

5. Remove the bird from the killing cone and place it in the scalder. Submerge the bird for 30 to 75 seconds in 138 to 140°F water. In some scalders several birds are placed in a basket that rotates in the water, keeping the birds submerged and the feathers agitated for the desired length of time. The lower temperatures will require longer periods of submergence. If the feather picker does not remove most of the feathers, the submergence time can be increased. Conversely, if the skin appears to be cooked too deeply, causing several layers of the skin to be removed during picking, the time can be decreased or the water temperature lowered. Experience with the first few birds through the processing line will tell you whether or not adjustments are necessary.

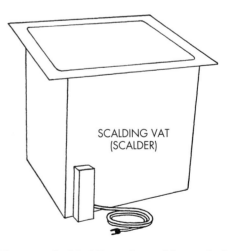

6. Remove the bird from the scalder and place it into a mechanical picker. This is a machine that looks like a large, old-fashioned washing machine. Inside the drum of the machine are rubber "fingers" that contact the bird as the tub or drum rapidly spins. Some mechanical pickers are large enough to process several birds at once, and the length of defeathering can be preset with a timer.

Another type of machine, more economical for small-scale producers, is a model that operates on the same principle as the drum type (i.e., spinning rubber "fingers"), except instead of placing the bird into the machine, you hold it in your hand and push it against the rubber fingers, which are attached to a revolving cylinder. With a little practice, this machine can do a very nice job of defeathering a bird.

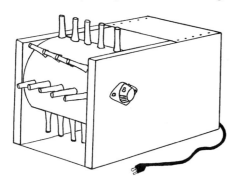

MECHANICAL PICKER
(DE-FEATHERER)

7. Take the bird from the picker and hang it by its feet from the shackles. Remove by hand all the feathers that the machine left.

8. Remove the pinfeathers (these are short, immature feathers in various stages of growth). Some may be so short that they cannot be grasped with the fingers. A pinning knife (a knife with a dull cutting edge) is excellent for removing the short pins. Grip the pinfeather between the edge of the blade and your thumb and pull outward. Do not scrape the pinfeathers with the knife, because a reddened, abraded area may result.

9. After pinning, singe the bird with an open flame from a gas torch to burn off any of the long hairs that were not removed in the picking process.

CAUTION: Pass the flame rapidly over the skin of the bird or rotate the bird rapidly over the flame to avoid cooking or burning the skin.

10. After singeing the carcass, spray it with water as it hangs from the shackle, washing all feather residue, blood, and dirt from the skin.

The bird is now ready for evisceration. There are two methods used to eviscerate broilers; the choice of method depends upon whether a whole carcass or halves for barbequeing is desired.

Ready-to-Cook Whole Broiler

1. Make a circular cut around the chicken's neck just behind the head, cutting through the skin and muscle to the vertebrae. Twist off the head.

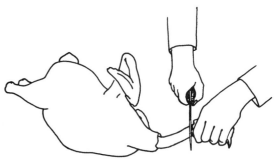

2. Remove the shanks by cutting through the hock joint. Remove one foot from the shackle, hold the leg straight out, and cut through the front of the hock joint, following the contour of the joint. Hang the part of the leg above the hock joint in the shackle. Do the other shank.

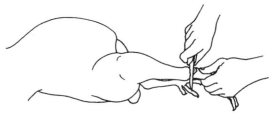

3. Slit the skin on the back of the neck from the point where the head was severed to a point on the

back between the shoulders. Peel the skin away from the neck and pull it down to the base of the neck.

4. Remove the esophagus, crop, and windpipe by pulling them away from the neck skin. Cut the esophagus at the point where it enters the body cavity just to the rear of the crop. The windpipe is pulled out at the point where it enters the body cavity.

> **CAUTION:** The esophagus and crop should contain no feed if the bird was starved for 12 hours before killing. If feed is present, do not rupture the crop and be careful when cutting the esophagus to avoid contaminating the neck and breast tissue.

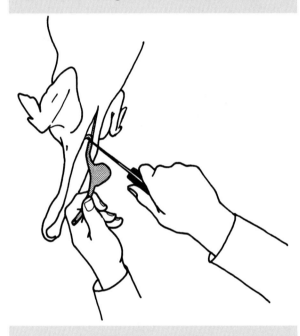

> The crop is a pouch of the esophagus located at the base of the neck in the V formed by the wishbone. It is used by the bird to store feed. The crop and the esophagus are very closely attached to the tissue under the skin of the neck. Be persistent in tearing and peeling them from the neck tissue until they are completely free.
> The esophagus narrows to its original tubular form as it disappears between the forks of the V in the breast into the body cavity.

5. Cut the neck from the body at the beginning of the back by making a circular cut around the neck as close to the body as possible. Make the cut at the junction of two vertebrae, if possible. Place the cutting edge of the knife in this cut on the back of the neck. Break the vertebrae apart by bending the neck back over the knife and cut through the remaining tissue to sever the neck.

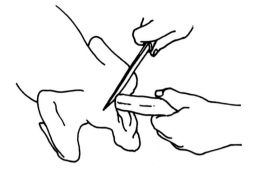

6. Remove the oil sac, located on the back at the base of the tail, by cutting under the sac to the backbone and up toward the tail. You can assist the operation by pulling up on the papilla of the oil sac with the fingers of your free hand.

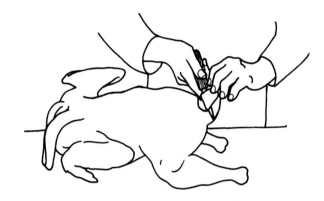

(The oil sac has two lobes and lies just under the skin at the base of the tail. The area of the oil sac is marked by a small papilla that projects about $\frac{1}{4}$ inch above the skin in front of the tail. The bird oils its beak by rubbing it over the papilla. The beak is then rubbed over each individual feather to deposit the oil. This whole process is called *preening,* and the gland often is called the *preen gland.*)

The content of the oil sac is a brownish orange, pastelike substance. All of the gland should be removed. A second, deeper cut may be necessary to remove it all.

7. Remove the viscera from the body cavity. Start by making a cut through the skin, underlying tissue, muscle, and fat into the abdominal cavity, beginning about an inch behind the end of the keel and continuing back to and around the vent. This cut should make an opening that is big enough to enable you to work the fingers of your right hand into the body cavity, freeing the intestines, gizzard, and liver from their attachments. Gradually work your fingers toward the front of the body cavity, freeing the heart and proventriculus (the glandular stomach attached to the esophagus, which was severed earlier). When the internal organs are fairly well worked free, pull them out through the opening in the abdomen.

CAUTION: Do not cut into or break an intestine, because the contents will spill into and contaminate the body cavity. Once the body cavity is contaminated it is almost impossible to cleanse it. A USDA inspector would condemn that carcass as unfit for human food.

Do not rupture the gall bladder, which lies between the two lobes of the liver. The greenish fluid of the gall bladder is the bile, which could impart a bitter taste to the meat.

---

8. Remove the gizzard, liver, and heart from the viscera. The heart and liver can easily be pulled away from the viscera. Cut off the intestines where they enter the gizzard.

9. Split the thick muscle of the gizzard lengthwise, cutting through the muscle down just to the lining. Peel out the lining, leaving the contents of the gizzard intact if possible.

10. Grasp the gall bladder at its base between the thumb and forefinger of one hand, holding the liver with the other, and pull the gall bladder away.

CAUTION: Grasp the gall bladder deep enough at its base, even if this means pinching a bit of the liver tissue, that there is no danger of breaking it. Hold the liver higher than the gall bladder so that if the gall bladder does burst, it will not contaminate the liver.

---

11. Trim the heart free from its sac and the large blood vessel that can be seen protruding from its large end. Wash it free of blood inside and out.

12. Wash the giblets (gizzard, liver, and heart). The giblets and neck can be chilled separately and tucked inside the body cavity of the bird at a later time.

13. Place the carcass and giblets in ice and water and chill them to 40°F as quickly as possible. The carcass should be chilled to 40°F or below in less than 4 hours if its weight is less than 4 pounds, in less than 6 hours if its weight is 4 to 8 pounds,

---

CAUTION: Enough clean, crushed ice should be used with clean water in a clean tank, vat, or other container to maintain the temperature under 40°F at all times.

Chilling poultry to 40°F or below immediately after processing and holding it as close to that temperature as possible until it is to be cooked is an important requirement for maintaining wholesomeness and top quality without freezing.

---

and in less than 8 hours if its weight is more than 8 pounds.

Halves, Ready to Barbecue

Broilers that are to be split or halved are often eviscerated in a manner different from that used for whole birds.

1. The head, shanks, and oil sac are removed as described above.

2. Leave the neck attached to the backbone.

3. Starting at the shoulder, cut along either side of the backbone to the tail. Continue your cut down to and around the vent. Return to the starting point at the shoulder and neck by cutting upward and along the other side of the backbone. Keep the knife shallow to avoid cutting into the intestinal tract.

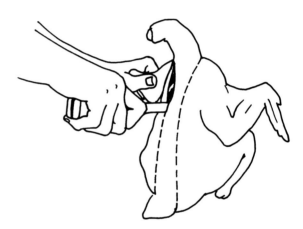

4. Strip out the backbone, tail, and neck.

5. Spread the sides of the carcass apart and remove the internal organs.

6. Wash the carcass thoroughly.

7. Trim the giblets as previously described.

8. The carcass may now be halved or quartered.

9. Chill as previously described.

10. If the carcass is to be cut into pieces, chill before cutting.

9.14 PROCESSING EGGS FOR FOOD

As with any perishable product, the initial quality of eggs cannot be improved upon. Therefore it is incumbent upon the producer to manage the flock to produce high-quality eggs and then maintain this high quality. Some of the factors affecting quality are as follows: (1) health and age of the flock; (2) space (square feet) allowed per bird; (3) environmental temperature; (4) litter and nest material management; (5) frequency of egg collection; (6) handling methods; (7) storage conditions; (8) cleaning, grading, and sizing. Each is discussed in turn.

1. The health and age of the flock can materially affect egg quality. Certain diseases cause not only a decrease in production, but dramatic quality deterioration as well. Pullet flocks produce much higher egg quality than 2-year (or older) flocks and higher quality than flocks that are force-molted.

2. Hens that are afforded proper space in which to move about and that have easy access to feed, water, and nests produce cleaner eggs with fewer checks (cracks).

3. Extremes in environmental temperature cause reductions in egg numbers and egg quality. Cold may result in frozen eggs, and temperatures of 90°F or above reduce exterior and interior quality, even when eggs are gathered frequently.

4. Litter (floor-covering material) management is essential to producing clean, sound eggs. Frequent stirring of litter (shavings, straw, peanut hulls, peat moss, etc.) and the addition or replacement of soiled material is mandatory in conditions of high environmental moisture and poor ventilation. This can also be true of nesting material.

CAUTION: Overcrowding of nests results in dirty and broken eggs, even when nesting material is clean and adequate. One nest per five to seven hens for small breeds, and one per three to five hens for heavy breeds are the recommendations. The material selected for nesting is often the same as the producer has selected for floor litter.

5. Eggs should be collected no less than three times per day, and even more frequently when the weather is extremely hot or cold. Approximately 80% of a flock's eggs are laid before noon, so more frequent A.M. gathering is recommended.

6. Handling eggs can help preserve initial quality or destroy it. Containers used to collect eggs range from hats, coat pockets, buckets, and boxes to plastic-coated wire baskets and fiber egg flats. Of these, the 30-egg fiber flats are ideal. They provide individual egg protection and an automatic counting

advantage. A cushion of three or four empty flats as a base provides support for up to 10 dozen eggs gathered in one stack. Because air movement is restricted when flats are stacked, cooldown is facilitated in the storage area if flats are separated.

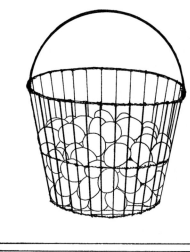

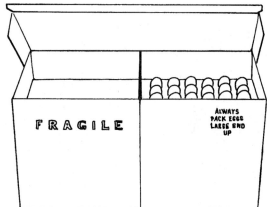

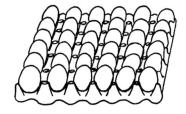

CAUTION: Regardless of egg-gathering methods, never feed broken eggs to the flock. This practice will result in egg-eating habits, which can be corrected only by removing the offenders—if they can be identified.

7. Storage areas for eggs vary widely and include the back porch, behind the kitchen stove, in the basement, in the root cellar, and in the refrigerator. The ideal place should provide temperatures between 45 and 60°F, with a relative humidity of 70%.

CAUTION: Keep eggs from pungent odors caused by products such as fruits, garlic, onions, and kerosene. Eggs stored nearby can absorb these odors and thus develop objectionable flavors.

8. Processing eggs for consumers includes sorting, cleaning, grading, and sizing. Sort eggs for cleanliness and clean only those that are soiled. "Dry cleaning" for a small number of eggs can be done with fine sand paper, steel wool, or emery cloth. Washing dirty eggs is the fastest method to clean large numbers. Water should be 115°F, and eggs should not be immersed longer than 2 to 3 minutes.

CAUTION: Never wash eggs with household detergents that have a "flowery" odor that can cause "off" flavors. An industry-recommended egg-washing detergent should be used. Never wash eggs in water colder than the temperature of the eggs, as this will cause contamination through the shell pores.

Grading eggs is the process of sorting them into quality classes by differences in interior and exterior factors. These include cleanliness, shell shape and soundness (exterior), size of the air cell, shape and size of the yolk, presence of foreign material (meat or blood spots), and the condition of the albumen (interior). The latter must be determined by use of a candling light as mentioned in Section 9.6.

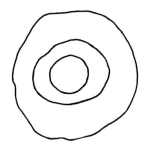

"AA" QUALITY

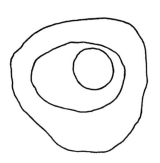

"A" QUALITY

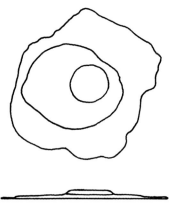

"B" QUALITY

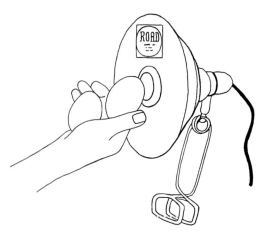

CANDLING LAMP

The egg grades are "AA," "A," "B," and "Dirty" and "Chicks." These grades are explained in detail in "Shell Eggs," USDA-AMS 56, USDA-AMS Poultry Programs Standardization Branch, (202) 690-0941.

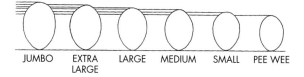

JUMBO   EXTRA LARGE   LARGE   MEDIUM   SMALL   PEE WEE

EGG WEIGHT CLASSES
(MINIMUM WEIGHTS PER DOZEN)

Table eggs sold commercially are sorted by weight in terms of ounces per dozen. These classes are as follows:

Class	Oz. Per Dozen
Peewee	15, 16, or 17
Small	18, 19, or 20
Medium	21, 22, or 23
Large	24, 25, or 26
X-large	27, 28, or 29
Jumbo	30 or greater

9.15 PLANNING POULTRY HEALTH PROGRAMS

The health of a flock of poultry is critical to the efficient production of meat and eggs or birds for show. It doesn't matter whether the flock size is 10 or 12 or numbers in the hundreds of thousands—disease prevention is paramount to the success of any poultry enterprise. A good poultryman should not be taken by surprise by an outbreak of disease; he manages his flock in such a manner that an outbreak is unlikely.

One of the basic management tools for disease prevention is cleanliness. This means clean feed, uncontaminated with disease organisms, pesticides, insecticides, or rodent feces; clean water; clean water troughs; clean houses, especially for starting day-old chicks or for housing pullets just starting to lay; and clean chicks, poults, or started pullets that are free of disease when delivered to the poultry house.

Another basic management practice is isolation. Isolate the flock from other flocks. Do not allow unauthorized people to enter the poultry houses. The reason is simple—to stop the transmission of disease by shoes, clothing, hands, or equipment from other poultry houses.

It is also accepted as a good management practice to limit the birds in a flock to one cage and to maintain a closed flock. Once the flock is housed, no new birds should be placed in that house. This eliminates the introduction of diseases by new chickens or turkeys.

There are methods of disease prevention that recognize the fact that no matter what a poultryman does, certain diseases will either be in his house already or eventually gain entry. Some of these methods are discussed below.

Feed Additives

Some diseases can be controlled by feeding chemicals in the diet. Coccidiosis and blackhead are two serious diseases that are handled in this manner. There are a variety of commercial products available for this purpose. The use of additives such as these is regulated by the U.S. Food and Drug Administration (FDA). Companies that add chemicals to their poultry feeds must apply to the FDA for a license to do so.

If turkeys or chickens are grown on the floor or on dirt, a coccidiostat added to the feed is a must. In addition, turkeys will need a blackhead preventative in the feed. Coccidiosis and blackhead are spread from one bird to another by their droppings. Poultry raised in wire cages or on wire floors where the droppings fall through the wire and are not within reach need not be fed a coccidiostat or a blackhead preventative. Check with your poultry disease specialist or feed dealer for recommendations about when and for how long these drugs should be fed.

Blood Testing

Certain diseases are passed from the hen to the chick or poult by way of the egg. Pullorum (*Salmonella pullorum*), certain paratyphoids, and M.G. (*Mycoplasma gallisepticum*) are three examples of diseases capable of causing serious economic losses in turkeys and chickens. Losses may be in the form of mortality or the reduction of growth and egg production.

Research has shown that dams and sires who carry these diseases form antibodies in their blood against the organisms and can be identified by mixing a sample of their blood with an antigen made from the killed disease organism that they carry. Blood from a disease carrier will "clump" or agglutinate the antigen. This bird can then be identified as a reactor. Blood from a noncarrier will not clump the antigen and can be identified as coming from a nonreactor or noncarrier. In the case of *Salmonella pullorum*, the reactors are removed from the breeding flock. Periodic blood testing of a breeding flock will eventually result in a flock free of this disease.

Chicks purchased from a commercial hatchery rarely are infected with pullorum disease. A voluntary blood-testing program supervised by the U.S. Department of Agriculture (USDA) has resulted in the virtual elimination of this disease from commercial flocks, but pockets of infection are believed still to exist in some breeding flocks of show stock and such wild birds as pheasants and quail.

State blood-testing schools are held throughout the United States each year to train and certify blood testers. Because of the critical nature of the blood test and the skill needed to interpret the test results, it is recommended that those interested in learning these techniques attend one of these schools. Dates and places can be obtained from your county agent or state poultry extension specialist.

Vaccination

Certain viral diseases such as Marek's disease, fowl pox, laryngotracheitis, Newcastle disease, epidemic tremors, infectious bronchitis, and infectious bursal disease are controlled by vaccination. Any one of these diseases can have a serious economic impact on a flock of chickens.

Vaccinations, in most cases with a live virus, produce an immunity in the chicken that protects it from contracting the disease at a later age. In some instances, the immunity from vaccination will last for the productive life of the chicken. Such is the case with Marek's disease, fowl pox, laryngotracheitis, and epidemic tremors. In other vaccinations, such as those for Newcastle disease and infectious bronchitis, the immunity is short-lived and pullets raised for layers and breeders must be revaccinated several times to strengthen their immunity.

Most pullets grown commercially as table-egg or hatching-egg producers are vaccinated routinely for five of the seven diseases just listed. It has recently been recommended that infectious bursal disease be added to the list of diseases against which pullets should be routinely vaccinated. Laryngotracheitis vaccination is done only if the farm or area has a history of previous outbreaks.

With the need to vaccinate for this many diseases, a well-planned schedule of vaccinations is a must. It is recommended that the advice and counsel of a trained poultry disease specialist be used in putting together a pullet vaccination program and that the program be reviewed yearly. Advice should be sought on products to be used, methods of applying the vaccine, age or ages at which to vaccinate, and methods of checking immunity levels.

Because recommendations on vaccination programs vary with location and change frequently, a discussion of specific vaccination techniques is not undertaken here. A general discussion of timing and methods of application follows.

Marek's Disease. Vaccinate as soon as chicks are removed from the incubator. The longer the lag time, the greater is the risk of an infection with Marek's virus. Vaccination is made with a herpes virus that has the ability to immunize the chick against the Marek's virus. The vaccine is administered subcutaneously with a hypodermic needle. There are a variety of automatic vaccinators available.

CAUTION: Read and follow the manufacturer's directions on preparing and administering the vaccine.

Newcastle Disease (N.D.) and Infectious Bronchitis (I.B.). Both vaccines are administered in the drinking water, generally at the same time, when the chicks are 10 days, 4 weeks, and 14 weeks of age. A spray mist can be used to administer the vaccine if desired. N.D. and I.B. immunizations are sometimes administered separately on the third vaccination.

CAUTION: Follow the manufacturer's directions in administering the vaccine. In preparing water-carried vaccines, chlorinated water or water containing heavy metals could be detrimental to the effectiveness of the vaccine. Birds should be checked after vaccination for level of immunity. Check with a poultry disease specialist for details.

Epidemic Tremors. The scientific name for this disease is avian encephalomyelitis, and the disease is frequently referred to simply as A.E. The vaccine is administered once only, in the drinking water, when the birds are 12 to 14 weeks of age.

CAUTION: Follow the manufacturer's directions on preparing and administering the vaccine.

Fowl Pox. The vaccine is administered once in the wing web by the "stick" method. A suspension of the live virus is prepared, a device consisting of a handle and twin needles is dipped into the vaccine, and the needles are then jabbed into the underside of the wing web, delivering a fixed amount of the vaccine to the punctured areas. Recommendations on the age at which to vaccinate vary from 7 days to 14 weeks.

CAUTION: Check for "takes" at the site of vaccination. In 7 to 10 days a reddened welt should appear if the vaccination was successful. Vaccinate all birds on the same wing to make it easier to check for takes.

Laryngotracheitis. This vaccine is administered only once, after 10 weeks of age, either by way of the drinking water or by dropping it into the eye.

Infectious Bursal Disease (I.B.D.). Vaccine is given one time only, in the drinking water, between 10 and 16 weeks of age.

CAUTION: All of the vaccines except the one for Marek's disease are made from live viruses that are capable of producing disease outbreaks in nonvaccinated chickens. Clean up carefully after vaccinating. Wash your hands, change your clothes, and destroy vaccine bottles and unused vaccine. Take every precaution to be sure that you do not inadvertently carry the vaccine to susceptible flocks.

There are also vaccination protocols for cholera, chronic respiratory disease, coccidiosis, erysipelas, infectious coryza, infectious synovitis, and viral arthritis. Check with your veterinarian or drug representative for details.

HEALTH MANAGEMENT TEAM: VETERINARIAN–LIVESTOCK PRODUCER

chapter |TEN|

Animal Health Management

10.1 INTRODUCTION

Most of the livestock losses that result from poor health can be prevented by proper management. The farmer, rancher, herdsman, or owner is responsible for the health of his animals. It is most important that good management practices and a herd health program be implemented to minimize or prevent disease in the herd or flock.

There is a need for close communication between the livestock producer and his veterinarian. The veterinarian's expertise is needed in the planning of a preventative health program and in making a definitive diagnosis, and he must be available when emergencies arise.

The objective of a herd health management program is to produce the most product of the highest quality at least cost. It is necessary for your veterinarian to make periodic visits to your premises, at whatever frequency is agreed upon, for discussing and formulating immediate and future needs for the program. Goals must be set, evaluated at given

intervals, and revised as the program grows or as new problems develop.

The mechanics of the program should be determined by the livestock producer and the veterinarian. They may agree to have fertility tests, pregnancy exams, vaccinations, castrations, wormings, and other management chores performed at prearranged times of the year. The veterinarian may help to train the farm or ranch personnel to perform many of these procedures themselves. These arrangements can prevent many predictable problems from occurring. Management of potential problems can be discussed and methods of prevention established at regularly scheduled meetings of the people involved in the day-to-day chores or management.

Each farm or ranch is different, and each should have its own program. The geographic location, the type of operation, the existing disease conditions, and the financial risks involved are factors to consider in making decisions about a health management program.

10.2 CONCEPTS OF HEALTH AND DISEASE

Health

Health is the state of the individual living in complete harmony with his or her environment. One needs to remember that an animal may appear healthy but still undergo physiological, anatomical, and chemical changes in response to stress from the environment. The basic difference between a healthy and a diseased animal is that the healthy animal has not yet exhausted its normal adaptive powers.

Disease

Disease is a condition in which the individual overtly shows physiological, anatomical, or chemical changes from the normal (symptoms). Diseases are recognized by different symptoms and differences in the changes produced in the body fluids and cells. There are two types of diseases: noninfectious and infectious.

Noninfectious diseases result from injury, improper nutrition, genetic abnormality, uncomfortable environmental conditions, or exposure to toxic materials. Examples are vitamin and mineral deficiencies, overeating, toxemia, azoturia, and wild cherry and sweet clover poisoning. Toxic or poisonous materials include *bacterial toxins, metallic and chemical poisons* (lead, arsenic, copper), *phytotoxins* (poisonous substances produced by plants), *zootoxins* (toxins from snakes or insect bites), and *mycotoxins* (toxins from fungi or mold).

Infectious diseases are produced by microorganisms that gain entrance to the body in sufficient numbers and virulence to result in symptoms and changes in body fluids and cells. Examples are brucellosis, swine erysipelas, ascarid infections, pseudorabies, bovine virus diarrhea, and ringworm.

Contagious disease is the term used to describe an infectious disease that is transmitted by the passage of an infectious agent from animal to animal. Contagious diseases are all infectious, but not all infectious diseases are contagious. For example, tetanus (lockjaw) is not spread directly from animal to animal, so it is not considered a contagious disease. The bacteria are in the soil and gain entrance through a wound. Pseudorabies, on the other hand, is a contagious disease that is spread from animal to animal by body discharges.

Disease Transmission

Infectious agents are shed (potentially spread) from an infected animal to the environment (including another animal) in a variety of ways.

Many types of microorganisms can survive for only short periods of time outside the body of a live animal.

Shedding Route	How Shed	Types of Diseases
Body surface	Pus, hair, scabs	Abscesses, ringworm, cowpox
Mouth	Saliva, phlegm	Foot and mouth, rabies, tuberculosis, erysipelas
Nose	Snot, blood	Influenza, strangles (distemper), anthrax
Eyes	Tears	Pink eye
Anus	Feces	Johne's disease, enteritis, septicemia
Mammary gland	Milk	Mastitis
Urinary and reproductive tracts	Urine, semen, eggs (poultry)	Leptosporosis, infertility, pullorum
Wounds	Blood	Q-fever

Other types can survive for a long time outside the animal's body if they are protected from sunlight and putrefaction. Still other organisms multiply rapidly outside the animal's body. Carrier animals may recover from symptoms of an infection and become immune to further effects of the organism but still harbor and shed the live organism in large numbers.

Microorganisms gain entrance to the animal's body in a variety of ways.

Entry Site	Method of Entry	Types of Diseases
Mouth	Water, licking, sucking, food (spoiled/infected)	Salmonellosis, brucellosis, leptospirosos
Nose	Breathing (dust, droplets)	Influenza, Newcastle disease, shipping fever
Eyes	Rubbing, swishing	Pink eye, conjunctivitis
Urinary and reproductive tracts	Placenta, urethra, eggs	Nephritis, pullorum, leucosis
Body surface	Breeding, rubbing, wounds	Vibrio (abortion), ringworm, tetanus, rabies, anthrax, myxomatosis
Vectors	Fly, flea, tick, bat, or animal bites, birds	Salmonellosis, foot and mouth, rabies, Q-fever

Most organisms have a marked preference for a certain type of tissue. For example, the rabies organism prefers nervous tissue; salmonella, the digestive tract; and pasteurella, lung tissue.

An animal's body possesses such protective structures as skin, hair, and feathers. Other protection is provided by the mucous membranes. The mucous membranes of the digestive tract are further protected by secretions of saliva, gastric juices, and bile. Membranes of the respiratory tract secret mucus, have cilia on special epithelia, and are further protected by the coughing reflex. The genitourinary tract is protected by mucous membranes, the washing action of urine, and the formation of a cervical plug in the pregnant animal. Eyes are protected by the eyelashes, eyelids, and tears; ears are protected by their structure and by hair and wax.

Immunity and Immune Responses

Immunity is an animal's resistance to the effects of a foreign substance. *Antigens* are the substances that are recognized as foreign to the body. They may be any type of substance, but typically are proteins, and may be a separate substance or part of another organism. Some common antigens are microorganisms, venom of animals and insects, and animal serum proteins. With reference to disease, immunity refers to the lack of susceptibility to an infectious agent, to an antigen, on the part of the animal.

Antibodies are proteins that are produced in plasma cells (a special type of white blood cell) in response to a specific antigen. Antibodies are secreted by the plasma cells to bind with those specific antigens. For each antibody produced there is an individual plasma cell line (type) that produces that antibody. These cell lines are determined while the animal is still an embryo. Note the specificity—a specific antigen elicits a specific antibody from a specific plasma cell line. Occasionally some cross-reaction does occur, with the antibody from one antigen reacting with a second, closely related antigen.

Types of Immunity

Animals have immunity (are immune to a disease) for one of three reasons: it can be an innate result of the animal's species (*natural immunity*); it can be *passively received;* or it can be *actively developed* following exposure to an antigen. This exposure can be a natural infection or a challenge from a vaccine.

Passive immunity is the result of antibodies produced by one animal being placed into a second animal. The second animal will be temporarily protected by this passive immunity derived from the antibodies from the first animal. Passive immunity found in very young animals is achieved by antibodies from the animal's mother, which either pass through the placenta, are consumed in the colostrum during the first few hours of life, or are artificially given via an intravenous transfusion of immune plasma.

In domestic animals, the majority of the antibodies an animal receives are passed in the *colostrum*, the first antibody-rich milk of the dam. The antibodies are absorbed through the intestinal wall. This ability to absorb antibodies across the intestinal wall is lost after the first 3 to 48 hours following birth, depending upon the species. Because of the large amount of antibodies absorbed during this time, it is essential that a newborn animal nurse within the first few hours of life. Without these colostral antibodies, the infant's only protection against disease is the immune response it can muster on its own. This is seldom adequate and usually results in death from bacterial or viral infections.

Any type of passive immunity is lost over time and must be replaced by an active immunity that is long lasting.

Active immunity can be produced by antibodies that the animal produces as a result of a natural infection (followed by some degree of disease progression, followed by recovery from the disease). Vaccination with a less dangerous type of the infectious antigen will also cause antibodies to be produced and active immunity to be received by the animal. The length of time a vaccination lasts (actually, the length of time the immunity provided by the antibodies produced as a result of the vaccination lasts) varies, depending upon the disease, the health and immunological status of the animal at the time of vaccination, and the vaccine used.

Resistance (Natural Immunity). Fortunately, natural resistance is a frequent occurrence. Horses are resistant to the virus that causes foot-and-mouth disease. Animals are resistant to plant viruses. Hog cholera virus will not infect man. This is a typical natural resistance that often is called *species resistance.* Cattle are much more susceptible to anthrax than are swine. Cattle die quickly from a generalized septicemia when infected with the anthrax organism, while hogs usually have a localized infection with very few deaths. This resistance is not a result of specific antibodies but is related to the physiological and biochemical functions of both the animal and microorganism.

In addition to the species-related differences in resistance, there are differences between individual animals within a species. These differences may be influenced by age, environment, nutrition, fatigue, and genetic makeup. There are also several examples of resistance being associated with genetics. Certain strains of laboratory mice are resistant to mouse typhoid, and some races of human beings are resistant to tuberculosis.

Immune Responses

An *immune response* is the process of recognizing and destroying antigens. There are three types of immune responses: (1) a humoral immune response, (2) cell-mediated immunity, and (3) tolerance.

A *humoral immune response* is the production of an antibody in response to exposure to an antigen. This occurs the first time an animal is presented with an antigen. The animal's response, however, may not be sufficient in degree or fast enough to prevent disease. Second and subsequent exposures to the same antigen result in a more rapid and stronger response than the first. This is known as an *anamnestic response* ("recollection"). If the second exposure to the antigen is too close to the first, there is actually a drop in the amount of antibodies in the blood for a few days until the secondary immune response gets under way. This is called a *negative phase* and is the result of the antigens binding to antibodies and removing them from circulation. Repeated injections of antigen don't indefinitely lead to greater levels of immunity. This is one of the most important reasons to pay attention to manufacturer's recommendations regarding the timing and frequency of booster shots.

Cell-mediated immunity (CMI) is similar to humoral immunity in that a second exposure to the same antigen causes a faster and stronger response. However, cell-mediated immunity cannot be transferred with serum. It can be transferred from one animal to another by transferring the cellular portion of the blood. The cells that are involved in the cell-mediated immune response are lymphocytes derived from the spleen, lymph nodes, or peripheral blood.

In addition to developing an immune response to an antigen by a humoral or cell-mediated response, the immune system is *tolerant* to some substances. That is, the immune system does not recognize the substance as foreign and therefore does not try to get rid of the antigen. This is how an animal keeps from attacking its own body and is a normal immune response. If this tolerance breaks down, then disease will occur (autoimmune disease) as either antibodies or lymphocytes destroy normal cells in an attempt to eliminate the offending antigen.

Tolerance can be induced, at least temporarily, by administering an overwhelming dose of the antigen. Tolerance is specific for the antigen administered and may be boosted by reexposure to the identical antigen. If the animal is not reexposed to the antigen, then tolerance is gradually lost. This is the principle behind allergy shots. The individual is desensitized to a specific antigen or group of antigens.

Some diseases are better prevented by stimulating a CMI response, while others are defended via a humoral immune response. The route of ad-

ministration of vaccines, in addition to the type of vaccine itself, has a profound effect on the type of immune response elicited. Consult your veterinarian for selecting appropriate vaccines for your herd.

10.3 SANITATION

Many microorganisms live and even multiply outside the host, infesting buildings, lots, and pens and constantly challenging the animals therein. The number of organisms in the environment and the incidence of disease outbreaks can be reduced by the implementation of sanitation practices. These include providing the animals with clean, dry, well-ventilated housing and the intelligent use of antiseptics and disinfectants.

The first step in a comprehensive sanitation program is a thorough cleaning, including the mechanical removal of waste, manure, and organic material. Organic matter furnishes nutrients for some microorganisms and protects most of them from destruction caused by temperature, drying, and disinfectants.

Such cleaning agents as soap and detergents, which are added to water, are helpful in removing the remaining organic material after the bulk of it has been shoveled, scraped, or scooped away. Sprayers capable of applying soap and water at a pressure of 1,000 psi will remove manure and dirt with great efficiency. Steam cleaning is also useful in removing waste, but because the steam nozzle must be held within 6 to 8 inches of the surface being cleaned to kill microorganisms, it is not a practical method of disinfection.

The intelligent selection and application of antiseptics and disinfectants is the second step in a complete sanitation program. It becomes possible only after one achieves a thorough understanding of the terminology involved; this terminology is discussed next.

Antiseptics are substances that kill or prevent the growth of microorganisms. The term *antiseptic* refers to preparations that may be applied to living tissues. The original meaning of the term was a substance that opposes sepsis (putrefaction or decay).

An ideal antiseptic should (1) have a high degree of germicidal potency, (2) have a broad antimicrobial spectrum, (3) have a low surface tension, (4) have a long-lasting germicidal activity, (5) have a rapid and sustained action, and (6) not be harmful or toxic to animal tissue.

Disinfectants are products that prevent infection by the destruction of pathogenic microorganisms.

The term is commonly used in reference to preparations applied to inanimate objects.

An ideal disinfectant (1) is stable and does not lose its effectiveness on contact with organic materials, (2) dissolves readily in water, (3) readily destroys many forms of microorganisms, (4) has minimal toxicity in that it is capable of killing lower forms of life with little or no effect on man and animals, (5) is noncorrosive, (6) has a minimal odor which will not contaminate products intended for human consumption, (7) has the power of penetrating cell walls so as to be effective on certain microorganisms such as mycobacteria and spore-forming bacteria, (8) is economical, and (9) is compatible with and effective in the presence of soaps and other chemical substances that are likely to be encountered.

A *sanitizer* is a special form of disinfectant that is capable of reducing the numbers of microbial contaminants to levels considered safe by public health requirements.

Sterilization refers to the complete destruction of all forms of life, especially microorganisms, by some chemical or physical means.

Germicide, in its broad and most useful meaning, is an agent that destroys microorganisms. It may be further defined by the specific terms *amebicide, bactericide, fungicide,* and *viricide*.

Disinfectants are defined as products that prevent infection by destroying pathogenic microorganisms. One of the characteristics of an ideal disinfectant is its ability to readily destroy many forms of microorganisms. As you might expect, microorganisms vary widely in their resistance to disinfectants. Below is a ranking of microorganisms according to their ability to resist the action of commonly used veterinary disinfectants.

Types of Disinfectants and Antiseptics

Disinfectants and antiseptics are conveniently divided into several groups. Most are readily available under various trade names.

Alkalies

Alkalies have a pH of 9 or higher, which is detrimental to the survival of most microorganisms. This group of compounds is very irritating to living tissues,

requiring rubber gloves, eye goggles, and protective clothing (such as coveralls) when being applied.

> **CAUTION:** Always follow directions for the use of antiseptics and disinfectants. Too little will not kill the microorganisms, and too much can result in a buildup of toxic materials. After they have been disinfected, it is advisable to rinse feeding and watering equipment before they are used again.

Sodium hydroxide is the most widely used of this group. In a 0.5% concentration it can be used as a germicide for mechanical milking equipment. In stronger solutions, it is used to kill many of the more resistant bacteria and viruses, such as the anthrax bacillus and hog cholera virus, on floors, stalls, and pens.

Unslaked lime (quicklime) is added to water, with which it reacts to form calcium hydroxide. It has excellent disinfecting action when used as a whitewash on walls, floors, and certain types of equipment. Quicklime is also used to cover carcasses when burying them.

Acids

Acids are most frequently used as preservatives for food or feed because they prohibit growth of many microorganisms. Examples are acetic acid (vinegar) and lactic acid. Boric acid is freqently used as an antiseptic wash (in low dilution) for the eye.

Phenols and Related Compounds

Phenol (carbolic acid) is one of the most efficient disinfectants; most are not inactivated by organic matter, hard water, or soaps, which makes them useful for surface disinfectants in difficult-to-sanitize areas. However, phenols are expensive, must be applied at temperatures greater than 60°F, and are toxic to tissues (especially to cats and other *Felidae*).

Cresol is frequently used to disinfect trucks, pens, and railway cars. It is more active then phenol and is toxic to living tissue.

Orthophenylphenol is an odorless disinfectant that frequently is used around dairy barns. It is also effective against the tubercular organisms.

	Bacteria						
Mycoplasmas	Gram–	Rickettsias	Chlamydia	Fungal spores	Parvo viruses	Coccidia	
	Bacteria	Pseudomonas	Enveloped	Nonenveloped	Picoma viruses	Bacterial spores	Prions
	Gram+		viruses	viruses			

Least Resistant   Most Resistant

Resistance to Disinfectants

Halogens

Chlorine is an excellent disinfectant but is readily destroyed by organic material. It is used in the sterilization of water supplies, treatment of sewage water, and sterilization of dairy equipment, dishes, and glassware. Chlorine has the added disadvantages that it is corrosive to some metals, tarnishes silver, bleaches fabric, is unstable in hard water, and must be prepared fresh for each use.

Iodine is one of the most effective antiseptics for use on the skin. Two forms are used: tincture of iodine (2 or 7%) and iodophors. Tinctures stain tissues reddish orange while iodophors do not. Because of this staining effect, 2% tinctures have largely been replaced by iodophores for topical application. Seven percent tinctures are used for dipping navel cords. Both forms make excellent washes in which to soak instruments between their use on successive animals. Iodophors may be inactivated by hard water.

Alcohols

Alcohols require prolonged contact (up to 22 minutes) to kill many bacteria. They may be used in combination with 0.2 to 1% iodine for disinfecting thermometers. By themselves, alcohols are poor disinfectants.

Ethanol. Ethyl rubbing alcohol contains 70% ethanol by weight. It is not very effective as an antimicrobial drug.

Isopropanol. Isopropyl rubbing alcohol contains 70% isopropanol by weight. It is slightly more germicidal than ethanol in undiluted form. It causes dilatation of surface blood vessels, thereby creating a greater tendency for bleeding.

Oxidizing Agents

Hydrogen peroxide is generally available as a 3% aqueous solution. It is used mostly for washing and cleansing wounds.

Permanganates are antiseptic and antifungal. The most common of these compounds is potassium permanganate, which is used in conjunction with formaldehyde to fumigate buildings.

Other Disinfectants and Antiseptics

Quaternary ammonium compounds are good disinfectants for use against non-spore-forming bacteria. Their primary use is as an equipment rinse and spray disinfectant. Quaternary ammonium compounds are inactivated by organic matter and soaps, making them a poor choice for footbaths.

Detergents, wetting agents, and *emulsifiers* are used extensively as cleaning agents because they are nontoxic to living tissues.

Soaps are used to remove oily compounds and organic matter. They are often used to remove bacteria from the skin.

Chlorhexidene (*Nolvasan*) is used as a general disinfectant and serves as an excellent bath in which to soak instruments between their use on successive animals.

Fumigation

At times, it is advisable to use fumigation in place of disinfection. Fumigation is the most effective method of killing microorganisms, but one must be able to make the area airtight for 8 hours during the fumigation period. Buildings and equipment must be thoroughly cleaned before fumigation and the area allowed to air out for 48 hours before it can be reentered. It is important that the temperature not be below 65°F (18°C), and there must be an adequate concentration of the toxic gas in the presence of sufficient humidity if the fumigation is to be effective.

The most common method of fumigation employs wide-bottom pans or buckets placed every 10 feet throughout the building so that each container fumigates 1,000 cubic feet of space. Place 10 tablespoonfuls of potassium permanganate into each bucket. Then place widemouthed jars containing 1½ cups of formaldehyde next to each bucket. Starting at the back of the building, pour the formaldehyde into the pails, progressing with reasonable haste to the front of the building and to the outside, securing the door after leaving.

CAUTION: It is advisable to have at least two persons present at the time of mixing the ingredients in case of an accident. Place a danger sign on the door and lock or nail the door shut.

A second method of generating gaseous formaldehyde is to heat paraformaldehyde powder. The same precautions are to be followed.

Miscellaneous Sanitation Measures

Dispose of dead animals and refuse quickly and properly. If available, rendering plants offer the best solution for removal of large animals. If dead animals are buried, the trench should be deep enough that at least 6 feet of soil will cover the carcasses. The carcasses should be covered with quicklime and then with soil. Burning is most difficult and expensive for large animals. Fumes and smoke may be a problem when birds or small animals are cremated.

Mudholes should be frequently filled or livestock fenced away from them. This is important to prevent the spread of such diseases as foot rot and mastitis.

An adequate supply of clean, cool, fresh water is very important in control of disease. Stagnant, contaminated water is frequently the route of spread of such diseases as salmonellosis and leptospirosis.

The control of internal parasitic diseases is also dependent upon sanitation procedures. Livestock stalls and pens must be cleaned on a regular basis, with the manure and other material discarded in such a way that animals will not be infected. Excessive moisture in bedding of stalls and pens often results in the accelerated incubation of such parasites as coccidia and strongyloides, thus causing an infestation. Certain developmental stages of many internal parasites take place in paddocks, lots, or pastures. Larval development must be minimized by collecting manure and exposing it to the sun by harrowing or dragging pastures, or by plowing fields. The rays of the sun are detrimental to most parasites and microorganisms. Rotation of pastures can be of some help, and composting of manure is advisable if one cannot spread it on fields *and plow it under.*

10.4 DRUGS

A drug is a compound administered to provide a therapeutic benefit. Drugs used by veterinarians and animal owners may be classed as either *pharmaceuticals* or *biologicals.* Pharmaceuticals or chemotherapeutics are used mainly for *treatment* of an animal disease or infection, while biologicals are primarily given to *prevent* disease. Both are necessary in implementing a herd or flock health program.

Pharmaceuticals

Pharmaceuticals should be used only for their approved and intended purposes because some are very specific in their action and effectiveness. They are available in a variety of forms, including drenches, boluses, liquids, feed additives, and powders. Keep in mind that a diagnosis of a problem is necessary in order to determine the effective pharmaceutical to use. They should not be recommended or used without good reason.

Products vary greatly in their ingredients and in the way the manufacturer puts them together. Adequate dosage levels must be reached and maintained for a sufficient length of time if satisfactory results are to be obtained. The route of administration, dosage, and frequency of dosage are all important factors in achieving and maintaining adequate levels of the active drug in the various body systems.

The ideal chemotherapeutic agent should (1) have selective and effective antimicrobial activity, (2) be bactericidal (capable of killing bacteria) and not merely bacteriostatic (capable of preventing the multiplication of bacteria), (3) not result in a buildup of resistance in target microorganisms, and (4) be capable of providing adequate tissue levels quickly and maintaining these levels for a sustained period of time.

Bacteria are often more sensitive to one drug or pharmaceutical than to another. In the laboratory, this sensitivity is determined by growing the bacteria in the presence of the various drugs. After an incubation period, the ability of the drug to inhibit the growth of the organism is evaluated. At times this can be misleading because the method of growth of the organism in the animal's body differs from the laboratory test conditions. When the drug is administered, the body may protect the organism from being exposed to a sufficient level (action level) of the antibiotic. A sensitivity test can, however, be a valuable tool in helping to determine treatment measures to be followed.

In many cases, not all bacteria are destroyed when an animal is treated with a selected drug. When this happens, resistant strains of the organism may develop and the drug become ineffective. Much remains to be learned about this phenomenon.

Some animals may develop allergic reactions to drugs. These can vary from a mild skin irritation to hives, difficult breathing, and even sudden death.

Be aware that many drugs have withdrawal times. Withdrawal time is the time required between the administration of a drug and the slaughter of the animals or use of milk for human consumption. This time period allows for the clearing of the drug from animal tissue.

CAUTION: There are several considerations that must be carefully considered before any animal health product is purchased and used in your herd or flock. Failure to do so can result in needless expenditure of scarce financial and time resources, violation of federal and state regulations and laws (and possible conviction and punishment), poor-performing animals as a result of product misuse, and possible illness or death from contamination of yourself or nontargeted animals. The guideline that will prevent all of these untoward occurrences from happening is a simple one: **Read the manufacturer's product label!**

Current FDA regulation limits the use of (i.e., FDA does not "approve" the use of or "withholds approval" of) a growing list of chemicals, drugs, and feed additives for certain categories of animals or for all animals under certain conditions.

Examples of nonapproval categories include lactating animals, reproducing animals or animals intended for reproduction, and animals producing meat or milk intended for human consumption. There are also products that are approved only for use in animals after a certain age or weight is attained.

There are drugs that can be used in certain situations only if they are ordered for use on the basis of a veterinary prescription. For example, there are drugs that can be purchased on the open market, by producers, that are not "labeled" for use in lactating dairy animals. These same "nonlabeled" products can be legally used in lactating dairy herds if a veterinarian so orders it. *Labeled* and *nonlabeled* means that the use to which you want to put the product either is or is not listed on the manufacturer's product label. If it is labeled for lactating dairy cows, then the label will say so. If it is not listed on the label, it is a nonlabeled product.

There are products that can be used for animals up to a certain point or for a period of time in their lives, after which they must be withheld or withdrawn from use in those animals. For example, there are biologicals, pharmaceuticals, feed additives, implants, and other products that can be used as necessary, or as ordered by a veterinarian, only if managers agree to withdraw the drug from use or withhold the animals from slaughter or their products (such as milk) from the marketplace for a predetermined and labeled amount of time. These **withdrawal** or **withholding** times must be adhered to for the public safety and to protect the continued use of the products.

Producers and their veterinarians should implement the practices outlined in the Food Animal Residue Avoidance Databank (FARAD), sponsored by the Food and Drug Administration (FDA) branch of the USDA. This program is an expert-mediated, decision-support system for producers to help reduce the chance of antibiotic and other pharmaceutical residues occuring in meat and milk products. A constantly updated database of available pharmaceuticals and withdrawal periods is available on the Internet at http://sulaco.oes.orst.edu:70/1/ext/farad. Further information on FARAD may be obtained from:

University of California at Davis
North Carolina State University
University of Florida

There is a sampling and screening process in place to identify chemical residues in meat and milk products. Failure to comply with FDA regulations can result in criminal prosecution, heavy fines, and jail sentencing.

There are insecticides, dewormers, and drugs that contain certain kinds of chemical compounds that, if used in combination with certain other chemical compounds, will cause a serious and potentially fatal reaction. For instance, organophosphates (which are one kind of cholinesterase inhibitor) are used in some dewormers, insecticides, and drugs. If you administer one of these organophosphate-containing products with another product that uses another type of cholinesterase inhibitor, you are headed for a serious reaction. The animal's nervous system can react violently, respiration failure can occur, and the animal can die. A double dose of either of these product types can cause the same reaction to occur. Manufacturers attempt to build safety margins into their products, but all animals are different, and what is safe for one animal will kill another. Avoid all of the problems: **Read the manufacturer's product label!**

Reading manufacturers' directions, visiting with other trusted producers, contacting your extension faculty, and consulting with your veterinarian for the wisest and safest use of animal health products in your management program is the wise manager's safety margin.

Antibiotics

Antibiotics, chemical substances produced by various species of microorganisms including bacteria, fungi, and actinomycetes, are pharmaceuticals that suppress the growth of other microorganisms.

The word *antibiotic* means "against life," and as applied in veterinary medicine it means "against bacteria." Some antibiotics are specific for the bacteria that they can destroy, while other antibiotics are capable of controlling a variety of different bacterial organisms. The latter are called *broad-spectrum antibiotics.*

Penicillin, ampicillin, ceftiofor, tetracyclines, florfenicol, and cephalosporin are a few examples of the more commonly used antibiotics.

Sulfa Compounds

The sulfa compounds are effective against bacterial infections of animals. They are bacteriostatic (capable of preventing the multiplication of bacteria), which implies that they are effective because they help the host's body to overcome the infection. They do not destroy the bacteria by their own action. They are considered to have a broad-spectrum activity. The sulfa or sulfanomid products most often used are sulfanilamide, sulfamethazine, sulfathiazole, sulfamerazine, and sulfapyridine. Two or more sulfa drugs are often used in combination to get a preparation that is more effective than one sulfa drug alone. They have been used for most types of bacterial infections that cause diarrhea, pneumonia, and mastitis. Bacteria can develop resistance to these compounds. An animal must have an adequate water intake when receiving these drugs. When sulfa drugs are used, one must *allow for an adequate withdrawal time.* Prolonged usage may result in toxicity. Sulfas may also cause hypersensitivity to light.

Nitrofurans

Nitrofurans are broad-spectrum drugs whose primary action is to inhibit the growth of the disease-causing organism. They are not very toxic. Excellent formulations are available for topical use in skin, eye, ear, and genital infections. An occasional animal may be allergic to the drug. Nitrofurans may only be used topically in food animals.

Steroids

Steroids such as cortisone are used in combination with other drugs to reduce inflammation. It must be realized, however, that the exogenous steroids inhibit the body's own ability to fight infection and to develop immunity. They have been most helpful in certain arthritic problems and to relieve lameness. Their total influence on the body is very complex, and they should be used only under the direct supervision of your veterinarian.

Hormones

Hormones are produced by the endocrine glands of the body (e.g., the thyroids, adrenals, testicles, pituitary, and ovaries), which have specific effects on the activity of other organs in the body. The many interactive effects of the hormones on body organs and systems are complicated, and it is quite easy to do more harm than good by the indiscriminate use of hormone therapy.

Oxytocin is a hormone used by swine producers during the farrowing process. It causes milk letdown and contractions in the uterus. It should never be given routinely to all animals, but only when there is a diagnosed problem.

Biologicals

Many biological products are available to stimulate immunity against specific diseases and provide the most reliable and effective form of livestock health management. A list of these products is given in Appendix R. Modern biologicals have reduced the incidence of viral and bacterial diseases and have prevented economic losses while improving performance.

Biologicals are effective in preventing disease because they elicit the *antigen–antibody reaction* in the animal's body. This is an involved process and beyond the scope of this text, but basically the biological is the antigen (reaction-causing agent) that causes the body's own defense mechanisms to react by producing antibodies against the antigen. The livestock manager and veterinarian select the disease organism (antigen) against which the animal is to be protected and inject a small amount of this into the animal. The animal responds by producing antibodies to the disease organism (and to the disease) and in essence protects (immunizes) itself against the time when it will be exposed naturally.

Biological products are carefully manufactured and tested for safety by the manufacturers and the federal government. Their ability to stimulate adequate antibody production varies greatly, however, and is not guaranteed.

The route of administration, quantity, and frequency of administration vary with the antigen and with the type of product. The amount of antigen given will influence the level of antibody response obtained.

Live immunizing products depend for their effectiveness upon some multiplication of the organisms within the animal. An ideal antigen would stimulate a high level of protection in each and every animal that would last for life after one exposure to the antigen. This does not occur, so a vaccination program must be developed for each farm or ranch and for each disease.

Biologicals must be stored, handled, and administered according to the manufacturer's recommendations. Ignoring or not following label directions can ultimately result in animal loss.

CAUTION: The following guidelines are recommended for handling biologicals.

Keep biologicals refrigerated, but not frozen.

Reconstitute vaccines at the time they are to be used and keep them refrigerated or in the shade when using them. Vaccines are very sensitive and can be destroyed by exposure to direct sunlight.

Never use needles or syringes that have been sterilized by chemical disinfectants. Residual disinfectant can kill the organisms in live vaccines. Use boiling water, steam, or an autoclave to sterilize needles and syringes.

Never mix a biological with any other substance that is not specified on the label. Some combinations will kill the living organisms.

Never use a product after the expiration date printed on the label. The efficacy of many products is dependent upon the number of living organisms available to stimulate adequate immunity, and this number may diminish with age.

All products on the market are not equal in their ability to stimulate adequate immunity. Be certain that the products being used come from a reputable manufacturer and distributor.

Vaccines

Vaccine is a term that implies or is used to mean all types of biological agents that produce active immunity.

Types of Vaccines. *Bacterin* is a killed culture of bacteria. It usually requires a series of injections of a bacterin to stimulate adequate protection against a field exposure. Subsequent (for example, yearly or every 6 months) booster injections against many bacterial infections are needed if a satisfactory level of protection is to be maintained. Bacterins are excellent antigens for certain diseases and organisms. Blackleg, leptospirosis, streptococcus, salmonella, pasteurella, bordetella, and staphylococcus are examples.

A *mixed bacterin* contains more than one type of organism.

An *autogenous bacterin* is made from an organism isolated from an infection in animals on a given premise. The bacterin is then used to protect other animals on the same premise. It usually gives better protection than a stock bacterin.

Stock bacterin is made from a bacterial culture maintained in a laboratory. Be aware that many are effective, but others will not stimulate satisfactory immunity.

Live bacterial spore vaccines are effective against bacteria that have the ability to form spores that are

very resistant to environmental conditions. The cultures for the vaccine are grown under adverse conditions (usually increased temperature) and are then tested to determine the dosage for use in different species of animals in different geographical locations. An example is the vaccine for anthrax, which is caused by the *Bacillus anthracis* organism.

Live bacterial vaccines of reduced virulence are made from certain strains of bacteria that are less virulent (less capable of causing disease) than others but are still able to stimulate adequate antibody production. Such bacteria are found in nature or developed in the laboratory; an example is the *Brucella abortus* strain 19 vaccine. These vaccines usually give better protection than the killed cultures of bacteria.

Killed virus vaccines contain virus organisms that have been killed and are then used as antigens. Because the organism is dead, there is no danger of producing the disease. Most of these vaccinations must be repeated at regular intervals to maintain adequate protection. An example is equine encephalomyelitis, for which horses should be vaccinated each year.

Modified live virus vaccines are produced by growing the virus in such a manner that its ability to produce disease is diminished while it still retains its ability to stimulate an active immunity and protection. Infectious bovine rhinotracheitis and equine rhinopneumonitis are examples of diseases that are prevented by the use of a modified live virus vaccine. Adequate protection against the disease is dependent upon the multiplication of the virus in the body of the recipient. For most diseases for which live virus vaccines can be used, good protection can be produced. It may be necessary to repeat the immunization at frequent intervals.

Antiserum is blood serum containing specific antibodies against a disease. The antiserum is obtained from an animal that has been immunized against the disease. It is then used to treat another animal that has contracted the same disease, or for producing immediate protection in a susceptible animal. Antiserums are short-lasting. The erysipelas antiserum is an example. Other vaccines must be used to provide longer-lasting immunity.

Antitoxins are antiserums that contain antibodies to bacterial toxins. Tetanus antitoxin is an example.

Toxoids are detoxified toxins that are used as antigens to stimulate production of an antibody against the toxin (e.g., tetanus toxoid). Toxoids are used for protection against diseases in which the toxin produced by a bacteria causes the sickness.

Adjuvants are substances added to a biological that enhance the antigen's ability to stimulate antibody production. Adjuvants perform this enhancement by delaying absorption of the antigen and by themselves stimulating the immune system to some extent.

The *type of vaccine and route of administration used* affect the immune response of the animal. In general, killed vaccines are safer than modified live vaccines; however, modified live vaccines provide a better stimulation of the immune system. For example, modified live, intranasal vaccines for infectious bovine rhinotracheitis and bovine parainfluenza provide better immunity than injectable killed products. These agents require stimulation of the cell-mediated immune system for protection, which occurs to a greater extent via the intranasal route. Your veterinarian can help select the appropriate vaccines and routes of administration for your situation.

10.5 ADMINISTERING PHARMACEUTICALS AND BIOLOGICALS

The application or administration of medicine to an animal, whether it is administered topically, orally, or parenterally, is a technique that must be approached with caution. If the application is topical, the animal may react very quickly to your touch because of the sensitive lesion. If the medicine is administered orally, the animal may resist the drenching bottle, dose syringe, or balling gun. Injections with a needle (parenteral administration) almost always elicit a strong, quick reaction from the animal. Be aware of these predictable reactions and take steps to cope with them. Restrain the animal properly, both for its safety and your own. Position yourself so that your safety will not be compromised by any reaction of the animal. The injection of a medicine should always be preceded by a study of the anatomy of the injection site and by a review of the proper procedures for the type of injection necessary.

Topical Route

A *topical application* is one in which the medicine is applied to the skin or to the mucous membranes of the eye, ear, or nasal passages. Such medications are available as ointments, aqueous solutions, powders, and aerosols. Remember that absorption does occur when a preparation is applied topically, so the medication must be nontoxic.

> **CAUTION:** Rubber or latex gloves should be worn when applying topical compounds. Always apply sprays downwind, that is, with the wind blowing the spray away from you instead of back into your face or onto your body or clothing.

Oral Route

Many preparations are given by mouth, and to a large extent the results obtained are dependent upon the condition of the digestive system at the time of administration and the tolerance of the animal's system to the product. These preparations are available in tablet, pill, capsule, and liquid form. Depending upon the form, one or more of the following methods of oral administration should be selected.

The *feeding* of drugs is sometimes practical. For this method to work, however, the animals must be eating and the drug acceptable or palatable to them. Feeding space and methods must be such that all individuals will get the proper dosage within the proper time if the treatment is to be successful.

Balling guns are used to give boluses, capsules, and tablets. The instrument is placed on top of the tongue, just back of the highest portion when the head is being held in a normal position. The plunger is then pushed to deposit the drug onto the back of the tongue. For the complete step-by-step procedures, refer to the appropriate chapters of this book.

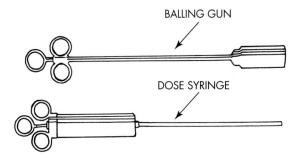

Drenching is a method that is satisfactory for giving moderate amounts (up to about 1 quart) of liquids or suspensions to cattle, sheep, and goats. A bottle with a long neck can be used for the procedure. Stanchion or headgate restraint is required for cattle. Calves, sheep, and goats must be held or restrained so that they will not be able to back away. Stand at the right side of the animal, reach over the head with your left hand, place your thumb into the animal's mouth (interdental space), and allow the fingers of your left hand to fall under the animal's lower jaw. Then lift the animal's head until the nose is slightly above a plane parallel to the ground. The attendant's left hip should be placed just back of the animal's head, allowing for extra help in holding the animal. The neck of the bottle should be placed into the right side of the mouth and on top of the tongue. Allow the animal to slowly swallow the contents (medicine). For the complete step-by-step procedure employed for a given animal, refer to the appropriate chapter of this book.

CAUTION: Do not tie the animal so that its head cannot be quickly released in case of choking.

Dose syringes are used to administer small amounts of liquids and suspensions, especially to horses and hogs. The syringe should be small enough that the attendant can hold it comfortably in his hand while giving the medicine. Most people can handle a 6-ounce syringe. For the horse, the nozzle should be at least 6 inches long. A 4-inch nozzle is best for calves, sheep, goats, and swine.

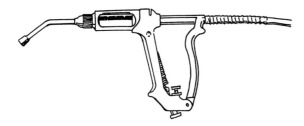

The animal's head should be held so that its mouth is parallel to the ground. The nozzle is placed into the corner of the mouth, oriented over the base of the tongue, and the plunger slowly depressed to deposit the drug into the back of the mouth.

CAUTION: Do not tie the animal so that its head cannot be quickly released in case of choking.

Injectable (Parenteral) Route

For all practical purposes, the terms *injectable* and *parenteral* refer to the same thing—the administration of a drug directly into an animal's body by the use of a needle and syringe. Drugs given parenterally act rapidly, are utilized more efficiently, and frequently act longer than when they are given orally or applied topically. Frequently, the parenteral route must be used because a drug may produce gastrointestinal irritations or be destroyed in or not absorbed from the digestive tract if it is given orally.

Injectable drugs may be more expensive than oral or topical medications. Their administration requires a sterile technique, they may cause local irritations, and there is a greater chance of a systemic allergic reaction, particularly with intravenous injections. Care must be taken in the preparation of the injection site, equipment, and product if complications are to be minimized.

Syringes and needles are used for injecting drugs. Syringes are constructed of either glass or plastic, and are available in many sizes. Plastic syringes have a big advantage over glass in that they won't break.

Higher-quality plastic syringes can be sterilized and used several times, while the lower-cost plastic syringes should be discarded after one use. In purchasing syringes, make certain that the plunger fits the barrel properly. If it does not, a part of the dose of medicine may be lost. A 1-cc syringe should be used when administering a small amount of a drug to ensure accurate dosage. Other convenient sizes of syringe to have available are 1, 5, 10, and 20 cc. The larger sizes can be used in administering large doses or for multiple doses.

Needles also come in many diameters (14 gauge is larger in diameter than 22 gauge) and in variable lengths. The length and diameter of the needle must be considered when preparing to give injections. Twenty-two-gauge, 1-inch and 18-gauge, 1½-inch needles will be adequate for giving most injections. A needle with a Luer-Lok hub can be screwed onto the syringe, minimizing the possibility of drug or needle loss.

Filling the Syringe

Step-by-Step Procedure

1. Pull back on the plunger and fill the syringe with an amount of air equal to the amount of medication to be placed in the syringe.

CAUTION: The use of a large-diameter needle (18 gauge or larger) for filling the syringe will be helpful; many drugs thicken when they are kept in a refrigerator.

---

2. Pass the needle through the cleansed rubber stopper of the medicine bottle (use alcohol to cleanse the stopper) and slowly inject the air into the bottle. (It may be necessary to withdraw some of drug before injecting all of the air.)

CAUTION: Always use a clean, sterile needle in a multidose bottle. Leave this needle in place if the syringe must be filled several times.

---

3. Make certain that only the bevel of the needle is through the stopper so that you can remove the last of the drug from the bottle.

4. Remove all air bubbles from the syringe by holding it upright, tapping on it with your finger, and allowing time for air bubbles to move upward. While holding it in this position, push upward on the plunger, pushing the air bubbles and extra drug back into the drug bottle.

5. Pull the needle straight out to remove it from the stopper. Hold the syringe and plunger in a manner that will not allow spillage or contamination of the needle.

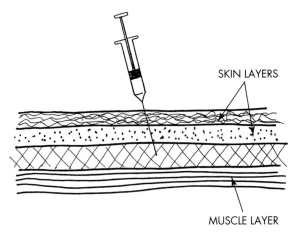

CAUTION: You must be careful not to carry foreign material into the injection site with the needle. Alcohol is effective in removing oil and dirt from the hair. It is a poor disinfectant, however, and requires several minutes to be effective against those bacteria that it can kill. When necessary, remove mud and manure from the injection site with soap and water, rinse well, and dry. It is not necessary to clip or shave the injection site before making an injection.

Make certain that the material injected does not leak from the opening in the skin when the needle is removed. If this should begin to occur, place your finger over the injection site and pinch firmly, then massage gently for a few seconds. Some tricks for giving injections: (1) pinch or pull on the skin 2" to 3" from the injection site; (2) slap firmly two or three times on the injection site, then insert the needle with a quick thrust; (3) pull loose skin down over the needle when making a subcutaneous (SC) injection.

CAUTION: The route, or location within the body, by which a drug is administered affects how long it takes for the drug to be absorbed and eliminated from the body. Medications should always be administered by the route on the label unless directed otherwise by your veterinarian. Administration by an incorrect route may render a drug useless, at best, or result in severe injury or death to the animal and/or yourself (from violent reactions).

Types of Injections

Subcutaneous (SC) injections are made beneath the skin but on top of (not into) the muscle layer. The side of the neck is a good area in which to make the injection in horses and cattle. The axillary space (under the front leg), flank, and abdomen are the usual sites for swine, sheep, and goats. Depending upon the product and the size of the animal, 2 to 30 cc can usually be injected into a site. The area should be massaged to disperse the drug when larger amounts are given. Needles of 18 to 20 ga and 1" to 1½" in length usually are used. They should be 20 to 22 ga and ¾" to 1" long for pigs, lambs, and kids.

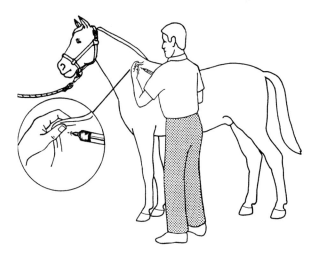

To properly administer the injection, lift the skin away from the underlying tissue with the thumb and index finger of your free hand. Into this raised fold of skin, insert the needle with a sharp, quick, yet controlled jab. Push the needle in until you are holding a fold of skin over the hub of the needle. Lightly pull out on the plunger and observe the syringe barrel for the appearance of blood. If blood should appear, withdraw the syringe slightly before making the injection.

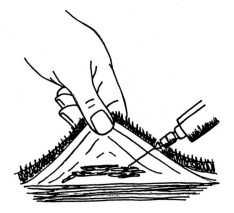

With the needle properly positioned, administer the injection. Massage the area to aid in dispersing the drug for absorption. Any enlargement of the injection site should disappear in a short time.

The active agent is absorbed more slowly when it is administered subcutaneously than when it is administered intravenously, but it is available over a longer period of time. Irritating products cannot be given by this route because of tissue damage. The implantation of growth-promoting pellets in the ears of calves is another form of subcutaneous administration of a drug.

Ear injections, more properly **base-of-the-ear injections,** are a type of subcutaneous injection. This route of administration was developed by Boehringer-Ingelheim Vetmedica, Inc., for use with clostridial injections. Typically, clostridial vaccines cause a substantial lesion at the injection site. Injection-site lesions are blemishes that cause the value of the beef carcass to be reduced at the packing plant. Very few ear injections cause an injection-site lesion, and those that do are of little or no consequence (and do not result in a value-reducing blemish) because the ear is discarded at slaughter.

Step-by-Step Procedure

1. Determine whether the animal has been implanted with some type of growth stimulant. Check for this at the base of the ears. If the animal has been implanted, select the opposite ear for the injection.

2. Locate the actual injection site at the base of the ear. This site will be mid-ear, from top to bottom, and lie just outside of (distal to) the auricular cartilage of the ear.

3. Securely grasp the animal's ear with one hand. Your hold should be near the end of the ear.

4. Use a 16-gauge needle, $\frac{3}{4}$" to 1" long. Begin inserting this needle into the injection site, as though you were going to give a subcutaneous injection (as just explained).

5. The precise injection site location is where the skin of the ear becomes loosened from the auricular cartilage.

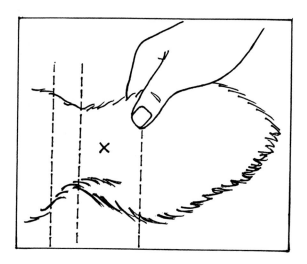

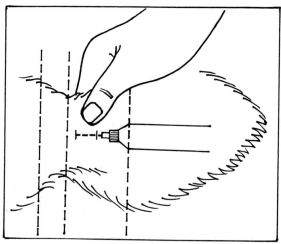

6. Insert the needle fully to the hub (as far as it will go).

7. Inject the vaccine with a deliberate, but not hurried, push on the plunger of the syringe.

8. Remove the syringe and needle from the base of the ear while continuing to hold some pressure on the plunger of the syringe.

9. As the needle is being removed, place your thumb on the injection site. The application of

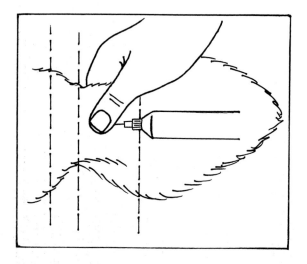

some pressure at the injection site will help to seal the opening caused by the needle.

10. This method of injection may not work well with other than certain clostridial vaccines. Check with your veterinarian before using this technique.

Intramuscular (IM) injections are made directly into a major muscle mass with an 18- or 20-ga needle that is 1" to ½" long. Absorption is usually rather

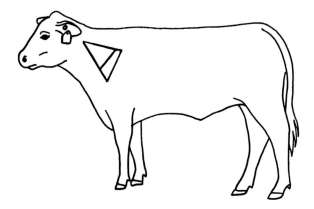

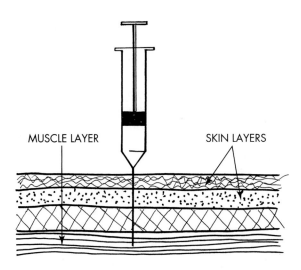

rapid because of the good blood supply to muscle tissue. Only small amounts of medication should be deposited in one site because too much drug in one area may result in muscle necrosis. Pain and lameness frequently occur when this route of administration is used, even if small amounts are injected in a sterile and proper manner.

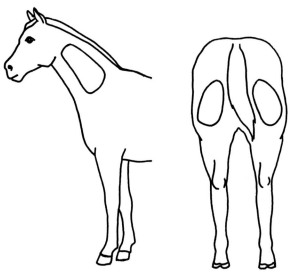

INTRAMASCULAR INJECTION SITES—HORSES

The proper administration of an intramuscular injection begins with the selection of an appropriate muscle mass. This selection should be based upon the animal's temperament, its primary use,

and the presence or lack of an injury. In meat-producing animals, intramuscular injections should be administered into an area of the body involving the lowest-quality meat cuts first . . . providing the injection site selected is satisfactory for the product to be administered. Only after all these sites are exhausted should the higher-priced-meat areas of rear leg and back muscles be used. The muscles of the neck and upper forelimb are the first choice.

After the site is selected, distract the animal's attention by pinching the skin next to the intended injection site or by slapping the injection site firmly. Immediately following the slap or during the pinching, insert the needle with a sharp, quick thrust. The angle is not critical, but it should be nearly straight inward.

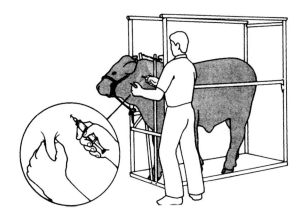

After inserting the needle, pull back on the plunger of the syringe to see if any blood appears in the lumen of the syringe. If no blood is present, inject the drug with a slow depression of the plunger. If blood is aspirated, slightly withdraw the syringe until the point of the needle is removed from the blood vessel. With the needle properly positioned, slowly depress the plunger of the syringe and administer the drug. Do not administer more than 12 to 15 cc per injection site.

Intravenous (IV) injections are made directly into the lumen of a blood vessel. Because of the extensive knowledge of anatomy necessary to be able to locate and control a blood vessel, and because of the extensive experience required to insert a needle accurately and efficiently into the blood vessel after it is isolated, it is recommended that this injection route be used only by an experienced herdsman following the specific recommendations and instructions of a veterinarian. There is a very real possibility that the animal may experience a strong allergic reaction to the drug. The veterinarian would have appropriate remedies for the reaction.

Intravenous injections are particularly useful when large volumes of drug must be given, when the drug must be rapidly available to the body, and when a drug that must be given is irritating to tissues.

CAUTION: Intravenous injections should always be administered with the needle pointing in the direction of the blood flow in the vein.

Intradermal (ID) injections are made with a 22-ga, ¾" needle inserted between the layers of the skin. Only 1 to 2 cc of medication can be deposited at a site. A small bump can be felt in the skin after the injection is made. ID injections are particularly useful whenever slow absorption is desired.

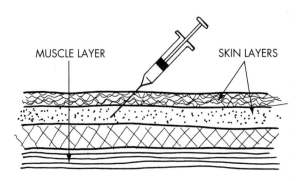

MUSCLE LAYER   SKIN LAYERS

Intraperitoneal (IP) injections involve the administration of a drug into the peritoneal cavity. The drug is made available at a more rapid rate than with the SC route, but much more slowly than with IV therapy. It is necessary to avoid placing the needle and product in the lumen of the intestine. This route is not recommended for the horse.

Intramammary infusions (IMI) are made by inserting a cannula into the teat canal and injecting the drug into the milk cistern. It is most important to clean and disinfect the teat orifice before inserting a sterile teat cannula. Massage the gland well to disperse the drug. Irritating materials cannot be infused.

Subconjunctival injections involve the injection of a drug beneath the conjunctiva of the eye.

Nebulization and **inhalation** refer to the administration of a drug by the respiratory route.

Calculating Dosages

Livestock producers very often must administer medications to animals. In most cases, dosages will have to be calculated. An example is given below. Be sure to follow label directions for the medication used.

EXAMPLE:

A 300-pound sick animal requires an injection of antibiotic at a dosage rate of 5,000 units/kg. The antibiotic to be used contains 200,000 units/ml. How much antibiotic should the livestock producer administer to the animal? Calculate the dosage as follows:

1. Change 300 lbs to kilograms (kg).
   2.2 lbs = 1 kg
   300 lbs ÷ 2.2 lbs/kg = 136 kg
   300 lbs = 136 kg

2. Determine how many units of the antibiotic the 136-kg animal will require.
   Dosage rate = units of antibiotic per kg of body weight
   5,000 units × 136 kg = 680,000 units
   Dosage for the 300-lb animal = 680,000 units of antibiotic

3. Calculate how many ml of the drug are needed to provide the required dose. The antibiotic product contains 200,000 units per ml, so
   680,000 units ÷ 200,000 units/ml = 3.4 ml
   Dosage = 3.4 ml, or 3.4 cc (1 ml = 1 cc)
   Sometimes dosage rates are given in mg per kg of body weight instead of units of drug per kilogram of body weight.

EXAMPLE:

A 300-pound animal requires an injection of antibiotic at the rate of 6 mg/kg of body weight. The antibiotic product contains 200 mg/ml. The dosage is calculated as follows:

$$300 \text{ lb} \div 2.2/\text{kg} = 136 \text{ kg animal}$$
$$136 \text{ kg} \times 6 \text{ mg/kg} = 816 \text{ mg antibiotic needed}$$
$$816 \text{ mg} \div 200 \text{ mg/ml} = 4 \text{ ml}$$

So the correct dosage is 4 ml, or 4 cc (1 ml = 1 cc).

Managing Drugs and Drugs Administration

Following are 10 steps vital to safe and efficient use of your drug supply:

1. When your animals are showing distress, be certain to consult with your veterinarian to determine the cause. Only when the cause and effect are known, can you select the proper drug to administer.

2. Administer drugs according to the manufacturer's directions, unless expressly directed to do differently by your veterinarian. To follow the manufacturer's directions exactly, you must read them thoroughly and carefully.

3. Leam how to administer the drugs properly. The desired action of some drugs is directly dependent upon giving them in the proper manner.

4. Use "sterile" technique when giving an injection drug. This caution is both for the drug bottle and for the injection site on the animal.

5. Be aware of the expiration date on the bottle labels. Efficacy of the drug, as well as the safety of the drug for the animal, depends upon the drug being current.

6. Do not mix different products in the same bottle or in the same syringes, unless directed to do so by your veterinarian.

7. Do not save partially used bottles of vaccine for later use. It is normal to save antibiotics in partially used bottles as long as the dates are current and the condition of the product in the bottle appears normal.

8. Dispose of used containers, needles, and syringes in an appropriate manner. Do not leave them lying around thinking that you can throw them away later—do it now!

9. Know the side effects and untoward reactions of the drug. Modern drugs are carefully tested, but adverse reactions do occur. They are easier to cope with if you are aware of what they might look like.

10. Store drugs in a safe place—far away from inquisitive children and untrained workers. Also be certain that the drugs are stored under the required conditions as specified by the manufacturer.

10.6 THE VITAL SIGNS: TEMPERATURE, PULSE, AND RESPIRATION

The body temperature, pulse rate, and respiratory rate of an animal are called its *vital signs*. There is a change from normal in one or more of these vital signs as a result of the body's response to an infection, and often to noninfectious problems. By noting these changes along with other symptoms, the problem can be identified early and appropriate treatment can be started, often preventing severe losses.

Body temperature is a reflection of the balance between the production and the dissipation of heat by the body. The animal's body has the ability to control the body temperature under conditions for which it is adapted. In domestic animals, there is very little fluctuation of the body temperature (about 1 to 2°F). The body temperature of the camel, however, can vary by as much as 11°F. In very small animals, the balance can easily be disturbed by 5 to 10°F. Severe environmental stress can influence the body temperature by as much as 3°F in domestic animals.

The animal can only report its feelings by its actions. When symptoms—abnormal actions—are noticed in an animal, a thermometer can give you valuable information. An abnormal temperature, either higher or lower than normal, is a good indication of a problem and should cause you to communicate with your veterinarian.

Fever is an increase in body temperature. There is a rise in body temperature with most infectious diseases, and an animal that is overheated also has an elevated temperature.

A subnormal temperature indicates a severe chilling or a grave prognosis—it is likely that the animal is in critical condition. The subnormal temperature is telling us that the body is no longer able to function adequately in response to the problem and is unable to maintain body temperature.

In animals, the temperature is always taken rectally. A veterinary thermometer should be used; it has an eye at the end opposite the bulb in which to tie a 10- to 12-inch length of string and a clip.

Table 10.1 contains the normal pulses and respiration rates for domestic farm animals.

Taking the Temperature

1. The animal should be placed in suitable restraint.

2. Shake the mercury in the thermometer down to a level below 96°F. This is done by grasping the

TABLE 10.1 **The Vital Functions of Domestic Livestock**

Animal	Rectal Temperature		Respiration Rate (per Minute)	Heart Rate (per Minute)
	°F	°C		
Cattle	101.5	38.6	30	50
	(100.4–102.8)	(38.0–39.3)		(40–70)
Goat	102.3	39.1	15	90
	(101.3–103.5)	(38.5–39.7)		(70–135)
Sheep	102.3	39.1	19	75
	(100.9–103.8)	(38.3–39.9)		(60–120)
Horse	100.0	37.8	12	45
	(99.1–100.8)	(37.3–38.2)		(25–70)
Swine	102.5	39.2	16	60
	(101.6–103.6)	(38.7–39.8)		(55–85)
Chicken	107.1	41.7	12–36	275
	(105.0–109.4)	(40.6–43.0)		(250–300)
Turkey	105	40.5	28–49	165
				(160–175)

ring end firmly between your thumb and index finger and, with a flipping motion of arm and wrist, shaking the thermometer vigorously. Be careful not to strike anything when shaking the thermometer.

3. Moisten the end of the thermometer with water or cover it with Vaseline, and then carefully insert about three-fourths of the length of the instrument into the rectum with a slow rotation and slight inward pressure. Attach the the clip to the hair of the animal. The thermometer should be left in the rectum for at least 3 minutes.

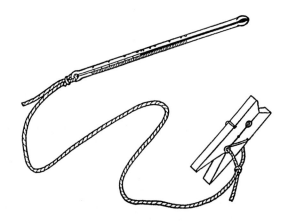

4. Remove the thermometer by pulling on the string. To read the thermometer, wipe it clean and hold it so that the numbers are almost facing your chest; rotate it back and forth until you see the mercury line and make your reading. If available, digital thermometers may be used instead of glass thermometers.

Taking the Pulse

Pulse is the surge of blood through the arteries following the contraction of the heart ventricles. The pulse rate is a close reflection of the heart rate and is usually defined as the number of heartbeats that occur in 1 minute.

In beef and dairy cattle, the pulse is most easily detected by palpating the artery (middle coccygeal) on the underside of the tail about 6 inches down from the head of the tail. In horses, the best pulse can be felt by palpating the facial artery where it crosses the underside of the jawbone, immediately in front of where the large cheek muscle attaches. In small ruminants, goats and sheep, the heartbeat can be felt directly by placing the fingertips between the ribs just behind the elbow area of the foreleg. The pulse can be felt in sheep and goats by palpating the femoral artery about one-third of the way down on the inside of the rear leg (up high on the thigh). In most cases, the pulse cannot be felt in pigs. However, for young or thin pigs, the femoral artery can be palpated, just as in sheep and goats.

Once the pulse can be felt, count the number of heartbeats felt in a timed minute. The pulse cannot be felt on hogs. Adult animals have a slower heart rate than do immature animals.

Counting Respirations

Respiration is the act of breathing. It is important to note the rate and the actions associated with breathing. These include sneezing, wheezing, groaning, rattling, head extension, and the expansion and relaxation of the rib cage and abdominal wall. The rate of respiration can be determined by counting the contractions of the rib cage or by holding the back of your hand near the nostril where you can feel the exhaled air. Count the number of times that the animal breathes in 1 minute.

10.7 RECOGNIZING THE HEALTHY ANIMAL

It is necessary that you know the behavior of normal animals if you are to recognize changes or symptoms that are indications of disease or sickness. A thorough understanding of animal behavior is also necessary if one is to make sound decisions about administering first aid and about when to solicit help from a veterinarian. The ability to be almost subconsciously yet completely aware of the normal behavior and activity patterns of your livestock ("livestock sense") enables you to detect a problem with only a glance at the herd or flock or one of its members. This trait is developed and sharpened by the experience gained in working and caring for livestock.

Until you have gained this "livestock sense," use the following items as a basis for characterizing the normal or healthy animal.

Learn the Normal Eating Behavior of the Animal. It is important to understand the general eating habits of a species, including the normal patterns of grazing, drinking, and ruminating, because in a large portion of animal health problems the animals will very quickly alter their eating habits or go "off feed" completely. If you are observant, many problems can be spotted early and animal and production losses minimized. Some normal behavioral patterns are described next.

Hogs root and move food around with their snouts. They bite food with their front teeth and then chew the food before swallowing it. They like to take a few bites of grain and then a swallow of water, then more grain, and then water again. They require a high-energy diet. When its stomach is full, a hog likes to go to a comfortable spot to lie down while digestion takes place.

Cattle, sheep, and goats are ruminants and need a large volume of forage. Cattle bring feed into place with the tongue and bite it off and take it into the mouth with the lower teeth and dental pad in conjunction with a unique head movement. The food is swallowed without chewing and enters the rumen. When its rumen is full, a cow will find a comfortable spot, lie down, and ruminate for several hours. (Rumination is the process by which the feedstuff is regurgitated for chewing. After it is chewed, the foodstuff is swallowed and moves through the omasum, the abomasum, and the intestine for additional digestion and absorption.) Sheep and goats also graze for a period of time and then lie down for rumination.

Horses are browsers, grazing continuously. They bite grass with their incisors and tear it off with a jerk of their head. Large mouthfuls of grain may also be consumed, necessitating large, shallow feeders for some horses (so they are less able to take huge quantities of grain with each bite).

For all animals, one should notice the way the head is held during chewing and swallowing and whether there appears to be a normal fill in the flanks. Are the animals selective in what plants they choose to eat? Will they eat less-palatable plants when food is scarce?

Learn the Normal Fecal Pattern and Consistency. Amount, consistency, odor, and color of stool are important in making an evaluation of the health of an animal. Types and amounts of feed ingredients will influence stool characteristics. Observe the defecation patterns and locations. Hogs, for example, usually develop a definite pattern of location and time for defecation and urination and can be toilet-trained.

Learn the Normal Stance, Movement, Posture, and Activity Patterns of the Animal. Understanding the normal stance, movement, and recumbency of an animal species is most important if one is to recognize the abnormal as an aid in making a diagnosis. Normal breathing is done easily and in rhythm. Livestock have definite patterns of sleeping, resting, and bedding down during the day or night. Activity patterns and the way animals group often are altered when weather conditions change, especially when extremes in temperature occur.

Recognize That Some Variation in Behavior Is Normal. The stage of gestation, stage of lactation, and type of work being performed are all factors that must be considered when evaluating the health status of an animal. These will vary between species and within a species, but each will evoke pronounced changes in the way an animal eats, sleeps, and moves about.

Observe the behavior of the animal in estrus, or "heat." Know the signs that the animal exhibits to indicate that parturition will soon occur.

Learn to Observe the Entire Herd or Flock. Note how the healthy herd or flock acts as a group while grazing, eating, drinking, and resting. When the opportunity arises, note the normal reaction of animals to visitors, dogs, wild animals, or additions to the herd. Social organization is present in most herds and flocks. The boss animal can usually be easily recognized by its dominance over other animals. Observe how animals act when it rains or snows. Note how animals act in the presence and absence of flies and certain insects.

Learn the Sounds or Communication Skills of Animals. All livestock have specific grunts, bellows, clucks, and gobbles by which they communicate with each other and even with the manager. A herd of very hungry cattle waiting for hay can be very vocal in letting the herdsman know that it is time for feeding. The lost baby lamb will make a specific concerned bleating sound which both the mother sheep and an experienced shepherd can identify. The mother sow has a grunting song that she sings to her pigs as she nurses them. Baby chicks will cheep loudly when they are cold, and an adult flock will cackle when something unusual is taking place. With experience, these sounds become meaningful and the herdsman knows automatically when the sound is normal or when it is one about which to be concerned.

10.8 RECOGNIZING THE ABNORMAL OR SICK ANIMAL

The efficient livestock manager has the ability to observe his animals with a knowing eye. This implies that while he is gazing across the herd or flock, he is in reality doing more than merely looking at them. He is noting generalities as well as details; he is noting which animals stick out because they do not fit the activity and behavioral patterns he has come to expect. He is noting which animals appear abnormal.

Section 10.7 deals with learning how to observe and recognize the normal animals. It is noted there that it is a more important first order of business to be able to notice that something is wrong than to be able to identify the exact problem.

It is essential that the manager have access to sufficient information to determine which physiological system in the animal's body is causing it to appear abnormal. It is not the intent of this text to provide you with all of the information necessary to diagnose the exact problem. That is the role of the veterinary diagnostic laboratory. The role of the manager is to manage all aspects of his animals, including their health. Being able to spot which body system is causing the animals to perform abnormally will provide you with the information necessary to efficiently manage the herd health program.

Many common animal health problems are described in Appendix S. The problems are organized according to the body systems—digestive, respiratory, genitourinary, nervous, integumentary (skin), locomotor, and circulatory—and then subdivided into noninfectious and infectious problems.

Several representative diseases and most of their symptoms are presented under each heading.

CAUTION: Zoonoses are diseases that can be transmitted from animals to man. Some of these are quite serious, and appropriate care must be taken to not contract these diseases. Throughout this book we have discussed management techniques that bring you into very close contact with animals, their body fluids, and their excreta. In every case, we have presented appropriate cautions involving adequate restraint and proper equipment, such as rubber gloves for rectal palpation for pregnancy status.

In Section 10.7, we have discussed ways of observing the behavioral patterns of your animals to help in determining if all is well with the herd or flock. In Section 10.8, we will alert you to ways to determine when the health status of your animals is in jeopardy. In the confines of this text, we cannot attempt to list all the symptoms of the diseases you are likely to encounter.

In the material following this **CAUTION,** we will present a listing of the major zoonoses. It is not our intent to frighten the animal manager with this information. However, it is to your benefit to be aware that zoonoses exist, what some of them are, and that they can be serious diseases if you contract them. Your veterinarian will be able to brief you regarding additional safeguards, beyond following careful management technique, that you should put in place should he or she suspect a possible zoonotic situation on your operation.

Humans contract these diseases in a variety of ways: inhalation of fecal or feather dust, ingestion of infected animal products, direct contact with animal secretions or excreta, infection of open wounds, contaminated soil or water, handling aborted fetuses, insect bites, animal bites, and simply handling infected animals.

The symptom list should be used to help identify the type of problem, and the disease list used as a starting point for follow-up reading from a veterinary manual if that is your wish.

As you use the information in Appendix S, keep in mind some biological principles:

1. A noninfectious problem will not move through the herd from animal to animal. Usually only a few animals will be affected.

2. An infectious problem usually involves the entire herd and moves through it quite rapidly.

3. If a fever is present, the problem is usually infectious.

4. More than one body system can be affected, and consequently more than one set of symptoms can be present at one time.

Zoonotic Diseases

Disease (Zoonose)	Causative Organism	Animal Reservoirs
Anthrax	Bacillus anthracis	Ruminants, equine, swine
Brucellosis	Brucella abortus, brucella spp	Ruminants, equine, swine
Campylobacteriosis	Campylobacter spp	All domestic animals
Clostridial disease	Clostridium spp	All domestic animals and fish
Erysipeloid (erysipelas)	Erysipelothrix rhusiopathiae	Swine, poultry, pigeons, fish, sea mammals
Eschericia coli	E. coli 0157-H7	Cattle, sheep, swine, poultry
Leptospirosis	Leptospira spp	Domestic animals, wild animals, especially rodents
Listeriosis	Listeria spp	Cattle, sheep, swine, rodents, birds
Plague	Yersinia pestis	Rodents, cats, dogs
Pseudotuberculosis (Yersiniosis)	Yersinia pseudotuberculosis	Rodents, felines, fowl
Psittacosis (ornithosis)	Chlamydia psittaci	Poultry, ducks, geese, pigeons, parakeets, parrots
Rat bite fever	Actinobacillus moniliormis Spirillum minus	Rodents
Salmonellosis	Salmonella spp	Mammals, birds, rodents
Tetanus	Clostridium tetani	Cattle, sheep, horses
Tuberculosis	Mycobacterium bovis Mycobacterium tuberculosis	Cattle, goats, swine, domestic animals, wild ungulates
Tularemia	Francisella tularensis	Sheep, rabbits, hares, wild rodents
Arbovirus infections	Var. arboviruses	Sheep, goats, swine, equines, rodents, birds
Bovine spongiform encephalopathy (BSE)	Prion	Cattle
Contagious ecthyma (Orf)	Parapoxvirus	Sheep, goats
Cowpox	Poxvirus	Cattle
Foot and mouth disease	Rhinovirus	Cattle, pigs
Herpes B virus disease (simian herpesvirus)	Herpes B virus	Monkeys
Influenza	Influenza virus type A	Swine, birds
Newcastle disease	Paramyxovirus	Fowl
Rabies	Rabies virus (lyssavirus)	Carnivores, bats, wild animals
Murine (endemic typhus) (flea-borne)	Rickettsia mooseri	Rats, mice
Rickettsial pox (mite-borne)	Rickettsia akari	Mice
Scrub typhus (mite-borne)	Rickettsia tsutsugamushi	Rodents
Spotted fever (tick-borne)	Rickettsia rickettsiae	Dogs, rodents, other animals
Q-fever	Rickettsia burnetti	Cattle, sheep, goats, wild and domestic mammals, birds
Ringworm	Microsporum spp Trichophyton spp	Cattle, horses, poultry, cats, dogs, small mammals
Amoebiasis	Entamoeba histolytica	Dogs
Balantidiasis	Balantidium coli	Swine
Coccidiosis	Isospora spp	Dogs
Malaria	Plasmodium spp	Monkeys
Sarcosporidiosis	Sarcocystis spp	Cattle, sheep, swine, ducks
Toxoplasmosis	Toxoplasma gondii	Mammals (especially cats), birds
Trypanosomiasis	Trypanosoma spp	Cattle, dogs, antelope, small mammals
Beef tapeworm (cysticercosis)	Taenia saginata	Cattle, buffalo, llama
Echinocosis, hydatid	Echinococcus granulosis	Cattle, sheep, cats
Pork tapeworm (cystcercosis)	Taenia solium	Swine

10.9 MANAGEMENT AIDS FOR PROTECTING THE HERD FROM INFECTIONS

The use of biological agents as management tools in preventing disease and of pharmaceuticals as management tools to combat disease once it has erupted are discussed in Sections 10.4 and 10.5. These medications should not, however, be relied upon as the sole basis for your herd health program. The list of potential diseases and their causative agents is far too formidable for you to depend upon the needle and syringe alone. The following guidelines are additional measures that the astute manager can utilize to reduce the chances of introducing infectious diseases into his herd or flock.

1. Purchase animals from reputable producers. Be inquisitive; ask questions that might uncover disease problems. Ask specifically: (1) to see production records and reproductive records; (2) about the disease program being followed; and (3) to look at the entire herd, in addition to the animal(s) being considered for purchase.

2. Ask for the appropriate health papers.

3. Isolate the animal(s) for a suitable length of time before exposing them to or adding them to your herd. Sufficient time should be allowed for symptoms of disease to appear or for measurable levels of antibodies to be produced (in the event that the infection is in the incubation stage). Additions to a swine breeding herd should be introduced to the base herd 2 to 3 weeks before breeding time. Parvovirus infections in swine are being controlled in this way.

4. Control outside traffic, both people and vehicles. Particularly watch for feed delivery, rendering trucks, and livestock haulers. Construct load-out facilities near the public road. Do not allow trucks that have been hired for grain or livestock hauling onto your premises unless they have been adequately cleaned and disinfected.

5. Have coveralls and boots available for people who must enter your livestock buildings, lots, and pastures. Street clothes should not be worn into these areas. Have and use footbaths with fresh disinfectant.

6. Control birds, rodents, and stray animals.

7. Double fencing along line fences may be necessary to protect your animals from those of a poorly managed neighboring herd.

8. Consider the location of your buildings and lots in order to protect them from a neighbor's drainage. Realize that you owe the same consideration to your downstream neighbors.

10.10 EVALUATING HERD HEALTH

The following are tools that the livestock manager should use in the ongoing evaluation of the health status of his herd or flock. Each of these tools should be monitored carefully; the combined package can be used as a powerful deterrent against disease.

1. The general condition of the herd—hair coat, amount of flesh, attitude.

2. Freedom from specific symptoms of sickness.

3. Production records—feed conversion, rate of gain, milk production, egg production.

4. Breeding records—conception rate, number of animals born alive and dead, number of animals weaned.

5. Fecal egg counts as indicators of internal parasitism.

6. Evaluation of the mastitis status of the milking herd by reviewing somatic cell summaries on DHI record summaries. Establish a routine, bulk-tank, milk-analysis program including milk culture and standard milk-quality test. Selectively sample and culture cows with clinical mastitis. Use the California Mastitis Test to verify subclinical mastitis in cows.

7. Blood tests to determine red blood cell profile (anemia) and antibody levels, where applicable, as indicators of infection or protection from an infectious disease.

8. Routine evaluation of tissues at slaughter for pneumonia, digestive problems, atrophic rhinitis, abscesses, joint problems, or other lesions of disease.

9. Percent mortality in the herd.

10. Autopsy and evaluation of all animals that die.

10.11 WOUNDS

Healing of Wounds

The healing of a wound ideally occurs without infection. The edges of the wound remain in apposition (touching one another) with a minimum of movement, and there is minimal disturbance to the local circulation.

The healing process begins with a joining of the wound edges by a blood clot and an invasion of the site by leukocytes (white blood cells). These are soon followed by the multiplication of connective tissue cells. These join with small capillary buds (blood vessels) and infiltrate the blood clot, gradually replacing it. Simultaneously, the epithelial cells (skin) multiply and cover the wound. This series of events forms the scar tissue, the amount of which is dependent upon the size of the wound, the degree of infection, and the amount of irritation that occurs during healing.

There should be no discharge during this healing process unless irritating (harmful) medicinal agents are applied or infection occurs. These complications slow the healing process and cause more scar tissue to be laid down in the wound.

Wounds that become infected, that are treated with irritating drugs or overtreated, or that have the local circulation severely damaged will heal with granulation tissue ("proud flesh"). Unfortunately, most wounds of animals heal in this manner. The new tissue consists mainly of capillary vessels and fibroblasts. This tissue is reddish, granular in appearance, free of nerves, and bleeds very easily. It grows inward and upward from the bottom of the wound. The granulation tissue gradually fills the wound, and the surface becomes resistant to bacterial infection. The epithelial cells slowly multiply and attempt to cover the granulation tissue. The latter gradually becomes less vascular and more firm (fibrous) as the healing process is completed.

Immobilization of the wound is most important if granulation tissue is to be kept to a minimum and healing time shortened. Wound irritation by insects will also delay the healing process.

The use of irritating disinfectants and harsh medicinal agents prevents proper wound healing. These products cause additional damage to the sensitive new cells that are growing to fill in the wound.

Classification of Wounds

Abrasions are wounds in which only the surface of the skin is damaged, exposing the sensitive layer of the skin. Irritation from tack or poor adjustment of tack, rope burns, falls on rough surfaces, and scraping of legs during loading and unloading or hauling are common causes of abrasions. The blood supply is frequently damaged, so such wounds heal very slowly and frequently have excessive granulation formation.

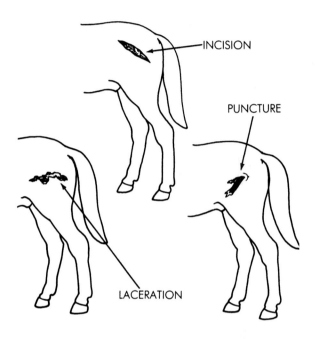

Incised wounds are made by sharp objects (scalpels during surgery, tin, field discs, glass). The wound has clean-cut edges and very little tissue damage. Accidental lacerations may do serious damage to deeper tissues, such as blood vessels, nerves, tendons, joints, and muscles, or penetrate the thoracic or abdominal cavities. Hemorrhage is frequently severe. These wounds can often be sutured and will heal quickly (7 to 10 days) and leave a minimal scar.

Lacerations differ from incisions in that the wound edges are irregular, the tissue is more severely damaged, hemorrhage is usually minimal, and they are slow to heal. Wire fences are the most frequent cause of lacerations. Having a foot cut while loading and getting a leg caught under a partition or other object when rolling on the ground are also common causes. These wounds heal slowly with the formation of a substantial amount of scar tissue.

Puncture wounds are a common occurrence with horses. Nails, splinters, and pipes are objects that frequently penetrate the feet, joints, tendon sheaths, muscles, and sometimes thoracic or abdominal cavities. All or part of the penetrating object, plus dirt, manure, and other debris are carried into the wound. These wounds must be cleaned, foreign material removed, and damaged tissue trimmed for adequate healing to occur. Considerable scar tissue is formed and the time required for recovery from this type of injury is lengthy.

Contusions result in very little damage to the skin while bruising the soft tissue below it. Their importance is dependent upon the area damaged and the extent of the damage. Sides of the abdomen and hips are frequent sites for these injuries. Kicking, butting, mistreatment, and hauling accidents are often the causes of these wounds. Bleeding sometimes occurs within the damaged tissues (hematoma), indicating the breakage of a blood vessel. The hematoma must not be opened until the blood vessel has healed. Drainage must be established and any foreign material removed in order to get healing. These wounds may take two or more weeks to heal. The livestock marketplace penalizes carcasses that show evidence of contusions.

Treatment of Wounds by Location

Wounds of the head should be sutured, and it is best that this be done as soon as possible. Lacerations of the eyelid should also be sutured. Lacerations of the ear frequently require extra surgery because of the cartilage present there. Because of the equipment needed, it is best to seek the help of your veterinarian. Healing should take place within a week.

Wounds of the tongue are frequent in horses and should be sutured for good healing.

Neck wounds are often serious because of deeper structures that may be damaged. These include the jugular vein, carotid artery, vagas nerve, esophagus, and trachea.

Wounds of the axilla (area between front limb and body) frequently occur when an animal attempts to jump a fence and lands on top of the wire or boards. Movement by the animal in walking often compounds this problem by admitting air to the subcutaneous tissue. One must make certain to remove all foreign material from these wounds after bleeding is stopped. These usually heal as well as open wounds.

Wounds over the thorax and abdomen must be carefully evaluated. It is necessary to determine if the thoracic or abdominal cavity has been penetrated. If it has not, the wound can be treated in the same manner as those occurring on other body surfaces. If the wound penetrates the thorax, or causes a fractured rib to do so, air can be aspirated (sucked) into the chest cavity, resulting in collapse of the lung. These wounds are often followed by pleuritis (inflammation of the pleural membranes). Wounds that penetrate the abdominal cavity can result in damage to the digestive tracts, followed by peritonitis (inflammation of the peritoneal membranes).

Wounds of the withers, back, croup, and tail can be difficult to heal because of problems in establishing adequate drainage. It is also important to check for fractures of the vertebrae.

Wounds of the limbs are common in all domestic livestock, and particularly so in horses.

1. Wire lacerations of the forearm are common and can be extensive enough to allow protrusion of the extensor muscles. In these cases, there is often a noticeable dragging of the leg or difficulty in taking a step for some time after the wound has healed. This animal should be placed in a tie stall for several weeks during healing.

2. Wounds of the knee must be evaluated for damage to blood vessels, tendons, sheaths, and joint structures, and for fractures or chips. These wounds should always be bandaged and movement of the animal minimized. This is done to minimize granulation tissue and bony spicule formation, with the resultant loss of mobility.

3. Wounds of the metacarpal (lower forelimb) and metatarsal (lower hindlimb) regions may involve the blood vessels, nerves, tendons, ligaments, and bones of the area. All of these wounds must be bandaged to minimize the formation of granulation tissue. Special treatment is necessary for damaged tendons and ligaments.

4. Wounds of the fetlock and pastern usually heal with minimal problems if they are kept under a pressure bandage and the animal is kept quiet.

5. Wounds of the feet are very common in livestock. Nail punctures of the sole and frog result in severe lameness. The puncture must be located and reamed out so that the area will drain. This wound must be kept open and clean for adequate healing. If the wound fails to respond to good treatment, radiographs will be needed to evaluate damage to deeper structures (third phalanx, navicular bone, or joint capsule).

6. Wounds of the heels and the coronary band are not uncommon and may do damage to cartilage, joint capsule, and tendon sheath. These must be thoroughly cleaned and put under a pressure bandage. Stall rest is most important.

7. Wounds of the thigh can result in extensive muscle damage. After good drainage is established, the wounds usually heal with minimal problems.

8. Wounds of the stifle area are often penetrating wounds and involve deeper structures. These should be cleaned and the animal given stall rest.

9. Wire injuries are common over the front surface of the hock. Pressure bandages must be applied, along with stall rest, if satisfactory healing is to be accomplished.

10.12 FIRST AID TECHNIQUES

First aid procedures are commonly thought of as being performed in an emergency situation, such as in treating an injury to an animal. This interpretation of first aid is an important one, but it is altogether too narrow in its scope. First aid should be interpreted as being the first procedures employed to alleviate the distress of any animal exhibiting symptoms of abnormal health—whether the cause of the distress be a laceration, a case of ketosis, or colic.

First aid is just that—the first help that an animal should receive so that it can be kept alive until the veterinarian arrives. It is not meant, nor should it ever be used, to replace proper veterinary care. It is well within the ability and should be considered the responsibility of every herdsman to become proficient in the administration of first aid to his animals.

The following procedures are broken into two parts: (1) treating the injured animal—the traditional approach, and (2) treating the sick animal—the expanded approach.

Treating the Injured Animal
Step-by-Step Procedure

1. *Properly restrain the animal.* All large animals are potentially dangerous to handle. Their physical size and weight and instinctive defensive behavior are usually more than a herdsman can safely handle. Because the animal is sick or injured, he may be unpredictable and doubly dangerous. Select the restraint method that allows you safely to control the animal and efficiently perform necessary first aid procedures.

2. *Stop the blood loss.* It is frightening to see blood gushing from a wound. But, as frightening as it may be, swift, correct attention to the situation can usually stop the blood flow before a life-threatening amount is lost. A mature horse can lose up to 6 quarts before he is in danger.

On the limb of an animal, use a pressure bandage to stop the bleeding. Hopefully, you have two first aid boxes in the barn—one for humans, the other for your animals. Quickly, but without panic, take a piece of absorbent material sufficiently large to make a small to medium first-sized ball. Place this "ball" into the wound area and hold it there with pressure. If the bleeding slows, the pressure is likely adequate, if it does not, push harder. Your goal is to stop all of the bleeding, including the oozing blood. It should stop bleeding in 2 or 3 minutes. When that happens, you can attempt to place a wrap over the pressure ball to hold it in place instead of your hand. Any movement of the ball could very well start the subdued bleeding. If you are going to wrap a leg that tightly, be certain to place a soft towel of

sheet cotton below the wrapping material and allow it to protrude below the wrap so that the tendons of the leg are not damaged. Next apply the bandage (wrapping) material, using enough pressure to stop the flow of blood from the wound.

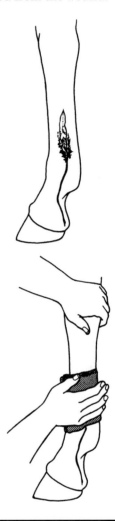

CAUTION: Always carry the bandage down the limb sufficiently far, usually down around the hoof, so as not to interfere with the blood circulation. Always allow an inch or more of the second layer of cotton to stick out above and below the bandaging material. This also helps to prevent cutting off the blood circulation.

When an area does not allow the use of a pressure bandage to stop blood loss, you have little choice but to continue holding the bandage there with hand pressure until professional help arrives.

Wounds should be cleaned and evaluated for further treatment after the bleeding has been controlled. The use for which the animal is maintained often dictates the need for suturing a wound. Wounds on horses are most frequently sutured because of cosmetic considerations. Most incisions should be sutured, and many lacerations will heal better if they are sutured as well. Wounds of the lower limbs seldom respond well to suturing.

3. *Clean the wound site.* Clip the hair from around and below the wound, keeping the clipped hair out of the wound by first coating the wound with Vaseline or by covering the wound with clean, damp gauze. The wound can be cleaned of dirt or foreign material by rinsing it with physiological saline solution. The damaged tissue may have to be trimmed to shorten the time required for healing.

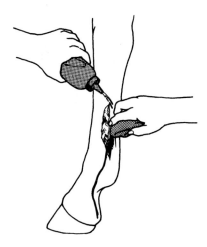

Preparation of Physiological Saline Solution

Physiological saline solution is used for cleaning, washing, or rinsing a wound. This saline solution is isotonic; that is, it has the same osmotic strength as such body fluids as blood and tears. It is nonirritating and will not cause cellular edema or dehydration, and thus it is preferable to water for washing wounds.

To prepare physiological saline, mix 1 pint of warm water with 1 teaspoonful of sodium chloride (table salt). Use noniodized salt. One teaspoonful of sodium chloride weighs between 4 and 5 grams.

4. *Identify the type of injury.* This is an important step in the first aid treatment of livestock because proper identification of the type of injury will assist you in determining the subsequent course of action.

Lacerations of the head, neck, and upper body will usually respond well to suturing. It is more difficult to get good healing after the suturing of limb wounds because of movement and pressure. Wounds to be sutured should be taken care of as soon as possible after they are incurred.

Puncture wounds must be examined carefully to make certain that all foreign material is removed and that adequate drainage is established. Puncture wounds of the foot should be protected by bandaging or with a pack.

Wounds into joints or tendons need special attention by a veterinarian to minimize structural problems.

Fractures in domestic animals are most difficult to handle because of the difficulty of keeping the bones in proper position, without movement, during healing. The size (weight) of the animal and its temperament also contribute to problems in handling fracture cases.

When you recognize a fracture, minimize all movement of the animal. The limb should be immobilized with temporary splints to prevent further damage. Pillows and strips from a bed sheet are frequently used for a temporary splint. A suitable vehicle should be used in transporting the animal. The fracture should be repaired as soon as possible. Euthanasia is advisable in many fracture cases. Meat animals may be slaughtered if this is done before inflammation or secondary problems occur.

5. *Medicate the wound.* Medicinal agents used on wounds should be nonirritating, mildly antiseptic or bactericidal, and provide some physical protection. It is wise to visit with your veterinarian about medicinal agents for wounds before you need them.

The area below the wound should be protected from the wound exudate. A solution of baking soda is good for removing this exudate. After it is cleaned, the area can be protected by a thin covering of Vaseline.

6. *Bandage the wound.*

   a. Select a piece of gauze or nonadhering pad large enough to cover the wound.

   b. Place a thin coating of nonirritating (Furacin) ointment over the wound or onto the gauze; then place the gauze over the wound.

   c. Wrap the leg with cotton, going well above and below the wound. It is best to carry the cotton down over the coronary band.

   d. Use 3-inch gauze bandage to wrap the area, pulling the bandage snugly as it is applied. Be sure to leave cotton extending above and below the gauze bandage.

   e. A covering of elastic gauze is helpful in keeping the bandage in place. This frequently can be reused two or three times.

   f. Bandages should be checked twice daily to see that they are staying in place. The frequency with which the bandage should be changed varies with the wound. After wounds start to heal and there is a minimal amount of exudate, bandages should be changed every 2 or 3 days. Before that, bandages must be changed daily, or at least every other day.

   g. Wounds below the knee and hock should always be placed under bandage. When applying the bandage, one must be certain not to cut off circulation. This is accomplished by

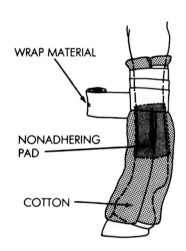

WRAP MATERIAL

NONADHERING PAD

COTTON

keeping plenty of cotton under the bandage and by carrying the bandage down over the top of the hoof.

h. The growth of an excessive amount of granulation tissue (proud flesh) frequently occurs in the process of healing. This prevents proper healing, and if it occurs, you should get help from the veterinarian. Surgical removal followed by pressure bandaging is usually the best procedure.

7. *Provide proper nursing care.*

a. When an animal is injured, always evaluate the need for tetanus antitoxin or a tetanus toxoid booster.

b. Good management of stables and pens is a must in keeping injured animals out of dust, mud, and manure.

c. Does the wound type indicate a need for parenteral antibiotics to minimize the possibility of systemic infection?

d. The location of the wound, the stage of healing, and the condition of the bandage will indicate the frequency with which the wound must be dressed. New, highly contaminated wounds must be checked more frequently than ones that are nearly healed. Bandages that become wet, soiled, or displaced must be changed.

Treating the Sick (Abnormal) Animal Step-by-Step Procedure

1. *Observe the abnormal animal from a distance.* Exactly what is it that is abnormal or unusual about the animal? Is it off-feed? Does it stray from the herd and isolate itself? Does it move reluctantly, and then with a stilted, staggering gait? Is it standing swaybacked and spraddle-legged or humpbacked and tucked in?

This list could be endless, but the point is that you must decide what it is that is signaling to you that one or more of your animals is in distress. This is why the skill of observation must be developed in a herdsman and why it is so important to know how the normal animal should act. It really is not nearly as important that you be able to identify what is distressing your animals as it is that you be able to identify the fact that they are distressed!

2. *Catch and restrain the abnormal animal.* Detailed discussions of approaching, catching, handling, and restraint are given elsewhere in this handbook.

> **CAUTION:** The animal is distressed and perhaps out of sorts. It may not react predictably. Take precautions to cope with this eventuality.
> Move the animal without stressing it unnecessarily. It is already in distress, and thoughtless handling could aggravate the situation.

3. *Observe the abnormal animal up close.* Try to verify your earlier observations made from a distance. Is the animal really dull-eyed and droopy-eared? Is it distended in the left flank? Is it showing signs of diarrhea about the tail and buttocks? Does it stand with weight removed from one of the front legs? Is the nose moist or dry? Is it panting? Is its breathing labored?

4. *Check the vital signs.* Take the temperature, pulse, and respiration and compare them to the norms. Realize that the hassle of catching and restraining is going to cause a slight elevation in each of these parameters. See Section 10.6 for a discussion of the vital signs in domestic livestock.

5. *Think in terms of body systems.* The body's anatomical and physiological machinery can be divided into the digestive, respiratory, genitourinary, circulatory, locomotor (muscles, bones, tendons, joints), integumentary (skin), and nervous systems. Each of these has a normal manner of functioning and each responds in a somewhat predictable manner under stress. It is important that you begin to interpret the animal's symptoms in terms of which system is affected. For example, raspy breathing indicates involvement of the respiratory system; a grossly distended upper left abdomen probably indicates a digestive system problem; and so on.

6. *Determine whether the problem is infectious or noninfectious.* In the majority of cases, if the cause of the animal's distress is infectious, an increase in body temperature (a fever) will be present. Distress from noninfectious causes is seldom accompanied by fever.

The occurrence of infectious disease means that pathogenic organisms are present on your livestock premises. This fact should dictate to you that an entirely different set of first aid procedures is indicated than if the animal were suffering from a lame left foreleg.

7. *Consider the first aid possibilities for the affected system.* A checklist for digestive system problems follows.

Noninfectious Digestive Problems
Ruminants

1. Relief of gas. If severe, use trocar and a cannula in left flank.

2. Oral administration of laxative(s), antiflatulants, stimulants, vegetable oils (as defoaming agents in bloat).

3. Enemas are often beneficial in young animals.

4. Make available loose salt and clean fresh water. Correct any feeding problems.

Swine

1. Enemas, soapy water, and saline laxatives are helpful for constipation.

2. Correct feed, water, and exercise deficiencies.

Horses

1. Prevent self-inflicted injury by moving to a large, well-bedded stall. The horse will roll and toss in an attempt to alleviate his distress. If possible, keep the horse on his feet.

2. Oral administration of laxatives such as mineral oil, saline, or a colic mixture.

3. Call veterinarian if improvement is not noticed in an hour.

4. Correct feeding problems.

5. Evaluate the worming program.

All Species

1. When animals, especially ruminants and horses, overeat foodstuffs to which they are not accustomed, they should be treated with mineral oil or milk of magnesia as soon as possible. This is to prevent bloating, digestive intoxication, and perhaps laminitis (founder).

2. If animals continue to show more severe symptoms or fail to respond to treatment in 24 hours, you should seek help from a veterinarian. Minor ailments may need a more definitive diagnosis and other treatment initiated after surgery. This is the case for such conditions as rumenitis, displaced abomasum, and also such conditions as intussesception, torsion, and volvulus, which can occur in all species.

Infectious Digestive Problems
All Species

1. Laboratory support is usually needed to make the correct diagnosis. Appropriate specimens should be taken before treatment, because the treatment itself may interfere with isolation of the causative agent.

A similar checklist can be created for each of the bodily systems. (This is not done here, because it is not the purpose of this section to provide an encyclopedic coverage of first aid. The intent is to expand the usually limited interpretation of first aid procedures.) This is not a written list but rather a mind search, calling upon the sum of knowledge you have acquired from formal instruction, reading, and experience. As your knowledge grows, so will your ability to recognize the need for and administer first aid.

2. *Summon the veterinarian if necessary.* For infectious problems, this is always recommended procedure. Your skill, your rapport with your veterinarian, and the value you place upon your animals should be your guide in treating the noninfectious problems.

3. *Provide the proper follow-up care.*

10.13 FIRST AID KITS AND MEDICINE CHEST

Every manager of livestock should have at his disposal the necessary equipment and medications to provide first aid treatment to his animals as the need may arise. It is impossible to anticipate all potential emergencies, but with the assistance of your veterinarian, a comprehensive first aid kit and equipment box can be developed for each species. (See Tables 10.2a and 10.2b. Table 10.3 lists suggested programs indicating disease and timing protocols.)

TABLE 10.2a   Manager's Equipment Boxes and First Aid Kits (Cattle, Sheep, and Goats)

Cattle	Sheep	Goats
Antiseptic	Antiseptic	Antiseptic
Balling gun	Balling gun	Balling gun
Bloat medicine	Blood stopper/clotter	Bloat medicine
Blood stopper/clotter	Boric acid	Bloat needle (10-ga, 3")
Boric acid powder	(Epsom salts)	Blood stopper / clotter
Bottle for milk replacer	Bottle for milk replacer	Boric acid
Electrolytes for calves	Castrating tools	(Epsom salts)
Calf puller	Cotton roll	Bottle for milk replacer
California Mastitis Test	Dehorning tools	California Mastitis Test
Castrating tools	Disinfectant	Castrating tools
Clippers—electric	Dose syringe	Cotton roll
Cotton roll	Drench bottle	Disbudding iron—electric
Dehorning tools	Feeding tube	Disinfectant
Disinfectant	Foot trimmers	Dose syringe
Dose syringe	Forceps/hemostat	Drench bottle
Drench bottle	Gauze rolls	Feeding tube
Forceps/hemostat	Halter and lead rope	Hoof trimmers
Gauze rolls	Heat lamp	Forceps/hemostat
Halter and lead rope	Lamb nipples	Gauze rolls
Hoof trimmers	Lubricant jelly	Halter and lead rope
Intravenous outfit	Mineral oil	Heat lamp
Lubricant jelly	Needles	Lamb nipples
Milk of magnesia	Obstetrical chains	Lubricant jelly
Mineral oil	Ointment—bactericidal	Mineral oil
Needles	Penicillin-streptomycin	Needles
Nose lead	(injectable)	Obstetrical chains
Obstetrical chains	Scarlet oil	Ointment—bactericidal
Ointment—bactericidal	Shears—electric	Penicillin-streptomycin
Ointment—eye	Stomach tube	(injectable)
Penicillin-streptomycin	Syringes	Scarlet oil
(injectable)	Thermometer—rectal	Shears—electric
Rope—casting / restraint	Tincture of iodine	Stomach tube
Scarlet oil		Syringes
Stomach tube		Teat dip
Syringes		Thermometer—rectal
Teat cannula		Tincture of iodine
Teat dip		
Thermometer—rectal		
Tincture of iodine		
Trocar and cannula		
Udder infusion antibiotic		

TABLE 10.2b   Manager's Equipment Boxes and First Aid Kits (Horses, Swine, and Poultry)

Horses	Swine	Poultry
Antiseptic	Antiseptic	Blood stopper/clotter
Balling gun	Boric acid	Caponizing tools
Bot scraper/knife	(Epsom salts)	Disinfectant
Bottle for milk replacer	Castrating tools	Clippers—canine toe
Breeding hobbles	Cotton roll	Forceps/hemostat
Clippers—electric	Disinfectant	Heat lamp
Colic mixture	Dose syringe	Mineral oil
Cotton roll	Ear notcher	Rotenone powder
Dental float—teeth	Forceps/hemostat	
Disinfectant	Heat lamp	
Dose syringe	Lubricant jelly	
Drench bottle	Mineral oil	
Fly repellant	Needles	
Forceps/hemostat	Nose rings and ringing pliers	
Gauze rolls	Obstetrical forceps	
Halter and lead rope	Oxytocin	
Hoof knife and hoof pick	Penicillin-streptomycin	
Hoof-trimming tools	(injectable)	
Lubricant jelly	Scarlet oil	
Mineral oil	Syringes	
Needles	Thermometer—rectal	
Ointment—bactericidal	Tincture of iodine	
Penicillin-streptomycin (injectable)	Tooth nippers	
Rope—casting/restraint		
Scarlet oil		
Stomach tube		
Syringes		
Thermometer—rectal		
Tincture of iodine		
Twitch		

TABLE 10.3   Vaccination Program Models

Following are suggested vaccination programs, indicating the diseases being vaccinated against as well as a protocol for timing the vaccinations. Each of the diseases selected for the model is probably important enough for producers in all regions of the country to include in their health management program. However, it is costly to vaccinate when it is not necessary, and it is even more costly to not vaccinate when it is necessary. The good manager will consult with the attending veterinarian to plan a comprehensive health program, specific to the unique needs of the operation.

Cattle	Disease	Schedule
Heifers	Brucellosis	Calfhood
	Bovine virus diarrhea (BVD)	Preweaning + Prebreeding
	Infectious bovine rhinotracheitis (IBR)	Preweaning + Prebreeding
	Parainfluenza-3(PI3)	Preweaning + Prebreeding
	Bovine respiratory syncytial virus (BRSV)	Preweaning + Prebreeding
	Bovine respiratory syncytial virus (BRSV)	Prebeeding
	Leptospirosis	Prebreeding
Mother cows	Bovine virus diarrhea (BVD)	Prebreeding
	Infectious bovine rhinotracheitis (IBR)	Prebreeding
	Parainfluenza-3(PI3)	Prebeeding
	Leptospirosis	Prebreeding
Steers	Bovine virus diarrhea (BVD)	Preweaning & Weaning
	Infectious bovine rhinotracheitis (IBR)	Preweaning & Weaning
	Parainfluenza-3(PI3)	Preweaning & Weaning
	Bovine respiratory syncytial virus (BRSV)	Preweaning & Weaning
	Leptospirosis	Preweaning & Weaning

TABLE 10-3   Vaccination Program Models (continued)

Sheep	*Disease*	*Schedule*
Lambs	Tetanus toxoid	Docking time
	Enterotoxemia (clostridium perfringens types C and D)	10 weeks—initial 12 weeks—booster
Ewes	Tetanus toxoid	Yearly booster
	Enterotoxemia (clostridium perfringens types C and D)	Yearly booster
	Caseous lymphadenitis	Yearly
Rams	Tetanus toxoid	Yearly booster
	Enterotoxemia (clostridium perfringens types C and D)	Yearly booster
	Caseous lymphadenitis	Yearly

Poultry	*Disease*	*Schedule*
	Marek's disease	1 day
	Newcastle disease	10–14 days
	Infectious bronchitis	2 weeks and 15 weeks
	Fowl pox	3–15 (8) weeks
	Laryngotracheitis	6–20 weeks
	Avian encephalomyeletis	10–14 weeks
	Fowl cholera	12–16 weeks

Goats	*Disease*	*Schedule*
Kids	Tetanus toxoid	8 weeks
	Enterotoxemia (clostridium perfringens types C and D)	10 weeks
		12 weeks—booster
Does	Tetanus toxoid	Yearly booster
	Enterotoxemia (clostridium perfringens types C and D)	2 doses annually, 14–21 days apart
	Caseous lymphadenitis	Yearly
Bucks	Tetanus toxoid	Yearly booster
	Enterotoxemia (clostridium perfringens types C and D)	Yearly booster
	Caseous lymphadenitis	Yearly

Swine	*Disease*	*Schedule*
Baby pigs	Escherichia coli	First 12 hours
	Rhinitis	7 days—initial
		14 days—booster
Gilts and sows	Transmissible gastroenteritis (TGE)	5 weeks pre-farrow—initial
		2 weeks pre-farrow—booster
	Rhinitis	4 weeks pre-farrow—initial
		2 weeks pre-farrow—booster
Feeder pigs	Pneumonia	45 lbs—initial
		45 lbs + 3 weeks—booster

Horses	*Disease*	*Schedule*
Foals	Tetanus (antitoxin)	At birth
	Tetanus (toxoid)	3 months
	Rhinopneumonitis	2–4 months
	Equine encephalomyelitis (eastern and western)	3 months
	Influenza	3 months
Mares	Tetanus (toxoid)	Annually: 4–6 weeks prefoaling
	Rhinopneumonitis	Annually: 5,7, 9 months of pregnancy
	Equine encephalomyelitis (eastern and western)	Annually—pre–mosquito season
	Strangles (distemper)	Annually
	Equine viral arteritis (EVA)	Annually—3 weeks prebreeding
Geldings and stallions	Tetanus (toxoid)	Annually
	Rhinopneumonitis	Annually
	Equine encephalomyelitis (eastern and western)	Annually—pre-mosquito season
	Strangles (distemper)	Annually
	Equine viral arteritis (EVA)	Annually—3 weeks prebreeding

10.14 COMMONLY USED VETERINARY MEDICAL ABBREVIATIONS

At some point in your livestock management life, you will be called upon to administer a pharmaceutical or biological to an individual animal or to each of the animals in your herd or flock. Most of the time, manufacturer's label directions are clearly written in layperson's terms. If you read and follow those directions, you will know clearly which method to use to administer the correct amount of the product, and you will know how often to administer it.

However, as your relationship with your veterinarian progresses, you may be given an animal health product with a label affixed that is written in veterinary medical prescription language. Once you have learned the "alphabet," you will find the shorthand-like language quite clear and actually easier to follow than plain English. In fact, you will find yourself wondering why the product manufacturer doesn't label products in that manner directly from the factory! For example,

(Name of the pharmaceutical, such as an antibiotic)
   (Give) 10 cc b.i.d. IM q 5d

is an abbreviated and clearly stated substitute for:

*Administer the **pharmaceutical** by giving **10 cc, twice each day**, using an **intramuscular injection** as the route of administration, and continue giving this amount **for 5 days.***

The following are commonly used veterinary medical abbreviations that you will encounter as your experience with herd health matters increases.

Frequency of Administration

ad lib (ad libitum)	= freely, as much as necessary
a.m. (ante meridiem)	= morning
p.m. (post meridiem)	= evening
d or in d	= daily
b.i.d. (bis in die)	= twice in a day; normally 12-hour intervals
t.i.d. (ter in die)	= three times in a day; normally every 8 hours
q.i.d. (quater in die)	= four time a day; usually every 6 hours
qd (quaque die)	= every day, each day
dieb alt (diebus alternis)	= every other day
qh (quaque hora)	= every hour, each hour
alt hor (alternis horis)	= every other hour
s.o.s. (si opus sit)	= if it is needed

Blood Terminology

CBC	= complete blood cell count
RBC	= red blood cell count
WBC	= white blood cell count
Hgb or Hb	= hemoglobin
HCT or HT	= hematocrit
PCV	= packed cell volume

Routes of Administration

IC (intracardial)	= in the heart
ID (intradermal)	= within the skin
IM (intramuscular)	= in the muscle
IP (intraperitoneal)	= in the peritoneal cavity
IV (intravenous)	= in a blood vessel
SC or SQ (subcutaneous)	= below the skin
PO (per os)	= by mouth; orally
AD (auris dexter)	= right ear
AS (auris sinistra)	= left ear
AU (auris uterque)	= both ears, each ear
OD (oculus dexter)	= right eye
OS (oculus sinister)	= left eye
OU (oculus uterque)	= both eyes

Units of Measure

cc	= cubic centimeter
g or gm	= gram
mg	= milligram
ug	= microgram
kg	= kilogram
gr	= grain
l (L)	= liter
ml	= milliliter
lb	= pound
oz	= ounce
pt	= pint
qt	= quart
gal	= gallon
IU	= international units; a measure of activity or potency

CAUTION: It is enjoyable to learn something new, especially when it involves the health and health care program for your animals. However, the well-being of your herds or flocks depends upon careful vigilance and proper use of biologicals and therapeutics. If you do not understand the directions written on a label or those verbally given to you, *ask* for reexplanation.

For a more complete listing of abbreviations, metric equivalents, and conversions, refer to Appendix T.

Diluting Solutions

When it is necessary to dilute a solution, the following method can be used to make the necessary calculations.

1. In the upper left corner of a square put the percentage concentration of the known solution.

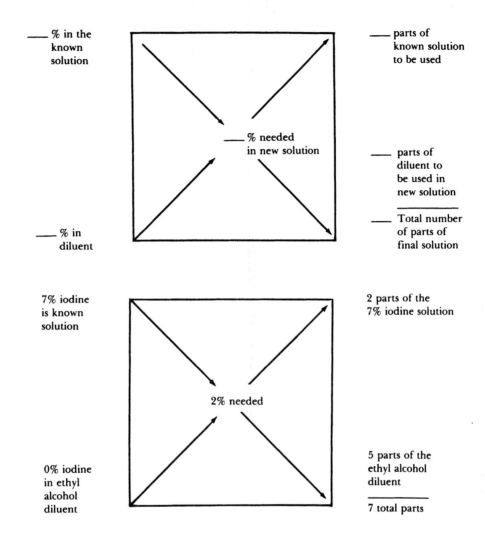

2. In the lower left corner put the percentage concentration of the diluent (material used to dilute the known solution).

3. In the center of the square put the percentage concentration needed in the new solution.

4. Calculate the difference between the percentage concentration needed in the new solution (center of the square) and the percentage concentration of the known solution (upper left corner) to obtain the parts of the diluent to make the new solution (lower right corner). Record this answer in the lower right corner.

5. Calculate the difference between the percentage concentration of the diluent (lower left corner) and the percentage concentration needed (center of the square) to obtain the parts of the known solution to be used in the new solution (upper right corner). Record this number in the upper right corner.

6. Add the number of parts of the known solution to be used to the number of parts of diluent to be used in the new solution to obtain the total parts.

EXAMPLE:

If a 7% iodine solution must be diluted to a milder 2% solution, calculate the amount of the 7% iodine solution that must be mixed with ethyl alcohol (diluent) to make the 2% iodine solution.

1. In the upper left corner of the square put the percent iodine in the known solution (7%).
2. In the lower left corner put the concentration of iodine in the ethyl alcohol diluent (0%).
3. Subtract the 2% needed (center of the square) from the 7% in the known solution (upper left corner of the square) to obtain the 5 parts of the ethyl alcohol diluent (lower right corner of the square) to be used to make the 2% solution.
4. Subtract the 0% in the ethyl alcohol diluent (lower left corner) from the 2% needed (center of the square) to obtain the 2 parts of the 7% iodine solution.
5. Add the 2 parts of the 7% iodine solution (upper right corner) to the 5 parts of ethyl alcohol (diluent) to get the total of 7 parts.

6. Prepare the quantity of new solution needed. For example, if 70 ml of a 2% iodine solution is needed, mix 20 ml of the 7% iodine solution (70 ml × $\frac{2}{7}$ part = 20 ml) with 50 ml of the ethyl alcohol diluent (70 ml × $\frac{5}{7}$ part = 50 ml).

10.15 ESTABLISHING AND MANAGING THE HERD HEALTH PROGRAM

A herd health program involves more than a series of vaccinations to administer to your young stock and breeding animals and a list of products to use to control parasites, moderate metabolic upsets, and repair injuries. If you remain with this narrow view of a herd health program, the amount of progress you will be able to make in the quest to raise the highest-quality product per lowest unit of input will be severely restricted.

A complete herd health program involves, at a minimum, the following *management practices*:

• *Monitor and adjust the nutritional program so that the specific requirements of each class of animal under your charge is met.* Each category of animal—mature breeding animals, replacement breeding stock, young preweaning animals, feeders, or the very old and infirm animals—has a specific set of nutritional requirements that must be met, and these will vary for each class through the management year. Monitoring the nutritional program and adjusting it as necessary to meet these requirements will help ensure healthy animals, quality meat and milk, and the best possible bottom line.

• *Design, construct, and maintain working and handling facilities that are appropriate to your operation's needs.* Inadequate facilities will result in added stress for the animals, extreme frustration for you

as the manager, and added bruises and hide damage for the animals. And very likely a reduced amount of attention will be paid to animal health, reproductive management, and product quality . . . because it is such a hassle to work the animals through the inadequate facilities that it is easier to skip the program details and hope for the best.

• *Learn to keep an excellent set of animal records.* At the very least, the records should contain the following: permanent animal identification; breeding, birthing, and weaning information; disease prevention and control measures; disease-incidence information; and a dated history of biologicals and pharmaceuticals used, including dosages and routes of administration.

• *Assemble and integrate your herd management resources.* These must include your veterinarian, your nutrition consultant (whether that be a paid consultant, feed company representative, or extension service faculty), your banker, the local buyer/packer representative, and your management consultant (usually one of the local or state extension faculty members). These men and women are available and provide a resource far too valuable not to use.

• *Train your workers to properly use animal health products, to keep good and complete records, to use and maintain your facilities properly, to take enough time to do the job correctly the first time, and to care for the animals' well-being as if they were their own.* The very best designed herd health and total quality management program is of minimal value if it is not carried out on your operation. It is impossible to do everything yourself, so proper training of those folks, upon whom you must depend, is essential. A commitment to doing things correctly the first time will save money, time (because you will not have to retreat the "misses" who later turn up in the sick pen), and animal stress. It will also lead to the production of top-quality consumer-satisfying products.

- *Custom-design a vaccination program for your region of the country and for your herd.* The first stop in building this program is, naturally, your veterinarian. Review with him or her which diseases are likely to cause problems in your region, vaccines available to help prevent these diseases, the efficacy of the vaccines, and the proper timing of their administration.

The following is a list of beef cattle diseases for which vaccines are available. Similar lists of diseases could be prepared for each species of livestock and poultry.

Anaplasmosis
Anthrax
Black disease
Blackleg
Bovine respiratory syncytial virus (BRSV)
Bovine virus diarrhea (BVD)
Brucellosis (Bang's)
Enterotoxemia (C&D)
E. coli
Hemophilus somnus
Infectious bovine rhinotracheitis (IBR)
Leptospirosis (1–5 strains)
Malignant edema (ME)
Parainfluenza-3 (PI3)
Pasteurella hemolytica
Pinkeye
Redwater
Rota-coronavirus
Trichomoniasis
Salmonellosis
Vibriosis (campylobacter)
Warts

After your health management team has decided which diseases to attempt to control, the next step involves the *proper timing of the administration* of the products. Normal times to administer animal health products are at preweaning time (when the young animals are still nursing the mother), weaning time, and when the animal is entering the feedlot. Each time has its strong points and drawbacks. Make the decisions with your veterinarian to best fit your management scheme.

When you have decided which diseases to vaccinate against and at what time of your management year to vaccinate, there are still critical steps to be followed.

Proper Handling and Administration of Selected Animal Health Products

1. Select products that have been manufactured by fully licensed, reputable companies. Animal health costs represent a very small portion of a total quality management program and an even smaller portion of your total animal investment. It is truly penny-wise and pound-foolish to purchase products to protect your animal investment from companies that do not enjoy the reputation of manufacturing top-quality products.

2. Always read and reread the labels on the animal health packages.

3. Use animal health products only at the dosages recommended and for the purposes intended.

4. Use transfer needles to reconstitute the dehydrated or freeze-dried animal health products.

5. Never combine vaccines into the same syringe or into any other container. They may wind up canceling one another out and you could lose the efficacy of both.

6. As you administer doses of a vaccine from a large bottle, shake the bottle well before initial use, and then shake the bottle well before each refill of the syringe.

7. If you use a vaccine that you reconstitute on site, do not mix more than you will be able to use in an hour or so. It will begin to degrade if left standing longer than that. Also, keep the vaccine out of the sunlight and heat. A styrofoam cooler works very nicely.

8. Use separate syringes for different types of products. For example, modified live vaccines should not be used in the same syringes as killed vaccines or bacterins. It is very likely that a bacterin will destroy the modified live vaccine.

9. Use nothing more than hot water to clean the syringes used for modified live vaccines. If you use a disinfectant, you run the risk of any disinfectant residue destroying the modified live virus that will be used in the syringe the next time.

10. Be certain that all of the air is evacuated from the syringe before using it on the animal. This will ensure that the animal is getting the intended dose of health product and not a partial dose of product and a partial dose of air.

11. Always restrain the animal properly before administering the product. Basically, you are trying to avoid all bruising.

12. Select the best route for administration of the product. Intramuscular and subcutaneous are the usual routes. Intramuscular is normal for modified live vaccines, and subcutaneous is preferred for bacterins or killed products. There are a variety of products on the market today that are meant to be given intraruminally. Follow the manufacturer's directions.

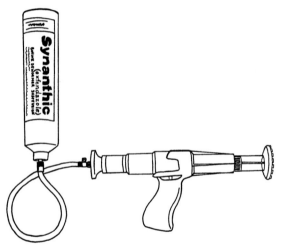

INTRARUMINAL DEWORMING INJECTOR

13. Choose the site for injection that will damage the least amount of muscle tissue that is used for expensive retail cuts at the supermarket.

14. Use the correct needle size. Subcutaneous injections normally use a 16- or 18-gauge needle from $\frac{1}{2}$ to $\frac{3}{4}$ inch long. Intramuscular injections use the same 16- to 18-gauge needle, but they are usually 1 to $1\frac{1}{2}$ inches long.

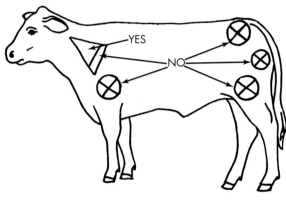

INJECTION SITES
NEW & OLD

15. Use good sanitation procedures when vaccinating your animals. This will include using a needle for the vaccine bottle that is separate from the needle used for injecting the animals, discarding bent or dulled needles, changing needles every 15 to 20 animals, making certain that the injection site is clean, and wiping down the needle with alcohol between each animal when using a bacterin or killed vaccine. Do not wipe the needle with any disinfectant when using modified live vaccines because the disinfectant may destroy the vaccine.

16. Use the proper site to implant your animals with growth stimulants. If implants are not properly placed, their efficacy is affected, there is residue present at slaughter, the packer will have to trim excessively, and consumer acceptance and confidence will suffer when the knowledge of poorly used implant technology "hits the press."

| APPENDICES |

External Parts of the Cow

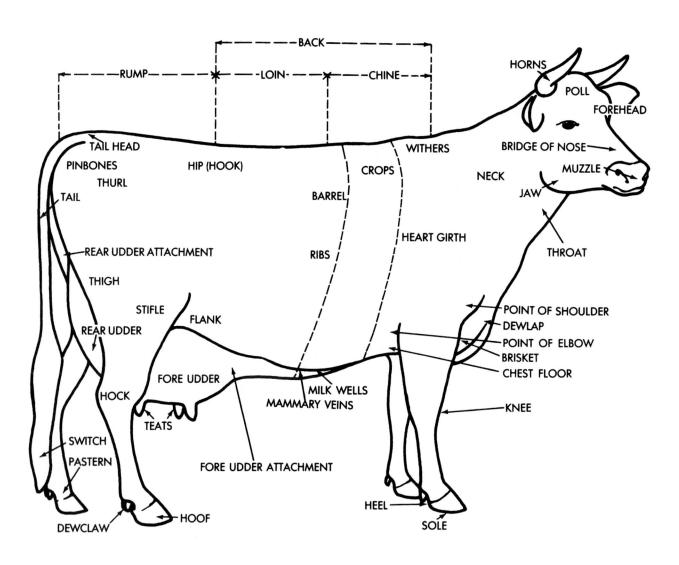

Skeleton of the Cow

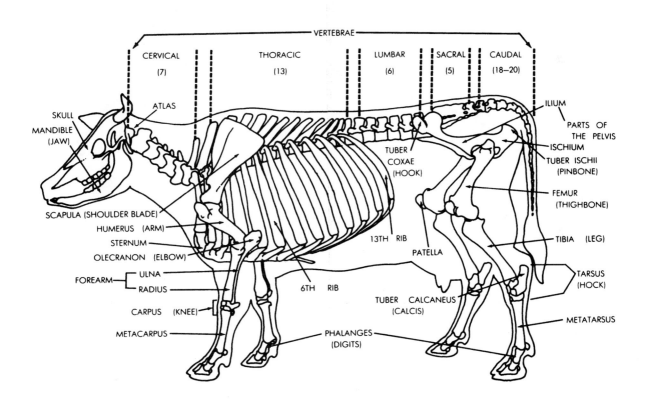

External Parts of the Pig

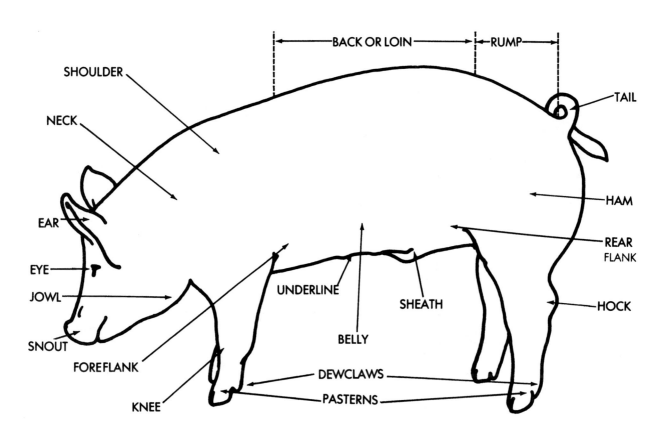

Skeleton of the Pig

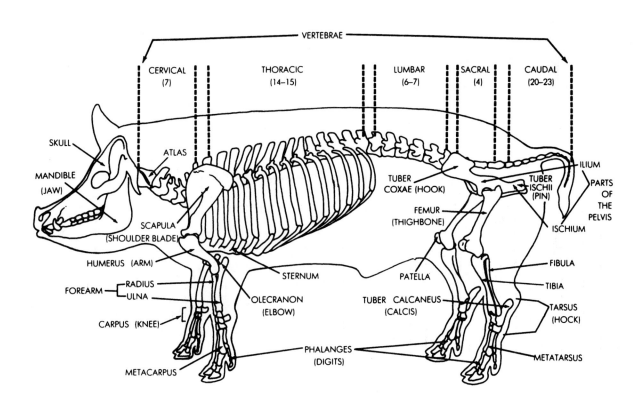

External Parts of the Horse

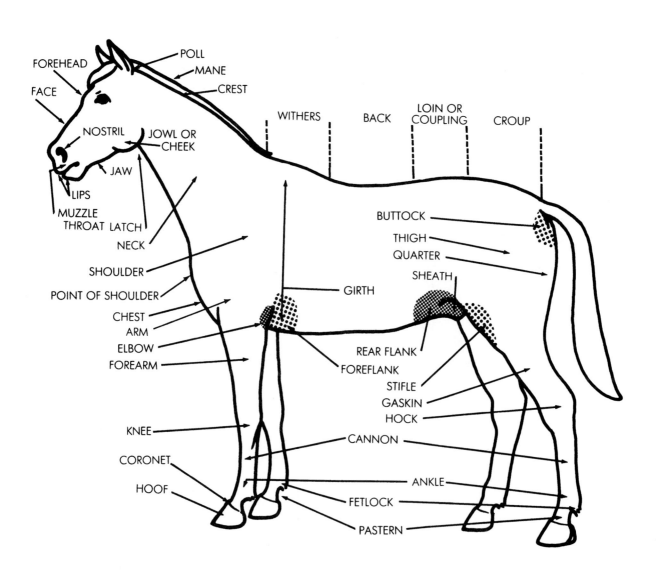

Skeleton of the Horse

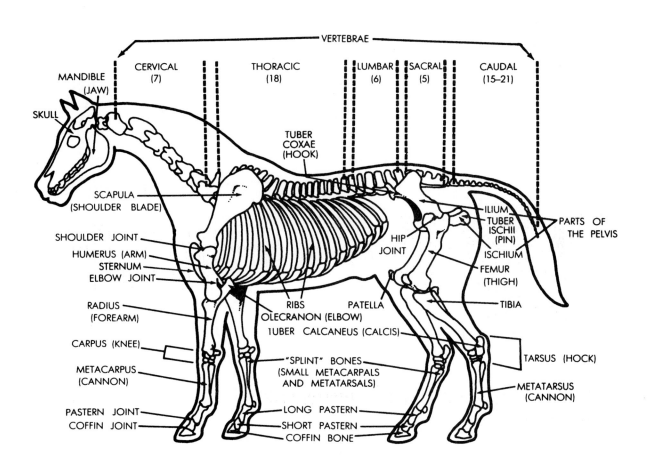

External Parts of the Sheep

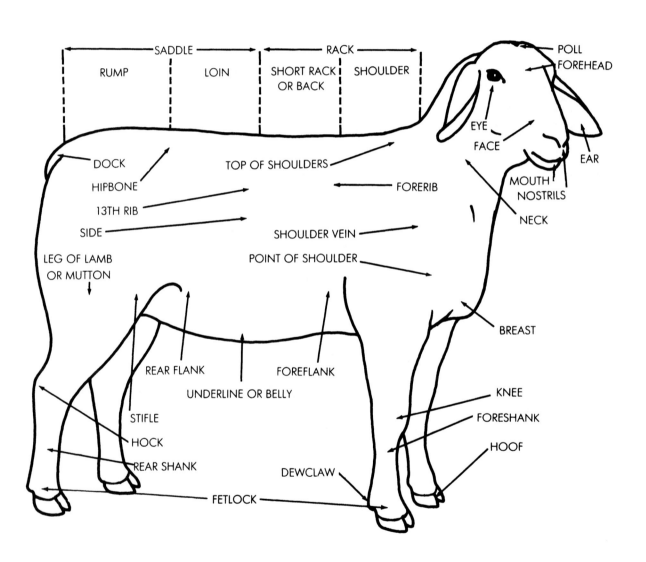

Skeleton of the Sheep

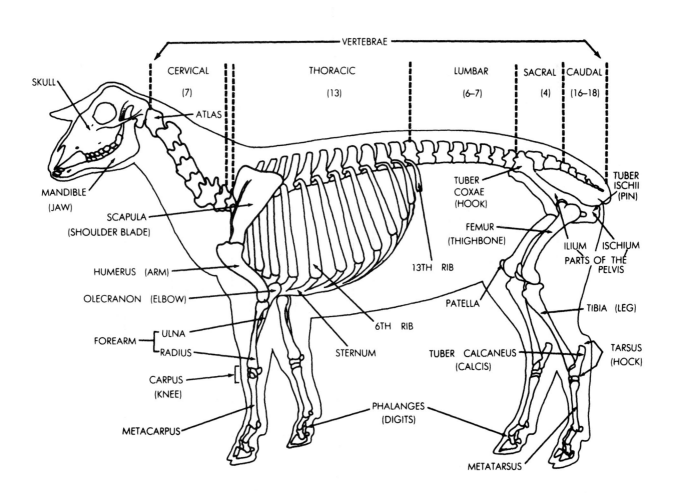

External Parts of the Goat

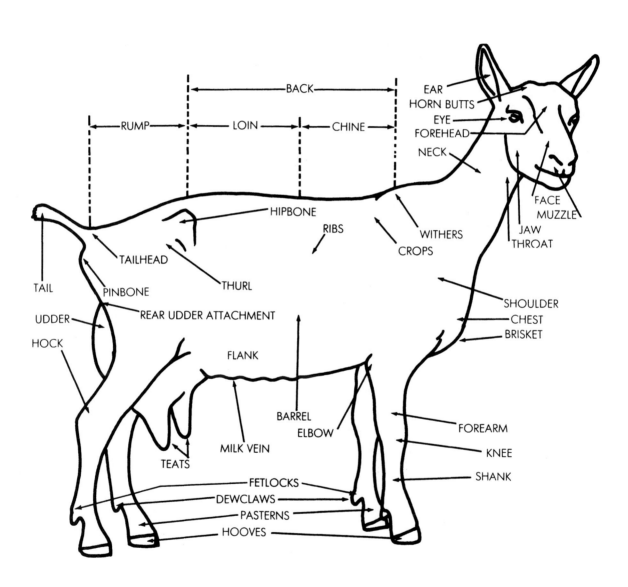

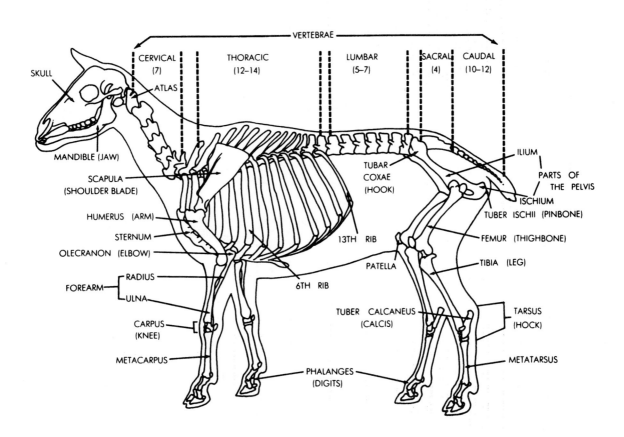

Skeleton of the Goat

External Parts of the Rooster

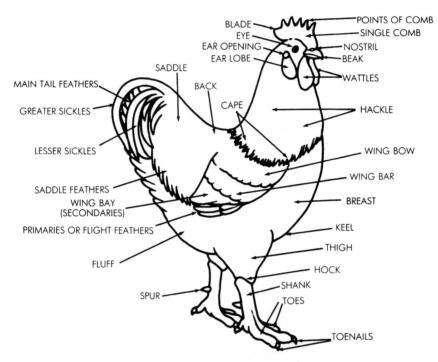

BLADE
EYE
EAR OPENING
EAR LOBE
POINTS OF COMB
SINGLE COMB
NOSTRIL
BEAK
WATTLES
SADDLE
BACK
CAPE
HACKLE
MAIN TAIL FEATHERS
GREATER SICKLES
LESSER SICKLES
WING BOW
WING BAR
BREAST
SADDLE FEATHERS
WING BAY
(SECONDARIES)
PRIMARIES OR FLIGHT FEATHERS
KEEL
THIGH
FLUFF
HOCK
SHANK
SPUR
TOES
TOENAILS

PARTS OF THE ROOSTER CHICKEN (MALE)

External Parts of the Hen

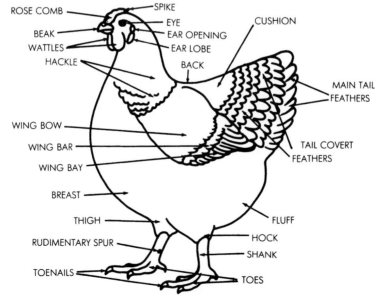

ROSE COMB
SPIKE
CUSHION
BEAK
EYE
EAR OPENING
WATTLES
EAR LOBE
HACKLE
BACK
MAIN TAIL
FEATHERS
WING BOW
WING BAR
TAIL COVERT
FEATHERS
WING BAY
BREAST
THIGH
FLUFF
RUDIMENTARY SPUR
HOCK
SHANK
TOENAILS
TOES

PARTS OF THE HEN CHICKEN (FEMALE)

Skeleton of the Fowl

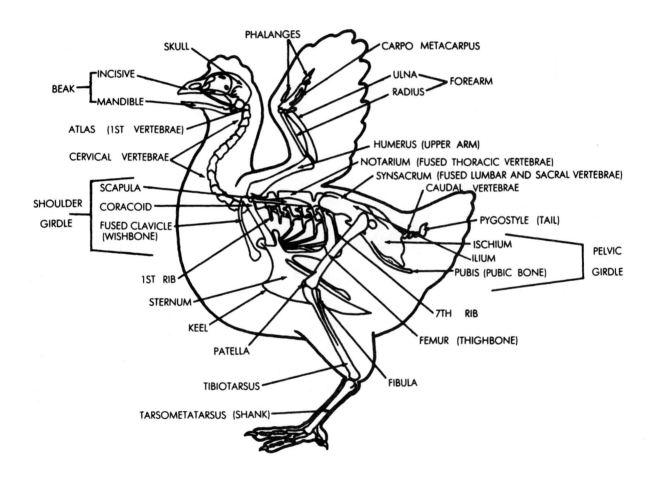

Parts of the Turkey Tom and Hen

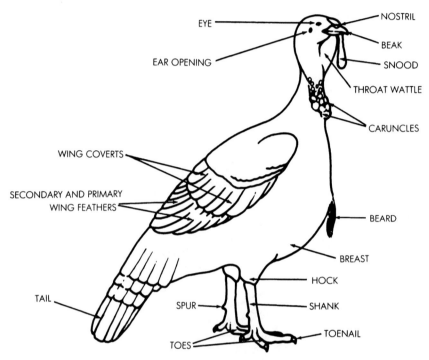

PARTS OF THE TURKEY TOM (MALE)

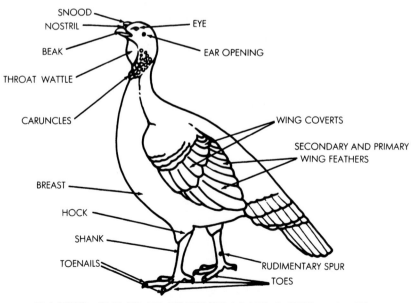

PARTS OF THE TURKEY HEN (FEMALE)

Age–Weight Relationships for Domestic Livestock

Growth in livestock is an increase in the size and the weight of an animal over a specific period of time. With most farm species, fast and efficient growth is important to the success of a livestock operation. The rate at which an animal grows is determined by its genetic potential and the environment provided.

Growth curves of all animals are S-shaped, but this characteristic shape will not be observed when animals are slaughtered before reaching maturity, when animals are weighed infrequently, or when abnormal weight changes occur because of environmental conditions or the physiological status of the animals (e.g., pregnancy).

The purpose of these curves is to provide the reader with an idea of the age–weight relationships of livestock. It is important to keep in mind that these are not absolute and that growth rates do vary with species, breeds, genetic potential, and the environment and management provided. Thus the curves may not necessarily be completely accurate in predicting expected growth in specific situations. Unless it is stated otherwise, the curves represent average to excellent performance for the animals indicated.

Figure N.1 The age–weight relationship for dairy cows.

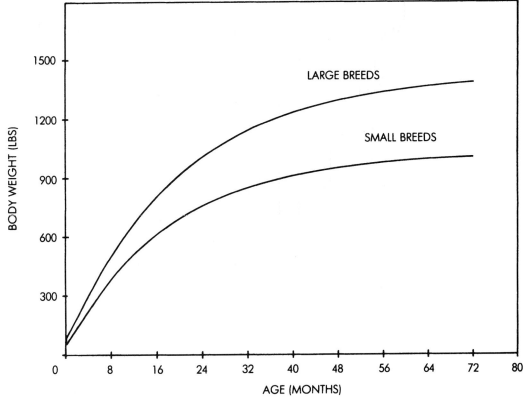

Figure N.2 The age–weight relationship for beef cattle.

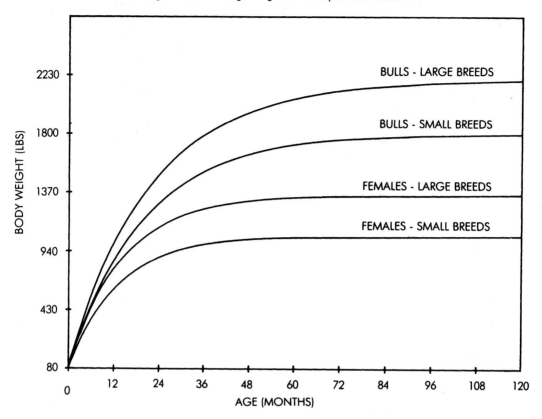

Figure N.3 The age–weight relationship for market swine.

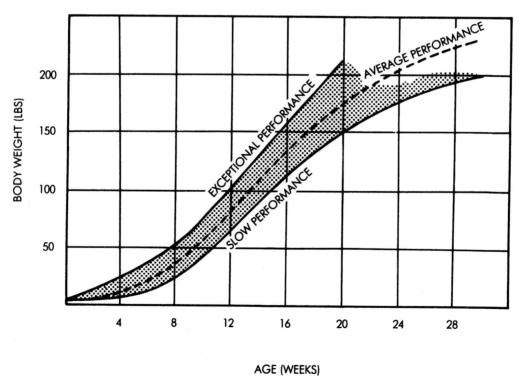

Figure N.4 The age–weight relationship for horses.

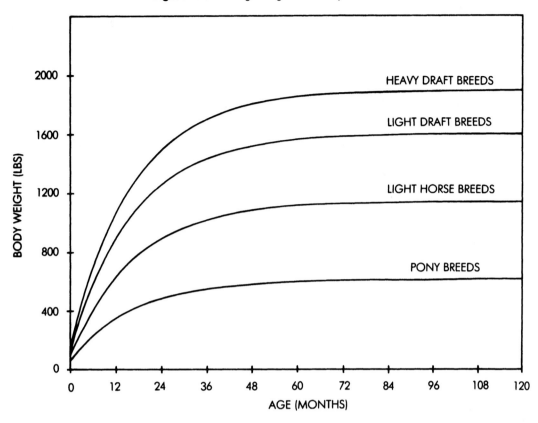

Figure N.5 The age–weight relationship for sheep.

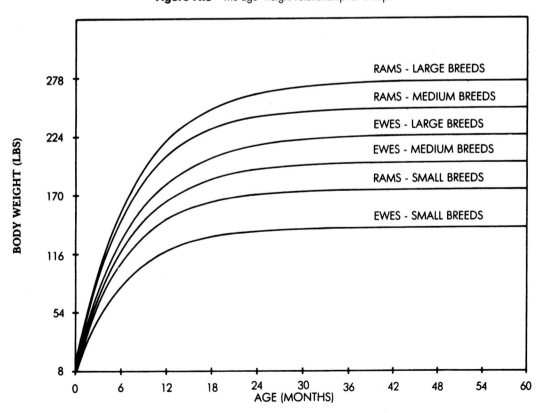

Figure N.6 The age–weight relationship for goats.

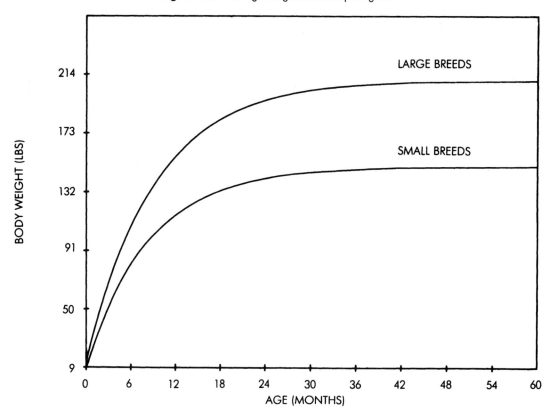

Figure N.7 The age–weight relationship for broilers.

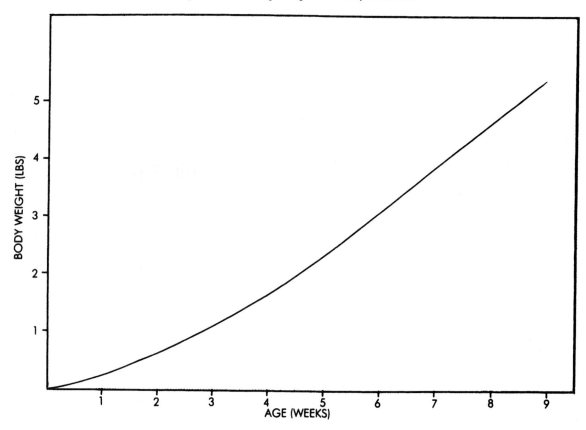

Figure N.8 The age–weight relationship for turkeys.

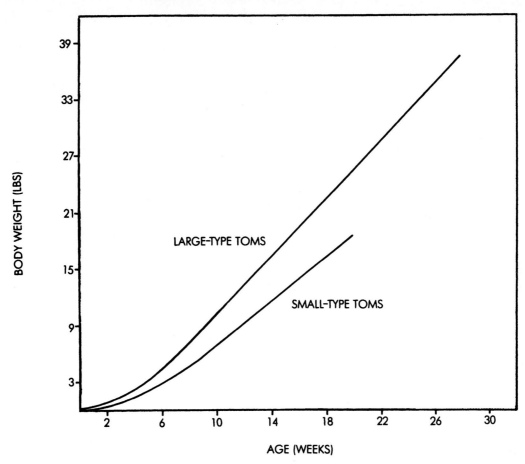

Gestation and Incubation Table

Date Bred or Date Eggs Set	COW 283 days Date due	DOE 150 days Date due	EWE 148 days Date due	MARE 336 days Date due	SOW 114 days Date due	CHICKEN 21 days Hatch date	TURKEY 28 days Hatch date
Jan. 1	Oct. 11	May 31	May 29	Dec. 3	Apr. 25	Jan. 22	Jan. 29
Jan. 6	Oct. 16	June 5	June 3	Dec. 8	Apr. 30	Jan. 27	Feb. 3
Jan. 11	Oct. 21	June 10	June 8	Dec. 13	May 5	Feb. 1	Feb. 8
Jan. 16	Oct. 26	June 15	June 13	Dec. 18	May 10	Feb. 6	Feb. 13
Jan. 21	Oct. 31	June 20	June 18	Dec. 23	May 15	Feb. 11	Feb. 18
Jan. 26	Nov. 5	June 25	June 23	Dec. 28	May 20	Feb. 16	Feb. 23
Jan. 31	Nov. 10	June 30	June 28	Jan. 2	May 25	Feb. 21	Feb. 28
Feb. 5	Nov. 15	July 5	July 3	Jan. 7	May 30	Feb. 26	Mar. 5
Feb. 10	Nov. 20	July 10	July 8	Jan. 12	June 4	Mar. 3	Mar. 10
Feb. 15	Nov. 25	July 15	July 13	Jan. 17	June 9	Mar. 8	Mar. 15
Feb. 20	Nov. 30	July 20	July 18	Jan. 22	June 14	Mar. 13	Mar. 20
Feb. 25	Dec. 5	July 25	July 23	Jan. 27	June 19	Mar. 18	Mar. 25
Mar. 2	Dec. 10	July 30	July 28	Feb. 1	June 24	Mar. 23	Mar. 30
Mar. 7	Dec. 15	Aug. 4	Aug. 2	Feb. 6	June 29	Mar. 28	Apr. 4
Mar. 12	Dec. 20	Aug. 9	Aug. 7	Feb. 11	July 4	Apr. 2	Apr. 9
Mar. 17	Dec. 25	Aug. 14	Aug. 12	Feb. 16	July 9	Apr. 7	Apr. 14
Mar. 22	Dec. 30	Aug. 19	Aug. 17	Feb. 21	July 14	Apr. 12	Apr. 19
Mar. 27	Jan. 4	Aug. 24	Aug. 22	Feb. 26	July 19	Apr. 17	Apr. 24
Apr. 1	Jan. 9	Aug. 29	Aug. 27	Mar. 3	July 24	Apr. 22	Apr. 29
Apr. 6	Jan. 14	Sept. 3	Sept. 1	Mar. 8	July 29	Apr. 27	May 4
Apr. 11	Jan. 19	Sept. 8	Sept. 6	Mar. 13	Aug. 3	May 2	May 9
Apr. 16	Jan. 24	Sept. 13	Sept. 11	Mar. 18	Aug. 8	May 7	May 14
Apr. 21	Jan. 29	Sept. 18	Sept. 14	Mar. 23	Aug. 13	May 12	May 19
Apr. 26	Feb. 3	Sept. 23	Sept. 21	Mar. 28	Aug. 18	May 17	May 24
May 1	Feb. 8	Sept. 28	Sept. 26	Apr. 2	Aug. 23	May 22	May 29
May 6	Feb. 13	Oct. 3	Oct. 1	Apr. 7	Aug. 28	May 27	June 3
May 11	Feb. 18	Oct. 8	Oct. 6	Apr. 12	Sept. 2	June 1	June 8
May 16	Feb. 23	Oct. 13	Oct. 11	Apr. 17	Sept. 7	June 6	June 13
May 21	Feb. 28	Oct. 18	Oct. 16	Apr. 22	Sept. 12	June 11	June 18
May 26	Mar. 5	Oct. 23	Oct. 21	Apr. 27	Sept. 17	June 16	June 23
May 31	Mar. 10	Oct. 28	Oct. 26	May 2	Sept. 22	June 21	June 28
June 5	Mar. 15	Nov. 2	Oct. 31	May 7	Sept. 27	June 26	July 3
June 10	Mar. 20	Nov. 7	Nov. 5	May 12	Oct. 2	July 1	July 8
June 15	Mar. 25	Nov. 12	Nov. 10	May 17	Oct. 7	July 6	July 13
June 20	Mar. 30	Nov. 17	Nov. 15	May 22	Oct. 12	July 11	July 18

Date Bred or Date Eggs Set	COW 283 days Date due	DOE 150 days Date due	EWE 148 days Date due	MARE 336 days Date due	SOW 114 days Date due	CHICKEN 21 days Hatch date	TURKEY 28 days Hatch date
June 25	Apr. 4	Nov. 22	Nov. 20	May 27	Oct. 17	July 16	July 23
June 30	Apr. 9	Nov. 27	Nov. 25	June 1	Oct. 22	July 21	July 28
July 5	Apr. 14	Dec. 2	Nov. 30	June 6	Oct. 27	July 26	Aug. 2
July 10	Apr. 19	Dec. 7	Dec. 5	June 11	Nov. 1	July 31	Aug. 7
July 15	Apr. 24	Dec. 12	Dec. 10	June 16	Nov. 6	Aug. 5	Aug. 12
July 20	Apr. 29	Dec. 17	Dec. 15	June 21	Nov. 11	Aug. 10	Aug. 17
July 25	May 4	Dec. 22	Dec. 20	June 26	Nov. 16	Aug. 15	Aug. 22
July 30	May 9	Dec. 27	Dec. 25	July 1	Nov. 21	Aug. 20	Aug. 27
Aug. 4	May 14	Jan. 1	Dec. 30	July 6	Nov. 26	Aug. 25	Sept. 1
Aug. 9	May 19	Jan. 6	Jan. 4	July 11	Dec. 1	Aug. 30	Sept. 6
Aug. 14	May 24	Jan. 11	Jan. 9	July 16	Dec. 6	Sept. 4	Sept. 11
Aug. 19	May 29	Jan. 16	Jan. 14	July 21	Dec. 11	Sept. 9	Sept. 14
Aug. 24	June 3	Jan. 21	Jan. 19	July 26	Dec. 16	Sept. 14	Sept. 21
Aug. 29	June 8	Jan. 26	Jan. 24	July 31	Dec. 21	Sept. 19	Sept. 26
Sept. 3	June 13	Jan. 31	Jan. 29	Aug. 5	Dec. 26	Sept. 24	Oct. 1
Sept. 8	June 18	Feb. 5	Feb. 3	Aug. 10	Dec. 31	Sept. 29	Oct. 6
Sept. 13	June 23	Feb. 10	Feb. 8	Aug. 15	Jan. 5	Oct. 4	Oct. 11
Sept. 18	June 28	Feb. 15	Feb. 13	Aug. 20	Jan. 10	Oct. 9	Oct. 16
Sept. 23	July 3	Feb. 20	Feb. 18	Aug. 25	Jan. 15	Oct. 14	Oct. 21
Sept. 28	July 8	Feb. 25	Feb. 23	Aug. 30	Jan. 20	Oct. 19	Oct. 26
Oct. 3	July 13	Mar. 2	Feb. 28	Sept. 4	Jan. 25	Oct. 24	Oct. 31
Oct. 8	July 18	Mar. 7	Mar. 5	Sept. 9	Jan. 30	Oct. 29	Nov. 5
Oct. 13	July 23	Mar. 12	Mar. 10	Sept. 14	Feb. 4	Nov. 3	Nov. 10
Oct. 18	July 28	Mar. 17	Mar. 15	Sept. 19	Feb. 9	Nov. 8	Nov. 15
Oct. 23	Aug. 2	Mar. 22	Mar. 20	Sept. 24	Feb. 14	Nov. 13	Nov. 20
Oct. 28	Aug. 7	Mar. 27	Mar. 25	Sept. 29	Feb. 19	Nov. 18	Nov. 25
Nov. 2	Aug. 12	Apr. 1	Mar. 30	Oct. 4	Feb. 24	Nov. 23	Nov. 30
Nov. 7	Aug. 17	Apr. 6	Apr. 4	Oct. 9	Mar. 1	Nov. 28	Dec. 5
Nov. 12	Aug. 22	Apr. 11	Apr. 9	Oct. 14	Mar. 6	Dec. 3	Dec. 10
Nov. 17	Aug. 27	Apr. 16	Apr. 14	Oct. 19	Mar. 11	Dec. 8	Dec. 15
Nov. 22	Sept. 1	Apr. 21	Apr. 19	Oct. 24	Mar. 16	Dec. 13	Dec. 20
Nov. 27	Sept. 6	Apr. 26	Apr. 24	Oct. 29	Mar. 21	Dec. 18	Dec. 25
Dec. 2	Sept. 11	May 1	Apr. 29	Nov. 3	Mar. 26	Dec. 23	Dec. 30
Dec. 7	Sept. 14	May 6	May 4	Nov. 8	Mar. 31	Dec. 28	Jan. 4
Dec. 12	Sept. 21	May 11	May 9	Nov. 13	Apr. 5	Jan. 2	Jan. 9
Dec. 17	Sept. 26	May 16	May 14	Nov. 18	Apr. 10	Jan. 7	Jan. 14
Dec. 22	Oct. 1	May 21	May 19	Nov. 23	Apr. 15	Jan. 12	Jan. 19
Dec. 27	Oct. 6	May 26	May 24	Nov. 28	Apr. 20	Jan. 17	Jan. 24

Normal Range of Reproductive Behavior

Animal	Age of Puberty (Months)	Length of Estrous Cycle (Days)		Duration of Estrus		Type of Estrous Cycle	Length of Gestation or Incubation Period (Days)		Time of Ovulation	Optimum Time to Breed
		Range	Avg.	Range	Avg.		Range	Avg.		
Cattle	10–15	19–23	21	6–30 hrs.	18 hrs.	Polyestrous all year	279–290	283	12 hours after the end of estrus	Mid-point to end of estrus
Goat	4–8	12–24	20	1–4 days	39 hrs.	Seasonally polyestrous from early fall to late winter	140–160	150	30–36 hours after the start of estrus	24 hours after estrus begins
Sheep	5–7	14–20	16–17	20–42 hrs.	30 hrs.	Seasonally polyestrous early fall to winter	144–152	148	At or near the end of estrus	24 hours after estrus begins
Horse	12–15	10–37	21	1–9 days	4–6 days	Seasonally polyestrous spring	332–340	336	24 to 48 hours before to 24 hours after the end of estrus	2nd–3rd day of estrus
Swine	4–7	18–24	21	1–5 days	2–3 days	Polyestrous all year	112–116	114	8–12 hrs. before end of standing estrus, or 37–40 hrs. after start of estrus	18–30 hours after estrus begins
Chicken	5–7	N/A	N/A	N/A	N/A	N/A		21*	Possible to ovulate daily*	
Turkey	5–7	N/A	N/A	N/A	N/A	N/A		28*	Possible to ovulate daily*	

N/A = not applicable
*Poultry are oviparous; the embryo develops outside the body of the mother.

Absorptive Capacities of Various Bedding Materials

Material	Pounds of Water Absorbed Material per cwt of Dry Bedding
Peat moss	1,000
Chopped oat straw	375
Oat straw, long	280
Vermiculite	350
Wood chips (pine)	300
Wood chips (hardwood)	150
Wheat straw, chopped	295
Wheat straw, long	220
Sawdust (pine)	250
Sawdust (hardwood)	150
Cornstalks, shredded	250
Newspaper, shredded	400
Corncobs, ground	210
Broadleaf leaves	200
Sand	25

Diseases of and Immunizing Agents for Domestic Livestock

Disease and Microorganism	Product(s)	Methods and Comments

CATTLE

Disease and Microorganism	Product(s)	Methods and Comments
1. Anaplasmosis *Anaplasma marginale*	Inactivated vaccine	Requires 2 doses, 4 weeks apart prior to vector season, then a single yearly booster vaccination.
	Infected blood. Not permitted in USA	Given to calves 6 mo. of age during winter months. Use only in areas that have anaplasma problems. No booster.
2. Anthrax *Bacillus anthracis*	Attenuated spore vaccine; avirulent culture; bacterin	Vaccinate each spring. Use only in anthrax areas. The extent of the problem and the animal species will dictate which product should be used.
3. Blackleg *Clostridium chauvoei*	Bacterin	Vaccinate animals between 4 and 18 months of age. If younger calves are vaccinated a second dose should be given at 5–6 months of age. Calves should be revaccinated prior to going to pasture as yearlings. This infection seldom occurs in cattle over 18 months of age. It is necessary to give two initial injections 3 weeks apart to get adequate protection on some premises.
4. Botulism *Clostridium botulinum*	Toxoid—C and D	The disease is present on some farms so animals are vaccinated 2 weeks prior to turning on pasture.
5. Enterotoxemia *Clostridium perfringens*	Bacterin—usually B and C types mixed	Vaccinate dams 4–6 weeks prior to calving on some premises for passive immunity and then vaccinate calves at 4 weeks of age.
	Antiserum B, C and D	Administered for immediate protection of newborn calves.
6. Other *Clostridium— septicum, novyi, sordelli*	Bacterins	Used on premises where these soilborne bacteria are a problem.
7. Brucellosis *Brucella abortus*	Strain 19 vaccine— lyophilized	Administer to heifer calves, usually 4 to 6 months of age. Vaccinated animals must be identified. Check with State Veterinarian.
8. Contagious pleuro- pneumonia *Mycoplasma mycoides*	Vaccine	Give to young cattle in infected areas.
9. Foot and mouth disease virus	Killed vaccine(s)	There are many strains of this virus. Used only in the foot-and-mouth disease–infected areas.
10. Infectious bovine rhinotracheitis (IBR) virus	Modified live virus vaccine; Killed vaccine	This vaccine gives good immunity for about 12 months. Requires 2 doses 21 days apart. Often give first dose at birth. Then give booster at 6–10 months of age.
11. Leptospirosis *Leptospira* spp.	Bacterin containing one or more serotypes	For good immunity, start immunity with 2 doses 21 days apart and then give a booster prior to the pasture season or more often. May have to give booster to bulls every 60 days throughout breeding season.
12. Pasteurellosis *Pasteurella multocida* *Pasteurella hemolytica*	Bacterin (stock and autogenous)	Administer 2 doses 21 days apart prior to known stress such as the movement or mixing of animals. Must give booster every 6 months on some premises.

Disease and Microorganism	Product(s)	Methods and Comments
13. Parainfluenza-3 virus	Killed vaccine	Administer 2 doses, 21 days apart and 3 weeks prior to stress. Should be repeated every 6–12 months for adequate protection. It often pays to give first dose at birth and repeat at 6 months.
	Modified intranasal	Spray nasal cavity 3 weeks prior to stress.
	Antiserum	Provides treatment or immediate protection.
14. Babesiosis (piroplasmosis) *Babesia bigemina*	Infected blood or by infected ticks	Infect calves under 6 months of age during cool months and under close supervision.
15. Rinderpest virus	Modified vaccine	Requires yearly immunization.
	Antiserum	Treatment for immediate protection.
16. Salmonellosis *Salmonella* spp.	Autogenous bacterin of species causing infection	Requires at least 2 doses 3 weeks apart with the last dose administered at least 2 weeks prior to calving to provide passive immunity to the calf. Can start calf injections at 2 weeks of age, repeating every 3 weeks until 4 doses are given in problem herds.
17. Warts (papillomas) virus	Autogenous killed vaccine	Autogenous vaccines provide the best control. The removal of some of the warts may be all that is required to control the infection.
18. Bovine virus diarrhea (BVD) virus	Modified vaccines	Modified live products are more effective than killed, but cannot be given to pregnant cows.
19. *Staphylococcus aureus* mastitis organism	Autogenous bacterin	Bacteria must have phage type causing the herd infection; thus the autogenous bacterin must be used. Requires injections; give first 2 doses 3 weeks apart and then every 6 months.
20. Rabies virus	Modified vaccine	Vaccine used in areas where infection is common. Make sure it is cleared for cattle.
21. *Dictyocaulus viviparus* (lungworm)	Irradiated larvae	Give 2 doses, 6 weeks apart as an aid in control of the infection.
22. Johne's disease *Mycobacterium paratuberculosis*	Bacterin	Shows promise as an aid in eradication of the disease in an infected herd. Vaccination of replacement breeding animals at 6–7 months of age has been helpful in infected herds.
23. Calf scours	Oral	Given to calves immediately following birth.
Rio virus, Corona virus, and Rotavirus	Modified live virus	Inject cows 6 weeks and 3 weeks prior to calving for passive immunity for calves.
24. Vibriosis *Campylobacter fetus*	Bacterin	Requires annual vaccination, 30 days prior to breeding season on premises where disease is a problem.

SHEEP AND GOATS

1. *Bacillus anthracis*	Attenuated spore vaccine; avirulent culture; bacterin	Vaccinate each spring. Use only in anthrax areas. The extent of problem and animal species will dictate which product should be used.
2. *Clostridium* spp.	See recommendations for cattle	
3. Blue tongue virus	Modified vaccine	Vaccinate nursing lambs prior to vector season and ewes prior to breeding.
4. Contagious ecthyma virus	Live vaccine	Vaccinate lambs prior to pasture season. It is necessary to vaccinate adults on initial outbreak and all replacement animals.
5. Enzootic ovine abortion *Chlamydia* sp.	Live vaccine	Use in infected areas. One injection gives adequate protection for 1 year and perhaps for life.
6. Epididymitis *Brucella ovis*	*Br. abortus* strain 19 plus bacterin of *Br. ovis.*	Vaccinate rams at 9 to 12 months of age or 3 weeks prior to breeding season.
7. Johne's disease *Mycobacterium paratuberculosis*	Bacterin	Vaccination of replacement breeding animals at 6–7 months of age has been helpful in infected flocks.
8. Louping ill virus	Killed vaccine	Vaccinate lambs at weaning time.
9. *Pasteurella* spp.	See recommendations for cattle	

Disease and Microorganism	Product(s)	Methods and Comments
10. Rift valley fever virus	Modified vaccine	Vaccinate ewes while open before breeding.
11. Tetanus *Clostridium tetani*	Tetanus antitoxin, tetanus toxoid	Used as directed on animals on an infected premise.
12. Vibriosis *Campylobacter intestinatis*	Bacterin	Vaccinate shortly after mating and again in 8 weeks and then annually before breeding season.
13. Leptospirosis	See recommendations for cattle	

SWINE

1. Hog cholera virus	Modified vaccines, antiserum	Available in countries where the disease exists.
2. Leptospirosis	See recommendations for cattle	
3. Swine erysipelas *Erysipelothrix insidiosa*	Vaccine, bacterin Antiserum	Usually administered to weanling pigs. Valuable for immediate protection and treatment.
4. Jowl abscesses *Streptococcus* Group E	Bacterin Autogenous bacterin	Healthy swine should be vaccinated. Animals should be at least 8–10 weeks of age. Autogenous bacterin may be more effective.
5. Pseudorabies virus	Modified vaccine Killed vaccine	Important to use product frequently enough (6-mo. intervals) to maintain adequate antibody level. Check regulations on the use of the vaccine.
6. Salmonellosis *Salmonella* spp.	Autogenous bacterin of species causing infection	Give at least 2 doses, 3 weeks apart with the last dose at least 2 weeks prior to farrowing for passive immunity for baby pigs. Can start pig injections at 2 weeks of age and then repeat every 3 weeks. Give 4 doses in problem herds.
7. Transmissible gastroenteritis (TGE) virus	Modified vaccine Planned exposure	The problem in each herd must be carefully evaluated and then a program of control planned and implemented.
8. Coliform diarrhea *Escherichia coli*	Live culture	Give oral culture to sows on each of three successive days at least 3 weeks prior to farrowing for passive antibody protection in baby pigs when *E. coli* has been a problem in pigs under 10 days of age.
9. Atrophic rhinitis *Bordetella bronchiseptica*	Autogenous bacterin Bacterin	Autogenous bacterin has been of help in some herds. Follow the manufacturer's directions.

HORSES

1. African horse sickness virus	Modified vaccine	Requires annual vaccination in infected areas.
2. Contagious equine metritis (CEM)	No immunization	Acute infection is self-limiting in 10–14 days. Antibiotic therapy can hasten recovery. Wash stallions with chlorhexidine.
3. Equine encephalomyelitis virus	Killed vaccine	Give 2 to 3 doses each year depending on length of mosquito season and the antibody level needed. Modified vaccine may not provide the protection needed.
Eastern and western Venezuelan strain	Modified vaccine	This vaccine was developed for human protection and was found to stimulate good protection in horses.
4. Equine influenza virus	Killed vaccine—multiple strains	Has proven to be beneficial in minimizing respiratory infections in some herds. Requires frequent boosters (60–90 days) in exhibition horses.
5. Equine Rhinopneumonitis virus (EHV-I and EHV-IV)	Modified vaccine	Vaccine must be given at frequent intervals (at least every 4 to 6 months) so that adequate antibody level can be maintained to protect and minimize respiratory problems and abortions.
6. Equine viral arteritis (EVA)	Modified live virus	EVA MLV produces antibody that is not distinguishable from disease-causing antibody. Results in importation problems.
7. Leptospirosis *Leptospira* spp.	Bacterin	Very few herds have a problem with this infection.

Disease and Microorganism	Product(s)	Methods and Comments
8. Lockjaw (tetanus) *Clostridium tetani*	Toxoid	Excellent protection by giving 2 doses 3 months apart and then yearly booster.
	Antitoxin	Antitoxin used for immediate protection and for treatment.
9. Malignant edema *Clostridium septicum*	Bacterin	Vaccinate animals if infection is a problem on the premises.
10. Salmonellosis *Salmonella* spp.	Autogenous bacterin	Give at least 2 doses, 3 weeks apart with the last dose at least 2 weeks prior to foaling for passive antibody immunity for the foal. Then start foal injections at 3–4 weeks of age, repeating every 3 weeks. Give 4 doses in problem herds.
11. Strangles *Streptococcus equi*	Bacterin	Vaccinate the foal at about 3 months of age, then repeat in 3 weeks. May cause soreness, and at times abscesses.
12. Viral papilloma	Autogenous vaccine	Does not speed regression of existing warts, but can prevent occurrence of new lesions.

POULTRY

Disease and Microorganism	Product(s)	Methods and Comments
1. Infectious bronchitis virus	Vaccine—many strains	Given to chicks via drinking water, nasal passages, or eye at 7 to 14 days of age; repeated at 4 to 5 weeks and at 14 to 16 weeks.
2. Fowl pox virus	Chicken pox vaccine	Given by feather follicle or skin puncture at least one month prior to coming into production.
	Pigeon pox vaccine	Less virulent than chicken pox so sometimes used when birds are in production.
3. Marek's disease virus	Live vaccine	Given at 1 day of age.
4. Erysipelas *Erysipelothrix* spp.	Bacterin	Given to turkey poults before turning out to range.
5. Fowl cholera (Pasteurellosis) *Pasteurella multocida*	Bacterin There are many strains	Used as an aid in control of fowl cholera.
6. Newcastle disease virus	Vaccine Many strains	Programs must meet the need of the area and premises.
7. Laryngotracheitis virus	Vaccine	Given via the eye route to birds over 6 weeks of age.
8. Encephalomyelitis virus	Vaccine	As necessary, depending upon area of the country.

Common Diseases of Domestic Livestock and Their Symptoms

Digestive System

Common Noninfectious Problems

Impaction—all species
Overeating—all species
Rumenitis—ruminants
Displaced abomasum—ruminants
Constipation—sows at farrowing
Gastric ulcers—swine
Colic—horses

Common Symptoms of Noninfectious Problems

Animals not eating
Listlessness
Straying from the flock or herd
Increased respiration in response to pain
Dry, hard, mucus-covered feces
Normal temperature
Rumen may be distended
Obvious signs of discomfort
Bloody stool

Common Infectious Problems

Salmonellosis—all species
Bovine virus diarrhea—cattle
Overeating (enterotoxemia)—ruminants, swine
Johne's disease—ruminants
Internal parasites—all species
Transmissible gastroenteritis (TGE)—swine
Rotavirus—cattle, swine
Bloody dysentery—swine
Bacterial diarrhea—all species

Common Symptoms of Infectious Problems

Loose stool
Loss of appetite or off-feed completely
Fever
Roughened hair coat
Dehydration
Bloody stool

Respiratory System

Common Noninfectious Problems

Pulmonary emphysema—ruminants;
    heaves—horses
Heavy raspy breathing—all species
Roaring—horses
Aspiration pneumonia—all species
Nose bleed—all species

Common Symptoms of Noninfectious Problems

Rib cage movement forced and abnormal
Rate of respiration is increased
Extra noises on inhalation and exhalation
Obvious signs of distress upon
    inhalation/exhalation
Symptoms worsen following exercise
Temperature usually normal

Respiratory System (continued)

Common Infectious Problems

Infectious bovine rhinopneumonitis—ruminants
Influenza—all species
Pneumonia(s)—all species
Catarrhal fever—ruminants
Pseudorabies—swine
Atrophic rhinitis—swine
Equine rhinopneumonitis—horse
Equine arteritis—horse
Distemper (strangles)—horse
Pasteurellosis—all species

Common Symptoms of Infectious Problems

Labored, increased, noisy respiration
Extended head to allow for freer air movement
Fever
Decreased appetite
Increased heart rate (pulse rate)
Catarrhal discharge from nose
Secondary problems may arise (abortions)
Sneezing, coughing

Genitourinary System

Common Noninfectious Problems

Low conception rates
Urinary calculi
Anatomical abnormality

Common Symptoms of Noninfectious Problems

Straining to urinate
Frequent urination
Repeated estrous cycles
Failure to show estrus
Sterility

Common Infectious Problems

Brucellosis—ruminants, swine
Vibriosis—cattle, sheep
Trichomoniasis—cattle
Chronic bacterial infections—all species
Leptospirosis—all species
Infectious bovine rhinotracheitis—cattle
Parvovirus infection—swine
Pseudorabies—swine
Equine rhinopneumonitis—horse
Contagious equine metritis—horse
Equine viral arteritis—horse

Common Symptoms of Infectious Problems

Low conception rates
Abortions
30-, 60-, 90-day returns to estrus
Retained placentas
Irregular estrous cycles
Fever

Nervous System

Common Noninfectious Problems

Birth (hereditary and developmental) defects—all
 species
Plant toxicities—all species
Nutritional deficiencies—all species
Salt poisoning—swine
Wobbles—horse
Obturator nerve damage—cattle

Common Symptoms of Noninfectious Problems

Abnormal disposition
Pushing of the head against an immovable object
Circling
Head held in tilted position
Spasms, convulsions, inability to rise
Sudden unexplained deaths
Incoordination, muscle rigidity
Paralysis, complete or partial

Common Infectious Problems

Meningitis—all species
Listeriosis—ruminants
Rabies—all species
Parasitic damage—all species
Pseudorabies—swine, ruminants
Encephalomyelitis—horse

Common Symptoms of Infectious Problems

The symptoms of infectious problems of the
 nervous system are indistinguishable from those
 of noninfectious problems.
Fever is often present.

Integumentary System (Skin)

Common Noninfectious Problems

Allergic dermatitis—all species
Anhydrosis (inability to sweat)—horse
Neoplasms—all species

Common Symptoms of Noninfectious Problems

Reddening, edema, itching, and some necrosis
Lack of sweating, dull hair coat, hair loss

Common Infectious Problems

External parasites: lice, mites, ticks—all species
Tick paralysis—sheep
Maggots—all species
Warbles or grubs—cattle
Sore mouth—sheep, goats
Pox—all species
Warts—all species

Common Symptoms of Infectious Problems

Scratching, rubbing, self-biting
Hair loss, scab formation
Inability to rise following tick infestation
Presence of blowfly larvae
Warble or grub bumps
Scabbing, crusting around mouth and udder
Vesicle, pustule, and scab formation
Wart lesions

Locomotor System

Common Noninfectious Problems

Fractures—all species
Azoturia—horse
Joint or sesamoid bone inflammation—all species
Bruises—all species
Strains, sprains—all species
Calcification of soft tissue—all species
Laminitis—ruminants, horses

Common Symptoms of Noninfectious Problems

Lameness
Stiffness
Hardened muscles, especially of lumbar region
    and hind limbs
Muscle tremors
Malformation of hooves

Common Infectious Problems

Foot rot—all species
Joint ill or navel ill
Secondary infectious following injury

Common Symptoms of Infectious Problems

It is impossible to distinguish between the
    symptoms of the infectious locomotor and
    noninfectious locomotor problems.
Fever is frequently present.

Circulatory System

Common Noninfectious Problems

Nutritional anemia

Common Symptoms of Noninfectious Problems

"Poor doers"
Low conception rate
Sudden death

Common Infectious Problems

Anthrax—ruminants
Anaplasmosis—ruminants
Foot-and-mouth—ruminants
Leptospirosis—all species
Erysipelas—swine
Hog cholera—swine
African swine fever—swine
Equine infectious anemia—horse

Common Symptoms of Infectious Problems

Fever
Listlessness, weakness
Loss of appetite
Constipation, followed by diarrhea
Nasal discharge
Labored breathing
Watery, pasty eyes
Incoordination
Sudden deaths

Poultry Health Problems

Common Noninfectious Problems

The majority of the noninfectious problems in poultry are associated with nutritional deficiencies. It is a matter of obtaining a correct diagnosis and then adjusting the ration. An inadequate water supply can result in serious losses especially during hot weather.

Breast blisters

Common Symptoms of Noninfectious Problems

"Poor doers"
Lowered egg production or growth
Lowered fertility
Large, fluid-filled blister on breast

Common Infectious Problems

Erysipelas
Mycoplasmosis
Necrotic dermatitis
Navel ill
Fowl cholera
Salmonellosis
Avian encephalomyelitis
Hemorrhagic enteritis
Infectious bronchitis
Newcastle disease
Ectoparasites
Internal parasites
Coccidiosis
Blackhead
Laryngotracheitis
Infectious bursal disease
Fowl pox
Marek's disease

Common Symptoms of Infectious Problems

Off-feed
Diarrhea
Swollen caruncle—reddish/blue in color
Reluctance to move, depression
Lameness
Swelling of joints
Necrosis of skin over thighs and breasts
Ruffled feathers
Coughing
Sneezing

Useful Abbreviations, Conversion Factors, and Mathematical Formulae

Abbreviations

U.S.

teaspoon	= tsp
tablespoon	= tbsp
cup	= c
pint	= pt
quart	= qt
gallon	= gal
fluid ounce	= fl oz
peck	= pk
bushel	= bu
grain	= gr
ounce	= oz
pound	= lb
hundred weight	= cwt
inch	= in
foot	= ft
yard	= yd
rod	= rd
mile	= mi

acre	= ac
degree Fahrenheit	= °F
parts per million	= ppm
square	= sq

Metric

microgram	= ug
milligram	= mg
gram	= gm
kilogram	= kg
milliliter	= ml
liter	= l
kiloliter	= kl
millimeter	= mm
centimeter	= cm
meter	= m
kilometer	= km
cubic centimeter	= ha
degree centigrade	= °C
square	= sq

Length Equivalents

U.S.

1 ft	= 12 in
1 yd	= 36 in
	= 3 ft
1 rod	= 16.5 ft
1 hand	= 4 in
1 furlong	= 220 yd
	= 1/8 mile
1 mile	= 5280 ft
	= 1760 yd

U.S. to Metric

1 in	= 2.54 cm
1 ft	= 30.48 cm
	= .3048 cm
1 yd	= .9144 m

1 mi	= 1609.34 m
	= 1.609 km

Metric

1 cm	= 10 mm
1 m	= 100 cm
1 km	= 100 m

Metric to U.S.

1 mm	= 0.03937 in
1 cm	= 0.3937 in
1 m	= 39.37 in
1 m	= 3.281 ft
1 m	= 1.094 yd
1 km	= 0.6214 mi

Length (continued)

Length Conversion Examples

Convert 13 inches to centimeters.

From the table: 1 in = 2.54 cm

Multiply: 2.54 cm/in × 13 in = 33.02 cm

13 in = 33.02 cm

Convert 33.02 centimeters to inches.

From the table: 1 inch = 2.54 cm

Divide: 33.02 cm ÷ 2.54 cm/in = 13 in

33.02 cm = 13 in

Area Equivalents

U.S.

1 sq ft	= 144 sq in
1 sq yd	= 1296 sq in
	= 9 sq ft
1 sq rd	= 272.25 sq ft
	= 30.25 sq yds
1 ac	= 43,560 sq ft
	= 4840 sq yds
	= 160 sq rd
1 sq mi	= 640 ac

U.S. to Metric

1 sq in	= 6.452 sq cm
1 sq ft	= 0.0929 sq m
1 sq yd	= 0.8361 sq m
1 ac	= 0.4047 ha
1 sq m	= 259.0 ha

Metric

1 sq mm	= 0.000001 sq m
1 sq cm	= 0.001 sq m
1 sq ha	= 10,000 sq m
1 sq km	= 1,000,000 sq m

Metric to U.S.

1 sq cm	= 0.155 sq in
1 sq m	= 1.196 sq yd
	= 10.764 sq ft
1 ha	= 2.471 ac
1 sq km	= 0.386 sq mi
	= 247.1 ac

Area Conversion Examples

Convert 18 square inches to square centimeters.

From the table: 1 sq in = 6.452 sq cm

Multiply: 18 sq in × 6.452 sq cm/in = 116.136 sq cm

18 sq in = 116.136 sq cm

Convert 116.136 square centimeters to square inches.

From the table: 1 sq in = 6.452 sq cm

Divide: 116.136 sq cm ÷ 6.452 sq cm/sq in = 18 sq in

116.136 sq cm = 18 sq in

Volume Equivalents

U.S.

1 cu ft = 1728 cu in
1 cu yd = 27 cu ft
3 tsp = 1 tbs
1 tbs = ½ fl oz
2 tbs = 1 fl oz
4 tbs = ¼ c
8 tbs = ½ c
16 tbs = 1 c
1 c = 8 fl oz
= ½ pt
1 pt = 16 fl oz
4 c = 32 fl oz
= 1 qt
4 qt = 1 gal
1 gal = 128 fl oz
8 qt = 1 pk
4 pk = 1 bu

U.S. to Metric

1 cu in = 16.387 cc
1 cu ft = 0.0283 cu m
= 28.316 l
1 cu yd = 0.7646 cu m
1 fl oz = 29.573 ml

1 pt = 473.166 ml
= 0.4732 l
1 qt = 0.9463 l
1 gal = 3.7853 l
1 tsp = 5 ml
1 tbs = 15 ml

Metric

1 cc = 1 ml
= 1 ccm
1 ml = 0.001 l
1 l = 1000 ml
= 1000cc

Metric to U.S.

1 cc = 0.061 cu in
1 cu m = 35.315 cu ft
= 1.308 cu yd
1 ml = 0.0338 fl oz
30 ml = 1 fl oz
1 l = 33.81 fl oz
= 2.1134 pt
= 1.057 qt
= 0.2642 qt
1 kl = 264.18 gal

Volume Conversion Examples

Convert 20 cubic inches to cubic centimeters.

From the table: 1 cu in = 16.387 cu cm
Multiply: 16.387 cu cm/cu in × 20 cu in = 327.74 cu cm
20 cu in = 327.74 cu cm

Convert 327.74 cubic centimeters to cubic inches.

From the table: 1 cu in = 16.387 cu cm
Divide: 327.74 cu cm ÷ 16.387 cu cm/cu in = 20 cu in
327.74 cu cm = 20 cu in

Weight Equivalents

U.S.

1 lb	= 16 oz
	= 7000 gr
1 cwt	= 100 lb
1 ton	= 2000 lb

Metric

1 mg	= 0.001 g
1 g	= 1000 mg
1 kg	= 1000 g
1 metric ton	= 1000 kg

U.S. to Metric

1 gr	= 0.065 g
1 oz	= 28.35 g
	= 437.5 gr
1 lb	= 453.592 g
	= 0.4536 kg
1 short ton	= 907.18 kg
1 long ton	= 1016.05 kg
	= 1.016 metric ton
1 ppm	= 1 ug/g
	= 1 mg/l
	= 1 mg/kg

Metric to U.S.

1 g	= 0.03527 oz
	= 15.43 gr
1 kg	= 35.274 oz
	= 2.205 lb
1 metric ton	= 1.102 short ton
	= 0.984 long ton
	= 2204.6 lb

Weight Conversion Examples

Convert 60 ounces to kilograms.

From the table: 1 oz = 28.35 g or 0.02835 kg

Multiply: 60 oz × 0.02835 kg/oz = 1.701 kg

60 oz = 1.701 kg

Convert 1.701 kilograms to ounces

From the table: 1 oz = 28.35 g or 0.02835 kg

Divide: 1.701 kg + 0.02835 kg/oz = 60 oz

1.701 kg = 60 oz

Temperature

Centigrade (Celsius) = 5/9 (Fahrenheit − 32)
Fahrenheit = 9/5 centigrade (Celsius) + 32

Temperature Conversion Examples

Change 104° Fahrenheit to centigrade.

Centigrade = 5/9 (Fahrenheit − 32)

= 5/9 (104 − 32)

= 5/9 (72)

= 360/9

104°F = 40°C

Change 40° centigrade to Fahrenheit.

Fahrenheit = 9/5 centigrade + 32

= 9/5 × 40°C + 32

= (360 ÷ 5) + 32

= 72 + 32

40°C = 104°F

Some Useful Formulae

1. **To calculate the circumference of a circle:**

    Circumference = π × diameter

    $C = 3.14 \times d$

    Example: Calculate the circumference of a circle with a diameter of 100 feet.

    $C = \pi d$

    $C = 3.14 \times 100$

    $C = 314.00$ feet

2. **To calculate the area of a circle:**

$$\text{Area of a circle} = \pi \times \text{radius}^2$$
$$A = \pi r^2$$

Example: Calculate the area of a circle with a diameter of 100 feet.

$A = 3.14 \times 50^2$

$A = 3.14 \times (50 \times 50)$

$A = 3.14 \times 2500$

$A = 7850$ square feet

3. **To calculate the area of a rectangle:**

$$\text{Area of a rectangle} = \text{length} \times \text{width}$$
$$A = LW$$

Example: Calculate the area of a lot 150 feet by 300 feet.

$A = LW$

$A = 150 \text{ ft} \times 300 \text{ ft}$

$A = 45{,}000$ square feet

4. **To calculate the volume of a rectangular container:**

$$\text{Volume is equal to length} \times \text{width} \times \text{height}$$

Example: Calculate the volume of a container 2.5 feet long by 2 feet wide by 1.5 feet high.

$V = LWH$

$V = 2.5 \text{ ft} \times 2 \text{ ft} \times 1.5 \text{ ft}$

$V = 7.5$ cubic feet

5. **To calculate the number of board feet:**

1 board foot = a board 1 inch thick, 1 foot wide, and 1 foot long

1 board foot = 144 cubic inches

Example: Calculate the number of board feet in a 1- by 6-inch board 16 feet long.

No. of board feet = 1 in × 6 in × (16 ft × 12 in/ft) ÷ 144 cu in/ft

No. of board feet = 6 sq in × 192 in ÷ 144 cu in/ft

No. of board feet = 1152 cu in ÷ 144 cu in/ft

No. of board feet = 8

Shortcut method

No. of board feet = thickness (in) × width (in) × length (ft) ÷ 12

No. of board feet = 1 × 6 × 16 ÷ 12

No. of board feet = 96 ÷ 12

No. of board feet = 8

Convenient Multipliers

1 inch	×	0.08333	=	feet
	×	0.02778	=	yards
	×	0.00001578	=	miles
1 square inch	×	0.00695	=	square feet
	×	0.0007716	=	square yards
1 cubic inch	=	0.00058	=	cubic feet
	×	0.0000214	=	cubic yards
1 foot	×	0.3334	=	yards
	×	0.00019	=	miles

1 square foot	×	144	=	square inches
	×	0.1112	=	square yards
1 gallon	×	8.33	=	pounds
	×	0.13368	=	cubic feet
	×	231	=	cubic inches
1 cubic inch	×	0.036024	=	pounds
of water	×	0.004329	=	gallons
1 cubic foot	×	62.425	=	pounds
of water	×	7.48	=	gallons

Weight per Ton—Percent per Ton—Parts per Million per Ton

Weight of Additive	Percent per Ton	Parts per Million (ppm)
10 grams	0.00110	11.0
50	0.01101	55.1
100	0.01101	110.1
200	0.02203	220.3
400	0.04405	440.5
500	0.05507	550.7
1000	0.11013	1101.3
1 pound	0.05	500
2	0.10	1000
4	0.20	2000
5	0.25	2500
10	0.50	5000
20	1.00	10000
100	5.00	50000

Nutrient Functions and Deficiency Symptoms

Nutrient	Function	Major Deficiency Symptoms
Energy (Fats, fatty acids, carbohydrates)	Growth, fattening, milk production, muscle and nerve function	Reduced growth rate, diminished milk production, poor body condition, silent heats, increased tendency to metabolic diseases such as fatty liver syndrome and ketosis
Fiber	Stimulates rumen function, stimulates salivation, helps keep rumen pH near neutral, fermented to short-chain fatty acids for milk synthesis and energy	Inflammation of the rumen, rumen shutdown, founder, low milk fat test, tendency toward displaced abomasum, poor muscle contractility
Protein	Cell formation, muscle and blood proteins, enzymes, milk synthesis	Retarded growth rate, poor body condition, reduced milk production, poor feed conversion, reduced immune response, reduced reproductive performance
Salt NaCl	Nerve and muscle function, acid-base balance, water retention	Reduced appetite, unthrifty condition, craving for salt with the resulting depraved appetite manifestations
Calcium Ca	Skeletal growth and strength, milk production, muscle integrity	Weakened bones and teeth, low calcium content in bones
Phosphorus P	Energy metabolism, skeletal growth and strength, milk production	Lack of appetite, depraved appetite, irregular heat periods
Vitamin D	Ca and P absorption, reduced excretion of P, mobilization of Ca and P from the skeleton	Rickets, lack of appetite, wobbly gait, enlarged joints, lameness, arched backs, deficiency of Ca or P
Magnesium Mg	Muscle response, electrolyte balance, enzyme structure	Grass tetany, skin twitching, staggers, downer cattle
Potassium K	Enzyme activation, acid-base balance, heart contractility, muscle tone	Muscle weakness, intestinal distension, loss of appetite, cardiac and respiratory muscle weakness
Iron Fe	Blood hemoglobin	Anemia
Iodine I	Metabolic rate, thyroid hormone synthesis	Diminished signs of heat, retained placentas, abnormal calves (enlarged necks, hairless, dead)
Fluorine F	Teeth	Dental caries
Manganese Mn	Enzyme synthesis	Deformed calves, weak calves, incoordination (ataxia)
Copper Cu	Enzyme synthesis, blood pigment for respiration	Anemia, young delivered dead, wool loss, hind leg incoordination, heart attacks in cattle, baby pig "thumps"
Cobalt Co	Vitamin B12 synthesis	Anemia, depressed appetite, emaciation, poor-doing calves

(continued)

Nutrient	Function	Major Deficiency Symptoms
Vitamin B12	Energy metabolism, red blood cell development	Anemia, depressed appetite, emaciation, poor-doing calves
Selenium Se	Integrity of muscle	Calf and lamb mortality, nutritional muscular dystrophy, retained placenta, liver damage in swine
Sulfur S	Amino acid synthesis, formation of coenzyme A	Poor nitrogen utilization, decreased production
Zinc	Enzyme synthesis	Stiff gait, itching, joint swelling, dermatitis, depressed female fertility, small testicle size
Vitamin A	Epithelial tissue, integrity of respiratory tract, reproductive tract, growth	Muscle incoordination, night blindness, respiratory diseases, rough hair coat, edema, abortions, weak calves at birth, eye abnormalities
Niacin	Enzyme formation	Dermatitis, mouth ulcers
Thiamin B1	Energy metabolism, nerve activity	Beriberi, brain tissue degeneration, muscular incoordination, grinding of teeth
Riboflavin B2	Enzyme synthesis, fetal growth and development, epithelial integrity	Seborrheic dermatitis, mouth, nose, and eye lesions, hair loss, excessive salivation and tearing
Biotin	Coenzyme system, growth, nervous system integrity	Hindquarter paralysis
Pantothenic acid B3	Coenzyme system, growth, epithelial integrity	Depressed growth, emaciation, dermatitis, respiratory infections
Folic acid	Nucleic acid synthesis, growth	Depressed growth, low white blood cell count
Pyridoxine B6	Metabolic processes, coenzyme systems, red blood cell formation, growth	Demyelination of nerves, epilepsy-like seizures, depressed growth, lack of appetite
Choline	Metabolic processes, muscle integrity	Weakness, labored breathing
Vitamin C Ascorbic acid	Antioxidant, amino acid, metabolism, growth, wound healing, iron absorption, collagen formation	Scurvy, anemia, depressed growth, slow wound healing, below-skin hemorrhaging
Vitamin E Tocopherols	Antioxidant, growth, fat metabolism, muscle metabolism and integrity	White muscle disease in calves, stiff lamb disease, heart muscle degeneration, nutritional muscular dystrophy

Male, Female, Castrate Livestock Terminology

Species	Intact Male		Castrate Male		Intact Female	
	Pre-puberty	Post-puberty	Young	Mature	Pre-puberty	Post-puberty
Cow	Bull calf	Bull	Steer	Stag/steer	Heifer	Cow
Sheep	Ram lamb	Ram	Wether	Stag/steer	Ewe lamb	Ewe
Swine	Boar pig	Boar	Barrow	Stag/barrow	Gilt	Sow
Horse	Colt	Stallion	Gelding	Gelding	Filly	Mare
Goat	Buck	Buck	Wether	Wether	Doe	Doe
Chicken	Chick/cockerel	Rooster/cock	Capon	—	Chick/pullet	Hen
Turkey	Poult	Tom	—	—	Poult	Hen
Donkey	Colt	Jack	Gelding	Gelding	Filly	Jennet

Mule	Offspring from male donkey (or jackass) bred to mare
Hinny	Offspring from stallion bred to female donkey (or jennet)

Digestive System Sizes and Capacities

	Cow	*Sheep*	*Swine*	*Horse*	*Goat*
Stomach	20+ Gallons	2–3 Gallons	1.5 Gallons	3 Gallons	2–3 Gallons
Small Intestine	120' Length 2" Diameter	90' Length 1" Diameter	60' Length 1"–2" Diameter	80' Length 3" Diameter	90' Length
Cecum	1 Gallon	1 Quart	1 Gallon	5 Gallons	1 Quart
Large Intestine	35' Length 3" Diameter	15' Length 2" Diameter	12' Length 2" Diameter	25' Length 3"–4" Diameter	15' Length 2" Diameter

GLOSSARY

Abomasum. Often referred to as the "true stomach," the abomasum is the fourth compartment of the ruminant stomach. It is the site of enzymatic digestion and is comparable to the nonruminant stomach.

Abscess. A collection of pus in a closed or semi-closed pocket or cavity. Pressure from the buildup of pus may cause pain and diminished function of nearby limbs.

Absorption. The process of being passed into or through, as by osmosis or other processes. For example, nutrients are absorbed into the bloodstream through the wall of the gastrointestinal tract.

Acute. Refers to a disease condition or sickness that is quick in its onset and often severe. Compare to a chronic condition that is long term, continuous, and not acute.

Adjuvant. A product given along with an antigen to enhance the animal's immune response to the antigen.

Adsorption. The process of being taken up and held on the surface of a substance.

Afterbirth. The placenta and fetal membranes expelled from the uterus after parturition. The expulsion of these membranes is often referred to by livestock producers as "cleaning."

Agglutination. A process or reaction in which particulate antigens (for example, bacteria) that are suspended in a liquid are caused to collect into clumps. This occurs when the cell suspension of a given antigen is treated with serum from animals immunized against the antigen. Classic example of the antigen–antibody reaction.

Agonist. A drug that elicits a specific reaction by binding with a targeted and appropriate receptor.

Albumen. Sometimes spelled *albumin,* it most often refers to the runny, clear fluid surrounding the yolk of an uncooked egg. The albumen of an egg turns white upon cooking. Albumin also refers to water-soluble proteins found in blood.

Allergy. A hypersensitivity to a particular substance which causes a body reaction such as hives, sneezing, labored breathing, etc., upon reexposure to the substance.

Ambient temperature. Environmental temperature. The temperature surrounding the animal at a given point and time.

Anatomy. The structure and arrangement of the internal parts of an animal. Skeletal anatomy refers to the structure and arrangement of the animal's bones. See Appendixes A through M.

Anemia. A deficiency in the quantity of blood or one of its constituents—specifically, red blood cell count, hemoglobin concentration, or packed cell volume.

Anthelmintic. A drug or chemical agent used in a commercial worming preparation.

Antibiotic. A substance that destroys or inhibits the growth or action of microorganisms.

Antibody. Protein produced by the body and carried in the blood that provides protection against specific diseases by interacting with the disease-causing agent and neutralizing it.

Antigen. An enzyme, toxin, or other "foreign" protein to which an animal body reacts by producing antibodies specific to the invading antigen.

Antihistamine. A drug that blocks the effects of histamine. Antihistamines are effective in reducing the bronchoconstriction that results from an allergic response.

Antimicrobial. An agent that destroys or inhibits the growth or action of microorganisms. These agents include antibiotics and synthetic agents. Lonophores and arsenicals are examples of synthetic antimicrobial agents.

Antiseptic. A chemical agent used on living tissue to control the growth and development of microorganisms.

Antiserum. A serum (blood fluid) that contains a rich complement of antibodies. The enriched serum resulting from an animal that has been hyperimmunized against one or more infectious agents.

Antitoxin. An antibody to a toxin that has been produced by a microorganism. The antitoxin will combine with the specific toxin to neutralize it.

Antitussive. A drug that suppresses the cough reflex.

Arrhythmia. A disruption or variation in the normal beating rhythm of the heart.

Artificial insemination. The depositing of semen into the female reproductive tract by means other than natural.

Ascarids. Large roundworms. Found in the intestinal tract of animals and man. Can completely block the intestine. Causes unthriftiness in spite of what should be an adequate diet.

Aspirate. To remove by suction. After the needle of a syringe is inserted into an animal, the syringe handle is pulled back slightly to create a suction (aspirated) before the injection to confirm that the needle is not in a blood vessel.

Astringent. A substance that causes shrinking or contraction in localized blood vessels, thereby reducing blood flow and any discharge.

Atresia. The absence of a normal body opening or passage, as in *atresia ani.*

Atrophy. A wasting away or decrease in size of a part of the body.

Attenuated. Weakened or made less virulent. Used to describe a type of vaccine that will cause antibody production and protection against a disease without causing the disease itself.

Avirulent. Incapable of causing disease.

Axillary space. The armpit. The medial or ventral space or cavity at the junction of the arm or front leg and the shoulder.

Azoturia. An acute and noninfectious disease of horses and sometimes cattle that causes severe pain, perspiration, muscular rigidity in the hindquarters, cellular membrane destruction, and discoloration of the urine.

Bacteria. Tiny single-celled microorganisms, most of which can be seen only through a microscope. Some bacteria are capable of causing disease.

Bactericidal. Having the capability to destroy bacteria.

Bacterin. A type of vaccine that has been prepared from bacterial growth.

Band. A small group of mares and other subordinate horses that a stallion has claimed as his own and will fight all challengers to protect. A band may contain 30 or more animals. Most commonly used in reference to a breeding group.

Barrow. A male hog that has been castrated before reaching sexual maturity.

Bars. *Referring to the horse's teeth:* the interdental space or gum area lying between the incisors and the premolars. This area is devoid of teeth, with the exception of the tusks, or canine teeth. *Referring to the horse's feet:* the extensions of the wall of the hoof that extend from the wall itself near the heel of the foot toward the frog. Sometimes called the *buttresses* of the hoof.

Bellwether. The animal, trained to the routine of the livestock operation, that is used to lead the other members of its flock to the catch pen for some management procedure or for marketing. This lead animal, usually a wether, is outfitted with a neck strap and bell. Because the lead animal carries over from year to year, newer flock members are easily conditioned to following the bell. The term is synonymous with *Judas goat.*

Bight. The temporary loop of a rope, formed by laying one end of it over the other.

Bile. A yellowish-green fluid secreted by the liver into the small intestine that aids in digestion, particularly the emulsification of fat. In most species, bile is stored in the gallbladder, which is located in the lobes of the liver.

Biologicals. Antigens or reaction-causing agents used to stimulate immunity against specific diseases. A vaccine is an example of a biological.

Bitting. The process and procedures involved with accustoming the horse to the bit and conditioning him to yield to pressure transmitted through it.

Bleat. The sound or cry made by sheep or goats. Sheep and goats bleat as part of normal vocalization as well as when lost, frightened, or endangered.

Bloat. An abnormal condition in ruminants characterized by a distention of the rumen, usually seen on the animal's upper left side, due to an accumulation of gases.

Block. This is a specialized form of clipping that involves removing the tips of hair or wool on the body in a systematic manner. The final result makes the animal appear smooth, trim, and youthful.

Blotch. To produce a brand that is illegible or nondescript. Movement of the animal while the iron is applied or the use of too hot an iron, which produces a heavy scar, are the main causes of blotched brands.

Boar. An uncastrated male hog.

Body condition scoring. A management tool that helps determine whether cattle are being fed and managed adequately. Body condition scoring allows the livestock manager to periodically evaluate the "flesh" cattle are carrying and, if necessary, to adjust their feed intake so that they are receiving neither too much (wastes money) nor too little (wastes productivity) feed.

Bolus. A mass of food or medicine that is given orally. This is usually administered with an instrument called a *bolusing* or *balling gun*.

Bolt. To suddenly start and run away as in response to fear or anger. Also used to describe a rapid and careless eating of feed.

Bone. This term is used to describe the process by which hairs on the legs of a beef animal are pulled upward. "Boning" makes the animal appear larger-boned and is generally accomplished by using a bar of glycerine-base soap.

Botulism. A type of food poisoning. Caused by the neurotoxin produced by *Clostridium botulinum*. Produces vomiting, diarrhea, weakness, and death.

Bradycardia. A slower than normal heart rate.

Breech birth or delivery. The birth of an animal with the buttocks or rear feet first rather than the front feet and head first.

Bridle path. The closely clipped area of the horse's mane (3" to 6") lying directly behind the poll and ears. The area is clipped to allow the pollstrap or crownpiece of the halter or bridle to lie smoothly in place without ruffling the mane.

Buck. Male goat. Male goats are at times disparagingly called "billy goats."

Buckling. A yearling buck, as compared to a doeling, or yearling doe.

Bull. A sexually mature, uncastrated male bovine.

Cannibalism. The habit of some animals of biting at or eating the body parts of penmates. For example, tail biting in pigs.

Capon. A male chicken castrated before reaching sexual maturity.

Caruncles. The maternal "buttons" of the uterus that attach to the cotyledons or buttons of the fetal placenta in ruminant animals.

Cast. To "throw" (down) or lay an animal upon the ground on its side and restrain it. Usually performed in lieu of more elaborate restraint facilities. Animals are cast so that management techniques can be performed or emergency situations coped with. An animal may cast itself (lie down) and be unable to rise because of its nearness to a wall or corner.

Castrate. As a verb it means to deprive of or remove the testicles. Other terms are *emasculate* and, in horses, to *geld*. When ovaries are removed, the process is called *spaying*. The castrated animal is sometimes referred to as the *castrate*.

Catarrhal. A catarrhal condition is one in which the inflammation of the mucous membranes of the nose or throat causes the afflicted animal to suffer a mucous discharge from the nostrils.

Catch pen. A small, fenced-in area, usually in the corner or end of a much larger pasture, into which livestock can be driven or attracted (by food and/or water) and held for the performance of management practices.

Cattle guard. A structure used in lanes, driveways, or roads that prevents cattle from crossing but allows vehicles to move over it. Cattle guards are made of concrete, wood, or metal with rails spaced over a pit, so as to discourage livestock from crossing.

Caudal. Toward the tail or posterior part of the animal. Compare to *cranial*, or toward the head or front end (anterior) of the animal.

Cauterize. To burn away or seal with a hot instrument or caustic substance. Proliferated tissue and blood vessels can be cauterized to remove them or to stop a flow of blood.

Cervix. The thick-walled structure of the reproductive tract located between the vagina and the uterus. The cervix is considered the opening to or the neck of the uterus.

Cesarean section. The surgical procedure of taking an unborn animal, at or near parturition, from the uterus by cutting through the abdominal and uterine wall. Cesarean sections are required when the pelvic opening is too small to accommodate the passage of the fetus or when the fetus is hopelessly malpositioned.

Chestnuts. The hornlike growth on the inside of the legs of the horse. They occur above the knees and below the hocks. They are thought to be the remnants of a fourth or fifth toe from the ancestral horse. Also called *gallosities* or *night eyes.*

Chevon. The meat of goats. At its best, it should be the meat from a weanling or yearling animal.

Chronic. Refers to a disease condition that is continuous and lasts for a long time. Usually, a chronic disease does not cause the severity of stress to the animal that an acute disease does.

Cilia. Very small hairs or hairlike structures. For example, cilia are a part of the cells that line the respiratory tract and aid in the removal of dust, pollen, mucus, and other foreign particles.

Clitoris. A reproductive organ in the female that is the homologue of the penis of the male. Basically without function, its enlarged presence often indicates a hormonal problem in the animal.

Cleaning. See *Afterbirth.*

Cloaca. A common chamber in fowl into which the reproductive, urinary, and intestinal tracts discharge.

Cockerel. A young rooster not more than a year old.

Cod. The small bag or pouch that remains in a steer after it is castrated. The remnants of the scrotum.

Colic. Acute abdominal pain. May be due to overeating, muscle spasm, torsion of the intestine, mechanical obstruction, or overproduction and accumulation of gas. Occurs mostly in the colon and cecum.

Colostrum. The first milk secreted by the mammary gland shortly before and for a few days after parturition. Contains large amounts of antibodies.

Colt. A male horse or pony, 4 years of age or younger, that has not been castrated.

Commissure. A joint or seam. In reference to the horse's feet, the grooves on the sides of the frog.

Conception. The process of becoming pregnant. The union of sperm and ovum to form a zygote.

Conjunctiva. The mucous membrane that lines the inner surface of the eyelid. One of the mucous membranes examined when observing the horse for anemia.

Conjunctivitis. Inflammation or infection of the mucous membrane that lines the inner surface of the eyelid. Also refers to inflammation of any of the other tissue surrounding the eye.

Contagious. Refers to a disease that is capable of being transmitted from one animal to another.

Corpus luteum. The reddish yellow mass that fills the cavity of the site where the ovum was released from the ovary. The corpus luteum secretes the hormone progesterone. It is sometimes called the *yellow body.*

Cotyledons. Buttonlike structures on the fetal placenta that attach to the maternal caruncles in the uterus of ruminants.

Cow. A mature female of the bovine species, usually having had at least one calf.

Cow-hocked. Condition of horses and cattle typified by the points of the hocks being oriented inward toward one another instead of being directed to the rear, and the rear hooves being oriented outward (splayfooted). Accompanied by the points of the hocks being more closely positioned to one another than are the fetlocks.

Creep area. An area within a livestock pen, corral, or pasture so constructed that baby animals can enter and leave, but adult animals do not have access. For example, farrowing crates and pens usually have a creep area where the baby pigs sleep and are creep-fed.

Cribber. A horse or pony that has the bad habit (vice) of biting or bracing his front teeth against some object while sucking in air. Also known as *wind sucking.* Not to be confused with wood chewing.

Crop. A saclike enlargement of the gullet of birds that serves as a receptacle and storage site for food.

Crossbreeding. The practice of mating parents of two or more different breeds. The offspring are called *crossbreds.*

Crutching or "crotching." The clipping or shearing of wool from the dock, udder, and vulva of female sheep prior to breeding and lambing. Also referred to as *tagging.*

Cryptorchid. An animal in which one or both testicles have failed to descend into the scrotum; also referred to as a *ridgling, original,* or "cryp." The failure of testicular descent can involve one testis only (unilateral) or both testes (bilateral).

Currycomb. A circular metal or hard rubber comb with teeth for rubbing, cleaning, and combing an animal.

Cyst. An enclosed smooth mass with a liquid or solid center that is produced by the cells lining the cyst walls.

Dam. The mother or female parent of an animal.

Debeaking. The removal of the tip of the beak of chickens to prevent cannibalism. Usually chicks are debeaked at a very young age and while still at the hatchery.

Decongestant. A drug that reduces the swelling of mucous membranes.

Dehorning. The removal of the horns of an animal. This can be done at any age, but is easier on the animal and on the livestock manager if done at an early age, while the horn is still in the bud stage.

Dehydration. Refers to the loss of water from the animal's body as a result of sickness or lack of drinking water.

Deodorizing. The removal of the scent glands from the horn area of the head of a goat. Usually done while the goat is a kid.

Dermatitis. An inflammation of the skin of the animal.

Desnooding. The removal of the fleshy tubelike appendage from the head, immediately behind the beak, of the turkey poult.

Dewattling. The removal of the two wattles from chickens.

Diarrhea. Frequent bowel movements of increased liquidity. Scours.

Disbudding. The removal of the horn buds from the very young goat.

Disease. Literally, "dis-ease," or not at ease. Disease is a deviation from the normal state of health.

Disinfectant. A chemical agent used on nonliving objects to kill microorganisms. Its high toxicity precludes its use on live animals.

Docking. Refers to the removal of all or part of the tail of an animal. Normally, this is a procedure that is applied to young sheep. It was also common with draft horses.

Doe. A female of the goat, deer, or rabbit family. Female goats with kids are sometimes referred to as "nanny goats."

Doeling. A yearling doe, as compared to a buckling, or yearling buck.

Dorsal. Pertaining to or located on the back of an animal, as opposed to ventral, which pertains to the lower part or belly of the animal.

Downer cow. A cow that is "down" and cannot get up and stand by itself. The problem results from injuries caused by slipping, being bumped by other cows, by paralysis as a result of calving, or from a disease such as milk fever.

Drenching. The administration of a fluid product for the control of parasites or disease control. Normally, a dose syringe is used to administer a drench.

Drying off (up). The process of using certain management practices to stop milk production. Dairy cattle are usually dried off for a rest period of 60 days before calving. A dry cow is not lactating.

Dubbing. The removal of the comb from chickens.

Dystocia. Difficult parturition. Sometimes used to describe the pain involved with a difficult birth.

Ectoparasite. A parasite that lives on the outside of the body of the host animal (e.g., face flies, mites, ticks, and mosquitoes). Compare to internal parasites such as the roundworms, lungworms, and bots.

Edema. Swelling caused by the accumulation of fluid in the tissues of the body. The excess fluid is trapped between rather than within the cells.

Elastrator. A device used in bloodless castration and in the docking of tails of livestock. The elastrator is used to apply a strong rubber ring over the scrotum or tail, which then cuts off the circulation of blood and ultimately causes the death and subsequent sloughing of the tissue.

Electrolytes. Body acids, bases, or salts. Necessary for transmission of nerve impulses, oxygen and carbon dioxide transfer, digestive processes, and muscle contraction.

Emaciation. Severe thinness. A loss of conditioning (fat) and flesh (muscling) so that the underlying bones are prominent and easily seen.

Emasculatome. An instrument used for the bloodless castration of livestock by severing the spermatic cord without injury to the scrotum. Commonly referred to as the *Burdizzo* in livestock operations.

Emasculator. An instrument used for castration of livestock. It contains both a pair of cutting blades used to remove the testes or tail and crushing jaws to help prevent bleeding.

-emia. Referring to the blood—e.g., toxemia is the presence of a toxin in the blood.

Encephalitis. Referring to an inflammation of the brain.

Endocrine. Refers to the endocrine glands. These glands secrete their products (hormones) directly into the bloodstream or lymph. Examples of endocrine glands are the thyroid and pituitary.

Endometrium. The membrane lining of the uterus.

Endoparasite. Parasite that lives inside the host animal body (roundworms, strongyles, bots). Compare to ectoparasites that live on the outside of the host animal.

Endotoxin. Toxin produced by and contained within an organism. Endotoxin can be released when the organism's integrity is destroyed.

Enterotoxemia. Overeating disease. A toxemia or blood poisoning caused by the toxin of *Clostridium perfringens*. Unless vaccinated for, it is a highly lethal disease.

Entropion. An abnormality of the eye in which the eyelid and eyelashes are turned inward instead of outward and come in contact with the eyeball.

Emphysema. Heaves. An abnormal lung condition resulting from the rupture of pulmonary alveoli. Likely caused by allergies to dust, pollen, and mold. The symptoms are difficult breathing (dyspnea) and a forced, often double, expiratory effort.

Epithelium. The covering of the surfaces of the body, both internal and external. The skin consists of epithelium, as do the linings of the respiratory and reproductive tracts.

Eructation. Belching. The sudden ejection or release of gas from the stomach via the esophagus and mouth.

Estrous cycle. The period of time between one estrus or heat period and the next, or between two successive ovulations.

Estrus (noun); estrous (adjective). The time during which the female is sexually receptive to the male. Also referred to as the *heat period* or *estrous period.*

Euthanasia. The painless killing of an animal.

Eviscerate. To remove the viscera as in slaughtering or butchering an animal.

Ewe. A female sheep.

Exogenous. That which is put into or introduced into the body from the outside. Compare to endogenous—that which is produced naturally by the body's own mechanisms.

Exotoxin. Toxin produced and released by an organism, e.g., a bacterium, and absorbed by the tissues of the host animal.

Expected progeny difference (EPD). The difference in performance that can be expected from the future progeny of a sire compared to the performance that can be expected from the future progeny of an average animal of the breed.

Expectorant. A medicine that causes the release and expulsion of the thick mucous discharge from the nose, throat, and lungs that usually accompanies a cold.

External anatomy. The morphology or shape of an animal by section. Also referred to as the *parts* of an animal. The sections or parts are referred to in livestock-producer terminology. See individual species chapter headings.

Extra-label. Use of a drug in an animal in a manner that is not in accordance with the approved labeling. For example, use in a species not approved on the label, use for conditions not listed on the label, use at dosage levels not in accordance with label directions, or use of a route of administration not approved on the label.

Faired. Rounded or made less oddly shaped by rolling between the hands or between a shoe and the floor. Certain splices in a rope are faired after formation to make them more attractive.

Farrier. A specialist in the care and shoeing of horses' feet.

Farrow. To give birth to a litter of pigs.

Feral. Wild and untamed. A free-roaming, "wild" animal that was once domesticated. An escapee.

Fertilization. The union of sperm with the ovum or egg to form a new life.

Filly. A female horse or pony up to 4 years of age that has not foaled.

Floating. The filing or rasping away of the sharp edges that develop on the premolar and molar teeth of the horse. Because of the rotating motion of the horse's jaws, the sharp edges develop on the buccal (cheek) side of the teeth in the upper jaw and on the lingual (tongue) side of the teeth in the lower jaw.

Fly strike. A situation or seasonal occurrence in livestock operations in which there is a large number of flies irritating livestock or in which they attack and infest open wounds, sores, or incisions.

Fornix. An arch formed by the cervix projecting into the reproductive tract. It is not present in all species.

Founder. An inflammation in the foot of a horse, cow, sheep, or goat. Also known as *laminitis.* It can be caused by overeating grain, green grass, or cold water, severe concussion to the hooves, and prolonged trailering.

Freshen. With reference to cattle, to give birth to or calve (parturition) and thereby begin a new lactation period. A fresh cow has recently calved and is actively lactating.

Frog. The flexible, elastic, triangular-shaped cushion in the rear center of the sole of the horse's foot. Assists the horse by acting as a shock absorber and as a pump to return fluids from the lower leg to the upper portion.

Flushing. The practice of feeding a higher than normal level of energy at breeding time to increase ovulation rate, especially in sheep and gilts.

Fumigation. The process of fumigating or disinfecting, usually done to buildings to kill all living organisms and vermin for the purpose of controlling disease transmission. Chemical fumes are allowed to penetrate throughout an airtight livestock building for a day or longer. Precautions must be taken with this procedure to protect human and animal life.

Gangrene. The decay of body tissue caused by an interruption of the blood supply to, and consequent death of, the part affected.

Gastric ulcer. Gastric refers to the stomach. A gastric ulcer is an open sore on the inner lining of the stomach.

Gelding. A castrated male horse.

Generation interval. The average age of the parents within a species when the offspring are born, or the time period between the same stage in the life cycle of two successive generations.

Gestation. Pregnancy. Refers to the carrying of the products of conception in the uterus from fertilization to parturition. Gestation period is the time between mating (conception) and parturition.

Gilt. Female hog, usually less than 15 months of age, that has not produced a litter. Sometimes used to include females that have produced one but not two litters.

Gizzard. The second stomach of poultry. It has thick, muscular walls that assist the digestive process by mechanically grinding the food.

Globulin. Family of proteins found in the plasma portion of the blood, part of which are the immunoglobulins (antibodies), associated with the immune response of the animal.

Goatherd. The person who takes care of goats. Compare to *shepherd* for sheep.

Grafted. The process of transferring an orphan baby animal to a foster mother. Term usually used in the sheep and beef cattle industries.

Gram negative. Refers to the classification of bacteria based on the bacteria's ability or inability to retain a purple stain used in Gram's method of staining. Gram-negative bacteria appear rose-colored or pink following the process, because they are unable to retain the purple Gram's stain.

Gram positive. Refers to the classification of bacteria based on the bacteria's ability or inability to retain a purple stain used in Gram's method of staining. Gram-positive bacteria appear purple following the process, because they retain the purple Gram's stain.

Granulation tissue. Small, rounded clumps of tissue that grow outward to heal or fill a wound site. Sometimes referred to as *proud flesh*, granulation tissue is richly vascularized and full of connective tissue cells.

Gregarious. Fond of or liking to live in flocks, herds, or other groups.

Hand mating. A mating procedure in which the herdsman controls the pairings by allowing males to mate with females in heat, but pens them separately before and after mating occurs.

Heavy metals. Metals of high specific gravity. They are of concern to livestock producers because their presence in food and water causes heavy metal poisoning. Lead and mercury are examples.

Heifer. A female bovine that has not given birth to a calf. Sometimes used to denote females until their second calving.

Hematoma. A swelling or tumor that contains blood. *Heme* refers to blood, *oma* denotes a tumor or swelling.

Hemoglobin. The pigment of the red blood cell that is involved in the transport of oxygen and carbon dioxide in the blood.

Hemorrhage. Bleeding. Loss of blood from the blood vessels.

Hen. An adult female chicken or turkey.

Hepatitis. Inflammation of the liver.

Hermaphrodite. An animal that has the reproductive organs, and very often the behavioral patterns, of both sexes.

Hernia. The protrusion of an organ or a part of the intestine through a break (rupture) in its surrounding structures (e.g., muscle walls, mesentery).

Hobble. To restrain an animal by tying two or more of its legs together.

Homeotherm. An animal that maintains a constant internal temperature regardless of environmental temperature.

Honda. A small loop or ring affixed to or formed into the end of a lariat through which the catch loop passes. Can be made of rope, leather, or metal.

Hormone. A substance secreted or released by a gland and carried by the circulatory system to a target organ, cell, or tissue in other parts of the body. The function of hormones is to regulate the activity of the target organs.

Hyperimmunize. Repeatedly challenge an animal with (expose an animal to) an antigen so that it will produce a larger quantity of antibodies than it would have after a normal single challenge or exposure.

-icide. Suffix denoting something that destroys or kills. A bactericide kills or destroys bacteria. A fungicide kills or destroys fungi. A viricide kills or destroys viruses.

Immunity. The animal's resistance to the effects of a foreign substance such as a toxin or disease organism. With reference to disease, immunity is the animal's lack of susceptibility to an infectious agent or other disease-causing antigen.

Immunization. The process and procedures involved in creating immunity (resistance to disease) in an animal. Vaccination is a form of immunization.

Impaction. A condition in which partially digested food material has been firmly wedged into a segment of the digestive system causing a stoppage of movement through the tract.

Implant(ing). The placement of a growth-stimulating compound, usually in the form of a pellet, just below the surface of the skin of an animal. The slow but steady absorption of the compound causes the animal to experience accelerated growth, up to the limits of the animal's genetic potential.

Incisors. The sharp cutting teeth located in the front part of the mouth.

Incubator. An appliance, varying in size from 10" × 18" × 6" boxes to 10' × 12' walk-in rooms, constructed for the purpose of providing the proper environmental conditions to allow for the development of poultry young while still in the egg. Fertilized poultry eggs are placed into the incubator for the entire incubation period.

Infection. The contamination of healthy tissues with a disease-producing microorganism or "germ."

Inguinal (ring). *Inguin* means groin, or in the groin region of the body. The inguinal ring in the male is the space or opening through which the spermatic cord passes into the scrotum from the body.

Instinct. A natural ability, very likely inherited, for doing or accomplishing a task that is not based on learning through habit. A knack or instinct for doing certain things.

Intramuscular. Refers to the injection route in which the product is deposited directly into a major muscle mass.

Intraperitoneal. Refers to the injection route in which the product is deposited into the peritoneal (abdominal) cavity of the animal.

Intravenous. Within (*intra*) the vein. Usually used in conjunction with injection, as in an intravenous injection. Contrast with *intramuscular* and *subcutaneous* injections.

Intussusception. The telescoping of one section of intestine into the lumen of an adjoining section, thereby causing blockage.

Involution. The process in which an organ returns to its normal size, as, for example, the uterus does after parturition.

Irradiated. Having been treated with some form of radiant energy.

-itis. Referring to inflammation—e.g., tendonitis is inflammation of a tendon.

Join. The area where two letters of a branding iron touch or where two parts of a single letter touch or cross. The letter *T* has one join; the letter *E* has three.

Joint ill. Also referred to as *navel ill,* this malady results in a variety of problems ranging from a localized umbilical abcess to diarrhea, peritonitis, convulsions, lameness, and death. It is the result of the invasion of the raw umbilical stump of the newborn by any of several species of bacteria. The infection can be prevented by treating the navel with iodine.

Judas goat. See *Bellwether.*

Ketosis. A metabolic disease characterized by hypoglycemia (low blood sugar). It is a disease of lactating ruminants that occurs within a few days after parturition. It is caused by an imbalance between nutrient intake and the nutrient requirements of the animal.

Kid. A young goat.

Kidding. To give birth (parturition) in the goat. Compare to *foaling, lambing, calving, farrowing.*

Lactation. The secretion of milk. The period between birth and weaning when the dam nurses her young.

Lamb. A young sheep. A ewe lamb or ram lamb, depending upon the sex.

Lambing jug. The small penlike enclosures into which the ewes and their lambs are placed immediately after birth. This allows the mother and lambs time to imprint on one another.

Lanolin. The fatlike substance contained in wool. It imparts the greaselike or smooth feeling present in wool and on your hands after handling wool.

Lesion. The change in the structure or form of an animal's body caused by a disease or injury.

Libido. Sexual drive or sexual desire. Usually reserved to describe the male of the species.

Liniment. Medicine carried in an oil-, soap-, or alcohol-based vehicle to be massaged into the skin of the animal to relieve pain or act as a counterirritant.

Lumen. The space within a tubular organ; for example, the lumen of the intestine or blood vessel.

Lyophilized. To remove water by freezing a material for the purpose of storing.

Mare. A mature female horse or pony.

Mastitis. Inflammation of the mammary gland.

Meconium. The dark-colored, semisolid fecal matter in the digestive tract of the fetus that forms the first bowel movement of the newborn.

Metastasis. Transfer of cancer cells from one location in the animal's body to another.

Metritis. Inflammation of the uterus.

Molars. The flattened back (jaw) teeth in animals that are used for grinding.

Molting. Used in reference to poultry, molting refers to the seasonal loss of feathers.

Mucolytic. A drug that has the ability to cause the breakdown of mucus.

Mucous membranes. The soft membranes that line the passages and cavities of the body such as the respiratory, alimentary, and genitourinary systems.

Mycosis. A disease caused by a fungus.

Mycotoxin. A toxin compound produced by a fungus.

Navel cord. Umbilical cord. The umbilical cord connects the fetus with the placenta and conveys food and removes waste products. It is severed or broken at birth.

Nearside. The animal's left side. Used primarily when referring to horses. Horses are customarily mounted and dismounted from the nearside. Compare to the horse's *offside* or right side.

Neat's-foot oil. A light, yellow oil extracted from the leg bones of cattle by boiling. Used primarily as a leather preservative.

Necropsy. Examination of the organs of a dead body in order to determine the cause of death or to study the pathologic changes present.

Necrosis. Death and decay of a body tissue that is still in contact with living tissue.

Needle teeth. The eight sharp teeth present in baby pigs at birth. Four are incisors and four are canine teeth. Sometimes referred to as *wolf teeth*.

Neonatal. That period of time immediately before and after birth of the young. From a management point of view, these are especially critical times in the lives of mother and young.

Neoplasm. A new and abnormal growth of tissue; a tumor.

Neurotoxin. A toxin that is destructive or poisonous to the nerve tissues of the animal.

Nonprotein nitrogen (NPN). An economical source of nitrogen from other than typical protein sources that can be fed to ruminant animals to partially satisfy the daily protein requirement. The microbes in the rumen utilize carbon chains for energy and NPN to synthesize microbial protein that can be digested by the animal.

Nutraceutical. Nontoxic food component that has been proven to have a beneficial effect on the health of an animal.

Obstetrical. Having to do with parturition.

Offside. See *Nearside.*

Ointment. Medicine carried in a semisolid vehicle for application to the eyes or skin of the animal.

Omasum. The third compartment of the ruminant stomach. It is located between the reticulum and the abomasum.

Omnivore. An animal that eats food from both animal and vegetable sources.

Ophthalmic. Of or pertaining to the eye.

Ovary. The reproductive organ in the female where eggs are produced and where the hormones estrogen and progesterone are synthesized.

Overbrand. This term refers to leaving a freeze-branding iron on the hide of an animal long enough to produce a bald brand instead of white hair. This is usually done on white or cream-colored cattle.

Ovulation. The release of the ovum or egg from the ovary of the female.

Oxytocin. A hormone secreted by the pituitary gland that causes milk letdown and contractions in the uterus by causing the contraction of smooth muscle.

Palpate. To investigate or examine by feeling or touching with the hands. An animal can be palpated to determine the state of pregnancy.

Papilla. A small projection that resembles a nipple or wart.

Parasite. A living organism that lives on or in another animal at that animal's expense. From a livestock management perspective, there are *internal* and *external* parasites.

Pasture breeding. Breeding system where the herd, flock, or band of females is placed into a pasture setting and allowed to roam freely with the breeding bulls, bucks, rams, or stallions. Usually, no attempt is made to control the onset of estrus of the females or to determine which males are mated to which females. It is as close to a natural breeding setting as modern management systems allow. Contrast to artificial insemination and hand-mating systems.

Pathogen. An agent, primarily a microorganism, that is capable of causing disease.

Paunch. See *Rumen*.

Pen mating. The management technique used to breed livestock wherein the male animal and one or more females are placed in a corral, paddock, or other type of pen for breeding.

Peritoneum. The membrane that lines the abdominal wall of an animal. It can also be viewed as the membrane that contains the viscera (large internal organs) of the animal.

Peritonitis. Inflammation of the peritoneum, which is the thin membrane that covers the organs in the abdomen. Characterized by an accumulation of serum, fibrin, cells, and pus in the peritoneum.

Phalanx. A part of the distal appendages on an animal's fore- and rear limbs. The phalanges of a horse are the long pastern bone (P1), the short pastern bone (P2), and the coffin bone (P3).

Pharmaceutical. A drug or other preparation used as a treatment in medicine.

Placenta. A reproductive organ containing the fetus that attaches to the inner wall of the uterus, thereby forming the route by which the fetus is nourished.

Pleura; pleural cavity. Pleura are delicate membranes that line the thorax. The thorax is the region between the neck and abdomen. The space within the pleura is called the *pleural cavity*.

Pneumonia. A disease of the lungs.

Poll. The top, crown, or back of the head of an animal. More specifically, it refers to that area directly between the ears of the animal.

Polyestrous. Having a regularly recurring estrous cycle throughout the year, as in female cattle and hogs. The mare and the ewe show regular recurrence of estrus only during certain seasons of the year and so are referred to as *seasonally polyestrous*.

Poult. A young turkey of either sex, usually not more than 2 months of age.

Poultice. A soft, moist mass of material such as mustard that is applied to a part of the body to promote healing.

Preconditioning. Completing one or more management techniques such as castration, dehorning, vaccination for the shipping fever complex, and training cattle to eat feed from a bunk. Preconditioning is usually done shortly before or at weaning time to reduce stress upon arrival at the feedlot.

Prognosis. A prediction or estimate of the progression and final outcome of a disease.

Prolapse. To slip or fall out of place through a natural body opening. For example, a prolapse of the uterus results when the uterus slips backward and is expelled from the reproductive tract through the vagina.

Proud flesh. See *Granulation tissue*.

Proventriculus. In the digestive system of poultry, the glandular or true stomach. Located between the crop and the gizzard.

Pruritis. An itching sensation.

Puberty. The time in the life cycle when an individual reaches sexual maturity. Ovum production in the ovary of the female and sperm formation in the male's testes are initiated.

Pullet. A female chicken less than a year old, or a young female chicken before she begins to lay.

Purebred. An animal whose parents are of the same breeds, and which is eligible to be recorded with a registry association.

Purulent. Referring to an injured area of an animal that contains or is exuding pus.

Pus. The liquid product of an inflammation process consisting of white blood cells and a liquid portion (*liquor puris*).

Putrefaction. Decay. Occurs as a result of enzymatic and/or bacterial action.

Quarter (udder). Refers to one of the four sections of the cattle udder. For example, the right-rear quarter could be spoiled or infected with mastitis.

Quick. To cut too deeply when trimming the horse's foot or to orient a shoeing nail into the sensitive laminae of the foot.

Ram. A male sheep that has not been castrated.

Rectum. The last, lowest, or caudal-most part of the large intestine.

Regimen. The protocol for administering a drug. A regimen includes the dosage, the route, the frequency, and the duration of administration.

Reticulum. The function of the reticulum is similar to that of the rumen. It is one compartment of the ruminant stomach where pregastric fermentation takes place. It is also the compartment involved in hardware disease.

Rooster. An adult male chicken. Also called a *cock*.

Rumen. The rumen is the pregastric fermentation pouch and is often called the *paunch*. It is the largest compartment of the ruminant stomach and is located on the left side of the animal. The rumen and reticulum are often referred to as the *reticulorumen*. They are the structures involved with bloat.

Rumenitis. Inflammation of the wall of the rumen.

Scrotum. A part of the male reproductive system. The pouch or bag that contains the testicles.

Scur. The small, round remnant or regrowth of horn tissue attached to the skin at the site of the horn in a mechanically dehorned or polled animal.

Sensitivity test. A test used to detect the sensitivity of microorganisms to various antibiotics. Paper discs impregnated with antibiotics are placed on agar plates that have been seeded with some organism. The presence of organism growth around the disc indicates an absence of sensitivity to the antibiotic. The absence of growth around the disc indicates that the antibiotic may be effective against the organism.

Septicemia. A blood poisoning caused by the growth or multiplication of bacteria in the circulatory system of the body.

Serotyping. A classification of microorganisms based on the way an animal's immune system (antibodies) reacts to unique antigens on the surface of microorganisms.

Serum. Normally referring to the clear liquid portion of an animal's blood remaining after the clot and any extraneous cells are removed.

Shoat (shote). A young pig of either sex that has been weaned; usually of medium weight (60 to 160 lbs).

Shock. A physical or mental disturbance that is very often the result of a major physical injury or some major stress. Also, acute failure of peripheral circulation. Can be caused by loss of blood or damage to the circulation control center.

Shod. The past tense of the verb "to shoe." For example, the horse was shod.

Shorn. The past tense of the verb "to shear." For example, the sheep were shorn.

Shot. An injection.

Signalment. The method of identifying or describing an animal that uses obvious, naturally occurring, distinguishing characteristics such as hair color, scars, and white markings.

Sire. The father or male parent of an animal.

Sow. Female hog usually over 15 months of age that has produced at least one litter.

Species. A group of animals possessing common characteristics that distinguish them from other animals.

Spermatic cord. The cord that passes from the scrotum to the body and includes the vas deferens, muscle, artery, vein, nerves, and lymphatics leading to and from the testicles.

Spermatogenesis. The formation and development of sperm in the testes.

Splayfooted. Feet pointed outward. Compare to its opposite, which is *pigeon-toed*, or feet turned inward. The splayfooted horse will typically wing, or wing-in, as it travels.

Spook. To scare or frighten. Animals, especially those not accustomed to being handled, are easily spooked by unusual noises or the quick movements of their handlers.

Spore. A bacterial cell that has an extra-thick wall, making it resistant to most sterilization processes.

Stag. An animal castrated after reaching sexual maturity or acquiring secondary sex characteristics.

Stallion. A mature, noncastrated male horse or pony.

Standing heat. That part of estrus (estrous period) that is characterized by the highest level of female receptivity to the male. A few hours before or just after this highest point of receptivity, the female may not stand still or as easily allow the advances of the male, hence the term *standing heat*.

Steer. A male bovine castrated before reaching sexual maturity.

Stencil. Term used in sheep production; refers to paint-branding numbers or letters on the backs or sides of sheep.

Sterilization. The process of sterilizing by boiling or by use of chemicals to make free of germs or microorganisms that could cause infections or disease. Primarily used on cutting or surgical tools used in livestock management.

Stress. Any environmental factor causing a change in an animal's internal functions.

Striker. A horse or pony that has the bad habit of reaching out quickly with one of its forelegs with the intent of striking its handler.

Subclinical. Referring to a disease condition that exists at a low enough level so as not to cause the animal to exhibit symptoms. A good example is subclinical mastitis, which costs the dairy industry hundreds of millions of dollars annually.

Subcutaneous. Refers to a route of injection in which the product is deposited beneath the skin but not into the muscle layer below it.

Supernumerary. Exceeding the regular number. Sheep, goats, and mares usually have two teats on their udders; cattle have four. A third teat on sheep or a fifth on cattle is considered supernumerary.

Surcingle. A belt or girth encircling the horse about the withers and foreflank. Can be used by itself or with rings attached as a training aid or to hold a saddle or pad in position.

Surfactant. A chemical substance such as a detergent that has the ability to reduce surface tension of water or other liquid. A surfactant may be given to bloated cattle and sheep to break up the accumulations of "foam" in their digestive systems that are causing the animals' discomfort.

Suture. The act of sewing a wound, the material used, or one of the stitches employed in sewing up a wound.

Synchronization. Verb is *synchronize.* To cause events to occur at the same time. In a management sense, it refers to the synchronization of the estrous cycles or heat periods of the females. This allows for breeding and the subsequent birthing processes to occur in a shorter period of time than if the herd were allowed to come into heat and breed at natural intervals.

Tachycardia. A faster than normal heart rate.

Tail rubber. A horse or pony that persistently leans backward against a fence or wall and rubs his tail area by moving his rear end from side to side.

Teasing. Technique used to determine when a mare is in standing heat and has, therefore, a follicle at maximum development, containing an ovum that is ready to be ovulated. The technique involves bringing a stallion to the mare and observing her reactions to him. Receptivity indicates heat; rejection—lack of heat. Whatever the reaction, the stallion is not allowed to mount, his purpose being only to "tease" so that the handler can observe the mare.

Test mating. The process of mating a male animal to a female in heat and observing the male for libido and the ability to penetrate and breed a female. To provide an additional test, a semen sample is sometimes collected to check for sperm quantity and quality. Test matings are conducted prior to the breeding season.

Testosterone. Male sex hormone. Produced by the Leydig cells in the testis.

Tetanus. A bacterial disease that results in spasms and stiffness of the muscles. The prognosis for a *tetanic* animal (one showing symptoms of tetanus) is grave.

Tether. To fasten or tie an animal with a rope or chain so that it can move or graze only within a specific area.

Tie bow. An enlargement or "bow" in the section of the tendon behind the cannon bone over which a string or cord was tied. Can be caused by tying too tightly, by tying without adequate padding, or by tying the knot or bow directly upon the tendon.

Tom. A male turkey.

Toxemia. A general systemic toxification or poisoning caused by the absorption of toxic substances (toxins) produced by bacteria. The toxins are usually produced at and absorbed from a localized source of infection. The presence of toxins found in the blood of an animal.

Toxic. Poisonous.

Toxoid. A toxin that has been treated so that it has lost its toxicity but is still capable of causing the production of antitoxin when injected into an animal.

Tranquilizer. A drug used to reduce physical or nervous tension. Used to quiet and calm nervous animals and thereby make them more receptive to handling attempts.

Trapnesting. A management method in poultry production for ensuring the identification of the parents. Eggs from trapnested hens are kept separate from those of other hens.

Trichoglyph. A swirled area (whorl) or natural disturbance in the lay of an animal's hair coat. These occur on the face, side of the neck near the crest, the front (underside) of the neck, the rear flank, and at several other locations. They are usable for identification because no two animals have identical whorled areas. See *Whorls*.

Tusk. The four greatly elongated teeth that project from the mouth of a mature boar. They are the canine teeth and arise, one from each side, top and bottom, from behind the incisors and in front of the premolars. Also found in horses.

Twitch. A tool used to restrain or control a horse by its application to the horse's nose. Consists of a loop of rope or chain attached to a handle.

-uria. Pertaining to urine. Glucouria or glycouria refers to the presence of sugar in the urine.

Urinary bladder. An organ of the body having the function of collection and storage of urine until it is excreted from the body.

Urinary calculi. "Stones" formed in any part of the urinary tract.

Uterus. Part of the reproductive tract in which the fertilized ovum attaches and develops into an embryo and fetus.

Vaccine. Any biological agent that produces live immunity. See *Antigen*.

Vagina. The part of the female reproductive tract between the cervix and the external opening.

Vasectomy. The surgical removal of a part of the vas deferens. The *vas* is the portion of the reproductive tract in the male through which the sperm must travel from epididymis to the penis.

Veal. Meat from a young calf, usually less than 3 months of age. In its strictest sense, veal production implies that the calves have been fed milk as their only source of nutrients.

Vector. A carrier, usually an insect, of a disease or disease organism from one animal to another.

Vent. The anus of the fowl.

Ventral. Defines location, and on an animal refers to the lower part or surface of the body, near or on a level with the belly or abdominal area.

Virulent. Pathogenic or capable of causing disease.

Virus. Infectious organism smaller than bacteria that is dependent upon other living tissues for its nutrients and replication sites. Does not respond to antibiotic treatment. Difficult to identify and control.

Viscera. The large internal organs of the body—includes the organs of the digestive, respiratory, urogenital, and endocrine systems, as well as the spleen, heart, and great vessels.

Volvulus. An obstruction within the intestines resulting from or caused by the twisting of the intestines.

Weaning. Literally, getting a young animal accustomed to eating food other than its mother's milk. In a livestock management sense, it refers to the separation of the young animals from their mothers. This is done at weaning time.

Weaver. A horse that stands in the stall in one spot and rhythmically swings his head and neck from one side to the other. The likely cause is boredom.

Wether. A male sheep or goat castrated before reaching sexual maturity.

Whipping. A method of finishing the cut end of a length of rope to prevent the strands from unlaying and fraying. Accomplished by tightly wrapping the cut end with a small-diameter string or cord.

Whorl. A small swirled or twisted area of an animal's hair coat having an arrangement or alignment of hair that is different from the surrounding pattern. An animal typically has a whorl on its forehead and on its rear flank. These whorls are unique and can be used for animal identification. See *Trichoglyph*.

Winking. The opening and closing of the vulva of the mare indicating that she is coming into estrus or "heat."

Withdrawal time. Time required between the application or feeding of a drug or additive and the slaughter of the animal to prevent any residue of the drug from remaining in the carcass. Withdrawal times are legally specified by the Food and Drug Administration (FDA).

Wolf teeth. The rudimentary premolar teeth sometimes found immediately in front of the normal upper premolar teeth of each jaw. In the horse, these teeth can become infected and sensitive and cause behavioral problems. In such cases, they must be removed. See *Needle teeth*.

Wormer. A commercial product containing an anthelmintic administered by stomach tube, in the feed or water, or by syringe to control internal parasites. Sometimes referred to as a *dewormer*.

Zoonoses. Diseases or parasites that can be transmitted between man and animals.

| INDEX |